OKB YAKOVLEV

A HISTORY OF THE DESIGN BUREAU AND ITS AIRCRAFT

Yefim Gordon
Dmitriy Komissarov
and Sergey Komissarov

MIDLAND
An imprint of
Ian Allan Publishing

OKB Yakovlev
A History of the Design Bureau and its Aircraft

ISBN 1 85780 203 9

Published by Midland Publishing
4 Watling Drive, Hinckley, LE10 3EY, England
Tel: 01455 254 490 Fax: 01455 254 495
E-mail: midlandbooks@compuserve.com
www.midlandcountiessuperstore.com

Midland Publishing is an imprint of
Ian Allan Publishing Ltd

Worldwide distribution (except North America):
Midland Counties Publications
4 Watling Drive, Hinckley, LE10 3EY, England
Telephone: 01455 254 450 Fax: 01455 233 737
E-mail: midlandbooks@compuserve.com
www.midlandcountiessuperstore.com

North American trade distribution:
Specialty Press Publishers & Wholesalers Inc.
39966 Grand Avenue, North Branch, MN 55056, USA
Tel: 651 277 1400 Fax: 651 277 1203
Toll free telephone: 800 895 4585
www.specialtypress.com

Design concept and layout by
Polygon Press Ltd. (Moscow, Russia)
Line drawings by Yakovlev OKB, PolygonPress, Aleksey Alyoshin, the late Vladimir Klimov, Vladimir Tootikov, *Kryl'ya Rodiny*, *Modelist-Konstrooktor* and via RART.

Printed in England by Ian Allan Printing Ltd
Riverdene Business Park, Molesey Road,
Hersham, Surrey, KT12 4RG

Acknowledgements

The authors wish to express their gratitude to the persons who assisted in and contributed to the making of this book:

Yuriy V Zasypkin; Yuriy I. Yankevich; Konstantin Udalov and AVICO-Press; the Russian Air Force's press department; the Russian Aviation Research Trust which kindly supplied valuable photos; the magazines *Kryl'ya Rodiny* and *Modelist-Konstrooktor*; the CIS Interstate Aviation Committee; and East Line Aviation Security and Vnukovo Airport Aviation Security.

Below: **Russian Air Force Commander-in-Chief Vladimir M. Mikhaïlov (right) shakes the hand of Yakovlev OKB chief test pilot Roman P. Taskayev after a flight in the second production-standard Yak-130 combat trainer in Nizhniy Novgorod. The Russian Air Force has recently ordered the first batch of the type.**

Opposite page: **A formation of UT-2 basic trainers operated by Osoaviakhim, the pre-war Soviet paramilitary sports society.**

Vladimir Drobyshevskiy

Sergey and Dmitriy Komissarov archive

CONTENTS

Aleksandr Sergeyevich Yakovlev, the founder of the OKB-115 aircraft design bureau.

A BRIEF HISTORY OF THE YAKOVLEV OKB

The design bureau, or OKB (***op**ytno-kon**strook**torskoye byu**ro*** – Experimental Design Bureau), founded by Aleksandr Sergeyevich Yakovlev is one of the most prominent and versatile Russian (Soviet) aircraft design bureaux. This book covers the activities of Yakovlev, his design team and his successors from 1924 to the present day, that is to say, in the course of 80 years. The aircraft created by this organisation are described in the main body of the book. This account is intended to show the origins of the Yakovlev OKB and its progressive transformations, detailing the general scope of its activities and its way to prominence. Naturally, the personality of Aleksandr Yakovlev stands in the centre of this brief outline. Due attention is also given to those who have succeeded Yakovlev and are carrying on with the firm bearing his name.

Aleksandr Yakovlev. First Steps to a Career in Aviation

Aleksandr Yakovlev was born on 1st April 1906 in a prosperous family in Moscow, where his father was the chief representative of the Nobel Oil Co. As a schoolboy Yakovlev took an interest in all matters technical, but gradually his interests began to narrow down to the aeronautical field. In 1923, his final school year, Yakovlev formed the first ODVF section at a Moscow school (ODVF = ***Ob**shchestvo droo**zey** voz**doosh**novo **flo**ta* – Air Fleet Friends Society). Among many other activities Yakovlev and a friend obtained permission to take a crashed Nieuport fighter from the dump at Moscow's Central airfield (Khodynka), bring it to the school and take it completely to pieces. This provided the young enthusiast with a valuable grounding in aircraft design.

Yakovlev obtained access to the world of aviation thanks to Konstantin K. Artseulov who was in charge of the first All-Union Glider Meet. This was to be held in November 1923 at Koktebel' near Feodosiya in the Crimea. Artseulov appointed Yakovlev assistant to N. D. Anoshchenko, one of the participants of the competition. Yakovlev soon proved his worth as a conscientious person and a skilled woodworker. He played an increasingly responsible part in building Anoshchenko's strap-on glider dubbed *Ma**ka**ka* (Macaque); as a reward, he went with its designer to the Crimea. On his way there he met some glider builders and, notably, such people as Sergey V. Il'yushin and Vladimir Pyshnov who were students of the Air Fleet Academy (AVF – *Aka**de**miya voz**doosh**novo **flo**ta*). In 1925 it was renamed VVA, *Voy**en**no-voz**doosh**naya aka**de**miya*, or Air Force Academy (college). Yakovlev made up his mind to gain admission to the Academy, but he lacked the necessary service in the Red Army.

At the 1923 glider meet the rather primitive Makaka glider had no luck, crashing after stalling during its very first take-off. As the Makaka was beyond repair, Yakovlev had time to watch the other gliders. He decided to do his utmost to design and build his own glider for the 1924 All-Union Glider Meet. This wish was backed up by Yakovlev's conscientious self-education in the course of which he diligently studied numerous books on aircraft design.

In 1924 Yakovlev gained admittance to the AVF as a humble shop-floor worker. He did not shun work of any kind and, thanks to his perseverance, in two years he acquired enough skills to earn him the position of hangar keeper at the airfield.

Enlisting support from his school friends and from the ODVF section, Yakovlev embarked on the construction of his first glider. The result was the AVF-10 (that is, the tenth glider designed at the Air Fleet Academy). It was a capably designed monoplane which was taken to the 1924 All-Union Glider Meet at Koktebel' and tested there. The glider behaved faultlessly and impressed the pilots who had flown it. Yakovlev received 200 roubles and a diploma.

With his award Yakovlev was able to build a second glider, the AVF-20, which was a refined version of the first. The AVF-20 again flew superbly and was described by the Aviakhim adjudicator I. S. Unshlikht as 'the best training glider design' (Aviakhim was an acronym for the voluntary society for the support of Soviet aircraft and chemical industry, a forerunner of the Osoaviakhim, DOSAV and DOSAAF). This glider remained at Koktebel' where it became a valued trainer in a group set up by Sergey Pavlovich Korolyov (later renowned as the mastermind behind Soviet rocketry and space achievements).

In September 1925, before he had taken the AVF-20 to Koktebel', Yakovlev had completed the first, preliminary three-view drawing of his first aeroplane. It was designated VVA-3 (as noted above, the AVF was renamed VVA in April 1925), the figure three denoting his third design. Despite the harshness of his long working days, young Yakovlev found time and resources to prepare the numerous detail drawings of his aeroplane and to undertake all the stressing calculations. The aircraft was built and made its first flight at the hands of Yulian I. Piontkovskiy who at that time commanded the VVA squadron and thus was Yakovlev's commanding officer. Later he became Yakovlev's test pilot.

The AIR-1 (as the VVA-3 was soon renamed) proved to be an outstanding success; the immediate consequence of this was that Yakovlev was granted admission as a full student to the VVA.

Over the years 1927-31 Yakovlev not only worked hard at his studies but also managed the construction of a succession of aircraft to his design. All were successful, and in his graduation year (1931) he ambitiously produced the AIR-5 five-seat cabin transport. Powered by a Wright Whirlwind engine, it might well have gone into series production, had a suitable Soviet engine been available. When he produced the smaller AIR-6, matched to the available 100-hp engine, this did indeed go into production for the Civil Air Fleet (GVF – *Grazh**dahn**skiy voz**doosh**nyy flot*), as the Soviet airline Aeroflot was then known.

Concerning the AIR designations

The AIR designations worn by Yakovlev's early aircraft call for some explanation. These letters are the initials of Aleksey Ivanovich Rykov, the then head of the Council of People's Commissars (that is, the Soviet Government). Yakovlev's first aircraft was called AIR-1 in honour of A. I. Rykov, and subsequent designs continued to bear this designation with consecutive numbers – AIR-2, AIR-3 and so on. This was quite in conformity with the practice of that time, when new hardware was given brands named after

Above: The Soviet leader Iosif V. Stalin, Kliment Ye. Voroshilov (centre) and Aleksandr S. Yakovlev (right) at an air event held at Moscow-Tushino airfield in 1935.

Soviet statesmen. Notable examples are the IS and KV series of heavy tanks (the initials stand for Iosif V. Stalin and Kliment Ye. Voroshilov respectively; Marshal Voroshilov headed the Defence Committee), the Class SO17 and Class FD20 freight steam engines (for Sergo Ordzhonikidze, People's Commissar of Heavy Industry, and Felix E. Dzerzhinskiy, the founder of the Soviet secret service) and so on.

This was not so much a case of 'toadying' to the authorities but rather an expression of gratitude for the support which Yakovlev, as a non-professional (at that time) designer, was receiving from the ODVF and its successor, Aviakhim. Rykov was an honorary chairman of these organisations; considering that Yakovlev's early designs were mostly trainers and sports aircraft intended for operation by the air clubs of ODVF/Aviakhim, Rykov's support was very welcome indeed. Contrary to persistent rumours, Yakovlev was in no way a relative of Rykov.

Rykov was arrested on 27th February 1937 and executed, becoming one of the victims of Stalin's purges. Of course, the AIR abbreviation immediately became 'taboo'. It was replaced with Ya – the last letter of the Russian alphabet and the first letter of Yakovlev's last name – for the aircraft already built (Ya-1, Ya-2 and so on). New designs likewise received the Ya designator (Ya-20, Ya-21 and so on) or were referred to simply as 'No.20' (or 'aircraft No.20'), 'No.21' and so forth.

There is some uncertainty as to which of Yakovlev's design was the last to bear the AIR abbreviation. According to the well-known Soviet aviation historian and aircraft

The members of Aleksandr S. Yakovlev's original design team in 1932. Seated, left to right, are Vadim V. Barsookov, Sergey D. Trefilov, Ivan S. Ivankovich and Nikolay D. Savitskiy. Standing behind them are Viktor V. Alekseyev, Gheorgiy S. Lekanov, Vasiliy I. Chubukov, Gleb V. Sedel'nikov, Yevgeniy G. Adler, Viktor N. Yefimov, Pyotr A. Belyayev, Konstantin V. Sinel'shchikov and Viktor A. Staurin.

designer Vadim B. Shavrov, it was the AIR-19 (a passenger aircraft derived from the AIR-17/UT-3 trainer), but this name must have been abandoned at a very early stage, since the aircraft was not completed before late 1938. Subsequent types could not possibly have borne the AIR prefix, and designations such as AIR-20, -21, -23, -24 that crop up in some publications are wrong.

However, 'ya' is not only the last letter of the Russian alphabet but also the pronoun 'I' in the Russian language. Perhaps someone decided eventually that using this 'oh marvellous me' designation prefix was unbecoming; or, equally possibly, someone decided to borrow the German system of using the first two letters of the company name for model designations (Ju for Junkers, Do for Dornier and so on). Anyway, in 1941 the Soviet Union switched from the previously used principle of designating aircraft by their function to a new system which is still in use today, namely using the first two letters of the Chief Designer (or the OKB's founder) as the designation prefix. Thus the letter 'Ya' (Я) gave place to the well-known 'Yak' (Як) brand.

Upon leaving the VVA Yakovlev gained employment at Plant No.39 as engineering supervisor on the staff of Nikolay Nikolayevich Polikarpov, a famous Soviet designer who had been arrested by OGPU (Soviet security) in 1929 and continued his aircraft design activities in detention (the now famous Hangar 7 at this plant was turned into an 'internal prison'). Having gained an acquaintance with the Polikarpov I-5 fighter powered by the M-22 (licence-built Bristol Jupiter) engine, Yakovlev and his team decided to make a much faster aircraft fitted with the same engine. The resulting AIR-7 two-seat monoplane, first flown in November 1932, outpaced the I-5 fighter, exceeding the magic figure of 300 km/h (186 mph). Soon thereafter it set a national sped record at 332 km/h (206 mph).

However, this was followed by an unexpected setback. During a demonstration flight the AIR-7 suffered aileron flutter, and one aileron nearly came off. Thanks to Piontkovskiy's skill the flight ended in a textbook forced landing, but for Yakovlev this event started off a period of disgrace which might well have ended in a permanent ban on further work as aircraft designer. He and his team became virtual outcasts, being deprived even of access to the factory where they worked. Yakovlev fought hard to clear himself and his team and regain the right to engage in design activities. In this he eventually succeeded, aided in no small part by the favourable impression that his AIR-6 cabin four-seater had produced on one of the Communist Party senior officials,

Yakovlev OKB

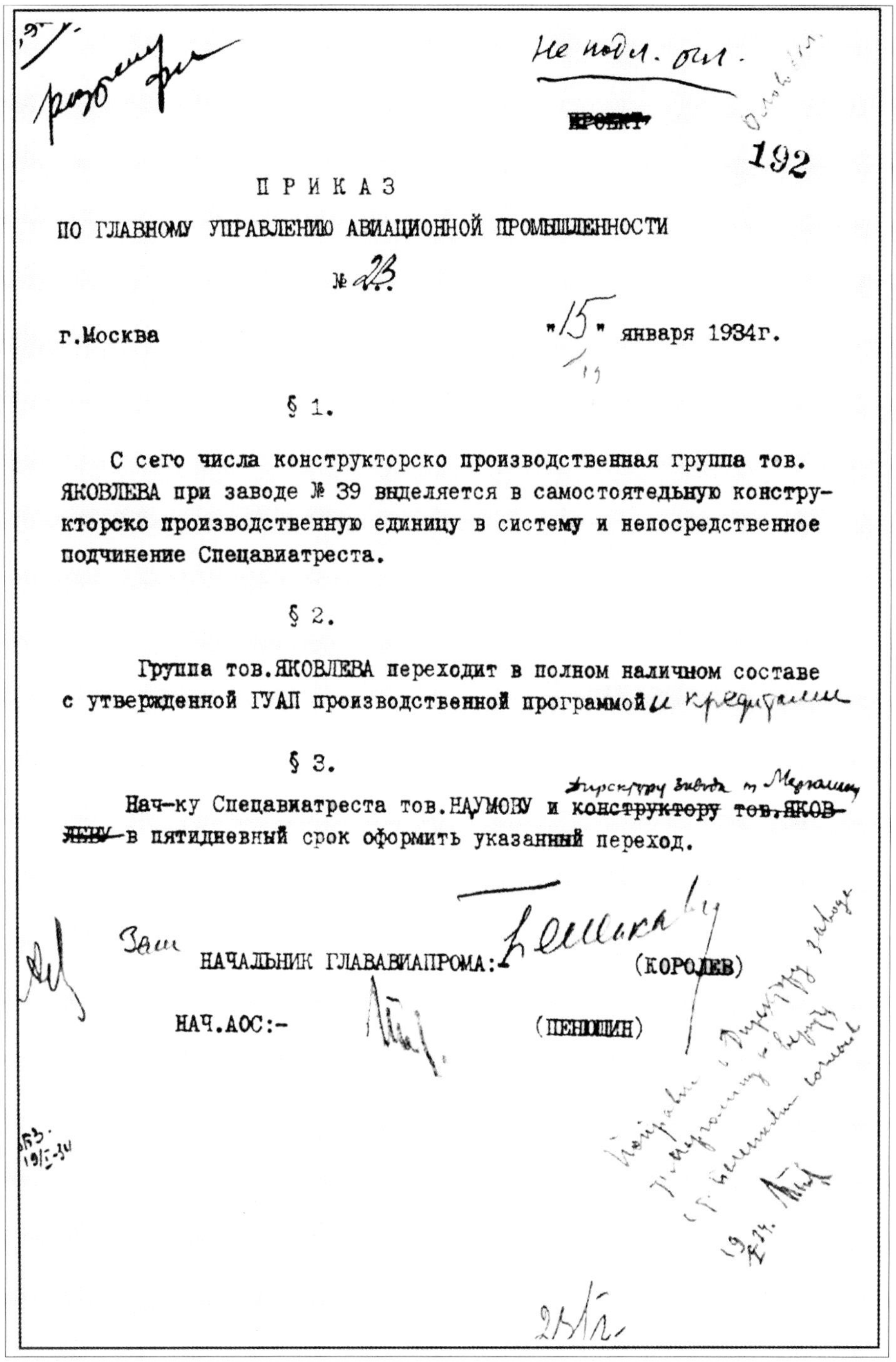

ПРОЕКТ

192

ПРИКАЗ

ПО ГЛАВНОМУ УПРАВЛЕНИЮ АВИАЦИОННОЙ ПРОМЫШЛЕННОСТИ

№ 23

г.Москва «15» января 1934г.

§ 1.

С сего числа конструкторско производственная группа тов. ЯКОВЛЕВА при заводе № 39 выделяется в самостоятельную конструкторско производственную единицу в систему и непосредственное подчинение Спецавиатреста.

§ 2.

Группа тов.ЯКОВЛЕВА переходит в полном наличном составе с утвержденной ГУАП производственной программой и кредитами

§ 3.

Нач-ку Спецавиатреста тов.НАУМОВУ и директору завода т. Моргалину ~~конструктору тов.ЯКОВЛЕВУ~~ в пятидневный срок оформить указанный переход.

Зам НАЧАЛЬНИК ГЛАВАВИАПРОМА: (КОРОЛЕВ)

НАЧ.АОС:- (ПЕНЮШИН)

Order No.23 of 15th January 1934, signed by Chief Directorate of Aircraft Industry (GUAP) head Sergey P. Korolyov and Experimental Aircraft Construction Department chief Penyushin. '1. Comrade Yakovlev's design/production group at plant No.39 is forthwith transformed into a separate design team reporting directly to the Special Aviation Trust. 2. The group is transferred in full in accordance with the production programme [*and credits*, added in handwriting] endorsed by GUAP. 3. Special Aviation Trust chief Comrade Naumov and plant director Comrade Morgalin shall fill out all required paperwork within five days.'

Yan E. Rudzutak who headed the Party's Central Control Commission. On 15th January 1934 the Chief Directorate of Aircraft Industry (GUAP – ***Gla**h**v**noye oopravleniye aviatsionnoy promyshlennosti*) issued an order permitting Yakovlev to set up a shop in a derelict bed factory on Leningradskiy Prospect avenue in Moscow. His team was given the status of a separate design bureau subordinated to the Special Aviation Trust.

The OKB is born. From Peace to War.

The new premises were hardly suitable for aircraft design and construction, and getting them into shape cost much effort, but Yakovlev and his growing team went on producing new aircraft. This was duly recognised by the Aviation Trust which in 1935 voted funds to build a real aircraft factory and a design office. The old sheds and huts were pulled down and a new building with a floor area of 1,500 m^2 (16,130 sq ft) was

Above: The first group of Yakovlev OKB employees to receive Soviet government awards in the Kremlin in April 1939 when they received their orders and medals.

erected. Yakovlev found himself the head of an officially named OKB (Experimental Design Bureau) with a small associated State Aviation Factory No.115 (hence the name OKB-115 under which Yakovlev's design bureau was subsequently known).

For some time light sporting, trainer and utility aircraft remained the new OKB's predominant line of activity. Yakovlev scored notable success in this field in 1937-38 in producing two types which were put into series production and built in large numbers. These were the AIR-14/UT-1 single-seat trainer for fighter pilots and the 'aircraft No.20', or Ya-20, which entered production as the UT-2, a standard pilot trainer. Emboldened by his growing experience, Yakovlev ventured into the domain of combat aircraft design. His first attempt at producing a combat aircraft was unorthodox enough. In 1939 he designed a twin-engined aircraft designated 'aircraft No.22', or Ya-22, the concept of which centred around the idea of attaining a speed record. Indeed, the prototype tested in February 1939 reached an outstanding speed of 567 km/h (367 mph), far in excess of what had been previously attained in the Soviet Union. However, the all too one-sided emphasis on speed, coupled with neglecting the purely operational aspects of the machine, made the aircraft ill-suited for the planned combat roles of a short-range bomber, escort fighter and reconnaissance aircraft. Albeit placed into series production as the Yak-2 and, in its later guise, as the Yak-4, the aircraft proved disappointing and was manufactured only in small numbers, making virtually no impact on aerial warfare on the Eastern Front after the outbreak of the war with Germany.

In 1939-40 Yakovlev was able to expand his OKB by adding a new 1,300-m^2 (13,980 sq ft) building for the assembly of prototypes. Several important designers and engineers joined him at this time, but his OKB was still modest in size. It comprised 45 designers and draftsmen and 152 construction workers with 44 machine tools. With these rather limited resources Yakovlev embarked on another task in the field of combat aircraft design. This time it was a fighter project. Initially known as the I-26 and later redesignated Yak-1, it proved to be a basically sound design which started a long line of progressively improved fighters. The Yak-1 gave birth to such derivatives as the Yak-7, Yak-9 and Yak-3 forming a distinguished family of aircraft that made Yakovlev famous the world over. Overall production of these fighters exceeded 37,000 – more than any other single basic design.

The war years were for Yakovlev the years of hectic work associated with an incessant race for improving his fighters so as to make them capable of meeting their adversaries on equal terms. At the same time he had to ensure a continuous and growing output from production factories – a formidable task, bearing in mind the dislocation caused by the German invasion and the resulting redeployment of enterprises eastwards. Finally, the enormous burden he had to shoulder was compounded by the fact that in January 1940 Yakovlev was appointed Deputy Commis-

The premises of the Yakovlev Design Bureau in Moscow as they looked in 1940, shortly after Yakovlev's design team became an independent entity (OKB-115). The outbreak of the war and the German advance on Moscow soon forced the OKB to vacate these premises, evacuating to Novosibirsk in September/October 1941, but the OKB returned home in 1944.

sar of Aviation Industry and Head of the Department of Experimental Aircraft Construction. He was thus responsible not only for his own design and production problems but also for those of the aircraft industry as a whole. Yakovlev had to manage much of the planning for moving aircraft plants from threatened areas to safer locations. An added element of urgency was to try to get everything relocated before the onset of the Siberian winter.

Yakovlev's own OKB was evacuated in September/October 1941 to distant Novosibirsk. The huge NKAP plant No.153 erected there in the pre-war years became the OKB's new home. Only a skeleton staff was left at Leningradskiy Prospekt avenue in Moscow where it dealt mainly with the rebuilding of battle-damaged Polikarpov I-15 and I-16 fighters and the modification of Hawker Hurricanes supplied under the Lend-Lease Agreement with heavier armament.

On 18th September 1942 the Yakovlev OKB was collectively awarded the Order of Lenin. In the autumn of 1944, the threat of Moscow's occupation having receded, the evacuated OKB progressively returned, and the Leningradskiy Prospect site was back at full strength by late November of that year.

In 1946 Yakovlev was promoted to the rank of Colonel-General of Aviation, and also made a Deputy of the Supreme Soviet (roughly corresponds to Member of Parliament). He was able to resign his post of Deputy Commissar at the People's Commissariat of the Aircraft Industry (NKAP – *Na**rod**nyy komissari**aht** aviatsi**on**noy pro**mysh**lennosti*; soon to be renamed MAP – *Mini**ster**stvo aviatsi**on**noy pro**mysh**lennosti*, Ministry of the Aircraft Industry) and return full-time to his now very large OKB. In June of that year he signed an order for the organisation of a new department for Science and Research. Later that year the original two-storey main building was reconstructed and a third storey added, increasing the floor area of this building to 2,000 m^2 (21,500 sq ft).

Yakovlev OKB

Aleksandr S. Yakovlev in the immediate post-war years in his Colonel-General's uniform. Note the Gold Star Order that went with his Hero of Socialist Labour title.

The Jet Age Sets In

The advent of the jet age was marked by the participation of the Yakovlev OKB in the efforts aimed at creating the Soviet first-generation jet fighters. Several design bureaux were engaged in this task, OKB-155 headed by Artyom Ivanovich Mikoyan and Mikhail Iosifovich Gurevich and OKB-301 led by Semyon Alekseyevich Lavochkin being among the contenders. Yakovlev chose a cautious and conservative approach, producing his first jet fighter, the Yak-15, as a more or less straightforward adaptation of his piston-engined Yak-3 fighter to take a captured German turbojet. This machine and its derivative, the Yak-17, played their useful role in making easier the conversion of Soviet Air Force pilots to jet aircraft. These two fighters, as well as the subsequent Yak-19, Yak-23 and Yak-25 (1947; the first fighter to be thus designated), featured the traditional straight wings, reflecting Yakovlev's initially cautious attitude towards swept wings. Yakovlev's first swept-wing fighter, the Yak-30 of 1948 (again the first to be this designated), lost out to Mikoyan's successful MiG-15; this placed the design bureau led by Mikoyan and Gurevich in an advantageous position which it succeeded in preserving subsequently. In the 1950s the Yakovlev OKB produced several prototypes and projects of single-engine front-line jet fighters, notably the Yak-50, but failed to gain production orders, losing again to Mikoyan's designs.

The OKB's activities were not confined to developing fixed-wing aircraft. In 1946 Yakovlev formed a special team to take on the problem of helicopter design. That was a hard nut to crack. Two small machines (the EG featuring contra-rotating rotors and the single-rotor Yak-100) were built in the late 1940s, providing the OKB with some experience in this field. This enabled the OKB to take up, in response to a government order, the challenging task of creating a tandem-rotor helicopter intended to carry 24 troops. As a result of arduous work the OKB eventually succeeded in tackling the numerous problems associated with the development of this machine. However, the Yak-24 was built in relatively small numbers, being eclipsed by the Mi-6 turboshaft-powered giant. The last helicopter project of the OKB was that of a huge twin-rotor machine, a tandem-layout counterpart of Mil's V-12.

In the late 1950s and the 1960s the Yakovlev OKB's activities in the field of combat aircraft design centred around the development of the Yak-25 twin-engined interceptor and its derivatives. The result was a large family of interceptors, tactical bombers and reconnaissance aircraft, many of which went into production and squadron service. Among these, the numerous versions of the Yak-28 supersonic aircraft occupy a prominent place.

A special chapter in the OKB's activities was opened in 1960 when it embarked on the development of vertical/short take-off and landing (V/STOL) attack aircraft. The Yak-36 technology demonstrator was followed by the production Yak-38 carrier-based aircraft. This subsonic aircraft had limited success, but it was to be superseded by a far more potent machine, the supersonic Yak-41M (better known as the Yak-141) which was created by Yakovlev's successors. Prototypes of this machine were successfully tested, but its development was brought to a halt by the sweeping changes in Russia's political life and economic situation in the early 1990s.

Throughout the post-war period, the OKB continued the design and development of sports, training and light utility aircraft drawing on the experience gained during the pre-war years. In this particular field the Yakovlev OKB established itself in a pre-eminent position with a long line of trainers for the Air Force and the Civil Air Fleet, supplemented by several successful aerobatic aircraft designs. Here mention must be made of the Yak-18 basic trainer, with its Yak-18U and Yak-18A tricycle-gear versions, and of the Yak-18P, 'PM and 'PS and Yak-55 aerobatic aircraft. They were superseded in due course of time by the equally successful Yak-50 and Yak-52 machines, of which the latter remains in operation to this day. The Yak-18T cabin monoplane was a successful trainer/utility aircraft, and the Yak-30 and Yak-32 marked the OKB's venture into the field of jet trainer design.

Yet another direction in the activities of the OKB was represented by passenger aircraft. The first attempt by Yakovlev to create a twin-engined airliner dates back to 1938 with the Ya-19 six-seater. The Yak-8 and Yak-16 followed up this line, albeit these basically sound machines did not go into production. A revival of this direction came with the decision to embark on the development of a three-turbofan airliner for local services. Designated Yak-40, this machine, seating anything from 24 to 32 passengers in airline configuration, was built in considerable numbers and made Yakovlev's name popular among Aeroflot's passengers on short-haul routes; in addition, it enjoyed some success with export orders. The Yak-40 became a starting point for more ambitious programmes of the same kind. Next in the line was the Yak-42 medium-haul airliner which looked like a scaled-up Yak-40. Despite some vicissitudes, this machine achieved a measure of success and remains in operation to this day both in Russia and in a number of other countries.

In 1982 the OKB ventured into a completely new field of activity by establishing a research and design team for unmanned aerial vehicles. They were intended for use as targets or, in most cases, as tactical reconnaissance platforms. This effort was initially managed by Sergey Aleksandrovich Yakovlev, the General Designer's elder son. Later he was transferred to other programmes, and Yuriy I. Yankevich took over as UAV programme manager.

In 1956 Chief Designer Aleksandr S. Yakovlev, as well as several of his colleagues, was conferred the title of General Designer. He held this post until 1984.

In 1966 the OKB-115 headed by General Designer A. S. Yakovlev was renamed MMZ ***Sko**rost'* (Speed; MMZ = *Mos**kov**skiy ma**shin**ostro**itel**'nyy za**vod***, Moscow Machinery Plant). On 21st August 1984 Aleksandr S. Yakovlev retired; in the official capacity of a consultant he continued for some time to take part in the work of the Design Bureau. He died on 22nd August 1989 in retirement in the age of 83. After his demise the enterprise came to be officially known as MMZ Skorost' named after A. S. Yakovlev. Aircraft created by this enterprise continue to bear the Yak brand. This was the final tribute to the man whose achievements had been richly rewarded in his lifetime. Suffice it to say that he had been awarded the Order of Lenin, the Soviet Union's highest civic order, ten times, to say nothing of the numerous state prizes and other high awards.

Yakovlev as a person

Aleksandr S. Yakovlev was a multi-faceted personality. Undoubtedly, he was one of the most prominent figures in the history of the Soviet/Russian aircraft industry. His wartime series of piston-engined Yak fighters brought him world renown; after the war his name was associated above all with the successful series of sports and training aircraft and the popular Yak-40 airliner. During his lifetime and long afterwards he was given much publicity (to which he had contributed through the numerous editions of his memoirs titled first *A Designer's Notes* and then *A Life's Goal*. Needless to say, this publicity created a somewhat idealised image of the designer with an emphasis on his merits and achievements. In recent years, some of his associates in their memoirs have portrayed Yakovlev as a personality endowed both with great talents and with certain human failings. These frank and realistic accounts contribute to a better understanding of the man whose work has set an indelible imprint on an important branch of technology in Russia. To quote Yevgeniy G. Adler, a designer who had worked with Yakovlev for several decades, Yakovlev had to his credit:

a sober, clear mind aided by a wealth of expert knowledge in the most diverse fields;

faultless, discerning taste as regards shape and colour;

a strong, indomitable will in reaching the goals that he had set for himself, a will to

Yakovlev OKB

Aleksandr S. Yakovlev with the legendary Marshal Semyon M. Budyonnyy (left) and Air Marshal K. A. Vershinin at a Kremlin reception in the 1980s.

surmount the inevitable obstacles and bring a project to fruition;

the ability to combine a fine aesthetic taste with a technically and scientifically sound and reasonable approach;

a disposition to order in everything and no mean abilities as an organiser;

charisma and sense of humour compensating for stern attitude to faults.

On the minus side, one notes Yakovlev's temper which sometimes prompted him to be overly harsh, even brutal in criticising mistakes made by his subordinates. In his later years, Yakovlev displayed a tendency to rely on the advice and opinions of a narrow circle of persons whose motives were not always unselfish.

Yet, in making the overall appraisal, one must recognise Yakovlev's indisputable ability to organise the work of his design staff, which won him great authority among the OKB employees. He is remembered as an exacting person, making succinct remarks always to the point, intolerant of incompetence and carelessness, capable of promptly taking the necessary decisions in a difficult situation.

It was characteristic of Yakovlev that he always exercised tight control over his Design Bureau and resisted suggestions that the organisation be 'departmentalised' into sections dealing with specific directions of design (he was obviously wary of the possibility of such sections eventually gaining independence). Yet, the success of his activities is in no small degree due to the fact that he had been able to attract enthusiastic and competent designers to form a united team. Among these, mention must be made of Leon M. Shekhter who was responsible for evolving the general layout of many of Yakovlev's aircraft. Yakovlev's close associates included, to name just a few, G. K. Sinel'shchikov, Oleg K. Antonov (the future General Designer of GSOKB-473), Yevgeniy Adler, Aleksandr Sinitsyn, Longin Lis, Igor' A. Erlikh, Leonid L. Selyakov, Yuriy I. Yankevich, Dmitriy K. Drach, Vyacheslav P. Kondrat'yev and others.

Yakovlev had two sons both of whom became aircraft engineers and worked on the staff of the OKB led by their father. Sergey, the elder son, was eventually promoted to the rank of Deputy General Designer and was directly responsible for the work on the Yak-40 and Yak-42 airliners; Aleksandr, the younger son, directed the work on the Yak-52 trainer.

Yakovlev OKB

Aleksandr S. Yakovlev during an informal session with high-ranking Soviet Air Force officers in the 1980s.

The OKB with Yakovlev's successors at the helm

In 1989-92 the Soviet aircraft industry underwent a process of reorganisation which was effected through orders issued by the Government's Department of Aircraft Industry. Design bureaux were again renamed; this was followed by a process of transforming these state enterprises into joint-stock companies (a process that swept over the whole of the Russian industry at that time). As a result, the MMZ named after A. S. Yakovlev was transformed into the 'OKB named after A. S. Yakovlev' Joint-Stock Co. (for the sake of convenience it will still be referred to hereinafter as the Yakovlev OKB).

Further changes in the aircraft industry included the formation of associations comprising design bureaux and production plants. As a part of this process, in 1992 the Yak Aircraft Corporation was formed as a management structure. In addition to the Yakovlev OKB, it included the Saratov and Smolensk production factories (outside the corporation were production factories in Tbilisi, Ulan-Ude and Irkutsk which were also engaged in the manufacture of the Yakovlev-designed aircraft).

In the period between 1985 and December 1990 the Yakovlev OKB was headed by A. A. Levinskikh as acting General Designer. He had taken part, among other things, in the development of the Yak-38 and Yak-141 V/STOL aircraft. The work on the latter project became the OKB's main task under Levinskikh.

In December 1990 A. A. Levinskikh was succeeded in the capacity of General Designer by Aleksandr N. Dondukov who thus became head of the 'Yakovlev OKB' JSC.

A few words about Aleksandr Dondukov

Aleksandr N. Dondukov was appointed General Designer of the Yakovlev OKB at the age of 36. Dondukov had started his career in aviation after graduating from the engine faculty of the Moscow Aviation Institute, whereupon he worked for several years at the Mikoyan OKB, dealing with the development and testing of the MiG-29 fighter. In December 1984 he was transferred to the Yakovlev OKB where a deputy Chief Designer well versed in engine technology was urgently needed. Three years later Dondukov became project manager for the Yak-141 VTOL aircraft which at that time accounted for about 70% of the OKB's work. That played its part in the decision to place him at the head of the OKB in December 1990.

For the following ten years Dondukov led this enterprise under the turbulent and harsh economic conditions associated with the sweeping political and economic changes in Russia. Apparently he displayed to advantage his qualities and capabilities as industrial leader. When in May 2000 the Ministry of Industry, Science and Technology was set up in Russia, he was appointed the new Minister, while retaining for some time his posts as General Designer and Chairman of the Board of Directors of the Yakovlev OKB JSC.

Somewhat later Dondukov left the Yakovlev OKB, concentrating on his job as minister. New people came to lead the Yakovlev Company. As of 2001, Nikolay N. Dolzhenkov held the post of technical manager of the enterprise, Oleg F. Demchenko being General Director.

In April 2004 the Yakovlev Design Bureau became a part of the Irkut Corporation (until 2002 known as the Irkutsk Aircraft Production Association, renamed Irkut Science and Production Corporation in December 2002). The Irkut Corporation acquired a

Yakovlev OKB

A. A. Levinskikh, acting General Designer of the Yakovlev OKB in 1985-1990

75.4% stake in the Yakovlev OKB which, thus, retains its identity. It is to be hoped that this merger will enhance the Yakovlev Design Bureau's financial stability and ensure favourable conditions for its design activities and series manufacture and marketing of its aircraft.

The merger of the two organisations within the framework of a holding was reflected in appointments to their top management posts. In June 2004 Aleksey Fyodorov, President of the Irkut Corporation, was elected Chairman of the Board of Directors of the Yakovlev Design Bureau JSC. In the same month Oleg Demchenko, General Director of the Yakovlev Design Bureau JSC, was elected Chairman of the Board of Directors of the Irkut Corporation.

Yakovlev OKB

A. N. Dondukov, who became General Designer of the Yakovlev OKB in December 1990.

Some facts on Oleg Demchenko

Oleg Fyodorovich Demchenko, General Director of the Joint Stock Society 'OKB named after A. S. Yakovlev' was born on 13th October 1944 in Kazakhstan, then part of the USSR. In 1968 he graduated from the Kuibyshev Aviation Institute (KuAI; at present the Samara State Aerospace University named after Andrey N. Tupolev) and started his work at various enterprises of the aircraft industry. In 1981 he was promoted to a post in the Ministry of Aircraft Industry where he was appointed head of a Chief Directorate and became a member of the Board – the consultative body under the Minister.

In 1987 Demchenko graduated from the Academy of National Economy. In 1992 he left the Ministry to take up a job in the Yakovlev OKB where he held the posts of First Deputy General Designer and General Director of the production facility in 1992-94.

In the period between 1994 and 2001 Oleg F. Demchenko held the posts of President and General Director of the 'OKB named after A. S. Yakovlev' JSC. Since 2001 he occupies the posts of President and Chairman of the Board of Directors. Since 2003 Demchenko is General Director of the Joint Stock Society 'OKB named after A. S. Yakovlev'. Since 26th June 2004 he concurrently holds the post of Chairman of Board of Directors of the Irkut Science and Production Corporation.

During his work in the Yakovlev OKB Demchenko took part in the projecting and development of such aircraft as the Yak-42D (2001 model), Yak-42A, Yak-112, Yak-48, Yak-152, Yak-52M, Yak-58, Yak-77, Yak-242, Yak-130, unmanned aerial vehicles.

Oleg F. Demchenko took an active part in evolving the Federal development programmes for civil aviation and defence branches of industry.

As noted above, in the period after Aleksandr S. Yakovlev's retirement the activities of the OKB were concentrated to a large extent on the programme of the Yak-141 V/STOL aircraft until it had to be abandoned in the early 1990s. An important programme in the military field was the Yak-130 jet advanced trainer, under development since 1990. It remains one of the OKB's main assets at present. The drastic reduction of budget funding of the OKB's work and state-financed purchases of military aircraft compelled the Yakovlev OKB, as well as other Russian design bureaux and production aircraft factories, to embark on the programmes of the so-called *kon**ver**siya*. This Russian term denoted a reorientation of defence industry enterprises towards commercial products,

Yakovlev OKB

O. F. Demchenko, General Director of the Yakovlev OKB Joint-Stock Co. in 2001

not necessarily aircraft. The Yakovlev OKB strove to retain to the utmost possible degree its aviation specialisation and well-established positions in the field of aircraft design by shifting the emphasis to civil aircraft – not an easy task, given the limited resources of Russian air carriers and the ever-increasing competition from abroad. In addition to upgrading the Yak-40 and Yak-42 airliners, the OKB set about designing a whole series of new civil aircraft intended for medium- and long-haul routes, such as the Yak-42M, Yak-242, Yak-46 and, lately, the MS-21. Concurrently, several new types in the utility, business and short-haul airliner class were developed, including the Yak-58, Yak-112, Yak-48 and Yak-77. Piston-engined trainers and sports aircraft, such as the Yak-54, Yak-52M and Yak-152, are also part of the OKB's activities. Further work is being done on UAVs.

Lastly, one must mention the OKB's involvement in the recent years' discussions around the fifth-generation fighter development programme. It was officially named PAK FA (*Perspek**tiv**nyy aviatsi**on**nyy **kompleks** fronto**voy** avi**aht**sii*, Advanced airborne system for the front-line aviation). The AVPK Sukhoi (the Sukhoi Corporation) was chosen as the chief contractor for the implementation of this programme. There was some talk about a possible co-operation with the Mikoyan OKB and the Yakovlev OKB in this project. In 2002 Mikhail Pogosyan, the then head of the Sukhoi enterprise, stated that 'there was a complete mutual understanding with the Yakovlev OKB as a potential participant in the joint development of the aircraft'. However, the possible role of the Yakovlev OKB in this project remains uncertain.

THE LIGHTPLANES OF 1924-1943

Gliders, trainers, sports and utility aircraft

Makaka glider

In October 1923 Aleksandr S. Yakovlev, then a youth of 17, took part in the construction of the *Makaka* (Macaque) glider designed by Nikolai Anoshchenko for the first All-Union Glider Meet held in November 1923 in Koktebel'. This biplane glider was of an antiquated design akin to Otto Lilienthal's gliders, controlled by moving the pilot's body in much the same way as today's hang glider. During the tests conducted by Anoshchenko the Makaka made only a few short hops in tow; during one of these hops the glider stalled at a height of a few feet and overturned, suffering irreparable damage – luckily with no injuries to the pilot. The accident prevented the glider from participating in the competition.

AVF-10 conversion training glider

Impressions from his participation in the first glider meet at Koktebel' prompted Yakovlev to embark on the construction of his own glider. The work was done under his guidance by a group of his friends, aviation enthusiasts from his school, supported by the local section of the ODVF. The glider was intended for training flights as an intermediate stage between the initial training gliders and record gliders. It was a wire-braced monoplane with the two-spar constant-chord wings mounted atop the fuselage. The bracing wires were supported by a V-shaped tubular cabane above the wing centre section. The fuselage was a wooden truss formed by four longerons and rectangular frames with internal wire bracing, the whole being covered with fabric. The pilot sat ahead of the wing leading edge. The undercarriage comprised two V-shaped struts joined by a common axle with two lightweight wheels.

In September 1924 the glider took part at the second All-Union Glider Meet at Koktebel'. Its first flight took place on 15th September and was eminently successful. The glider was flown by many pilots and drew praise from all of them. It won recognition as one of the best conversion training gliders.

AVF-10*bis* initial training glider

This version of the AVF-10 was not completed. A photo published in the *Kryl'ya Rodiny* (Wings of the Motherland) magazine shows the machine under construction. Surprisingly, the layout was changed, the wings being perched on struts above the fuselage and braced by twin lateral struts.

AVF-20 conversion training glider

This glider was designed by Yakovlev for participation in the third All-Union Glider Meet at Koktebel' in 1925. It was based on the AVF-10 and again served for conversion from ab initio training gliders to record gliders. Compared to its predecessor, the AVF-20 incorporated some structural refinements. This time the wing outer panels were joined to a small centre section (on the AVF-10 the two wing halves were joined directly at the centreline and placed atop the fuselage). The wing spars were redesigned; the two-strut cabane was replaced by a three-strut one. Changes were made to the control system, thicker-section longerons were used in the fuselage structure.

The glider was built in the AVF workshops by Yakovlev's colleagues, engine mechanics of the AVF air squadron. It was flown at Koktebel' in September-October 1925 and, again, displayed good handling, became very popular with the pilots. After the contest Yakovlev was awarded a prize by

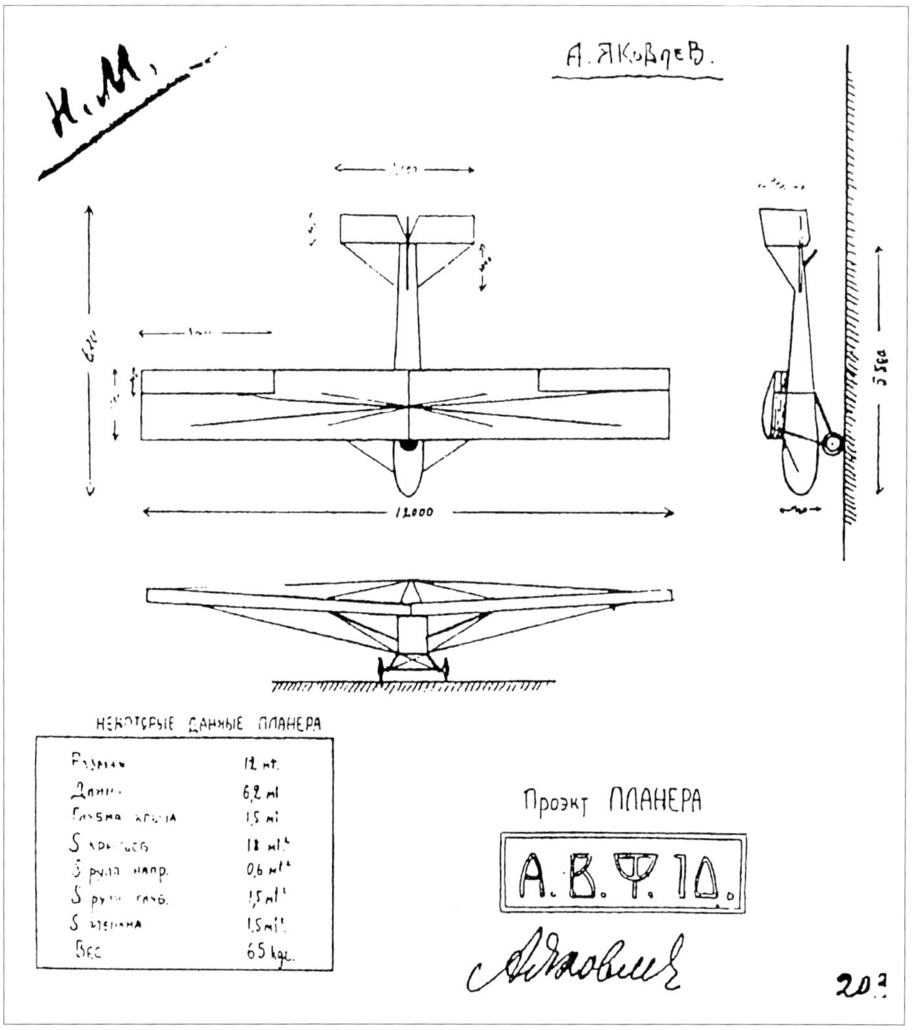

A project drawing of the AVF-10 signed by A. S. Yakovlev. The specifications include a wing span of 12 m (39 ft 4 29/64 in), a length of 6.2 m (20 ft 4 in), a wing area of 18 m^2 (193 sq ft) and a weight of 65 kg (143 lb).

Yakovlev OKB

Yakovlev OKB

Top and above: The AVF-10, Yakovlev's first glider, in the ground and in flight.
Below: The broadly similar AVF-20 glider as originally completed and flown with a short, blunt nose.

Yakovlev OKB

Yakovlev OKB

The AVF-20 flying in modified form with a longer and more pointed nosecone.

Aviakhim 'For the best training glider design'.

In 1927 the AVF-20 was used for training during the 4th All-Union Glider Competitions. It was damaged in a rough landing and had to be repaired; in the course of the repair work the glider received a more pointed nose.

AVF-34 record glider (project)

This hitherto unknown project of a record glider tentatively designated AVF-34 was prepared by Aleksandr S. Yakovlev in 1926 together with his AVF colleague A. A. Sen'kov and pilot G. A. Shmelyov. Recent research by Yuriy Zasypkin has unearthed some documents on this strut-braced high-wing glider which had a wing span of 13.5 m (44 ft 3½ in) and a length of 6 m (19 ft 8¼ in). In the initial project version the wings comprised a 5-m (16 ft 5 in) constant-chord centre section and tapering outer panels. In a revised version the outer wing panels were also made rectangular, with slightly cropped wingtips reducing the wing span to 13.0 m (42 ft 8 in). The glider was to have an aspect ratio of 8.6, a wing area of 19.5 m^2 (210 sq ft) and an AUW of 183 kg (403 lb).

AIR-1 (VVA-3, Ya-1) two-seat light aircraft prototype

This aircraft started its life under the designation VVA-3 (*Voy**enno**-voz**dooshnaya*** *aka**dem**iya*, Air Force Academy). At the time of its inception Aleksandr S. Yakovlev worked as assistant engine mechanic of the Academy's training air squadron. Encouraged by his success in designing and building gliders, Yakovlev decided to try his hand at the construction of a powered aeroplane and take part in a contest of light aircraft projects which had been announced by the Aviakhim voluntary society. Yakovlev was supported by a section of the OSO (***Ob**shchestvo **so**deystviya obo**ron**e* – Defence Enhancement Society) which had been set up at VVA; in 1927 it was transformed into a section of Osoaviakhim after the merger of the two voluntary organisations.

The VVA-3 project was developed to meet a specification issued by the Osoaviakhim section; it envisaged a military liaison light aircraft powered by a 40- to 60-hp engine, with a landing speed of 50 km/h (31 mph), a maximum speed of not less than 120 km/h (75 mph) and an endurance of three hours. The aircraft was built to accept a 60-hp Airdisco Cirrus M-1 four-cylinder in-line engine; its construction was performed by a group of enthusiasts headed by Yakovlev and it took eight months to complete the machine. By 1st May 1927 the finished aircraft was transferred to the Central airfield, where its first flight took place on

12th May with Yulian I. Piontkovskiy at the controls.

The VVA-3 was a single-bay biplane of wooden construction, with emphasis on the simplicity of design and weight saving; the airframe weighed only 335 kg (739 lb). Flight tests showed that the aircraft had good stability and controllability, presented no difficulties during take-off and landing and the performance exceeded the stipulated figures. The VVA-3 attained a maximum speed of 150 km/h (93 mph) and had a landing speed of 60 km/h (37 mph); the endurance was four hours.

In July 1927, at Yakovlev's initiative, the VVA-3 was sent on a long-range flight that spanned over the Moscow-Khar'kov-Sevastopol'-Moscow route. Before the flight, the machine received the civil registration R-RAIR. Under the then-current system, Soviet civil aircraft were registered in the R-Rxxx series (or, much less often, had the RR- prefix followed by three individual letters); in both cases the two R's denoted Russian Republic. Later, when new designs by Yakovlev came into being, this individual letter combination (AIR) used for the VVA-3 was chosen as their common designator as explained in the Introduction and the VVA-3 was renamed AIR-1.

The southbound leg of the flight was covered on 12th July with Piontkovsky at the controls and Yakovlev as a passenger, with an intermediate landing in Khar'kov. The return flight from Sevastopol' to Moscow was made non-stop by Piontkovskiy alone because the forward cockpit was occupied by an extra fuel tank. Struggling against a stiff headwind and rain, Piontkovskiy covered a distance of 1,420 km (883 miles) in the course of 15 hours 30 minutes. This was a world record as regards non-stop distance and endurance for the light aircraft category (the world distance record registered by FAI at that time was 868 km/539 miles). This world record remained unofficial because the Soviet Union became an FAI member only in 1935.

In September 1927 the AIR-1 took part in the role of a liaison machine in a Red Army exercise near Odessa. Despite inclement autumn weather, the aircraft faithfully performed liaison services between the headquarters and the units in the field, often making several flights a day to unprepared airstrips. The machine proved capable of sustaining prolonged operation under harsh conditions. Commissions set up at Osoaviakhim and the Military Air Academy came to the conclusion that the VVA-3 was the best among Soviet light aircraft and could be recommended for series production (which, however, failed to materialise). After the Odessa exercise the AIR-1 took part in various demonstration and propaganda flights, as well as training flights. Sadly, like most Soviet aircraft types of the pre-war period, it did not survive.

Yakovlev OKB

Yakovlev OKB

Top and above: The AIR-1, Yakovlev's first powered aircraft, as originally completed (with no markings whatever). The aircraft's clean lines are evident. Note the exposed wheel spokes.

AIR-1 replica

In 1977 the staff of the Yakovlev Design Bureau commemorated the 50th anniversary of the AIR-1's construction. To mark the occasion, enthusiasts from the staff built a replica of the historic aircraft, using photographs of the real thing and information from aeronautical publications of that period as the starting point. The pristine-looking aircraft without insignia (this was one of the AIR-1's guises) became a part of the Yakovlev OKB museum and was shown, together with other veteran Yakovlev aircraft, at Moscow-Tushino in August 1977.

AIR-2 (Ya-2) prototype two-seat lightplane with ADC Cirrus engine

The AIR-2 was built in 1928 and also bore the name *Pioner* (Pioneer) because its construction was partly financed by money raised by Leningrad's Young Pioneers youth organisation. The first prototype of the AIR-2 made its first flight in early October 1928.

Specifications of the AIR-1

Engine type	ADC Cirrus
Engine power, hp	60
Length overall	6.9 m (22 ft 7⅝ in)
Wing span	8.8 m (28 ft 10½ in)
Wing area, m² (sq ft)	18.7 (201)
Empty weight, kg (lb)	335 (739)
All-up weight, kg (lb)	535 (1,179)
Maximum speed at sea level, km/h (mph)	140 (87)
Landing speed, km/h (mph)	60 (37)
Time to 1,000 m (3,280 ft), min	8.0
Range, km (miles):	
normal	500 (311)
maximum	1,240 (770)
Take-off run, m (ft)	80 (260)
Landing run, m (ft)	60 (200)

Note: Some sources state a length of 6.99 m (22 ft 11¼ in) and a wing span of 8.85 m (29 ft 0½ in).

Yakovlev OKB

Above: The still-unmarked AIR-1 on skis during trials, with test pilot Aleksandr S. Yakovlev (wearing a pointed Red Army cap known as a *budyonnovka*) standing in front. The colour scheme has been revised.

Yakovlev OKB

The AIR-1 on wheels again, wearing a different colour scheme and the symbolic registration R-RAIR; in keeping with the international standard of the day the nationality prefix is repeated on the rudder. Note the addition of hub caps on the wheels.

Like its predecessor, it was powered by an ADC Cirrus engine. The AIR-2 was structurally and dimensionally identical to the AIR-1 (VVA-3), differing from the latter outwardly in having a revised engine cowling which was less bulbous in side view. Another point of difference was the deletion of wire braces between the front and rear pairs of cabane struts; they impaired access to the front cockpit, and it was decided to replace them with additional bracing struts between the wing and the fuselage forward of the front cockpit. Other changes included deletion of ailerons on the upper wings and a tailplane of revised planform. Several more examples of the AIR-2 were built, featuring different powerplant options.

Yakovlev OKB

The beautiful replica of the AIR-1 built to commemorate the 50th anniversary of the aircraft's first flight.

AIR-2 (Ya-2) prototype two-seat lightplane with M-23 engine

An example of the AIR-2 was fitted with a 65-hp M-23 (alias NAMI-65) three-cylinder radial engine developed at NAMI (*Naoochnyy* ***avtomotornyy*** *institoot* – Automobile and Engine Research Institute); its testing commenced in July 1931. This machine featured several aerodynamic improvements. One of these was the provision (for the first time in Soviet aircraft design practice) of automatic leading-edge slats on the upper wings; the lower wings were equipped with large-area slotted ailerons. Sometimes the cockpits of this machine were enclosed by a transparent sideways-opening canopy. The use of this canopy, however, had no marked effect on performance; moreover, the cockpit canopy was not popular with the pilots, all the more so since it could be opened only on the ground from outside.

This particular example of the AIR-2 was successfully tested for aerobatics and spinning; it showed good controllability and prompt spin recovery.

The M-23 engine suffered from severe vibrations; this prompted the designers to try other powerplants.

AIR-2 with Walter NZ-60 engine

A version of the AIR-2 powered by a 60-hp Walter NZ-60 five-cylinder radial engine possessed slightly lower performance characteristics as compared to the AIR-1 (VVA-3). A small batch of four machines was built. A floatplane version powered by the Walter engine was studied, but not built, the 60-hp output being obviously insufficient.

AIR-2S with Siemens engine (landplane version)

In 1931 a single AIR-2 was fitted with an 85-hp Siemens five-cylinder radial engine; this version was designated AIR-2S, the suffix letter standing for Siemens. The additional engine power led to a noticeable improvement in performance, making it possible to convert it into a floatplane (see below).

AIR-2S floatplane prototype

The floatplane version of the Siemens-powered AIR-2S was equipped with floats designed by Vadim B. Shavrov. The tests were conducted on the Moskva River on 18th May 1931. Pilot B. L. Bukhgol'ts, with V. B. Shavrov as a passenger, performed a take-off and then a 10-minute flight around the area. This was followed by several flights with Piontkovsky at the controls, Yakovlev being among the passengers. All flights proceeded without a hitch. On one occasion Bukhgol'ts ventured to fly under the Krymskiy bridge not far from which the aircraft had its mooring place.

AIR-2S replica from Nevinnomyssk

A replica representing the AIR-2S was built by a group of aviation enthusiasts from the ***Kryl'**ya* (Wings) technical club in the town of Nevinnomyssk, Stavropol' Region, in southern Russia. It was presented at a general aviation fly-in held at Myachkovo airfield near Moscow in June 2001. The aircraft was painstakingly built, using original technical drawings kindly supplied by the Yakovlev OKB. However, the original Siemens engine was unobtainable and had to be replaced by a restored Shvetsov M-11 five-cylinder radial that happened to be available. With its 160 hp, the M-11 was 50% heavier and had twice the power compared to the Siemens, but the designers managed to mate it to the airframe. The replica was successfully flown in Nevinnomyssk, attaining a maximum speed of 145 km/h (90 mph) at an AUW of 665 kg (1,470 lb). It was displayed again at the *Le**ta**yushchiye le**ghen**dy* (Flying Legends) veteran and vintage aircraft show held at Monino south of Moscow on 13th-15th August 2004 – unfortunately only statically, as beastly weather and a soggy grass runway made flying the replica too risky.

AIR-3 'Pionerskaya Pravda' long-range light aircraft

This two-seat parasol monoplane powered by a 60-hp Walter NZ engine was designed by Yakovlev, then a second-year student of the Air Force Academy, in early 1929. The work was undertaken in response to an order from Osoaviakhim which wanted to have a lightweight machine with a maximum possible range and endurance; the aircraft was to be used for long-distance flights and propaganda flights.

In designing the AIR-3, Yakovlev made use of some components from his preceding AIR-1 and AIR-2 machines. The fuselage, tail unit and undercarriage of the new machine were basically similar to the AIR-1 and AIR-2, but Yakovlev abandoned the biplane layout in favour of the monoplane configuration and stuck to it consistently in all his subsequent designs, the only exception being the experimental Yak-12B biplane version of the Yak-12A lightplane. The two-spar wings of the AIR-3 were placed above the fuselage on a cabane consisting of six wire-braced struts. On each side of the fuselage the wing was supported by a pair of bracing struts. Instead of the usual wire braces the wing was stiffened by 1.5-mm (0.059-in) plywood skinning extending from the top of the rear spar over the leading edge to the lower side of the front spar. The rectangular wing centre section housed three fuel tanks holding a total of 176 kg (388 lb) of fuel with gravity feed to the engine. The fuel load ensured a 14-hour endurance and a range of 1,700-2,000 km (1,057-1,243 miles). The outer wing panels (outboard of the wing strut attachment points) had a slight taper and rounded wingtips and their the trailing edge was occupied by ailerons.

The fuselage of rectangular cross-section was formed by four wooden longerons and housed two cockpits, the rear one being fitted with flight controls; the other cockpit was used by a passenger (observer).

The undercarriage comprising two mainwheels and a tailskid made use of the wheels from an old Hanriot aircraft. The wheels were mounted on a common axle supported by steel tube struts with wooden fairings.

Construction of the AIR-3 started in April 1929 and was effected partly by Plant No.39 named after V. R. Menzhinskiy (then located

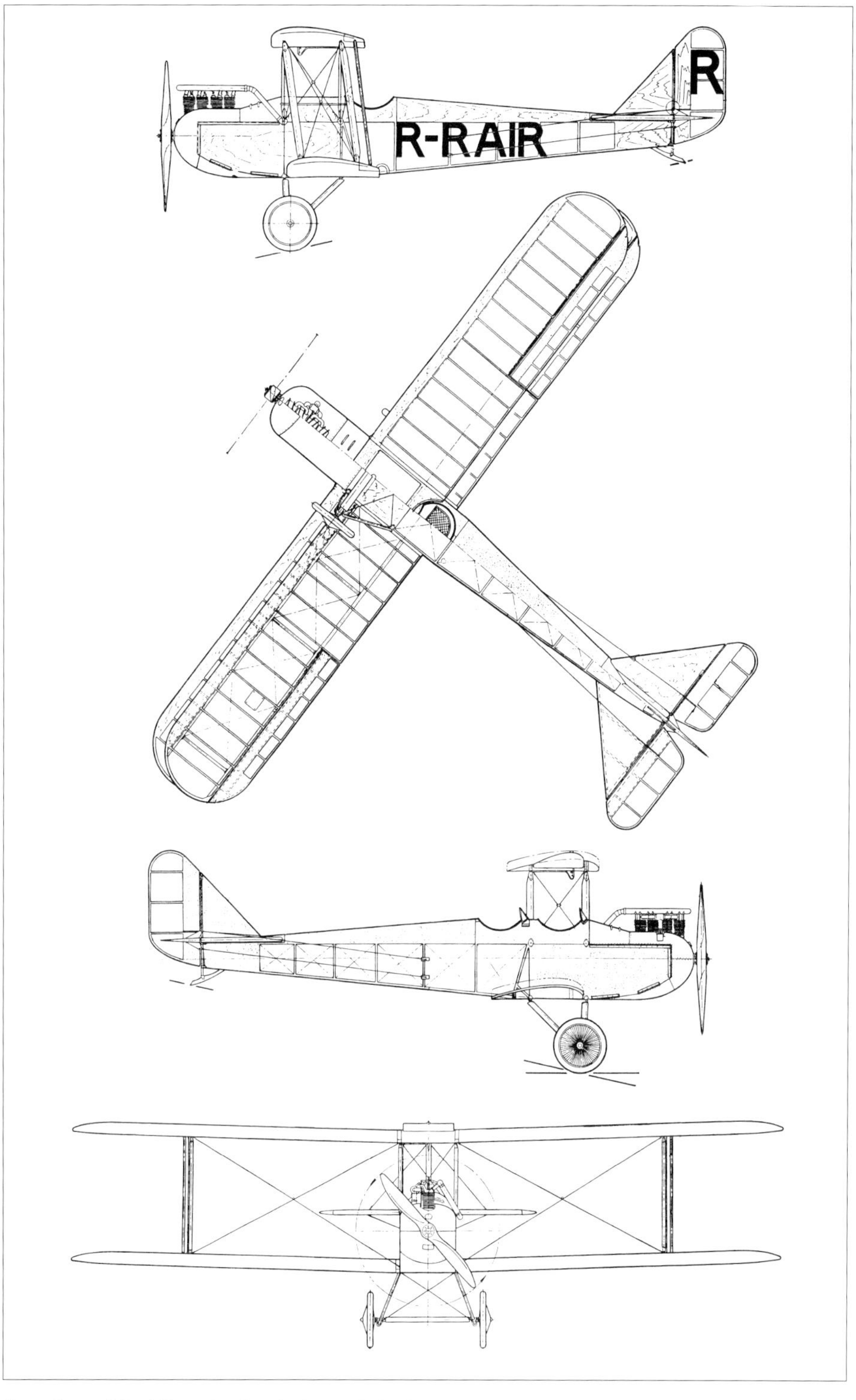

Four views of the AIR-1, showing the internal wire bracing of the fuselage and wing structure.

Yakovlev OKB

Yakovlev OKB

Top and above: The ADC Cirrus-engined AIR-2 prototype during tests. These views show clearly the slimmer engine cowling and the additional sloping cabane struts allowing the bracing wires to be deleted.

Yakovlev OKB

This AIR-2 has been fitted with an M-23 (NAMI-65) three-cylinder radial and an enclosed cockpit. Note the different landing gear design and the red/white striped tail that was to become a Yakovlev trademark.

Specifications of the AIR-2 versions

	AIR-2 with NAMI engine	AIR-2S (landplane)	AIR-2S (floatplane)
Engine type	M-23 (NAMI-65)	Siemens	Siemens
Engine power	65	85	85
Length overall	7.0 m (22 ft 11½ in)	7.0 m (22 ft 11½ in)	7.7 m (25 ft 3 in)
Wing span	8.8 m (28 ft 10½ in)	8.8 m (28 ft 10½ in)	8.8 m (28 ft 10½ in)
Empty weight, kg (lb)	403 (888)	420 (926)	Approx 470 (1,040)
All-up weight, kg (lb)	646 (1,424)	660 (1,455)	Approx 710 (1,565)
Maximum speed, km/h (mph)	141 (88)	150 (93)	140 (87)
Landing speed, km/h (mph)	60 (37)	60 (37)	65 (40)
Service ceiling, m (ft)	3,534 (11,594)	n.a.	n.a.
Range, km (miles)	540 (336)	n.a.	n.a.
Take-off run, m (ft)	80 (262)	n.a.	n.a.

in Moscow) and partly at the workshops of the Air Force Academy. It was financed by money raised by the Young Pioneers youth organisation; hence the aircraft was christened *Pionerskaya **Pravda*** (Pioneers' Truth), the name of a newspaper (a kids' version of the *Pravda* daily). The work was completed in a matter of three months, and flight testing began in July, in parallel with preparations for a long-range flight. Test pilots who flew the machine spoke highly of its handling qualities, noting in particular its excellent stability; the aircraft could be flown with hands off the stick. Quick reaction to all control inputs made the aircraft easy and pleasant to fly.

The flight tests were successfully completed on 17th August 1929. On 26th August the AIR-3 piloted by A. I. Filin, with A. F. Koval'kov as observer, started on a non-stop flight from Moscow to Mineral'nyye Vody, the route spanning over 1,750 km (1,087 miles). This distance was covered within 12 hours of flying time, the aircraft arriving at Mineral'nyye Vody on 1st September (an intermediate landing had to be made in Rostov because of an unexpected problem: the fuel was sucked out of the wing-mounted tanks through venting holes by the slipstream). The return flight took place on 6th September; this time the distance of 1,750 km was covered within 10 hours 23 minutes at an average speed of some 170 km/h (106 mph). Thereby Filin and Koval'kov established two unofficial world records: a distance record and an average speed record for Category I light aircraft (two-seat aircraft with an empty weight of up to 400 kg).

In September and October 1929 the AIR-3 bearing the alphanumeric registration CCCP-310 under the recently introduced new system was used in the Crimea as a utility aircraft for rendering all sorts of services (liaison, search and rescue and so on) during the glider competitions held there. Within two months the aircraft logged 70 flying hours, covering some 10,000 km (6,200 miles) in all, which was regarded as an excellent job performed under difficult field conditions.

AIR-3 with M-23 (NAMI-65) engine

In 1930 the AIR-3 was re-engined with a 65-hp M-23 (NAMI-65) engine developed by V. A. Dollezhal' at NAMI. This was the first Soviet light aircraft powered by an engine of indigenous design.

AIR-4 two-seat lightplane with Walter engine

The success of the AIR-3 prompted a decision to build a small batch of these aircraft, and five machines were laid down at plant

No.39. They featured a number of changes dictated by the conditions of series manufacture; in the new configuration the aircraft received the designation AIR-4. This aircraft was basically identical to the AIR-3, retaining all its aerodynamic features. The changes included a wider fuselage with more comfortable cockpits; access to them was made easier by providing doors in the sides. The ailerons had four hinges instead of three. The aircraft was provided with dual controls and a twin set of instruments. The modified undercarriage featured separate three-strut main units without a common axle. This undercarriage layout was typical for braced high-wing monoplanes of the period because it ensured a wide wheel track; it was also adopted for Yakovlev's subsequent high-wing aircraft.

The first two AIR-4s registered CCCP-311 and CCCP-312 and powered by Walter engines passed their initial tests in September 1930. Their performance was very close to that of the AIR-3. In order to test the aircraft under tough operational conditions, in October 1930 the two mentioned machines were sent on long-range flights from Moscow to the Crimea and back, in which they proved their robustness and dependability.

AIR-4 with M-23 engine

In the course of the abovementioned long-range flights of the AIR-4s problems were experienced with the Walter engines which proved capricious during the season's cold and rainy weather. In 1930 first the AIR-3 (see above) and then one of the AIR-4s was equipped with the new M-23 engine. With its higher rating (65-70 hp) this engine ensured improved performance, raising the speed by some 20 km/h (12.5 mph). However, this engine failed to reach production status.

Yakovlev OKB

Above: Aleksandr S. Yakovlev stands on the starboard float of the AIR-2S powered by a five-cylinder Siemens radial and fitted out as a floatplane, with B. L. Bukhgol'ts in the pilot's seat.

Yakovlev OKB

Above: Another view of the AIR-2S floatplane moored on the Moskva River during trials. Note the 'hammered effect' on the forward fuselage/engine cowling panels.

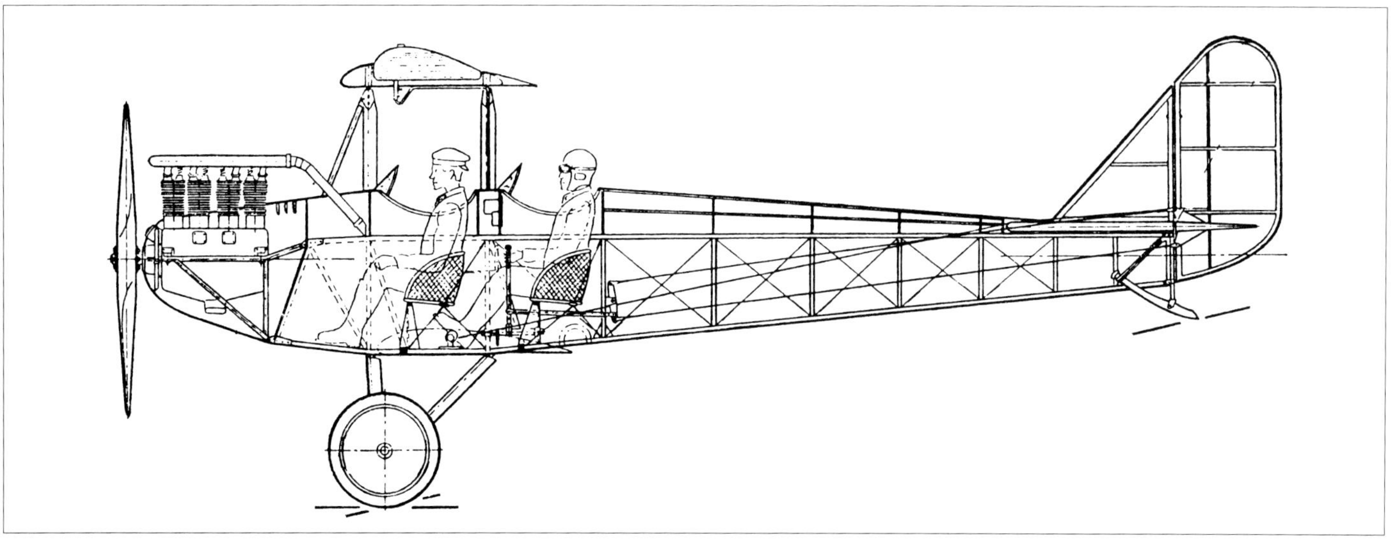

A cutaway drawing of the AIR-2 in Cirrus-engined configuration.

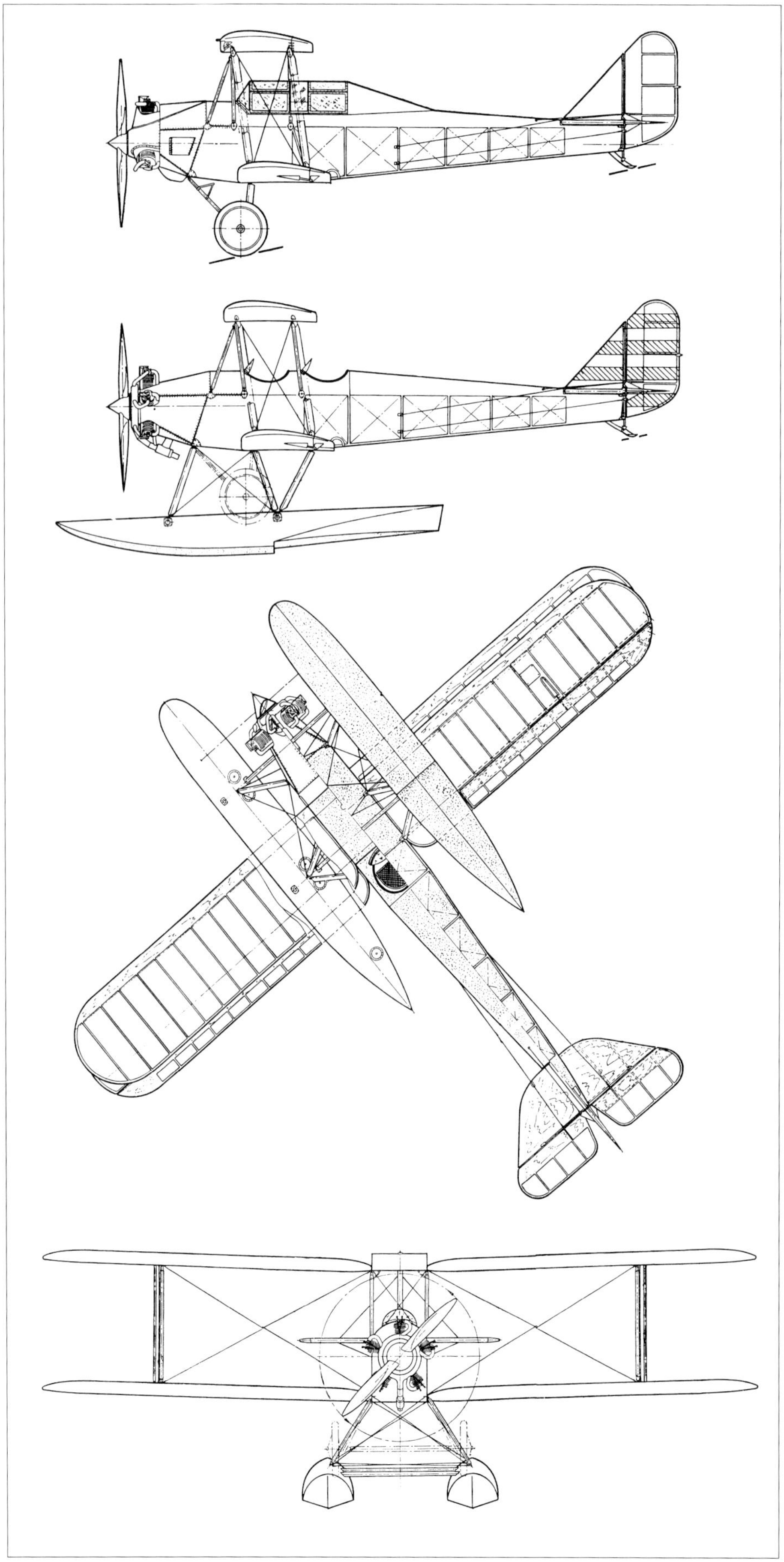

A four-view of the AIR-2S in floatplane configuration, the wheel undercarriage being shown by hatched lines, with an additional side view (top) of the experimental M-23-powered version with enclosed cockpits.

AIR-4MK (E-31) experimental aircraft

An AIR-4 was used by the Civil Air Fleet Research Institute (NII GVF – *Naoochno-issledovatel'skiy institoot Grazhdahnskovo vozdooshnovo flota*) for experiments aimed at reducing aircraft's landing speed through the use of various high-lift devices. Among the means envisaged was the use of full-span landing flaps. To make the greater part of the wing trailing edge available for these flaps, the normal ailerons were to be replaced by so-called 'floating' ailerons which were, in fact, the wingtips tilting around transverse axles.

To test this idea, NII GVF researchers modified an AIR-4 which is sometimes referred to as AIR-4MK (*mekhanizeerovannoye krylo* – 'mechanised wings', the Russian term of that period for wings with high-lift devices). The new wings were developed by B. N. Zalivatskiy and Leon M. Shekhter. The aircraft was also known as E-31, after its registration CCCP-Э-31 (that is, SSSR-E-31 in Cyrillic characters).

(Note: In 1932 the Soviet Union changed its civil aircraft registration system again. Under the new system used up to 1958, instead of the three digits used hitherto the CCCP- country prefix was followed by a letter designating the aircraft's operator plus up to four figures. In this case the operator designator is an E (for eksperimental'nyy – experimental) denoting NII GVF. Cf. CCCP-Лxxxx (that is, SSSR-L in Cyrillic characters, derived from *leeneynyy* [*samolyot*] – aircraft in airline service) denoting the Main Directorate of the Civil Air Fleet (GU GVF – *Glahvnoye oopravleniye grazhdahnskovo vozdooshnovo flota*) which operated scheduled passenger/cargo services. CCCP-Hxxx (the Cyrillic N) indicated the Main Directorate of the Northern Sea Route (GU SMP – *Glahvnoye oopravleniye severnovo morskovo putee*) which included the Polar Aviation branch, CCCP-Cxxxx (the Cyrillic S) stood for Osoaviakhim, CCCP-Axxxx for the agricultural division, CCCP-Kxxx (derived from *Krahsnyy Krest* – Red Cross) for the People's Commissariat of Health and so on. The rendering of the registrations as actually applied is used throughout.)

The wing span of the AIR-4MK was increased from 11.0 m (36 ft 1⁵⁄₆₄ in) to 12.5 m (41 ft 0⅛ in). Of these, 10.3 m (33 ft 9½ in) were occupied by the flaps. Outboard of the flaps, the wingtips, each of them having a span of 1.1 m (3 ft 7⁵⁄₁₆ in), were transformed into 'floating ailerons' with the axis of rotation close to their leading edges. The ailerons could be deflected differentially within ±15° for roll control; upon completion of the manoeuvre they switched to 'floating' mode, aligning themselves with the airflow. With these ailerons, the wing area was increased from 16.5 m² (177.6 sq ft) to 18 m² (193.8 sq ft). The aircraft's empty weight increased by 40 kg (88 lb) and the AUW by 45 kg (99 lb).

The tests undertaken from 1933 showed that the use of 'floating ailerons' was a basically sound idea. Thanks to the large flap area the landing speed was reduced from 66 km/h (41 mph) to an unprecedented 34 km/h (21 mph) at the cost of an insignificant drop in the maximum speed, which was reduced from 150 to 145 km/h (from 93 to 90 mph). One source claims, though, that the cruising speed was increased by 7%.

PS experimental aircraft ('transparent' version of the AIR-4)

The AIR-4 became involved in an experiment associated with the idea of making aircraft less noticeable in the air. An unorthodox solution to this problem was suggested by engineer Sergey G. Kozlov who headed the chair of aircraft design in the Air Force Academy. Kozlov believed that an aircraft might be made 'invisible' if the usual skinning were replaced by a transparent film. A transparent material called 'cellon' was proposed for this purpose (Shavrov mistakenly wrote about the use of a French material called *rhodoide*). The idea was applicable primarily to light aircraft with truss-type fuselages.

For starters, in 1934 Kozlov suitably modified an example of the Polikarpov U-2 trainer, replacing part of its fuselage skinning with cellon. The experiments were continued on an AIR-4, the testing of which started on 25th July 1935. The aircraft was subjected to considerable alterations. Not only the fabric wing skinning but also part of the plywood fuselage skin was replaced by the new material, which necessitated some structural changes. To make the structural members (wing ribs and spars) less noticeable, they received a coat of white paint containing aluminium powder. After the modification the AIR-4 was renamed PS – *pro**zrach**nyy samo**lyot*** (transparent aircraft).

Visual observations from the ground and from another aircraft confirmed the expectations, the 'transparent aircraft' becoming considerably less noticeable or even completely unnoticeable at some distance. Additional bonus was the excellent visibility for the crew afforded by the transparent skin of the cabins and the wing. Encouraged by the results, Kozlov proposed the idea of designing and building a special single-seat high-speed reconnaissance aircraft featuring cellon skinning. However, the use of this material on high-speed machines was considered dubious for reasons of strength. The Experimental institute headed by Pyotr I. Grokhovskiy was tasked with continuing the experiments in 1936, but no 'transparent aircraft' was built there.

Yakovlev OKB

Above: Aleksandr S. Yakovlev in military uniform beside the AIR-3 light aircraft.

Yakovlev OKB

Yakovlev OKB

Two views of the AIR-3. The parasol wings and the pointed nose give the aircraft a racy look.

Specifications of the AIR-3 and AIR-4 utility aircraft

	AIR-3	AIR-4	AIR-4MK
Engine type	Walter NZ-60	Walter NZ-60	Walter NZ-60
Engine power, hp	60	60	60
Length overall	7.1 m (23 ft 3½ in)	7.1 m (23 ft 3½ in)	7.1 m (23 ft 3½ in)
Wing span	11.0 m (36 ft 1 in)	11.0 m (36 ft 1 in)	12.5 m (41 ft 0⅛ in)
Empty weight, kg (lb)	392 (864)	395 (871)	440 (970)
All-up weight, kg (lb)	762 (1,680)	630 (1,390)	670 (1,480)
Maximum speed, km/h (mph)	146 (90)	150 (93)	145 (90)
Landing speed, km/h (mph)	66 (41)	66 (41)	34 (21)
Range, km (miles)	1,835 (1,139)	500 (311)	450 (280)
Take-off run, m (ft)	n.a.	n.a.	50 (164)
Landing run, m (ft)	n.a.	100 (330)	40 (130)

Above: An AIR-4 with the pre-1932 registration CCCP-310 as originally flown, with clear-doped tail surfaces. Note the fuel tank venting air pipes above the wing leading edge.

Above: The same aircraft following a repaint, with a red/white striped vertical tail, a lighter-coloured fuselage with a stripe and the registration in smaller digits.

Above: An AIR-4 'buzzes' the flight line at an air club's airfield with a VERY provisional hangar on the right.

A smartly painted and as yet unregistered AIR-4.

Some sources claim that the cellon fairly quickly lost its properties and became opaque under the influence of the sun's rays. Compounded by the dust and oil traces that soiled the film in the course of operation, this nullified the 'invisibility' effect.

'P' six-seat passenger aircraft with M-26 engine (project) (AIR-6, first use of designation)

In 1930 Yakovlev developed a project of a six-seat passenger aircraft designated 'P', powered by Bessonov's 300-hp M-26 radial engine. It also bore the designation AIR-6, which was later re-used for a different aircraft powered by an M-11 engine (see below). The aircraft was intended to carry two crew and four passengers plus 40 kg (90 lb) of luggage. The design performance included a speed of 230 km/h (143 mph), a ceiling of 5,250 m (17,225 ft) and a range of 700-750 km (435-466 miles) at an AUW of 1,810 kg (3,990 lb).

Four project versions were developed, but the aircraft remained on the drawing board, probably as a result of losing out to a competitor – A. I. Putilov's Stal'-2 passenger aircraft with a similarly rated engine.

'P' four-seat passenger aircraft with a 165-hp engine (project)

The 'P' project was re-worked to accept a 165-hp engine of Soviet manufacture. The number of seats was reduced to four, endurance was reduced to five hours as against six in the original version. The aircraft would have a comfortable automobile-style cabin and would feature detachable wings for ease of transportation by rail. This project became a stepping stone on the way towards the AIR-5.

AIR-5 passenger aircraft prototype with Wright J-4A engine (VVA-5, 'aircraft No.49')

In 1931 Yakovlev graduated from the Air Force Academy and was posted as engineer to plant No.39 where his first aircraft had been built. Yakovlev's first major accomplishment at this plant in his capacity as an engineer became the construction of the AIR-5 four-seat passenger aircraft. In common with the preceding 'P' designs, it had a comfortable enclosed cabin reminiscent of contemporary Ford cars; this earned it the sobriquet *voz**doosh**nyy Ford* ('aerial Ford' or 'flying Ford').

The machine was designed and built as a voluntary undertaking in addition to the plant's official production plans; it was known as 'aircraft No.49'. The designation VVA-5 occasionally encountered in documents of that period presumably also refers to this machine.

The mock-up of the AIR-5 was built in May 1931, and prototype construction began in July. By early November the machine was completed, making its first flight that month. The AIR-5 was a braced high-wing monoplane with a fixed undercarriage; it was designed around a 165-hp engine of Soviet manufacture, but provision was made for the installation of other types of air-cooled engines delivering 175 to 250 hp. Eventually it was fitted with a 200-hp Wright J-4A radial, one of the two examples that had been purchased at that time.

In the course of the testing the AIR-5 displayed excellent qualities. The State acceptance trials were completed in April 1932; the aircraft was pronounced suitable for passenger and mail transportation, as well as for liaison and aerial photography, and was recommended for series production.

Specifications of the AIR-5 with the Wright J-4A engine

Engine power, hp	200
Length overall	8.0 m (26 ft 3 in)
Height, tail up	2.46 m (8 ft 0 55/64 in)
Wing span	12.8 m (42 ft 0 in)
Wing area, m^2 (sq ft)	23.0 (248)
Empty weight, kg (lb)	812 (1,790)
All-up weight, kg (lb)	1,390 (3,064)
Maximum speed, km/h (mph)	193 (120)
Cruising speed, km/h (mph)	152 (94.5)
Landing speed, km/h (mph)	75 (46.6)
Time to 1,000 m (3,280 ft)	6.5 min
Service ceiling, m (ft)	4,275 (14,000)
Range, km (miles)	1,000 (621)
Take-off run, m (ft)	100 (330)
Landing run, m (ft)	100 (330)

AIR-5 with M-48 engine (production version prototype)

The AIR-5 proved its worth during operational trials in the Civil Air Fleet for which it was to be built in series. There were also plans for producing up to 50 machines for Osoaviakhim. The imported Wright J-4A engine was to be replaced on production machines by the indigenous M-48 engine rated at 220-240 hp. This seven-cylinder radial was developed at aero engine factory No.29 under the direction of A. S. Nazarov in 1930-32. An example of the AIR-5, fitted with an M-48 engine and intended to serve as a pattern aircraft for series manufacture, was publicly demonstrated in August 1934 and was flight-tested in November of that year, attaining a speed of 210 km/h (130 mph). However, this engine failed to reach production status, only a few examples being built. In consequence, series production of the AIR-5 had to be cancelled at its very start.

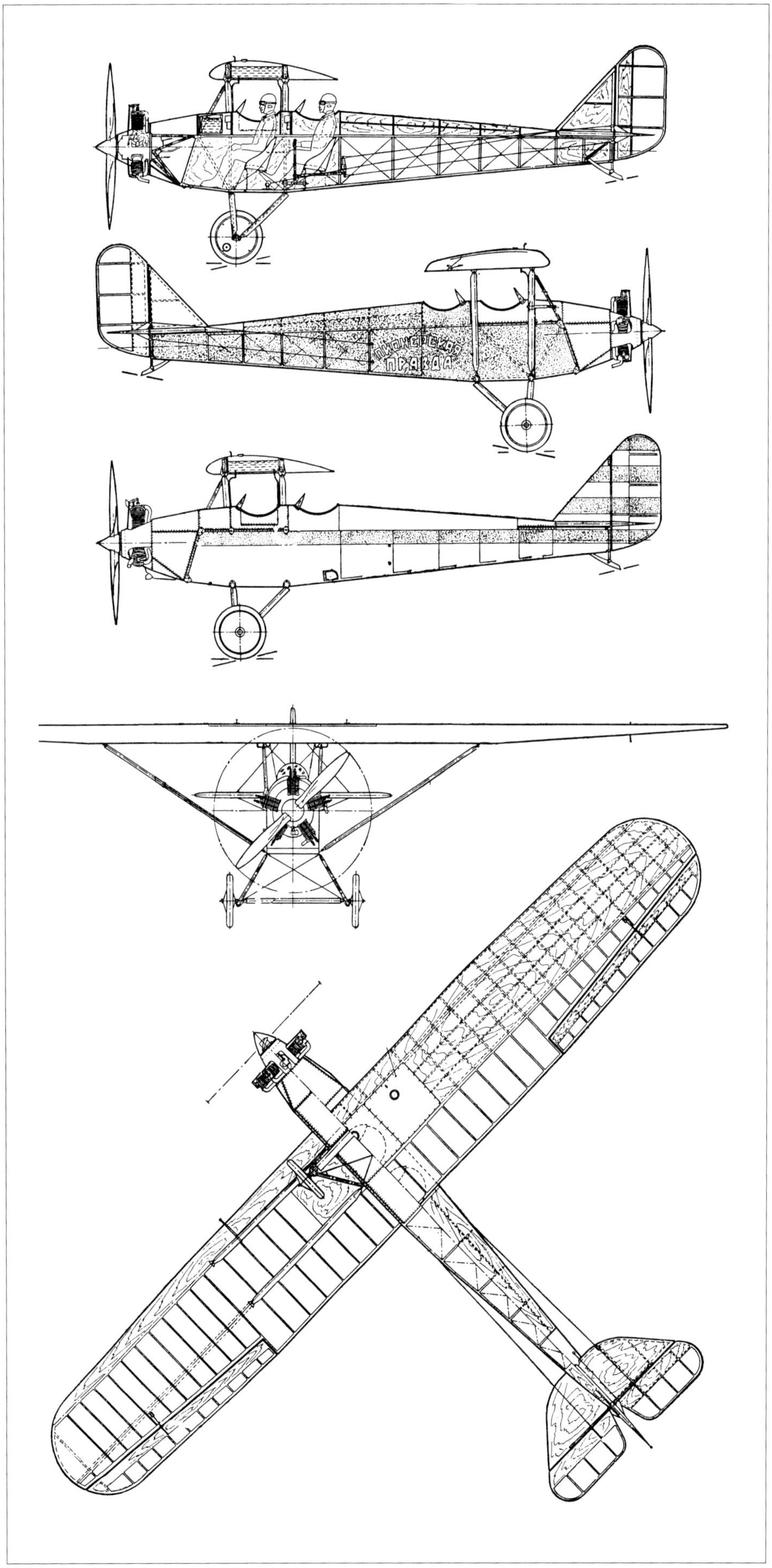

The AIR-4 light aircraft.

Yakovlev OKB

Above: Another AIR-4 with a pre-1932 registration, CCCP-312, carries 'Osoaviakhim SSSR' (Soviet Society for the Support of Defence, Aviation and the Chemical Industry) titles under the front cockpit.

Yakovlev OKB

Yakovlev OKB

Yakovlev OKB

Three aspects of the highly modified AIR-4MK, CCCP-Э-31 (SSSR-E-31), showing the almost full-span flaps and wingtip ailerons. Note the balancing booms on the ailerons.

AIR-10 (AIR-5 with M-48 engine) (project, first use of designation)

Design work was conducted on an improved version of the AIR-5 powered by the 220-hp M-48 engine and designated AIR-10 (first use of designation). It carried the same number of passengers as the preceding version, but the maximum speed was raised from 210 km/h (130 mph) to 240 km/h (149 mph). Other performance figures included a range of 1,000 km (620 miles), an endurance of five hours and a landing speed of 70 km/h (43 mph). No prototype was built, and in 1935 the designation AIR-10 was reused for a two-seat trainer prototype which was later developed into the production UT-2 trainer.

AIR-6 three-seat utility aircraft with M-11 engine (second use of designation)

The unavailability of the M-48 engine for the planned AIR-5 production prompted Yakovlev to develop the two- or three-seat AIR-6 utility aircraft fitted with the less powerful M-11 engine. Its general layout was modelled on that of the AIR-5. A group of designers under Yakovlev's direction undertook design work on this machine in 1932 in response to the wishes of Osoaviakhim. This aircraft was intended for liaison, pilot training, sports flying and touring. Its design philosophy was based on such requirements as the use of small unprepared airstrips, ease of handling, simple take-off and landing procedures, low manufacturing costs and simplicity of maintenance and repair.

AIR-6 prototypes

The first prototype of the AIR-6 was completed in May 1932, four months after the start of the design work. To save time and costs, it made use of undercarriage struts from the Polikarpov U-2 primary trainer/utility aircraft and the tail surfaces from the Tupolev I-5 fighter. The pilot and a passenger sat in tandem, but the cockpit was roomy enough to accommodate a third person sitting further aft. The aircraft had an empty weight of 582.8 kg (1,285 lb) and a normal AUW of 853.3 kg (1,182 lb), the maximum take-off weight being 965 kg (2,128 lb). The prototype was not yet fitted with the wheel brakes and high-lift devices specified by technical requirements.

In the autumn of 1932 the AIR-6 passed its official manufacturer's tests at the flight test facility of plant No.39. Despite numerous manufacturing defects and some design faults revealed during the testing, the pilots' general impression of the aircraft was quite good. It was stated that the aircraft basically met the technical requirements, was undemanding in operation and was suitable for the role of an aircraft intended for mass operation.

In October 1933 state acceptance trials of the AIR-6 with a wheeled under-carriage were completed. In the winter of 1932-33 the AIR-6 was flown with a ski landing gear, and a second aircraft was fitted with skis during the following winter. The first prototype was later converted into a floatplane (see AIR-6A below).

The AIR-6 had bright prospects of entering production when an accident with the AIR-7 (see below) placed Yakovlev in disfavour with the nation's aviation authorities and led to cancellation of the AIR-6 programme (for the time being, as it turned out). In this situation Yakovlev decided to build two more prototypes; they were completed in June and September respectively. Yevgeniy G. Adler, who was appointed project engineer for these machines, introduced a number of alterations on them, including a revision of the bungee cord shock absorbers and local lightening of the structure. The reduction of empty weight permitted four fuel tanks to be installed on the third machine instead of two. The third prototype was fitted experimentally with a Townend ring and wheel spats, but this aerodynamic refinement proved to give little advantage at the aircraft's moderate speeds.

AIR-6 production aircraft

Eventually problems with the AIR-7 were sorted out and the AIR-6 again became 'eligible' for series production. The Civil Air Fleet's aircraft repair factory No.47 in Leningrad was tasked with building 60 machines in 1934. By May 1934 the first production machine was rolled out, but then the short supply of basic raw materials led to a temporary halt, the production beginning in earnest only in the autumn of that year. Difficulties with launching series production at Plant No.47 prompted Yakovlev to propose that the AIR-6 be built also at other factories. As a result, the aircraft was also placed in production at plant No.23 in Fili, then a western suburb of Moscow (now long since part of the city). Machines manufactured by that plant had their cabin widened by 150 mm (5$\frac{29}{32}$ in) to improve passenger comfort.

There are conflicting reports as to the overall number of AIR-6s built; relatively recent research (in 1999) suggests that 128 machines were built at plants No.39, 47 and 23. Several production versions existed as detailed below.

One AIR-6 had an unusual registration, CCCP-N1002 (*sic*) – with a Cyrillic prefix but a Roman N as an operator designator.

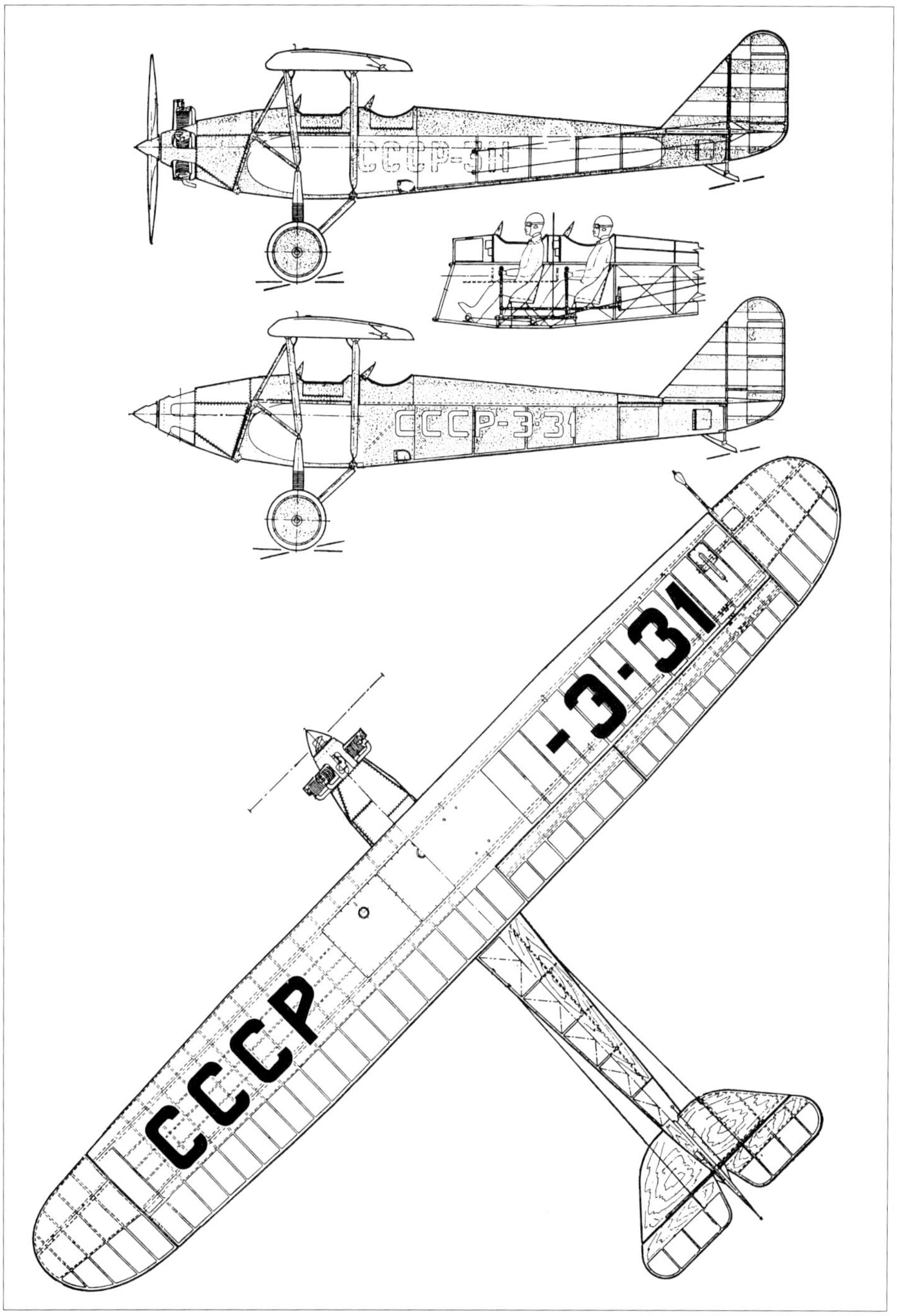

Above: The AIR-4MK (E-31), with a side view of a standard AIR-4 at the top for comparison. The engine is not shown in the side view.

Yakovlev OKB

The prototype AIR-5 passenger aircraft, alias VVA-5 or 'aircraft No.49', on skis.

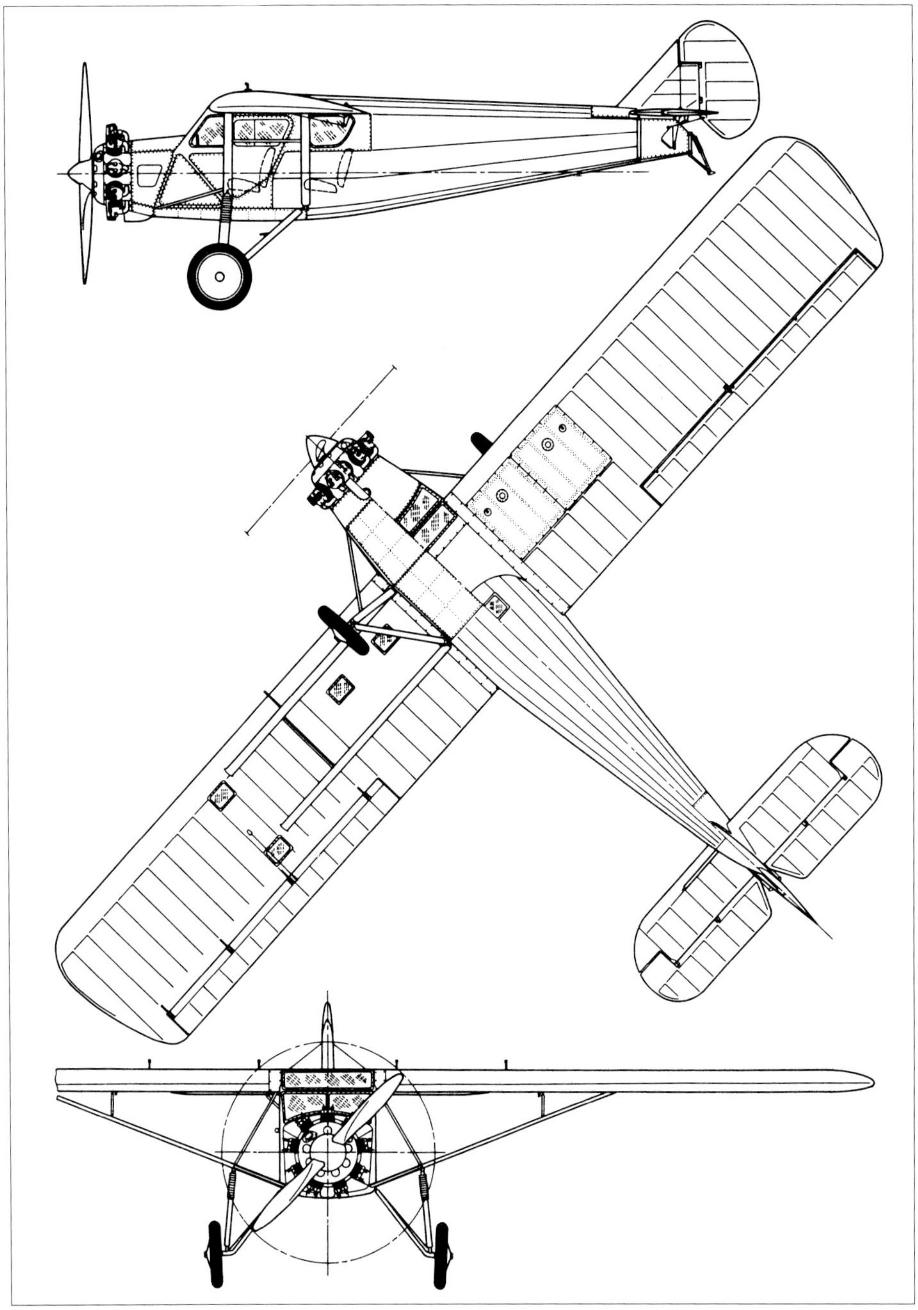

A three-view of the AIR-5. Note the rudder cut away at the base to provide room for tailskid shock absorption travel.

Specifications of the AIR-6 versions

	AIR-6 prototype	AIR-6A floatplane
Length overall	7.8 m (25 ft 7 in)	8.5 m (27 ft 11½ in)
Wing span	12.08 m (39 ft 7½ in)	12.0 (39 ft 4½ in)
Wing area, m² (sq ft)	19.8 (213)	19.8 (213)
Empty weight, kg (lb)	616 (1,358)	668 (1,473)
All-up weight, kg (lb)	961 (2,119)	958 (2,112)
Maximum speed, km/h (mph)	168.5 (104.7)	150 (93)
Cruising speed, km/h (mph)	130 (81)	n.a.
Landing speed, km/h (mph)	80 (49.7)	70 (43.5)
Service ceiling, m (ft)	4,600 (15,090)	3,000 (9,840)
Range, km (miles)	715 (444)	600 (373)
Take-off run, m (ft)	85 (280)	n.a.
Landing run, m (ft)	165 (541)	150 (492)

Before the Second World War, Soviet aircraft flying international services usually had a different prefix, URSS (*Union des Républiques Sovietiques Socialistes*) or USSR. This aircraft was named *Ivan Mikheyev*.

AIR-6A (AIR-6 'Ghidro') floatplane

In the summer of 1932 the first prototype of the AIR-6 was equipped with floats designed by Vadim B. Shavrov. These were single-step floats basically similar to those of the AIR-2 but slightly scaled-up.

Designated AIR-6A or AIR-6 ***Ghid**ro* (Hydro), this aircraft made its first flight from the Moskva River with Yulian I. Piontkovskiy at the controls in June 1933. It passed its State acceptance trials at NII GVF, demonstrating just as good handling qualities as the landplane version. Later several production machines were also fitted with floats.

AIR-6 for use in the High North ('Arctic version')

In 1934 a special 'Arctic' version of the AIR-6 was produced. It could be operated on wheels, skis or floats, all types of undercarriage being interchangeable in field conditions. Special measures were taken to ensure trouble-free running of the M-11 engine in low ambient temperatures. The first machine for the High North was ready in October 1934; in February 1935, upon completion of the testing, it was handed over to the Northern Sea Route authorities.

AIR-6 ambulance version

A number of AIR-6s (20 machines, according to some reports) were built in an ambulance version featuring a triangular door on the port side of the fuselage for loading the stretcher. They could be identified by their registrations containing the letter K as an operator designator denoting ambulance aircraft – for example, CCCP-K181. These machines were operated in Kirghizia and other mountainous areas

AIR-7 high-speed courier monoplane

The concept of this aircraft – a high-speed low-wing monoplane – owed much to a series of sports monoplanes built in the USA in the late 1920s. These included the Travel Air Texaco 13, or Mystery Ship, the Gee Bee Senior Sportster, and similar machines built by Weddell Williams and Lockheed. Yakovlev was inspired by the idea of creating an aircraft powered by an M-22 radial (the Soviet licence-built version of the Bristol/Gnome-Rhône Jupiter) that would match the high-speed performance of these aircraft and become the fastest aircraft in the Soviet Union thanks to the use of such aerodynamic refinements as the low-wing monoplane layout, a Townend ring on the engine and 'trouser' fairings on the undercarriage wheels. Design work on the AIR-7, funded by Osoaviakhim, started in late 1931; seven months later, in August 1932, the completed machine was rolled out from the assembly workshop of Plant No.39. This was a two-seat aircraft, the first low-wing machine in the series of Yakovlev's sports, training and light utility aircraft.

The AIR-7 was a sleek aircraft; the two seats arranged in tandem were covered by a common canopy featuring separate hoods which folded to the right, with a fixed section between them. The wings were braced by overwing struts and bracing wires (this permitted the use of relatively thin wings having a constant chord and thickness/chord ratio of 8% over the whole span). The wings made use of the Göttingen-436 airfoil; the structure was based on two spars of spruce laminates (outboard of the overwing struts) and ribs of plywood. The covering was plywood up to the front spar and fabric elsewhere. The wing centre section had metal spars and was mostly skinned in D1 duralumin sheet. The ailerons of rectangular planform had deeply inset hinges and large slots ahead of the forward balance area.

The fuselage was based on a square-section truss of welded steel tube. Ahead of the front cockpit the skin was aluminium, mainly in removable panels. The rest of the fuselage was streamlined by adding a light wooden secondary structure of stringers supporting fabric covering. The tail surfaces were made of light alloy with fabric covering, the stabilisers being wire-braced

Each main undercarriage unit was an assembly of steel tubes attached to the ends of the centre section spars. Two tubes joined at the bottom formed a rigid triangle to which was pivoted a fork carrying the wheel. The weight of the aircraft was supported by a further fork with a vertical strut incorporating shock absorption by a stack of rubber discs. The entire unit was enclosed by an aluminium trouser fairing. The tailskid comprised multiple steel leaf springs.

The first flight of the AIR-7 took place on 19th November 1932 with Yulian I. Piontkovskiy at the controls. On the following day, with Aleksandr S. Yakovlev as a passenger, Piontkovskiy made another flight in which he attained a speed of 325 km/h (202 mph) – a national speed record at that time. The AIR-7 proved to be faster than the I-5 single-seat fighter and became one of the world's fastest two-seat aircraft among those powered by similarly rated engines. The AIR-7 also boasted a very favourable ratio of 3.6 between the maximum speed and the landing speed – 325 and 90 km/h (202 and 56 mph) respectively.

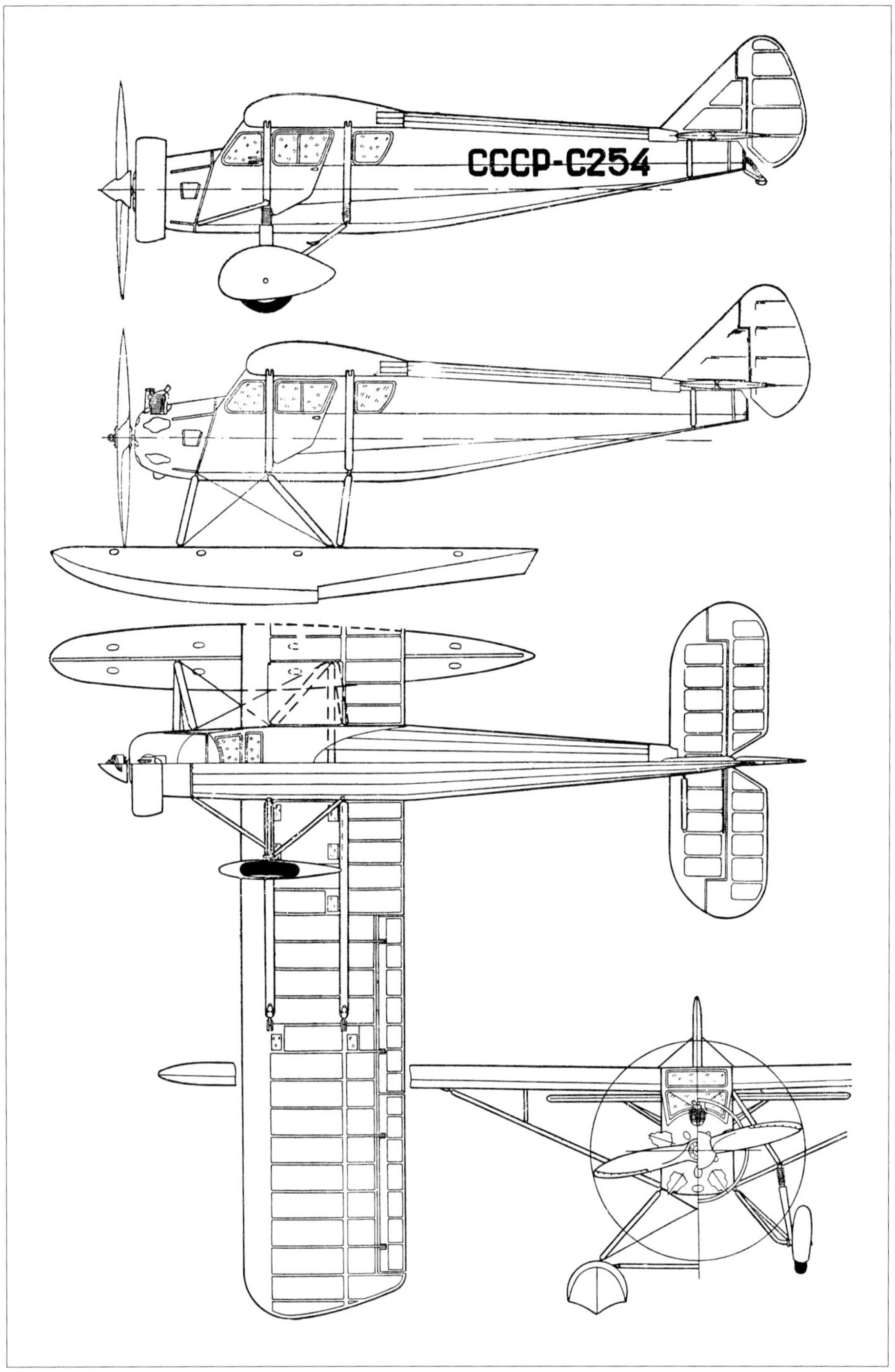

Above: A three-view of the AIR-6 in alternative landplane and seaplane (AIR-6A) configurations, with an additional side view of AIR-6 CCCP-C254 (that is, SSSR-S254) with a Townend ring and wheel spats.

Yakovlev OKB

AIR-6 CCCP-C254, an Osoaviakhim aircraft, in standard configuration. The aircraft is named *Latyshskiy strelok* (Latvian rifleman).

Yakovlev OKB

Above: Two Osoaviakhim AIR-6s, CCCP-C254 and CCCP-C283, fly over the Moskva River in the suburbs of Moscow.

Yakovlev OKB

Yakovlev OKB

Yakovlev OKB

Three views of the AIR-7 sports aircraft as originally flown with almost constant-chord wheel 'trousers'. The machine was finished in red and silver dope, with the trademark Yakovlev red/white striped tail.

The first two flights of the AIR-7 were truly impressive, and the Soviet Air Force command organised a demonstration of the aircraft on 23rd November. Piloted by Piontkovskiy, the AIR-7 took off from Moscow's Central airfield (Khodynka) and made a low pass over the airfield. It was to make two or three circuits of the airfield and make a landing. Suddenly, when the aircraft was above the southern edge of the airfield, a glittering strip parted company with the aircraft and started falling to the ground. The aircraft started a gradual descent and disappeared from view behind some trees. To everybody's relief, it soon became known that the aircraft had landed safely. The accident was caused by a failure of one of the starboard aileron hinges, whereupon the aileron assumed a vertical position. The aircraft was saved thanks to Piontkovskiy's excellent airmanship and the robustness of the wing structure, which did not disintegrate, despite the violent vibrations caused by the aileron failure.

Yakovlev assumed full responsibility for the accident, citing his own mistake in calculating the strength of the aileron hinge, albeit it was the members of his group who had made the calculations. An investigating panel was set up; as it transpired, the panel was intent not so much on discovering the causes of the incident as on victimising the young designer. The incident became a pretext for heaping all sorts of blame and ill-founded accusations on Yakovlev in an effort to discredit all his preceding work which had been marked by some indisputable successes. His future as a designer was called into question at that time. Yakovlev displayed no mean courage in rebuffing these attacks. In the end, he succeeded in defending his right to create new aircraft.

AIR-7 as rebuilt after the incident

The AIR-7 was repaired, undergoing some modification in the process. In addition to strengthening the aileron hinges, the view from the cockpit was improved by adding new glazing panels in the fuselage sides below the canopy; the undercarriage was reworked so as to make it simpler and more reliable. The 'trousers' gave place to wheel spats of a different shape.

On 25th September 1933 Piontkovskiy made two flights in the AIR-7 and established a new national speed record, attaining 332 km/h (206 mph). In October and December of that year the AIR-7 underwent tests at NII GVF. The rate of climb attained during these tests was also a record high. The AIR-7 left its mark on the development of aircraft design in the USSR by stimulating the transition to the aerodynamically clean monoplane layout.

This machine ended its days as a target on a firing range, as was the case with many aircraft withdrawn from use. Sadly, no one cared much about preservation of historic aircraft at that time.

Specifications of the AIR-7

Length overall	7.80 m (25 ft 7 in)
Height	3.10 m (10 ft 2 in)
Wing span	11.00 m (36 ft 1 in)
Wing area, m^2 (sq ft)	19.40 (209)
Empty weight, kg (lb)	900 (1,980)
All-up weight, kg (lb)	1,400 (3,090)
Maximum speed, km/h (mph)	332 (206)
Landing speed, km/h (mph)	90 (56)
Range, normal/max, km (miles)	270/1,300 (167/808)
Take-off run, m (ft)	150 (490)
Landing run, m (ft)	150 (490)

AIR-8 two-seat trainer prototype

This parasol monoplane seating two pilots in tandem in open cockpits was built in 1933 as a further development of the AIR-4. It featured redesigned wings of greater area with no taper and with two-section slotted ailerons. The wings were mounted above the fuselage on a cabane and braced by struts running from mid-span to the lower edge of the fuselage sides. Both the wings and the main gear units were slightly moved forward The shock struts of the pyramid-type main undercarriage were linked to the forward pair of the wing bracing struts, like those of the AIR-4, but an additional strut from that junction went to the wing, not to the fuselage. The shock-absorption stroke of the legs was increased. The horizontal tail

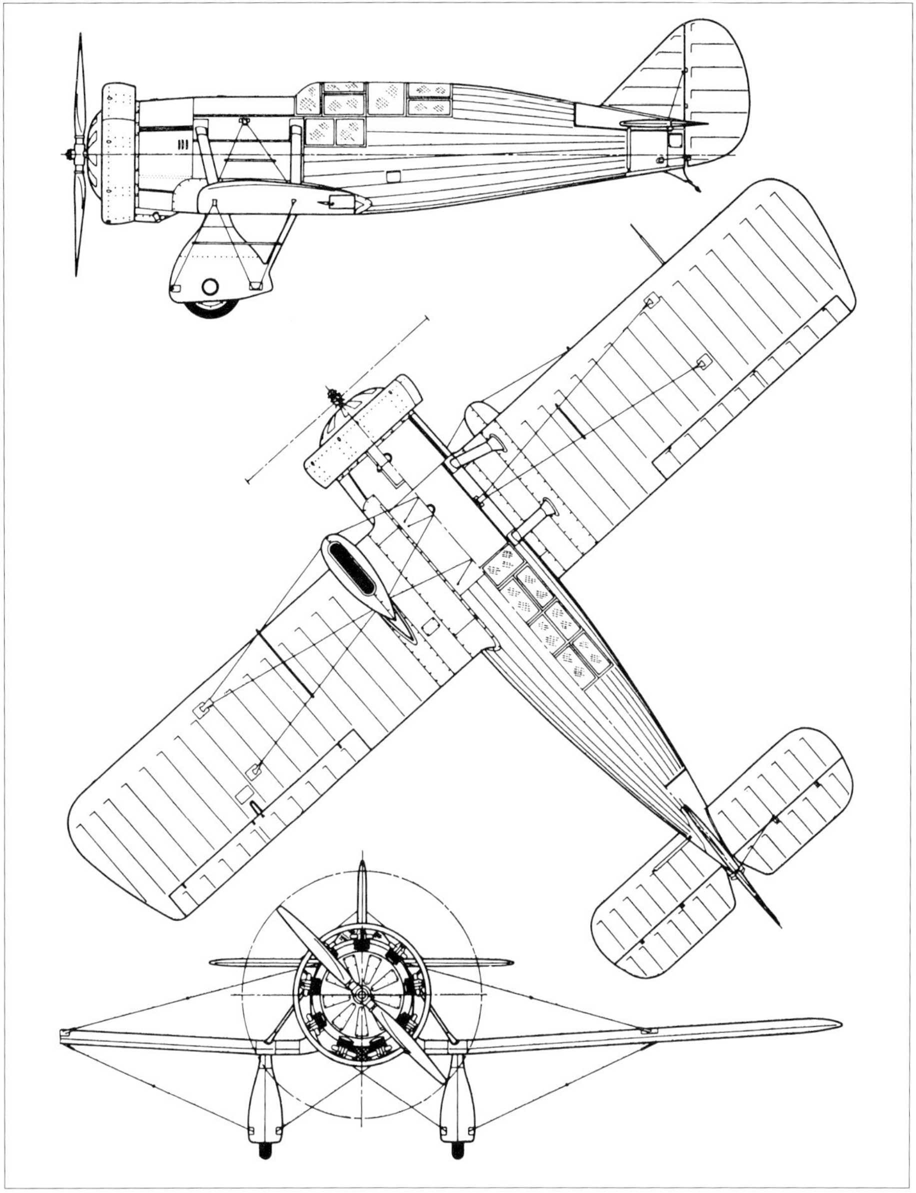

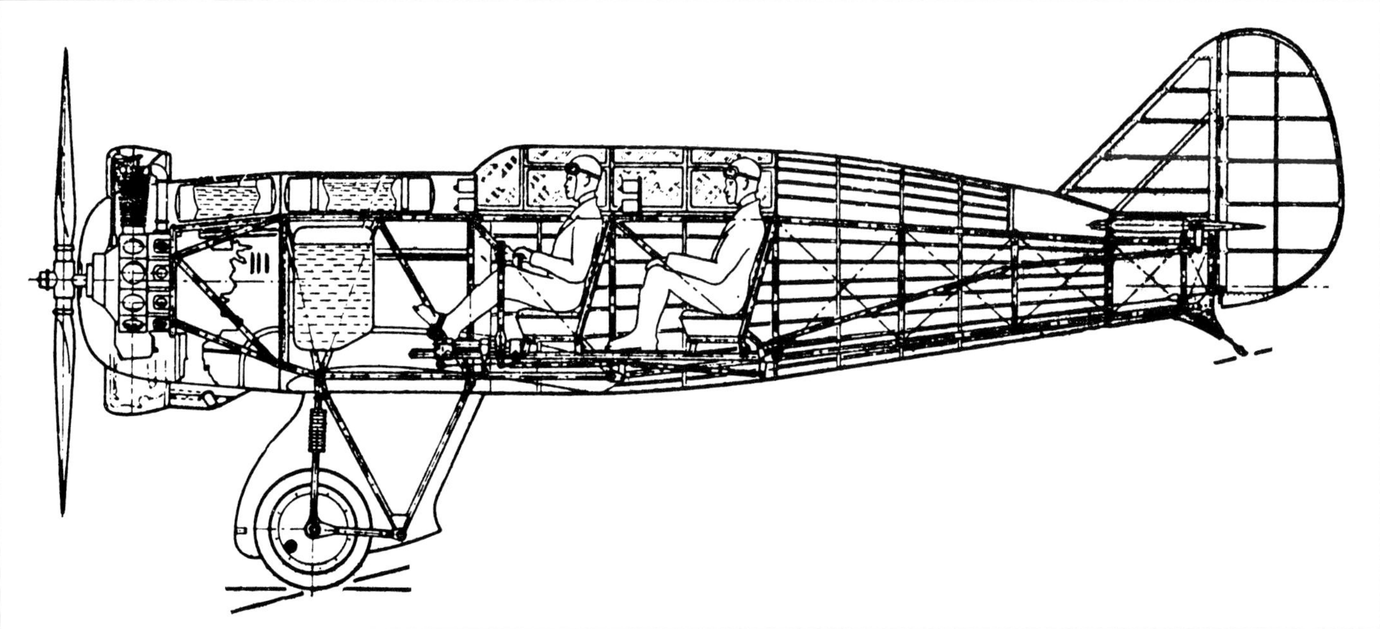

Top: Three views of the AIR-7 as modified after the 23rd November 1932 incident, showing the modified wheel 'trousers' and canopy.
Above: A cutaway drawing of the AIR-7. Note the main gear design with leading arms.

Yakovlev OKB

Yakovlev OKB

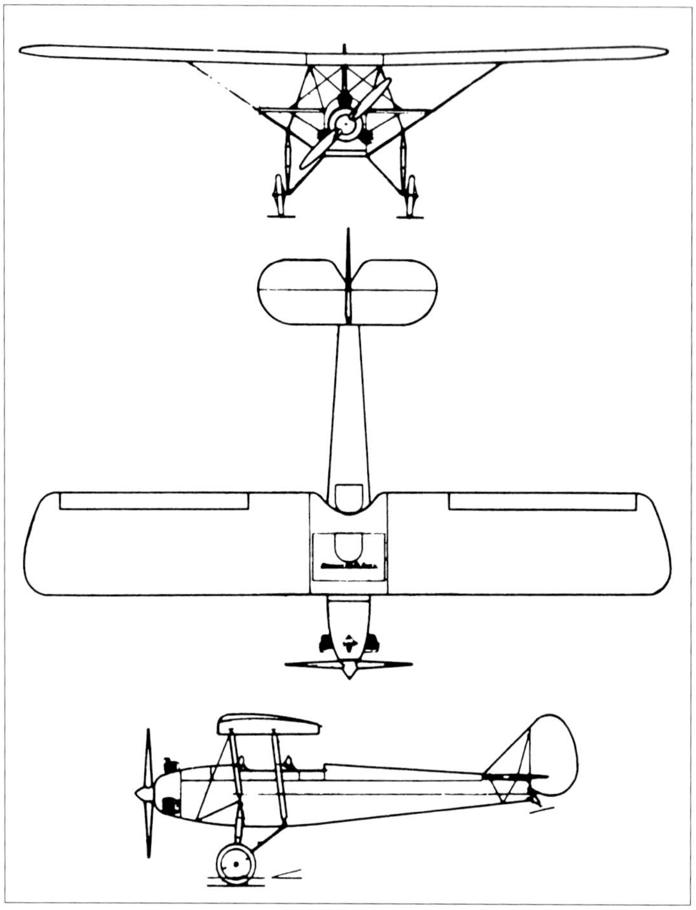

Top and above: Two views of the AIR-8 trainer, showing the fat low-pressure tyres known as *sooperballony* or, in American parlance, 'airwheels'.

Right: A three-view of the AIR-8.

was redesigned, receiving parallel leading and trailing edges, and a rudder with a curved outline and increased area replaced the previously used rudder with a straight trailing edge. Fat low-pressure tyres (so-called 'superballoons' or 'airwheels') were introduced for use on soft ground.

The aircraft was originally powered by an M-23 three-cylinder engine delivering 70 hp. This engine was soon replaced by a 60-hp Walter NZ-60 engine. Finally, in 1934 the AIR-8 was re-engined with a 85-hp Siemens engine.

The aircraft was developed in response to an Air Force (VVS) requirement for a front-line observation and liaison aircraft, but personnel changes in the VVS staff led that service to lose interest in the aircraft.

Specifications of the AIR-8 with the 85-hp Siemens engine

Length overall	7.1 m (23 ft 3½ in)
Wing span	11.1 m (36 ft 5 in)
Wing area, m² (sq ft)	18.0 (194)
Empty weight, kg (lb)	430 (950)
All-up weight, kg (lb)	675 (1,490)
Maximum speed, km/h (mph)	150 (93)
Landing speed, km/h (mph)	65 (40)
Range, km (miles)	750 (466)
Take-off run, m (ft)	50 (164)

Two-seat sports aircraft with a monowheel undercarriage (project)

An archive drawing dated 1933 reveals the existence of this project of a sports aircraft seating two pilots in tandem in an enclosed cockpit. It was a monoplane with a low-set wing braced by wires both on the upper side and underneath. The lower bracing wires converged on a keel-like centreline fairing which housed the unorthodox monowheel undercarriage. There was no tailwheel or tailskid, their role being played by a reinforced bulge beneath the rudder axle. The aircraft was to be powered by an air-cooled in-line engine.

Single-engined reconnaissance aircraft (project)

This project dating back to 1933-34 remained unknown until recently; evidence of its existence appeared in 1999 in a publication based on the OKB archive which contained an artist's impression of this aircraft. It shows a sleek cantilever shoulder-wing single-engined monoplane with cockpits fore and aft of the wing joined by a long common canopy. The aircraft had a conventional tail unit and a retractable undercarriage.

The shape of the engine cowling reveals the planned use of a liquid-cooled V-12 engine (possibly the Mikulin M-34 rated at 750/800 hp). This, to the best of the authors' knowledge, is the only Yakovlev project from the mid-1930s which represented a departure from the prevalent line of his activities at that time (light training, sports and utility aircraft).

AIR-9 two-seat sports aircraft (project of 1933)

In 1933 Yakovlev and his team developed a project of a two-seat low-wing sports monoplane powered by the M-11 engine. The primary fuselage structure was a welded steel tubular truss. Tail surfaces had a duralumin framework; the wings were of wooden construction with plywood skins from the leading edge to the rear spar on the upper surface and to the front spar on the lower surface, the rest of the skinning being fabric. The wings were equipped with landing flaps. The cockpits were open. The project was submitted for a contest of projects of safe aircraft; therefore the AIR-9 featured automatic leading-edge slats.

This project did not reach the hardware stage. However, in 1934 it was reworked to feature enclosed cockpits as described below.

AIR-9 sports aircraft prototype

This two-seat sports aircraft, powered by a 100-hp M-11 engine, was the first machine to be built in the former bed factory to which the Yakovlev design team moved after obtaining the status of an independent organisation in January 1934. The AIR-9 was

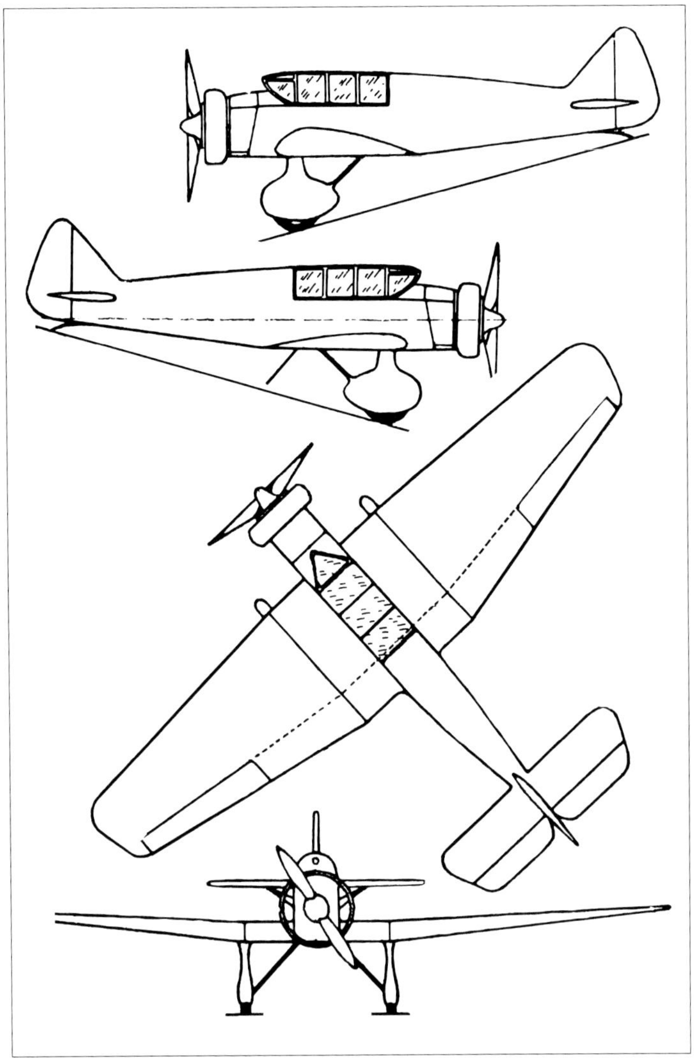

Yakovlev OKB

Yakovlev OKB

Yakovlev OKB

Left: Four views of the AIR-9*bis*.
Right above, centre and below: The AIR-9 during various stages of its flying career; all three photos show different paint jobs. Note the racing number in the centre photo. The windshield design changed in the course of time, too; the upper view shows a frame, while the other two guises have a one-piece visor.

completed in 1934, in time to be exhibited at that year's Paris Air Show.

It was a neat low-wing cantilever monoplane with two seats arranged in tandem under a common canopy with the front hood sliding rearwards over the fixed central portion and the rear hood apparently sliding forward under the fixed section (on some pictures the rear hood has additional framing which suggests that it was sideways-hinged and folding). In its original form this aircraft had a traditional windshield with a rearward-sloping windscreen. Photos show two kinds of this traditional windshield used on the AIR-9. A machine wearing the number '31' (probably a racing number) on the forward fuselage can be seen with a one-piece rounded frameless windshield (it also featured an exhaust collector ring); a differently painted machine carrying no number was shown with a V-shaped windshield formed by flat panels.

The AIR-9's airframe featured wooden two-spar wings attached to a welded-steel centre section built integrally with the fuselage. The tail featured the usual duralumin structure with fabric covering; the rudder had a curved trailing edge. The outer wings were tapered and were covered with plywood over the entire upper surface. The wings were fitted with slotted ailerons and large manually operated split flaps. The fuselage was entirely fabric-covered, with the exception of the forward portion where a removable hemispherical dome of aluminium covered most of the engine (apart from the cylinder heads). The main undercarriage units had lateral and longitudinal bracing struts, the wheels being provided with spats.

The AIR-9 was fitted with a standard M-11 driving a two-blade wooden propeller.

Yakovlev OKB

The AIR-9*bis* in a silver dope colour scheme with red trim, as originally flown with a Townend ring on the engine. Note the characteristic undercut windshield.

Yakovlev OKB

Yakovlev OKB

Top and above: The AIR-9*bis* with the front cockpit canopy slid open. Note the exhaust collector ring and port side exhaust pipe. The wheel spats are still of the early type, with exposed struts.

Boris Vdovenko

A ground crewman in Red Army uniform prepares to swing the prop of the AIR-9*bis*, now lacking the Townend ring and exhaust collector but adorned with the racing number 32. Note the different wheel spats.

Initially the AIR-9 had no Townend ring but, when exhibited in Paris in 1934, it was equipped with a Townend ring while retaining the rearward-sloping windshield formed by flat panels.

AIR-9*bis* (modified AIR-9 prototype?)

In 1935 the AIR-9 underwent further modifications. The traditional cockpit windshield was replaced with a V-shaped forward-sloping windscreen. In this configuration (with the Townend ring and the new windscreen) it was designated AIR-9*bis*. Photos show it to be fitted with an exhaust collector ring. Some sources believe the AIR-9*bis* to be a conversion of the original prototype, while others assert that this designation applies to a second machine that was allegedly built.

Originally the AIR-9*bis* had a tailskid and wheel spats leaving the undercarriage struts exposed. The aircraft was exhibited at the Milan airshow in 1935; there it was shown equipped (temporarily?) with a tailwheel instead of the original leaf-spring tailskid.

In 1937 the AIR-9*bis* was used for a record flight. On 4th July 1937 the aviatrices I. N. Vishnevskaya and Ye. M. Mednikova set a women's FAI Class C altitude record of 6,518 m (21,385 ft). For this flight the aircraft was lightened by removing the Townend ring and wheel spats. The record-setting machine had the number '32' on the forward fuselage sides, suggesting its participation in some kind of races. There are photos showing this aircraft, again with the number '32' (before or after the record-breaking flight?), without the Townend ring and, interestingly, with individual exhaust stubs, but with wheel spats and, additionally, the undercarriage struts faired by 'trousers' of the kind that became standard on the UT-1 and UT-2 trainers. In all, available pictures show the AIR-9/9*bis* in no fewer than six different configurations with regard to powerplant, cockpit details, undercarriage fairings and colour scheme/number, which may well support the theory that more than one example was built.

Specifications of the AIR-9*bis*

Length overall	6.97 m (22 ft 10⅜ in)
Wing span	10.2 m (33 ft 5⅝ in)
Wing area, m^2 (sq ft)	16.87 (182)
Empty weight, kg (lb)	495 (1,090)
All-up weight, kg (lb)	768 (1,693)
Maximum speed, km/h (mph)	215 (134)
Landing speed, km/h (mph)	65 (40)
Service ceiling, m (ft)	6,080 (19,950)
Range, km (miles)	695 (432)
Take-off run, m (ft)	80 (262)
Landing run, m (ft)	90 (295)

AIR-10 trainer prototype (second use of designation)

In 1935 Yakovlev designed and built a prototype of a tandem two-seat trainer featuring a low-wing cantilever monoplane layout. Designated AIR-10, this machine was powered by an M-11 engine. In its project configuration it featured slats and split flaps. These were omitted in the prototype which made its first flight on 11th June 1935.

Yakovlev OKB

Above: The AIR-10 trainer prototype, the precursor of the UT-2, as originally flown. Note the exhaust collectors serving two cylinders to port and three to starboard.

The AIR-10 had much in common with the AIR-9; in fact, it might be regarded as an AIR-9 with open cockpits. The airframe structure was basically the same, with a welded steel-tube fuselage truss, wooden two-spar wings and light-alloy tail surfaces. The wing structure differed from that of the AIR-9 in having the ply covering only up to the front spar on the upper side. The open cockpits had dual controls; they were fitted with small windscreens formed by two flat panels joined at the centreline and with fold-down doors on the port side to facilitate access. The main undercarriage units were fitted with spats and fairings enclosing both the main leg and the rear bracing strut.

The AIR-10 was among the Soviet sporting aircraft that took part in an All-Union competitive long-range flight organised by Osoaviakhim in 1935. The flight took place between 2nd and 9th September on a circuit spanning over 5,500 km (3,418 miles). Piloted by Yulian I. Piontkovskiy and wearing the racing number '30', the AIR-10 won first place in this contest; prizes were awarded to the designer, the Design Bureau, the pilot and the aircraft itself. In 1936 the AIR-10 repeated this victory.

After the 1935 race the AIR-10 was subjected to a prolonged evaluation by the Air Force. Military test pilots gave a very favourable report on the aircraft, which was deemed to be a good prospective primary trainer, as well as an intermediate stage facilitating transition from the current U-2 primary trainer to combat aircraft. Minor improvements were requested, including the installation of flaps, trim tabs and a crash pylon to protect the occupants in the event of a nose-over. The design load factor was expected to be raised from 8 to 10, the landing speed was to be brought down to 55 km/h (34 mph). With these alterations, the

Boris Vdovenko

Above: With a ground crewman supporting the starboard wingtip, the still unmarked AIR-10 taxies during trials.

Boris Vdovenko

Boris Vdovenko

Two views of the AIR-10 at a later date, now sporting the side number '1 White'. The forward fuselage appears to have been lengthened. One can only guess what the man standing on the wing is hollering to the pilot. Note AIR-6 CCCP-C283 in the background with the racing number '22'.

Yakovlev OKB

Above: The AIR-10 with the racing number 30 with which it participated in the air races on 2nd-9th September 1935.

Yakovlev OKB

Still wearing its racing number, the AIR-10 is seen here in floatplane configuration being tested on the Moskva River in 1937.

aircraft was to enter series production. However, it was deemed advisable to rework the airframe structure, making it all-wooden so as to simplify production methods. The result was the Ya-20, which was put into production as the UT-2 (described below) some three years later.

AIR-10 on floats

In 1937 the AIR-10 was tested as a seaplane, fitted with floats taken from the AIR-6A. Some sources claim that in this configuration it was designated VT-2 (unconfirmed).

AIR-11 (LT-1) touring aircraft prototype

Powered by a 120-hp de Havilland Gipsy Major four-cylinder inverted in-line engine, this three-seat touring aircraft with an enclosed cockpit was completed in late 1936. It was a low-wing cantilever monoplane with spatted non-retractable mainwheels. The airframe comprised two-spar wooden wings, a steel-tube fuselage and tail surfaces with duralumin frames and fabric covering. The fuselage was wide enough for a passenger to sit beside the pilot, with a third seat at the rear. The canopy incorporated an upward-folding hood between the conventional aft-sloping windscreen and the fixed rear portion with side passenger windows. The undercarriage was similar to that of the AIR-10, with trousered mainwheels.

The aircraft successfully passed manufacturer's tests and State acceptance trials, but the absence of a suitable indigenous engine prevented it from entering series production. The AIR-11 was also tested on skis. It took part in the sporting aircraft races held in the Soviet Union on 24th July 1937 and won second place in the two-seat section in that event.

The AIR-11 immediately attracted the attention of Sergey V. Il'yushin who took a fancy to it and used all his influence to come into possession of this aircraft, despite Yakovlev's reluctance to part with it. Finally, Yakovlev yielded to this pressure and deemed it right to present the aircraft to Il'yushin who, being a pilot himself, made use of the machine for his business trips from Moscow to his production plant in Voronezh. All went well for some time, until one of these flights ended in a crash landing. The accident was caused by the negligence of a mechanic who had forgotten to fill the oil tank before the flight. After this episode Il'yushin was strictly forbidden by his superiors to take the pilot's seat in an aircraft.

Specifications of the AIR-10

	AIR-10 landplane	AIR-10 floatplane
Engine type	M-11	M-11
Engine power, hp	100	100
Length overall	6.80 m (22ft 3¾ in)	7.65 m (25 ft $1\frac{3}{16}$ in)
Wing span	10.2 m (33 ft $5\frac{37}{64}$ in)	10.2 m (33 ft $5\frac{37}{64}$ in)
Wing area, m² (sq ft)	16.87 (182)	16.87 (182)
Empty weight, kg (lb)	510 (1,124)	624 (1,376)
All-up weight, kg (lb)	820 (1,810)	896 (1,975)
Maximum speed, km/h (mph)	217 (135)	200 (124)
Landing speed, km/h (mph)	70 (43.5)	70 (43.5)
Service ceiling, m (ft)	5,700 (18,700)	3,200 (10,500)
Range, km (miles)	950 (590)	700 (435)
Take-off run, m (ft)	100 (330)	120 (390)
Landing run, m (ft)	n.a.	n.a.

Specifications of the AIR-11

Length overall	7.32 m (24 ft 0¼ in)
Wing span (m (ft)	10.2 m (33 ft 5⅝ in)
Wing area, m² (sq ft)	16.8 (181)
Empty weight, kg (lb)	566 (1,248)
All-up weight, kg (lb)	891 (1,964)
Maximum speed, km/h (mph)	209 (130)
Landing speed, km/h (mph)	82 (51)
Service ceiling, m (ft)	4,480 (14,700)
Range, km (miles)	720 (447)
Take-off run, m (ft)	200 (660)
Landing run, m (ft)	340 (1,115)

AIR-12 racing aircraft prototype

This two-seat sporting aircraft was designed and built by Yakovlev in 1936 with a view to undertaking long-range flights. It was a cantilever low-wing monoplane with a retractable undercarriage, powered by a mass-produced M-11 engine which had by then established a reputation for reliability – no small advantage for a long-range aircraft. Other design features were also dictated by the aircraft's mission. In particular, the wooden wings had a fairly high aspect ratio, with very narrow pointed tips; they had marked taper, with pronounced leading-edge sweep and a straight trailing edge. No slats or flaps were fitted. Each main gear unit was braced in the extended position by a rear strut and a lateral strut; the undercarriage retracted inwards by means of cables wound on to a drum, the pilot using a hand crank in his cockpit. The tailskid was a faired leaf spring, as on Yakovlev's earlier aircraft.

The fuselage featuring a steel-tube primary structure was covered by removable aluminium panels at the extreme nose, with plywood skin up to the wing trailing edge and fabric further aft. The pilot's cockpit was placed aft of the fuselage behind the wing trailing edge; it was enclosed by glazed side panels folding down on each side and a forward-sliding hood with an integral windscreen which fitted quite closely around the pilot's head when closed. The second cockpit, intended for a passenger, was placed considerably further forward, above the wings, and was covered with a transparent hood flush with the upper decking.

The engine drove a fixed-pitch wooden propeller. The fuel tank was located in the forward fuselage; it could be supplemented by an auxiliary tank installed in the passenger cockpit if necessary.

The AIR-12 commenced its manufacturer's testing before the end of August 1936. The test programme included a non-stop long-range flight which was performed by Yulian I. Piontkovskiy on 21st September. Flying the Moscow-Khar'kov-Sevastopol'-Khar'kov route, he covered a distance of about 2,000 km (1,240 miles) in 10 hours 45 minutes. The AIR-12 was then re-engined with an M-11Ye rated at 150 hp. On 24th October 1937 a woman crew comprising pilot Valentina Grizodoobova and navigator Marina Raskova flew the AIR-12 over 1,444 km

Yakovlev OKB

Yakovlev OKB

Yakovlev OKB

Top, centre and above: The AIR-11 was a fairly elegant aeroplane whose lines echoed those of contemporary British light aircraft – mainly due to the inverted in-line de Havilland engine.

Yakovlev OKB

The AIR-12 racing aircraft might well have been used as a fast mailplane, the cargo being housed in the forward cockpit above the wings. This cockpit was definitely claustrophobic.

Boris Vdovenko

Above: A day at the races. The AIR-12 heads a long line of Yakovlev aircraft – mostly UT-1 trainers. Note the strong wing leading-edge sweep and the revised forward cockpit canopy with increased glazing area.

Boris Vdovenko

Above: This view of the AIR-12 in original configuration (with a small glazing area shows the twin Venturi tubes to starboard and the open pilot's cockpit canopy.

Yakovlev OKB

The AIR-12 in flight with the landing gear retracted, now wearing the racing number 2. The pilot's cockpit canopy is left open, as it fitted uncomfortably closely around the pilot's head; in this instance the pilot even appears to be too tall for the canopy to be closed!

Specifications of the AIR-12

Engine type	M-11
Engine power, hp	100
Length	7.17 m (23 ft 6 9/32 in); also reported as 8.0 m (26 ft 3 in)
Wing span (m (ft)	11.0 m (36 ft 1 in); also reported as 12.0 m (39 ft 4½ in)
Wing area, m² (sq ft)	15.6 (167.9); also reported as 17.0 (183)
Empty weight, kg (lb)	558 (1,230)
All-up weight, kg (lb)	1,204 (2,654)
Maximum speed, km/h (mph)	235 (146)
Landing speed, km/h (mph)	93 (58)
Range, km (miles)	2,990 (1,860)
Take-off run, m (ft)	220 (722)
Landing run, m (ft)	n.a.

(897 miles) from Moscow to Akhtoobinsk (Saratov Region, southern Russia). Both flights were record-setting, albeit that the FAI refused to recognise the performance figures achieved by Grizodoobova and Raskova as a Ladies' Class record because no official observers were present.

AIR-13 racing aircraft (project)

This project of a single-seat aircraft with an enclosed cockpit and a retractable undercarriage powered by two 240-hp engines was developed in 1935 specially for participation in air races that were planned for 1936. It did not progress further than the drawing board. A sketch unearthed and published by Yuriy Zasypkin shows that the AIR-13 bore a close similarity to the de Havilland DH.88 Comet racing aircraft.

AIR-14 (UT-1) trainer and aerobatic aircraft

The AIR-14, renamed UT-1 in its production version (see below), was a small single-seat low-wing cantilever monoplane intended for use as an aerobatic sporting aircraft and a conversion trainer for fighter pilots. It was designed by Yakovlev and his team in early 1936, Longin Lis acting as project manager.

The basic airframe followed Yakovlev's previous practice. The two-spar wings, with dihedral from the roots, had a pronounced taper. The skin was thin plywood inboard of the ailerons on both the upper and lower surfaces, except aft of the rear spar. Each aileron was divided into inner and outer sections, with duralumin frames and fabric covering. There were no flaps. The fuselage was based on a welded steel truss, with each bay braced by OVS grade piano wires tightened by turnbuckles. The fabric covering of the fuselage was supported by a light secondary structure of ply and stringers, which a narrow dorsal fairing behind the pilot's head. The tail surfaces had the usual duralumin structure with flush-riveted skin back to the front spar, the rest being fabric. The fixed tailplane was braced by a strut below and single wires above.

Each main undercarriage unit comprised a slightly forward-sloping main leg braced by a rear strut and a lateral strut inboard. The main leg with a non-braking wheel in a fork fitting was pinned at the top to the front spar; the landing shocks were absorbed by a stack of eleven rubber discs with steel spacers. The rear strut also had a fork fitting and was pinned at the upper end to the rear spar. Except for the lateral brace, the entire undercarriage was faired inside a spat and trouser-leg of duralumin sheet. A leaf-spring tailskid was provided.

The powerplant was a 100-hp M-11 engine driving a two-blade wooden pro-

Yakovlev OKB

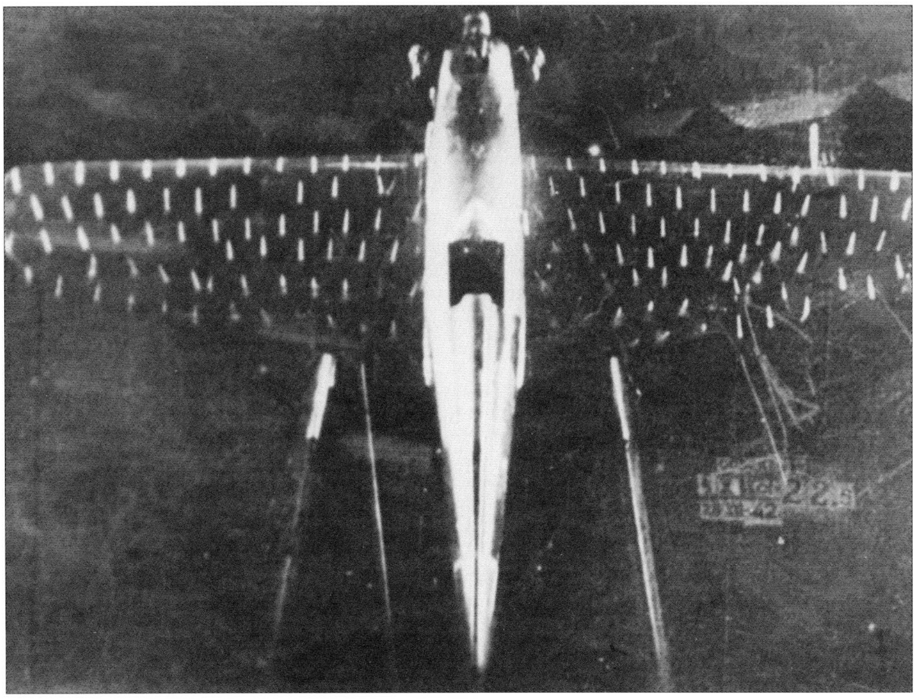

Above: A UT-1 with wool tufts for airflow visualisation in one of TsAGI's wind tunnels on 28th December 1942. The aircraft has rounded wingtips; the horizontal tail has been removed for some reason.

Yakovlev OKB

Another UT-1 in the wind tunnel; the date of the experiment is illegible in this case. This aircraft has clipped wings with angular tips. The wool tufts clearly show a problem, with whirlpool-like vortices arising at the trailing edge near the wing roots.

peller. The oil tank was immediately behind the engine; the main and upper (auxiliary) fuel tanks were accommodated between the firewall and cockpit.

The open cockpit had a small windscreen, usually V-shaped. On each side was a downwards-hinging panel for access.

AIR-14 prototype with M-11 engine

The AIR-14 prototype was powered by a standard 100-hp M-11 engine which had an exhaust collector ring. The latter feature was omitted on subsequent prototypes and production machines in favour of individual exhaust stubs on each cylinder. (One source claims that the prototype had individual exhausts, but there is photo evidence to the contrary.) A photo of the prototype also shows it to have a horn balance on the rudder, which gave place to a rudder with a straight hinge line on subsequent machines. This particular example was painted red with white stripes on the tail and red/white undercarriage fairings. (The red/white striped tail was generally characteristic of Yakovlev's early aircraft.)

The prototype made its first flight in 1936 and, upon completion of the manufacturer's tests, was submitted for State acceptance trials in the same year. The AIR-14 displayed good performance characteristics but was demanding in handling, requiring skill and a steady hand. This was considered by many as a merit rather than a drawback, because the aircraft was intended for pilots converting to the even more demanding Polikarpov I-16 fighter. Yet, the need to make the handling of the UT-1 less capricious and more suitable for average-skilled pilots was also recognised. Tackling this problem proved to be a fairly difficult and time-consuming task.

AIR-14 prototype and early production UT-1 with M-11G engine

In 1937 the AIR-14 was re-engined with an M-11G rated at 115 hp and resubmitted for State acceptance trials with improvements in response to the critical remarks made by the military test pilots. The trials results were satisfactory; the report prepared by the NII VVS pilots stated that the aircraft had excellent manoeuvrability, could easily perform aerobatics and created no problems in servicing. On the other hand, it remained as demanding as ever.

The M-11G-powered version of the AIR-14 entered production under the service designation UT-1 (*oo**chebno**-treni**rov**och-nyy* [*samo**lyot***] – trainer) and was promptly dubbed *Oo**tyo**nok* (Duckling) by witty airmen as a play on the UT designator. Series manufacture had been started (prior to the completion of the second stage of the State acceptance trials) in the autumn of 1936 at two plants. The overall production run between 1937 and 1940 totalled 1,241 machines.

The UT-1s were supplied to military flying schools and first-line units where they served for proficiency training of fighter pilots; additionally, they were often used as personal 'hacks' by commanders of various ranks who had good piloting skills. The number of UT-1s manufactured by the end of 1940 was clearly in excess of what the Air Force really needed, and attempts were made to transfer a sizeable number of these machines to Osoaviakhim. This met with some reluctance on the part of this organisation. The UT-1s were successfully used by skilled aerobatic pilots in Osoaviakhim air clubs, but the number of such pilots was relatively limited. As for the large-scale *ab initio* training of pilots, which was the main task of the air clubs, the tricky and demanding UT-1 was ill-suited for this role. A special directive

Yakovlev OKB

Yakovlev OKB

Top and above: The immaculate AIR-14 prototype during trials. The aircraft had a butch and businesslike 'baby fighter' look. Note the exhaust collector ring and the cooling air shutters around the propeller hub.

Boris Vdovenko

A production UT-1 in standard white finish with red trim. Note the straight rudder hinge line (unlike the prototype, there is no horn balance).

Specifications of the UT-1 trainer

	UT-1 with M-11G	UT-1 with M-11Ye
Length overall	5.75 m (18 ft 10⅜ in)	5.75 m (18 ft 10⅜ in)
Wing span	7.3 m (23 ft 11⅜ in)	7.3 m (23 ft 11⅜ in)
Engine power, hp	110	160
Empty weight, kg (lb)	434 (957)	429 (946)
All-up weight, kg (lb)	506 (1,116)	597.5 (1,317.5)
Maximum speed, km/h (mph)	234 (145)	257 (160)
Rate of climb, m/sec (ft/min)	6.0 (1,181)	7.4 (1,457)
Service ceiling, m (ft)	4,600 (15,090)	7,120 (23,360)
Take-off run, m (ft)	120 (390)	90 (295)
Landing run, m (ft)	190 (620)	n.a.

issued by the Defence Committee in October 1940 obliged the Osoaviakhim to accept 300 UT-1s, but considerably fewer machines were actually transferred before the outbreak of the war, when all aircraft belonging to the Osoaviakhim were turned over to the Air Force.

UT-1 with M-11Ye engine

As a mid-life upgrade (in present-day parlance), the UT-1 was fitted with the M-11Ye engine rated at 150 hp. The extra power of this engine proved to be exactly what was needed to achieve a perfect match between the engine power and the airframe weight and size. This version was manufactured in considerable numbers

UT-1 prototype with M-11F engine

The existence of this version is open to question. A Soviet publication dating back to 1961 asserts that in July 1937 a UT-1 powered by a 150-hp M-11F won an air race between Soviet sports aircraft on the Moscow-Sevastopol'-Moscow route, having attained an average speed of 238 km/h (148 mph). A later Western publication mentions the same aircraft as being powered by a 150-hp M-11Ye engine.

UT-1 with lengthened engine mount

In 1939 a UT-1 was fitted with a new engine mount whose length was increased by 260 mm (10¹⁵⁄₆₄ in). The resulting forward shift of the CG led to a considerable improvement in longitudinal stability and better handling. The powerplant was modified to permit inverted flight.

UT-1 floatplane

In 1937 an example of the UT-1 powered by a 150-hp M-11Ye was fitted with floats. The empty weight rose from 430 to 505 kg (from 948 to 1,114 lb), the AUW from 597 to 673 kg (from 1,316 to 1,484 lb); maximum speed at sea level dropped from 257 to 218 km/h (from 160 to 135 mph).

The aircraft was used by pilots Yulian I. Piontkovskiy, Valentina S. Grizodoobova, N. D. Fedoseyev and Yekaterina M. Mednikova for record-setting flights in which seven world records in the appropriate category were established. These included a speed of 218 km/h (135 mph), a range of 1,174.8 km (730 miles) and an altitude of 4,086 m (13,406 ft). Some sources credit this version with designations VT-1 and UT-1V (***vod**nyy* – in this case, waterborne), both of which are unconfirmed.

UT-1 M-11 armed fighter trainer prototypes

In 1940 the Yakovlev OKB converted an example of the UT-1 into a fighter trainer; this machine was subjected to factory tests between 19th May and 23rd August 1940. The aircraft was fitted with a synchronised 7.62-mm (.30 calibre) Shpital'nyy/Komarnitskiy ShKAS machine-gun, a PBP gunsight and a PAU-22 gun camera. This was the first armed version of the UT-1 which was developed in response to a directive of the Defence Committee under the Council of People's Commissars (= Ministers) of the USSR dated 4th March 1940. The conversion was effected under the direction of Yevgeniy G. Adler, Chief Designer of Plant No.47 named after Maksim Gor'kiy (then located in Leningrad; it was evacuated to Orenburg in 1941). During tests this aircraft showed insufficient stability and was rejected.

Somewhat later two examples of the UT-1 manufactured by Plant No.47 (c/ns 0/01 and 0/02 – that is, Batch Zero, first and second aircraft in the batch) were converted into armed fighter trainers by equipping them with a synchronised machine-gun. At the same time modifications were made to the tail unit to cure the stability problem: the tailplane area was increased by 6% and the vertical tail area by 26%. The engine mount was lengthened by 150 mm ($5\frac{29}{32}$ in).

A test report concerning these two aircraft was endorsed by General Pavel Rychagov, Deputy People's Commissar of Defence and Chief of the Air Force Directorate, on 27th March 1941. The report stated that the increased area of the tail surfaces and the change in the CG position (it was moved forward to 27.5% MAC) resulted in some improvement of stability, yet the aircraft still remained very demanding in its handling, which hampered its wide use for training service pilots. Note was also made of the cramped and uncomfortable cockpit. Hence the aircraft was not accepted for production.

Apparently it was one of these two machines (c/n 0/02) that was preserved in the Yakovlev OKB museum and demonstrated to the public on several occasions in the 1960s. This UT-1 was fitted with a ShKAS machine-gun located in the upper decking of the forward fuselage, offset to starboard. It also featured a fin of enlarged area with a blunt, rather than pointed, top of the vertical tail. Surprisingly, it also featured non-standard main gear struts which were straight instead of being forked, the wheels being mounted outboard of the struts (like those of the IL-10, for example). Later this machine was converted to represent a production-standard UT-1 with a pointed vertical tail and a normal 'trousered' undercarriage.

In some cases the UT-1s intended for air gunnery training were equipped only with a gunsight and a gun camera.

Yakovlev OKB

Yakovlev OKB

Top and above: This UT-1 in Red Army Air Force insignia was modified as a fighter trainer. Note the optical gunsight and the anti-slip rubber strip on the port wing root.

Yakovlev OKB

Above: A UT-1 (again in Red Army Air Force insignia) on skis; the skis' attachment points are hidden inside the wheel spats.

Yakovlev OKB

This UT-1 in military markings carries the Cyrillic letters 'IYe' on the tail to signify ownership by the People's Commissariat of Aircraft Industry (NKAP) and the aircraft's development status.

Yefim Gordon archive

Above: A trio of red/silver dope UT-1s operated by one of the Osoaviakhim air clubs The aircraft are civil-registered (CCCP-C5158, CCCP-C5159 and CCCP-C5256) but also wear tail numbers which are probably their fleet numbers within the club.

Yakovlev OKB

Above: This overall silver dope (except for the striped rudder) UT-1 has been fitted with floats. No water rudders are provided, directional control when afloat being provided by the standard rudder only.

Yefim Gordon archive

This UT-1 (c/n 0/02) preserved at the Yakovlev OKB museum is depicted here at Moscow-Tushino in the 1970s. Note the modified UT-2M style vertical tail. By the way, the car is a 1956-model GAZ-12.

UT-1 modified by Sutugin

In 1943 a production example of the UT-1 was modified by engineer L. I. Sutugin who re-engined it with a 200-hp M-12 five-cylinder radial and installed a machine-gun; the laminated rubber shock absorbers of the undercarriage were replaced by oleo-pneumatic shock absorbers. A small batch was produced in this configuration, but with standard M-11 engines.

UT-1b (UT-1B) wartime attack version

In 1942 an armed version of the UT-1, designated UT-1b (sometimes rendered as UT-1B), was produced as a field modification without participation of the OKB. This work was effected during the defence of Sevastopol' at the initiative of Air Major General Ostryakov, Commander of the Black Sea Fleet Air Force. The armament comprised two ShKAS machine-guns with 420 rpg mounted above the wings and two or four guide rails for 82-mm (3.22-in) RS-82 rocket projectiles (*ra**ket**nyy sna**ryad***) mounted under the wings outboard of the undercarriage legs. The wheel spats were deleted.

A batch of UT-1s was modified in this fashion and was operated by the Black Sea Fleet Air Arm with various armament options, including bombs. The UT-1Bs made their contribution to the defence of Sevastopol' and combat activities in the North Caucasus area, flying more than 2,000 close air support sorties. Their missions included suppression of enemy flak, destruction of floodlights, attacks against railway trains and other targets.

Armed UT-1 version as modified by Volkov

During the war, in 1941, armament specialist A. I. Volkov fitted a UT-1 with two 7.62-mm ShKAS machine-guns which were mounted under the wings near the undercarriage legs. Each machine-gun had a complement of 200 rounds.

Armed UT-1 version as modified by Moskatov (UT-1 Moskit)

During the war engineer K. A. Moskatov developed his own version of armament fit for the UT-1 which included two ShKAS machine-guns and rocket projectiles. It was dubbed Moskit (Mosquito).

UT-1 with two BK machine-guns (project)

In 1941 there was a proposal which envisaged equipping the UT-1 with two 12.7-mm (.50 calibre) Berezin BK machine-guns in underwing pods in a fashion similar to the BK installation on the Mikoyan/Gurevich MiG-3 fighter. One of the two machine-guns was to be mounted in the upright position and the other one in the inverted position. The idea was not put into effect.

UT-1 with Renault Bengali 4 (MV-4) engine

In 1939 an example of the UT-1 was fitted with a four-cylinder Renault Bengali 4 inverted in-line engine rated at 140 hp. The aircraft was tested with a ski undercarriage.

One source states that three UT-1s were fitted with 140-hp MV-4 (licence-built Renault Bengali 4) engines for the 1937 Osoaviakhim race.

UT-1E, alias UT-1 (15) experimental aircraft

In 1938 a modified example of the UT-1, known as UT-1E (the E presumably stands for *eksperimen**tahl'**nyy* – experimental), or UT-1 (15), was used for experiments at TsAGI. It was fitted with the so-called floating ailerons at the wingtips (such ailerons had already been tested on the AIR-4MK high-wing aircraft, see above).

The wings of this machine had a momentless airfoil section developed by F. G. Glass which was the subject of studies and experiments at TsAGI. Vadim B. Shavrov erroneously refers to this aircraft as the AIR-15 or UT-15 in his reference book on Soviet aircraft, describing it as a racing aircraft with the airfoil developed by F. G. Glass which was 'tested in early 1938 and transferred to TsAGI for investigations' (he mentions separately a UT-1 version with floating ailerons, as though this were a different aircraft).

Yefim Gordon archive

Above: Another aspect of the same aircraft at a much earlier date. This photo illustrates the non-standard landing gear with the wheels mounted outboard of the struts rather than in forked fittings.

Yefim Gordon archive

Above: Close-up of the synchronised ShKAS machine-gun on a UT-1 modified as an armed fighter trainer with the cowling panels removed. The engine mount is visible on the right.

Yefim Gordon archive

Yefim Gordon archive

Two views of a UT-1 fitted experimentally with a GAZ-Avia (GAZ-11) in-line water-cooled engine. Note the exhaust ports on the port side of the cowling and the lateral cooling air scoops aft of the wings for the radiator buried in the rear fuselage; the air outlet was presumably located ventrally.

Yakovlev OKB

Above: This UT-1 was re-engined experimentally with an MV-4 (licence-built Renault Bengali 4) inverted in-line engine.

Yakovlev OKB

Above: The UT-1E at the OKB's premises, showing the 'floating' ailerons. Curiously, both ailerons are deflected fully up.

UT-1 aerodynamics testbed

An example of the UT-1 was experimentally fitted with wings incorporating a VVS (Air Force) airfoil. The modification proved to be disappointing because the landing speed increased.

UT-1 development aircraft with retractable undercarriage and enclosed cockpit

In 1938 an example of the UT-1 powered by an unidentified version of the M-11 engine was equipped with a retractable undercarriage (the main units retracted inwards) and an enclosed cockpit. It makes an interesting comparison with the later AIR-18 (see below) from which it differed only in being fitted with a radial engine and in having no wheelwell doors on the undercarriage legs.

UT-1 development aircraft with automatic flaps (1941 and 1944)

In 1941 an example of the UT-1 was fitted with TsAGI-type automatic flaps. Another source mentions the testing of this aircraft or a similarly modified UT-1 at the Flight Research Institute of the People's Commissariat of Aircraft Industry (LII NKAP – ***Lyot**no-is**sled**ovatel'skiy insti**toot***) in 1944.

UT-1 development aircraft with GAZ-11 (GAZ-Avia) engine

In 1939 one example of the UT-1 was experimentally fitted with a modified GAZ-11 liquid-cooled six-cylinder in-line automobile engine with a displacement of 3,485 cc developed by the Gor'kiy Automobile Factory (GAZ – ***Gor'**kovskiy avtomo**bil'**nyy za**vod***). The aircraft version of this engine was designated GAZ-Avia and rated at 92 hp (some sources quote 85 hp). The aircraft had a characteristic pointed nose with an angular cowling and a large propeller spinner. Cooling air was routed to the water radiator buried in the rear fuselage by two shallow intakes aft of the wing trailing edge.

This adaptation was made by Ye. V. Agitov who experimented with using automobile engines on several types of light aircraft. The engine could run on low-octane petrol and was cheaper than the M-11 but had a poorer power/weight ratio (automobile engines are typically heavier than their purpose-built aviation counterparts).

UT-1s fitted experimentally with different propellers

Several types of propellers were tested on the UT-1. No details are available.

UT-1 remote-controlled target drone conversion

According to one account, a radio-controlled version of the UT-1 was tested before the war. The UT-1 and Polikarpov's U-2 biplane trainer were the first radio-controlled unmanned target drones developed in the Soviet Union. A team led by Prof. Nikol'skiy at the NII-20 research institute undertook the development effort.

There are reasons to suspect that the mentioned account contains an error, the radio-controlled aircraft being in fact a UT-2. Compare with the description of the radio-controlled UT-2 below, where, incidentally, the research institution is cited as NII-22.

Yakovlev OKB

A close-up of the UT-1's wing flaps.

'AIR-15 (UT-15) (UT-1E, UT-1 (15), UT-1 No.15)'

Shavrov attributes the AIR-15 designation to an experimental version of the UT-1 described above as the UT-1E. In the light of recent research by Yuriy Zasypkin, Shavrov appears to be wrong; the AIR-15 designation belongs to a different aircraft (see below).

AIR-15 single seat-aircraft (project)

This was a project of a single-seat fighter trainer powered by an MM-1 air-cooled inverted in-line engine rated at 230 hp. An archive drawing shows this machine to be a low-wing cantilever monoplane with an enclosed cockpit and a retractable undercarriage; the aircraft bore a superficial resemblance to the Caudron C-690 and C-713 light fighters. The project was not proceeded with.

AIR-16 (LT-2) passenger aircraft prototype

The AIR-16 (also known as LT-2, or Aircraft No.16) was a four-seat passenger aircraft powered (according to the project) by a 220-hp Renault Bengali engine. It was a low-wing cantilever monoplane with fixed spatted mainwheels, externally similar to the AIR-11.

via RART

via RART

Top and above: The UT-1B light strike version developed at the initiative of Maj. Gen. Ostryakov was armed with two overwing ShKAS machine-guns and two launch rails for RS-82 rockets.
Below: A very similar attack version called UT-1 Moskit was developed by engineer K. Moskatov.
Bottom left: Close-up of the twin rocket launch rails and the port main gear unit on the UT-1 Moskit.
Bottom right: Close-up of the starboard ShKAS machine-gun on the UT-1 Moskit.

via RART

via RART

via RART

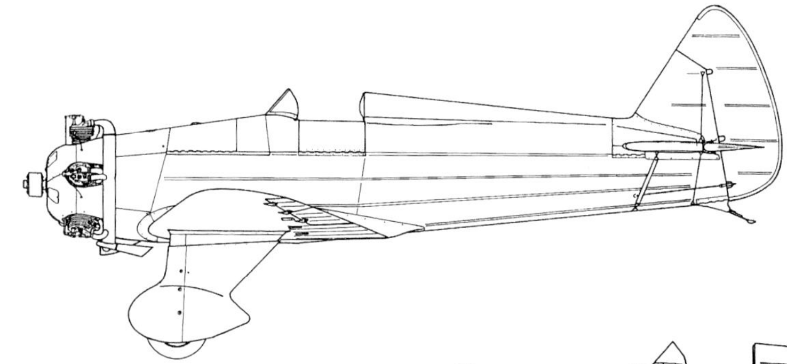

The AIR-14 prototype

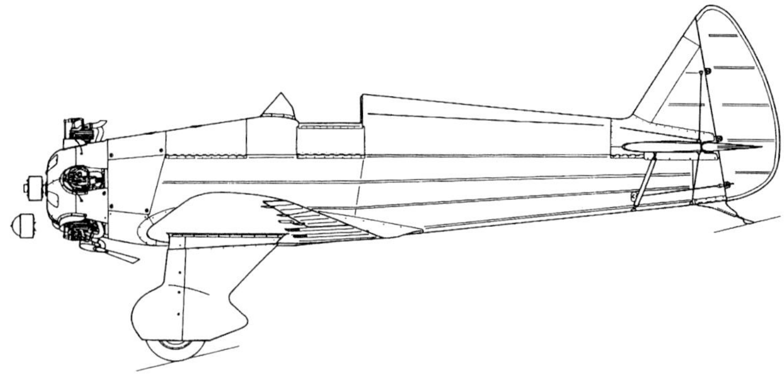

A standard production UT-1
(c/n 150038)

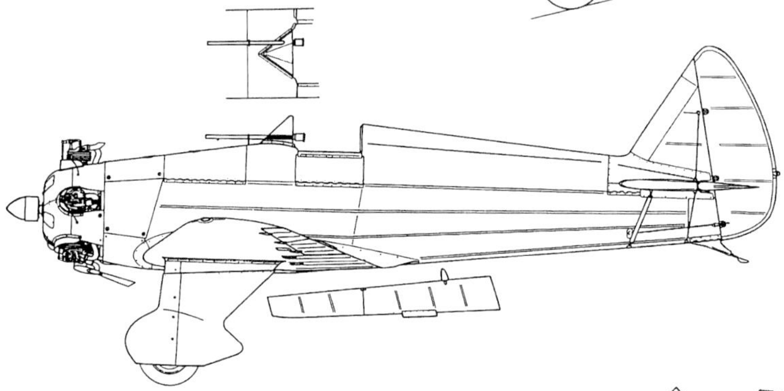

A UT-1 with an OP-1 gunsight

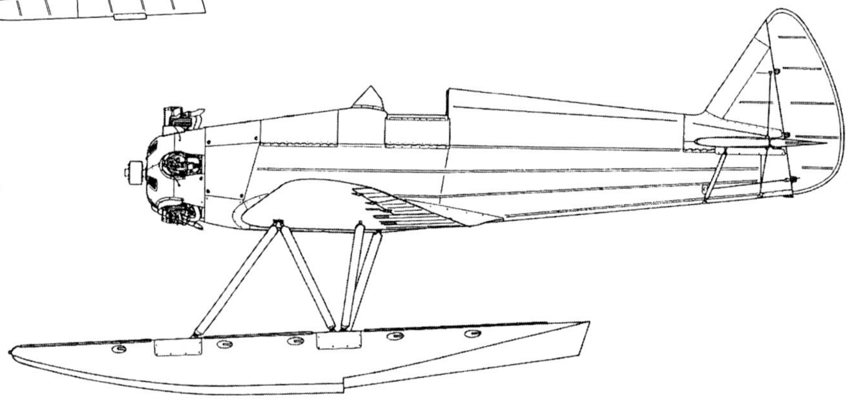

A UT-1 fitted experimentally
with floats

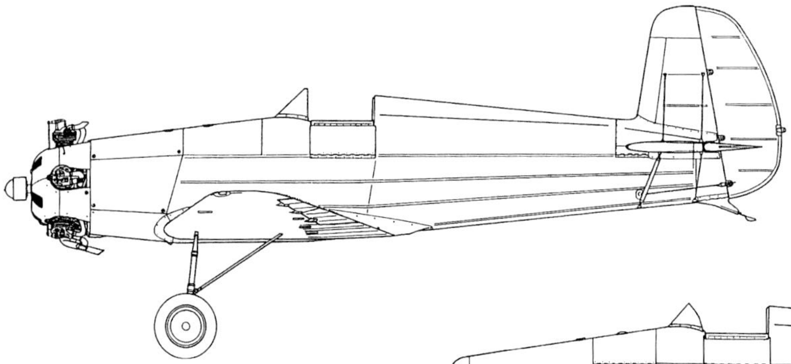

Modified UT-1 c/n 0/02

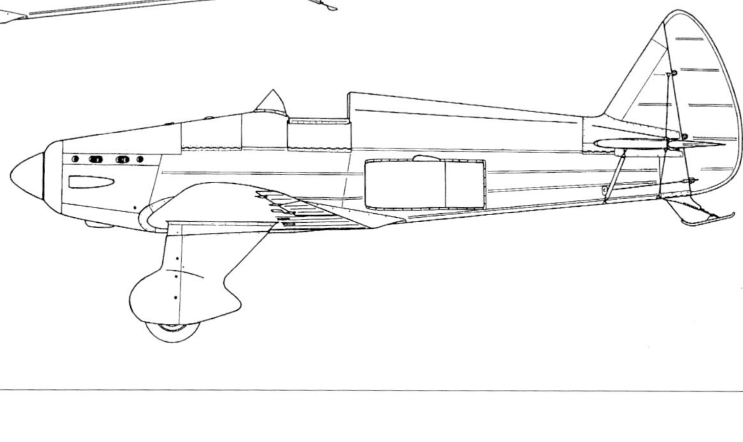

The UT-1 fitted experimentally
with a GAZ-11 engine

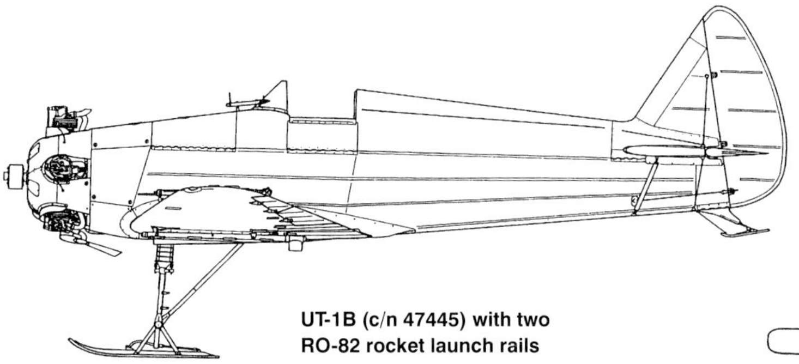

UT-1B (c/n 47445) with two RO-82 rocket launch rails

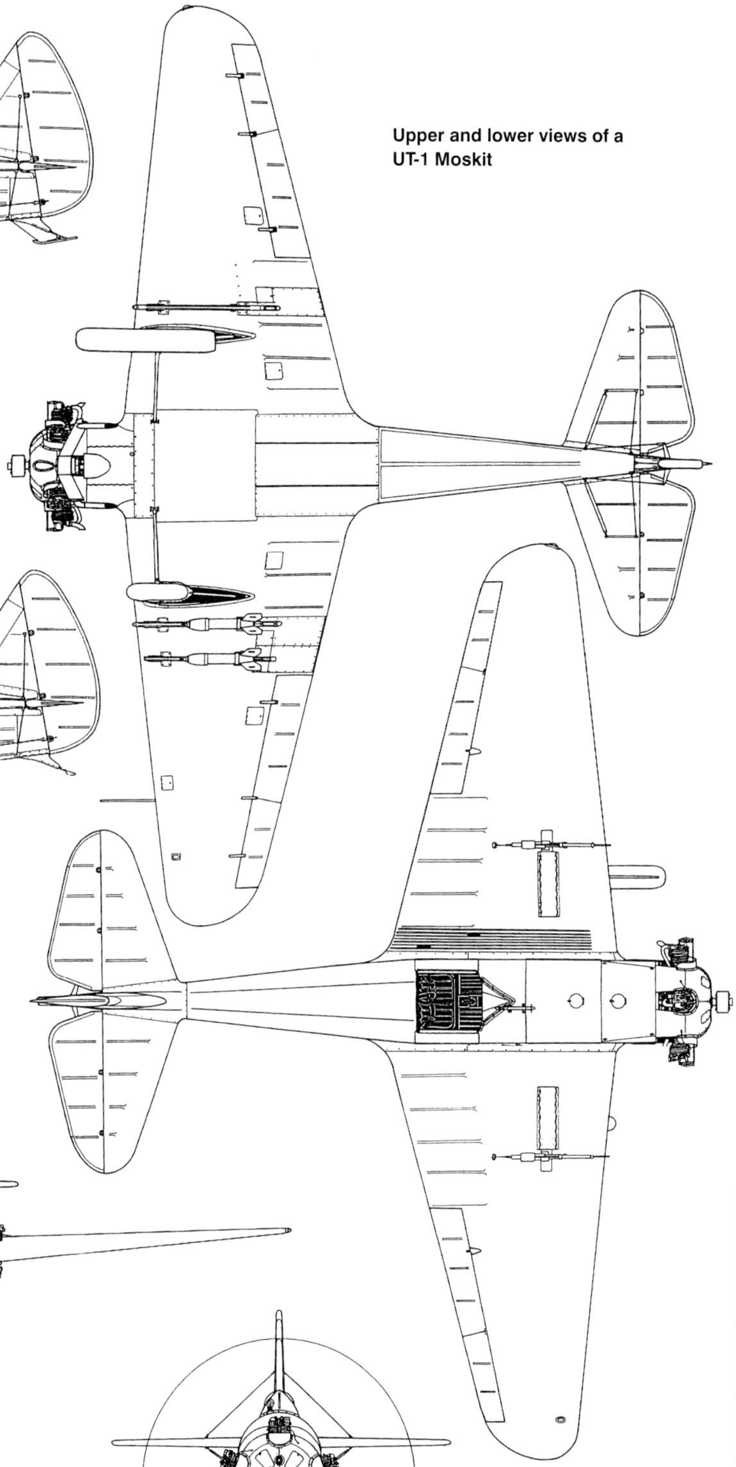

Upper and lower views of a UT-1 Moskit

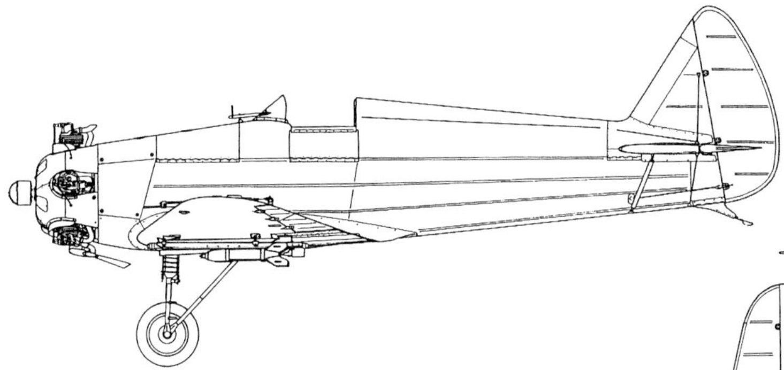

A UT-1 Moskit with four RO-82 launch rails

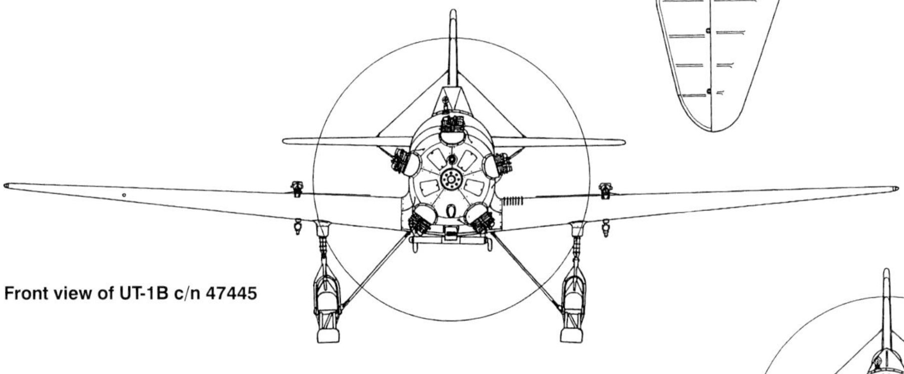

Front view of UT-1B c/n 47445

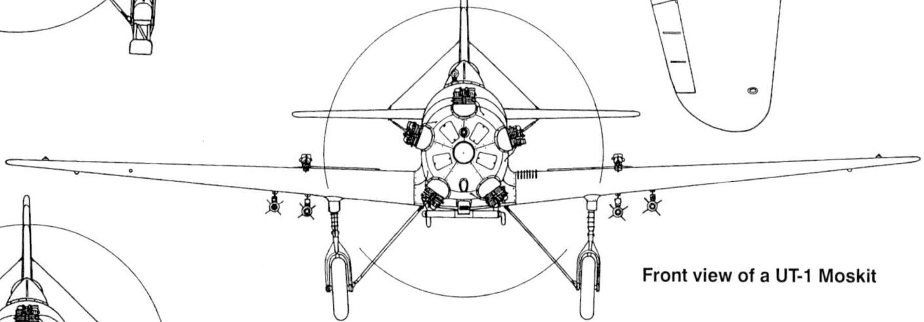

Front view of a UT-1 Moskit

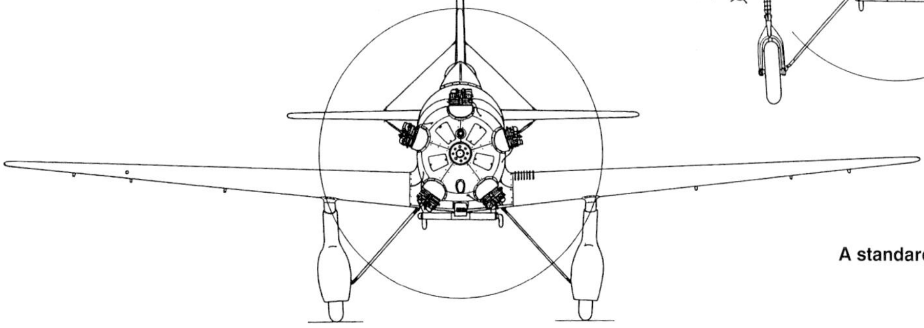

A standard production UT-1

Boris Vdovenko

Boris Vdovenko

Top and above: A day at the races, Part II. Unfortunately, despite its appealing looks, the sole prototype AIR-16 cabin monoplane (seen here beside the AIR-12 racer) never flew, much less raced.

The single prototype built in 1937 was never flown and had to be relegated to the role of a static exhibit. This is usually explained by the absence of the engine.

A different explanation is given by Yevgeniy Adler in his memoirs. Due to the inexperience of Vladimir Kotov, project engineer for this machine, it came to incorporate so many inherent weaknesses that all efforts to rectify them proved useless and there was no point in flight testing (Adler, as Kotov's successor, shared responsibility for this failure).

Yakovlev OKB

Above: The AIR-17 (Ya-17) bomber trainer prototype on wheels. Note the rudder horn balance, the opaque nosecone ahead of the navigator's station and the open glazing of the dorsal gunner's cockpit.

Yakovlev OKB

A Ya-17 (UT-3) trainer on skis, showing the greatly increased area of the nose glazing for the benefit of the navigator trainee. No armament is fitted to this one.

AIR-17 ('aircraft No.17', S-17, Ya-17, UT-3, M-17) bomber trainer

Prototype with Renault engines and French propellers

In 1938 Yakovlev and his team built the prototype of their first twin-engined aircraft intended to fill the bomber trainer role. It was the seventeenth aircraft in the succession of Yakovlev's machines, and apparently at its inception it bore the designation AIR-17. However, by the time the first prototype was rolled out in the spring of 1938, the AIR initials were already taboo after Aleksey I. Rykov's arrest, and the machine was referred to as '*samo**lyot*** No.17' (Aircraft No.17), often shortened to S-17. In the OKB documents it was also designated Ya-17.

This was a cantilever low-wing monoplane of mixed construction. The wings were built as a one-piece structure featuring two box-section wooden spars tapering towards the tips and truss-type wooden ribs. The skin was plywood up to the rear spar and fabric aft of it. The slotted ailerons had wooden frames and fabric skinning. Duralumin split flaps divided by engine nacelles occupied the entire trailing edge between the centreline and the ailerons. The fuselage was a welded steel-tube truss reinforced with bracing wires. The forward fuselage had duralumin skinning, the aft fuselage was covered with fabric.

The crew of three comprised a navigator, a pilot and a gunner/radio operator. The latter two crewmembers sat above the wings in separate cockpits joined by a fairing, while the navigator was accommodated in the glazed nose; his place was occupied by an instructor pilot when the aircraft was used for pilot training. The lower glazing of the nose cabin was used for bomb aiming.

Placed aft of the pilot's cockpit was a small bomb bay. The tail surfaces had duralumin frames covered with fabric. The tailplane was reinforced with bracing struts.

The S-17 prototype was powered by two Renault 6Q-01 Bengali engines of French manufacture, rated at 220 hp. The Soviet licence-built version of this six-cylinder inline air-cooled engine was designated MV-6. The engines drove French Ratier 1363 two-blade variable-pitch metal propellers. Two 310-litre (68.2 Imp gal) fuel tanks were housed in the wing roots.

The main undercarriage units retracted aft into the engine nacelles and had hydraulic shock absorbers. The castoring tailwheel was fixed. Provision was made for non-retractable skis for winter operation. The aircraft had dual controls, the second set of controls being fitted in the navigator's cockpit.

The bomb armament comprised such options as six 50-kg (110-lb) FAB-50 high-explosive bombs or three 100-kg (220-lb) FAB-100s, the normal bomb load being 300 kg (660 lb). For self-defence the aircraft was to be armed with two flexible 7.62-mm ShKAS machine-guns, one of which was placed in the navigator's cockpit and the other one in the rear cockpit on a TUR-8 mount.

Upon completion of the manufacturer's tests the S-17 was submitted to NII VVS for State acceptance tests which began on 16th May 1938. Military test pilots gave a generally positive appraisal to the aircraft but noted some drawbacks and recommended a few modifications, including the introduction of detachable wing panels to simplify transportation and repairs. The overall conclusion tasked the OKB (plant No.115) with both rectifying the defects and starting preparations for series manufacture.

On 26th August 1938 the prototype was redelivered to NII VVS for a second round of testing. By then the machine had received the official designation UT-3; occasionally the designation M-17 also cropped up in documents. The prototype had undergone a number of modifications: the tail structure was stiffened, an electro-hydraulic system was installed for operating the flaps and undercarriage, for greater reliability the French propellers were replaced by Soviet wooden two-blade propellers without spinners. The engine cowlings were slightly altered for better cooling and the mainwheels were fitted with pneumatically-operated disc brakes. A full complement of armament was installed (initial testing was conducted with only the gun mounts fitted).

The new round of testing confirmed the previous positive impression of the aircraft. The UT-3 was deemed suitable for the conversion of pilots to high-speed bombers and for combat training of pilots, navigators, observers and gunner/radio operators under conditions coming close to those of a high-speed bomber.

Some deficiencies persisted, however, and, concurrently with preparations for series production, the OKB continued its efforts aimed at perfecting the prototype which was subjected to service trials during September and October 1938.

UT-3-2MV-6 three-seat initial production version

UT-3 production was launched initially at plant No.81 at Moscow-Tushino. The first batch of ten was to be produced by 1st September 1939, but due to some delays in the tooling-up the first production example was not completed until 4th September. It made use of some parts manufactured by plant No.115 and was fitted with locally-built MV-6 engines driving Soviet-made AV-3 variable-pitch propellers. By that time the plant had produced fuselages and some other components for the remaining nine machines.

Somewhat later, production drawings were also transferred to plant No.301 in Khimki, a northern suburb of Moscow, which had likewise been tasked with manufacturing the UT-3. In August 1939 the decision was taken to concentrate all the production of the UT-3 at plant No.301 and the incomplete airframes from plant No.81 were transferred thereto. After that, plant No.301 set about manufacturing the first batch of five machines, to be followed by a second batch

Yakovlev OKB

Above: An early 'notch-nose' three-seat UT-3 2MV-6 in flight. The ski landing gear was non-retractable.

Yakovlev OKB

Yakovlev OKB

Centre and above: Later UT-3s were two-seaters with an extended parabolic nose and a redesigned rudder. This one stood on its nose when landing on a snow-covered airfield but suffered almost no damage.

Yakovlev OKB

Above: This two-seat UT-3 2MV-6 built by plant No.47 appears to be much the worse for wear, having shorn off the landing gear in a heavy landing. Note the bay replacing the dorsal gunner's cockpit.

of five aircraft. The first machine (c/n 301101 – that is, plant No.301, Batch 1, 01st aircraft in the batch) was completed on 1st September 1939. It was submitted for testing at NII VVS in December 1939 and January 1940. The production machine turned out to be much heavier than the prototype and, importantly, had a CG located considerably farther aft. All this resulted in markedly worse performance, stability and controllability compared to the prototype. One of the test pilots, Pyotr M. Stefanovskiy, dismissed the production UT-3 as unsuitable for the trainer role, citing poor stability and controllability, as well as the fact that its handling characteristics were very far from those of a combat aircraft.

UT-3 2MV-6 – two-seat unarmed production version

As soon as the first flights of the UT-3 c/n 301101 revealed its vices, plant No.301 took a decision to assemble the rest of the production batch in a two-seat version without armament. (The aircraft's handling proved acceptable only when used in the pilot training configuration – that is, without the dorsal gunner, his machine-gun and its mount and without bombs and bomb racks.) It is in this configuration that all subsequent machines were completed at this plant (externally the aircraft remained basically similar to the prototype).

Nine UT-3s from the first production batches were sent to Air Force units for service tests which were conducted between 15th May and 15th August 1940. The nine machines had c/ns 102, 103, 201, 202, 203, 204, 301, 302 and 303. In the course of the test various design faults came to light, compounded by poor manufacturing standard and the resulting numerous failures and accidents. Not surprisingly, test personnel reached the conclusion that the UT-3 2MV-6 aircraft, in the version presented, had failed to pass the service tests. The negative results of the tests played their part in the decision to discontinue series production of the UT-3 at plant No.301, which switched over to the manufacture of the new Yak-1 fighter.

However, in early 1940 one more factory, the Leningrad plant No.47, started tooling up for UT-3 production. Some of the incomplete airframes were transferred there from plant No.301. Yakovlev, in his capacity of Deputy People's Commissar of Aircraft Industry, appointed one his associates, Yevgeniy G. Adler, as Chief Designer of plant 47, despite the fact that the plant already had a Chief Designer in the person of G. I. Bakshayev who headed a resident OKB set up in 1939. Development of the UT-3 at plant No.47 is detailed below.

UT-3 2MV-6 two-seat production version developed by plant No.47

Plant No.47 built its own version of the UT-3 incorporating changes made to address the complaints voiced during State acceptance trials. It was a two-seater with the rear cockpit faired over and the front cockpit (navigator's station) featuring revised contours without the characteristic step. The upper part of the nose became fully glazed. The propeller hubs were fitted with spinners. There was no armament, no radio was fitted. The first production Leningrad-built UT-3 (c/n 01147 – that is, Batch 01, first aircraft in the batch, plant No.47) was completed on 26th July 1940.

The introduction of the UT-3 at plant No.47 was accompanied by an NKAP decision to commit all the staff of the Bakshayev-led OKB to this task, to the detriment of the OKB's own projects (Bakshayev had been working on the project of the RK-I variable-wing fighter which was terminated in March 1940). Naturally, this did not endear the UT-3 to Bakshayev and his staff; Bakshayev voiced strong criticism of the UT-3 which, in his opinion, did not meet the Air Force requirements and had some inherent drawbacks that could be rectified only at the cost of a drastic redesign. Small wonder that UT-3 production at plant No.47 progressed at a snail's pace (only eight machines had been built by 1st November 1940); further work on improving the production machines was inspired chiefly by Ye. G. Adler who was responsible for development of the version described below.

UT-3 'standard-setter for 1941' (S-17A, UT-3 2MV-6A, UT-3M)

(standard-setter No.1/plant No.47)

This version of the UT-3 was developed at production plant No.47 under the direction of Yevgeniy Adler, Yakovlev's engineer seconded to this plant. Designated S-17A, or 'standard-setter for 1941', it was optimised for the conversion trainer role, differing considerably from the original S-17. It was a two-seat aircraft in which the trainee's and instructor's seats were accommodated under a common canopy close to each other. The seat of the trainee was

Yakovlev OKB

A UT-3 (S-17A) nearing completion at plant No.47. This close-up gives a good view of the port MV-6 engine.

placed forward, offset to port, and the instructor's seat behind it was offset to starboard to provide forward view. The navigator's station in the forward fuselage was deleted, together with its glazing, and the nose became 'solid'.

The undercarriage was now fixed, the mainwheel legs being provided with bracing struts. The tailwheel had a hydraulic shock absorber. The hydraulic system was deleted; the flaps were now operated pneumatically. The wing incidence was changed from zero to 3° and the stabiliser incidence was increased from –0°30' to –2°.

Many improvements were introduced into the powerplant. The machine was fitted with two MV-6A engines which featured cooling fins of greater area on the cylinder heads, bigger oil coolers and cartridge starters. The latter proved too dangerous in operation to win favour with the technical staff. The cooling of the cylinders was also improved by providing additional openings and air outlet gills in the engine cowlings. The length of the engine mounts was increased by 150 mm ($5\frac{29}{32}$ in) to shift the CG forward. This made sure that the modified machine retained the benign handling of the original UT-3 prototype.

Upon completion of initial factory tests the S-17A was turned over to NII VVS for State acceptance trials which took place in Moscow in April-May 1941. The tests revealed a noticeable improvement in performance, including better longitudinal stability and an increase in speed, despite the fixed undercarriage. This was due partly to the reduction of the AUW by some 450 kg (990 lb). The aircraft passed the State acceptance trials successfully and was recommended for series production as the UT-3 2MV-6A (Ye. G. Adler refers to it in his memoirs as the UT-3M). Some deficiencies remained, however, and a number of improvements was planned to iron them out (some of the projected versions are detailed below). The aircraft seemed, at last, to have bright prospects for series manufacture, but the outbreak of the war radically changed the situation, shifting priorities to combat aircraft. In addition, series production of both the MV-6 and the MG-31 radial engine was halted. Thus, only a very limited number of production UT-3s were completed, and the type played no operational role.

UT-3 with MG-31F engines (project)

In March 1941 the Soviet Air Force command endorsed a new specific operational requirement (SOR) for a two-seat trainer. One of the projects developed to meet these requirements was a project for a UT-3 powered by two Kossov MG-31F nine-cylinder radials rated at 350 hp. It was a two-seat machine with a 'solid' nose, externally distinguishable from the S-17A by its 'fatter' cylindrical engine cowlings.

Yakovlev OKB

Yakovlev OKB

Yakovlev OKB

Three views of a much-modified UT-3 version developed by plant No.47 and designated UT-3 2MV-6A or S-17A. The 'solid' nose and the common canopy for the trainee and the instructor are readily visible.

UT-3 'standard-setter No.2' (plant No.47) (project)

One more project prepared by plant No.47 was the UT-3 'standard-setter No.2'. It differed from the 'standard-setter No.1' in having a revised tail unit with twin fins and rudders. The wings were redesigned, the outermost parts being slightly swept back in relation to the inboard wing section (presumably for CG reasons). This resulted in a characteristic kink on the leading and trailing edges.

Specifications of the UT-3

	UT-3 2MV-6 Three-seat production version	UT-3 2MV-6A (S-17A, standard-setter for 1941)
Engine type	MV-6	MV-6A
Engine power, hp	2 x 220	2 x 220
Length overall	10.70 m (35 ft $1\frac{17}{64}$ in)	10.83 m (35 ft $6\frac{3}{8}$ in)
Wing span	15.0 m (49 ft $2\frac{1}{2}$ in)	15.0 m (49 ft $2\frac{1}{2}$ in)
Wing area, m^2 (sq ft)	33.42 (360)	33.42 (360)
Empty weight, kg (lb)	2,040 (4,497)	2,042 (4,502)
All-up weight, kg (lb)	2,900 (6,390)	2,627 (5,791)
Maximum speed, km/h (mph)	273 (170)	260 (161.5)
Landing speed, km/h (mph)	90 (56)	95 (59)
Service ceiling, m (ft)	4,000 (13,123)	6,200 (20,340)
Range, km (miles)	1,000	1,050 (652)
Take-off run, m (ft)	319 (1,046)	245 (800)
Landing run, m (ft)	230 (755)	115 (377)

Yakovlev OKB

Above: The AIR-18 racing aircraft powered by a Renault Bengali 4 engine. Note the large glazing area and the landing gear retraction actuators mounted outboard of the struts.

Yakovlev OKB

No, this is not an AIR-18 powered by an M-11 radial but an earlier experimental version of the UT-1.

UTPB dive-bombing trainer (project)

The projected derivatives of the UT-3 included the UTPB (*oo**cheb**no-treni**rov**och-nyy pi**kee**ruyushchiy bombardi**rov**shchik* – dive-bomber trainer). This machine was to be equipped with dive brakes under the wings.

Design work on this version was conducted in 1940 under the direction of G. I. Bakshayev, Chief Designer at the production plant No.47. He had proposed his own project of a twin-engined dive-bomber trainer designated UPB 2MV-6, but on 10th August 1940 his project was rejected by a standing commission of NKAP in favour of modifying the UT-3 for this role. Apparently Bakshayev did some work pursuant to this decision, but the UTPB prototype was never built.

AIR-18 racing aircraft prototype

The AIR-18 was a single-seat racing monoplane built in 1937. Based on the UT-1 trainer's airframe, it was fitted with a Renault Bengali 4 engine, a retractable undercarriage and an enclosed cockpit. This machine achieved a speed in excess of 300 km/h (186 mph).

AIR-18 on floats

According to some accounts, the AIR-18 was also tested on floats. No pictorial evidence to confirm this is available.

AIR-19 (S-19, Ya-19) passenger aircraft prototype

In 1938, concurrently with the work on the UT-3 trainer, the Yakovlev OKB undertook development of its civil version designated S-19, or Ya-19. (Shavrov also cites the AIR-19 designation as the last in the AIR series, but it could have existed only inside the OKB at the earliest stage of the project.) The airframe of the Ya-19 combined the wings, tail surfaces, powerplant and undercarriage of the UT-3 with a new, roomier fuselage. The pilot's cockpit was moved forward and the navigator's cockpit was deleted and replaced by a shorter nose fairing. The steel-tube truss forming the passenger cabin received higher sides with plywood skinning and a ceiling. There were five passenger seats, two to port and three to starboard; the passenger cabin was accessed through a door on the port side. Placed aft of the cabin was a small baggage hold. The aircraft was powered by Soviet-manufactured MV-6 engines; the electric equipment included a landing light.

After the manufacturer's tests conducted by OKB pilot Yulian I. Piontkovskiy, in October 1939 the aircraft was transferred to NII GVF for State acceptance trials which it passed with good results.

Specifications of the AIR-18

Length overall	5.99 m (19 ft 7⅞ in)
Wing span	7.3 m (23 ft 11⅜ in)
Wing area, m^2 (sq ft)	9.58 (103)
Empty weight, kg (lb)	475 (1,047)
All-up weight, kg (lb)	645 (1,422)
Maximum speed, km/h (mph)	310 (193)
Landing speed, km/h (mph)	85 (53)
Service ceiling, m (ft)	6,500 (21,325)
Range, km (miles)	600 (373)
Take-off run, m (ft)	120 (394)
Landing run, m (ft)	160 (525)

Ya-19 production version (project)

The proposed production version of the Ya-19 was to be fitted with variable-pitch propellers and a V-3 radio intended for radio beacon navigation. The Civil Air Fleet showed much interest for the Yak-19 which was regarded as a replacement for the Putilov Stal'-3 and the Kalinin K-5 on local services, and plans were made to introduce the new aircraft on some routes. However, a prerequisite for the series manufacture of the Ya-19 was the launching of mass production of the baseline UT-3 which, as stated above, failed to materialise due to the outbreak of the war.

Ya-19 ambulance version (project)

The OKB proposed to the VVS an ambulance version of the Ya-19. It was intended to take on board two stretcher cases, one walking wounded and a medical attendant. The stretchers were placed in two tiers replacing the three seats on the starboard side. The door was enlarged and given a trapezoidal shape. A full-size mock-up of this version was built in October 1939 and approved by the mock-up review commission which recommended that a standardised version easily convertible from passenger to ambulance configuration and back. For reasons stated above this recommendation could not be implemented.

UT-2 (Ya-20) trainer

In 1936-37 Yakovlev reworked the AIR-10 trainer to meet the requirements of series production and some of the wishes expressed by the military. Notably, they wanted the airframe to be stressed to a design load factor of 10, which would clear the aircraft for unlimited aerobatics. This requirement was met, whereas some of the

Specifications of the AIR-19 (Ya-19)

Length overall	10.83 m (35 ft 6⅝ in)
Wing span	15.0 m (49 ft 2½ in)
Wing area, m² (sq ft)	33.42 (360)
Empty weight, kg (lb)	2,134 (4,705)
All-up weight, kg (lb)	2,950 (6,500)
Maximum speed, km/h (mph)	271 (168)
Landing speed, km/h (mph)	86 (53.5)
Service ceiling, m (ft)	5,600 (18,375)
Range, km (miles)	783 (487)
Take-off run, m (ft)	410 (1,345)
Landing run, m (ft)	365 (1,200)

requested features (such as a crash pylon to protect the crew in the even of a nose-over) had to be waived for reasons of production simplicity.

The new machine was designated Ya-20, or simply 'aircraft No.20' (the AIR designations having been abandoned by then). The most important new feature of the aircraft was the introduction of an all-wooden fuselage structure instead of the welded tubular truss. Thus, practically the whole of the airframe was made of wood, with the exception of joints and some highly stressed parts where steel and duralumin were used. The wing structure was revised, the wooden box-beam spars being replaced by spars made of solid machined spruce planks. There were also numerous detail changes in the spar/rib structure. There were no flaps, and he ailerons were the usual duralumin/fabric two-piece assemblies with slotted profiles and plywood leading edges.

The fuselage was based on a frame with four continuous longerons. Two large bays above the wing accommodated the tandem cockpits. As on the AIR-10, the cockpits were fitted with fold-down doors on each side. Above the main truss was a deep secondary structure giving a rounded top. The tail was a duralumin structure with flush-riveted leading edges over which the fabric was wrapped. The horizontal tail was untapered and the vertical tail, with its pointed top, was typical for Yakovlev's aircraft of that period. Unlike the UT-1, the rudder had a large balance horn at the top. The rudder and elevators were operated by cables. The main undercarriage was almost identical to that of the AIR-10, featuring unbraked wheels and rear struts faired inside a spat and a trouser fairing.

On the whole, in comparison with the AIR-10 the Ya-20 was an almost completely new airframe, albeit virtually identical externally to its predecessor.

The AIR-10 with its 100-hp M-11 engine was considered underpowered; an engine delivering 140 to 150 hp was required. No

Yakovlev OKB

Yakovlev OKB

Top and above: The sleek AIR-19 (Ya-19) passenger aircraft was developed as commercial derivative of the UT-3 and would have undoubtedly gone into production, had not the outbreak of the Great Patriotic War ruined these plans.

Yakovlev OKB

Above: The second prototype Ya-20 trainer, the prototype of the future UT-2, with an M-11Ye radial. Note the number '3 Red' on the underside of the starboard wing.

Yakovlev OKB

The first prototype of the Ya-20, seen here at the premises of the Yakovlev OKB, represented the 'competing' version powered by Renault Bengali 4 in-line engine.

Yefim Gordon archive

Above: A further M-11Ye-powered Ya-20 prototype (or perhaps an initial-production UT-2) fitted out with test equipment; note the sensors on the tail recording control surface deflection.

Yefim Gordon archive

The same aircraft on the NII VVS hardstand at Chkalovskaya (note the characteristic hexagonal concrete pavement slabs).

Soviet engine with that rating was in production at that time, available options being the French series-produced Renault Bengali 4 rated at 140 hp and the Soviet M-11Ye rated at 150 hp that was still at a test stage. Both options were used in the construction of prototypes. There is some ambiguity in the available sources as to which of the two prototypes (described below) was the first to appear, but most accounts state that the first prototype was powered by the Renault engine.

Ya-20 (No.20) Renault-powered trainer prototype

The prototype powered by the inverted in-line air-cooled Renault Bengali 4 engine featured a neatly fitted tapered cowling with the cooling air inlet on the left and the carburettor inlet on the right. It was pained white overall with a red-striped rudder, red wing leading edges and front parts of the undercarriage spats. Initially there was no number on the fuselage, but later the number '3 Red' was writ large on each side of the fuselage, on the port wing's upper surface and on the starboard wing undersurface.

The aircraft was promptly subjected to manufacturer's tests in the course of which it took part in one of the regularly held air races. In July 1937 the Renault-powered prototype took second place in an air race on the Moscow-Sevastopol'-Moscow route (another report states that test pilots Pyotr M. Stefanovskiy and P. I. Nikitin flying a UT-2 prototype with a Renault Bengali 4 engine won first place among two-seat aircraft in a racing flight of Soviet sports aircraft which was held on 24th July 1937). Contrary to expectations, the French engine proved to have some deficiencies, and attention was turned to the M-11Ye radial.

Ya-20 (No.20) M-11Ye-powered trainer prototype

The second prototype fitted with the 150-hp M-11Ye development engine was built in 1937 and, together with the Renault-powered prototype, passed factory testing in the same year. The two machines differed in fuel tankage: the first Ya-20 had only a single 75-litre (16.5 Imp gal) fuel tank, while the second (M-11Ye-powered) example had two welded tanks, each of 90 litres (19.8 mp gal) capacity, in the wing roots between the spars. The two initial Ya-20s, together with a third example featuring increased tankage,

Yefim Gordon archive

An impressive formation of UT-2s operated by Osoaviakhim, including CCCP-C5728, CCCP-C5713/'7 Black', CCCP-C5733 and CCCP-C5727.

were tested in the summer of 1937, and in September it was decided to put the aircraft into series production as the standard conversion trainer coming between the U-2 and the combat types, as the UT-2.

UT-2 trainer (initial production version)

The initial production version of the UT-2 (sometimes unofficially referred to as 'standard-setter for 1938') was normally powered by the M-11G engine with a maximum rating of 110 hp at 1,650 rpm and a nominal rating of 100 hp at 1,600 rpm, or the similarly rated M-11M. The set of instruments fitted to this version comprised an altimeter, a compass, a three-pointer engine gauge unit, an airspeed indicator, a turn and bank indicator and a vertical speed indicator.

The production UT-2s behaved excellently in the hands of experienced pilots, but their operation in first-line Air Force units where they were often flown by average-skilled or inexperienced pilots brought to the fore some problems. The most serious one was the aircraft's spinning characteristics. The UT-2 was prone to entering a flat spin and recovered reluctantly from it due to the aft CG position. The OKB, aided by TsAGI specialists and test pilots A. A. Sinitsyn and V. L. Rastorguyev, undertook serious research in an effort to rectify the problem. For the first time in Soviet practice a spin recovery parachute was used during test flights to ensure safe recovery. This speeded up the work, and eventually the problem was cured. The research results were incorporated in the production version described below.

UT-2 production trainer (standard-setter for 1940)

To improve the UT-2's spinning characteristics, the following modifications were introduced. The forward fuselage was lengthened by 150 mm ($5\frac{29}{32}$ in), which produced a 3% forward shift of the CG. Stabiliser incidence was changed from –0°30' to –1.5°. The fin was set at an angle of 30' to port relative to the aircraft's centreline to counteract the tendency towards the left turn. The machine thus modified was dubbed 'standard-setter for 1940'; it also incorporated a number of other improvements unrelated to spin.

UT-2 trainers manufactured to the 1940 standard were powered by M-11D engines with a maximum rating of 125 hp at 1,760 rpm and a nominal rating of 115 rpm at 1,700 rpm. The installation of this engine necessitated strengthening of the engine mount. A number of other airframe members were re-stressed to enhance airframe sturdiness. The flight instrument set of this version was

Yakovlev OKB

Above: A single UT-2 was manufactured with a Soviet-built MV-4 engine after all. Here it appears to bear the number '2 Red'.

Yefim Gordon archive

Yefim Gordon archive

Two views of the same MV-4 powered UT-2 in Red Army Air Force markings undergoing tests at NII VVS.

Yefim Gordon archive

Yefim Gordon archive

Two views of a production UT-2 in a camouflage scheme. In-service aircraft were often flown without wheel spats which tended to get clogged on muddy airstrips.

Yefim Gordon archive

Above: This UT-2 has been fitted out to carry a small torpedo (obviously a practice weapon) and equipped with an appropriate sight in the front cockpit.

Yakovlev OKB

This aircraft is the UT-2N (alias SEN) testbed for an experimental air cushion landing gear developed by A. D. Nadiradze. The fan forcing air into the cushion is driven by a separate motor.

nearly identical to that of the initial production version, the only difference being the engine three-pointer gauge unit which in this case was divided into three separate instruments (oil thermometer, oil manometer, fuel manometer). The instruments were slightly rearranged. A new feature was the canvas hood for blind flying training. The cockpit windshields could be either semi-circular, or V-shaped, or composed of three triangular panels.

The modifications effected on the UT-2 turned it into a thoroughly reliable machine which for many subsequent years remained the main trainer type in Soviet aviation. In the period between 1938 and 1946, before the advent of the Yak-18, the production of different versions of the UT-2 (including the UT-2M described below) totalled 7,243 aircraft. It was exported to Poland, Bulgaria, Romania and Hungary.

Specifications of the UT-2 production versions

	UT-2 (standard for 1938)	UT-2 (standard for 1940)
Engine type	M-11G	M-11D
Engine power, maximum, hp	110	125
Length overall	7.0 m (22 ft 11⅝ in)	7.15 m (23 ft 5½ in)
Height, tail up	2.99 m (9 ft $9\frac{23}{32}$ in)	2.99 m (9 ft $9\frac{23}{32}$ in)
Wing span	10.2 m (33 ft 5⅝ in)	10.2 m (33 ft 5⅝ in)
Wing area, m^2 (sq ft)	17.12 (184.3)	17.12 (184.3)
Empty weight, kg (lb)	616 (1,358)	628 (1,385)
All-up weight, kg (lb)	938 (2,068)	940 (2,073)
Max speed at 1,000 m (3,280 ft), km/h (mph)	205 (127)	210 (131)
Cruising speed, km/h (mph)	160 (99)	160 (99)
Landing speed, km/h (mph)	90 (56)	90 (56)
Climb to 1,000 m, minutes	4.8	5.0
Service ceiling, m (ft)	3,500 (11,480)	5,000 (16,400)
Maximum range, km (miles)	1,130 (702)	1,130 (702)

UT-2 Renault-powered floatplane trainer prototype

In 1937 a UT-2 powered by an in-line Renault engine (presumably the UT-2 prototype with the Renault Bengali-4 engine mentioned above) was equipped with floats. It was used for record-setting flights. In October 1937 test pilots P. M. Stefanovskiy and P. I. Nikitin established on it a world speed record for two-seat light hydroplanes of the 1st category: 210 km/h (131 mph) on a 100-km (62-mile) closed-circuit route. It was also used by women pilots V. S. Grizodoobova and Ye. Slobozhenko for setting records: a speed record of 200 km/h (124 mph) and an altitude record of 3,267 m (10,719 ft).

MV-4-powered UT-2 trainer prototype (standard-setter)

In 1939 Plant No.301 manufactured an example of the UT-2 fitted with the MV-4 engine (a Soviet-built licence-built version of the Renault Bengali 4). It was tested as a standard-setter for series manufacture (this version was intended to supplant the M-11-powered UT-2). However, this version failed to enter production.

UT-2V production bomber trainer

This was an armed version of the UT-2 (the V stands for *vo'oru**zhon**nyy* – armed). In 1942 Yakovlev's OKB developed the UT-2V which was intended for practice bombing in flight schools. The aircraft was manufactured in quantity.

UT-2 with machine-gun armament

According to Vadim B. Shavrov, during the war some examples of the UT-2 were fitted with one or two 7.62-mm ShKAS machine guns and launch rails for rocket projectiles.

UT-2M production trainer

In 1941 the Yakovlev OKB produced an upgraded version of the UT-2 which was designated UT-2M (*modifit**see**rovannyy* – modified). It had new wings featuring leading-edge sweepback, a straight trailing edge and stronger dihedral. The tail surfaces were enlarged, the vertical tail being given a rounded contour; the aft fuselage was extended by 350 mm (1 ft $1\frac{25}{32}$ in), bigger cockpit visors with a rounded shape were fitted and the wheel spats were deleted. This version was manufactured in series at Plant No.168 in Rostov-on-Don (and possibly other plants as well).

UT-2MV production light bomber

Experience gained with the UT-2V was used by Plant No.47 in 1942 to produce the UT-2MV, an armed version of the UT-2M (again, V stands for *vo'oruzhonnyy* – armed). This aircraft, powered by a 140-hp M-11F engine, was intended for use as a front-line light bomber. Four DER-31 bomb shackles mounted under the wings enabled the aircraft to carry four 50-kg (110-lb) bombs with one pilot and two 50-kg bombs with two crew (pilot and observer). The ordnance load could also comprise eight RS-82 unguided rocket projectiles and two 50-kg bombs in pilot-only configuration; the rocket armament made it possible to use the UT-2MV as a light attack aircraft.

The aircraft was tested at NII VVS, where the military test pilots noted its low bombing accuracy. Nevertheless, in September 1942 a decision was taken to put the new version into production, and Plant No.600 started manufacturing UT-2MVs, initially without rocket launch rails.

Yefim Gordon archive

Above: A UT-2M trainer during checkout tests at NII VVS, Chkalovskaya AB. The reshaped tail and the lack of wheel spats are evident; the redesigned wings are less obvious at this angle.

Yefim Gordon archive

A production UT-2M in flight.

UT-2 modified by K. A. Moskatov

In 1942 a single UT-2 was modified by engineer K. A. Moskatov in front-line field conditions. He covered the seats with a canopy and added an extension to a centre fuselage frame to act as a crash pylon protecting the crew in the event of a nose-over. Trim tabs were added to rudder, elevators and ailerons.

UT-2 torpedo-bomber trainer

In 1946 at least one production example of the UT-2 was converted into a torpedo-bomber trainer carrying a small dummy torpedo beneath the fuselage. It was equipped with a sight and an aerial camera. No information is available on its eventual operational use.

UT-2N (SEN) experimental aircraft

A production UT-2 was converted into a testbed or, one might say, a technology demonstrator, in experiments with an air cushion undercarriage conducted by Nikolay Ivanovich Yefremov and Aleksandr Davidovich Nadiradze at TsAGI and LII in 1939-1941. The aircraft, designated UT-2N, was also known as SEN (or, more exactly, SYeN, *samo**lyot** Ye**frem**ova i Nadi**ra**dze* – Yefremov and Nadiradze's aircraft). Nadiradze had previously worked on air cushion vehicles at a different OKB headed by V. I. Levkov.

The air cushion undercarriage was intended to replace the normal wheel or ski undercarriage and enable the aircraft to take off from any surface along a downhill path with next to no take-off run. Mounted under the aircraft on a centreline pylon and lateral N-struts was a big annular rubber balloon or skirt resembling an inflatable dinghy turned upside down. The size of the skirt was calculated to ensure the aircraft's stability in the event of sideways movement during the take-off. Compressed air was fed into the skirt by a fan driven by a 25-hp motorcycle engine mounted on a platform atop the skirt.

During the tests, which were conducted in 1940 by test pilots Igor' I. Shelest, Mikhail

Yefim Gordon archive

A UT-2MV light attack aircraft with four 100-kg bombs under the wings.

M. Gromov, A. B. Yumashev and A. P. Chernavskiy, the aircraft behaved quite normally, demonstrating the practicability of the concept. Encouraged by the results, NKAP tasked the two engineers with developing a similar but retractable undercarriage for the Petlyakov Pe-2 dive-bomber. This was duly done in 1941 and the suitably modified Pe-2 began initial taxying runs until the outbreak of the war led to further work on the project being halted.

UT-2 as a radio-controlled unmanned aircraft

In the pre-war years Soviet engineers conducted experiments with radio-controlled aircraft (known as 'telemechanical aircraft' in the terminology of the time). The UT-2 was one of the aircraft types involved in these experiments. In January 1940 the Defence Committee issued a special directive tasking NKAP with the development and testing of prototype 'telemechanical' versions of the TB-3 and SB bombers and the UT-2 trainer.

In September 1940 People's Commissar of the Aircraft Industry Akeksey I. Shakhoorin asked the Defence Committee to postpone the deadlines for completion of this work; in particular, he requested five more months for the development of a UT-2 version capable of radio-controlled take-off and landing (this work was entrusted to the NII-22 research institute where a team led by engineer Nikol'skiy was responsible for the task). This request was granted. The remote-controlled UT-2 was undergoing manufacturer's tests in Leningrad in early 1941, and there were plans for submitting it for State acceptance trials in July-August 1941.

UT-2 equipped for intercepting flare bombs (project)

In 1941 the TsAGI Design Bureau (KB TsAGI) headed by A. A. Sen'kov came up with a proposal envisaging the use of the UT-2, as well as the Polikarpov U-2 and R-5 aircraft, for intercepting flare bombs dropped by the enemy during night bombing raids. For this purpose the mentioned aircraft were to be equipped with devices called paravane (presumably wires mounted on some kind of outriggers at the wingtips). There is no evidence to suggest that any practical work had been done on the proposed UT-2 version.

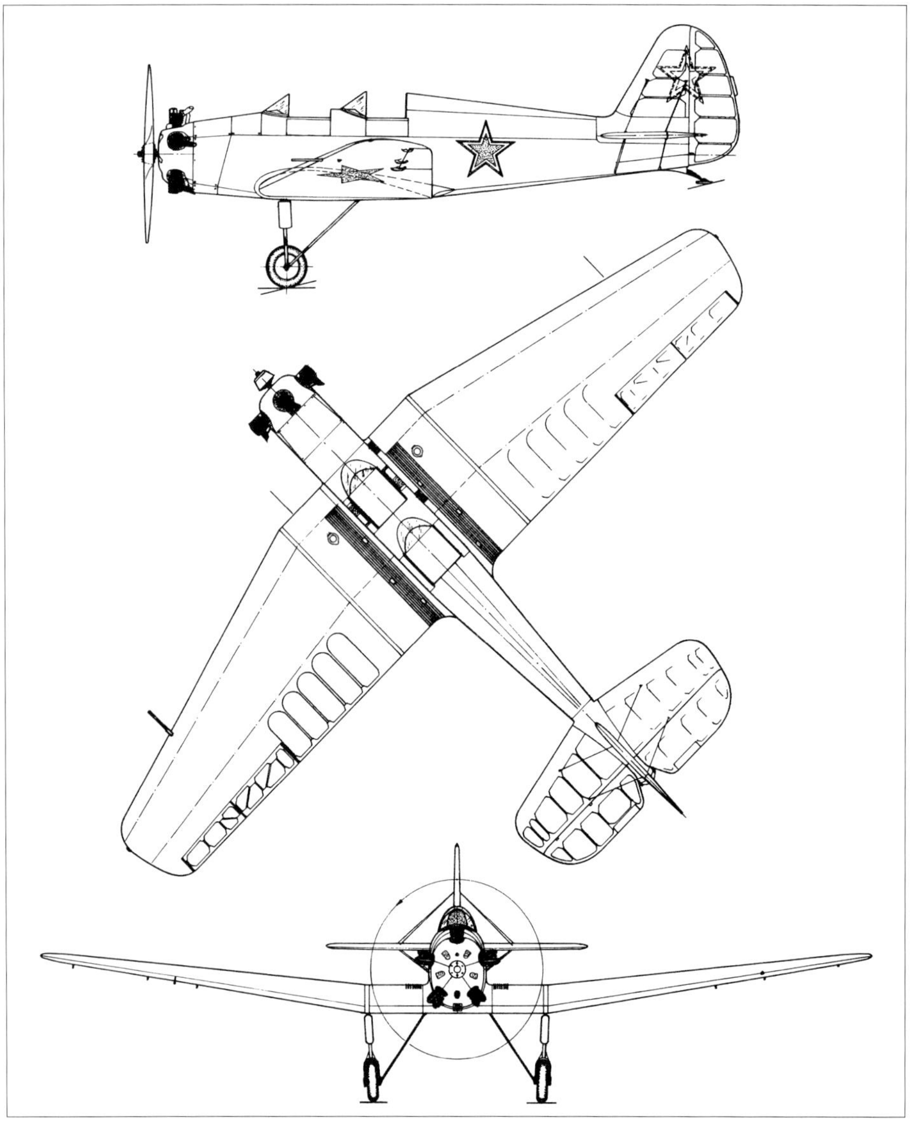

A three-view of the production-standard UT-2M trainer.

Primary trainer (project)

An OKB drawing dated 11th December 1940 testifies to the fact that the Yakovlev OKB was working on the project of a primary trainer in which the trainee and the instructor were to be seated side-by side. This low-wing monoplane would have wings with a marked leading-edge sweep and no trailing-edge sweep. The aircraft was to be powered by the 72-hp M-16, an obscure engine type which was never built in quantity. Interestingly, the drawing shows the aircraft to be fitted with a flat-four or flat-six engine, judging by the shape of the cowling. Design performance included a maximum speed of 180 km/h (112 mph), a cruising speed of 140 km/h (87 mph), a service ceiling of 4,000 m (13,120 ft) and a range of 300 km (186 miles).

UT-2L ('Limousine') courier aircraft of 1942 (first use of designation)

In 1942 the first version of the UT-2 with enclosed cockpits made its appearance. Dubbed UT-2L 'Limousine' to reflect its 'executive transport' role, it featured an extensively glazed cockpit canopy with a solid fairing behind the rear seat; in other respects the aircraft seemed to be identical to the standard version. Thirteen examples of this UT-2L were built.

UT-2L standard setter for 1944 (UT-2L 'Limousine', second use of designation)

This version produced in prototype form in 1944 was quite different from the aircraft described above. It strongly resembled the future Yak-18 (the initial 'taildragger' version) and may be regarded as its immediate predecessor. The features that distinguished the 'second-generation' UT-2L from the standard UT-2 and at the same time presaged the future Yak-18 were: an enclosed five-piece cockpit with a canopy modelled on that of the Yak-9V conversion trainer (the windshield, centre and rear portions were fixed, with two aft-sliding portions in between); a helmeted engine cowling with individual cooling air inlets and exhaust stubs for each cylinder; a rounded vertical tail; and a tailwheel instead of a tailskid.

The engine mount was extended forward by 150 mm ($5\frac{29}{32}$ in); the main under-

Yakovlev OKB

Above: A fine in-flight study of the UT-2L 'Mk.2'. The second aircraft to bear this designation was remarkably similar in appearance to the future Yak-18, the fixed and spatted landing gear being the most obvious external difference.

carriage units had spatted wheels equipped with brakes, the mainwheel legs proper and their inboard bracing struts being provided with close-fitting fairings. The wings were inherited from the standard UT-2 (not from the UT-2M) and fitted with a centre section landing flap. The UT-2L featured a wooden construction with two-spar wings skinned with plywood; the fuselage was a wooden truss with plywood upper decking. The tail surfaces and ailerons had duralumin frames with fabric skinning. The aircraft was powered by an M-11D rated at 145 hp. At an AUW of 950 kg (2,095 lb) it had a maximum speed of 225 km/h (140 mph), a landing speed of 75 km/h (47 mph) and a range of 700 km (435 miles).

Together with its single-seat derivative, the Yak-5 (see below), the UT-2L was to enter production as soon as the demand for combat aircraft at the frontlines became less acute. However, by that time the Soviet aircraft industry was producing mostly aircraft of metal construction, and it was found inexpedient to start manufacturing in series a wooden aircraft. Still, some features introduced on the UT-2L and the Yak-5 later went into the design of the all-metal Yak-18 trainer.

Yak-5 single-seat fighter trainer prototype

In the same year of 1944 the Yakovlev OKB produced a single-seat derivative of the UT-2L described above. Designated Yak-5 (second use of designation), it was intended for the fighter trainer role. The Yak-5 was similar to its two-seat progenitor in outlines and dimensions and retained its wooden construction, but differed in having a retractable undercarriage and a variable-pitch propeller. The main gear units retracted aft into the wings, leaving the wheels partially exposed to lessen the damage in the event of a wheels-up landing. This feature was later incorporated into the design of several training and aerobatic aircraft developed by the Yakovlev OKB.

The Yak-5 featured a more comprehensive flight and navigation equipment suite which included a strong landing light and enabled the machine to be used for training flights under adverse weather conditions and at night. The aircraft was equipped with a radio, a radio compass and was armed with a 7.62-mm ShKAS machine-gun. As stated above, the Yak-5 was not put into production, but its concept was later reflected in the Yak-11 fighter trainer.

UT-21 (Ya-21, No.21) single-seat fighter trainer prototype

This fighter trainer was developed by Yakovlev in response to wishes from the VVS in 1937. It was a derivative of the UT-1 single-seat trainer in which the M-11 radial engine was replaced by a Renault 6Q-01 six-cylinder in-line air cooled engine. This engine change entailed a replacement of the engine mount and necessitated changes in the fuel and oil piping and tanks. Some changes had to be introduced into the first two sections of the fuselage truss. The cockpit was enclosed by a sliding canopy, with large transparencies in the fuselage sides below the canopy. The undercarriage wheels were provided with brakes. In accordance with the machine's mission it was fitted with armament consisting of a fixed synchronised 7.62-mm ShKAS machine-gun.

The first flight of this machine is presumed to have taken place in the second half of September 1937, possibly between 19th and 22nd September. During manufacturer's tests the UT-1 attained a speed of 322 km/h (200 mph). In 1938 the UT-21 was mentioned in the Soviet press as 'an aircraft for training fighter pilots' and as 'a high-speed sports and training aircraft'.

Two UT-21 prototypes were expected to be submitted for the State acceptance trials to NII VVS (Research Institute of the Air Force) in October and December 1939 respectively. However, the second UT-1

Yakovlev OKB

The Ya-21 (UT-21) fighter trainer was yet another Renault-engined derivative of the UT-1.

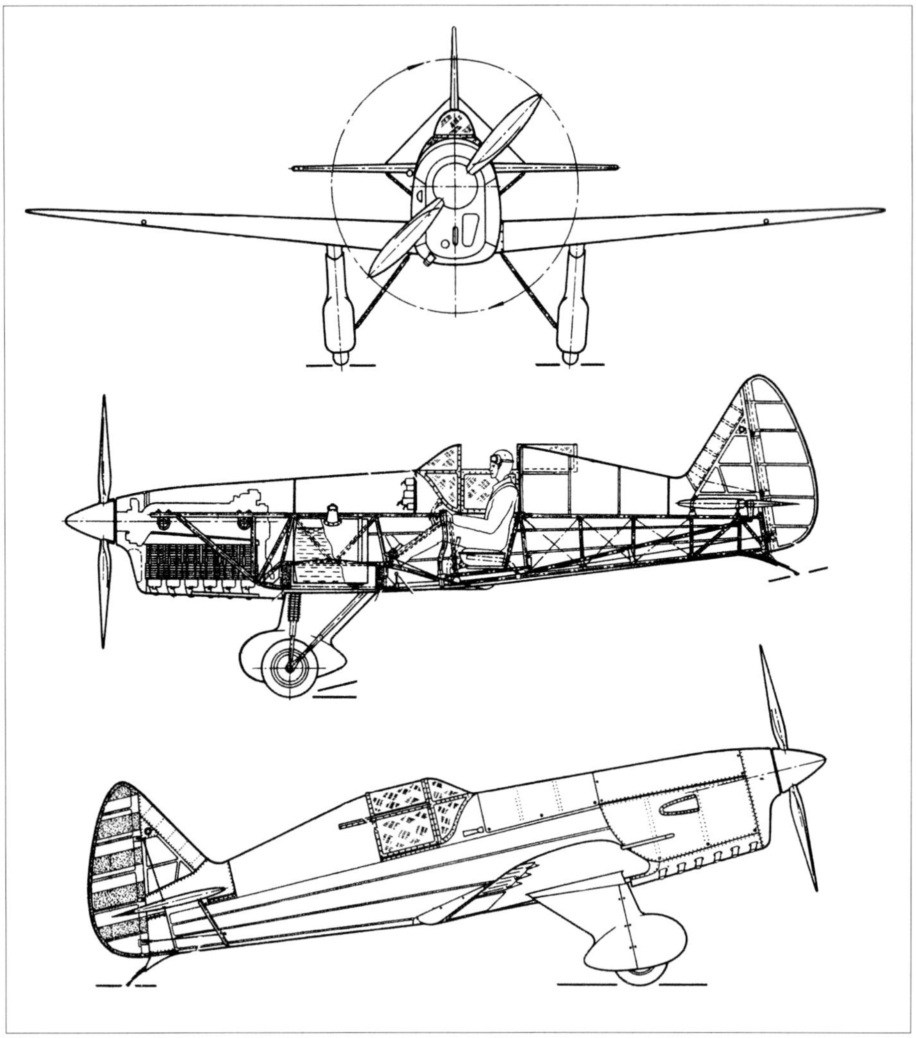

Two views of the UT-21 (Ya-21) trainer.

Specifications for the Ya-21 (UT-21)

Engine type	Renault Bengali 6Q
Engine power, max hp	220
Length, tail down	6.4 m (21 ft 0 in)
Height	2.05 m (6 ft 8⅔ in)
Wing span	7.8 m (25 ft 7 in)*
Wing area, m^2 (sq ft)	9.58 (103)
Empty weight, kg (lb)	611 (1,347)
All-up weight, kg (lb)	831 (1,832)
Maximum speed	
at sea level, km/h (mph)	322 (200)
Cruising speed, km/h (mph)	290 (180)
Landing speed,	
without flaps, km/h (mph)	90 (56)
Time to 1,000 m (3,280 ft), min	2.7
Service ceiling, m (ft)	7,000 (22,960)
Range	
at cruising speed, km (miles)	715 (444)
Take-off run, m (ft)	150 (492)
Landing run, m (ft)	190 (620)

* Some sources give the wing span as 7.3 m (23 ft 11⅜ in)

prototype was never built, and the testing in NII VVS was confined to the first and only machine. The testing revealed some serious faults calling for structural changes and partial airframe redesign. Installation of the inline Renault engine led, in the opinion of the military test pilots, to a deterioration of handling qualities because of greater distribution of masses along the longitudinal axis of the aircraft. A wish was expressed to have the Renault engine replaced by the MG-31 nine-cylinder radial rated at 270-320 hp.

The UT-21 was to be accordingly re-engined, but this plan was not implemented because of a change in the intentions of the customer (the Air Force). Priority was given to the project of converting the UT-21 into 'aircraft No.25' (see below).

'Aircraft No.25' single-seat fighter trainer prototype

As related above, the aircraft No.21 (UT-21) was converted into a machine designated 'aircraft No.25' and powered by the 220-hp MV-6 engine (the Soviet-built version of the French Renault Bengali 6). Available drawing shows that 'aircraft No.25' differed from 'aircraft No.21' in having a revised cockpit. Instead of being fully enclosed, with extensive side glazing, the cockpit was open and featured a deeper visor providing room for a gunsight. Opaque side panels hinging down for access were installed instead of the glazed side panels.

Neither the UT-21 nor its derivative, 'aircraft No.25', had a chance to be fully developed and reach production and service status. Yakovlev's OKB was in the process of switching over to combat aircraft design, with 'aircraft No 22' (the BB-22) bomber and the I-26 fighter in the focus of its attention, and the work on a fighter trainer had to be given up for the time being.

UT-23 ('aircraft No.23', Ya-23) reconnaissance trainer (project)

In the period between 10th November and late December 1938 a team of engineers within the Yakovlev OKB was engaged in design work on 'aircraft No.23'. The work was aimed at investigating the feasibility of developing a reconnaissance trainer broadly similar to the UT-2. Before the end of 1938 the assignment was re-designated from No.23 to No.24, and was termed 'UT-23, or 'aircraft 24' (*sic*). The UT-23 was to be armed with two forward-firing synchronised machine-guns operated from the front cockpit and a flexible machine-gun in the rear cockpit.

According to calculations, the aircraft was to have a maximum speed of 250-280 km/h (155-174 mph) and a landing speed of 75-80 km/h (47-50 mph). But this machine was never built: pursuant to a government directive, the work on the UT-23 was eliminated from the prototype construction plan for 1940.

UT-24 ('aircraft No.24') reconnaissance trainer (project)

There is some confusion about this designation and the foregoing one in the available sources. It is not quite clear whether the UT-24 (this is the designation accompanying a published drawing) is just a different name for the UT-23 described above or a further development of this type. The mentioned drawing depicts an aircraft which in its general outline fully corresponds to the Renault-powered UT-2 and differed from it primarily in being armed, as described under the UT-23 heading. The visor of the front cockpit resembled those of the future Yak-1/ Yak-7/Yak-9 fighter family, being shaped so as to provide place for a gunsight; the visor of the rear cockpit remained the same as on the UT-2 trainer. The aft cockpit's equipment included a vertically mounted telescopic bombing sight. The aircraft was shown with the characteristic fairings on the main undercarriage units that were normally sported by the UT-2s.

THE COMBAT AIRCRAFT OF 1940-1945

2

The Ya-22/BB-22/Yak-2/Yak-4 family

Ya-22 high-speed aircraft prototype

A new twin-engine monoplane appeared at a test airfield near Moscow in the spring of 1939. Its red/white colour scheme and streamlined contours attracted everybody's attention. It was clear that the aircraft was capable of high performance. Indeed, in the very first tests flights the aircraft reached a speed of 560 km/h (348 mph), which was more than more than most of the contemporary fighters could boast.

The new aircraft was designated No.22 ('aircraft No.22') or Ya-22; in its prototype form it had no armament and was intended primarily for demonstrating its ability to attain a record high speed. But Aleksandr S. Yakovlev did envisage a military role for it, namely that of a reconnaissance aircraft. Subsequently it was called upon to fulfil the role of a bomber and was accordingly redesignated BB-22 (***blizh**niy bombardi**rov**shchik* – short-range bomber). The Ya-22 was the first military aircraft designed by the Yakovlev OKB which had earlier engaged in developing and building only light aircraft.

The aircraft was a low-wing monoplane of mixed construction. The one-piece wooden wings had a two-spar structure with plywood skin. The wooden centre fuselage was integrated with the wings. The forward fuselage was made of duralumin, while the detachable rear fuselage was a welded steel-tube truss with fabric skin. The tail unit featured endplate fins and rudders. The pilot and the navigator sat in separate cockpits, that of the navigator being placed behind the wing trailing edge.

The aircraft was powered by two 930-hp Klimov M-103 Vee-12 liquid-cooled engines driving three-bladed tractor propellers; the engines were housed in nacelles adhering to the wing undersurface so that the propellers' rotation plane was ahead of the fuselage nose. All three landing gear struts featured single wheels in forked fittings and retracted aft, the main units stowing in the rear portions of the engine nacelles which also accommodated the water radiators.

The Ya-22's factory test programme was conducted by the OKB's chief test pilot Yulian I. Piontkovsky. He demonstrated the aircraft to Red Army Air Force Commander Yakov Smooshkevich, who was impressed by the machine. Soon Iosif V. Stalin learned about its existence; he always attached

Yakovlev OKB

Yakovlev OKB

Top and above: The sleek 'aircraft No.22' was optimised for high-speed performance, and the cherry-red and white prototype showed extra special attention to surface finish. Note the spaced cockpits and the stalky landing gear with small mainwheels and a skid on the tail strut only.

Yakovlev OKB

Yakovlev OKB

Top and above: Two photos of the BB-22 bomber prototype. Note the repositioned gunner's cockpit and the larger mainwheels. The white panels on the fins are for access to the rudder bellcranks.

Yakovlev OKB

Above: A production BB-22 with white (or silver-doped) fuselage and fins.

Yakovlev OKB

A production BB-22 in temporary winter camouflage and on skis. Note the aerial mast on the starboard engine nacelle.

paramount importance to speed and was fascinated by the Ya-22's performance. The aircraft made its public debut on 1st May 1939 when it participated in the May Day flypast over the Red Square. On 29th May it entered tests at NII VVS, with project engineer Kholopov, pilot Shevaryov and navigator Tret'yakov as members of the test team. The machine was tested in the reconnaissance aircraft role, albeit no mission equipment or armament was fitted. The State acceptance trials were completed on 9th June 1939.

The 'aircraft No.22' attained a maximum speed of 567 km/h (352 mph) at 9,900 m (32,480 ft), and a climb to 5,000 m (16,400 ft) took 5.75 min. Such performance made the machine one of the best among the aircraft of this class both in the USSR and abroad. It compared favourably with the Italian Breda-88, the French Bréguet 690 and other foreign aircraft. The Ya-22 outran the production Tupolev SB 2M-103 bomber powered by the same engines, with a top speed almost 100 km/h (62 mph) higher, and was considered capable of attaining 600km/h (373 mph), provided the engine cooling system was improved, the exhaust system modified and advanced propellers installed.

The flight tests also revealed several shortcomings. The engine cooling system fell short of the requirements; the engines overheated, especially during climb. For example, the estimated time required to reach 7,000 m (22,965 ft) was 8.7 minutes, but in fact it took 34 minutes to climb to that altitude because the pilot repeatedly had to level off in order to bring down the coolant temperature. The design of the braked mainwheels needed refinement and the fuel supply system was not reliable enough.

Now we know from the recollections of Aleksandr S. Yakovlev that on 17th August 1939, soon after the completion of the aircraft's State acceptance trials programme, Stalin summoned the Chief Designer to the Kremlin. Stalin wondered very much how such a speed could be reached with the same M-103 engines as fitted to the SB bomber. Yakovlev wrote in his book *My Life's Ambition*: *'Stalin was still walking around the office, expressing his surprise and saying: "Miracles, veritable miracles; this is a revolution in aviation".'* Indeed, the Ya-22 outclassed Soviet and foreign bombers, as well as many fighters, as far as speed was concerned. Nevertheless, this aircraft was doomed soon to be forgotten and even the Yakovlev OKB specialists later recalled it reluctantly. The main reason was that too big hopes were pinned upon it – hopes that it failed to justify during the Second World War.

The Ya-22 could not justify these hopes, in particular, because it was capable of such a speed to the detriment of some other qualities. The State acceptance trials report pointed out that the aircraft had no armour plating. The cramped cockpit hampered piloting and made emergency escape virtually impossible. In order to transform the 'aircraft 22' into a short-range bomber, it was necessary to install both machine-guns and bomb armament, a communications radio and other equipment.

Thus, the miracles noted by Stalin did not exist in the perception of such Soviet specialists as Andrey N. Tupolev, Vladimir M. Petlyakov and Vladimir M. Myasishchev. Yet, the Ya-22 (BB-22) is not to be dismissed altogether. In the words of test pilot Igor' I. Shelest, *'...the concept put forward by A. Yakovlev proved to be an incentive to the majority of Soviet aircraft designers, prompting them to design new high-speed aircraft. The result was the development of the '100' (Pe-2) and later the '103' (Tu-2).'* These aircraft could reach approximately the same speed as the BB-22, but they had powerful armament comprising cannons and heavy machine guns, armour plating, sufficient range and endurance.

BB-22 initial production tactical bomber

However, the cognisance of this situation came later, and in the summer of 1939 the Powers That Be and the Air Force were not yet aware of it. Simultaneously with the programme of State acceptance trials of the Ya-22, the mock-up review commission was working at plant No.115 where the project of converting the aircraft into a short-range bomber was discussed in the presence of designers and representatives of NII VVS. On 7th June 1939 the commission took the decision to allocate the BB-22 designation to the aircraft. Problems associated with the bomb and gun armament, as well as navigational equipment, were discussed. It was decided to equip the machine with bomb racks for carrying a 400- to 700-kg (880- to 1,540-lb) bomb load; an aft-firing machine gun was to be installed. At the insistence of the military, the navigator's cockpit was repositioned to provide better view and permit direct communication with the pilot. The navigator/gunner was now seated close behind the pilot; this arrangement was accompanied by a redesign of the canopy over the back-seater, so that when the whole assembly was raised a hydraulic jack could pivot the ShKAS machine-gun mounting upwards to give a good field of fire. This arrangement was tested on the second prototype of the Ya-22 and introduced on production machines after the completion of an initial service evaluation batch based largely on the original prototype configuration.

Yakovlev OKB

Yakovlev OKB

Top and above: This Yak-2 manufactured by plant No.81 and pictured during State acceptance trials represents the late version with twin mainwheels. a cutback fuselage spine and a revised canopy. The position of the water radiator inlets on the sides of the engine nacelles is clearly visible.

Yefim Gordon archive

A fine air-to-air of a production Yak-2 with pre-war style black-outlined insignia. The position of the propeller rotation plane ahead of the nose is clearly visible.

Yefim Gordon archive

Above: A trio of late-production Yak-2s with twin-wheel main gear units heads on a bombing mission in the early days of the Great Patriotic War.

RART

German soldiers and officers examine the wreckage of a Yak-2 destroyed on the ground by a Luftwaffe air raid. Many Red Army Air Force aircraft shared this fate in the opening days of the war.

Simultaneously, the preparations for series production of the aircraft were going on. As early as March-April 1939 – even before the BB-22's test programme was launched – Aleksandr S. Yakovlev began to transfer photocopies of the manufacturing drawings to plant No.1 where the bomber was to be produced.

On 20th June 1939 the Defence Committee issued directive No.171ss ordering the BB-22 into production at two plants. Plant No.1 was tasked with manufacturing 50 BB-22s in 1939, and plant No.81 was to tool up for the production of this aircraft and manufacture 300 machines in 1940.

The abovementioned plans proved to be overly optimistic. The first production aircraft was built on 31st December 1939 and test flown in February of the following year. As was often the case, the quality of production aircraft left a lot to be desired. Despite the reduction of fuel capacity from 1,000 kg (2,200 lb) to 600 kg (1,320 lb), the all-up weight of the production machines increased to 5,380 kg (11,860 lb) as compared to the prototype's 5,023 kg (11,073 lb). The speed decreased markedly to 515 km/h (320 mph) at 5,000m (16,400 ft), 52 km/h (32 mph) lower than the prototype's.

The service evaluation batch was built soon and was tested in February-April 1940. These first ten machines were built by plant No.1 to the original standard with the navigator/gunner sitting aft of the wings, with a ShKAS machine-gun installed in his cockpit. To improve this gun's field of fire, Yakovlev redesigned the secondary structure above the steel truss of the rear fuselage so that it could retract in flight. (According to a document, these ten machines featured as many as four different types of the rear machine-gun installation.) The aircraft had an internal bomb bay between the wing spars which was moved behind the rear spar in the subsequent production versions.

The initial stage of the service tests revealed that the powerplant and undercarriage were underdeveloped. Due to tail unit vibrations the tests had to be suspended at the very beginning. Also, the BB-22 had unsatisfactory longitudinal and lateral stability, due to which it could be flown only by well-trained pilots. The engines often overheated during climb, water leaking from the cooling system joints.

A further deterioration of flight performance was noted. The speed of the 5,660-kg (12,480-lb) aircraft was only 399 km/h (248 mph) at sea level as compared to 455 km/h (283 mph) attained by the prototype, and 478 km/h (296 mph) at 4,600 m (15,090 ft). Climb time to 5,000 m was 9.5 minutes. The service ceiling was 8,100 m (26,576 ft) – 2,700m (8,860 ft) lower than that of the prototype. Thus, the BB-22 certainly did not outperform the SB built two years earlier.

The report on the evaluation results stated that the aircraft was not combat-capable and reliable; even with a 400-kg (220-lb) bomb load the flights were dangerous for the crew. The final conclusion was that the BB-22 2M-103 manufactured by plant No.1 in the configuration submitted had failed the service tests.

BB-22 2M-103 (Yak-2) production short-range bomber (*izdeliye* 70)

On 9th May 1940 the Technical Council at NII VVS reviewed the service test results and stated that the BB-22 was unfit for operational use, unless the numerous faults revealed were rectified. Among the military a sentiment arose in favour of halting the series production of the BB-22.

This played its part in the decision taken by the Defence Committee to stop production of the BB-22 at plant No.1 and concentrate the manufacture and further development of this aircraft at plant No.81. Issued on 25th May 1940, directive No.224ss to this effect stipulated that plant No.1 was to complete 100 machines; the remaining parts and subassemblies, together with the jigs and tooling, were to be turned over to plant No.81.

A special design bureau had been set up at the latter plant to tackle the introduction and development problems. Headed by engineer L. P. Koorbala, it was known as KB-70 (the '70' denoted the number of personnel). Hence the production BB-22s built by plant No.81 had the factory designation *izdeliye* 70 and were allocated c/ns commencing 70. (*Izdeliye* (product) such and such was, and still is, a common way of coding Soviet/ Russian military hardware items.)

Interestingly, the machines from the remaining deliveries by plant No.1 were to correspond to the standard reached by 20th May 1940. It included a modified undercarriage with twin-wheel main units (first introduced on the BB-22*bis*/Yak-4 prototype, see below), a rear gun mount of the type fitted to BB-22 c/n 1020 (that is, the 20th example built by plant No.1), improved flap actuators and modified bomb bays.

The aircraft produced at plant No.81 turned out to be somewhat better. First of all it concerned the production methods: the quality of the surface finish and painting was higher, and the engine cowlings and doors were a better fit. This reduced drag and increased the top speed by 10-20 km/h (6.2-12.4 mph) as compared to production aircraft built at plant No.1.

The test flights of the aircraft in maximum TOW configuration showed that take-off at short-term engine boost with a 400-kg (880-lb) bomb load carried internally and two 100-kg (220-lb) bombs on external racks improved the aircraft's combat capabilities. The defects of the armament and the stability problems noted earlier were not eliminated.

Having passed the test programme in November 1940, the production BB-22 was soon renamed Yak-2, reflecting the last name of the Chief Designer in keeping with the new policy

Yak-2 dual-control trainer prototype

One Yak-2 (BB-22 2M-103), c/n 1010, was fitted with dual controls at an Air Force experimental plant. It underwent State acceptance trials between 6th March and 5th June 1941, making 13 flights. After the flights performed on 7th-10th March the aircraft was pronounced unsuitable for piloting from the second cockpit.

To rectify the faults, the cockpit canopy was thoroughly redesigned, and on 23rd May 1941 the aircraft was resubmitted for testing with a view to passing a final judgement as to its suitability for conversion training. The conclusions at which the test pilots arrived were as follows. The dual-control version of the Yak-2 presents considerable difficulties for piloting due to high rudder pedal forces, cramped cockpits and strong draughts into the cockpits, and its handling characteristics are dissimilar to those of the most up-to-date aircraft. For these reasons there is no point in modifying the Yak-2 into a conversion/proficiency trainer.

No special designation for this version is mentioned in documents.

Specifications of the No.22 prototype and the production BB-22 (Yak-2)

	Aircraft No.22 (Ya-22)	Production BB-22 (Yak-2) manufactured by plant No.81
Length	9.34 m (30 ft 7¾ in)	9.34 m (30 ft 7¾ in)
Wing span	14.0 m (45 ft 11⅛ in)	14.0 m (45 ft 11⅛ in)
Wing area, m^2 (sq ft)	29.4 (316)	29.4 (316)
Empty weight, kg (lb)	3,796 (8,369)	4,258 (9,387)
All-up weight, kg (lb)	5,123 (11,294)	5,630 (12,412)
Max speed, km/h (mph)	455 (283) at sea level	410 (255) at sea level
	567 (352) at 4,900 m (16,080 ft)	498 (309.5) at 4,800 m (15,750 ft)
Climb time to altitude	4.6 min to 4,000 m	8 min to 5,000 m
Landing speed, km/h (mph)	160 (99)	130 (81)
Service ceiling, m (ft)	10,800 (35,435)	8,700 (28,545)
Range, km (miles)	1,000 (621)	n.a.
Take-off run, m (ft)	375 (1,230)	390 (1,280)
Landing run, m (ft)	855 (2,805)	298 (978)

Yak-2KABB attack aircraft prototype

A single M-103-powered Yak-2 was fitted with the KABB-MV gun/bombing installation (*kombi**nee**rovannaya artille**reey**sko-bombardi**rov**ochnaya bata**rey**a Mozha**rov**skovo i Vene**vid**ova*, combined gun/bomb battery designed by Mozharovskiy and Venevidov). It enabled the aircraft to deliver its bombs and strafe the target in a single pass.

The forward fuselage underwent extensive modification. The aircraft became a single-seater because the former navigator's cockpit housed the KABB-MV installation and its ammunition boxes. To improve forward visibility and the view to the sides, the fuselage nose was shortened and the opaque skin was partly replaced by Plexiglas panels.

Yakovlev OKB

Above: The Yak-2KABB low-altitude attack aircraft prototype featured a ventral twin-cannon pack and a redesigned extensively glazed nose with a shape akin to the Messerschmitt Me 210/Me 410.

Yakovlev OKB

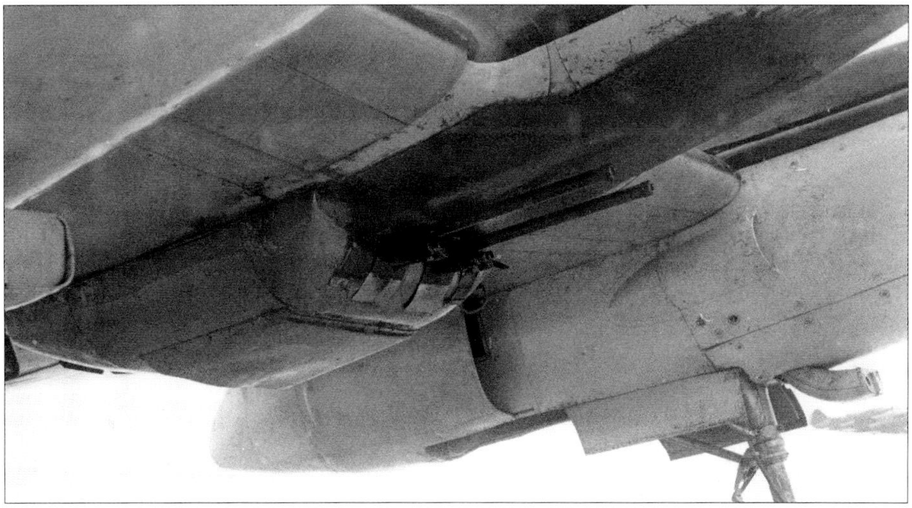

Close-up of the KABB-MV depressable installation on the Yak-2KABB with two ShVAK cannons and two ShKAS machine-guns. The underside of the nose is liberally covered in soot, a result of firing the cannons.

Yakovlev OKB

Yakovlev OKB

Top and above: This unmarked aircraft (c/n 1002) was the prototype of the BB-22*bis* (later redesignated Yak-4), introducing the new rear fuselage and twin mainwheels. Note the dog-leg exhaust pipes.

Yakovlev OKB

The BB-22*bis* prototype after the ground accident on 23rd May 1940 when it hit two parked aircraft. Due to the one-piece wing design the aircraft was a write-off.

The KABB-MV installation beneath the fuselage was shielded by a special fairing. The design of the control stick was also changed.

The bomb release switches were relocated to the pilot's cockpit. Two 20-mm (.78 calibre) Shpital'nyy/Vladimirov ShVAK cannons and two ShKAS machine-guns were mounted on a flexible frame. The guns could be fired either in a fixed position or while being tilted by electric motors within an angle between 0° and 30°. The usual bomb complement could be carried – 20 AO-8 or AO-20 fragmentation bomblets in two cassettes, or four 50-kg (110-lb) FAB-50 bombs, or four 100-kg (220-lb) FAB-100 bombs.

Designated Yak-2KABB, the aircraft thus modified was subjected to testing at NII VVS in March-April 1941. Nineteen flights were performed, of which one flight was not accompanied by the use of weapons, nine flights were made with the use of guns only, two flights with bombing only and in seven flights both types of armament were used.

The KABB-MV installation on the Yak-2 received a generally positive appraisal; the concluding part of the test report stated that the installation could be recommended for new types of twin-engined attack aircraft or single-engined aircraft with pusher propellers. Yet, after the outbreak of the war all work on the KABB-MV was terminated.

BPB-22 (BPB-22 2M-105, BB-22PB) dive-bomber prototype (*izdeliye* 31)

The BPB-22 (factory code '*izdeliye* 31') was a short-range dive-bomber fitted with the newly available 1,050-hp Klimov M-105 Vee-12 engines (BPB stands for ***blizh****niy* *pi****kee****ruyushchiy bombardi****rov****shchik*, short-range dive bomber). It was also known as BPB-22 2M-105 or BB-22PB.

Development of this version began in response to a government directive and the appropriate NKAP order No.317 dated 27th June 1939 as a back-up for Petlyakov's PB-100 (the future Pe-2). The mentioned order tasked Klimovitsky (Director of plant No.81) and engineer L. P. Koorbala (head of the plant's design office) with undertaking the projecting and construction of the machine. The stipulated design performance included a maximum speed of 570 km/h (354 mph) at 5,000 m (16,400 ft) and a range of 1,200 km (746 miles).

The advanced development project of this version submitted by Koorbala was discussed by the NKAP project commission on 21st September 1940. Apart from the new engines, this version differed from the baseline bomber in being fitted with airbrakes, an automatic dive entry and recovery system and an increased area of the cockpit transparencies. In the dive bomber role the aircraft could use only bombs carried externally, typical bomb loads comprising two FAB-250 high-explosive bombs, or four FAB-100s, or two BetAB-150DS concrete-piercing bombs, or two BrAB-200DS armour-piercing bombs. In overload configuration two FAB-250 and two FAB-100 bombs could be carried.

The project was not approved by the Air Force, but the first BPB-22 prototype was nevertheless completed in October. In the same month Deputy Commander of the Air Force Astakhov endorsed the conclusions on the project; these contained a recommendation that the already completed first prototype be used for the refinement of the main assemblies and units, and the second prototype, then under construction, be subjected to evaluation by a mock-up review commission with a view to making a final verdict on the project.

In late October 1940 factory test pilot Lipkin (some sources say Lapkin) flew the BPB-22 for the first time. Further testing was done by Ya. Paul' at Ramenskoye. In one of the flights the engines cut soon after take-off due to fuel starvation; the ensuing emergency landing resulted in the aircraft being damaged beyond repair. Further work on this version was halted.

Interestingly, a comparison was made by NKAP specialists between the PB-100, the BB-22, the Arkhangel'skiy SB-RK (alias Ar-2) and the Arkhangel'skiy 'B' in the dive-bomber role. The conclusion was that the PB-100 was superior to the other types in overall performance, but its all-metal airframe made it more expensive. Hence it was deemed advisable, in addition to starting series manufacture of the PB-100, to retain the BB-22 in production as a cheap aircraft made of readily available materials which could be used mostly for conversion and proficiency training.

Yakovlev OKB

Yakovlev OKB

Top and above: Two views of the first production Yak-4. Note the oil coolers relocated from the nacelles' sides to a chin position.

RART

This Yak-4 coded '3' was knocked out at an airfield by a German air raid together with several other aircraft, including two Polikarpov I-16 fighters (left) and an Arkhangel'skiy Ar-2 bomber (far centre).

Yak-4 2M-105 production short-range bomber (BB-22*bis*, *izdeliye* 70*bis*)

While the M-105-powered dive-bomber version of the BB-22 failed to gain acceptance, the standard (basic) bomber variant fitted with these engines instead of the M-103s received a go-ahead and was put into production at plant No.81 as the BB-22*bis*, or *izdeliye* 70*bis*. The VISh-2K (or VISh-22K?) propellers were replaced by VISh-22Ye units. In December 1940 this version was renamed Yak-4, the original BB-22 2M-103 receiving the designation Yak-2, as noted above.

In addition to new engines, the Yak-4 featured some airframe and equipment modifications. The rear fuselage was redesigned, its length being increased and the unwanted retractable decking eliminated. Various improvements were made to the engine mounts and the engine cooling and fuel systems. Each main undercarriage unit now had a single shock strut braced only by a welded triangular gusset on each side and carrying twin wheels which were semi-exposed in flight. The crew section was modified *again*, the pilot being given additional windows low down on each side and the back-seater operating a ShKAS on a TSS-1 mount under a low-drag pivoting canopy. Further modifications in the course of production included fitting stronger external bomb racks to carry FAB-250 bombs, giving a theoretical bomb load of 900 kg (1,980 lb), and adding a 250-litre (54.3 Imp gal) drop tank under each wing. Another modification was to relocate the oil coolers from the port side of the cowling to a chin duct under each engine.

Thanks to the extra power of the M-105s, the Yak-4's speed increased to 533 km/h (331 mph) at 5,050 m (16,570 ft) and 458 km/h (284 mph) at sea level. Rate of climb and service ceiling were also improved as compared to the Yak-2.

The prototype BB-22*bis* was produced in March 1940 at plant No.1 by converting the second production BB-22 (c/n 1002); it was the first of the BB-22/BB-22*bis* aircraft to be equipped with twin mainwheels. Manufacturer's tests lasted until May 1940. On 12th May BB-22*bis* c/n 1002 attained a speed of 574 km/h (356 mph). The testing was halted due to an accident on 23rd May; while taxying at a high speed, pilot

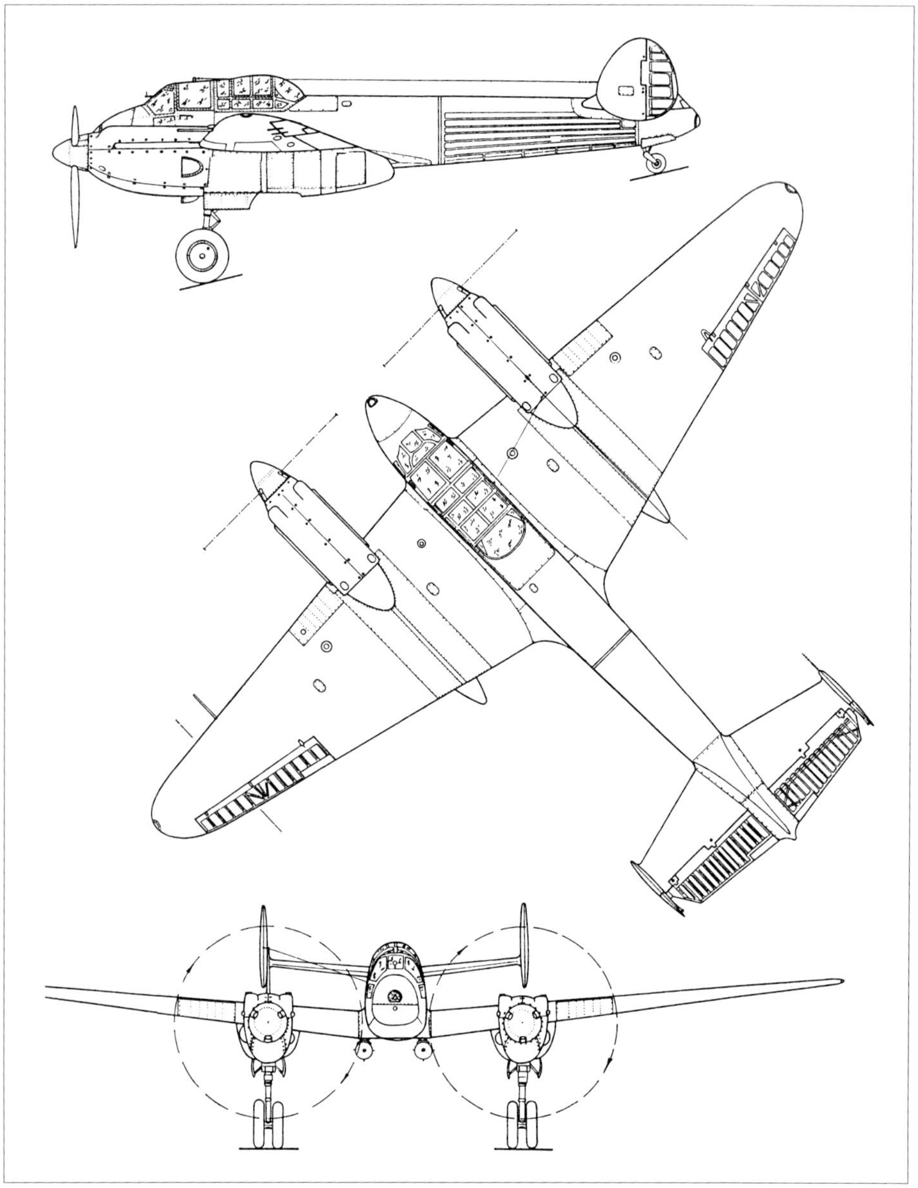

A three-view drawing of a late-production Yak-2 with a cut-down rear fuselage and twin-wheel main gear units. Note the external bomb carriage beneath the wing roots.

Specifications of the BB-22*bis* (Yak-4) versions

	BB-22*bis* prototype (c/n 1002)	Production BB-22*bis* (Yak-4) manufactured by plant No.81
Length	10.18 m (33 ft 4¾ in)	10.18 m (33 ft 4¾ in)
Wing span	14.0 m (45 ft 11⅛ in)	14.0 m (45 ft 11⅛ in)
Wing area, m² (sq ft)	29.4 (316)	29.4 (316)
Empty weight, kg (lb)	4,251 (9,373)	4,560 (10,050)
All-up weight, kg (lb)	5,845 (12,888)	6,115 (13,481) *
Max speed, km/h (mph)	574 (357) at 5,300 m (17,390 ft)	535 (332) at 5,050 m (16,570 ft)
Climb time to 5,000 m, minutes	5.45	6.5
Landing speed, km/h (mph)	160 (99)	140.5 (87)
Service ceiling, m (ft)	10,000 (32,810)	9,700 (31,824)
Range, km (miles)	900 (559)	925 (575)
Take-off run, m (ft)	300 (980)	492 (1,614)
Landing run, m (ft)	550 (1,800)	496 (1,627)

* normal

Moiseyenko lost control of the machine and crashed into two parked SB bombers, the BB-22*bis* suffering irreparable damage to the starboard wing.

A directive dated 4th March 1940 stipulated that two examples of this version were to be submitted by plant No.1 for State acceptance trials by 1st July 1940; the required performance included a maximum speed of 590 km/h (366 mph) at 5,000 m, a service ceiling of 11,000 m (36,090 ft) and a range of 1,200 km (745 miles). Then the task was transferred to plant No.81, which started series production of the new version in October-November 1940, prior to the State acceptance tests. On 10th December 1940 two Yak-4s (c/ns 70502 – that is, *izdeliye* 70, batch 5, 02nd aircraft in the batch – and 70603), were submitted for State acceptance trials which were completed on 29th January 1941 with disappointing results. The final test report contained the following conclusions:

'1. The Yak-4 aircraft submitted by plant No.81 are not combat-capable in their present condition and unfit for operational use (from the point of view of maintenance and servicing).

2. The Yak-4 does not meet Air Force requirements and should be withdrawn from series production.'

This recommendation was followed up by a corresponding Government directive. On 24th February 1941 Klimovitskiy, Director of plant No.81, reported to People's Commissar of the Aircraft Industry Aleksey I. Shakhoorin that, pursuant to the Government directive on halting production of the Yak-4 and introducing the Yak-3 (I-30 – *Auth.*) at plant No.81, further manufacture of subassemblies for the Yak-4 had been halted.

The last 22 Yak-4s were delivered in April 1941; the number of Yak-2s and Yak-4s built totalled 201, not 600 as Yakovlev wrote in the book *Soviet Aircraft*. Both versions took part in the fighting during the opening stage of the Great Patriotic War, seeing action with the 314th and 316th Reconnaissance Air Regiments, the 136th Bomber Regiment, the 10th, 44th, 48th and 53rd Medium-Range Bomber Regiments and the 225th Short-Range Bomber Regiment. In general the Yak-2/Yak-4 was not a major success in combat, partly due to the limited time available for mastering the types, and the few operational aircraft had been destroyed by enemy action or in accidents by the end of 1941.

BB-22 2M-107 (project)

Concurrently with the development of the M-105-powered version of the BB-22, plant No.1 and its chief designer Ya. N. Strongin were tasked with adapting the BB-22 to accept the even more powerful Klimov M-107 engine rated at 1,400 hp for take-off, which was under development at the time. The work on the M-107-powered version (sharing the designation BB-22*bis* with its M-105-powered stablemate) was under way as early as February 1940. It was formally sanctioned by a directive of the Defence Committee dated 4th March 1940, requiring the BB-22 2M-107 to be first flown before 15th December 1940 and submitted for State acceptance trials by 1st February 1941. This machine was to attain a maximum speed of 650 km/h (403 mph) versus the 590 km/h (367 mph) expected from the M-105-powered BB-22*bis*.

However, the work ran into difficulties. In mid-June 1940 it was stated that designers of plant No.1 were unable to complete the advanced development project (ADP) of this version; they came to the conclusion that fitting M-107s to the BB-2 was not feasible because the wing structure would be excessively stressed. The plant's management eventually succeeded in having this version cancelled. On 28th April 1941 NKAP issued order No.386 halting the work on the BB-22 2M-107 (the responsibility for which had by then passed from Strongin to Koorbala as chief designer of plant No.81).

R-12 reconnaissance aircraft prototype

R-12 (R stands for [*samo***lyot**-] *raz***ved***chik*, reconnaissance aircraft) was the most straightforward adaptation of the basic Ya-22 to practical uses because the necessary alterations were limited to installing the photographic equipment. Powered by M-103 engines driving VISh-22K propellers, the aircraft retained the original Ya-22 configuration with the gunner/navigator's cockpit placed amidships just aft of the wings. The navigator's station was fitted with an AFA-1 aerial camera (***a****ero***fot***oappa***raht**) and a ShKAS machine-gun. The canopy fitted flush into the fuselage contour; to enable the navigator/gunner to use his ShKAS, the rear upper decking of the fuselage retracted into the fuselage structure. An NAFA-19 camera (*noch***noy** ***a****ero***fot***oappa***raht**) was installed for night photography which was assisted by the use of flare bombs. Eight such 20-kg (44-lb) bombs were accommodated in a bay between the spars.

A single prototype of the R-12 was completed in June 1940 and passed manufacturer's tests only. A planned version of the R-12 with M-105 engines (the R-12 2M-105) never materialised, but Yak-4 bombers may have been adapted for reconnaissance duties in the field.

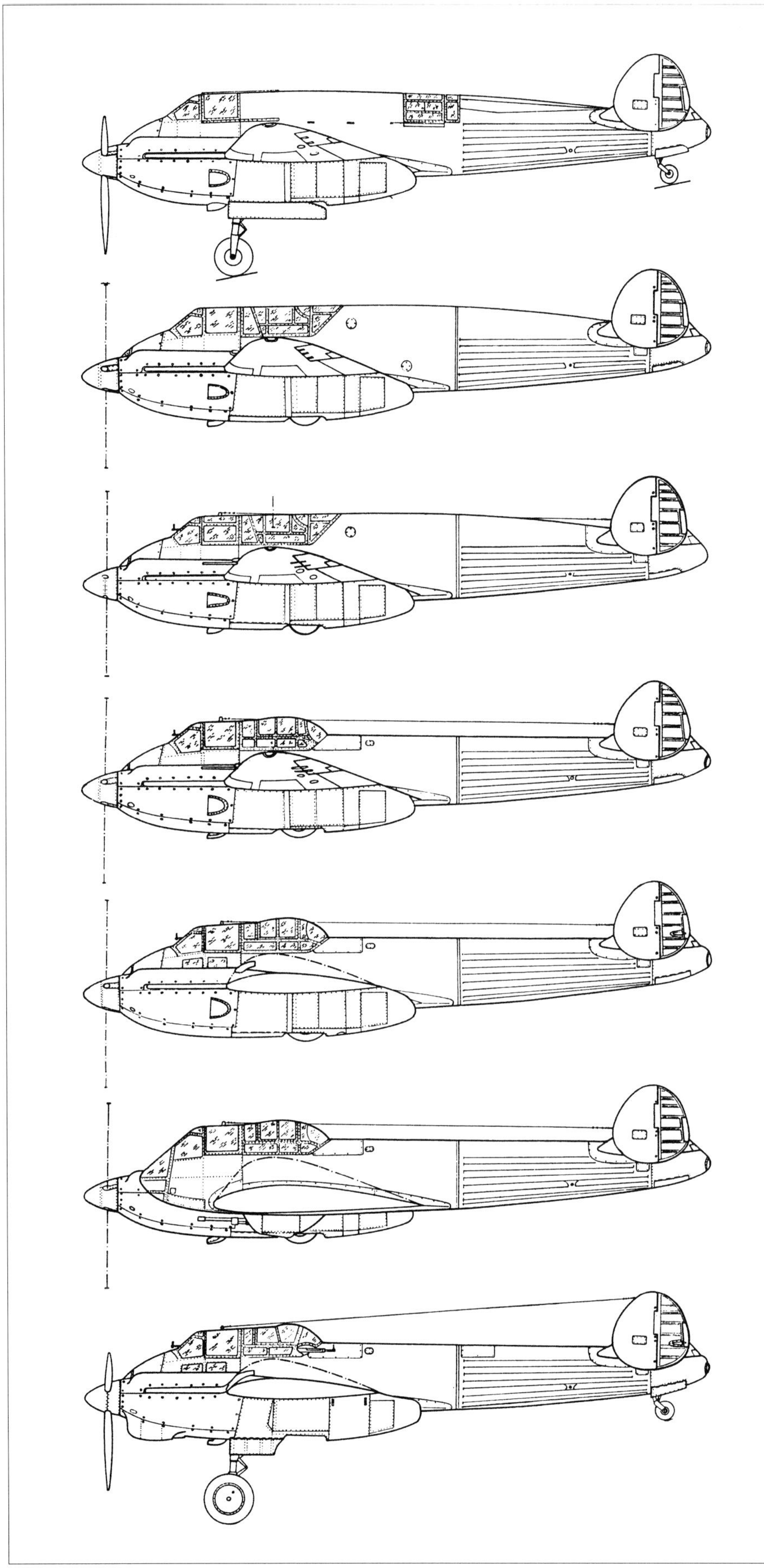

Top to bottom: The Ya-22 prototype; the BB-22 prototype; an early-production BB-22 2M-103A; a late-production BB-22 (Yak-2); a modified Yak-2 with extra glazing panels in the extreme nose; the Yak-2KABB prototype (the port wing and engine are omitted for clarity); and a production Yak-4 2M-105.

BB-22IS (BB-22DIS, I-29) escort fighter prototype

The R-12 served as the basis for one more version derived from the BB-22. Manufactured in early 1941, this was an escort fighter designated I-29 (*istrebitel'*, fighter), also known as BB-22IS (*istrebitel' soprovozhdeniya*, escort fighter) or BB-22DIS (D = ***dahl'niy*** – long-range). The BB-22DIS is mentioned in a document dated 15th October 1940 as being under design.

A government directive tasking the Yakovlev OKB with the development of the single-seat BB-22IS was issued on 25th November 1940; the BB-22IS powered by two M-105 engines was to be completed by 1st March 1941. A maximum speed of 585 km/h (364 mph) at 5,000 m (16,400 ft) was stipulated. However, the schedule was slipping – by the end of April the prototype was only 90% complete. The delay was due to urgent work on the I-28 and I-30 fighters.

Eventually the prototype was completed. Available drawings depict this machine as structurally identical to the original Ya-22, with the rear cockpit amidships; apart from the new engines, the only difference was the replacement of the inter-spar bomb bay by magazines for two ShVAK 20-mm cannon scabbed on under the fuselage. The armament was to include a further three ShKAS machine-guns. According to Vadim B. Shavrov, very little flight testing was undertaken on the machine, albeit the refinement work continued well into 1942.

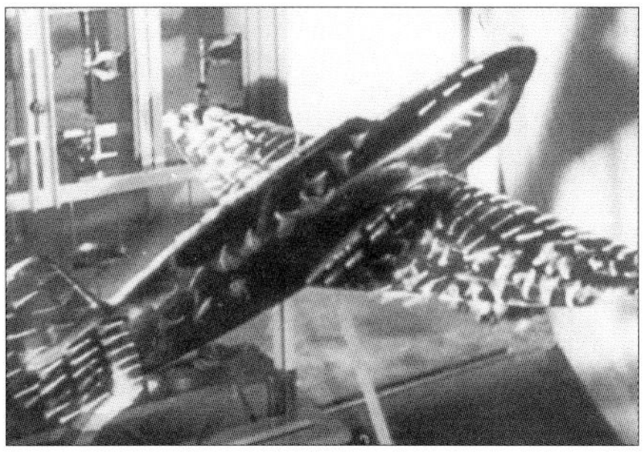

Yakovlev OKB

Yak-1 (I-26) fighter prototypes

The I-26 aircraft (later renamed Yak-1) occupies a special place in Aleksandr S. Yakovlev's creative activities as a designer. The military customers had good reasons to voice many complaints and criticisms concerning this machine, especially in 1940-42. Nevertheless, this fighter had a happy fate. It entered large-scale production and became the progenitor of a family of piston-engined fighters that achieved world renown. While retaining the baseline type's general appearance, the family included no fewer than 60 prototype and production versions.

The I-26 was proposed in several versions. While other design teams submitted one project each for a fighter contest, Yakovlev presented four projects at once. These were the I-26 front-line (tactical) fighter for low and medium altitudes; the I-27 (alias UTI-26) fighter trainer for conversion training; the I-28 high-altitude fighter for air defence; and the I-30 (alias I-26U) multi-cannon fighter. In contrast to other OKBs, the Yakovlev team submitted neither a preliminary design (PD) project nor a mock-up of the future machines. At that time the Chief Designer already had connections in high places that enabled him to skip such 'formalities'.

The work on the I-26 proceeded on a quicker pace compared to other projects, and all employees of the design bureau were fully aware of its importance. After all, the success or failure of the 'firstling' would seriously affect the destinies of the other machines. Yet, the I-26 was born in greater pains than its stablemates. This was quite natural: after all, the badly needed experience in matters pertaining to the construction of a modern high-speed fighter was being acquired in the course of the work.

Yakovlev appointed K. Sinel'shchikov, a young engineer, as project engineer for the I-26. Under his direction design work on the new machine started in May 1939; nearly eight months later, the prototype was completed and rolled out of the Experimental Plant's assembly shop.

The fighter that emerged served as a prototype for all the subsequent numerous versions. In its layout the I-26 was a single-seat mixed-construction monoplane. The fuselage featured a welded steel-tube truss. The forward fuselage housing the engine had duralumin skinning, while the aft fuselage had a fabric skin supported by a structure consisting of light planks. The wings featuring a Clark YH airfoil section were a one-piece unit, with no provision for detaching the outer panels, and were of wooden construction. Duralumin was used in the construction of the tail unit, ailerons (which had fabric skinning) and flaps.

Many parts and units featured a basically new design and had experimental status. These included the future fighter's 'core' – the engine. It was envisaged that the prototype would be powered by a Klimov M-106 high-altitude engine (an M-105 derivative uprated to 1,350 hp); this would enable the I-26 to attain a maximum speed of 620 km/h (385 mph) at altitude, the landing speed being 120 km/h (74.6 mph). The service ceiling was estimated at 11,000-12,000 m (36,090-39,370 ft), and it was expected that the fighter's manoeuvrability and rate of climb would be outstanding for its time. Installation of the supercharged M-105 engine on the second prototype was expected to enhance the fighter's high-altitude performance.

However, these plans proved to be wishful thinking – even the baseline M-105 engine existed only in prototype form when construction of the I-26 began. It should be noted that the engine developed under the direction of Vladimir Ya. Klimov was specially designed to permit installation of a cannon between the cylinder banks. The barrel of the 20-mm ShVAK cannon passed through the propeller hub, obviating the need for synchronisation and permitting the cannon to fire irrespective of the engine/propeller rpm. The cannon was supplemented by four synchronised 7.62-mm ShKAS machine-guns arranged in pairs above and below the engine.

The machine was built with no prior fighter design experience to fall back on. As

Yakovlev OKB

Top: A model of the I-26 in the TsAGI wind tunnel; note the wool tufts.
Above: The ill-fated I-26-1, the first prototype of the Yak-1, on skis (as originally flown). Note the red stripes on the propeller blades.

a result, the real fully-equipped airframe weighed 2,600 kg (5,730 lb) instead of the intended 2,300 kg (5,070 lb). To make the first flight possible, the fuel tanks of the I-26 were only partially filled and the cannon and machine-guns were not provided with ammunition. Yakovlev kept the promise he had given to Stalin; the first prototype (I-26-1) was ready to fly by the beginning of 1940. On 13th January the Design Bureau's chief test pilot Yulian I. Piontkovskiy took the machine, which was fitted with ski undercarriage for the occasion, into the air.

Piontkovskiy had a high opinion of the fighter. However, concluding his report after the maiden flight, he noted that the oil temperature had started to rise immediately after take-off; this alarmed him, compelling him to make an urgent landing.

The first series of flights showed that the machine was responsive, with low control forces, making the handling of the aircraft easy and simple. Its maximum speed at 5,000 m (16,400 ft) was 580 km/h (360 mph), which came close to the estimated figure. Only 5.2 minutes were needed to reach this altitude.

On the other hand, the I-26 suffered from a number of teething troubles caused both by design deficiencies and poor production standards. It took a long time before the oil overheating observed in the first flight could be eliminated. Changes were repeatedly introduced into the oil system but failed to produce a radical improvement. After a flight of only two or three minutes at full throttle the pilot had to abort his mission. The engine was replaced five times, the propeller had to be changed even more frequently. Very soon the VISh-52 propeller, which turned out to be ill-suited for the type, gave place to the more suitable VISh-61P.

It transpired that many elements of the aircraft's structure were not strong enough. In particular, static tests revealed that the wing leading edges disintegrated when the load reached two-thirds of the calculated maximum figure. Undercarriage retraction was troublesome and the wheels were not securely held by the locks, whether it be in the extended or in the retracted position. Apparently it was the defects of the main gear mechanism and insufficient strength that caused the machine to crash on 27th April 1940. On its 43rd flight, the aircraft made a flip roll at low altitude, entering a spin, and crashed in Bol'shaya Maslovka Street in Moscow near the Design Bureau's premises. Pilot Yulian Piontkovskiy was killed. Until then he had saved the Yakovlev fighter prototype 15 times thanks to his swift and faultless judgement in critical situations. Piontkovskiy was succeeded by test pilot Sergey Korzin'shchikov.

Above: the I-26-1 on wheels at a later stage of the manufacturer's flight tests; note the shape of the canopy. As per OKB custom, the fighter was painted cherry-red overall, except for the red/white striped rudder which was a trademark feature of Yakovlev aircraft in the pre-war years.

Above: The first prototype had a rather pointed fin top and a boxy fairing atop the engine cowling housing the carburettor air intake.

The wreckage of the I-26-1 after the crash in Moscow's Bol'shaya Maslovka Street on 27th April 1940. Yulian Piontkovskiy, Yakovlev's first chief test pilot, lost his life in the crash.

The accident investigation panel surmised that in the course of the roll the mainwheels had become unlocked sponta- neously, the abrupt load causing a failure of the wing structure which was not strong enough. This reasoning sounds convincing to this day.

By the time when the I-26-1 crashed, the second prototype (I-26-2) had already been completed. Its airframe structure was strengthened; in particular, the wing leading edges were reinforced and the plywood wing skinning was made thicker.

Above and below: the identically finished I-26-2, the second prototype, during manufacturer's flight tests. Outwardly it differed from the I-26-1 in lacking the oblong carburettor air intake fairing and having a more rounded fin top – the shape that was selected for the production model. Note the highly polished airframe finish

In his efforts to provide better cooling for the engine, Yakovlev had to relinquish the aerodynamically favourable position of the oil cooler. It was moved from its original location under the cylinder banks (that is, between the mainwheel wells) to a place under the engine sump (that is, under the extreme nose), which entailed an increase in drag. While the work on the machine was proceeding, the Government issued a directive stipulating that the M-107 engine – the most powerful of the liquid-cooled engines then available – be installed in the I-26. Yet this directive could not be implemented at that time because even the forerunner of this engine, the M-105, was still beset by teething troubles. It may be noted at this stage that more or less normal operation of the M-107 engine (which was renamed VK-107 by then to denote its designer Vladimir Klimov) on the Yak-9U fighter was achieved only after the end of the war.

The VISh-61P propeller which had been selected to suit the M-105 engine during the first flights was also installed on the second prototype. Changes were introduced into the armament of the I-26-2: of the four machine-guns only the upper pair was retained, chiefly for CG reasons. The angle at which the aircraft sat on the ground was insufficient, and the aircraft tended to be 'nose-heavy' during landing runs, which incurred considerable danger of nosing over; the omission of the lower machine-guns was expected to alleviate this problem. With the alterations listed above, the I-26-2 acquired the features that were characteristic for the subsequent prototype and production versions.

The manufacturer's tests of the I-26-2 were brief. Already on 1st June 1940 it was submitted for State acceptance trials at NII VVS. This was preceded in May by a Government directive requiring the I-26 to be put into large-scale production in Leningrad, Moscow and Saratov. This was an unprecedented occurrence: the aircraft still had many defects to be cured, Air Force engineers and pilots were only just beginning their part of the testing, yet large-scale series production was already envisaged. Presumably this was not solely due to A. Yakovlev's abilities in 'promoting' his progeny (in April 1940, in addition to his previous posts, he was appointed Vice People's Commissar of Aircraft Industry with responsibility for prototype aircraft construction). More probably, the top echelon of NKAP (Aleksey I. Shakhoorin, Pyotr V. Dement'yev and Pavel A. Voronin) became aware of the new fighter's potential and took the calculated risk in order to save time, since the Second World War was already raging in Europe.

The State acceptance trials conducted under the direction of engineer N. Maksimov went rapidly and brought no surprises. The machine demonstrated very high performance. For example, the maximum speed at sea level was 490 km/h (305 mph), and at 4,800 m (15,744 ft), which was the second rated altitude, the I-26 reached 585.5 km/h (363.9 mph). The service ceiling was 10,200 m (33,456 ft), and a climb to 5,000 m took six minutes flat.

The aircraft's structural strength remained insufficient, and it was tested at an all-up weight of 2,700 kg (5,954 lb), which was 100 kg (220 lb) lighter than in the fully equipped version. Despite this, the danger of structural failure was still there and had to be taken into account. A temporary ban had to be placed on most aerobatic manoeuvres. Due to G-load limitations, test pilots Pyotr M. Stefanovskiy and A. Nikolayev performed a full-circle turn at the altitude of 1,000 m (3,280 ft) during 24 seconds (they described it as a 'pancake turn', since they avoided banking the machine). Coolant and oil overheating forced the pilots to decrease the engine rpm during climb.

The State trials revealed 123 defects of varying seriousness (regrettably, about the same number of defects was found in the other prototype fighters – the Lavochkin/Gorboo-nov/Goodkov I-301 and the Mikoyan/Gurevich I-200 – which entered large-scale production before the war). Thus, the I-26 failed the State trials. However, it was recognised that the new fighter was about 100 km/h (62 mph) faster than the Polikarpov I-16 – the most widespread fighter in service at that time, and the I-26 was even easier to fly than Polikarpov's fighter. This promised great advantages.

When the test results of Yakovlev's first fighter were discussed by NKAP and Air Force representatives, two questions came to dominate the discussion. Firstly, the I-26 had no generator, nor a vertical speed indicator. The bulky PAN-23 telescopic sight fitted originally was replaced by a ring-and-bead sight; this finally brought home the awareness of how primitive the aircraft's equipment was. The aircraft did not even have a communications radio. This was a cause of concern for the leaders of NII VVS; Aleksandr Filin, head of the Institute and an experienced engineer, rightly pointed out that a transceiver had to be installed already on prototype examples of fighters. The Chief Designer objected, however, being of the opinion that the purpose of testing was to evaluate the aircraft's flight and tactical performance and a radio was not required for that. In the end, the commission ruled that the equipment of production fighters should be revised. Another point of doubt was the use of one-piece wings by Yakovlev. The absence of outer wing panel joints offered a weight saving, but quite obviously it greatly hampered transportation and repairs of the fighter under field conditions.

The aircraft was immediately put into series production at plant No.301 in Moscow. In the summer of 1940 18 machines were manufactured there for the purpose of service trials, using I-26-1 drawings. Changes had to be introduced into the new fighter's units and systems in the process of construction. For example, the design of the oil tank and its placement had to be changed three times; the shape of exhaust stubs was changed twice. Service trials of the machine were conducted at Kubinka airfield near Moscow under the direction of engineer K. Chasovikov. Pilots of the 11th IAP (*istre**bit**el'nyy aviatsi**on**nyy polk* – Fighter Regiment), who were the first in the Air Force to convert to the new fighter, could not begin flights until the wing leading edges had been reinforced.

The pilots took a liking to the machines and flew them confidently. Although some defects were noted, the impression was that the I-26 was simple, docile in handling and that the machine had a bright future in store. Yakovlev's aircraft was evaluated in the 11th IAP not only in the daytime but at night as well. Although the machines lacked landing lights (a serious drawback for night flights), the pilots made trouble-free landings with the landing strip illuminated by a floodlight.

A flight of five initial production I-26s took part in the traditional parade on 7th November 1940; its flypast over Moscow caused admiration of the public assembled in Red Square. Due attention was also paid to a comment made by test pilot A. Yakimov who, acting under orders from People's Commissar of Defence Kliment Ye. Voroshilov (succeeded by that time by Timoshenko), checked the flight performance of new fighters. He said, 'This is not a combat aircraft but rather a kind of a lightplane – so simple and docile is it in flight.'

However, prolonged operational service turned up defects which could not have been revealed during the expeditious State trials. It was necessary to enhance urgently the strength of the undercarriage wheels, as well as to eliminate the jamming of the sliding cockpit hood in a dive and the tendency of the tailwheel to collapse on landing.

In the meantime, the third prototype – the I-26-3 – was ready for State acceptance trials. On this machine many of the defects noted earlier had been eradicated. In particular, the airframe strength had reached the required level, enabling the fighter to attain an indicated airspeed of 635 km/h (395 mph) in a dive, which had been considered dangerous earlier, and perform aerobatics at maximum permissible G-loads. The anticipated good manoeuvrability was corroborated: a full-circle turn at the altitude of 1,000 m (3,280 ft) could be performed in 20-21 seconds, and the fighter gained 800-1,000 m (2,625-3,280 ft) in a climbing turn from that altitude.

The improvements, however, had a price: the all-up weight rose by 100 kg (220 lb). It was expected that the production fighter would be still heavier after the installation of the necessary equipment, in particular, the radio. However, this did not affect the I-26's ease and simplicity of handling. In

Yakovlev OKB

Yakovlev OKB

Top and above: Two more views of the I-26-2 during State acceptance trials. Note the radio set aft of the seat headrest, the lack of the gunsight and the characteristic hexagonal concrete slabs of the NII VVS hardstand at Shcholkovo airfield.

Yakovlev OKB

Above: The fifth production I-26 (c/n 0105) manufactured by the Moscow aircraft factory No.301 and delivered to NII VVS. The aircraft did not exactly conform to the manufacturing drawings.

October-November 1940 NII VVS test pilots Pyotr M. Stefanovskiy, A. Koobyshkin, A. Proshakov, K. Groozdev and A. Nikolayev checked the machine's behaviour in different flight modes. The work was conducted under the direction of project engineer A. Stepanets, who subsequently headed most of the teams testing Yakovlev's fighters and did much for improving the quality of prototype and production machines.

In early December a report was signed, stating that the I-26-3 had passed the tests with satisfactory results. However, the defects noted were still numerous. The main defects included imperfections in the design of the fighter (undercarriage, pneumatic system) and of the powerplant (oil overheating and unstable operation at cruising rpm) and the absence of the necessary equipment (radio, landing light, generator and so on).

Yak-1/M-105P production fighter

Series production of the fighter got under way in the late autumn of 1940. In addition to the small plant No.301 in Moscow, which had already gained some experience in manufacturing Yakovlev aircraft, production of the machine was started at plant No.292 in Saratov (former 'Sarcombine' plant producing agricultural machinery). It discontinued work on the I-28 fighter designed by Vladimir P. Yatsenko (not to be confused with Yakovlev's I-28 described below in this chapter) and completely switched over to mastering production of the promising I-26; in December 1940 the latter was allocated the designation Yak-1.

According to the Government's plans, the two abovementioned plants, supplemented by plant No.126 in Komsomol'sk-on-Amur which also joined in the Yak-1 programme, were to deliver 220 machines by the end of that year. However, the latter enterprise, being overburdened with work on the Il'yushin DB-3F bomber, was unable to start tooling up for Yak-1 production. A chronic shortage of propellers, engines, radiators and other units also hampered the expansion of production.

Launching Yak-1 production in Saratov was also accompanied by a multitude of problems. In accordance with the rules adopted in the Soviet Union at that time the plants engaged in aircraft production used a loft floor-and-template method. It involved providing the plant with big plywood sheets upon which full-size layouts of the main units and parts of the future machine were marked. Templates were made by copying the outlines on the loft floor panels; they were used for manufacturing the jigs and tools and, later, the parts themselves. When alterations were introduced into the drawings, this necessitated changes in the whole chain of the production process. As witnessed by the OKB reports, the Chief Designer approved 7,460 (!) changes in the drawings of its first fighter in the period between June and December 1940 alone. Besides, for a long time the plant's workers could not master the welding of the fuselage truss and the woodworking jobs associated with wing construction; this led to a big amount of defective products. Despite this, on 3rd October 1940 plant 292 test pilot Colonel P. Shootov took the first production Yak-1 into the air.

In the period between the end of 1940 and the beginning of 1941 the series production was marked by a gradual build-up of the monthly output; yet, the plan for 1940 was not fulfilled. By the end of 1940 military representatives took delivery of 48 Yak-1s in Moscow and 16 in Saratov. While in December 1940 one aircraft was rolled out each day, in April of the following year this figure rose to two per day, reaching three per day before the outbreak of the war, despite the cessation of Yak-1 production at plant No.301 since April.

Production flaws were gradually ironed out; still, critical remarks accompanied all the tests of production machines at the beginning of 1941. The defects noted included leaky oil lines, uneven fuel consumption from the port and starboard tanks, and water and oil overheating.

At that time almost all design changes were accompanied by an increase in weight; a Yak-1 tested in February-March 1941 (c/n 04-06 – that is, fourth aircraft in Batch 6) weighed as much as 2,858 kg (6,302 lb). Many defects of the aircraft were eliminated: the main gear doors were strengthened, the carburettor air inlet was redesigned, the

Yakovlev OKB

Yak-1 fuselages on the continuous flow assembly line. The line moves away from the camera, judging by the lack of the cowling attachment structure on the foremost machine.

wing root fairings and the numerous hatch covers became more tight-fitting. The oil cooler was made less protruding, which helped reduce drag. Analysis of materials of the numerous tests shows that the following performance figures were typical of the Yak-1 in 1941: speed at sea level, 480 km/h (298 mph); speed at the second rated altitude of 4,950 m (16,236 ft), 577.5 km/h (358.9 mph). Time to 5,000 m (16,400 ft) was now 5.7 minutes. The aircraft performed a 360° turn at 1,000 m (3,280 ft) within 20-21 seconds. The fighter's landing speed was 137 km/h (85 mph), and cruising range reached 700 km (435 miles).

Preparation for and participation in the 1941 May Day Parade became a crucial test for the newly produced machines; 40 Yak-1s were assigned for this mission. Pilots of the already mentioned 11th IAP, which was a part of Moscow's anti-aircraft defences, performed 162 missions in all. There were no fatal crashes or serious accidents with the Yaks, but all sorts of technical failures ran into dozens. Particularly disturbing were the cases of fuel filler cap disintegration – the fire hazard was great.

From mid-May 1941 onwards the 11th IAP was fully equipped with 62 Yak-1s (that was the regular complement at the time). The pilots of this regiment had made a worthy contribution to eradicating the defects of the first production Yaks.

Based in the vicinity of Baku, the 45th IAP commanded by Major I. Dzoosov became the second regiment in the Red Army Air Force to convert to Yakovlev's fighter. In April 1941, 78 Yaks were sent to the Transcaucasian region.

Small numbers of production Yak-1s supplemented the inventory of other units literally a few days before the outbreak of the war. The majority of flyable Yaks were stationed in the Moscow area, but 105 fighters of this type succeeded in reaching the five western military districts situated along the border.

As a result of delayed deliveries of new Yaks to Air Force service units in the western border area, in June 1941 the majority of these new machines were flown only by the commanders of some fighter regiments and smaller units. Among those who fully mastered the new fighter was Major B. Soorin, Commander of the 123rd IAP. On 22nd June, repulsing the Luftwaffe air raids against the 4th Army headquarters in Kobrin, Belorussia, he flew four sorties, claiming the destruction of three enemy aircraft, but was killed in action on the evening of that day.

Naval Aviation pilots also converted to the Yak-1. In the Black Sea Fleet Air Arm it was Lieutenant Yu. Shitov from the 9th IAP who scored the first 'kill' while flying this fighter type, shooting down a Romanian reconnaissance aircraft.

Most of the Yaks opposing the Luftwaffe invasion at the borders of the Soviet Union met a tragic fate. Unlike the I-16s and the SB bombers, the Yaks of the 91st and the 123rd IAP suffered losses mainly on the ground not only during the first day of the war, but until the end of June; if they had a choice, Soviet pilots preferred to fly their missions on older types which they had thoroughly mastered.

After just a few days of combat actions no Yaks remained in the inventory of the Air Force of the Western and South-Western fronts. Pilots of the 20th IAP gained fame in fierce combat against German airmen at the South-Western Front. During three days alone (10th-12th June), while committed to action at distant approaches to Kiev, the unit downed 26 enemy aircraft.

In the late summer and the autumn of 1941 Yakovlev's fighters saw action primarily near Moscow. As of 10th July 1941, 133 Yak-1s made up approximately 1/6th of the fighter aviation element which defended the capital in the ranks of the units of the 6th IAK

Yakovlev OKB

Above and below: Yak-1 '16 White' (c/n 0406) during checkout trials at NII VVS. Such trials were held from time to time to make sure that production aircraft conformed to the Air Force's specifications.

Yakovlev OKB

Yakovlev OKB

An early-production Yak-1 built by plant No.292; note the characteristic shape of the upper main gear door segments.

Yakovlev OKB

Above: Yak-1 '30 White' (c/n 0218) which was modified according to the decision of a joint NKAP and Air Force commission. The crudely hand-painted tail number was obviously temporary.

(*istre**bit**el'nyy aviatsionnyy **kor**poos* – Fighter Air Corps) of Air Defence.

The outbreak of the war posed an acute problem – it was necessary to organise a steady conversion of the flying personnel to new aircraft types, including the Yak-1. The conversion centres that had been set up at production factories and in the 11th IAP were already unable to cope with the needs of the Red Army Air Force. Hence Reserve Air Regiments, or ZAPs (*zapas**noy** aviatsi**on**nyy polk*) were set up. The 8th ZAP, which specialised in Yak-1 conversion training, was deployed in Bagai-Baranovka near Saratov in early July 1941.

The combat effectiveness of the Yak-1 would have been higher, had it not been for some annoying defects. Many troubles stemmed from the M-105P engines. They were more reliable than the 1,350-hp Mikulin AM-35A Vee-12 engine powering the MiG-3 fighter, but there were cases of magneto and speed governor failures and cases when oil was ejected through the output shaft of the reduction gearbox. Metal chips accumulated in the oil cooler of one of the engines during the second (!) sortie; the engine jammed and the pilot had to make an urgent forced landing in order to avoid more serious consequences.

Gradually, as the teething troubles were overcome, it became clear that the Yak-1 had greater chances of success in combat against German fighters compared to other aircraft of the Red Army Air Force. This is the appraisal given at that time by Luftwaffe experts: *'The Yak-1 is presumably the best Soviet fighter. It had a better speed and rate of climb compared to the MiG-3 and came close to the performance of the Bf 109F, but was inferior to the latter in speed. The Yak-1 was appreciably harder to hit from the rear than the MiG-3. It retained a good rate of climb up to 6,000 m* [19,680 ft]*, but manoeuvrability fell off at that altitude. For this reason the pilots dived from high altitudes to lower ones where they accepted combat.'*

Yet, there was no shortage of defects and complaints. They were not so dangerous in flight as was the case with the first MiG-3s, and did not cause such a serious deterioration of performance as in the case of the LaGG-3, but they did cause quite a few vexations.

The moving elements of the undercarriage remained unsatisfactory. Bumps and jerks occurred during retraction; sometimes the undercarriage legs jammed halfway through retraction. Frequently the undercarriage collapsed during landing; as for the tailwheel failures, they were due mainly to the insufficient castoring angle.

The Yak-1's armament also provided its share of troubles. Many pilots considered it inadequate and prone to jamming. The cockpit became excessively hot, and oil leaking from the engine breather sprayed onto the cockpit windshield, which hampered flying – the fighter had to be flown with the sliding canopy in the open position.

Yak-1/M-105PA production fighter

As early as in the summer of 1941, the M-105P engine installed in the Yaks was gradually supplanted by the improved M-105PA. The new model had a reinforced crankcase, stronger connecting rods and a floatless carburettor (the latter enabled the pilot to perform inverted flight for some time and enter a dive with negative G-loads). There were also other changes.

During the first month of the war the Soviet government adopted several important decisions calling for an increase in the Yak-1's production output. Thus, for example, on 24th June plant No.292 was prescribed to deliver to the front 1,350 combat machines, not 1,100 as stipulated earlier. It was envisaged that Yak-1 production should be mastered at aircraft factories in Leningrad (plant No.47) and Engels, Saratov Region (plant No.492). In addition, in the summer of 1941 the People's Commissar of Aircraft Industry approved a programme of Yak production at the new aircraft plant No.448 in Tbilisi, Georgia, and its branch in Kutaïsi (plant No.131). However, the exigencies of war prevented these plans from being implemented.

In the meantime, the staff of the Yakovlev OKB and plant No.292 in Saratov did not halt their efforts aimed at improving the fighter's airframe. In November 1941 a Yak-1 (c/n 2029) successfully passed testing at NII VVS. This fighter featured a supplementary hydraulic power cylinder which ensured smooth and bump-free landing gear cycling, and a simplified non-retractable tailwheel. A new easily detachable propeller spinner was introduced, as were a trim tab on the rudder, a landing light and a radio; the installation of the latter was accompanied by shielding and electrical bonding of the engine and the airframe. A special breather vent tank was added to eliminate the oil spill problem.

The reliability of the armament also improved. Unfortunately, this was accompanied by an increase of the all-up weight to 2,934 kg (6,469 lb) and some deterioration of performance. Maximum level speed was 468 km/h (291 mph) at sea level and 560 km/h (348 mph) at 4,800 m (15,740 ft) – that is, the aircraft proved to be 12-17 km/h (7.4-10.6 mph) slower than the early Yak-1s. The time required to reach 5,000 m rose to 6.8 minutes – an increase of more than one minute. The machine's manoeuvrability remained good – at low altitude the Yak-1 could make a full-circle turn in 19-20 seconds. In the course of a combat turn the fighter gained some 900 m (2,950 ft). The introduction of engine boost mode made it possible to reduce the take-off run by 5% and compensate for the Yak's increased take-off weight.

The abovementioned Yak-1 c/n 2029 became the first example of this type to be fitted with a transceiver radio at the production plant. Approximately 1,000 Yak-1s built earlier did not have even the simplest radio receiver. Subsequently, in the course of several months transceivers were installed in one Yak-1 out of every ten, the rest having only a receiver. This was due to the Chief Designer's firm conviction that, as long as the quality of radio communication was still unsatisfactory, the radio merely added extra weight to the airframe.

Yak-1 production fighter with rocket armament

Even prior to that, in October-November 1941, the Yaks were provided with rocket armament. Here note must be made of the initiative displayed by Major A. Negoda, Commander of the 562nd IAP. He performed four to five sorties after one refuelling, strafing the enemy's forward lines with the new 82-mm (3.22-in) RS-82 unguided rockets.

This was possible because the forward line of defence passed about 10 km (6 miles) from the regiment's airfield in Khimki north of Moscow. The German anti-aircraft defences were hard put to it to repulse effectively the attacks of Soviet fighters which made a surprise appearance at extremely low altitudes.

195 Yaks were fitted with rocket armament at the plant by the end of 1941; another 953 fighters were retrofitted with this armament by the late spring of 1942. The installation of six RS-82 projectiles on the machine found a positive response from the flying personnel; as a result, the units began fitting rocket armament to the fighters directly at the frontlines. Although the external stores increased the all-up weight by 65 kg (143 lb) and decreased the maximum speed by some 30 km/h (18mph), firing these projectiles against aerial targets (especially during head-on attacks) produced a strong psychological effect on the enemy. In the case of a direct hit (which, admittedly, happened extremely rarely) the enemy aircraft simply disintegrated in mid-air.

Yak-1/M-105PA production fighter, winter configuration

In the late autumn of 1941 the task of preparing a 'winterised' version of the Yak-1 came to the fore. It was a normal fighter in which the wheel landing gear was replaced by a retractable ski undercarriage. In addition, heating was provided for some units of the powerplant; a system for diluting the engine oil with gasoline to facilitate starting in winter was installed; the water radiator was filled with glycol solution; a thick cushioned cover was placed on the engine cowling to prevent the engine from getting overcooled when the aircraft was parked, especially on a frosty night. To make the aircraft less noticeable against a snowy background, a coat of washable white paint (a chalk and glue solution) was applied over the summer camouflage. In the winter version the Yak-1 became 70-80 kg (154-176 lb) heavier and 30-40 km/h (18.7-24.9 mph) slower at all altitudes.

The Yaks were intensively operated from snow-covered field airstrips without clearing away the snow. No other new fighter was used so extensively on a ski undercarriage.

Up to 25th February 1942, 830 Yak-1s were built in this version. One of the reasons for this was that ski undercarriage simplified landings on uneven airfields covered with a thick layer of snow. A fighter fitted with skis was more stable during a landing run and did not bounce. A strip of only 250 m (820 ft) was needed for the lift-off, and the landing run, with the use of brakes and landing flaps, was only 275 m (900 ft). The angle at which the aircraft sat on the ground was considerably increased, which reduced the danger of a nose-over in the case of a sharp decrease in speed.

Yak-1 – improvements on the production line

Efforts to provide plant No.292 'Sarcombine' with high-productivity jigs and tools and introduction of continuous flow-line assembly of the aircraft and its parts had their effect: even before the outbreak of the war the production of combat machines acquired a regular tempo. Whereas four fighters were delivered per day in June, at the end of October 1941 the daily output reached eight machines.

The concentration of efforts on the series production of the relatively well-developed Yak-1 fighter left no choice but to finally give up plans for switching over to manufacturing the improved Yak-3 (I-30) described below. The government directives on setting up series production of the Yak-1 in Tbilisi and Kutaïsi were not implemented either. All work on this fighter was concentrated in Saratov.

At first no steps were taken to organise sufficiently stringent control over structural weight. Almost all changes introduced into the design led to an increase of the airframe weight. From early August 1941, however, this drawback was partially remedied. Thanks to Yakovlev's initiative, designers and production engineers began to receive remuneration for literally every gram of weight that was saved. In mid-August 1941, starting with the 29th production batch, the weight was reduced to 2,917 kg/6,432 lb (without external stores and radio) which became characteristic for the Yak-1.

The first six months of the war showed that the Yak-1 had proved to be the most satisfactory among the fighter types put into series production before the war. However, the Yak-1 did not occupy a notable place numerically among other fighters. On the eve of the Soviet counter-offensive near Moscow, on 5th December 1941, service units of the Red Army Air Force had only 83 Yak-1s on strength (about 8% of all fighters), and of these only 47 were considered serviceable. This is due in part to the wide use of obsolete types and to serious difficulties in replenishing the heavy losses sustained

Yakovlev OKB

Yakovlev OKB

Top and above: A 'winterised' Yak-1 armed with six RS-82 unguided rockets. The latter were effective against both ground targets and enemy bomber formations. Note the modified cockpit canopy.

Yakovlev OKB

Yakovlev OKB

Yakovlev OKB

Top: This photo illustrates the standard shape of the oil cooler bath and supercharger air intake. Centre: The oil cooler and supercharger intake modified in accordance with TsAGI recommendations make an interesting comparison. Above: The fifth version of the oil cooler bath developed by TsAGI.

during the first six months of the Great patriotic War.

At the beginning of 1942 the Yak-1 was the best Soviet fighter as regards overall performance. New fighter units were being trained at an accelerated tempo in the rear areas. In January six regiments equipped with Yakovlev fighters could, for the first time, be simultaneously sent to the front, which was a sizeable contribution to enhancing the combat capabilities of the Red Army Air Force.

The number of pilots mastering the new machine grew steadily. While only 156 pilots could fly the Yak-1 as of 22nd June 1941, by 1st February 1942 their number increased to 637. Gradually Yak-1 pilots came to participate in combat on all fronts, with the exception of the Karelian front, and fought with considerable success.

The most famous Soviet fighter ace Aleksandr I. Pokryshkin flew combat sorties on a Yak-1 from mid-May until 9th August 1942. At that time he was Lieutenant Senior Grade with the 16th GvIAP (*Gvar**dey**skiy istre**bit**el'nyy aviatsi**on**nyy polk* – Guards Fighter Regiment; the Guards appellation was bestowed on units for special gallantry, and such units were the elite of the Armed Forces). This unit received well-used and fairly battered Yak-1s from other regiments fighting at neighbouring sections of the front; this was probably the reason for the very reserved comments from the pilots. Previously, Pokryshkin had thoroughly studied the MiG-3 and actually taken a liking to it; yet he noted that the Yak-1 was easier to fly, more forgiving and had more potent armament. Thanks to these advantages the famous ace achieved eight victories during a period of incessant air battles that lasted for almost three months, when he downed three Messerschmitt Bf 109s, three Junkers Ju 88s and two Messerschmitt Bf 110s.

The 16th GvIAP fought at the southern flank of the Soviet-German front. In the late spring the Luftwaffe considerably stepped up its activities there (especially fighter operations), and the Soviet airmen were in a tough situation. Yak fighters took part in combat in the areas of the Kuban' and Don rivers, and in the Crimea. Thus, airmen of the 45th IAP forming part of the 3rd Special Air Group commanded by Colonel I. Dzoosov fought especially hard battles repelling the German onslaught on Sevastopol'.

According to official records, in June 1942 Soviet airmen flew 186 combat sorties and scored 22 'kills', even though the enemy enjoyed air superiority (according to German sources, the destruction of six or seven German aircraft can be confirmed; according to Soviet sources, own losses amounted to eight Yak-1s and two pilots lost). Lieutenant N. Lavitskiy achieved special success, with seven 'kills' to his credit. When Wehrmacht troops stormed the blazing city of Sevastopol', eleven Yak-1s of the 45th and 247th IAPs left the Crimea on the night of 1st July, landing safely on airfields in the Krasnodar Area.

Yak-1 with a boosted version of the M-105PA

The staff of the Design Bureau and Chief Designer Aleksandr S. Yakovlev continued to seek possibilities of improving their machine. The Yak-1 was still inferior to its main opponent in air combat – the Bf 109F.

The task became all the more urgent when the Soviet intelligence reported that the Bf 109F-2 had been supplanted on the production line by the F-4 version featuring an uprated engine with better altitude performance, improved armour protection and heavier armament (in fact, the first Bf 109F-4s went into action on the Eastern front in August 1941). The new model further increased the advantages of the German machine over the Yak-1 with regard to rate of climb and other performance characteristics, while retaining a parity in manoeuvrability. The Bf 109F-4 also retained the superiority in speed.

While the German machines were progressively fitted with engines of steadily rising power, in the case of the Yak-1 substituting the M-105PA for the M-105P had not added a single new horsepower to the machine. Its designers were well aware of the fact that introduction of a new engine was an unrealistic option at that time. Therefore the idea came up of uprating the available engine.

The story of how the boosted version of the M-105PA came into being is very unusual. In April 1942 a mixed team comprising specialists from the engine factory and NII VVS and led by B. Nikitin, Military Engineer 3rd Rank, was sent on a mission to the 236th IAP of the Air Force of the Western Front (commanded by Major P. Antonets). At its own initiative the team made arrangements for seven Yak-1s powered by M-105PA engines to be subjected to trial operation with the boost pressure increased from 910 mm Hg to 1,050 mm Hg.

This resulted in a considerable improvement of performance. The Air Force was faced with the prospect of immediately enhancing the combat capabilities of all Yak-1 fighters then in service by simple and expeditious means. Introduction of the increased manifold pressure, it seemed initially, did not require any modification of the M-105PA engine and could be done in situ by the front-line units' own technical staff.

However, operational use of the boosted engines brought some unpleasant surprises. It turned out that in summer weather the fighter could fly with the radiator shutter fully closed (to achieve maximum speed) for no more than two minutes, after which the coolant and oil temperature exceeded the permissible limits. For the same reason it proved impossible to perform a continuous climb in boosted mode; it was necessary to level off several times in order to bring the temperature back to normal.

All this came to the knowledge of the authorities. The Commander of the Air Force issued an order requiring two of the seven modified Yaks to be transferred to NII VVS for research purposes. The test results were as follows: with the engine running at the same nominal rpm and under other conditions identical to those of the usual Yaks, the increased manifold pressure enabled the 'new' fighter to attain a 20-25 km/h (12.4-15.5 mph) increase in maximum level speed at altitudes up to 3,500 m (11,480 ft); time to 5,000 m (16,400 ft) was reduced by one minute, and the time needed for a banking turn at 1,000 m (3,280 ft) by one second. Take-off performance was also improved. On the down side, the water and oil in the boosted engine overheated. To keep their temperature within acceptable limits, the engine speed had to be reduced from the nominal 2,700 rpm to 2,400-2,500 rpm, which, in effect, nullified all the advantages afforded by increasing the boost pressure

On the basis of these results the People's Commissariat of Aircraft Industry tasked Chief Designer A. Yakovlev with redesigning the engine cooling system and taking measures to prevent oil spill from the breather, as well as introducing engine seals. Also, the engines themselves were to be perfected. The main improvements to be effected by engine designers included reinforcement of the crankcase and piston pins and increasing the diameter of borings in the carburettor nozzles.

Yakovlev OKB

A production 'razorback' Yak-1 with aerodynamic refinements based on TsAGI recommendations. Note the modified main gear doors and the faired engine exhaust stubs.

Yak-1/M-105PF production fighter

The production version of the boosted engine received the designation M-105PF (*for**see**rovannyy* – uprated, or boosted). Its readjustment, as compared to the M-105PA, entailed not only an increase of power but also a reduction in altitude performance. Thus, the output at the first rated altitude became 1,260 hp at 700 m (2,300 ft) instead of 1,100 hp at 2,000 m (6,560 ft), without taking into account the dynamic pressure; at the second rated altitude the power rose from 1,050 hp at 4,000 m (13,120 ft) to 1,180 hp at 2700 m (8,860 ft). At altitudes above 4,000 m the characteristics of boosted and unboosted engines proved to be identical.

In June 1942 a Yak-1 (c/n 15-69) with the boosted M-105PF engine and increased oil cooler area passed tests at NII VVS. Despite the modified oil cooler, the temperature condition of the powerplant deteriorated. At a take-off weight of 2,917 kg (6,430 lb) typical for a Yak-1 without a radio, the fighter attained 510 km/h (317 mph) at sea level and 571 km/h (355 mph) at the second rated altitude of 3,650 m (11,970 ft); it needed 6.4 minutes to climb to 5,000 m (16,400 ft), performed a full-circle turn at low altitude in 19-20 seconds and gained 980 m when climbing in a combat turn. However, to ensure normal operation of the aircraft the engine revs at low altitude had to be limited to 2,550 rpm.

It proved possible to reduce the Bf 109F's ascendancy at low altitude, as demonstrated by mock combat between the Yak-1 M-105PF and the Bf 109F conducted at NII VVS for training and test purposes. At 1,000 m (3,280 ft) the Bf 109F had a marginal advantage in vertical and horizontal manoeuvrability. The German fighter succeeded in getting on the tail of its 'adversary', but only after four or five turns. At the altitude of 3,000 m (9,840 ft) both fighters fought on equal terms; in effect, aerial combat became restricted to head-on attacks. It was to the Yak's advantage if its 'adversary' could be lured to higher altitudes. Already at the altitude of 5,000 m (16,400 ft) the Yak-1 showed better manoeuvrability and its pilot could impose his will on the enemy.

It should be remembered, though, that when the Bf 109F was evaluated at NII VVS the supercharger system of the Daimler-Benz DB 601N engine did not ensure a constant degree of supercharging. On the other hand, when the German engine ran normally, the Messerschmitt fighter having better engine performance at high altitude was considerably superior to the Yak-1 M-105PF at 5,000 m and higher. Besides, NII VVS had tested the Bf 109F-2, whereas, as noted earlier, it was the Bf 109F-4 with the more powerful DB 601E engine that had become Germany's main fighter type by the summer of 1942. The Yak-1 M-105PF was markedly inferior to this model on many counts.

Aerodynamically refined version of the Yak-1

Improving the aircraft's aerodynamics, along with measures designed to reduce the all-up weight, was one of the main areas of the OKB's activities during the whole period of the series production of the Yak-1. This work was usually conducted at the OKB's own initiative. But sometimes exigencies of life compelled the Design Bureau to resort to urgent measures.

One of such periods when the work acquired the character of a crash programme was the autumn and winter of 1942. A dire situation arose during the Battle of Stalingrad; it was caused, in particular, by the advent of new German fighters – Bf 109G-2s fitted with engines of greater power and better altitude performance which gave them considerable ascendancy in air combat. The Bf 109G-2 possessed only marginally greater speed compared to Soviet fighters but enjoyed a considerable advantage in rate of climb and, consequently, in the vertical manoeuvre.

The leaders of the nation demanded that the performance of Soviet fighters be enhanced. This applied, above all, to the Yak-1 fighter. It was envisaged that the work would be concentrated on improving aerodynamics and installing the more powerful M-106 engine with a single-speed supercharger. Aerodynamic improvements were effected in accordance with the Government decision adopted in December 1942.

The work conducted under Yakovlev's direct control comprised the following. Completely sealed bulkheads were installed in the fuselage; the retractable tailwheel was reinstated; the engine cowlings and wing root fairings were redesigned in accordance with the recommendations of theoretical aerodynamics; a debris guard wire mesh screen installed in the water radiator intake was deleted. The exhaust stubs were fitted with fairings and the shape of the stubs was altered so that they provided a measure of additional thrust; the shape of the water radiator and oil cooler ducts was altered. Finally, the aircraft's overall surface finish was improved.

Modified Yak-1 (Yak-1b) production fighter

Another direction of work in the Yakovlev OKB was concerned with improving the cockpit visibility, armour protection and armament of the production Yak-1. To achieve this, considerable alterations based on the available experience were incorporated in one example of the fighter in June 1942. The following design features were introduced.

To improve rear vision, the upper fuselage decking behind the cockpit was cut down and the cockpit canopy was given a

Yakovlev OKB

Above and below: Yak-1 c/n 3560 was converted into the *etalon* (production standard-setter) for 1942 with a teardrop canopy and cut-down fuselage spine. Note the moulded frameless windshield; the aircraft had a Yak-7 oil cooler, as it had been planned to install an M-105PF engine which ultimately was never fitted.

Yakovlev OKB

Yakovlev OKB

The Yak-1 *etalon* for 1943 during checkout trials at NII VVS.

streamlined teardrop shape. The moulded visor of that particular example gave place to a windshield formed by flat glazing panels which lessened the distortion of objects observed by the pilot.

To enhance the protection of the pilot's head, a bulletproof windscreen and a rear bulletproof glass panel were installed. The pilot was also protected by an armoured headrest and armoured armrest.

Changes in the armament involved replacing two 7.62-mm ShKAS machine-guns by one 12.7-mm (.50 calibre) Berezin UBS machine-gun (also synchronised), while the 20-mm ShVAK cannon was retained. The OPB-1 telescopic gunsight was replaced by a VV-1 ring-and-bead sight in response to insistent demands from front-line units dissatisfied with the poor quality of the OPB-1.

In addition, the modified fighter featured alterations to the control stick which was made similar to that of the German Bf 109. The stick had push-buttons actuating the armament; this permitted the pilots to fire their weapons without diverting their attention from piloting during violent combat manoeuvres.

On 1st July 1942 Aleksandr S. Yakovlev reported to the Government that the work on modifying the Yak-1 had been completed. The fighter was immediately submitted for State acceptance trials. NII VVS test pilots A. Proshakov, P. Stefanovskiy, A. Kochetkov, L. Koovshinov and V. Khomyakov assessed the new features as excellent. The trials report stated that *'the forward view and visibility to the sides and rearwards through the bulletproof glass and the transparent aft part of the canopy are good and can be regarded as the best compared to all Soviet fighters'*. It was recommended that these alterations be incorporated in production machines and the aircraft tested be regarded as a standard-setter.

Time was needed to implement in production the improvements described above. Not until September 1942 did plant No.292 begin producing Yak-1s modified to the new standard. This aircraft with the UBS-12.7 machine-gun and the bubble canopy was sometimes called Yak-1b, especially in the documents of front-line units (It should be noted that, in contrast to other Yak models, the versions of the Yak-1 were not assigned official suffix letters). Right from the start they were powered by M-105PF engines. By the end of 1942 the number of machines of this version reached 959 out of that year's total production (3,474) of the Yak-1.

In December 1942/January 1943 the Yak-1b passed operational service tests in the 32nd GvIAP of the 210th IAD (*istre**bi**tel'naya aviatsi**on**naya di**vee**ziya* – Fighter Air Division) at the Kalinin Front and in the 176th IAP of the 283rd IAD at the Stalingrad Front. During this time 58 new fighters flew up to 700 combat sorties, performing 38 dogfights and shooting down 25 enemy aircraft, as claimed by the units, for the loss of six Yak-1bs. The new fire control system and the cockpit canopy offering improved rearward visibility were appreciated by the flying personnel and were recommended for introduction on all fighters. The airmen also approved the other changes effected by the Design Bureau.

Lightened version of the Yak-1

The Messerschmitt fighter retained its main advantage – superior vertical manoeuvrability which was due to lower power loading. Soviet designers sought to strip the enemy of this advantage by further lightening the

Yak-1. In March 1942, in response to the Chief Designer's instructions, the OKB staff intended to achieve a radical lightening of the airframe. This work had experimental status at the time. Ten airframes were completed as interceptors. The list of measures aimed at reducing the all-up weight comprised 25 items. The most important of these included removing the ShKAS machine-guns with their ammunition supply, one of the two compressed air bottles and dispensing with the self-sealing layer on the fuel tanks.

The weight saving achieved totalled 162 kg (357 lb). One of the fighters was flight-tested by the OKB's chief test pilot Pavel Ya. Fedrovi who spoke highly of the aircraft. All ten lightened Yak-1s were delivered to the 12th GvIAP of the 6th IAK assigned to the Moscow Air Defence System and were used as interceptors.

The OKB made good use of this experience in September 1942. Twenty lightened Yak-1s were manufactured specially for the Stalingrad Front. At an all-up weight of 2,780 kg (6,130 lb) these machines were marginally faster than the usual Yaks. Importantly, vertical manoeuvrability was appreciably improved. The lightened Yak-1s formed part of the complement of the 512th IAP (commanded by Hero of the Soviet Union, Lieutenant-Colonel N. Gherasimov) and the 520th IAP (commanded by Major S. Chirva).

Combat experience showed that, when piloted by well-trained airmen, these aircraft were quite a match for the Bf 109F-4 and Bf 109G-2 at low and medium altitudes. While a standard production Yak-1 required an average 26 sorties per one German aircraft shot down, the lightened version required only 18 (according to Soviet reports). The latter version also boasted lower own losses. Lightened Yaks caught up with the Bf 109s during the climb, but it was in the banking turns that they were particularly superior to the adversary. The deletion of machine-guns did not seriously diminish the firepower because German aircraft, especially bombers, were destroyed primarily by cannon fire.

512th IAP pilots I. Motornyy and V. Makarov, who were sent to the Saratov plant to take delivery of the first lightened Yak-1s for their unit, engaged in a dogfight during their ferrying flight, claiming two Bf 109G-2s shot down.

The results of the work on lightening the Yak-1 were subsequently used by the Yakovlev OKB in developing other fighters, especially the Yak-3. But at that time, in September 1942, it was decided to refrain from committing the lightened Yak-1 to large-scale production. While the reduced weight of fire was not of crucial importance for experienced fighter pilots, the Soviet top command considered it too heavy a price for ordinary airmen and the lightened version of the Yak-1 was not built in quantity by plant No.292.

Yakovlev OKB

Above and below: Saratov-built 'bubbletop' Yak-1 c/n 04111 during tests at NII VVS. The aircraft wears winter camouflage, though this is not very obvious because the machine is covered all over in soot and oil from the engine!

Yakovlev OKB

Yak-1 – fighter-bomber version

In May 1942 the Yak-1 was once more subjected to alterations. In keeping with a Government decision, the rocket projectiles carried under the wings were dispensed with in order to regain the speed that was so necessary in combat against the Messerschmitt fighters. At the same time, it was decided that the Yaks be used in a new capacity – the fighter-bomber role. For this purpose two bomb racks (with sway braces and shackles) were mounted underwing. Each of them could carry one 25- to 100-kg (55- to 220-lb) bomb. A lever installed in the cockpit enabled the pilot to drop the bomb load.

Thus the aircraft could be used as a high-speed fighter-bomber – or as a tactical fighter without bombs. Installation of bomb racks and attachment of bombs increased the all-up weight and spoiled the machine's aerodynamics, adversely affecting its performance and handling. The speed reduction was especially serious, reaching 30 km/h (18.6 mph). The Yak-1 became heavy and sluggish. After the bombs had been dropped, the performance improved but, since the bomb racks were still there, the fighter-bomber remained slower than the pure fighter version.

In the opinion of the command of the Air Force, installation of bomb armament on the Yak-1 was unjustified because it had been done without taking into account the real capabilities of the M-105PA engine whose power was clearly inadequate for this task. Also, the pilots of front-line units were dissatisfied with the bomb racks; they often said they were carting around a useless load. As the best type of Soviet fighter aircraft at that time, the Yak-1 was intended primarily for the interception of enemy attack aircraft and for aerial combat, not for bombing and attack missions. Nevertheless, bomb racks on production machines were retained up to the end of series production.

Yakovlev OKB

Above: Yak-1 c/n 1047 was lightened and equipped with an all-round vision teardrop canopy.

Yak-1/M-106 fighter

It was presumed that a new, more powerful engine would afford still greater advantages. The Yakovlev OKB had thoroughly refined the airframe to adapt it to the installation of the M-106. For example, the wings had metal spars, the lightened tail unit also received a metal framework. Two fuel tanks holding 400 litres (88 Imp gal) were installed in the outer sections of the wings; two circular-section oil coolers patterned on those of the I-26-1 were installed in the wing centre section.

In its armament and equipment the experimental fighter differed little from its production stablemates. It also featured a similar engine cowling design.

The M-106-1sk was a modified version of the M-106 that had been recommended for installation on fighters as far back as the beginning of 1941. At that time the engine was supposed to develop 1,350 hp for take-off, to have the same rating at 2,000 m (6,560 ft) and deliver 1,250 hp at 4,000 m (13,120 ft). Now, the engine was redesigned under the direction of V. Ya. Klimov: it was made lighter; reliability was increased by installing a single-speed supercharger designed by V. Dollezhal' (hence the 1sk, the 'sk' being an abbreviation for ***sko**rost'* – speed). The increased boost pressure (1,175 mm Hg versus 910 mm Hg in the M-105) ensured a higher power output; it was approximately 150-200 hp greater than that of the M-105PA and 'PF at altitudes of 4,000 m (13,120 ft) and higher.

Importantly, the M-106 differed from its predecessor in having a lower compression rate, a greater output of the main oil pump, a stronger crankshaft and so on. Yet, the dry weight of the M-106 remained virtually the same as that of the M-105PF.

Thus, the main advantage of the M-106 over the M-105PF lay in its ability to deliver greater power while retaining the same weight and overall dimensions. Another advantage was associated with the character of variations in the power output depending on the altitude. The M-105 had a two-speed supercharger, and the need to switch the speeds entailed a fall in the power output at 1,800-2,000 m (5,900-6,560 ft). German pilots flying the Bf 109 knew this handicap of the Soviet engine and tried to force the Yakovlev fighters to wage battles exactly at these altitudes. The Messerschmitt fighters had turbo-couplings that switched over automatically; thus, they suffered no loss in speed. In the M-106 engine, thanks to the single-speed supercharger, the power output increased smoothly from sea level to the rated altitude, and then decreased depending on the changes in atmospheric pressure.

Tests of the Yak-1 M-106 (c/n 3299) conducted in January 1943 showed excellent performance. At a weight of 2,757 kg (6,078 lb), which was unprecedentedly low for aircraft of this type, the fighter attained a speed of 557 km/h (346 mph) at sea level and 630 km/h (391 mph) at 3,400 m (11,150 ft). The Yak-1 M-106 needed only 4.5 minutes to climb to 5,000 m (16,400 ft). It seemed that a brilliant future was in store for the machine. However, these hopes were shattered. Pilot A. Kokin noted that the water radiator and oil coolers on this aircraft ensured adequate cooling of the engine at nominal power only when the ambient temperature at sea level was not higher than 15°C. Otherwise, the machine could not be flown at the nominal engine power rating.

The testing was conducted by engineer K. Mkrtychan who had actively participated

Yakovlev OKB

Above: Yak-1 c/n 3299 served as the prototype for the M-106P engine installation.

Yefim Gordon archive

Production Yak-1s powered by the M-106P engine were outwardly identical to their M-105-powered 'brethren'

in the development of the Yak-1 M-106 prototype. Flight tests showed that the M-106 engine had not been developed to the standard required for normal operation: its running was accompanied by vibration, detonation, emission of smoke and ejection of oil through joints. It became clear that the aircraft could not stay airborne for a long time.

Almost simultaneously with this work, engineers of the Flight Research Institute (LII) installed the M-106 in a production Yak-1. As compared to the prototype fighter with the same engine, the speed dropped by approximately 20 km/h (12.4 mph) and the time to 5,000 m rose by 1.1 minute. However, engine operation proved unsatisfactory on this second modified aircraft as well.

Despite the setbacks, the Yak-1 M-106 was ordered into series production. By 18th February 1943 the Saratov plant had manufactured 47 aircraft. The work on the M-106-powered fighter was considered to be the most important in the first quarter of that year. Of the 47 machines completed, only 32 were officially accepted by the military customer. In spite of all the efforts, they could not be tested during the winter, and subsequently the engines on most of these machines were either replaced by M-105PFs or underwent readjustment. Several Yak-1 M-106s saw service with the 148th IAP (commanded by Lieutenant-Colonel G. Zaïtsev) with the supercharging pressure decreased. A basically sound concept ended in a failure in 1943, resulting in a considerable waste of work time, equipment and materials.

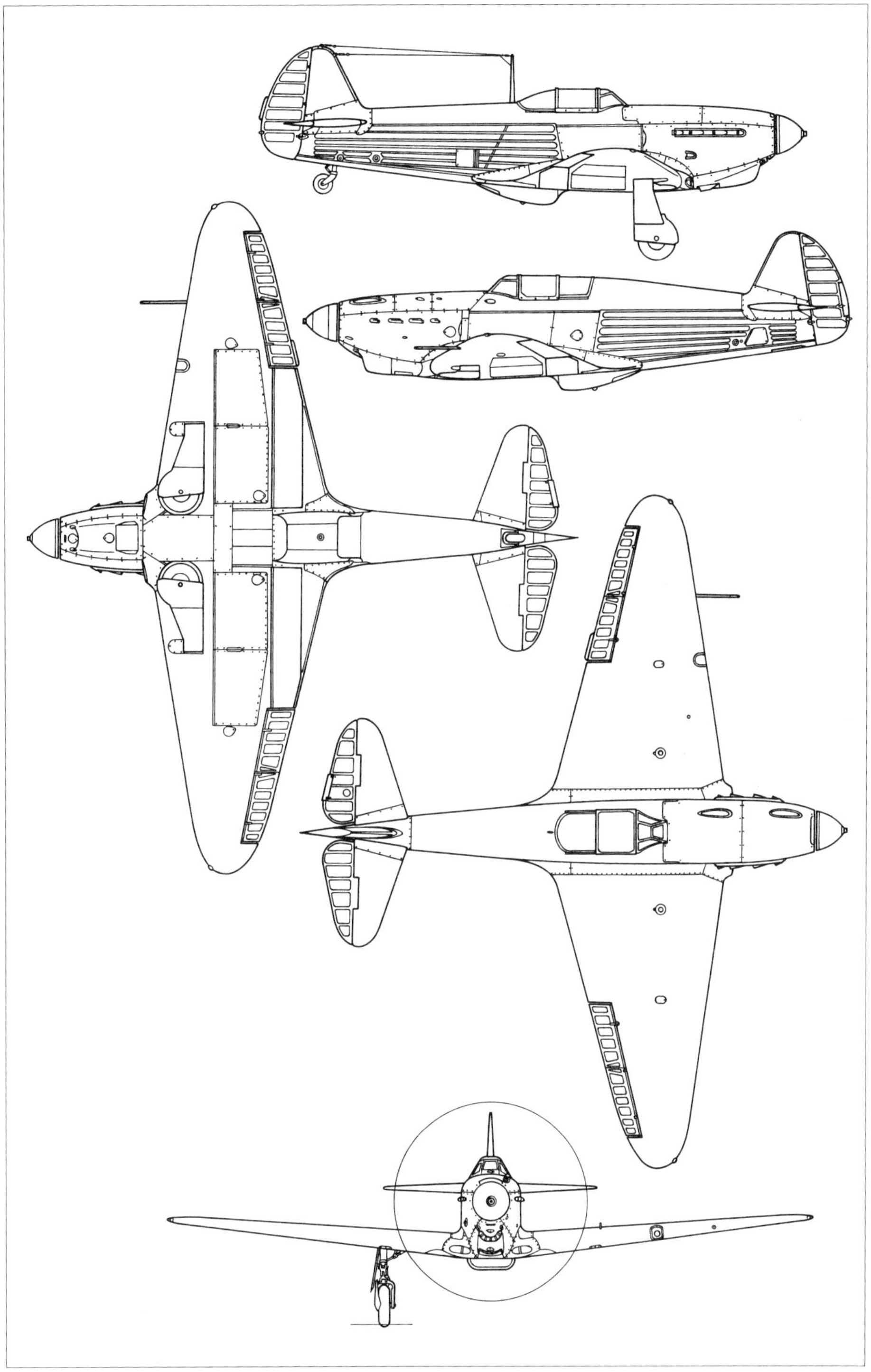

Four views of the 'bubbletop' Yak-1b, plus a side view of an early-production 'razorback' Yak-1.

Yakovlev's firstling soldiers on until the end of the war.

In the biggest air battles of 1943 the Yak-1 fighter, alongside its Yak-7B and Yak-9 stablemates and the Lavochkin La-5 fighter, constituted the main types of the Red Army Air Force's front-line machines. Yak-1s also made up a considerable part of the Soviet aviation reserves – the Air Corps of the Supreme Command.

The first blow dealt by the Luftwaffe during the battle at the Kursk Bulge was sustained by airmen of the 2nd and 16th Air Armies. By 5th July 1943 these formations had 659 Yak-1s and Yak-7Bs in their inventory. It was these formations that bore the brunt of the combat over the northern flank of the battle and suffered the greatest losses.

The Soviet airmen were up against not only all types of enemy aircraft, including the new Focke-Wulf Fw 190A-5 and Bf 109G-6 which were fitted with more powerful engines, heavier armament and increased armour protection, but also the most skilled and experienced Luftwaffe crews. But the Germans enjoyed air superiority for only a few days. Fresh units were committed to action in the area and corrections were made to combat tactics; this enabled the Soviet to check the German onslaught.

The test results obtained with Yak-1 c/n 07127 (seventh aircraft in Batch 127), which can be considered typical for mid-1943, make it possible to assess the machine's strengths and weaknesses. Many proposals aimed at perfecting the aerodynamics and armament and improving the handling had been implemented in this machine. At an AUW of 2,884 kg (6,359 lb), a top speed of 521 km/h (324 mph) was attained at sea level and 591 km/h (367 mph) at the second rated altitude of 4,100 m (13,450 ft). The climb to 5,000 m took 5.5 minutes. These figures matched those stipulated by the Government directive.

Some serious shortcomings were also noted. Fuel feed from the port and starboard tanks remained uneven. The range of two-way radio communication with the ground was improved – it surpassed 50 km (31 miles) at 1,000 m (3,280 ft), but the level of

Yefim Gordon archive

Above: A flight of Yak-1s operated by the Red Banner Baltic Fleet's 3rd GvIAP on quick-reaction alert.

Yefim Gordon archive

Above: Technicians push a late-production Yak-1 into a sheltered position after a sortie.

Specifications of Yak-1 variants

	I-26-2	Yak-1	Yak-1
Year of manufacture	1940	1942	1942
Powerplant	M-105P	M-105PA	M-105PF
Power at altitude, hp	1,050	1,050	1,180
Length	8.5 m (27 ft 10½ in)	8.48 m (27 ft 9¾ in)	8.48 m (27 ft 9¾ in)
Wing span	10.0 m (32 ft 9$\frac{45}{64}$ in)	10.0 m (32 ft 9$\frac{45}{64}$ in)	10.0 m (32 ft 9$\frac{45}{64}$ in)
Wing area, m^2 (sq ft)	17.15 (184.6)	17.15 (184.6)	17.15 (184.6)
Empty weight, kg (lb)	2,318 (5,110)	2,394 (5,277)	2,412 (5,317)
All-up weight, kg (lb)	2,700 (5,952)	2,883 (6,355)	2,917 (6,430)
Speed at sea level, km/h (mph)	490.0 (304.4)	478.0 (297.0)	510.0 (316.9)
Speed, km/h (mph)	585.0 (363.50	563.0 (349.8)	571.0 (354.8)
at altitude, m (ft)	4,800 (15,750)	4,850 (15,900)	3,850 (12,000)
Climb time to 5,000 m (16,400 ft), minutes	6.0	5.9	6.4
Service ceiling, m (ft)	10,200 (33,500)	10,400 (34,000)	10,000 (32,750)
Turn time, seconds	24	19	19-20
Operational range, km (miles)	700 (434)	650 (403)	650 (403)
Take-off run, m (ft)	300 (984)	320 (1,049)	320 (1,049)
Landing roll, m (ft)	540 (1,771)	530 (1,738)	520 (1,706)
Armament, mm	1 x 20-mm ShVAK 2 x 7.62-mm ShKAS	1 x 20-mm ShVAK 2 x 7.62-mm ShKAS	1 x 20-mm ShVAK 2 x 7.62-mm ShKAS

noise still remained very high. Stick forces were excessive, and pilots quickly got tired because of excessive heat in the cockpit.

Yakovlev continued the work on perfecting his 'firstling' up to the end of series production. For instance, excellent results were obtained during checkout tests of Yak-1 c/n 46139 in May 1943. At an AUW of 2,864 kg (6,315 lb) the aircraft attained a speed of 539 km/h (335 mph) at sea level and 605 km/h (376 mph) at the second rated altitude of 4,100 m (13,448 ft), a climb to 5,000 m took 5.7 minutes, and the Yak-1 gained 1,050 m (3,440 ft) during climb in a combat turn. Apparently, this was the limit of what could be obtained from this type of fighter. In this form this aircraft could successfully oppose all types of German production fighters at low and medium altitudes where most of the dogfights took place.

By the time when production of the Yak-1 was finally discontinued in July 1944, the production run totalled 8,666. The last machine was officially accepted by Air Force representatives in October 1944. Plant No.292 in Saratov manufactured 192 batches, while plant No.301 in Moscow managed to complete only six batches.

Owing to lack of attrition replacements the number of Yak-1s in front-line units gradually declined. Of the 5,810 fighters that were in the inventory of active Air Armies by the beginning of 1945, 735 were Yak-1s (the total attrition of Yak-1s caused by enemy action amounted to 3,336 machines in the period of 1941 to 1945).

Yak-1s soldiered on in the Naval Aviation and in the PVO (Air Defence) units throughout the war. Starting in the second half of 1944, however, they were gradually supplanted in the PVO units by Hawker Hurricanes, Supermarine Spitfires and Curtiss P-40 Kittihawks.

Mention must be made of the fact that Yak-1s fought in the ranks of the French 'Normandie' Air Regiment which received 14 fighters of this type in March 1943. Before re-equipping with Yak-9s in June of that year, the regiment destroyed nine German aircraft for the loss of four Yak-1s and four pilots. The Polish 1st Fighter Regiment 'Warszawa' (Warsaw) had 34 Yak-1s on strength by July 1944, but its participation in combat was limited. Furthermore, the Soviet 586th Air Regiment composed entirely of woman pilots fought in Yak-1s from the winter of 1942 onwards; initially it was assigned to the PVO (Air Defence) of Saratov.

Yak-1 MPVO air defence fighter

In 1943-44 Saratov's production included 385 aircraft modified for PVO units. (MPVO stands for ***mes**naya **pro**tivovoz**doosh**naya **obo**rona*, local air defence). Additional equipment for night-time operations included the FS-155 landing light, VR-2 VSI, AG-5 artificial horizon and RPK-10 direction finder.

Yak-1 with one ShVAK cannon and two UB machine-guns

In April 1941 Aleksandr S. Yakovlev, together with Levin, director of the production plant No.292, was tasked with organising production of a Yak-1 version carrying five guns and machine-guns. This complement was to include two wing-mounted 12.7-mm UBK machine-guns (the K stands for *kryl'ye***voy** – wing-mounted), presumably in addition to the standard fit of one ShVAK cannon and two ShKAS machine-guns. An order to this effect (No.348ss) was issued by NKAP on 21st April. The first machine of this variant was to be completed by 15th June 1941. However, this weapons arrangement would have led to a reduction of fuel load and range. Therefore, the plant's management decided instead to replace two synchronised ShKAS machine-guns by two synchronised UB machine-guns.

A prototype of this version was completed in May 1941. The Yak-1 with this weapon complement did not enter mass production, only a small batch for service evaluation being built in January-February 1942.

Yak-1 propeller testbeds

In 1941, before the outbreak of the war, the Yak-1/M-105P was used for testing new propellers under development. The types tested included the VEM-107 electromechanical propeller (*vint* ***elek****tromekha-***niche***skiy*), the AV-5LP, VISh-105P and VISh-61. Of these, the VISh-105P was found to be superior to other types and was recommended for series manufacture and installation on the Yak-1 and LaGG-3 fighters.

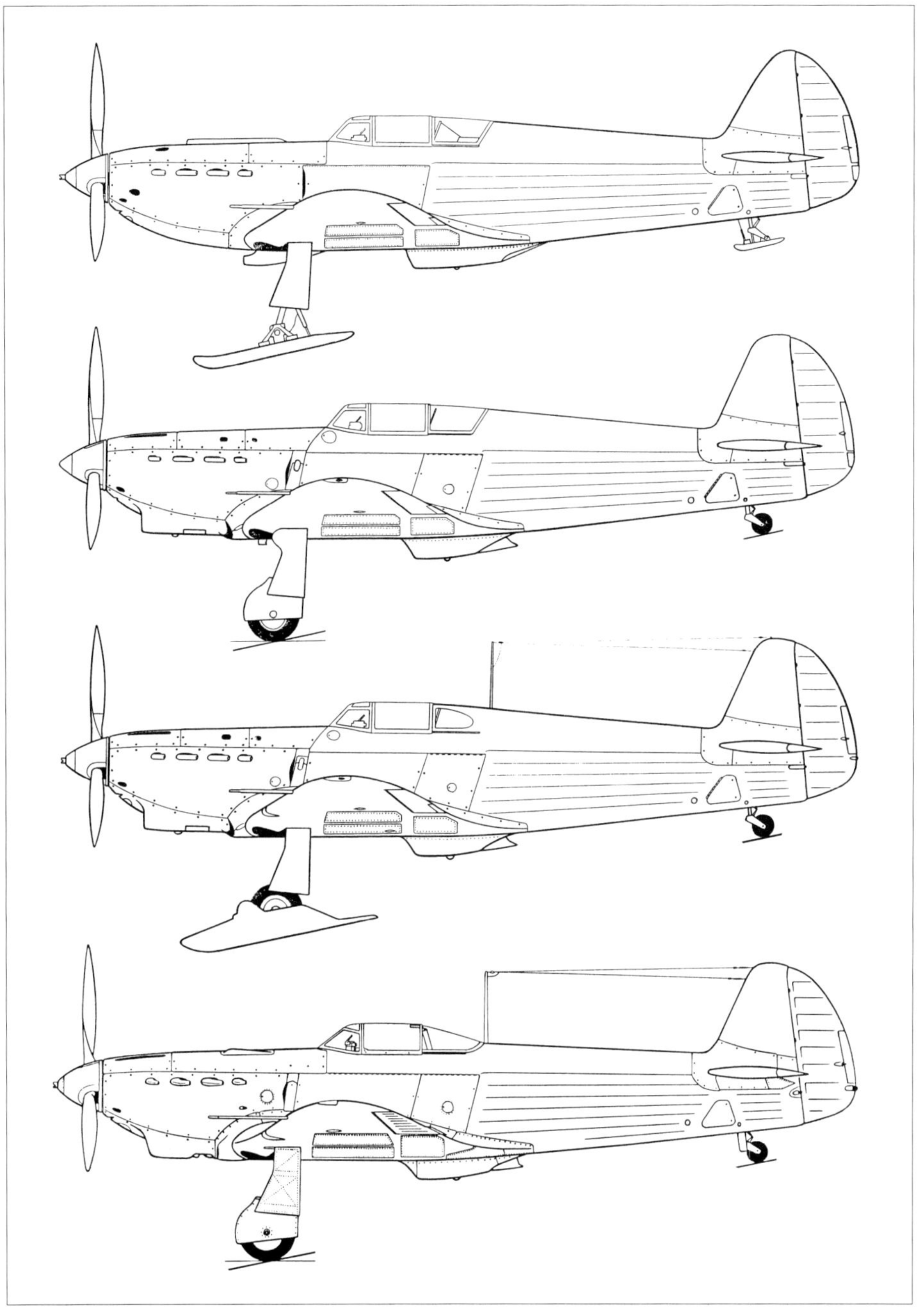

Top to bottom: The I-26-1 as first flown, an initial production 'razorback' Yak-1 *sans suffixe*, a late-model Yak-1 *sans suffixe* with 'slipper' skis falling away after take-off and a Yak-1b.

I-28 (Yak-5) high-altitude fighter prototype

Whereas previous prototype aircraft were created for operation at low and medium altitudes, the I-28 – also known as 'aircraft 28', I-26V (*vy***sot***nyy* – high-altitude), I-26N and I-28V – was intended for use by the PVO system. It was the first time Chief Designer A. Yakovlev embarked on the development of a high-altitude fighter.

Yakovlev placed his bets on the use of the prototype M-105PD engine fitted with the E-100 supercharger designed by V. Dollezhal'. The supercharger's impeller drive was provided with a hydraulic coupling which made it possible to change the rpm smoothly as the flight altitude increased. A similar device was incorporated in the Daimler-Benz DB 601E engine which had been designed in Germany at that time. But, most importantly, the German designers had created and fully developed an automatic supercharger speed control system, while in the Soviet engine the impeller speed was controlled manually. This caused much inconvenience for the pilot during the flight and prevented the engine power from being fully used.

The I-28 prototype made its first flight on 1st December 1940 with Pavel Ya. Fedrovi at the controls. The high-altitude fighter differed structurally from the I-26 in having an undercarriage borrowed from the UTI-26 (Yak-7UTI) and being provided with automatic slats of considerable span. The wing span was reduced from 10.0 m (32 ft 9¾ in) to 9.74 m (31 ft 11½ in), the wingtips being more angular.

The design performance of the I-28 included a speed of 515 km/h (320 mph) at sea level and 650 km/h (404 mph) at the altitude of 9,000 m (29,520 ft) with the engine running in boosted mode; it was to climb to 5,000 m (16,400 ft) within 5.2 minutes and reach a service ceiling of 12,000 m (39,360 ft). Since the power output changed smoothly between the first and the second supercharger speeds, the I-28 was expected to have a significant advantage in speed over the I-26 and I-301 fighters within the range of altitudes between 3,500 and 4,000 m (11,480-13,120 ft), because engines lacking

Above and below: Two views of the I-28 (Yak-5) fighter prototype. The fighter was powered by a supercharged M-105PD engine, hence the enlarged chin-mounted oil cooler.

the hydraulic coupling were afflicted by the so-called 'drops in power output'. Consequently, there was every reason to hope that that the I-28 would also prove to be a good tactical (front-line) fighter.

However, the engine proved troublesome from the outset. The very first flight ended in a forced landing. Attempts to establish the I-28's actual performance characteristics proved unsuccessful. The engine underwent modifications and was tested in February-March 1940, but this brought no success to the aircraft (which in the meantime was redesignated 'Yak-5, 1940 model').

Specifications of I-28 (Yak-5) and I-30 (Yak-3-1) fighters

	I-28 (Yak-5)	I-30 (Yak-3-1)
Year of manufacture	1941	1941
Powerplant	M-105PD	M-105P
Power at altitude, hp	1,160	1,050
Wing span	10.0 m (32 ft 9⁴⁵⁄₆₄ in)	10.0 m (32 ft 9⁴⁵⁄₆₄ in)
Wing area, m^2 (sq ft)	17.15 (184.6)	17.15 (184.6)
Empty weight, kg (lb)	2,450 (5,400)	2,550 (5,620)
All-up weight, kg (lb)	2,928 (6,455)	3,130 (6,900)
Speed at sea level, km/h (mph)	515.0 (320.0)	476.0 (295.7)
Speed, km/h (mph)	650 (403.9)	571 (354.8)
at altitude, m (ft)	9,000 (29,530)	4,900 (16,070)
Climb time to 5,000 m (16,400 ft)	5.2 min	7.0 min
Service ceiling, m (ft)	12,000 (39,370)	9,000 (29,530)
360° turn time, seconds	n.a.	19-20
Operational range, km (miles)	n.a.	975 (605)
Take-off run, m (ft)	n.a.	303 (994)
Landing roll, m (ft)	n.a.	525 (1,722)
Armament	1 x 20-mm ShVAK 1 x 7.62-mm ShKAS	3 x 20-mm ShVAK 2 x 7.62-mm ShKAS

In the first half of 1941 the OKB concentrated its efforts on the development of the Yak-1, resuming work on high-altitude aircraft in 1942 with the production Yak-7 fighter.

I-30 ('1941-model Yak-3') fighter prototype

The I-30 aircraft was built as the last of the prototype machines; in consequence, it incorporated all the experience accumulated by the OKB. Small wonder that it featured a large number of design improvements and operational advantages over its predecessors. These included an easily detachable propeller spinner, ejector exhaust stubs and a modified control stick patterned on that of the Bf 109. The I-30's wings had metal spars and detachable outer panels housing two additional ShVAK-20 cannons which fired outside the propeller disc. The overall weight of fire amounted to 4.28 kg/sec (9.43 lb/sec), which was considerably superior to the firepower of other Soviet front-line fighters.

The engine mount incorporated breaks which simplified engine replacement in field conditions. Some successful design features evolved on previous aircraft were also included, such as a forward shift of CG and an undercarriage patterned on that of the Yak-7, as well as slats of similar design to those of the Yak-5. This time the designers paid sufficient attention to the machine's special equipment. Pilots assessed it as meeting the specification requirements for a front-line fighter. Reliable two-way radio communication between the aircraft and the ground was ensured within a range of 200 km thanks to the extremely thorough electric bonding and shielding.

Piloted by Pavel Ya. Fedrovi, the I-30 fighter made its first flight on 12th April 1940, powered by an M-105PD engine featuring the Dollezhal' E-100 supercharger (hence the D). Unfortunately the high-altitude engine was not yet fully developed before the war, and an engine of the old M-105P model had to be installed in the aircraft. With this engine the aircraft passed its State trials on 3rd July 1941 (pilot A. Nikolayev, engineer A. Stepanets). At an all-up weight of 3,310 kg (7,300 lb) it attained a maximum speed of 476 km/h (296 mph) at sea level and 571 km/h (355 mph) at the second rated altitude of 4,900 m (16,070 ft); time to 5,000 m (16,400 ft) was 7.0 minutes and the service ceiling was 9,000 m (29,530 ft).

The high gross weight caused an increase of the landing speed to 142 km/h (88 mph). However, the landing run remained almost unchanged thanks to the possibility of using wheel brakes effectively. The provision of leading-edge slats brought down the I-30's minimum pre-stall speed and, in effect, precluded the possibility of a spin even in the case of grave piloting errors.

Increasing the internal fuel load from 305 to 383 kg (from 670 to 844 lb) and introducing a service tank improved the fuel feed system, enabling the fighter to cover a maximum distance of 975 km (606 miles) in high-speed flight mode. Summing up his impressions, pilot Stepan P. Sooproon, who had flown the fighter, told A. Yakovlev: 'With such fighters we need not fear any Messerschmitts!'. This conversation took place during the first days of the war when the renowned Soviet ace was leaving for the front as the CO of a fighter regiment.

As early as 8th April 1941 the Government issued a decision requiring the Moscow aircraft factory No.81 to halt production of the Yak-4 and switch to the I-30, which was renamed Yak-3 – the first aircraft to have this designation. By the time the war broke out a number of incomplete airframes had been manufactured.

On 21st April 1941 a decision was taken to launch production of the '1941-model Yak-3' (I-30) in Saratov as well. Chief Designer Aleksandr S. Yakovlev had no doubts that the Yak-1 would be superseded by its more refined stablemate. At pre-war conferences, when faced with numerous complaints concerning the Yak-1 (in particular, regarding the absence of a radio), the Chief Designer invariably reacted by saying that all the shortcomings would be eliminated on the Yak-3 (I-30).

In the meantime, the Yakovlev OKB continued the development of the I-30 – the Yak-3 *doob***lyor** (lit. 'understudy' – that is, second prototype, in the Soviet terminology of the time), or Yak-3-2, was completed in May 1941. The cockpit layout evolved on it was recommended as standard for all fighters of the Red Army Air Force. Obviously envisioning large-scale production, Yakovlev reverted to one-piece wooden wings without LE slats. Powered by an M-105PD engine, the *dooblyor* was damaged beyond repair in an accident at LII, turning turtle after a messed-up landing.

The work on the first prototype I-30 was not yet completed at the outbreak of the war. One of the documents stated: *'The Yak-3 powered by the M-105P engine possesses potent and well-developed armament; it is easy to fly and can easily be mastered by wartime average-skilled pilots. As such, it is recommended for introduction into service with the Red Army Air Force.'*

Many members of the OKB staff believed that the setback with the second prototype would not affect the fate of the promising new aircraft. Still, production of the Yak-3 (I-30) failed to materialise. At first its production entry was postponed to 1942 and then

Yakovlev OKB

Yakovlev OKB

Yakovlev OKB

Top, centre and above: The first prototype I-30 (I-30-1) fighter. Note the four-section flaps visible in the centre photo and the enlarged oil cooler air scoop; as was often the case with Yakovlev prototypes, the fighter was totally devoid of markings.

via Yefim Gordon

The second I-30 prototype (alias Yak--3-2) after a crash-landing. The aircraft was declared a write-off.

Yakovlev OKB

Above and below: The UTI-26-1, the first prototype of the future Yak-7 which was originally conceived as a trainer but ultimately evolved into a real combat aircraft. The aircraft is shown here during manufacturer's flight tests. Note the shape of the windshield and the small oil cooler.

Yakovlev OKB

the decision to launch series production was cancelled altogether. This was due to various reasons, notably the scarcity of aluminium caused by the industry's enforced redeployment to the eastern regions of the USSR and by the loss of several enterprises which remained in the territory occupied by the enemy. Also, the mass-produced Yak-1 had no special need for slats, showing sufficiently easy handling without them; finally, the phasing of a new aircraft into production would unavoidably have entailed a reduction in the output of series-manufactured machines – something that was ill-affordable in the course of the war.

Yak-3 M-500 fighter (project)

On 24th April 1941 the People's Commissariat of the Aircraft Industry issued order No.361 'On the designing and construction of the modified Yak-3M-500 aircraft' pursuant to the government directive dated 16th April 1941 and as a sequel to NKAP order No.88 of 28th January 1941. Under the terms of order No.361 A. S. Yakovlev was tasked with designing and building two examples of a Yak-3 (I-30) version powered by the M-500 engine; they were to be submitted for State acceptance trials by 1st September and 1st November 1941 respectively. The order stipulated a maximum speed of 640 km/h (398 mph) and an armament comprising two 20-mm cannons, one 12.7-mm BS machine-gun and two 7.62-mm ShKAS machine-guns.

The M-500 must have been a new engine under development which was presumably abandoned at an early stage. There is no evidence of any practical work undertaken by the Yakovlev OKB on the M-500-powered Yak-3 (I-30) version.

Yak-7 fighter and fighter trainer

UTI-26 fighter trainer prototype

As noted earlier, the Yakovlev OKB, unlike other design teams, prepared three other combat aircraft in parallel with the I-26 fighter. The first of them was the UTI-26 (*oochebno-trenirovochnyy istrebitel'* – fighter trainer), also known in various documents as UTI-26-4, UTI-1, UTI-27 and so on. The Design Bureau started work on this aircraft in early 1940. The Government authorised this work by adopting an appropriate directive. The assumption was that the new aircraft would have substantial commonality with the I-26 prototype, which lessened the risk of a failure.

Indeed, the first UTI-26 was basically similar to the I-26 fighter. It differed in having two cockpits (for the trainee and the instructor) placed in tandem under a common canopy and featuring dual controls. No intercom was fitted to at first, and the instructor could communicate with the trainee by means of a rubber hose. The provision of a second cockpit required the wings of the UTI-26 to be moved slightly aft.

On 23rd July 1940 pilot Pavel Ya. Fedrovi took the fighter trainer into the air. After a few development flights the machine was submitted for State trials. They were conducted by engineer A. Stepanets and pilots Pyotr M. Stefanovskiy and A. Koobyshkin between 28th August and 19th September.

On 30th August an incident occurred: the undercarriage collapsed during taxying due to faulty design. On the whole, most of the criticisms levelled at the UTI-26 were directed at this element of the airframe. The locks were unreliable, the wheels were not strong enough to cater for the aircraft's weight, and the mainwheel struts were set at an insufficient angle to prevent a nose-over; this, in effect, precluded the use of wheel brakes during the landing run, especially when the instructor was not in the rear seat. These complaints had already been voiced with regard to the single-seat I-26.

The presence of such defects appeared inadmissible in a fighter trainer intended for primary training in flying schools or for taking ill-trained pilots of service units on familiarisation flights. Also, the 11th IAP pilots who had flown the aircraft noted that in certain flight modes, including landing, the required elevator inputs were exceedingly small, which meant it was all too easy to apply excessive elevator input; this made using the aircraft as a trainer very problematic. Nevertheless, on 25th September the military customer recommended the UTI-26 for series production. It was decided that the work on perfecting the aircraft would be continued on the second prototype.

Known as the UTI-26-2, the second prototype differed from its predecessor primarily in having an altered ratio between the sizes of the elevators and stabilisers, the overall horizontal tail area remaining unchanged. It also featured changes in the undercarriage design; bigger wheels were fitted to match the machine's all-up weight and the main gear fulcrums were relocated, making it possible to increase the angle intended to prevent nose-overs. However, the wheels' plane of rotation noticeably deviated from the aircraft's direction of movement, thus creating an additional load on the undercarriage pivots.

Certain alterations in the fighter's structural layout caused a forward shift of the CG which was beneficial for stability. This was confirmed by the pilots who put the aircraft through an intensive flight test programme right from the first days of 1941. It makes sense to quote from the UTI-26-2's flight test report: *'The aircraft has good controllability and sufficient stability in flight, has pleasant handling and performs aerobatic figures easily. The visibility is quite satisfactory; the presence of a canopy obviates the need for using goggles. The aircraft enters a spin only in the case of a considerable loss of speed and recovers easily. It forgives even grave piloting errors with no serious consequences. The aircraft possesses a very wide range of safe speeds.'*

As for performance characteristics, they were close to those of the I-26. At a weight of 2,750 kg (6,060 lb) the fighter trainer attained a maximum speed of 500 km/h (296 mph) at sea level and 586 km/h (364 mph) at the altitude of 4,500 m (14,760 ft). The aircraft climbed to 5,000 m (16,400 ft) within 5.5 minutes. The machine needed 310 m (1,020 ft) for the take-off run and as many as 750 m (2,460 ft) for the landing run; the latter figure was due to insufficiently strong wheels making abrupt braking dangerous. The landing speed was 125 km/h (77.7 mph).

Production Yak-7UTI fighter trainer (Yak-7UTI M-105PA)

On 4th March 1941, exactly one year after the directive calling for the development of the machine, the UTI-26 entered series production under the designation Yak-7UTI at plant No.301 which had previously manufactured the Yak-1. Interestingly, in October 1940 there were plans to put the UTI-26 into production at new aircraft factories that were to be set up in Riga and Kaunas (Latvia and Lithuania had become Soviet republics earlier that year); these plans were presumably abandoned or overtaken by the German invasion in June 1941.

When the working drawings were transferred to the plant, some changes were introduced. Since a fighter trainer was supposed to perform short flights without retracting the undercarriage, the tailwheel was made non-retractable; the engine speed was limited to a figure that was substantially lower than the nominal rpm; and only the port ShKAS machine-gun was retained. The all-up weight of production machines was nearly 2,800 kg (6,174 lb), and performance was close to that of the production Yak-1s of the period.

On 18th May 1941 pilot Pavel Ya. Fedrovi test-flew the first production Yak-7UTI from the Central Airfield in Moscow (Khodynka). Deliveries of these aircraft to Air Force units commenced immediately before the outbreak of the war; the machines were operated mainly by the regiments deployed in the western regions. They played an important role in expediting the conversion of Soviet pilots to the new-generation fighters.

Taking into account the operational experience with Yak trainers, in early August 1941 engineer Cherepov (an employee of plant No.301) came up with an important proposal. Wishing to simplify production and thus increase the output, he suggested that production Yak-7s be manufactured with a fixed undercarriage. In his opinion, this was fully in conformity with the character of work performed by the aircraft and its pilots. Right away, in August, the suggestion was implemented on the production line.

On the 87 aircraft manufactured before 20th July that were operated by various Air Force units, the most serious complaints concerned the mainwheel leg pivots which failed by the dozens. The parts were reinforced, but even that did not immediately cure this annoying defect.

Yak-7R tactical reconnaissance aircraft prototype

One of the first adaptations of the Yak-7UTI was directed at adapting it for the tactical reconnaissance role. Following instructions from Chief Designer Aleksandr S. Yakovlev, two production aircraft were fitted with an AFA-I aerial camera and an RSI-4 radio set. To enable the latter to function, shielding and electric bonding were performed on the aircraft; the pilot's seat was provided with an armoured back and the canopy glazing was altered. The aircraft was armed with an engine-mounted 20-mm ShVAK cannon. When the Yak-7R ([*samo***lyot**-] *raz***ved***chik* – reconnaissance aircraft) had received an approval from the Research Institute of Specialised Services, a recommendation was issued to build a small batch of such machines.

Yak-7 front-line fighter

In August 1941 another transformation of the fighter trainer was effected under the direction of project engineer K. Sinel'shchikov at plant No.301. This time it was converted into a 'real' fighter. A production Yak-7UTI (c/n 04-11) was taken off the assembly line and subjected to the following alterations: an armoured seat back was installed in the rear cockpit, the gun camera was deleted, the fuel tanks were provided with self-sealing coating and an inert gas pressurisation system featuring a bottle with carbon dioxide. The fighter was armed with the engine-mounted 20-mm ShVAK cannon with 120

via Yefim Gordon

via Yefim Gordon

Two more views of the UTI-26-1 at NII VVS during State acceptance trials. These views show well the clean lines of the aircraft, an impression enhanced by the extensively glazed canopy.

Yefim Gordon archive

A lineup of Yak-7s at a tactical airfield. These fighters were paid for by donations from Young Communist League members of Novosibirsk; hence each aircraft carries the legend 'Novosibirskiy komsomol'.

rounds and two synchronised 7.62-mm ShKAS machine-guns with a total of 1,500 rounds, as well as six rails (three under each wing) for 82-mm (3.22-in) RS-82 unguided rockets. The rocket projectiles were fired by means of the electric bomb release mechanism placed in the pilot's cockpit. The canopy of the rear cockpit and the tailwheel remained unchanged, so that the fighter virtually did not differ outwardly from the two-seat Yak-7UTI.

The designers reported the results of their work to Yakovlev. The latter was sceptical about the idea, but then changed his mind and gave his approval. After the Chief Designer had informed the government that the conversion trainer had been evolved into a fully capable fighter, the work received support from NKAP. Two directives issued by the People's Commissariat in August stipulated that already from 15th September onwards the Yak-7UTI was to be manufactured *'to Yak-1 standard'*, referring to the identical armament. Here a curious occurrence took place: mistakes slipped into the texts of the directives. According to these texts, plant No.301 in Moscow and then, after evacuation, plant No.153 in Novosibirsk were to manufacture *Yak-1 fighters*!

Actually the Yak-1 was never built there, but fighters manufactured by these plants were entered into statistical reports under this designation. In reality, the collective of the plant built 186 Yak-7s in Moscow (51 of them with armament) and 21 in Novosibirsk (including 11 fighters).

Right from the outset, favourable comments came in concerning the Yak-7 fighter. For example, the document on the acceptance of the aircraft by the Design Bureau's technical commission stated that the Yak-7 single-seat fighter was a better machine than the Yak-1. The size of the undercarriage wheels was in full conformity with the all-up weight, which came close to three tons (6,610 lb). The detachable engine mount fitted to the Yak-7 facilitated repairs and made it possible to adapt the aircraft relatively easily to a different engine. The main undercarriage legs were set at a higher angle intended to counteract nose-overs, lessening the danger of the machine turning turtle after a resolute application of wheel brakes. The second cockpit, which was retained on the Yak-7, could be used for carrying maintenance personnel or cargo during redeployment to new bases, for bringing pilots back after an off-field forced landing or for installing a long-range fuel tank; almost all of these options were subsequently implemented.

As for performance, the Yak-7 proved to be very similar to the Yak-1 fighter. At an all-up weight of 2,960 kg (6,530 lb), which was 160 kg (350 lb) higher than the weight of the Yak-7UTI, the Yak-7 fighter attained a maximum speed of 471 km/h (293 mph) at sea level and 560 km/h (348 mph) at 5,000 m (16,400 ft). The latter altitude could be reached within 6.8 minutes. (All performance figures are quoted for the fighter without external stores and with the cockpit hood closed). Manoeuvrability turned out to be somewhat inferior to that of the Yak-1: 24 seconds were needed for a full-circle banking turn at 1,000 m (3,280 ft) and the Yak-7 gained 750 m (2,460 ft) during climb in a combat turn.

The fighter's handling qualities earned high praise. Referring to the Yak-7's spinning characteristics, A. Lazarev, a Novosibirsk aircraft factory test pilot, noted that spin recovery was 'exceedingly simple – no tricks, once the rudder pedals and the control stick in the transverse plane have been set neutral, the aircraft instantly ceases rotation and enters a dive'.

In September 1941 a Yak-7 armed with unguided rockets successfully passed trials. Thanks to the fighter's good stability, projectile dispersal during attacks against ground targets was somewhat less than in the case of the LaGG-3 and the Yak-1, and the rocket hits were much more closely grouped than in the case of the MiG-3.

In October 1941 all enterprises, including plant No.301, had to be evacuated from Moscow, which was detrimental to the deliveries of aircraft to front-line units. In the meantime, the work on perfecting the fighter continued unabated. Right before the transfer to Novosibirsk the designers succeeded in completing a 'winterised' example of the Yak-7 that was to serve as a standard for production; it incorporated features similar to those of the 'winterised' Yak-1 described

Yefim Gordon archive

Another view of the same unit's Yak-7s as the crews go to the flight line, preparing for the day's work. Not all of the fighters sported the red arrow near the engine exhaust stubs.

above. Work on fitting heavier armament to the Yak-7 was under way – it envisaged the installation of two synchronised ShVAK cannon; the evacuation prevented the completion of this work.

During the late autumn of 1941 a very strained situation arose in Novosibirsk. Plant No.153, which had been built there before the war, could not cope with the task of manufacturing LaGG-3 fighters; as if that weren't enough, after the transfer of the Yakovlev OKB to Siberia work started on setting up a second continuous flow assembly line there for the production of the Yak-7 fighter. At the same time trains coming from the USSR's western areas delivered personnel and equipment; barracks for housing the personnel and auxiliary premises were built. *Four* different aircraft enterprises happened to be sharing the same territory; each of them had its own director and chief engineer, and they wished to exercise control over their subordinates and the equipment. The government tasked A. S. Yakovlev with uniting everybody and everything into a single enterprise before too many cooks spoiled the broth. Being a capable organiser, Yakovlev coped with this task.

Simultaneously the Chief Designer succeeded in having the united plant (No.153) relieved of the LaGG-3 production programme so that all efforts could be concentrated on his Yak-7. Brigade engineer Leshukov, a representative of the State Defence Committee in Novosibirsk, advocated a different point of view: first of all, he maintained, it was necessary to improve the Yak-7 fighter because the latter had no tangible advantages over the LaGG-3 as yet, and it would be advisable to manufacture both types in parallel. However, Yakovlev used his influence on Stalin to secure control over the entire plant, which, after the evacuation, became one of the biggest in the Soviet aircraft industry.

Yak-7 fighters saw combat for the first time during the counteroffensive started by Soviet troops near Moscow. In early December 1941 a squadron of the 172nd IAP equipped with Yak-7s (the other two were equipped with LaGG-3s and Yak-1s) escorted a group of 65th ShAP (Attack Air Regiment) Il'yushin IL-2 attack aircraft and itself attacked enemy ground troops in the area of Teriayeva Sloboda and Rooza when the Soviet troops were liberating the town of Volokolamsk. Only eight Yak-7s were in the entire inventory of the Soviet aviation at that time, three of them being fully operational.

Yak-7M fighter prototype (Yak-7M M-105PA)

A very unusual modification of the Yak-7 was effected in August 1941. It was undertaken not in Moscow or Novosibirsk, where the Yak-7 was being manufactured at the time, but in Saratov. A group of Yakovlev OKB engineers was rendering assistance to plant No.292 in getting Yak-1 production under way. Today it is hard to say what prompted the initiation of this work, but many elements that had been evolved on the I-30 fighter prototype were incorporated into the design of the fighter trainer. As a result, a version emerged which was designated Yak-7M (*modifitseerovannyy* – modified).

The aircraft was stripped of all equipment in the rear cockpit and of the sole ShKAS machine-gun which was replaced by three 20-mm ShVAK cannons. One of them was a hub-mounted weapon, while the other two were mounted in the wings outside the propeller disc area and did not require synchronisation. Thanks to this weapons fit the Yak-7M was superior to all Soviet and German fighters in weight of fire.

The most serious alterations concerned the wings. The span was marginally reduced from 10.0 m (32 ft $9\frac{45}{64}$ in) to 9.74 m (31 ft $11\frac{15}{32}$ in), and the wingtips were made more angular because the leading edge incorporated slats of fairly large span. The installation of cannons and their ammunition boxes necessitated a reduction in the capacity of outer wing tanks; to compensate for this, an 80-litre (17.6 Imp gal) supplementary fuel tank was installed behind the pilot's armoured seat back. The wing structure was strengthened and flap area was increased.

In October 1941 the Yak-7M was transferred to NII VVS for trials; it immediately attracted the attention of leading engineer A. Stepanets and pilot V. Khomiakov who headed the team assigned to Yakovlev fighters at the time. They noted that, despite the all-up weight being increased to 3,160 kg (6,968 lb), the handling was appreciably simplified thanks to the LE slats. The danger of the aircraft flipping into a spin due to pilot error was virtually eliminated; stability, especially lateral stability, was improved. All this justified the conclusion that flight safety had been considerably enhanced. The minimum control speeds were reduced by 30-40 km/h (18.6-24.8 mph) as compared to production Yak-1s and Yak-7s, both when performing different aerobatic manoeuvres and during pancaking.

The cannons functioned reliably and faultlessly in all flight modes; shell scatter proved to be within acceptable limits. For the purpose of aiming the cannons were so adjusted that their fire converged in one point at the distance of 400 m (1,310 ft). After the necessary development and test firing it became clear that the Yak-7M's armament could be used very effectively against both aerial and ground targets. In overall performance the Yak-7M fighter was marginally inferior to the Yak-1 and Yak-7, mainly because this particular machine was a well-worn example which had suffered a minor accident and undergone the modification simultaneously with the necessary repairs to the fuselage and wings.

The Yak-7M fighter was recommended for series production. However, this enticing proposal made by NII VVS could not be put into effect during the winter of 1941-42 – the new machine would have necessitated too radical changes in the production process that were not feasible at that time.

Specifications of Yak-7 fighter trainer and fighter variants

	UTI-26	Yak-7A	Yak-7B
Year of manufacture	1941	1942	1942
Powerplant	M-105P	M-105PA	M-105PF
Power at altitude, hp	1,050	1,050	1,180
Length overall	8.5 m (27 ft $10\frac{1}{2}$ in)	8.48 m (27 ft $9\frac{3}{4}$ in)	8.48 m (27 ft $9\frac{3}{4}$ in)
Wing span, m (ft-in)	10.0 m (32 ft $9\frac{45}{64}$ in)	10.0 m (32 ft $9\frac{45}{64}$ in)	10.0 m (32 ft $9\frac{45}{64}$ in)
Wing area, m^2 (sq ft)	17.15 (184.6)	17.15 (184.6)	17.15 (184.6)
Empty weight, kg (lb)	2,181 (4,808)	2,450 (5,400)	2,490 (5,490)
All-up weight, kg (lb)	2,750 (6,060)	2,935 (6,470)	3,010 (6,635)
Speed at sea level, km/h (mph)	500.0 (310.6)	495.0 (307.5)	514.0 (319.3)
Speed, km/h (mph)	586.0 (364.1)	571.0 (354.8)	570.0 (354.1)
at altitude, m (ft)	4,500 (14,750)	5,000 (16,400)	3,650 (12,000)
Climb time to 5,000 m (16,400 ft)	5.5 min	6.4 min	5.8 min
Service ceiling, m (ft)	9,400 (30,750)	9,500 (31,000)	9,900 (32,500)
Turn time, seconds	22	21-22	19-20
Operational range, km (miles)	700 (434)	643 (399)	645 (400)
Take-off run, m (ft)	310 (1,020)	410 (1,345)	435 (1,430)
Landing roll, m (ft)	550 (1,804)	610 (2,001)	620 (2,034)
Armament, mm	2 x 7.62-mm ShKAS	1 x 20-mm ShVAK 2 x 7.62-mm ShKAS	1 x 20-mm ShVAK 2 x 12.7-mm UB

Yakovlev OKB

Above and below: A Batch 22 Yak-7B (c/n 22-03) during checkout tests at NII VVS. Note the one-piece main gear doors attached to the oleo struts that were characteristic of late-production Yak-1s and Yak-7s alike. The rear cockpit transparency has been deleted but the cockpit is still functional, allowing a passenger (usually the aircraft's technician) to be carried in case of need.

Yakovlev OKB

Yak-7A production fighter (Yak-7A M-105PA)

The Yak-7 underwent major alterations in 1942. Whereas at the beginning of the year it still retained all the features of the fighter trainer, by the end of the year it was transformed into one of the best fighters of the day. This was accompanied by an appreciable increase in the number of machines manufactured. While the monthly output of Yakovlev fighters totalled a mere six to seven machines during the winter months of 1942 (plant No.153 was still continuing production and deliveries of the LaGG-3), the production tempo reached six to seven Yak-7s per day in August of the same year.

In early January 1942 a Yak-7 (c/n 14-13) was transferred to NII VVS for joint manufacturer's/State acceptance trials. (The c/n – Batch 14, 13th aircraft in the batch – reveals that plant No.153 continued the sequence of batches initiated at plant No.301 after the amalgamation of the two plants.) The machine was test-flown with both wheels and skis. The tests revealed that the performance was still marginally inferior to that of the UTI-26. In the 'summer' version Yak-7 c/n 14-13 attained a maximum speed of 476 km/h (296 mph) at sea level, this figure rising to 550 km/h at 5,000 m (16,400 ft). This altitude could be reached within 6.8 minutes.

When fitted with skis and coated with the chalk-based winter paint over the summer camouflage, the Yak-7 lost some 30-40 km/h (18.6-24.8 mph) in speed. Rate of climb, range and field performance also deteriorated, while the landing run was reduced by approximately 150 m (490 ft) owing to the considerable friction of the skis on the snow.

The design of the skis proved to be successful. They were manufactured at plant No.153 entirely from non-strategic materials. The skis had a wooden construction and featured a framework consisting of spars, transverse frames, lugs and a runner. Wide use was made of the Siberian larch that grew in the forests of the Novosibirsk region. The framework was supplemented by plywood skinning, sheet aluminium being used for the edging on the sides. Mounted on the upper side of each ski was the so-called 'cabane' with 'paws' made from Cromansil alloy tubes. The ski mounted on the non-retractable tail strut featured a similar design. To evaluate operational qualities of the skis, test pilots made nearly 300 successful landings; no anomalies were revealed.

In accordance with A. Yakovlev's directions project engineers representing different organisations – A. Stepanets from NII VVS, P. Limar from LII, K. Tarootin from plant No.153 – joined forces to develop the fighter further as a standard for series production. As a result, with a wheeled undercarriage the fighter attained 495 km/h (308 mph) at sea level and 571 km/h (355 mph) at 5,000 m, climbing to this altitude within 6.4 minutes.

From the beginning of 1942 onwards the fighter was series-produced in Novosibirsk under the designation Yak-7A. Various improvements were envisaged and introduced into the design gradually, mainly in the spring of that year. The principal improvements included installation of a two-way radio with an aerial mast and wire aerial. A semi-retractable tailwheel was introduced and the fixed strut joining the upper part of the tail skid with the fuselage frame was replaced by a pneumatic actuator for tailwheel retraction; additional gear doors were introduced to close the mainwheel wells completely when the undercarriage was retracted. The sliding hood of the rear cockpit was replaced by a hinged plywood hood swinging open to starboard; when closed, it ensured a smooth transition between the canopy and the rear decking of the fuselage.

Ammunition belt link collectors were installed instead of link chutes; this made it possible to eliminate openings in the cowling and preserve the belt links for subsequent re-use. A lever controlling the engine boost was installed to shorten the take-off run. The fuel tanks' inert gas pressurisation system was altered and the instrument panel layout was changed.

When all the changes had been incorporated, the Yak-7A proved to be one of the fastest Soviet production fighters; even if it was somewhat inferior to the Yak-1 in manoeuvrability, the armament was equally capable.

In the early summer of 1942 several dozen Yak-7As were operated on the Volkhov and the Western Fronts. Citing the merits of the Yak-7A, Lieutenant Filatov nevertheless wrote in his letter addressed to Gheorgiy M. Malenkov, Secretary of the Communist Party Central Committee: *'The tactics employed by German pilots in air combat are based on making use of their advantages in speed and rate of climb – they try to force us to fight in the vertical plane. It is necessary by all means to increase the maximum speed of our fighters. Besides, Yak-7 pilots have poor rearward visibility; it is necessary to provide the fighter with a fully*

Yakovlev OKB

Above: Yak-7B c/n 22-41 was another example subjected to checkout tests at NII VVS.

Yakovlev OKB

Above: Another Yak-7B seen during checkout tests at NII VVS. The aircraft wears green/black summer camouflage, even though the tests were held in winter; but then, there was no enemy opposition there.

glazed all-round vision canopy because the metal framing reduces the observable zone by about 30%.'

The comments that came from the front were predominantly favourable. Here is an excerpt from a letter dated 13th July 1942 that was sent to A. Yakovlev by Major A. Morozov, commander of the 283rd IAP which had several Yak-7As on strength: *'My wingman and I were conducting a dogfight with eight Bf 109Fs. Despite his numerical superiority, the enemy could do nothing, because the Yak-7A enabled us to perform all manoeuvres at will, and the vaunted fascist carrion vultures* (that was a common allusion to Luftwaffe aircraft among Soviet people at the time – *Auth.*) *could not make a firing pass at us, no matter how they tried... [...] The pilots of my regiment who fly this aircraft have literally lost their hearts to it; it's no use trying to lure them to some other fighter type, and the pilots of the neighbouring units look at us with envy.'*

Yak-7B M-105PA production fighter

The letter to A. Yakovlev quoted above also noted some shortcomings of the Yak-7, however. Major Morozov noted the fighter's inadequate armament, insufficient speed and insufficiently powerful engine. It was these shortcomings that the Yakovlev OKB persistently sought to eradicate. The results of this work were embodied in the M-105PA-powered Yak-7B which differed from its predecessor in having heavier armament and some aerodynamic refinements.

The two 7.62-mm machine-guns were replaced by UBS-12.7 heavy machine-guns, while the engine-mounted 20-mm ShVAK cannon was retained. Additionally, in overload configuration six RS-82 rockets or two 25- to 100-kg (55- to 220-lb) bombs could be carried underwing.

TsAGI recommendations concerning aerodynamic refinements were fully incorporated in the Yak-7B. Improvements were also made to the carburettor air inlets with a view to using the ram air pressure to a fuller extent and achieving higher rated altitudes. The aircraft received a better surface finish; perfected radiator and oil cooler ducts were installed.

Despite the presence of an aerial and associated mast and the installation of new armament, which spoiled the clean contours of the upper part of the engine cowling, compounded by the increase of the all-up weight by nearly 100 kg (220 lb) over the Yak-7 and Yak-7A, the Yak-7B's performance proved to be somewhat higher than that of its predecessors. After the replacement of the ShKAS by the 12.7-mm UBS the Yak-7B's weight of fire fully met the requirements of front-line units, and this fighter could be used more effectively than the Yak-1 or Yak-7 *sans suffixe* against both air and ground targets. Its salvo weight of 2.72 kg/sec (6 lb/sec) was more than 1.5 times greater than the Yak-1's and Yak-7A's, 1.35 times greater than that of the late-production LaGG-3 and 3.5 times greater than that of the Bf 109F (as tested in the Soviet Union).

The State acceptance trials report noted: *'...In comparison with the Yak-1 aircraft which has built up a good service record at the front, the Yak-7B aircraft features a more perfected and promising design. During the period of its series production the Yak-7 has been progressively improved and fitted with new, more capable armament and special equipment with no adverse effect on its performance and combat capabilities.'*

Tests of the Yak-7B conducted by LII for the purpose of evaluating the balanced character of controls in different flight modes

Yakovlev OKB

This Yak-7B is armed with six 82-mm RS-82 unguided rockets on underwing racks. The picture was taken during trials at the Soviet MoD's Artillery Weapons Proving Ground (NIPAV).

Yakovlev OKB

Above and below: This late-production Moscow-built Yak-7B, c/n 41-01, was modified with a cut-down rear fuselage decking and bubble canopy. It is seen here during State acceptance trials.

Yakovlev OKB

Yakovlev OKB

Above and below: Another modified 'bubbletop' Yak-7B. This aircraft differs from Yak-7B c/n 4101 in having a recontoured water radiator bath.

Yakovlev OKB

gave excellent results. For example, in the opinion of test pilots, the La-5 was inferior to the Yak fighter: the former possessed insufficient directional stability, too 'light' ailerons and, on the contrary, too high stick forces from the elevators. The Bf 109E received a reasonably positive appraisal, yet its ailerons and tail surfaces were not quite so effective as those of the Soviet fighter.

Having compared the Yak-7B with indigenous and German fighters, as well as with British and American fighters delivered under the Lend-Lease arrangement, the institute arrived at the conclusion that this aircraft was among the best with regard to stability and controllability. Upon completion of the State trials the State Committee of Defence adopted a decision requiring plant No.153 to switch its assembly line from Yak-7A production to the Yak-7B. The first production Yak-7B was completed in Novosibirsk as early as April 1942.

Concurrently, plant No.21 in Gor'kiy and plant No.82 (which had returned to Moscow from evacuation) began tooling up for the production of this fighter. Yakovlev had succeeded in 'squeezing' Semyon A. Lavochkin and his LaGG-3 fighter out of plant No.21 (one of the biggest in the Soviet Union), but after the successful development of the promising La-5 powered by the Shvetsov M-82 radial Lavochkin regained control over the plant; only five Yak-7Bs assembled in Gor'kiy passed acceptance and were delivered to front-line units.

As early as May 1942 the first three Yak-7Bs were assembled at the Moscow plant from shipsets delivered from Novosibirsk. By the end of July of that year 27 fighters had been built; 11 of them were pronounced combat-ready and promptly sent to front-line units. They proved to be noticeably inferior in quality to their stablemates from Novosibirsk. Especially many complaints were voiced concerning the production standard of the armament. Thus, Yak-7B c/n 820307 (that is, plant No.82, Batch 03, 07th aircraft in the batch) which was operated by the 4th IAP proved to have a malfunctioning ShVAK-20 cannon.

Series-produced aircraft from all factories were immediately sent to the front. To improve aerodynamics, the rocket armament was deleted at the end of May 1942, but the bomb racks were retained. The first phenomenon encountered in the front-line units of the South-Western Front was an inadmissible forward shift of the CG caused by the greater weight of the armament. The use of wheel brakes on landing became dangerous. To prevent nosing-over, a supplementary 80-litre (17.6 Imp gal) fuel tank was installed in the rear cockpit from the end of May 1942.

Pilots often expressed their displeasure with having this tank, which lacked self-sealing into the bargain, behind their backs. Its presence increased the all-up weight and impaired the flight performance, especially vertical manoeuvrability; besides, it entailed an increased fire hazard. In consequence, the personnel of front-line units began removing the supplementary tanks without bothering about official approval.

The 434th IAP was among the first to receive the M-105PA-powered Yak-7B. This unit enjoyed a special status because it was subordinated not to the commander of the 8th VA (*voz**doosh**naya* ***ar**miya* – Air Army) in whose sector it was operating, but to Chief Inspector of the Red Army Air Force, Colonel Vasiliy I. Stalin, the Soviet leader's son.

Comparative trials of the Yak-1 and the Yak-7B were conducted in the 202nd IAD, but pilots failed to reach a consensus: each of the two machines had its adherents. Having analysed the combat experience accumulated in the 146th and 181th IAPs, 234th IAD Commander Yaroslavtsev came to the conclusion that the Yak-7B was superior to all other types flown by Soviet pilots earlier. However, as he noted, *'the Yaks have an excessively long landing run; the fighter is 'nose heavy' and, most importantly, it lacks some 30-50 km/h* [18.6 -31 mph] *to gain an advantage over the contemporary Bf 109F'.*

Yak-7B M-105PF production fighter

In the summer of 1942 an important direction of work on perfecting the Yak-7B lay in uprating its engine. The Yakovlev OKB actively participated in experiments with increasing the boost pressure from 910 to 950 mm Hg, and then to 1,000 and 1,050 mm Hg and with determining the influence of the engine boost on the maximum speed and other performance of the aircraft. The results of these experiments proved very promising; they formed the basis for a Government decision which tasked Vladimir Ya. Klimov with transferring the M-105 engine from the normal to the boosted mode within the shortest possible time.

To ensure reliable operation in boosted mode, the piston pins were reinforced and the carburettor was readjusted. At that time the production M-105PF engine had no other differences from the M-105PA.

In addition to the engine modification, a number of other measures were implemented on the Yak-7B from Batch 22 onwards in accordance with Yakovlev's instructions. They were aimed at improving the aerodynamics and reducing the all-up weight, mainly by lightening some elements of the structure without compromising the strength and impairing the aircraft's operational qualities.

Yakovlev CKB

This rare picture shows a 'bubbletop' Yak-7B in the TsAGI wind tunnel during tests of the canopy jettison system. Note that the engine is running and the wheels have been removed in order to secure the aircraft in the tunnel.

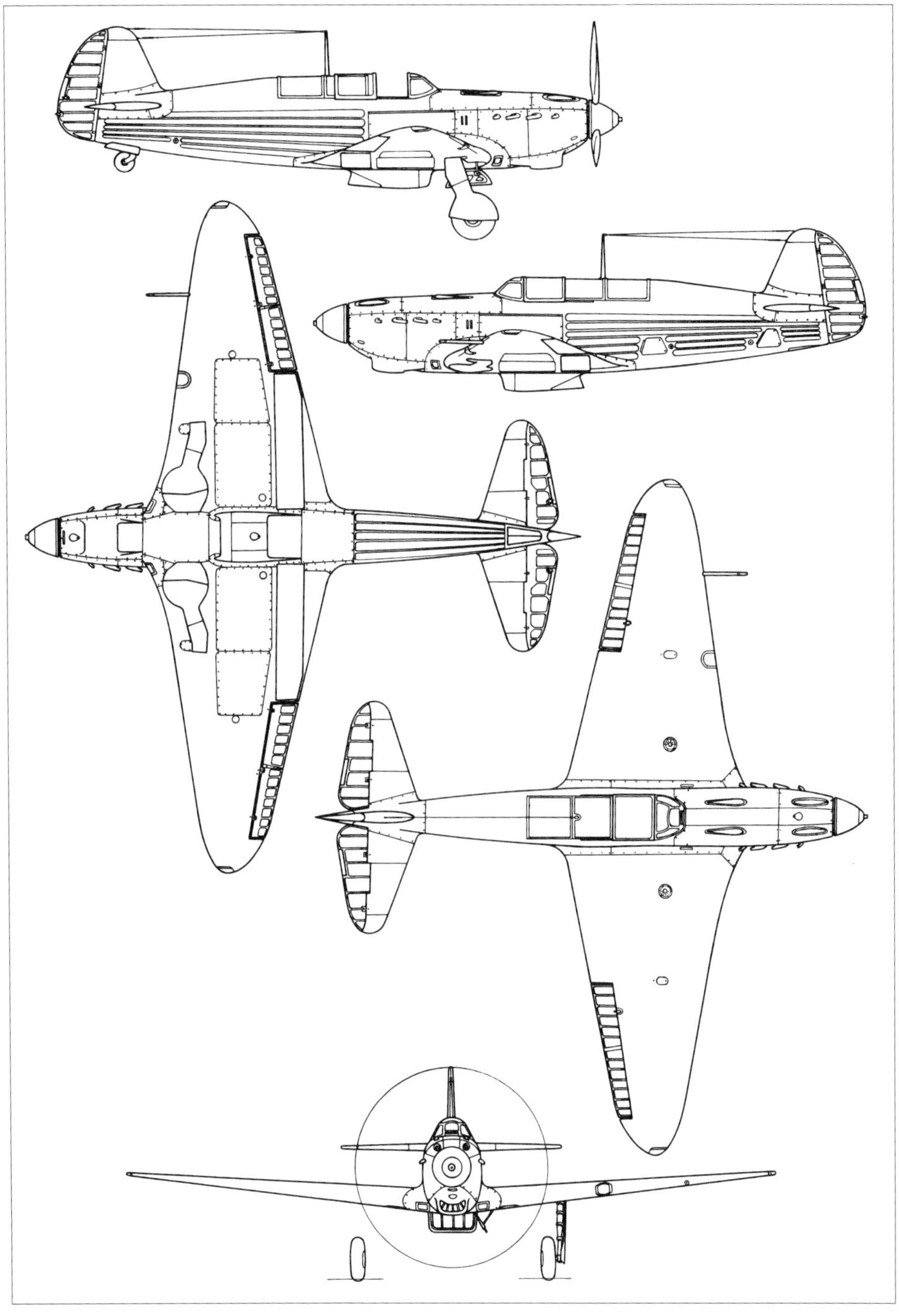

Five views of the Yak-7A.

Yakovlev OKB

Above: The Yak-7-37 was a derivative of the Yak-7B armed with a 37-mm Shpital'nyy Sh-37 cannon firing through the propeller hub. The long barrel of the Sh-37 is clearly visible in this view.

Yakovlev OKB

The Yak-7-37 prototype was totally devoid of insignia. Unusually, it had the designation painted on the tail instead of the construction number, as was customary on test aircraft.

The main alterations included lightening the fuselage framework, deleting the wiring to the rocket guide rails and altering the propeller blades' incidence at fine pitch. As a result, the M-105PF-powered Yak-7B (Yak-7B M-105PF) proved to be 30-35 kg (66-72 lb) lighter than its predecessor with the same armament and equipment. Tests conducted on Yak-7B c/n 2241 at NII VVS showed that a maximum speed of 514 km/h (319 mph) was attained at sea level, reaching 570 km/h (354 mph) at the second rated altitude of 3,650 m (11,970 ft). The time need to climb to 5,000 m (16,400 ft) was 5.8 minutes, and a banking turn could be performed in 19-20 seconds. In a combat turn beginning at 1,000 m (3,280 ft) the Yak-7B gained 950-1,000 m (3,120-3,280 ft) in altitude. In the opinion of NII VVS experts, such performance characteristics enabled the fighter to wage combat successfully with the Messerschmitts at low and medium altitudes.

In the summer of 1942, Yak-7B M-105PF fighters fought intensive battles with the Luftwaffe units near Stalingrad. On 20th August a whole fighter division (the 288th IAD) equipped with the last batches of Yakovlev's machines was committed to action. Making use of the Yak-7B's advantages, the 4th IAP staffed with the most experienced pilots shot down (according to pilot reports) 29 German aircraft, mostly Bf 109s, in the course of six days (7th-12th September) for the loss of only nine machines.

The results scored by the division's other regiments proved to be more modest, with higher own losses. 288th IAD Commander Lieutenant-Colonel Konovalov believed that the Yak fighters of his formation were inferior in performance to German fighters. He wrote to Stalin, noting that in field conditions it was impossible to obtain the characteristics obtained during tests at NII VVS, especially the maximum speeds stated in test reports. He asked that urgent measures be taken to ensure a tangible improvement in performance.

A commission was immediately sent to the division's regiments, establishing that the potential of the fighters was indeed not used to the full at the front. For the sake of convenience the factors causing a reduction of speed were divided into two groups: those depending on the pilot and those beyond the pilot's control. It transpired that many airmen flew their fighters in combat with the canopy hood open or even removed and with fully open radiator shutters; they piloted their machines at substantially lower engine speeds relative to the maximum revs (2,700 rpm) and so forth. As a natural result, the fighters flew 40-50 km/h (24.9-31 mph) slower than indicated in the reports of military test pilots and could not perform vigorous vertical manoeuvres.

One of the factors classed as being outside the pilot's control was the indifferent attitude of the tech staff to the condition of the aircraft's external surface. Technicians cut various openings in the engine cowling instead of carefully sealing all airframe joints. Poorly fitting wheel well doors, hatches and maintenance access panels (this was often the case after repairs in field conditions) led to a considerable increase of drag and hence a reduction of speed.

An urgent effort was undertaken in the 288th IAD to show the ground personnel how the Yaks should be properly maintained. One aircraft with no fewer that 60 hours on the clock (not many fighters lasted that long!) that had been repaired several times was 'treated' by specialists who eliminated some obvious causes of increased drag. After this, flown by fighter pilot A. Zaïtsev, the machine showed a top speed of 495 km/h (308 mph) at sea level instead of 467 km/h (290 mph) before the improvements! The results were on hand, but time was required to make them available to all front-line units.

Yak-7 tactical reconnaissance aircraft

During the period described here the Yakovlev OKB concentrated on perfecting the production machines. Yet, possibilities were sought for creating new prototype versions. One of the results of such work was a reconnaissance aircraft derived from the Yak-7B.

As far back as August 1941 two prototypes had been built at plant No.301 in Moscow on instructions from Chief Designer Aleksandr S. Yakovlev. These were Yak-7s in a reconnaissance version fitted with AFA-IM aerial cameras and an RSI-4 radio. At that time they were not ordered into production because of the acute need for the single-seat Yak-7 fighter.

The photo reconnaissance version of the Yak-7B was essentially similar to the abovementioned Yak-7 prototypes. It was intended for conducting reconnaissance from altitudes between 300 and 3,000 m (980-9,840 ft). A sample kit of aerial photography equipment was built at plant No.82 in 1943; this plant also manufactured nearly three hundred Yak-7s fitted with aerial cameras.

Provision was made on the production Yak-7B fighter for installing the AFA-IM camera and its mount. Therefore, conversion of the fighter for photographic reconnaissance missions could be performed *in situ* by front-line units in accordance with the technical manuals that followed the aircraft.

Yak-7-37 fighter

The Yak-7-37 was an important derivative of the Yak-7B M-105PA fighter, obtained by replacing the standard 20-mm ShVAK cannon with a 37-mm (1.45 calibre) cannon designed by Boris Shpital'nyy. This weapon had previously been fitted to the LaGG-3 for development purposes.

With a weight of fire reaching 4.15 kg/sec (9.15 lb/sec), the Yak-7-37 packed a considerably greater punch than all other known fighters used operationally. The new armament required some changes to the design of the Yak-7. The cockpit was moved 400 mm (1 ft 3¾ in) aft and the rear cockpit was deleted in order to provide accommodation for the cannon's breech in the engine bay and keep the CG within acceptable limits. The size of the modified fighter's tailwheel was increased. To reduce the danger of spin entry at low speeds and critical angles of attack, the Yak-7-37 was fitted with leading-edge slats.

Following brief trials conducted in April-May 1942, a decision was taken to build a small batch of Yak-7-37s, since the type was of some interest for the Red Army Air Force. In August 1942 a batch of 22 Yak-7-37s lacking LE slats was built in Novosibirsk; these machines were sent for operational trials to the 42nd IAP at the North-Western Front.

In 12 group air battles the Yak-7-37s shot down ten enemy aircraft at the cost of four own machines lost and another three damaged. The MPSh-37 cannon (*mo**tor**naya **push**ka Shpi**tal'**novo* – Shpital'nyy engine-mounted cannon) proved to be a reliable and deadly weapon. A shell hitting an enemy aircraft's fuel tanks caused an explosion, and a hit in the wing would tear a hole measuring more than one 1 m^2 (10.6 sq ft). Scoring a hit with just one shell was usually sufficient to destroy the enemy aircraft.

The MPSh-37 fired fragmentation/incendiary or armour-piercing/incendiary/tracer rounds with a muzzle velocity of 900 m/sec (2,950 ft/sec), making it possible to engage effectively not only aerial targets but also ground targets (armoured vehicles, light tanks) and even small ships. The tests were pronounced a success, and Yakovlev decided to continue the work on installing the 37-mm cannon between the cylinder banks on a fighter of improved design.

Yak-7 M-82 fighter prototype

The Yak-7 M-82 occupies a special place among the purely experimental versions of the Yak fighter. It was one of the first attempts to improve the aircraft, the programme being launched as early as August 1941. Yakovlev OKB documents show that the work on installing the Shvetsov M-82 14-cylinder air-cooled radial was undertaken largely for research purposes and was not necessary for Yakovlev in the same degree as for other designers (Semyon A. Lavochkin, Artyom I. Mikoyan, Mikhail I. Goodkov and others). This was due primarily to the fact that Yakovlev's designs were relatively lightweight and could also forthwith be successfully operated with the liquid-cooled M-105 engines.

When the design work on the Yak-7 M-82 commenced, the fighter was expected to have a maximum speed of 515 km/h (320 mph) at sea level with the engine boosted and 615 km/h (382 mph) at the rated altitude of 6,400 m (20,990 ft), climbing to 5,000 m (16,400 ft) in 5.5 minutes. This performance would enable the machine to compete comfortably with other Yakovlevs, as well as Lavochkin fighters. However, when the OKB's chief test pilot Pavel Ya. Fedrovi began flight tests of the new fighter, it became clear that the machine was a failure.

The M-82A engines that were available to the OKB were early-production examples; they had many production flaws and, in consequence, were plagued by malfunctions. Four engines had to be replaced successively on the prototype during the tests. However, it proved impossible to ensure satisfactory engine operation at the second supercharger speed.

Worse, the expected speed increase failed to materialise. The top speed at sea level obtained in the course of the tests was only 501 km/h (311 mph), falling short of the estimates. This was mainly because the propeller could not absorb the available engine power. The mainwheel legs were too short to allow a propeller of the required diameter (3.2 m/10 ft 6 in) to be installed, and the fighter barely avoided scraping the surface of the airfield with the propeller blade tips when a propeller of 2.8 m (9 ft 2 in) diameter was fitted.

Development of the Yak-7 M-82 was discontinued in May 1942.

Yak-7PD high-altitude fighter prototype

Development of the Yak-7PD represented a sequel to the work initiated in 1940 when the I-28 (Yak-5) was designed. This time it was not a new prototype machine, but rather a conversion of a production airframe which was mated to an experimental M-105PD engine provided with the E-100 supercharger

Yakovlev OKB

Above and below: Two views of the Yak-7 M-82 prototype during manufacturer's flight tests. The bulky radial engine ruined the sleek lines of the fighter. Note the wide chord of the propeller blades.

Yakovlev OKB

Yakovlev OKB

Above: The Yak-7 M-82 with virtually every panel open for inspection. Note that the cowling design is patterned on that of the Polikarpov I-185 and Lavochkin La-7, with a large one-piece cowling panel on each side rather than a two-piece folding panel as on the Polikarpov I-180 and La-5.

Yefim Gordon archive

The Yak-7PD was a high-altitude interceptor version with a turbocharged engine and a bubble canopy. These examples were operated by one of the PVO units defending Moscow. Note the different style and position of the tactical numbers.

Specifications of further Yak-7 fighter variants

	Yak-7 M-82	Yak-7-37	Yak-7PD
Year of manufacture	1942	1942	1942
Powerplant	M-82	M-105PA	M-105PD
Power at altitude, hp	1,330	1,050	1,160
Length	8.37 m (27 ft 5½ in)	n.a.	8.48 m (27 ft 9¾ in)
Wing span, m (ft-in)	9.74 (31 ft 11½ in)	10.0 m (32 ft 9 45/64 in)	10.0 m (32 ft 9 45/64 in)
Wing area, m^2 (sq ft)	17.15 (184.6)	17.15 (184.6)	17.15 (184.6)
Empty weight, kg (lb)	2,318 (6,051)	2,694 (5,939)	2,452 (5,405)
All-up weight, kg (lb)	3,370 (7,430)	3,235 (7,130)	2,904 (6,402)
Speed at sea level, km/h (mph)	515.0 (320.0)	485.0 (301.3)	500.0 (310.6)
Speed, km/h (mph)	615.0 (382.1	564.0 (350.4)	611.0 (379.6)
at altitude, m (ft)	6,400 (20,000)	4,730 (15,500)	7,600 (25,000)
Climb time to 5,000 m (16,400 ft)	5.6 min	7.2 min	5.4 min
Service ceiling, m (ft)	10,000 (33,000)	8,260 (27,000)	11,300 (37,000)
Turn time, seconds	24	23	19-20
Operational range, km (miles)	700 (434)	550 (341)	575 (357)
Take-off run, m (ft)	n.a.	n.a.	n.a.
Landing roll, m (ft)	n.a.	n.a.	n.a.
Armament, mm	2 x 20-mm ShVAK 1 x 12.7-mm UB	1 x 37-mm MPSh-37 2 x 12.7-mm UB	1 x 20-mm ShVAK

designed by V. Dollezhal'; the supercharger was developed at the Central Aero-Engine Institute (TsIAM – *Tsen**trahl'**nyy insti**toot** avi-atsi**on**novo mo**tor**ostro**yen**iya*). At the same time, in the late summer of 1942, the Yakovlev OKB fitted a Batch 22 airframe with an improved carburettor air intake and modified the upper fuselage decking and the cockpit canopy in a manner similar to the 'bubbletop' Yak-1; a bulletproof windscreen and a rear bulletproof glass panel were installed and some other alterations introduced.

To make the airframe as lightweight as possible, the armament was restricted to one 20-mm ShVAK cannon. Bearing in mind the low probability of a second firing pass against an enemy aircraft at altitudes close to the service ceiling, this armament's weight of fire was clearly inadequate, but the OKB had to choose between the aircraft's high-altitude performance and firepower. The Yak-7PD's service ceiling of 11,300 m (37,060 ft) was superior to that of any Soviet production fighter of the time (in 1942 the MiG-3 was no longer in production).

The performance of the Yak-7PD could have been much higher, had it been provided with automatic, not manual, control of the turbocouplings (that is, control of the supercharger's first-stage speed). The lack of automatic turbocoupling control made operation of the aircraft more complicated: the pilot was compelled to make constant corrections to the supercharger's impeller speed, which distracted his attention from flying the aircraft. Besides, he was simply physically unable to maintain the nominal value of the manifold pressure in accordance with the change in the atmospheric pressure; as a result, the aircraft's performance inevitably suffered.

In October-November 1942 the Yak-7PD was still used for development work on a device for automatically controlling the turbocouplings, but positive results could not be obtained at the time. The work on the M-105PD engine was continued at a later stage.

Yak-7V conversion trainer

At the end of 1941 plant No.153 completely switched to fighter production, and soon the resulting replenishment of the Red Army Air Force with first-class fighters had a noticeable effect on its combat capabilities. On the other hand, the output of fighter trainers was halted completely. The Yak-7UTI was the only such type in series production since the beginning of the war; by mid-1942 the Yak-7UTIs, which were produced in small numbers, were already fairly worn out. The Red Army Air Force's Directorate of Combat Training repeatedly urged upon the People's

Commissar of Aircraft Industry the need for creating an aircraft that would provide the transition from the Polikarpov U-2 and Yakovlev UT-2 primary trainers to a real fighter.

The Yak-7V (*vyvoz**noy*** – for familiarisation training) filled this role. Development of this version began at the end of 1941, making use of operational experience gained with the Yak-7UTI. As distinct from its predecessor, the undercarriage of the Yak-7V was fixed; this lessened its wear and precluded the possibility of the mainwheel legs collapsing because of malfunctions or the trainee's error. Now, in contrast to the 'combat' Yak-7s, the unfaired mainwheel legs were strictly vertical (when seen from the front) and the wheel rotation planes were not canted outwards. This measure also simplified production. The trainer was stripped of all armament, and the upper decking of the engine cowling had no recesses for the machine-gun barrel.

The Yak-7V proved to be a simple, reliable and rugged aircraft capable of almost all aerobatic manoeuvres. When the output reached 510 machines by the end of 1943, the Air Force command came to the conclusion that the needs of the Air Force had been fully met. Further production of the Yak-7V was stopped; however, shortly before the termination of hostilities 87 Yak-7Bs were converted into fighter trainers similar to the Yak-7V at the Air Force's repair shops.

The Yak-7 – various improvements

Work on perfecting of the Yak-7B went on uninterrupted in the middle of the war. In the winter of 1942-43 it proved possible to improve the quality standard of production machines. Tests of a Novosibirsk-built Batch 25 Yak-7B revealed that its maximum speeds at all altitudes were 22-25 km/h (13.7-15.5 mph) higher than those of Batch 22 machines thanks to a considerable improvement in production methods achieved in December 1942. The vertical manoeuvrability of combat machines was also improved.

Outstanding results were achieved when the OKB, LII and NII VVS jointly undertook an improvement programme on Yak-7B c/n 31-01. This fighter featured the following improvements. The radio antenna was buried in the fuselage and the aerial mast was deleted; the leading edges of the wings and tail surfaces were given a polish; ejector exhaust stubs were installed in keeping with TsAGI recommendations; all airframe joints were carefully sealed; wheel well doors and maintenance access panels were a better fit, and a fully retractable tailwheel was introduced.

Numerous other measures were taken, each of them affording a speed increment of

Yakovlev OKB

Yakovlev OKB

Above and below: The prototype of the Yak-7V conversion trainer (c/n 15-70) on skis during winter trials. The trainer could be distinguished from the combat versions by the glazed rear cockpit and the non-retractable main gear struts with faired-over wheel wells and no doors.

Yakovlev OKB

Above: A production Yak-7V (c/n 02-66) during checkout trials at NII VVS.
Below: A winter-camouflaged Yak-7UTI trainer with retractable landing gear at a frontline airstrip.

Yefim Gordon archive

Yakovlev OKB

This Yak-7 was used as a testbed for the Merkoolov DM-4S ramjets (it is often called Yak-7PVRD). The large size of the boosters is obvious in this view.

1-2 km/h (0.6-1.2 mph). As a result, the fighter attained 547 km/h (340 mph) at sea level and 615 km/h (382 mph) at 4,000 m (13,120 ft); no other Yak fighter had shown such high results before, or showed them later, with the exception of the Yak-3 versions. The Bf 109 and the Fw 190 had lower speeds at low and medium altitudes.

The quality of Moscow-built fighters proved to be much worse. They were 25-30 km/h (15.5-18.6 mph) slower than the Novosibirsk-built machines, had high control stick forces and were more sluggish when performing various manoeuvres.

Improvement of the fighter's airframe continued unabated – from the end of 1942 onwards the Yak-9 (described below) was series-produced on a mass scale. However, production of the Yak-7B had to be continued due to the shortage of duralumin from which the Yak-9's wing spars were manufactured. In the meantime, some features characteristic of the Yak-9's external appearance were incorporated into the design of the Yak-7B. Thus, starting with the 40th batch of plant No.153 in Novosibirsk and the 12th batch of plant No.82 at Moscow-Tushino, the upper decking of the rear fuselage was cut down and a tear-drop canopy provided with an emergency jettisoning system was installed.

When the 157th and 728th IAPs took delivery of new Yak-7Bs (aircraft from Novosibirsk-built batches 39 through 41) in both 'razorback' and 'bubbletop' versions, to use the analogy with the Republic P-47 Thunderbolt, the pilots noticed considerable discrepancies in the maximum speeds of different machines. The cause was traced to differences in the machines' production standard and in the nominal power of their engines. Liberal oil spillage from breathers during flight at maximum speeds, uneven fuel consumption from the port and starboard groups of tanks, the insufficient range of radio communication and various defects hampering the maintenance of the fighters led to the obvious conclusion: the results of the check-out tests were unsatisfactory.

In the late spring of 1943 dozens of Yak-7Bs were completely put out of action in the 7th VA operating on the Karelian Front. As it turned out, the large amount of water on the airfields resulting from the snow melting in the daytime and freezing again at night had created a devil's mixture of dirt, ice and water. This mixture clogged the Yaks' radiators, causing engine overheating, with frequent forced belly landings as a result.

Even more disturbing were the cases when the wing skins of Yak fighters ripped away in flight. Such facts had been noted earlier: four Yak-7Bs had crashed in the period between June and October 1942. In the spring of 1943 the number of accidents and incidents increased sharply due to defective bonding of the wing structure caused by breaches of the prescribed technology and by the use of substitutes of various chemical substances.

In spite of the frequently voiced complaints concerning the Yaks, they were popular with the pilots. There were many cases when these machines displayed high combat survivability. For example, during the Belgorod operation in July 1943 the starboard wing spar of one Yak-7B was pierced by a shell and the blast ripped away a piece of wing skin measuring 1 m^2 (10.6 sq ft). On another occasion a direct hit into the front spar by two shells destroyed 70% of the spar flanges in the most highly stressed part of the wing; on a third aircraft the fabric skin of the fuselage was torn off and the compressed air system was damaged. Yet all three aircraft belonging to the same unit (the 483th IAP) returned to base, landing safely.

Plant No.153 continued manufacturing Yak-7Bs until the end of 1943, and plant No.82 did so until July 1944. At the end of 1943, having studied 13 machines produced in Novosibirsk, a team of LII test pilots and engineers stated that the all-up weights of these machines were practically identical (around 3,000 kg/6,610 lb). Deviations of up to 15 kg (33 lb) from this figure arose from the presence or absence of bulletproof windscreens and to insignificant differences in production processes.

The maximum speeds of the Yak-7Bs showed more marked discrepancies amounting sometimes to 20 km/h (12.4 mph); this was due primarily to differences in the rated altitudes of the engines installed. Efforts to bring the quality of the Moscow-built fighters to the required level proved unsuccessful – Yaks from plant No.82 were, on the average, 30 kg (66 lb) heavier and 15 km/h (9.3 mph) slower than similar machines from plant No.153.

In all, the four production factories (Nos. 21, 82, 153 and 301) delivered 6,399 Yak-7s of different versions to the Air Force; of these, 5,210 were Yak-7Bs powered by the more powerful M-105PF engines. Yak-7 production at plant No.82 ended with the 28th batch, while the 50th batch was the last at plant No.153. In the final batches of this type built in Novosibirsk, several dozen in every batch consisting of 100 machines were fitted with direction finders, landing lights and undercarriage position indicators; they were produced specially for service in the PVO (Air Defence) units.

262 Yak-7Bs remained on strength in the 36th, 106th, 126th and other Air Divisions of the PVO system by the end of 1944; they continued to be operated fairly intensively. The role played by the Yak-7 in the front-line aviation at that time was somewhat more modest. According to reports from different Air Armies, 65 fighters of this type were engaged in operations against the Luftwaffe by 1st January 1945.

Yak-7 propulsion testbeds with DM-4S boosters

One of the unusual tasks tackled by the Yakovlev OKB was the construction of a flying testbed intended for verifying the experimental DM-4S ramjet boosters designed by I. A. Merkoolov. The DM-4S was 2.3 m (7 ft 6½ in) long, with a diameter of 500 mm (1 ft $7\frac{11}{16}$ in), and weighed 45 kg (99 lb). Two such engines were suspended under the fighter's wings outboard of the main gear pivots; they ran on petrol fed from the machine's standard fuel tanks.

The DM-4S engine was designed in 1941; this was the version that was flight-tested on the Polikarpov I-15*bis* and I-153 fighters before the war. After a period of bench tests two such engines manufactured in early 1942 were mounted under the wings of a UTI-26 operated by the 12th GvIAP. In

Yakovlev OKB

The Yak-7L development aircraft with laminar-flow wings, seen here in front of one of LII's hangars. Note the cranked leading edge wing.

the spring of 1942 the frontlines rolled back some distance to the west from Moscow; this made it possible to perform test flights without the risk of being intercepted by the Messerschmitts.

The testing proved abortive: installation of the boosters led to a significant forward shift of the CG, entailing the risk of a nose-over. Besides, no special fire protection measures had been taken, and vibrations generated by the ramjets caused frequent leakage of fuel from the tanks. The flights had to be suspended for almost two years.

In the middle of the war information started coming in about experiments with auxiliary powerplants conducted both by the Soviet Union's allies and the adversaries. For example, the British performed many experiments with ship-borne aircraft, while the Germans made use of jettisonable solid-fuel rocket boosters to assist the take-off of overweight bombers and assault gliders from small field airstrips.

In early 1944 changes were introduced into the design of the DM-4S engine with a view to enhancing its reliability. The engines were mounted under the wings of a production Yak-7B, a seat for the test engineer being fitted in the rear cockpit. The modified aircraft was known as the Yak-7VRD or Yak-7PVRD (*pryamo***toch***nyye vozdooshno-reak***tiv***nyye* ***dvig****ateli* – ramjets). The tests were conducted at LII NKAP, starting on 24th March 1944; on 15th May project test pilot Sergey N. Anokhin ignited the ramjets in flight for the first time.

The boosters consumed up to 20 kg (44 lb) of petrol per minute; therefore, the auxiliary powerplant could be operated only briefly – for example, to break off an engagement or to chase the adversary. The flights revealed that the increase in maximum speed afforded by the boosters amounted to 53 km/h (32.9 mph) as compared to 22 and 42 km/h (13.6 and 26.0 mph) respectively in the case of the I-15*bis* and I-153. However, owing to the considerable additional drag of the wings and the massive hollow cylinders attached to them the net speed increase did not exceed 19 km/h (11.8 mph).

These results were disappointing. The work went on until the end of 1944. It encompassed the development of boosted modes of ramjet operation, installation of the ramjets in an aerodynamically refined airframe, making the booster engines easily detachable and so on. Calculations showed that the speed increase could be as high as 100 km/h (62 mph) in the course of 10 seconds, but it proved impossible to corroborate these estimates by real test results.

Yak-7R experimental fighter with a mixed powerplant (project)

Several interceptor projects were evolved during the war. One of them, designated Yak-7R (*reak***tiv***nyy* – jet-powered) envisaged an aircraft featuring a combined powerplant comprising two Merkoolov DM-4S ramjets

Yakovlev OKB

Above: A Yak-7B development aircraft with a pressurised cockpit; note the non-standard sliding canopy. Below: Another Yak-7 equipped with a pressurised cockpit of a different design (note the shape of the windshield). Judging by the black-outlined insignia, this was an earlier version (compare with the white-outlined stars on the other Yak).

Yakovlev OKB

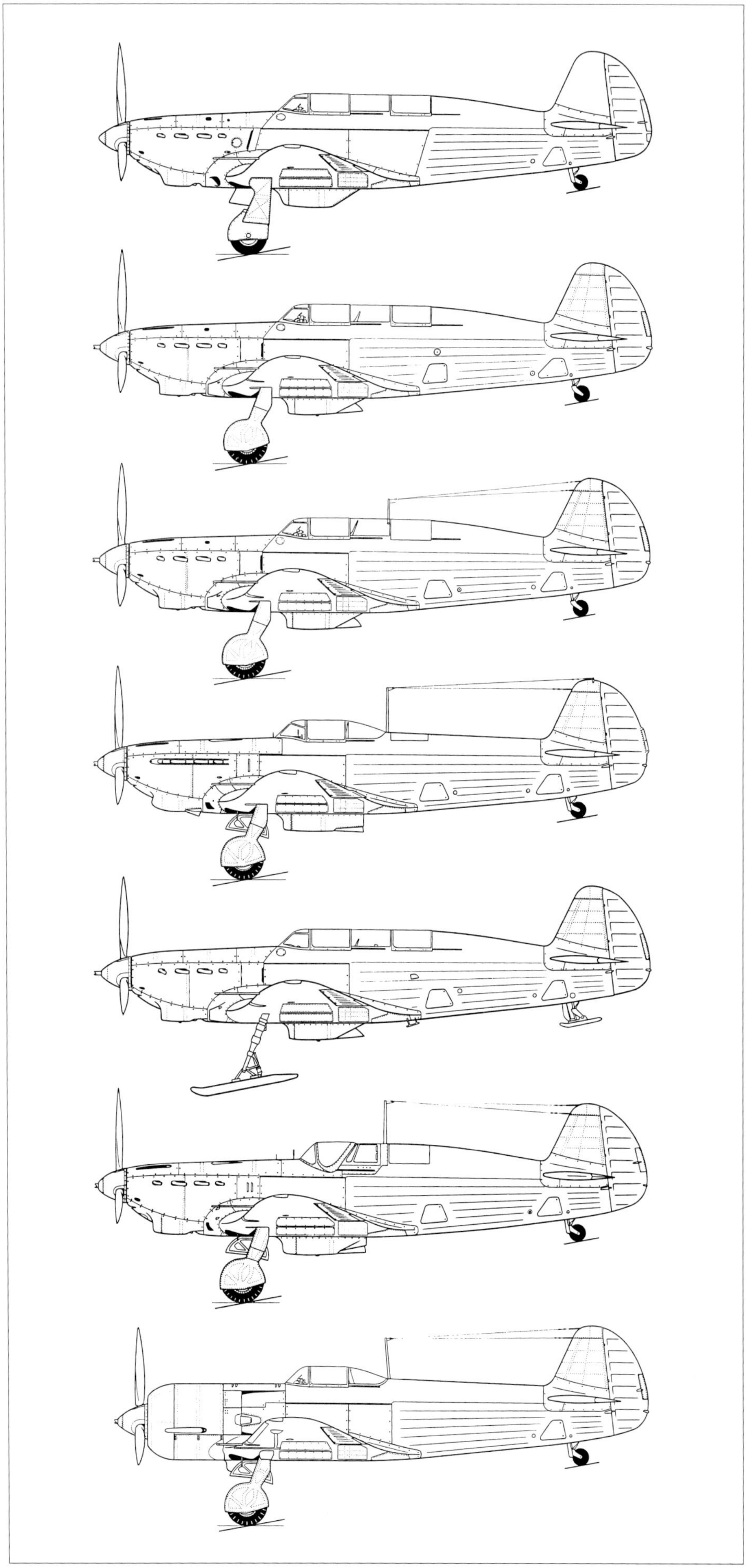

Top to bottom: The UTI-26-1 trainer prototype; a Batch 22 production Yak-7 trainer; a 'razorback' Yak-7B fighter; a 'bubbletop' Yak-7B; the Yak-7V conversion trainer; the Yak-7GK experimentally fitted with a pressurised cockpit; and the Yak-7 M-82 development aircraft.

and one D-1A liquid-propellant rocket motor designed by L. S. Dooshkin.

The ramjet boosters were mounted under the wings in the same way as on the previous model, and the D-1A was accommodated in the aft fuselage. The latter installation was intended for brief use during take-off and for imparting maximum acceleration. The rocket motor ran on petrol, using nitric acid as an oxidiser; this necessitated radical changes to the fuel system. Three additional tanks (two for petrol and one for the acid) were accommodated behind the cockpit. The piston engine was eliminated, and a fixed nosecone housing the armament replaced the cowling and propeller spinner.

Design work on the Yak-7R was completed on 27th August 1942. Many designers, including Yakovlev, were aware of the potential advantages offered by ramjets and liquid-propellant rocket motors; however, the enemy had forced its way to Stalingrad, and the bulk of the effort was concentrated on series production. Also, the flights of the rocket-powered BI fighter designed under Viktor F. Bolkhovitinov's direction showed that no reliable jet or rocket engines suitable for immediate installation in an aircraft were available in the USSR at that time. The experience gained in designing the Yak-7R was used later for developing the Yak-3RD mixed-power fighter.

Yak-7L experimental prototype

The Yak-7 aircraft were subsequently used probably more extensively than any other Yakovlev fighters for numerous factory and field modifications. The former is exemplified by the development of the Yak-7L fighter with laminar-flow wings (hence the L suffix for *lami**nar**noye kry**lo***) which afforded a maximum speed increase of 15-20 km/h (9.3-12.4 mph) thanks to the use of low-drag airfoil sections.

Yak-7GK fighter prototypes

Experiments were conducted with the installation of pressure cabins on the Yak-7. A production Yak-7B M-105PF was fitted with a Shcherbakov type cabin featuring a characteristic diagonal-type canopy. The machine was referred to as the Yak-7GK (*gher**metich**eskaya ka**bin**a* – pressurised cockpit); it was submitted for testing at a research institute of the Air Force (not NII VVS) on 1st May 1944. The pressurisation system had many deficiencies making the cabin unsuitable for introduction into service.

One more Yak-7B was fitted with a pressure cabin of a different type. Both installations had thick canopies with all joints sealed by rubber.

Yak-7P fighter with twin cannons

The field modifications included the Yak-7P (***push**echnyy* – cannon-armed). The heavier armament, which was installed in the workshops of the 1st VA in July 1943, comprised two synchronised ShVAK cannons instead of the two UBS machine-guns in the same fashion as on the La-5.

Yak-7 glider tug

Another version of the fighter was fitted with a special hook for towing gliders. It is not known which gliders were in question, but they were obviously not assault gliders.

Yak-7K liaison aircraft

According to Vadim B. Shavrov, an example of the Yak-7K liaison aircraft (*koor'**yer**skiy* – courier, used attributively) with a comfortable passenger cockpit at the rear was delivered in 1944 to meet an Air Force order. Several more machines were converted to this standard in the field (just like the later 'war weary' P-51B conversions in the USA).

Yak-9 fighter

The Yak-9 aircraft that followed the Yak-1 and the Yak-7 was the third basic type of the wartime Yakovlev fighter family. Structurally it was a further development of the Yak-7 and, though featuring only minor external differences from its predecessor, the Yak-9 incorporated many improvements that were not so obvious. This was natural because the design of this aircraft embodied the wealth of production and combat experience accumulated with the Yak-7. Also, duralumin was no longer as scarce in the Soviet Union as it had been at the beginning of the Great Patriotic War and could be used on a wider scale. The use of metal made it possible to reduce the airframe weight substantially. This weight saving was used for increasing the fuel capacity; alternatively, it could be used for installing heavier armament or fitting various equipment items.

Of all fighter types operated by the Red Army Air Force, the Yak-9 was built in the greatest numbers. In mid-1944 the aggregate number of Yak-9s, Yak-9Ts and Yak-9Ds surpassed the number of fighters of other types in the inventory of front-line units. They largely supplanted the Yak-1, Yak-7B and LaGG-3, to say nothing of the MiG-3s which had been manufactured in 1941, and the Polikarpov fighters built before the war.

The main special feature of the Yak-9 was its ability to be modified into a great variety of versions optimised for different roles. These included a front-line fighter with normal and heavy armament; a long-range escort fighter; a fighter-bomber; a photo-reconnaissance fighter; a high-altitude interceptor; a two-seat unarmed liaison aircraft for special duties; and a two-seat conversion training and familiarisation fighter. This basic type had 22 (!) main variants, including 15 production versions. The Yak-9 featured the installation of five different new engines or engine versions; six variants with regard to the number and capacity of fuel tanks; seven variants of armament and two variants of special equipment.

The Yak-9 was produced at three major aircraft factories – No.82 in Moscow (the M-107A-powered version only), No.166 in Omsk and No.153 in Novosibirsk. The production tempo at the latter enterprise reached 17 machines per day in mid-1944. In all, no fewer than 16,769 Yak-9s were built, including the 14,579 manufactured before the end of the war.

Yak-7D reconnaissance aircraft prototype

The Yak-9's predecessor was the Yak-7D (***dahl'**niy* – long-range). In the late spring of 1942 Aleksandr S. Yakovlev tasked engineer Nikolay K. Skrzhinskiy with developing a Yak-7 version intended for use as a long-range reconnaissance aircraft to suit the needs of the Air Force's fighter element. Built in May 1942, the aircraft combined the fuselage of a production Yak-7V conversion trainer with elements of the Yak-7B.

A novel design feature was the experimental wings which differed greatly from all the wing types built earlier. The wings' framework was formed by two continuous full-span duralumin spars joined together by six duralumin ribs and two wooden ribs (the latter placed at the tips). The wing skins were made of properly shaped Bakelite plywood and veneer which was bonded to the structure. While retaining the same area of 17.15 m^2 (184.6 sq ft), the wings had a shorter span – 9.74 m (31 ft 11½ in) instead of 10.0 m (32 ft $9\frac{45}{64}$ in) – and more angular tips introduced with a view to fitting leading-edge slats at a later stage.

Eight fuel tanks with a total capacity of 833 litres (183.26 Imp gal) were neatly 'squeezed' into the wings; a further 92 litres (20.24 Imp gal) of fuel were housed in the fuselage. The Yak-7D was unequalled in

Yakovlev OKB

Above: The Yak-7D long-range fighter, the first step towards the Yak-9. It is seen here in single-seat configuration revealing its origin from a Yak-7B.

Yakovlev OKB

The same aircraft in two-seat configuration. The position of the aerial mast remained unchanged.

Yakovlev OKB

Above and below: The Yak-7DI featuring a cut-back rear fuselage decking and a bubble canopy was, in effect, the prototype of the Yak-9.

Yakovlev OKB

range and endurance, covering a distance of 2,285 km (1,420 miles) in the course of six and a half hours.

Yak-7DI fighter prototype

In mid-June 1942, when Yak-7D development work was in full swing, A. S. Yakovlev issued directions requiring the machine to be reworked into a long-range fighter. The result was the Yak-7DI (***dahl'****niy istre****bi****tel'* – long-range fighter). In fact, it was the fuselage of a production M-105PF-powered Yak-7B mated to the experimental wings of the Yak-7D. This time the wings housed only four self-sealing tanks with a total capacity of 673 litres (148 Imp gal), or 500 kg (1,102 lb).

The Yak-7DI had its upper rear decking cut down and featured a teardrop-shaped cockpit canopy. The starboard Berezin UBS machine-gun was deleted for the sake of further reducing empty weight. This subsequently became a 'trademark' recognition feature of the 'nine' – the starboard heavy machine-gun was not installed on any of the Yak-9 versions powered by the M-105PF.

An important novel feature employed on the Yak-7DI was that all four fuel tanks were connected with a single service tank via check valves; this ensured uniform and complete consumption of fuel from the port and starboard groups of tanks – something that had not been possible on the Yak-1 and Yak-7.

The designers envisaged the possibility of the Yak-7DI being operated in two basic versions – the standard and the lightened one. In the latter case a 200-kg (441-lb) weight saving was achieved by reducing the amount of fuel and oil carried. In standard configuration the all-up weight reached 3,035 kg (6,692 lb) and the Yak-7DI possessed basic performance characteristics similar to those of the production Yak-7B manufactured in the summer and autumn of 1942, except that the range was appreciably greater. In this version the aircraft turned into a long-range fighter capable of lengthy patrolling, escorting bombers to the full stretch of their combat radius and fulfilling other missions that required the machine to stay airborne for a long time. In the lightened version the all-up weight was only 2,835 kg (6,251 lb), and the Yak-7DI became easy and tractable in handling, displaying very good manoeuvrability at low and medium altitudes. In mock combat with the Bf 109F conducted close to the ground the Yak-7DI gained an extra 250 m (820 ft) of altitude in a combat turn as compared to its opponent, latching onto the Messerschmitt's tail after three to four full-circle banked turns.

After a short stage of development work at the factory the Yak-7DI successfully passed State trials in August 1942. Leading engineer A. Stepanets wrote in his report that *'...thanks to the possibility of being operated in various versions and to its excellent performance the Yak-7DI arguably ranks first among our country's production fighters'*. It was recommended that the Yak-7DI be put into series production as quickly as possible. The aircraft industry implemented this recommendation without delay. In production the fighter was allocated the designation Yak-9.

Yak-9 production fighter (Yak-9 M-105PF)

Initially plant No.153 in Novosibirsk built the fighter in the lightened Yak-7DI version, deleting the two extra fuel tanks in the wings. The all-up weight rose by 35-40 kg (77.2-88.2 lb) compared to the prototype, which was due to the usually lower quality standard characteristic for mass production, but production machines suffered no deterioration in performance. Tests of the first production Yak-9s showed that the fighter had retained its excellent agility which enabled it to outmanoeuvre any adversary.

Plant No.153 began manufacturing the Yak-9 in October 1942, when the workers assembled the first 16 machines (in addition to 290 Yak-7Bs). Production was getting under way successfully – the military representatives at the plant accepted 15 Yak-9s. Subsequently, however, the plant ran into serious difficulties in manufacturing the wings: of the 44 Yak-9s built in the course of the following two months only four (!) machines could be sent to the front-line units before the end of the year.

Three new production fighters were delivered to the 434th IAP in mid-November 1942. The Yak-9s were committed to combat for the first time during the Soviet counteroffensive near Stalingrad in the second half of December 1942.

One machine (c/n 01-18 – that is, Batch 01, 18th aircraft in the batch) was transferred to NII VVS. The Yak-9 successfully passed tests there during February and March 1943. At the weight of 2,870 kg the fighter attained the speed of 520 km/h (323 mph) at sea level and 599 km/h (372 mph) at the second rated altitude of 4,300 m (14,100 ft). Only 5.1 minutes were needed to climb to 5,000 m (16,400 ft). Other important performance

Yakovlev OKB

Above and below: The fourth production Yak-9 (c/n 01-04) during manufacturer's flight tests.

Yakovlev OKB

characteristics of the machine were equally impressive: it had a service ceiling of 11,100 m (36,410 ft), a full-circle banking turn at 1,000 m (3,280 ft) took 16 to 17 seconds and the aircraft gained 1,120 m (3,670 ft) of altitude in a combat turn. For the first time a Soviet production fighter was virtually equal to the contemporary Messerschmitts in vertical manoeuvrability while surpassing them in ease of handling and horizontal manoeuvrability. Importantly, the improvement of performance was achieved without resorting to such radical measures as installation of a new engine.

Interestingly, testing of Yak-9 c/n 01-18 was not confined to determining its basic performance characteristics. Flights were made to determine the field performance, range at different engine running modes, gliding characteristics and even the engine's acceleration time in accordance with procedures adopted in the USSR. It was revealed that Yakovlev's new machine took more time to accelerate from cruising speed to a speed close to V_{NE} compared to the Yak-1 and the Bf 109G-2; on the other hand, it could decelerate quicker.

At the beginning of 1943 assembly of the Yak-9 was initiated at plant No.166 in Omsk, which had halted its work on the Tupolev Tu-2 bomber. The fighters of the first batch had a number of production defects, but already from the second batch onwards the plant's personnel, assisted by engineers from the Yakovlev OKB, achieved considerable success in improving the manufacturing standards. During tests conducted at LII and completed in April 1943, Yak-9 c/n 1102014 (that is, *izdeliye* 11, Batch 02, 014th aircraft in the batch) weighing 2,840 kg (5,822 lb) attained a speed of 539 km/h (334 mph) at sea level and 577 km/h (358 mph) at 3,810 m (12,500 ft); it climbed to 5,000 m (16,400 m) in five minutes flat. Consequently, in most parameters this aircraft was on a par with similar machines built in Novosibirsk.

In the first quarter of 1943 plant No.166 turned out 61 Yak-9s. A total of 135 such machines were built in January and February of that year, and the number of those accepted by the military representatives was even higher (140). Most of the 'nines' went to reserve regiments and training centres where these machines were assigned primarily to air formations of the Supreme Command's Reserve. Despite all the advantages of the Yak-9, series production of this type in its original form did not last long. This was due to the emergence of three versions developed by the Yakovlev OKB at the end of 1942; two of them subsequently became the baseline versions. They supplanted the Yak-9 *sans suffixe* (also known as the 'Yak-9 with two fuel tanks and standard armament'), of which nearly 500 copies had been built by then.

In early 1943 the Yak-9, like the other Yakovlev fighters, was plagued by numerous cases of the wing skin warping or even breaking away in flight, which often led to tragic results. This was mainly due to such reasons as breaches of bonding techniques in the wing manufacture, the use of poor-quality substitutes of various chemical substances (this refers primarily to plant No.153) and considerable fluctuations in the ambient temperature and humidity within 24 hours. The skinning was successfully reinforced on most of the fighters by July of the same year.

M-106-powered Yak-9 prototype (Yak-9 M-106, Yak-9 M-106-1sk)

Along with machines intended for large-scale production, the Yakovlev OKB built some experimental fighters based on the Yak-9. One of the first such versions was the Yak-9 M-106. Installation of the M-106-1sk engine with a single-speed supercharger was effected in the same way as on the

Yakovlev OKB

Yak-9 c/n 01-18 at NII VVS during State acceptance trials. Note the new three-tone grey camouflage, a departure from the dark green/black scheme used hitherto.

Yakovlev OKB

A Yak-9 fitted experimentally with an M-106-1sk engine. This one appears to have the dark green/black camouflage scheme.

similarly powered version of the Yak-1. The machine was built as the *dooblyor* (second prototype) of the Yak-7DI by mid-November 1942; shortly thereupon pilot Pavel Ya. Fedrovi made the first flight. Prior to the installation of the new engine the Yak-9 M-106 had passed operational tests, performing more that 1,000 different aerobatic figures and making 500 landings. Yet no signs of permanent deformation were noted on the fighter.

Testing of this aircraft at NII VVS showed that a speed of 602 km/h (374 mph) was attained at 3,250 m (10,660 ft), accompanied by an improvement in rate of climb and take-off performance compared to the usual Yak-9. The Yak-9 M-106 demonstrated its ascendancy in mock combat with the Bf 109G-2/R6, an example of which had been captured near Stalingrad. Test pilots noted that the Yak fighter was considerably easier in handling compared to the overweight Messerschmitt; consequently, a well-trained pilot on the Yak-9 had a better chance of imposing his will on the adversary in combat. The cockpit of the Soviet fighter afforded a considerably better view, especially in the rear hemisphere. At the same time the Yak-9 was clearly inferior to its opponent in firepower and performance at high altitudes.

The Government demanded that the Yak-9 M-106 be urgently put into large-scale production. One of the telegrams sent to Novosibirsk and addressed to director Lisitsyn stated: *'You are to regard switching completely to the production of the M-106P-powered Yak-9 fighters as your most important task for the first quarter of 1943 and to ensure the manufacture of 176 such machines.'* Huge resources were committed for the implementation of this task. Still, the immaturity of the M-106 engine prevented this machine from entering series production in Novosibirsk, just like the Saratov factory had failed to master production of the Yak-1 M-106.

Yak-9T production fighter (Yak-9-37)

It was the Yak-9T (***tahn**kovyy* – tank-busting) that became the first mass-produced version of the Yak-9. It differed from the prototype and the initial production Yak-9 in having the 20-mm (0.78 calibre) ShVAK cannon replaced by the 37-mm (1.45 calibre) 11P-37 cannon developed by the OKB-16 design bureau (subsequently it was redesignated NS-37). Hence the alternative designation Yak-9-37 has also been quoted.

Design work on this weapon was performed in 1941-42 by a group of engineers under the direction of A. Nudel'man and A. Sooranov. They continued the work on the development of a large-calibre aircraft cannon initiated by Yakov Taubin and M. Baboorin. By the time when it attracted A. Yakovlev's attention the 11P-37 cannon had successfully passed bench tests and a trial installation on the LaGG-3 fighter. In similar fashion to the LaGG-3, it was mounted between the cylinder banks with the barrel passing through the propeller hub; the cannon was attached at two points (to the engine and the airframe). The barrel protruded 160 mm (6¼ in) from the propeller spinner, increasing the fighter's length from 8.50 m (27 ft 10½ in) to 8.66 m (28 ft 4¹⁵⁄₁₆ in).

The installation of a big and heavy 37-mm cannon weighing 150 kg (331 lb) necessitated a number of substantial changes to the Yak-9's design. The fuselage structure was reinforced. To provide space for the cannon breach and keep the CG within acceptable limits, the cockpit was moved 400 mm (1 ft 3¾ in) aft, as on the Yak-7-37. This led to a marginal improvement in rearward visibility and made the fighter more agile.

The potent 37-mm cannon made it imperative for the staff of the Novosibirsk plant to raise the aircraft's manufacturing standard. Improving the quality of the assembly job was a matter of prime concern because the N-37's considerable recoil force (about 5,500 kg/12,127 lb) caused piping joints to leak and some elements of the airframe structure to develop fatigue cracks and fail.

The Yakovlev OKB built the Yak-9T prototype in January 1943; the aircraft passed special tests at the Aviation Armament Firing Range in February 1943 (L. Los' was the project engineer) and flight testing at NII VVS in March of that year (with A. T. Stepanets as project engineer). With an all-up weight of 3,025 kg (6,670 lb) the Yak-9T attained a speed of 533 km/h (331 mph) at sea level and 597 km/h at 3,930 m (12,890 ft). Manoeuvrability remained good, a full-circle turn being performed within 18-19 seconds and 1,100 m (3,610 ft) of altitude being gained in a combat turn. Immediately, in the same month of March, the Yak-9T entered series production. The Soviet Government demanded that a fighter regiment be fully equipped with Yak-9T fighters by mid-March. In actual fact, only three aircraft were built by the end of March; another 75 machines with the heavy cannon rolled off the assembly line in April. Shortly thereupon the monthly output reached approximately one hundred Yak-9Ts.

Thirty-four Yak-9Ts were used in combat evaluation tests which took place at the Central Front in the 16th VA between 5th July and 6th August 1943 during the battle of Kursk. According to official Soviet information, almost half of the 110 enemy aircraft destroyed there were shot down by Yak-9Ts (the rest were accounted for by Yak-1s, Yak-7Bs and Yak-9s). Own losses amounted to 12 Yak-9Ts, or a third of all the Yak fighters lost. On average, only 31 rounds of the NS-37 cannon were needed for the destruction of one enemy aircraft, while the corresponding figure for the 20-mm ShVAK cannon was 147 rounds.

The appearance of the Yak-9T at the front had a strong psychological effect on enemy aircrews. Until then, for example, the highly survivable and heavily armed Fw 190A had willingly undertaken head-on attacks against Soviet fighters. When the Yak-9T came into widespread use, the Fw 190As began to avoid head-on attacks.

The installation of the 37-mm cannon made it possible to increase substantially the distance at which fire could be opened. This proved very important because thereby the probability of the pilot being hit by return fire from the enemy bomber or reconnaissance aircraft was reduced. The fighter pilots could use this weapon effectively against twin-engined aircraft at a distance of 500-600 m (1,640-1,970 ft); in the case of enemy fighters being the target this distance did not exceed 400 m (1,310 ft). But if the purpose was to disrupt an enemy bomber formation, it was possible to open fire at a distance of 1,000-1,200 m (3,280-3,940 ft), providing the ammunition included fragmentation shells with time-delay fuses. Effective use of the Yak-9's heavy armament was severely hampered by imperfect design of gunsights and limited ammunition supply.

A pilot flying the Yak-9T was expected to fire his weapon only after aiming it carefully, and then only in short bursts of one or two rounds, three at the most. Firing long bursts meant spending ammunition to no avail (the fighter had only thirty or thirty-two 37-mm rounds on board); the strong recoil of the cannon caused the fighter to drop its nose and lose alignment with the target after a few shots, and the pilot had to take aim anew.

The flight stability when firing the N-37 depended on the flight speed and on the length of the burst (the greater the speed and the shorter the burst, the less was the influence of the cannon on the direction of flight). Lieutenant-Colonel Shinkarenko, CO of the 42nd IAP of the 240th Air Division, noted that the Yak-9T was appreciably more stable when firing than the LaGG-3-37 which his unit had operated earlier.

As the deliveries to front-line units increased, the Yak-9T came to be widely used and earned high praise from the flying personnel. As of 1st July 1943, only 14 machines out of the 153 Yak-9s in the front-line aviation had the 37-mm cannon, but shortly thereafter the 'tank-busting fighter' became one of the main types in the fighter element of the Soviet aviation – in all, 2,748 Yak-9T aircraft were built before the end of the war. Expeditious mastering of this machine in production and operational use played an important part in the winning of air superiority by the Soviet aviation.

Yak-9D long-range fighter

Even more widespread was the Yak-9D (***dahl'niy*** – long-range) – another version of the Yak-9 intended for large-scale production. This fighter featured as standard the fuel system borrowed from the Yak-7DI prototype. The installation of four fuel tanks holding a total of 650 litres/480 kg (143 Imp gal/1,058 lb) instead of 440 litres/320 kg (96.8 Imp gal/705 lb) on previous Yak-9 versions necessitated an increase in the amount of engine oil to 48 kg (109 lb).

The need for such a fighter arose in 1943 when Soviet ground troops began to undertake deep penetration of enemy defences. The rapid advance of the Red

Yakovlev OKB

Above: The experimental Yak-9-37 armed with a 37-mm cannon firing through the propeller hub.
Below: As this side view shows, the 37-mm cannon required the cockpit to be moved aft for CG reasons. These pictures were taken during the aircraft's State acceptance trials.

Yakovlev OKB

Yakovlev OKB

Above and below: The prototype of the Yak-9T production-standard 'tank killer' version (c/n 01-08) during State acceptance trials.

Yakovlev OKB

Specifications of Yak-9 fighter variants

	Yak-7DI	Yak-9	Yak-9T
Year of manufacture	1942	1943	1943
Powerplant	M-105PF	M-105PF	M-105PF
Power at altitude, hp	1,180	1,180	1,180
Length	8.48 m (27 ft 9¾ in)	8.5 m (27 ft 10½ in)	8.65 m (28 ft 4 in)
Wing span	9.74 m (31 ft 11½ in)	9.74 m (31 ft 11½ in)	9.74 m (31 ft 11½ in)
Wing area, m^2 (sq ft)	17.15 (184.6)	17.15 (184.6)	17.15 (184.6)
Empty weight, kg (lb)	2,360 (5,200)	2,277 (5,019)	2,298 (5,066)
All-up weight, kg (lb)	2,835 (6,250)	2,870 (6,327)	3,025 (6,668)
Speed at sea level, km/h (mph)	505 (313.8)	520 (323.1)	533 (331.1)
Speed, km/h (mph)	570 (354.1)	599 (372.2)	597 (370.9)
at altitude, m (ft)	3,900 (12,750)	4,300 (14,000)	3,930 (13,000)
Climb time to 5,000 m (16,400 ft)	5.5 min	5.1 min	5.5 min
Service ceiling, m (ft)	10,400 (34,000)	11,100 (36,500)	10,000 (32,750)
Turn time, seconds	17-18	16-17	18-19
Operational range, km (miles)	600 (372)	660 (410)	620 (385)
Take-off run, m (ft)	300 (980)	305 (1,000)	380 (1,250)
Landing roll, m (ft)	580 (1,900)	450 (1,480)	500 (1,640)
Armament, mm	1 x 20-mm ShVAK	1 x 20-mm ShVAK	1 x 37-mm N-37
	1 x 12.7-mm UB	1 x 12.7-mm UB	1 x 12.7-mm UB

Army formations and the absence of prepared airfields ahead of them created the danger of the Soviet aviation losing contact with the ground troops; the latter complained about not receiving sufficient air cover.

The Yakovlev OKB built the Yak-9D prototype in January 1943; by the end of February 1943 the machine had passed State trials at NII VVS (pilot V. Golofastov, leading engineer I. Rabkin). The tests showed that the Yak-9D had the same maximum speed as the other Yak-9s. However, the increase in all-up weight to 3,117 kg (6,873 lb) resulted in poorer manoeuvrability: the fighter needed 19-20 seconds to perform a banking turn and 6.1 minutes to climb to 5,000 m (16,400 ft); it gained 950 m (3,120 ft) in altitude during a combat turn.

Like the Yak-9T, this aircraft was immediately placed into large-scale production. As had often been the case previously, the decision to build the first 100 long-range fighters was taken before the State trials had even begun. These 100 Yak-9Ds were assembled before the end of the spring of 1943, whereupon the industry increased the output of such machines. It should be noted that in its directive dated 18th February the State Defence Committee had demanded that all fighters manufactured at plant 153 be fitted forthwith with four fuel tanks.

The first Yak-9Ds were assigned to the 20th and 18th Guards Fighter Air Regiments. Both units joined battle actively on 12th July, after the beginning of the Soviet counteroffensive against the German troops concentrated near Oryol (in the course of air battles over the Kursk Bulge). Low cloud and long spells of heavy rain hampered the operational activities of aviation. When the clouds lifted somewhat, the Yaks mostly participated in providing cover for attack aircraft. The fighters flew at altitudes of 200-400 m (660-1,310 ft), and the pilots often returned to base in machines riddled with bullets and shell fragments.

For example, between 12th and 27th July pilots of the 18th GvIAP commanded by Hero of the Soviet Union Lieutenant-Colonel A. Golubov flew 503 sorties (425 of them for the purpose of escorting Soviet attack aircraft) and waged 52 air battles in which only four Yak-9Ds were lost. The Soviet airmen claimed the destruction of 30 enemy aircraft; they came to know the strengths and weaknesses of the new Yaks very well. Therefore the command took a decision to conduct the Yak-9D's service tests in this unit.

During operations over the Western Front in August-September 1943 the regiment had, apart from three Yak-9Ds, 12 'regular' Yak-9s *sans suffixe* on strength. All types of aircraft were uniformly distributed among the three squadrons and flew roughly the same type of missions. Many sorties were flown by the airmen of the 18th GvIAP together with the 'Normandie' squadron which was deployed close at hand, forming part of the same 303rd IAD. At the end of August the French unit operated six Yak-9Ds, three Yak-9s (with two fuel tanks) and 11 Yak-9Ts. The pilots of the 'Normandie' squadron disapproved of having 'excess' fuel on board; at their own initiative they blanked off the fillers of outer wing fuel tanks, making it impossible to fill them with fuel, and made use only of the inboard fuel tanks.

In the course of service trials conducted in the 18th GvIAP, 58 flights were performed on the three Yak-9Ds (each of them with an average duration of 75 minutes); the regiment's pilots conducted seven dogfights, claiming two Junkers Ju 87 dive-bombers, two Fw 190s and one Heinkel He 111 bomber shot down. Of the three own machines one was shot down by the enemy, another one was damaged and sent for repairs. On 30th August 1943, Yak-9D c/n 0415317 (that is, Batch 04, plant No.153, 17th aircraft in the batch) received the following damage in the course of a dogfight with an Fw 190: a wing-root fuel tank was punctured by a cannon shell, the engine reduction gearbox casing was pierced by bullets and the fuselage was damaged in several places. Nevertheless, the pilot succeeded in bringing the machine back to base and landing on the airfield, which testified to the high survivability of the 'nine' in its 'four-tank' version.

Combat experience showed that it was inexpedient to use Yak-9Ds jointly with Yak-9s *sans suffixe* for fulfilling the same missions in fighter regiments where the vast

Yakovlev OKB

The prototype of the Yak-9-23 fighter armed with an engine-mounted 23-mm cannon. The aircraft type is painted on the tail in lieu of the construction number.

majority of aircraft had the standard two fuel tanks and were unable to stay airborne for several hours due to the limited amount of fuel. For instance, in the 303rd IAD the Yak-9Ds used up an average 270 litres (59.4 Imp gal) of fuel, or about 40% of the available amount, in each sortie. The remaining fuel merely increased the weight of the aircraft and made it more vulnerable compared to other Yaks.

Experience gained during service trials led to the conclusion that the Yak-9D would best be used for fulfilling special duties which could not be tackled by fighters carrying a limited amount of fuel (the Yak-1, Yak-7B and La-5). Yak-9Ds were widely used for escorting bombers on missions deep into the enemy's rear areas; they provided cover for tank and mechanised infantry groups making incursions into the enemy's rear. They were also used for patrolling flights of long duration. They proved useful in cases where an abrupt change in the weather while returning to one's own airfield proved difficult and it was necessary to divert to a remote alternate airfield which, for example, was not obscured by low clouds.

Undoubtedly the Yak-9D would have been more useful, had it not been for the absence of navigation equipment (in particular, the artificial horizon and the direction finder) which limited the possibility of poor-weather operations. The full use of the machine's potential was also hampered by the discrepancy between the aircraft's range in high-speed flight mode (905 km/562 miles) and the range of two-way radio communication with the ground (about 60 km/37.3 miles); in the course of virtually the entire flight the pilot could not receive assistance from a ground air control station.

In aerial combat with the German Fw 190A-4 and Bf 109G-6 fighters the Yak-9D had an advantage in horizontal manoeuvrability at altitudes up to 3,500 m (11,480 ft), especially when some of the fuel had been used up. On the other hand, the Soviet fighter was slower than the enemy machines, even at low altitudes.

In the summer of 1943 production of the Yak-9D began in earnest. While the workers of plant No.153 assembled 'four-tank' Yak-9Ds on one production line, another assembly line close at hand built the 'two-tank' Yak-9T. By the end of 1943 the Novosibirsk plant had built 706 long-range machines. In Omsk, 463 'four-tank' Yak-9Ds were built at plant No.166; the plant started providing the fighters with increased fuel capacity from September of that year on.

M-105PF-powered Yak-9s at war

At the end of 1943 the first machines of the 9th batches of the Yak-9D and Yak-9T

Yakovlev OKB

Above and below: Two Yak-9Ds pictured during checkout trials at NII VVS. Judging by the style of the national insignia, the aircraft on the upper photo is a later example.

Yakovlev OKB

respectively were tested at LII. As might be expected, the 'Ds proved to be somewhat heavier when fully equipped, the all-up weight being 3,070 kg (6,770 lb) versus 3,025 kg (6,670 lb). The Yak-9D was also a bit slower, the speed at sea level being 526 km/h (327 mph) and 532 km/h (331 mph) respectively, and the speed at the second rated altitude being 577 and 591 km/h (359 and 367 mph) respectively. The rate of climb was virtually identical, the fighters climbing to 5,000 m (16,400 ft) in 5.6 and 5.5 minutes respectively. On the whole, that kind of performance could not satisfy the Red Army command on the eve of major offensive operations.

In consequence, a modified Yak-9D (c/n 08-05) was submitted to NII VVS for tests; it had an all-up weight of 3,117 kg (6,873 lb) and featured improved aerodynamics in accordance with recommendations from TsAGI. The tests were conducted under the supervision of project engineer I. Rabkin. Flights performed in January and February 1944 by V. Golofastov showed that the fighter had become 20 km/h (12.4 mph) faster thanks to the sealing of fuselage and engine cowling joints, the generally improved finish of the wing surface and some other measures. The speed rose to 535 km/h (332 mph) at sea level and 591 km/h (367 mph) at 3,650 m (11,970 ft), while other performance characteristics remained virtually unchanged. Time to 5,000 m was 6.1 minutes, a 360° turn was performed within 19-20 seconds; the aircraft had a service ceiling of 9,100 m (29,850 ft) and a landing speed of 141 km/h (87.6 mph).

In early 1944 the Soviet command gave up the practice of sending front-line units to the rear for reorganisation; therefore, the majority of Yak-9s were sent to replenish the losses of other Yaks directly in front-line conditions. The combat activities of Soviet airmen during the Soviet summer offensive of 1944 in Belorussia (Operation *Bagration*)

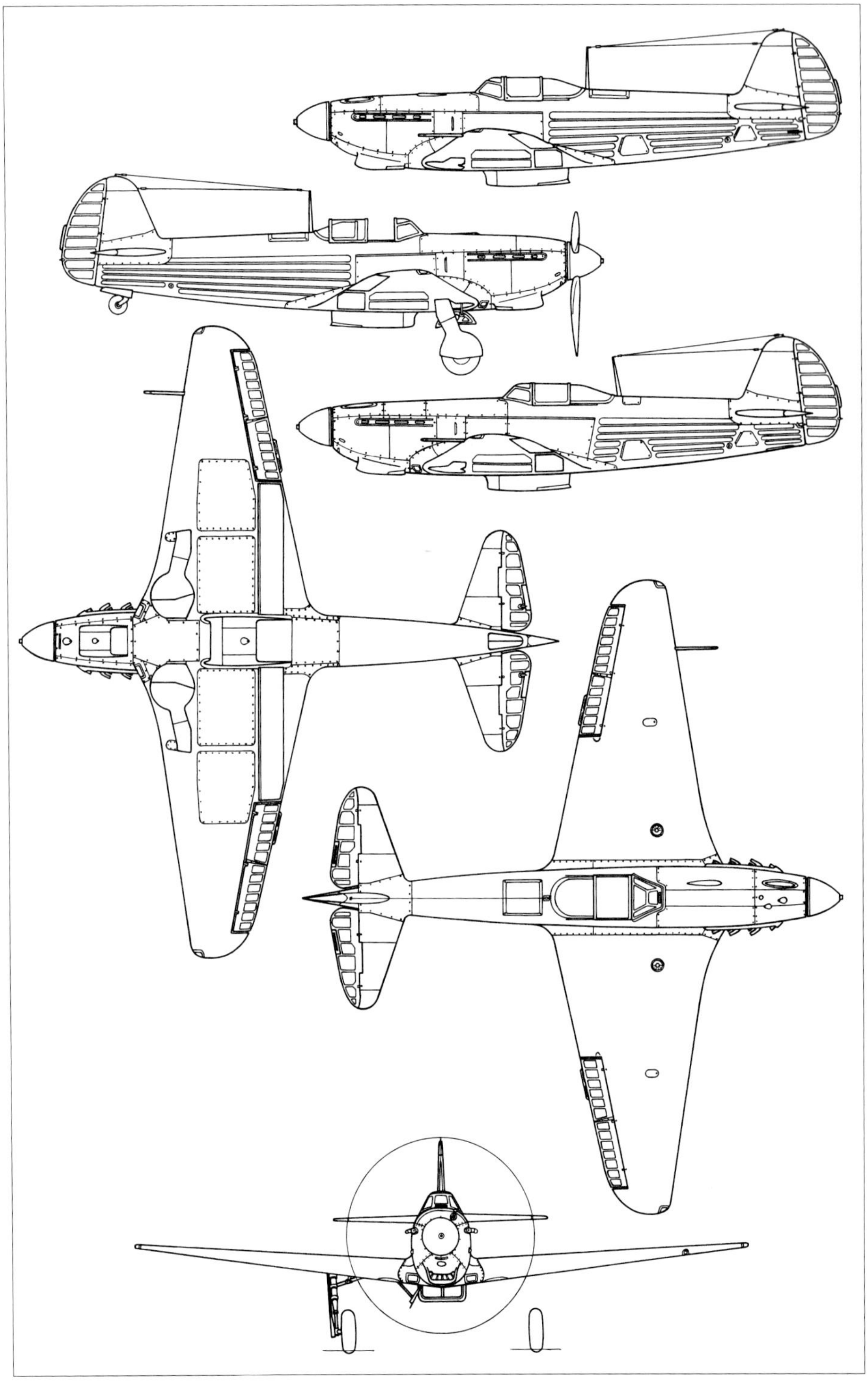

Four views of the Yak-9M long-range fighter with an RPK-10 DF, with an additional side view (third from the top) of the Yak-9M. Note the four fuel tank access panels in the lower view.

were conducted mainly in small groups. In contrast, during the L'vov offensive operation launched on 13th July 1944, the Soviet Air Force resorted to massed air strikes. The German opposition there appeared to be stronger, and a number of Yak-9 equipped air units and formations were waging continuous battles.

Numerous statistical reports dating back to 1944 make it possible to cite a few figures characterising the service rendered by different M-105PF-powered Yak-9 versions. Of the 6,289 Yak-9s manufactured that year, nearly 5,000 were sent to active Air Force units. During the same time 1,754 Yak-9s were written off for various reasons. 141 Yak-9s were shot down in air-to-air combat, 94 fell prey to anti-aircraft fire, 787 failed to return from a sortie for unspecified reasons, 12 were destroyed on the ground, 454 were lost in accidents and 266 were worn out completely. Thus, combat attrition of the Yak-9 amounted to 1,034 machines (the total combat losses of all basic Yakovlev types being 3,571 aircraft). The Yak-9's relative attrition rate was approximately one loss per 150 sorties, which was lower than that of the more advanced Yak-3 or La-7 – that was due to the latter two types being used at those sectors of the front where the fighting was more intense.

Yak-9TD and other Yak-9 upgrades

In the course of 1944 a number of substantial design changes were incorporated into the Yak-9 fighter on the production line. As an example, here are the upgrades introduced on the Yak-9T:

- from Batch 11 (February 1944) onwards all machines were built with four tanks with a total capacity of 480 litres (105.6 Imp gal). When the Nudel'man/Sooranov N-37 cannon was installed, the model was designated Yak-9TD;
- from Batch 13 (the end of March) onwards a two-way radio was fitted to all fighters, not to one in every two as before;
- from Batch 14 (April) onwards a push-button control was introduced for the radio transmitter. The switch was located on the throttle lever, enabling the pilot to easily maintain two-way communication in the air while keeping his attention on flying the aircraft;
- from the second half of Batch 14 (May) onwards the ballast installed in the aft fuselage was deleted (it had been introduced at an early stage for CG reasons) and the wing design of the Yak-9D and the Yak-9T was standardised;
- from Batch 15 (June) onwards an emergency canopy jettison system was introduced and dust filters were installed on the carburettor air intakes to ensure trouble-free operation from unpaved airfields;
- from Batch 16 (June) onwards the wire braces in the fuselage truss were altered, the wingtips were reshaped and the design of the identification lights was changed;
- provision was made on several Batch 17 machines (end of July) for the installation of the RPK-10 direction finder, for which purpose a hatch with a plywood cover was made in the upper fuselage decking;
- from Batch 19 (September) onwards a pneumatic engine starting system based on recommendations from LII was introduced;
- from Batch 20 (October) onwards the Huck starter dog intended for starting the engine by a truck-mounted starter was deleted and the spinner was give a more streamlined shape.

Yak-9M production fighter

The most significant alterations to the Yak-9 powered by the VK-105PF engine (as the M-105PF was redesignated in the spring of 1944) took place during the manufacture of Batch 25. The new fighter emerged as an upgrade of the Yak-9D, from which it differed mainly in having a fuselage patterned on the Yak-9T – that is, the cockpit was moved 400 mm (1 ft 3¾ in) aft. This version designated Yak-9M (*moderni**zee**rovannyy* – upgraded) was convenient in production, making it possible to standardise the design of fuselages and use them for manufacturing either the Yak-9T or the Yak-9D, depending on the needs of the Air Force.

In addition, on the Yak-9M the designers made major changes intended to enhance the fighter's combat efficiency and make it easier to operate. They succeeded in eliminating almost all the design flaws and manufacturing defects noted in various NKAP and Air Force reports. In particular, the wing strength problem was addressed; the Yak-9M's wings were reinforced by increasing the skin thickness, using Bakelite plywood and increasing the area of bonding between the skin and the wing framework etc. As a result, the required structural strength level was finally achieved. Furthermore, the engine cooling system featured an automatic device regulating the coolant temperature, and all machines with even construction numbers were fitted with direction finders.

Yefim Gordon archive

Above: A production Yak-9 with the tactical number '41 White' taxies out from an unpaved airfield in the second half of the war, creating a local dust storm. The yellow fin cap is a rapid recognition marling denoting that the aircraft belongs to a specific regiment.

Yakovlev OKB

Above: The prototype of the Yak-9TD which combined the Yak-9T's engine-mounted 37-mm cannon (and hence the aft location of the cockpit) with the increased fuel tankage of the Yak-9D. It is seen here during manufacturer's flight tests. Note that the width of the white border to the national insignia varied.

As regards fuel capacity and armament, the Yak-9M differed little from the Yak-9D, whereas the location of the cockpit was identical to that of the Yak-9T. Moving the cockpit aft in comparison with the Yak-9D had practically no adverse effect on forward visibility while being very useful in lessening the tendency to nose over. The Yak-9M did not differ from the Yak-9D and Yak-9T with regard to piloting techniques and basic performance. It may be mentioned that the modification was made directly at plant No.153 which was producing the Yak-9D and Yak-9T. The designation Yak-9M was allocated to all Yak-9s from Batch 25 onwards.

The first production Yak-9M underwent State acceptance trials at NII VVS in December 1944, with V. G. Ivanov as project test pilot and engineer G. A. Sedov in charge of the testing. At a weight of 3,095 kg (6,824 lb) the fighter attained a speed of only 528 km/h (285.4 mph) at sea level and 573 km/h (356 mph) at 3,750 m (12,300 ft). The rate of climb and manoeuvrability characteristics proved to be similar to those of other Yak-9s with the same all-up weight. Such performance could not be considered satisfactory for the end of 1944. Suffice it to say that the Yak-9M was some 50 km/h (31 mph) slower than a production Yak-3; it was also inferior to contemporary models of the Focke-Wulf and Messerschmitt fighters.

The Yak-9M was built in series at plant No.153 from May 1944 until June 1945, up to the moment when production of the Yak-9U powered by the VK-107A engine got under way there. The production run totalled 4,239 aircraft. In October 1944, when VK-105PF-2 engines (the -2 stands for 'boosted for the second time') with a take-off rating of 1,290 hp became available to plant No.153, Yak-9Ms began rolling off the production line with this more powerful engine. Until then, the entire output of the VK-105PF-2 had been reserved for installation on the Yak-3.

Yak-9M PVO (Yak-9 PVO) night interceptor

Designated in various sources as Yak-9M PVO or simply Yak-9PVO (***pro**tivovoz**doosh**-naya obo**ro**na* – air defence), this was a special version of the Yak-9M powered by the VK-105PF or 'PF-2 and intended for night intercept missions. Additional mission equipment included an RPK-10M radio compass, a gyro horizon and an FS-155

Yakovlev OKB

The Yak-9M featured an aft-positioned cockpit *à la* Yak-9T, even though it did not have the big cannon. This feature was introduced to ensure commonality between aircraft with different armament fits and simplify production at the Novosibirsk factory.

Above: Another Yak-9 with an aft-positioned cockpit pictured during checkout tests at NII VVS.

floodlight faired into the port wing leading edge, as well as an RSI-4M multi-channel radio and some other equipment items. This version was produced in limited numbers from early 1943 onwards.

Yak-9P fighter prototype

The next area on which the Yakovlev OKB concentrated was the Yak-9's armament, which needed to be reinforced. On one machine the designers installed a synchronised Shpital'nyy/Vladimirov ShVAK-20 cannon (also known as SP-20) with 175 rounds instead of the standard synchronised 12.7-mm UBS machine gun with 200 rounds. This did not entail any major modifications.

The performance of the new fighter designated Yak-9P (***push**echnyy* – cannon-armed) remained unchanged, while the weight of fire increased by 25% as compared to the Yak-9. The aircraft was regarded favourably by the specialists of NII VVS. Firing the synchronised cannon, as well as the synchronised machine-gun through the propeller disc was quite safe and the armament functioned reliably in all flight modes, regardless of the aircraft's attitude.

Initially the Government envisaged building 100 Yak-9Ps. In fact, however, this model never entered production because of the decision to use cannons of even greater calibre (for example, 37 and 45 mm).

Yak-9TK fighter prototype

The next step in the effort to enhance the fire-power resulted in the emergence of the Yak-9TK. It was intended specially for trial installation of different hub-mounted cannons. For the first time the possibility was provided to install alternative cannon types, depending on the demands of the Air Force: the 20-mm (0.78 calibre) ShVAK, the 23-mm (0.9 calibre) Volkov/Yartsev VYa, the 37-mm (1.45 calibre) Nudel'man/Sooranov NS-37 or the 45-mm (1.78 calibre) NS-45. The weight of fire in these cases was 1.997, 2.71, 3.74 and 4.07 kg (4.4, 5.97, 8.25 and 8.97 lb) respectively. To fit an alternative armament option, it was necessary merely to replace attachment fittings and ammunition feed devices suited to other types of rounds. Importantly, replacement of the weapon could be done under field conditions by the front-line units' personnel.

Pilot V. Khomiakov and armament engineer A. Aronov, who conducted the flight testing of the Yak-9TK in October 1943, came to the conclusion that the all-up weight, CG position, piloting techniques and performance depended wholly on the type of cannon armament installed.

When the ShVAK and VYa cannons were fired, their recoil had virtually no effect on the machine's behaviour, even at low speed. Firing the NS-37 had a more noticeable influence on piloting. At indicated airspeeds of 300-350 km/h (186-218 mph) the Yak-9TK rocked violently and accurate sighting was

The Yak-9TK development aircraft which incorporated provisions for installing engine-mounted cannons of various calibres. Here it is seen with a 20-mm cannon; this configuration was also known as the Yak-9T-20, hence the designation on the tail.

possible only during the first shot. Finally, firing the NS-45 cannon proved possible only in single shots at speeds close to the maximum.

Yak-9K production fighter

Versatile installations of cannon armament found their employment on the Yak-9 later, when the Yak-9U powered by the M-107A (VK-107A) engine entered mass production. Nevertheless, at the end of the war the NS-45 cannon attracted the attention of test pilots and engineers impressed by its exceptional capabilities. The work on installing this cannon on a fighter was continued on a machine which received the designation Yak-9K (***kroop****noka****lib****ernyy* – large-calibre). In this version the designers took into account the main peculiarity of the NS-45 cannon installation – the thin walls of its barrel and the very small radial gap between the cannon barrel and the hollow shaft of the engine gearbox, the shaft's bore being of a small diameter measuring merely 55 mm (2 11/64 in).

The recoil force of the 45-mm cannon exceeded that of the 37-mm cannon by 40%; to reduce it, for the first time in the Soviet Union the barrel was fitted with a large muzzle brake absorbing 85% of the recoil energy. The muzzle brake protruded noticeably from the propeller spinner, which increased the overall length of the Yak-9K to 8.87 m (29 ft 1⅛ in) as compared to 8.66 m (28 ft 4 61/64 in) for the Yak-9T and 8.50 m (27 ft 10½ in) for the Yak-9 *sans suffixe*.

Like the smaller NS-37, the NS-45 had continuous belt feed. It had an ammunition supply of 29 rounds and a counter in the cockpit showing the number of remaining rounds. As regards the weight of fire, the Yak-9K was superior to all Soviet fighters and the majority of foreign ones. Only such 'flying artillery batteries' as the Fw 190A-6/R1 or the Bf 109G-6/R6 boasted heavier armament. However, these two fighters had two or even four cannons mounted under the wings, which led to a substantial deterioration of performance and handling, while the Yak-9K was little different from the basic versions of the Yak-9: its all-up weight was only 3,028 kg (6,677 lb).

The Yak-9K prototype was completed by the Yakovlev OKB at the end of 1943; from the beginning of 1944 it underwent State trials at NII VVS. Building on the results of these tests, a batch of 53 machines was manufactured for the Air Force; almost all of them were assigned to the 3rd Fighter Air Corps commanded by Lieutenant-General Yevgeniy Ya. Savitskiy. These powerful fighters went into action for the first time in mid-August 1944, when German bombers (against which the Yak-9Ks were primarily

Yakovlev OKB

Yakovlev OKB

Above and below: Yak-9T c/n 0121 was converted into the prototype of the Yak-9K armed with a 45-mm NS-45 engine-mounted cannon. The large muzzle brake of the cannon was the chief recognition feature of this version.

Yakovlev OKB

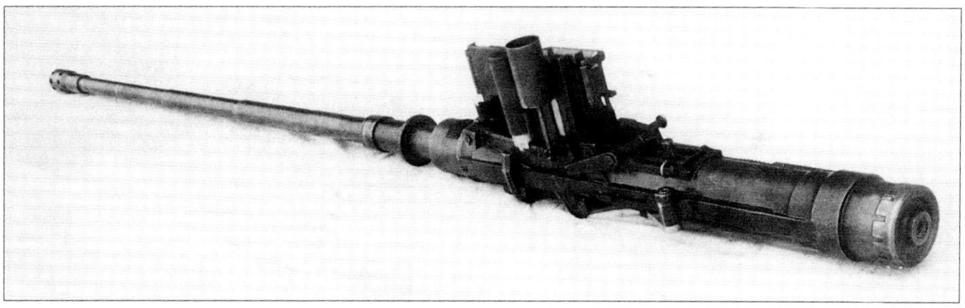

Above: The 45-mm Nudel'man/Sooranov NS-45 cannon was a potent weapon, allowing the Yak-9K to be used effectively against enemy tanks, but was difficult to use because of the strong recoil.

Yefim Gordon archive

A production Yak-9K. Note the badge of the 3rd IAP which operated this particular aircraft (a winged sword piercing a swastika).

Yakovlev OKB

Above: One of the prototypes of the Yak-9B (Yak-9L) fighter-bomber pictured during manufacturer's flight tests. At this angle it is identical to the standard Yak-9D.

Yakovlev OKB

Above: A Yak-9L on a snow-covered hardstand at NII VVS during State acceptance trials.

Yakovlev OKB

Another Yak-9L (c/n 0975), also photographed at NII VVS in the early spring of 1944.

intended to be pitted) had virtually ceased their activities on the Eastern front, their role being taken over by fighter-bombers (Fw 190s or, less often, Bf 109s).

In consequence, a large percentage of the Yak-9K sorties was accounted for by ground-attack missions. One well-aimed 45-mm shell was usually enough to set a truck, a wooden house, a locomotive or other similar targets ablaze. Major A. Nikashin from the 812th IAP, a participant of the service trials, defined the tactics of combating enemy aircraft in the Yak-9K as follows:

'Yak-9 fighters should be used in cooperation with Yak-3 light fighters (described below – *Auth.*) *flying top cover. Tangling with fighters is inadvisable for the Yak-9Ks because they are heavy and, owing to the insufficient engine power, have poor vertical component* (vertical manoeuvrability – *Auth.*). *On encountering bombers, the Yak-9K fighters should make a surprise attack from behind the clouds or out of the sun, trying to disrupt their formation. It is expedient to make the first attack from above at a distance of 400-600 m* [1,310-1,970 ft]. *If any part of a bomber is hit by one or two shells, this is enough for the bomber to be destroyed.'*

Pilots reported that the Yak's airframe could withstand a lot of punishment. One of the Yak-9Ks was hit by numerous shells from a 20-mm Oerlikon cannon; in addition, it lost a piece of skinning measuring nearly 1 m^2 (10.75 sq ft). Despite this serious damage, the Yak-9K covered a distance of more than 100 km (62 miles) and landed safely at its home base.

Along with its merits, the Yak-9K inevitably displayed some shortcomings as well. The powerful recoil of the NS-45 cannon seriously affected the aircraft's structure and caused leakage of water and oil through various seals and cracks in piping, radiators and the like. The cannon's insufficient reliability prevented the Yak-9K from entering large-scale production. Subsequently this work was continued with the improved Nudel'man N-45 cannon – its development was carried out on the Yak-9U.

Yak-9B (Yak-9L) fighter-bomber

Limited production was undertaken of a version designated Yak-9B (*bombardi**rov**shchik* – bomber). Bearing the factory designation Yak-9L, it was a modified production Yak-9D (c/n 1420). The modification consisted in fitting four bomb bays behind the cockpit (in the space previously occupied by the rear cockpit on the Yak-7). The bomb bays were arranged in two pairs in tandem, housing either four 100-kg (220-lb) FAB-100 high-explosive bombs or four cassettes with small PTAB 2.5-1.5 anti-armour bomblets weighing about 2 kg (4.4 lb) each. The cannon and machine-gun armament was identical to that of the Yak-9D.

The addition of bomb bays made it possible to extend considerably the range of combat missions performed by the 'nine'. Without bombs the Yak-9B could be used as a tactical fighter; with bombs it became a high-speed fighter-bomber intended for pinpoint attacks against well-protected targets.

For CG reasons and out of concern for the longitudinal static stability the bomb load had to be restricted to 200 kg (441 lb). In this configuration the all-up weight reached 3,356 kg (7,400 lb), and the aircraft could be flown by pilots possessing satisfactory piloting skills. A bomb load of 300-

400 kg (661-882 lb) was considered to be an overload configuration. Only highly skilled pilots were permitted to fly the Yak-9B with this bomb load, and then only in special circumstances.

The Yakovlev OKB managed to convert the tactical fighter into a fighter-bomber fairly quickly – by March 1944 the Yak-9B was prepared for flight-testing and subsequent thorough evaluation. State acceptance trials were conducted in four stages and lasted throughout the summer of 1944. The aircraft's stability and handling characteristics were determined and spinning trials were held during this period.

After that, plant No.153 manufactured a batch of 109 Yak-9Bs. The aircraft were delivered primarily to the 130th IAD (Fighter Division) commanded by Colonel Fyodor Shinkarenko in the autumn of 1944. Two of its regiments (the 168th and 909th IAPs) were equipped with Yak-9Bs, while the newly formed 409th IAP received Yak-9D and Yak-9M fighters.

The type was committed to action for the first time on 22nd October 1944 when 53 Yak-9Bs provided escort for attack aircraft of the 311th ShAP (*shtoormovoy aviatsionnyy polk* – Attack Air Regiment) in the area of Gumbinnen and themselves attacked targets of opportunity. At the end of December in the course of one sortie 120 Yaks from the 130th IAD took off on a mission; they encountered no visible opposition, yet two machines were lost and three more damaged. Later it transpired that the small PTAB bombs that had been loaded into the bomb bays collided in the air after being dropped and exploded under the aircraft. In the course of combat activities up to 20th February 1945 the Yaks performed nearly 2,500 individual sorties on attack and bombing missions. According to information from Soviet sources, they succeeded in destroying 29 tanks, 11 armoured personnel carriers, more than 1,000 soft-skinned vehicles, and many enemy depots and trains.

Despite these impressive results, the Yak-9B was adjudged as unsatisfactory on the basis of the service tests. The absence of a special sight for precision bombing prevented the Red Army Air Force from getting an efficient fighter-bomber, and many pilots noted that the bombed-up Yak-9B was difficult to fly. Much vexation was caused by the bombs refusing to leave the bomb bays when dropped in a shallow dive, as well as by difficulties in loading the bombs. Besides, fighter pilots lacked precision bombing skills, since they were not specially trained for bombing missions.

Air Marshal A. Novikov, Commander of the Red Army Air Force, considered it necessary to write when endorsing the service

Yakovlev OKB

Above: A production Yak-9B. Outwardly the fighter-bomber version differed from the Yak-9D in having a lengthened and reshaped rear canopy section associated with the provision of a bomb bay aft of the pilot's seat.

Yefim Gordon archive

Above: Despite the relatively small size of the Yak-9B's airframe, the capacity of its bomb bay was surprisingly large, as this photo testifies. Here the aircraft disgorges a load of 2-kg (4.4-lb) PTAB shaped-charge armour-piercing bomblets. In Soviet Air Force slang these munitions were dubbed *kapoostka* (Li'l Cabbage) because the smoke puffs created by the explosions resembled a head of cabbage when seen from the air. Note the open cockpit canopy.

Yefim Gordon archive

The Yak-9B's bomb bay seen from above, with two of the four FAB-100 bombs that could be carried by the aircraft peeping through holes in the upper stiffener of the bomb bay. Note the cables controlling the bomb release mechanisms

Specifications of Yak-9 fighter variants

	Yak-9D	Yak-9K	Yak-9B
Year of issue	1943	1944	1944
Powerplant	M-105PF	M-105PF	M-105PF
Power at altitude, hp	1,180	1,180	1,180
Length	8.5 m (27 ft 10½ in)	8.87 m (29 ft 1 in)	8.5 m (27 ft 10½ in)
Wing span	9.74 m (31 ft 11½ in)	9.74 m (31 ft 11½ in)	9.74 m (31 ft 11½ in)
Wing area, m² (sq ft)	17.15 (184.6)	17.15 (184.6)	17.15 (184.6)
Empty weight, kg (lb)	2,350 (5,180)	2,291 (5,050)	2,382 (5,251)
All-up weight, kg (lb)	3,117 (6,871)	3,028 (6,675)	3,356 7,398)
Speed at sea level, km/h (mph)	535.0 (332.4)	518.0 (321.8)	507.0 (315.0)
Speed, km/h (mph)	591.0 (348.5)	573.0 (356.0)	562.0 (349.2)
at altitude, m (ft)	3,650 (12,000)	3,900 (12,750)	3,750 (12,250)
Climb to 5,000 m (16,400 ft), min	6.1	5.7	6.5
Service ceiling, m (ft)	9,100 (29,750)	n.a.	8,600 (28,250)
Turn time, seconds	19-20	n.a.	25-26
Operational range, km (miles)	905 (562)	850 (528)	n.a.
Take-off run, m (ft)	370 (1,213)	305 (1,000)	440 (1,443)
Landing roll, m (ft)	550 (1,804)	450 (1,476)	580 (1,902)
Armament	1 x 20-mm B-20	1 x 45-mm NS-45	1 x 20-mm B-20
	1 x 12.7-mm UB	1 x 12.7-mm UB	1 x 12.7-mm UB

Yakovlev OKB

Above: Yak-9 c/n 0989 was effectively the prototype of the Yak-9DD long-range escort fighter. It is seen here at NII VVS during State acceptance trials.

Yakovlev OKB

A production Yak-9DD undergoing State acceptance trials. The taller aerial mast aft of the cockpit and the additional short aerial mast atop the fin were the main identification features of this version.

tests report: *'The Yak-9B aircraft is suitable only for very limited combat use; it is inexpedient to issue it to front-line units. The flying personnel flies this type of aircraft very reluctantly. It is necessary to ask the designer to improve the aircraft.'* After this appraisal further work on the machine was abandoned.

Yak-9DD (Yak-9Yu) long-range escort fighter

Yet another version of the Yak-9 was the Yak-9DD (***dahl'****nevo* ***dey****stviya* – long-range, the factory designation being Yak-9Yu). It embodied the concept initiated by the Yak-7D – creating a fighter with still greater range. The work was undertaken in response to a requirement for a fighter with a range exceeding 2,000 km (1,243 miles) issued in 1944. The special feature of the Yak-9DD was its wings, which incorporated eight metal fuel tanks and featured strengthened ribs. The fuel capacity of 845 litres (186 Imp gal), which equals a 630-kg (1,390-lb) fuel load, enabled the fighter to cover a distance of 1,325 km (768 miles) in high-speed flight and attain a range of 2,285 km (1,420 miles) in economic cruise.

The armament of the Yak-9DD was reduced to a single 20-mm ShVAK cannon. The Yakovlev OKB paid much attention to special equipment: the aircraft was fitted with a gyro horizon, an RPK-10 DF, an American SCR-274N radio set and oxygen bottles of increased capacity. The new radio equipment ensured reliable two-way communication at a distance of up to 150 km (93 miles) with the aircraft flying at 1,000 m (3,280 ft) and a reception range of 300 km (186 miles) at an altitude of 7,000 m (22,960 ft).

Despite its greater all-up weight of 3,387 kg (7,468 lb) accompanied by heavier handling and some deterioration of manoeuvrability, the Yak-9DD could still operate from unprepared airfields. The long range and endurance enabled it to be used as escort fighter, as well as for performing independently special duties in the enemy's rear.

The Yak-9DD passed State trials and was built in series from May 1944 onwards; a total of 399 were delivered. Production machines differed from the prototype mainly in navigation equipment, which had to be simplified, imported equipment items being replaced by locally-produced ones.

The first 40 production Yak-9DDs underwent service tests in the 386th IAP (commanded by Major M. Zhoolin) in the course of operations aimed at liquidating enemy troop concentrations in East Prussia. The Yak-9DDs were used mainly for escorting Pe-2 and Tu-2 bombers. Providing cover to the latter proved to be a fairly difficult task because the fast Tu-2s could rival the Yak-9DDs in speed.

Yakovlev OKB

Yakovlev OKB

Top and above: '29 White' was one of the five Yak-9PD interceptors powered by M-105PD engines. The long, deep oil cooler gave the aircraft a somewhat Fairey Firefly-ish appearance.

The high all-up weight, the considerable spanwise distribution of the weight and the inadequate engine output for this type of aircraft were the main causes for the deterioration of performance, handling and manoeuvrability. Therefore the pilots were very reserved in their appraisal of the Yak-9DD, dubbing it, rather unkindly, a 'flying cistern'. Maintenance of the overweight machine proved fairly troublesome, too: there were such nuisances as the short service life of wheel tyres and frequent failures of the tailwheel fork. Uneven consumption of fuel caused air ingestion into the carburettor, and there were cases of the engine cutting out in flight.

On the other hand, the Yak-9DD acquitted itself well as a long-range fighter in August 1944. A group of 12 Yak-9DDs commanded by Major I. Ovcharenko flew nonstop from Bel'tsy in Moldavia to Bari, Italy, in order to render assistance to the People's Liberation Army of Yugoslavia, covering a distance of 1,300 km (808 miles) without external tanks. Accompanying the group as a 'pathfinder' aircraft was a Douglas A-20G Boston bomber with a crew captained by M. A. Nyukhtikov, an experienced NII VVS test pilot. This successful long-distance flight and the subsequent missions associated with escorting Douglas C-47 transports to Yugoslav territory liberated by partisans demonstrated the high operational and flying qualities of the Yak-9DD.

Yak-9PD high-altitude fighter

The Yak-9PD signified the continuation of work on a high-altitude fighter initiated by the Yakovlev OKB with the I-28 and Yak-7PD. The OKB embarked on this work in November 1942, but, owing to delays with the delivery of M-105PD engines, the prototypes were not completed before April 1943. Structurally these were normal production Yak-9s with minor changes effected by the OKB and with the armament restricted to one 20-mm ShVAK cannon.

All five Yak-9PDs that were built underwent operational evaluation in the 12th GvIAP commanded by Major K. Marenkov which was part of the fighter element of Moscow's PVO system. By 25th June the high-altitude fighters had made 69 sorties; in 39 of them the machines climbed to altitudes in excess of 10,000 m (32,800 ft). The tests showed that the design was still immature. In particular, a continuous climb in optimum climb rate mode proved impossible: at 7,000 m (22,960 ft) the water and oil temperatures exceeded the maximum values permissible for this engine. In order to continue flight it was necessary to level off from time to time, allowing the engine to cool down. This greatly reduced the fighter's real rate of climb.

The report stated that the Yak-9PD had failed to pass the operational evaluation. This was due to powerplant deficiencies, the insufficient service ceiling for this class of aircraft (only 11,650 m/38,220 ft) and inadequate firepower.

It was high-altitude reconnaissance aircraft, primarily the Ju 88R-1 equipped with superchargers and a pressurised cockpit, that were the main opponents of Soviet fighters engaged in the air defence of Moscow in the summer of 1943. These aircraft repeatedly put in an appearance over the city. For example, on 2nd June 1943 Lieutenant-Colonel L. Sholokhov, an Air Corps piloting skills inspector, scrambled from the capital's Central airfield (Khodynka) to intercept an unidentified aircraft. The Soviet pilot quickly caught up with the adversary, while being some 1,000-1,500 m (3,280-4,920 ft) below the latter's flight level. He saw the yellow outer wing panels and the dimly visible Balkenkreuze painted on the wings of the Junkers machine. Sholokhov was on the point of opening fire when the oil pressure dropped all of a sudden and the cockpit windshield iced up. Having lost sight of the enemy, the Soviet pilot gave up the

Yakovlev OKB

Above and below: This Yak-9PD was fitted experimentally with am M-106PV engine optimised for high-altitude flight. Apart from the new powerplant, the aircraft had new wings of increased span and area.

Yakovlev OKB

chase and headed for home. Other attempts to intercept the Ju 88R-1 also proved abortive. On 23rd August G. Gromadin, Commander of the Western Front's air defences, reported to the Air Force command that no practical solution had yet been found in the question of creating a high-altitude fighter.

However, the Yakovlev OKB continued its quest for the stratosphere, working in concert with LII. One modified example of the fighter was fitted with an experimental M-105PD engine in which the gear ratio of the engine-driven centrifugal supercharger was increased from 8.48 to 9.72. In addition, steps were taken to raise the altitude efficiency of the engine cooling, oil and fuel systems. The prototype Yak was provided with wings of greater area, the span being increased by 1 m (3 ft 3⅜ in). This, coupled with an all-up weight reduced to 2,845 kg (5,479 lb), created favourable prerequisites for increasing the aircraft's service ceiling.

The aircraft attained an altitude of 12,500 m (41,000 ft), but many defects of the preceding model had not been rectified. Then the designers re-engined the machine with the prototype M-106PV (PV stands for *po**vysh**ennaya vy**sot**nost'* – enhanced high-altitude performance) which had the rated altitude increased from 8,500 to 9,500 m (from 27,880 to 31,160 ft). On 16th October 1943 a Yak-9 fitted with this engine reached an altitude of 13,100 m (42,970 ft).

In 1944 the Yakovlev OKB tried once more to develop a high-altitude interceptor, this attempt being the most successful so far. As before, the main means of raising the aircraft's service ceiling were to improve the powerplant's operation at high altitudes, as well as to further lighten the aircraft. The new version of the fighter was based on the airframe of the Yak-9U (described below) in which the VK-107A engine was replaced by the M-106PV. The latter was fitted with a device for cooling the air that had passed through the supercharger; this was not an intercooler – the desired effect was achieved by injecting a 50/50 mixture of alcohol and water. An additional pump in the fuel system ensured normal fuel feed at high altitude. Numerous weight-saving measures brought the airframe weight down to 2,500 kg (5,512 lb), which was an unprecedentedly low figure for the Yak-9.

The cooling systems ensured the possibility of a continuous climb up to the service ceiling of 13,500 m (45,280 ft) without intermediate levelling off. The success of this work made it possible to proceed to the manufacture of a series batch of 30 Yak-9 M-106PVs. (According to one account, they were to receive M-105PV engines and were based on Yak-9U airframes; only 12 examples were actually completed.) However, there was no opportunity to check their operational qualities because no overflights of Moscow by German aircraft were noted in the summer of 1944.

All high-altitude versions of the Yak-9 had one drawback in common – it was the lack of a pressurised cockpit. An experimental pressurised cockpit designed by A. Shcherbakov was not ready for installation in a fighter until May 1944, and even that proved to be faulty, as revealed by subsequent testing.

Yak-9R tactical reconnaissance aircraft

The Yak-9 served as a basis for the development of the Yak-9R reconnaissance fighter (R = *raz**ved**chik*). This version was produced both in the short-range configuration based on the 'two-tank' Yak-9 and in the long-range configuration based on the Yak-9D. In the latter case the machine-gun and its ammunition supply were often deleted to save weight. The photographic equipment comprised AFA-IM or AFA-3S/50 cameras installed aft of the cockpit.

The fighters were often converted into photo reconnaissance machines in the field, reflecting the need for relatively fast and manoeuvrable aircraft for fulfilling such missions in the enemy's rear areas adjoining the front-line. Additionally, 35 Yak-9Rs were built new at plant No.166 in Omsk in the summer of 1943. Some of them were delivered to the 48th GvRAP DD (*Gvar**dey**skiy raz**ved**yvatel'nyy **avia**polk **dahl'**nevo **dey**stviya* – Guards Long-Range Reconnaissance Air Regiment) of the Supreme Command (led by Lieutenant-Colonel Sadov) for operational evaluation.

On the whole, the results of the service trials were positive, but they showed that the Yak-9R, for all its high speed and manoeuvrability, could not supplant the Petlyakov Pe-2R in the reconnaissance air units; they could only supplement them when the enemy opposition in the air was particularly strong. After all, one airman whose attention was occupied by flying the aircraft could not make as many observations as a well-trained crew comprising three persons.

Yak-9R air defence version

In addition to the versions described above, at the end of the war the Yakovlev OKB developed a special air defence version intended for operations at altitudes of 10,000 m (32,800 ft); it featured a more substantial equipment fit, including a DF for navigation. This work drew on the experience gained in developing the Yak-1 and Yak-7B. This version was also built in series; no separate designation is known.

Yak-9S fighter

The Yak-9S emerged as one of the last versions developed in the process of updating the armament of the M-105-powered Yak-9. Apart from installing the VK-105PF-2 engine with additional boost and introducing the new VISh-105SV-01 propeller with blades featuring airfoil-section root cuffs, Yakovlev continued to develop promising new weapon installations. The Yak-9S had an engine-mounted 23-mm Nudel'man/Sooranov NS-23 cannon and two synchronised 20-mm Berezin B-20S cannons in the front upper decking.

This armament variant was considered to be promising and met the Air Force requirements for 1945. Two examples of the aircraft were completed literally on the eve of VE-Day; manufacturer's tests and State trials conducted in the summer of 1945 confirmed the soundness of the Chief Designer's concept. The weight of fire recorded during the tests was 4.23 kg (9.3 lb), which enabled the fighter to engage effectively both aerial targets and soft-skinned ground targets. Yet, since the basic flying characteristics of the Yak-9S were appreciably inferior to those of the Yak-3 and La-7 fighters, it was not put into series production. Yakovlev continued experimenting with this armament layout on fighters powered by the VK-107A engine.

Yak-9V two-seat conversion (familiarisation) trainer

The work on a two-seat conversion (familiarisation) trainer initiated before the war with the UTI-26 was continued with the Yak-9. In contrast to its predecessor, the Yak-7V, the new trainer designated Yak-9V retained the fighter's retractable undercarriage and was armed with a single engine-mounted 20-mm ShVAK-20 cannon. The new Yak featured a more substantial complement of equipment, which was important for training future pilots.

As distinct from the Yak-7V, the aircraft was provided with an intercom, a two-way radio which could be operated by the trainee and the instructor alike, an updated artificial horizon and some other instruments. True, the range of equipment was sufficient only for performing flights in daytime at altitudes up to 4,500 m (14,760 ft) because there was no provision for oxygen equipment. As regards the basic flight performance, the Yak-9V was little different from the Yak-9M, which was a distinct advantage for the training of flying personnel.

The Yak-9V trainer entered production after the end of the war. In all, 456 such machines were built new at plant No.153 and a further 337 trainers were converted from Yak-9Ms.

Yakovlev OKB

Above and below: The Yak-9S prototype during State acceptance trials at NII VVS. This was one of the last attempts to reinforce the fighter's armament without radical changes to the airframe and powerplant.

Yakovlev OKB

Yak-9V as a radiation reconnaissance aircraft

When the Soviet Union embarked on the development of nuclear weapons the need arose for aircraft capable of taking air samples during nuclear tests. In 1945 five Yak-9Vs were converted into radio-controlled unmanned aircraft intended for monitoring radiation levels. For taking air samples the Yak-9s were fitted with so-called 'barrels' (air sampling canisters with mesh filters and radiation measuring instruments). Two of these were mounted on the upper wing surface and one more above the fuse-

Yefim Gordon archive

The prototype of the two-seat Yak-9V at NII VVS. The aircraft had no aerial mast fitted at this stage.

Yefim Gordon archive

Above: A production Yak-9V trainer undergoing checkout tests; note the aerial mast installed on the starboard side.

Yakovlev OKB

Above and below: Reminiscent of 'war weary' North American P-51Ds converted into squadron hacks, the Yak-9 'Courier' was a custom-built high-speed VIP transport with comfortable cockpits. Note the stripes on the propeller blades, as applied to the I-26 prototypes before the war.

Yakovlev OKB

lage ahead of the cockpit. The Yak-9s were to be controlled from a Tu-2 bomber converted into an airborne command post.

The modified Yak-9Vs were transferred to the nuclear test range near Semipalatinsk (Kazakhstan). Two of the Yaks were lost during the preparations for the mission. The operational use of the remaining three machines is the subject of a controversy. One of the sources claims that a Yak-9V flew through the nuclear 'mushroom cloud' during the first Soviet nuclear test on 29th August 1949. Another source asserts that the Yak-9V radiation monitoring aircraft, together with their Tu-2 drone director aircraft, were ferried to the Semipalatinsk area as late as 1951. During the flights for simulated air sampling difficulties arose with the radio-controlled landing of the Yak-9s; a team of experts discovered deficiencies in the automated flight control system developed for these aircraft. Further work on unmanned aircraft for monitoring nuclear radiation was halted.

Yak-9 'Courier' liaison aircraft

This aircraft was intended for the transportation of one passenger over a considerable distance under the conditions of possible enemy fighter opposition. Structurally it was a Yak-9V fuselage mated to the wings of the Yak-9DD; it was a prototype VIP transport aircraft developed on the basis of production Yak-9 machines.

Yakovlev appointed N. Avtsyn as project engineer for this aircraft. Under his direction a certain amount of reworking was effected by July 1944; it was restricted primarily to ensuring a certain level of comfort for the pilot and the passenger during flights of long duration. Thus, the 'Courier' aircraft was fitted with comfortable seats, both cockpits were provided with urinals. The passenger's cockpit was stripped of its second set of controls and the instrument panel; instead, it was provided with cloth upholstery, a false floor, and pockets for maps and a map case on the sidewalls. In other words, measures were taken to ensure that the crew would feel very comfortable in this case, as distinct from the generally spartan style of crew accommodation in Soviet military aircraft.

The aircraft was not built in series and no documents were found testifying to its operational use.

Yak-9 control system testbed

At least one Yak-9 (the version is unknown) is on record as having been converted into a flying testbed which was used for testing new control systems with hydraulic actuators. The tests were successfully conducted by a design bureau headed by V. A. Kuznetsov at an experimental factory subordinated to TsAGI. As one may gather from the source, they took place in 1946-47.

VK-105PF-2-powered Yak-9U production fighter

The Yakovlev OKB was well aware of the fact that the Yak-9's performance, despite all its merits, was falling short of the demands that should be met by a fighter at the end of the war. From the outset of series production the Chief Designer endeavoured, as far as possible, to improve the design without introducing basic changes into production techniques in the manufacture of the Yak-9. It was with these thoughts in their minds that the designers embarked on the development of a new production model for the following year in November 1943.

This version of the Yak-9 marking the peak of its development received the 'U' suffix signifying '***ulooch**shennyy*' (improved). It embodied, first and foremost, the rich experience in the perfecting of aerodynamic design that had been accumulated by the design staff. Furthermore, the efforts were directed at reducing the all-up weight, making the aircraft more reliable in operation and improving the pilot's working conditions. Many of the novel features were borrowed from the successful Yak-1M which is described later.

The following main differences from the production Yak-9 can be noted. The wing centre section was altered to house an oil cooler in the same fashion as on the Yak-1M. The fabric skin of the fuselage was replaced by 2-mm ($^{45}/_{64}$ in) plywood; the sealing of the fuselage was improved and the aircraft was made more easily controllable while taxying.

The pilot's armoured seatback was cut down, a bulletproof windshield and a rear bulletproof glass panel were installed, plus an armoured headrest and armrest; all this provided excellent protection for the pilot in combat. The engine's output at low altitudes was increased by raising the boost pressure from 1,050 to 1,100 mm Hg. A further speed increase was obtained by fitting a propeller with blades featuring airfoil-section root cuffs.

The Yak-9U featured new, capacious water and oil radiators; the carburettor air intake design was changed in accordance with recommendations from TsAGI. The self-sealing coating of the fuel tanks had variable thickness so as to cater for both survivability and weight saving.

In Aleksandr S. Yakovlev's opinion, the optimum armament complement should include an engine-mounted 23-mm cannon and two 12.7-mm UBS machine-guns. The only 23-mm cannon available in November 1943 was the VYa-23; accordingly, it was installed between the cylinder banks. Installation of another cannon of the same calibre – the more lightweight and rapid-firing NS-23 – was done later on the Yak-9S fighter. Regrettably, this weapon came too late to be fired in anger by the Yaks.

However, the VYa-23 was also a potent and reliable weapon which had proved its worth on the IL-2 attack aircraft. The high muzzle velocity of its shells, coupled with their high destructive power, allowed the cannon to be used not only in air combat but against ground targets as well. A direct hit against an enemy aircraft's fuel tank usually led to an explosion, and a shell hitting a wing tore a large hole in it. The shells of the VYa pierced 25-mm ($^{63}/_{64}$ in) armour at up to 400 m (1,310 ft). Yet, despite all its merits, the Red Army Air Force found the VYa cannon to be too heavy and insufficiently rapid-firing to be

Yefim Gordon archive

Above and below: '40 White' was a stock Yak-9D converted into a trainer by a service unit. The rudder is painted yellow.

Yefim Gordon archive

Yakovlev OKB

Above and below: The Yak-9U VK-105PF-2 prototype. The lack of the chin-mounted oil cooler (the oil coolers are housed in the wing roots) and the aft position of the water radiator gave the fighter a much more 'racy' appearance.

Yakovlev OKB

Yakovlev OKB

Yakovlev OKB

Top and above: Another Yak-9U VK-105PF-2 prototype. This aircraft had a non-standard sharply raked windshield fitted to reduce drag.

Yakovlev OKB

Above and below: The ill-starred Yak-9U VK-107A prototype. The aircraft was damaged beyond repair in a forced landing on 23rd February 1943 when the engine caught fire in a test flight.

Yakovlev OKB

used on fighter aircraft, and production was restricted to a trial batch.

The designers paid much attention to making the pilot's work as convenient as possible. Thus, the radio set was controlled by push-buttons mounted on the throttle lever. The placement of the equipment was carefully chosen and came close to the contemporary standard. Outwardly the Yak-9U differed from the Yak-1M only in having shorter main undercarriage legs with one-piece gear doors.

Pilot V. Khomiakov, who test-flew the aircraft in January 1944, noted that the Yak-9U did not differ from production machines as regards the piloting techniques, retaining good stability, light controls and pleasant handling. The fighter's performance was appreciably higher thanks to improved aerodynamics and greater engine power. Maximum speed rose to 558 km/h (347 mph) at sea level and 620 km/h (385 mph) at 3,850 m (12,630 ft). The weight was reduced to 2,900 kg, which afforded better vertical manoeuvrability – a climb to 5,000 m (16,400 ft) took only 4.8 minutes, less than it took the opposing Messerschmitt fighters. In a climbing turn the Yak-9U gained 1,190 m (3,900 ft) versus 1,200 m (3,940 ft) for the best of the Bf 109 versions, the G-10, which testifies to the machines being roughly equal in this respect.

The protocol on the results of the State trials contained a recommendation to introduce in production all the changes associated with improving the fighter's aerodynamics and lightening its structure as described above. Whereas the Yak-9U powered by the VK-105PF-2 embodied an evolutionary line of development, a 'revolutionary' trend was associated with fitting a much more powerful liquid-cooled engine. Such an engine existed – it was the Klimov VK-107A. As early as at the end of 1942 this engine had passed 50-hour bench running tests, albeit with great difficulties.

VK-107A-powered Yak-9U production fighter

The work on the VK-107A (initially known as the M-107A) was arduous: a number of technological breakthroughs had to be scored, which was difficult to achieve during the war years. The first flight-cleared engine was installed in a Yak-7 in December 1942. At the beginning of January the machine received new wings with metal spars and was redesignated Yak-9. Manufacturer's trials of the Yak-9 M-107A were conducted by Pavel Ya. Fedrovi. On 23rd February 1943 the well-known test pilot Pyotr M. Stefanovskiy performed the last flight at Chkalovskaya airfield before handing the machine over for State trials. That flight ended in an accident.

'Having climbed to 1,000 m [3,280 ft], *I levelled off and flew some distance in level flight over the airfield in the direction of Shcholkovo to determine the maximum speed,* – Stefanovskiy was to recall later. – *The machine was surging forward at an increasing speed. Glancing at the engine cowling, I noticed that smoke was pouring out of slits between the cowling panels. Then a sheet of flame erupted. The engine was on fire...'*

While attempting to land the burning machine, the pilot was knocked out by the impact. As for the aircraft, the damage sustained in the crash-landing was so heavy that the fighter had to be written off.

Despite this setback, Fedrovi wrote to People's Commissar of Aircraft Industry Aleksey I. Shakhoorin that the M-107A-powered Yak-9 was one of the best fighters at the moment and that it could be compared only to Polikarpov's I-185 prototype. The latter offered roughly the same speeds but had heavier handling and poorer manoeuvrability due to the higher airframe weight. Fedrovi asked Shakhoorin to issue an order requiring M-107A engines to be installed on 10 to 15 production Yak-9 fighters for operational evaluation purposes. But the abovementioned accident and difficulties encountered in ironing out the bugs of the powerplant delayed the work by at least seven months.

In April 1943 eight Yak-9D airframes were fitted with M-107 engines. Attempts to eradicate the engine problems at the experimental plant proved unsuccessful, and four machines were transferred to LII for development work. Test flights resumed in early August, leading to the conclusion that the main reason for the unsatisfactory engine operation lay in the poor functioning of the ignition system. Between 4th August and 21st October LII test pilots performed a total of 64 flights and the technicians had to dismantle five engines out of seven because of serious defects. M-107A c/n 317-2 installed in one of the Yaks caught fire after 28 hours of running.

As noted earlier, many Soviet fighters were ordered into series production even before the testing began in earnest. This was exactly the case with the M-107A-powered Yak-9. The Government demanded that L. Sokolov, Director of plant No.166, should organise production of the new machine from June 1943 onwards and deliver 25 Yak-9 M-107As to the military by August.

It was no coincidence that the Omsk plant was chosen to fulfil this difficult and important task. In the opinion of Aleksandr Yakovlev, it was easier to get production under way in small workshops, provided that production routines were put under control and prescribed techniques strictly adhered to, coupled with seconding experi-

Above: Yak-9U c/n 0312 illustrates the dorsal air intake just aft of the spinner, the one-piece moulded windscreen and the seven exhaust stubs on each side characteristic of the VK-107A engine. The aircraft is seen here during checkout tests at NII VVS.

Above: Yak-9U c/n 25166021 was the first VK-107A-powered example built by factory No.166 in Omsk. It is seen here during State acceptance trials.

A standard production Yak-9U VK-107A pictured during checkout trials at NII VVS.

Boris Vdovenko

Above: A crew chief waves out a pair of Yak-9Us about to depart on a combat sortie. The nearest aircraft is equipped with a gun camera in front of the windshield.

Boris Vdovenko

A rare air-to-air shot of a pair of Yak-9Us equipped with gun cameras. Note the white fuselage band and the numeral 5 on the rudder of the example serialled '9 Yellow'.

enced engineers from the OKB to the plant. It would be easier to perfect the new machine under the conditions of limited production. However, this did not remove all the problems.

In the meantime, a new prototype was built in December 1943. Originally it was to be completed as a *dooblyor* of the Yak-9U VK-105PF-2 that had been transferred to NII VVS for testing. Yakovlev decided to make use of the experience gained with the new engine, albeit a negative one, and of the positive results shown by the latest machine. The fighter featured a modified engine mount, increased-area radiator cooling surfaces, and higher-capacity fuel and oil tanks. As distinct from the first Yak-9U prototype, the fighter was armed with the standard engine-mounted ShVAK-20 cannon. The all-up weight increased due mainly to the massive engine, reaching 3,150 kg (6,950 lb).

In the course of State trials which the aircraft passed in January-April 1944 (with A. T. Stepanets as project engineer and A. G. Proshakov as project test pilot) the aircraft attained a speed of 600 km/h (373 mph) at sea level and 700 km/h at 5,500 m (18,040 ft). The NII VVS report on the results of the State trials said: *'The Yak-9U powered by the VK-107A engine* (the M-107A had just been thus redesignated in March 1944 – *Auth.*) *is the best among the known Soviet and foreign fighters as regards basic performance characteristics in the range of altitudes from sea level to 6,000 m* [19,680 ft]'. However, it was also noted that it would be impossible to operate the fighter in the Air Force units unless the basic defects of the powerplant were rectified.

With regard to piloting techniques, the Yak-9 VK-107A remained just as simple in handling and easy to master for pilots of medium skill as the VK-105PF-powered Yaks.

Series production of the machine started at three factories as early as April 1944. Plant No.166, which by then had become one of the most advanced enterprises, was working at full capacity. It was there that the first production Yak-9U was built. It was not quite so refined aerodynamically as the prototype. At an all-up weight of 3,914 kg (8,630 lb) and at maximum engine speed (3,200 rpm) the fighter reached 562 km/h (349 mph) at sea level and 654 km/h (406 mph) at 5,150 m (16,892 ft). A climb to 5,000 m (16,400 ft) took 5.2 minutes. Thus, the characteristics proved to be sufficiently high, even though they fell short of the expectations.

However, production machines were plagued by the new engine's shortcomings to an even greater extent than the prototype. The main bugs of the VK-107A were the oil spill from the breather and from the front seal of the hollow gearbox shaft, oil pressure dropping below minimum permissible values as the altitude increased, vibration of the engine at low rpm and so on.

The main defect of the powerplant at that time was undoubtedly the propensity to water and oil temperatures exceeding the permissible limits. This occurred with the engine running in combat contingency mode (3,200 rpm) and, during hot summer months, also at nominal revs (3,000 rpm).

During level flight at maximum speed the oil temperature exceeded the limits even with the radiator shutters fully open. When gaining altitude in the optimum climb rate mode, the pilot had to level off repeatedly in order to bring the water and oil temperature down. Airmen in service units discovered that the operation of the aircraft was beset with great difficulties. It was far from always that the aircraft could demonstrate its performance to advantage.

The 42nd GvIAP commanded by Major Ya. Aleksandrovich was among the first to receive the Yak-9U VK-107A. Conversion of the unit proved to be a protracted affair which lasted until 25th August 1944. This was due in part to the fact that the Yak-9Us were powered by engines from the first batches manufactured at the end of 1943 and the beginning of 1944 – their cylinder banks were often damaged by the gases bursting through the seals. To enhance the airmen's confidence in the new materiel and provide a visual comparison of the manoeuvrability characteristics of the Yak-9U and the Bf 109G-4, the command arranged a kind of a contest. Captain L. M. Koovshinov, a NII VVS test pilot, arrived by air to a front-line airfield. It was he who piloted the captured German machine.

Despite all Koovshinov's skill, his opponents in mock combat – I. Gorboonov and G. Pavlov (both Heroes of the Soviet Union) from the 42nd GvIAP – emerged victorious from the single combat in vertical and horizontal manoeuvring. The Yak-9U gained 1,250-1,300 m (4,100-4,260 ft) of altitude in a combat turn and always proved to be positioned somewhat higher than the Bf 109G. These demonstration flights convinced the airmen of the entire regiment that they were flying the better aircraft.

The new fighters came to be used on a relatively wide scale in the late summer of 1944. The German command took note of these aircraft and their high operational qualities. On the other hand, maintenance of the Yak-9U in field conditions presented great difficulties. Continuous efforts were needed to overcome them at the least possible detriment to the fighter's performance. Thus, it was recommended that the use of the combat contingency rating be relinquished and flights be performed at not more than 3,000 rpm. This entailed a marginal decrease in speed and rate of climb but helped reduce the number of engine failures.

In October 1944 the command of the Red Army Air Force considered it possible to conduct combat evaluation of the Yak-9U. The 32 machines used for this were taken from the assembly line of plant No.82 in Moscow. The important task of conducting the service trials was entrusted to the airmen of the 163rd IAP in the 3rd VA (the regiment was commanded by Lieutenant-Colonel V. Ukhanov). The Yaks were mostly up against Fw 190As, 'Fs and 'Gs. In the course of the trials, which lasted nearly until the end of 1944, 398 individual Yak-9U sorties were put on record; they included 299 dogfights. According to the regiment's records, the Soviet airmen scored 28 victories for the loss of two aircraft (however, German documents do not confirm such high losses).

The Yak-9U demonstrated its complete ascendancy in combat with various versions of the Fw 190, especially during combat in the vertical plane. Commanders, pilots and ground crews of the 163rd IAP spoke highly of both performance and serviceability of the Yak-9U. They noted that the aircraft could fairly easily be mastered by young pilots and average-skilled technicians. Maintenance of the aircraft proved simple: preparation for a combat sortie did not exceed 30 minutes.

The enemy, too, became painfully aware of the Yak-9Us capabilities. Walter Wolfrum, a German ace who had flown the Bf 109G and scored 137 victories by the end of the war (which says a lot for his flying experience), noted: 'The best fighters that I have encountered in combat were the American P-51 Mustang and the Russian Yak-9U. Both of these types were clearly superior in their performance to all versions of the Bf 109, including the K. The Mustangs had unrivalled altitude performance, while the Yak-9U was a record-holder in rate of climb and manoeuvrability.'

It was the 4th VA that accumulated the most comprehensive experience with the Yak-9U. On 6th February 1945 I. Osipenko, the 4th VA's Chief Engineer, submitted a report enumerating the strengths and weaknesses of the new fighter. The merits included performance superior to all Bf 109 and Fw 190 versions at low and medium altitudes and the large speed envelope enabling the pilot to perform vertical aerobatic manoeuvres with ease. As for the shortcomings, he cited numerous cases of generator failures and breakdowns of the UBS machine-gun mounting brackets. The pilots of the 4th VA man-

Yakovlev OKB

Yakovlev OKB

This page: An Omsk-built Yak-9UT (c/n 40166022) seen during manufacturer's flight tests. The aircraft looked like a cross between the Yak-9U and the tank-busting Yak-9K. The muzzle brake of the 37-mm engine-mounted cannon is clearly visible, as is the dorsal carburettor air intake moved aft to a position about halfway between the spinner and the windscreen. Because of the Yak-9UT's strike role a bulletproof windshield was a must.

Yakovlev OKB

Above: Another Yak-9UT undergoing tests at NII VVS. Curiously, this example has a moulded windshield lacking the bulletproof windscreen.

Yefim Gordon archive

Yefim Gordon archive

Above: The Yak-9UV prototype built by plant No.82 (c/n 0000) seen during State acceptance trials. Like the Yak-9 'Courier', it is grey overall with a red stripe running the full length of the fuselage.

aged to keep the temperature of oil and water in the VK-107A engine within prescribed limits, but numerous cases were noted of the engines being overcooled in winter.

In the spring of 1945 the Red Army Air Force command embarked on a programme of re-equipping the active units with the Yak-9U. Up to then the machines had been assembled primarily at plant No.301 in Khimki north of Moscow. Now they were sent in kit form by rail to the west, and assembly workshops were established at Brest, Minsk, Kiev and L'vov. Shortly before VE-Day, assembly workshops were established in Insterburg, Bialostok, Poznań and other cities; they also served for distribution of the aircraft to the units and formations of the Red Army Air Force.

In all, nearly 2,500 VK-107A-powered fighters had rolled off assembly lines by the end of June 1945. As of 10th May 1945, approximately 750 Yak-9U fighters were on strength (out of the total of 6,267 fighters in the air arms of all the Western Fronts).

Yak-9U VK-107U fighter

According to Vadim B. Shavrov, in 1943 a single Yak-9U was fitted with the experimental VK-107U engine delivering 1,875 hp for take-off. This engine featuring a high degree of boost was prone to overheating, the water radiator becoming inadequate. Problems arose with the engine exhaust stubs, and the temperature in the cockpit rose alarmingly. The aircraft's AUW rose to 3,400 kg (7,497 lb); its performance included an impressive maximum speed of 720 km/h (447.5 mph) and just 4 minutes to climb to 5,000 m (16,400 ft), but the service ceiling was a mere 8,000 m (26,250 ft). It became obvious that the aircraft could not be used operationally unless the engine's teething troubles were rectified.

This information published in 1978 did not re-appear in later editions of Shavrov's works and has not been corroborated by other researchers. The little-known VK-107U engine (an uprated development of the VK-107A) likewise remains enigmatic, having been mentioned as 'a project'; in 1943 it was presumably known as M-107U.

Yak-9U versions with heavier armament (Yak-9UT)

Like many other basic subtypes of Yakovlev fighters, the Yak-9U served as a kind of proving ground for developing different armament options. In one machine provision was made for replacing the ShVAK-20 cannon in factory conditions by more potent cannon – an NS-23, an N-37 or even an N-45; in the latter case one of the two synchronised weapons in the front upper decking had to be deleted to lighten the machine. Synchronised machine-guns were replaced by new B-20 cannons (synchronised, of course).

The possibility of installing different engine-mounted cannons without any major redesign of the aircraft was a distinct advantage; it made it possible to switch series production quickly to this or that type of armament, depending on the needs of the Air Force. The NS-37 and its derivative, a 45-mm cannon, were experimental lightened weapons developed by OKB-16. A prototype fighter featuring this kind of armament (c/n 39-083, that is, Batch 39, 083rd aircraft in the batch) was built under A. S. Yakovlev's direct guidance in February and allocated the designation Yak-9UT (***tahn**kovyy* – tank-busting).

Naturally, each armament version had its own all-up weight, CG position and flying

Yakovlev OKB

Above: The same Yak-9UT c/n 40166022 on a snow-covered airfield during State acceptance trials. Note that the carburettor air intake has now reverted to its original location immediately aft of the spinner.

performance. However, the speeds were identical to those of the production Yak-9U. The handling qualities of the Yak-9UT were virtually the same as those of its predecessors, except for the control stick forces from the elevator, which proved to be too high. This was the Yak-9UT's most serious shortcoming.

On the credit side was the weight of fire: with the installation comprising one NS-37 and two B-20s, it amounted to 6.0 kg/sec (13.2 lb/sec) as compared to the Yak-9U's 2.81 kg/sec (6.2 lb/sec). In the final stages of the war such sizeable figures commanded respect even from the Germans, who also strove to increase the firepower of their fighters as much as possible in order to combat the sturdy and highly survivable Boeing B-17 Flying Fortresses and Consolidated B-24 Liberators.

The testing of this aircraft was conducted by engineer G. A. Sedov and pilot A. Manucharov at NII VVS in March 1945. The Yak-9UT proved to be considerably more stable under different manoeuvres compared to the Yak-9T and Yak-9K, owing primarily to the lesser recoil of the cannon and greater speed envelope. The aircraft was recommended for series production, and plant No.166 delivered 282 machines with the engine-mounted NS-23 cannon and two synchronised B-20S cannons.

Yak-9U standard-setter for 1945

An important stage in the history of the VK-107A-powered fighters was the emergence of the *etalon* (standard-setter, or pattern aircraft) for 1945. The main tasks facing the Yakovlev OKB were to eliminate the main defects of the Yak-9 VK-107A; to ensure a top speed of 600 km/h (373 mph) at sea level and 700 km/h (435 mph) at the second rated altitude with the engine running at its maximum rating; to ensure a climb to 5,000 m (16,400 ft) within 4.1 minutes and a service ceiling of 11,000 m (36,090 ft); to achieve a cruising range of 900 km (559 miles) and a range of 1,200 km (746 miles) in optimum cruise mode.

Yakovlev OKB

Yak-9UT c/n 40166074 is unusual in lacking the cannon muzzle brake.

The aircraft was built, using a production airframe (c/n 41-038). An important feature of this machine was its ability to accept alternative armament options in a manner similar to the Yak-9UT. The fighter was tested with three B-20s (two of them synchronised) and an all-up weight of 3,145 kg (6,935 lb). NII VVS specialists noted the advantages of the alternative armament options, the improved external finish and pointed out the need to use an antenna mast.

At the same time they stated that only some of the Yak-9U's shortcomings had been eliminated. For example, the installation of a bigger oil cooler with greater frontal area and introduction of an additional oil pump led to better engine operation at nominal revs only at high altitudes. As before, considerable oil spill from the breather occurred at maximum power.

In the course of 45 flights performed by the fighter, four (!) engine changes had to be made. In the last engine (c/n 527-21), which was part of an 'improved batch', the crankshaft main bearings broke down during the 14th hour of engine running. An extremely disappointed Aleksandr Yakovlev ordered Yak-9U c/n 41-038 to be withdrawn from testing and returned to the OKB so that development work could be resumed. All subsequent improvements introduced into the design came after the end of the war.

Yak-9UV conversion (familiarisation) trainer

The VK-107A-powered Yak-9UV (*u**looch**shennyy, vyvoz**noy*** – improved, familiarisation) created at plant No.82 embodied the final stage of the Yakovlev OKB's work on the development of a piston-engined fighter trainer. The armament of the Yak-9UV was restricted to one engine-mounted B-20M cannon, and the avionics/equipment fit ensured normal flight at altitudes up to 4,000-4,500 m (13,080-14,720 ft) in visual meteorological conditions for the trainee. As regards handling the aircraft did not differ from production Yak-9Us.

To ensure more effective engine cooling the designers installed new radiators. Even so, the engine suffered from overheating at nominal revs (3,000 rpm) due to the frequent take-offs and landings typical of the trainer's operational mode. Therefore the permissible rpm were limited to 2,800; at this power rating the speeds decreased by 35-90 km/h (21.7-56 mph), depending on the altitude.

Shortly after the end of the war the Yak-9UV was handed over for State trials, but it was clear that a different aircraft was needed. The Soviet Air Force started re-equipping with the first jet fighters; consequently, jet-powered conversion training machines were required. The Yak-9UV was not produced in quantity.

Yak-9UF reconnaissance fighter

This version was mentioned in 1975 in the Czechoslovak magazine *Létectví a Kosmonautika* (Aviation and Spaceflight) as a photo reconnaissance version of the Yak-9U fighter fitted with aerial cameras in the aft fuselage (F presumably stands for Foto).

VK-107A-powered Yak-9P production fighter

The war with Nazi Germany was over; with the onset of peacetime the aircraft industry was faced with new tasks. Soviet aircraft of mixed construction which had achieved a fairly good record at the front when a combat machine usually did not survive more than several dozen flying hours in combat were not expected to remain in operation for several years. Emphasis was placed on ease of production and low cost. The war 'devoured' military materiel in hundreds and thousands of examples, and the aircraft had no chance of growing old.

In peacetime there was no longer any need to build aircraft in such large numbers. However, one could no longer put up with such deficiencies as a short service life, susceptibility to the influence of the elements and other adverse natural conditions, deterioration of aerodynamic qualities and, as a consequence, a drop in performance, especially maximum speed. Research showed that it was impossible to solve these tasks without resorting to new structural materials.

The aircraft industry, including Chief Designer Aleksandr S. Yakovlev, was faced with the task of gradually switching to manufacturing aircraft of predominantly metal construction, to be followed by the creation of all-metal machines. Upgrading of the Yak-9 VK-107A fighter followed this pattern.

When the construction of the Yak-9P (unofficially dubbed *poslevo**yen**nyy* – postwar) was undertaken, the future aircraft had to meet fairly stringent demands concerning the quality of assembly work, production standards in the manufacture of airframe, powerplant and armament and thoroughness of surface finish. They were laid down in a Government directive dated 17th June 1946. Besides, the aircraft constructors were required to eliminate the main defects which had been noted in reports on the results of the Yak-9U's trials.

In 1946 most of the aircraft factories that had produced Yakovlev fighters (including the Yak-9U) during the war years were

Yakovlev OKB

Yakovlev OKB

Top and above: Yak-9UT c/n 40166074 was probably the most heavily armed example of all; it was converted to the one-off Yak-9-57 armed with a 57-mm cannon! Remarkably, the cannon barrel did not protrude beyond the spinner, so this was something of a wolf in sheep's clothing.

reoriented towards other work. For example, plant No.166 in Omsk switched to manufacturing Tu-2 bombers; plant No.82 in Moscow was completing the work on converting Yak-9s into trainers and tooling up for production of trolleybuses and trams. Only the giant plant No.153 in Novosibirsk persisted with perfecting the VK-107A-powered Yak-9. In the late spring of 1946 two initial production fighters featuring metal wings (c/ns 01-03 and 01-04) were built there; this was the beginning of a production batch bearing the P suffix.

After a brief period of adjustment work the two aircraft were handed over to NII VVS for State trials, which they passed under the direction of project engineer G. Sedov from 28th June until 23rd July 1946. Pilots Yu. Antipov and V. Ivanov performed 108 flights on these machines, logging over 60 flying hours. The aircraft were evaluated with regard to their performance and maintenance qualities.

Switching to the metal wings was accompanied by slight alterations of their planform – the wingtips were rounded, not angular as on the wooden version. The new version differed from the Yak-9U mainly in having ailerons with increased aerodynamic balancing, a dust filter in the engine air intake and additional uplocks for the flaps. Furthermore, the overall strength of the airframe was enhanced; this allowed the indicated airspeed in a dive to be increased from 650 to 720 km/h (from 404 to 447 mph), and the maximum G-load during recovery from a dive increased from 6.5 to 8.0.

The NII VVS test personnel also took a positive view of other alterations concerning the engine and some units and equipment items. Nevertheless, some shortcomings were noted, including leakage of engine coolant through sealed joints, spark plug failures after a mere eight or ten hours of operation, high temperatures in the cockpit... To make a long story short, the military test pilots and engineers placed the Yak-9P somewhere between the Yak-9U and the post-war demands to aircraft. Air Marshal K. Vershinin, who succeeded A. Novikov as Air Force Commander-in-Chief, acknowledged the results of the State acceptance trials as satisfactory and found it possible to begin service tests.

These tests were prepared especially thoroughly, since they were the first such tests of 29 production machines in peacetime. Between 4th October 1946 and 4th February 1947 hand-picked experienced pilots and engineers of the 246th IAD led by General Ye. Toorenko not only flight-tested the aircraft at Tolmachovo airfield near Novosibirsk (now the city's airport) but also determined the time limits for inspection and overhauls, inspection procedures and some other questions which had not been on the agenda of service tests during the war.

Alas, the results of this work were far from encouraging. While recognising the merits of these machines – above all, their ease of handling making them suitable even for pilots of less-than-average skill, – the military personnel noted a great number of defects precluding normal maintenance of the aircraft. They were also dissatisfied with the armament comprising one ShVAK-20 cannon and two UBS machine-guns (as on the Yak-9U) and the special equipment, which was very austere by the standard of 1946.

Equally unsatisfactory was the fact that the Yak-9P could wage aerial combat only at altitudes up to 7,000 m (22,960 ft). Above that altitude the aircraft was plagued by frequent jamming of the armament and deterioration of handling. Yakovlev's post-war fighters were assessed as unsatisfactory. For comparison purposes it may be noted that considerably fewer defects were noted

Yakovlev OKB

Yakovlev OKB

Top and above: These photos illustrate well the elegant lines of the Yak-9 VK-107A. Note the aft position of the carburettor air intake on this aircraft.

Yakovlev OKB

Yak-9 VK-107A c/n 52166082 was the *etalon* (production standard-setter) for 1946 and, in fact, the prototype of the post-war all-metal Yak-9P. Note the enlarged oil cooler air outlet immediately ahead of the radiator bath and the aerial mast.

Above: An early-production Yak-9P undergoing State acceptance trials. The aircraft has an unusually dark finish (probably medium grey).

This Yak-9P was equipped with a spin recovery parachute for the purpose of performing a spinning trials programme. Note the cutout in the fuselage spine closed by a Perspex panel, with the DF loop aerial visible underneath.

Specifications of VK-107A-powered Yak-9 fighter variants

	Yak-9U	Yak-9UT	Yak-9P
Year of issue	1944	1945	1947
Powerplant	VK-107A	VK-107A	VK-107A
Power at altitude, hp	1,500	1,500	1,500
Length	8.6 m (28 ft 2½ in)	n.a.	8.6 m (28 ft 2½ in)
Wing span	9.74 m (31 ft 11½ in)	9.74 m (31 ft 11½ in)	9.74 m (31 ft 11½ in)
Wing area, m² (sq ft)	17.15 (184.6)	17.15 (184.6)	17.15 (184.6)
Empty weight, kg (lb)	2,512 (5,537)	2,187 (4,821)	2,708 (5,970)
All-up weight, kg (lb)	3,204 (7,063)	3,260 (7,186)	3,550 (7,826)
Speed at sea level, km/h (mph)	575.0 (357.2)	578.0 (359.1)	590.0 (366.6)
Speed, km/h (mph)	672.0 (417.5)	671.0 (416.9)	660.0 (410.1)
at altitude, m (ft)	5,000 (16,500)	4,900 (16,000)	5,000 (16,500)
Climb to 5,000 m (16,400 ft)	5.0 min	5.2 min	5.8 min
Service ceiling, m (ft)	10,650 (35,000)	10,700 (35,000)	10,500 (34,500)
Turn time, seconds	20	20-21	21
Operational range, km (miles)	675 (419)	690 (428)	1,130 (702)
Take-off run, m (ft)	375 (1,230)	n.a.	540 (1,771)
Landing roll, m (ft)	530 (1,738)	n.a.	582 (1,909)
Armament	1 x 20-mm B-20 2 x 12.7-mm UB	1 x 37-mm NS-37 2 x 20-mm B-20	1 x 20-mm B-20 2 x 12.7-mm UB

during the service trials of the post-war La-9 fighter held in mid-1947; Semyon Lavochkin's fighter made the grade.

Yak-9P fighter, all-metal version

When the work on the service test batch of Yak-9Ps had been completed, the engine speed was limited to 3,000 rpm for the sake of reliability (a ban was imposed on the use of the 'combat' rating). Besides, it was stipulated that the industry should manufacture only all-metal aircraft featuring greater internal fuel tankage increased from 425 to 682 litres (from 93.5 to 150 Imp gal). The Novosibirsk plant met all these demands: after 29 Yak-9Ps with metal wings had rolled off the assembly line, all subsequent Yaks built by plant 153 were of all-metal construction.

In December 1947 one production machine – Yak-9P c/n 03-92 manufactured in July – successfully passed a series of State tests conducted by engineer V. P. Belodedenko and pilot L. M. Koovshinov. In comparison with the Yak-9Ps c/ns 01-03 and 01-04, the new fighter had somewhat inferior performance. Maximum speed amounted to 590 km/h (367 mph) at sea level and 660 km/h (410 mph) at 5,000 m (16,400 ft), which was 12-13 km/h (7.4-8.0 mph) less than before. The aircraft needed 5.8 minutes to reach 5,000 m versus 4.8 minutes required by its predecessor; this was due to the difference in the engine operation mode and to the weight increase from 3,227 to 3,550 kg (from 7,116 to 7,828 lb) caused by the extra fuel.

On the whole, the all-metal aircraft was virtually identical to its predecessor as regards handling; it was provided with special equipment which suited the machine's mission as all-weather fighter more fully and included such additional items as the RPKO-10M DF, an identification friend-or-foe (IFF) transponder, ultra-violet lamps lighting the instrument panel, a gun camera and the like. A better installation job and higher-quality electric bonding helped extend the range of radio communication to 118 km (73 miles) and the navigation instruments' operational range to 150 km (93 miles), which met the standards of the day. Unfortunately, like most of its predecessors, the production Yak was armed with one ShVAK cannon and two UBS-12.7 machine-guns.

In 1946-48 much effort was expended on perfecting the armament, including the Nudel'man N-37 and N-45 cannons and Berezin B-20s synchronised cannon. Additionally, some of the Yaks were equipped with reversible-pitch propellers (to reduce the landing run), the ASP-1N automatic gunsight, the 'Anschütz' artificial horizon and the like. Especially many flights on various Yak-9P examples were performed by Captain

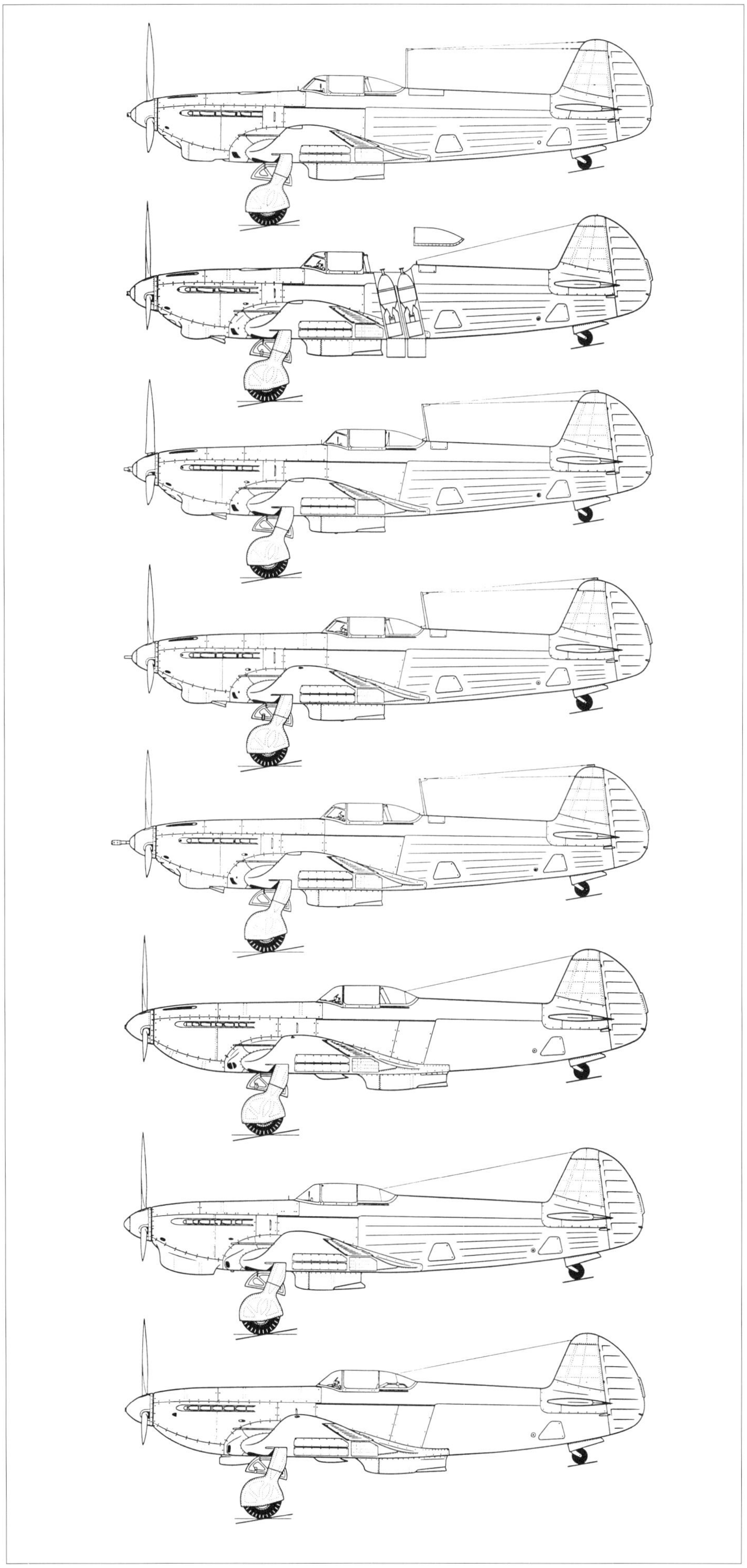

Top to bottom: The Yak-9D fighter; the Yak-9B fighter-bomber; the Yak-9-37 prototype; the Yak-9T; the Yak-9K; the Yak-9U; the Yak-9PD with an M-105PD engine; and the Yak-9PD with an M-106PV engine.

L. M. Koovshinov, in particular, for the purpose of perfecting methods of reducing the cockpit temperature in the and enhancing the aircraft's ease of handling.

In the meantime, the era of piston-engined fighters was coming to a close. Despite being beset by numerous teething troubles, the jet-powered machines of the late 1940s had a speed advantage of some 100 km/h (62 mph), enjoyed indisputable ascendancy in vertical manoeuvrability and carried more potent armament. In late March 1948 the Soviet Government took the decision to terminate production of the piston-engined Yak-9P at plant No.153 and start manufacturing Yak-23 jets instead. (Eventually the plant never built a single Yak-23, the type being manufactured by plant No.31 instead – but that's another story.) Production of the all-metal Yaks went on until the end of 1948; a total of 772 machines was delivered, supplemented by the 29 Yak-9Ps with metal wings. They stayed in service until the mid-1950s.

Yak-9P VK-107A for Bulgaria, modified

According to a report in *Air International* in November 1975, the VK-107A-powered Yak-9Ps delivered to the Bulgarian Air Force featured some alterations meeting the wishes of the customer. These alterations included the deletion of the outer fuel tanks in the wings and, importantly, the installation of two 12.7-mm UBS machine-guns in the wings.

Yak-9VRDK mixed-power fighter (project)

This project of a Yak-9 derivative featuring a mixed powerplant was developed in TsAGI in 1943. The work was undertaken by TsAGI's jet propulsion systems department headed by Prof. G. N. Abramovich. The powerplant comprised a piston engine plus a shaft-driven quasi-turbojet booster; the latter, in the parlance of that period, was dubbed 'compressor-type air-breathing jet engine' (*voz**doosh**no-reak**tiv**nyy **dvig**atel' kom**press**ornovo **tip**a*, or VRDK, hence the designation).

The piston engine required for the purpose already existed in the shape of the M-105RYeN which differed from the standard M-105F in having an auxiliary gearbox tapping 250 hp from it. This engine version had been intended to be used by A. D. Nadiradze on a Pe-2 bomber experimentally

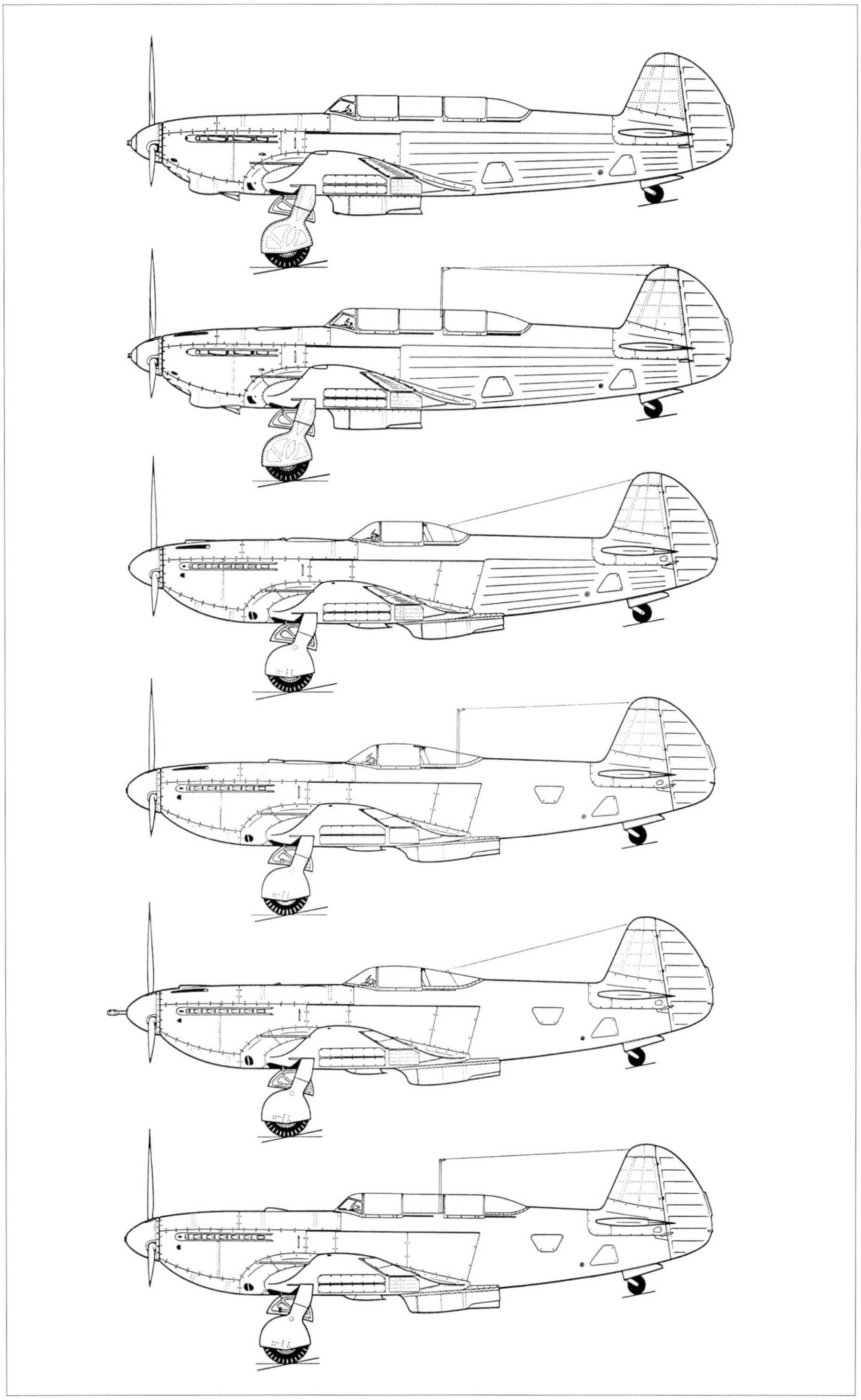

Top to bottom: A Yak-9D converted into a familiarisation trainer: the Yak-9 'Courier' liaison aircraft; the Yak-9U prototype; a production Yak-9U; the Yak-9UT; and the Yak-9UV prototype.

fitted with an air cushion undercarriage. The war prevented this plan from being implemented, and the modified engine remained unused.

Two years later TsAGI designers found a different application for it. The layout suggested by the institute required insignificant alterations to the Yak-9's fuselage truss in order to accommodate the combustion chamber of the VRDK; some elements of the armament had to be deleted. The original plan envisaged using a three-stage compressor measuring 600 mm (1 ft 11⅝ in) in diameter which was driven by two shafts via two intermediate gearboxes. The piston engine's coolant radiator was to be placed in the inlet duct forward of the compressor; later, this layout was deemed to provide unsatisfactory cooling and the radiator was relocated aft of the compressor. The rear part of the VRDK – an air-cooled combustion chamber with a pack of fuel injectors – was attached to the fuselage via a flexible mount because it had to be swivelled upwards 7° for take-off and landing to avoid scraping the ground.

The engine's power take-off gearbox and compressor drive shaft necessitated the deletion of the 20-mm ShVAK cannon, thereby reducing the fighter's firepower, and the extra weight of 115 kg (254 lb) associated with the VRDK booster and compressor drive made the fighter less agile in the vertical plane. The M-105RYeN delivered only 1,100 hp (as compared to the M-105PF's 1,250 hp), which was clearly insufficient for a mixed powerplant. As a result, the design top speed of the Yak-9VRDK exceeded that of the production Yak-9 by a mere 80 km/h (50 mph). The performance improvement would be insignificant, and the authors of the project admitted that its implementation was not worthwhile.

Yak-9UM (new-build Yak-9U with Allison engine)

The designation Yak-9UM was allocated to several new-build (replica) examples of a Yak-9U derivative that made its appearance in 1996. These aircraft were produced by the Orenburg-based *Strela* (Arrow) company (formerly plant No.47).

An example of the Yak-9UM was imported into the USA on 8th July 1996 for the aerobatic and display pilot Eddie Andreini of Half Moon Bay, California. Registered N900EA, painted bright red overall and named *Barbarossa* (!?), it was fitted with an Allison V-1710 Vee-12 engine driving a three-blade Hamilton Standard propeller. The cockpit and its canopy were modified: a second seat was placed behind the pilot's seat and the rear part of the cockpit canopy could be slid aft for access to the back seat. No armament was fitted, but the barrels of the engine-mounted N-37 cannon and two UBS 12.7-mm machine-guns were simulated for appearance's sake. The main gear doors were slightly altered in comparison with the real thing to accommodate the modern wheels.

Another Yak-9UM registered NX8250 was acquired by Jim Wickersham, the owner of a company in Nevada, who became a sales agent for the Yak-9UM in North America. The two machines took part in various air shows in the USA in late 1990s. Other known examples are N1157H (c/n 0470402 – that is, first digit unexplained, plant No.47, Batch 04, second aircraft in the batch), N908JW (c/n 0470408) and N9250 (later sold to France as F-AZYJ; c/n 0470403). The first digit in the c/n is always a zero but it obviously does not denote the year of manufacture (that would be 1990, which is much too early).

Yak-3 fighter (second use of designation, 1943)

The Yak-3 is the last and the most brilliant representative of Yakovlev's World War Two period fighter family. This fighter, based on the experience of the first two years of the war was, in effect, a new aircraft type. It embodied the enormous work performed by the Yakovlev OKB, the production plants and Soviet research establishments (first of all LII and TsAGI) with a view to improving the aerodynamic and structural design and enhancing the combat qualities of fighters.

The Yak-3 incorporated all the best features that had been embodied in the Yak-1, Yak-7 and Yak-9. When it had its baptism of fire in the summer of 1944, the Yak-3 powered by the VK-105PF-2 engine turned out to be the most lightweight and agile aircraft among the fighters of the USSR, the USA and Germany. Perhaps, only the British Supermarine Spitfire could rival this machine.

While possessing slightly lower performance than its stablemate (the VK-107A-powered Yak-9U), the Yak-3 VK-105PF-2 was superior to it as regards the proven powerplant which made the Yak-3 more reliable in operation and, consequently, more combat-ready. A prototype Yak-3 powered by the VK-108 engine reached a speed of 745 km/h (463 mph); this was the highest speed attained by Soviet fighters which came close to the limit for fighters of the 'pre-jet propulsion era'.

Yak-1M ('Moskit') fighter prototype

The Yak-3's predecessor was the Yak-1M fighter, the prototype of which was also dubbed 'Moskit' (Mosquito). The wings of the new fighter were structurally similar to those of the Yak-9, featuring metal spars, metal and wooden ribs and plywood skinning. However, wing area was reduced by 2.3 m^2 (24.7 sq ft) in comparison with the preceding Yaks and made up 14.85 m^2 (159.7 sq ft).

The fighter's wing span was reduced to 9.2 m (30 ft 2$\frac{13}{64}$ in). A production break coinciding with the fuselage axis was introduced; this afforded the possibility of replacing the port or starboard wing panels in field conditions in case of damage. The fuselage, undercarriage and control system were all borrowed from the production Yak-1. As distinct from the latter, the fuel system comprised three tanks: two main tanks in the wing outer panels and one service tank in the wing centre section. As regards the armament, equipment and armour protection, the Yak-1M had much in common with the Yak-1 in the 'bubbletop' version featuring improvements in visibility, armour plating and armament.

Yakovlev OKB

Above: A model of the Yak-1M (Yak-1 'Moskit') in the TsAGI wind tunnel. This aspect does not give a clear idea of the nose contour.

Yakovlev OKB

Above and below: Two views of the first prototype Yak-1 'Moskit' in front of a camouflaged hangar at LII. At this stage the fighter was powered by an M-105PF engine (note the three dual exhaust stubs on each side).

Yakovlev OKB

The engine's cooling system featured a more efficient radiator buried more deeply in the fuselage. Two circular-shaped oil coolers working in parallel were housed in the wing centre section under the cockpit floor. This made it possible to eliminate the characteristic chin fairing for the oil cooler, resulting in a smooth lower cowling contour and considerably reducing drag.

The Yak-1M had an all-up weight of 2,665 kg (5,876 lb) – some 245 kg (540 lb) less than that of a production Yak-1 manufactured in 1943. The wing loading rose marginally from 169 to 179 kg/m^2 (from 373 to 395 lb/sq ft), while the power loading was noticeably reduced – from 2.40 to 2.19 kg/hp (from 5.30 to 4.83 lb/hp). The weight reduction was achieved mainly by reducing the wing area and substituting duralumin alloy spars for wooden ones; these measures reduced the empty weight by 150 kg (331 lb).

Yakovlev OKB

Above: The Yak-1M during State acceptance trials at NII VVS which lasted from 27th June to 4th July 1943.

Yakovlev OKB

The first prototype Yak-1M after being re-engined with the M-107A (note the seven exhaust stubs). The photo shows the fighter as originally flown with this powerplant.

The prototype was completed in mid-February 1943. Throughout the spring months the aircraft was subjected to development work conducted under the direction of leading engineer M. Grigor'yev who had taken an active part in the design and construction of the Yak-1M. Yakovlev OKB chief test pilot Pavel Ya. Fedrovi made an uneventful first flight at the end of February 1943, whereupon the machine entered the manufacturer's test phase.

Testing of the aircraft at NII VVS lasted throughout June 1943 (with A. G. Proshakov as project test pilot and A. T. Stepanets as project engineer, both being the most experienced specialists of the Institute as regards Yakovlev aircraft). The tests revealed the new fighter had excellent performance. However, the OKB held the opinion that the fighter's improvement potential had not yet been exhausted. At Aleksandr S. Yakovlev's insistence additional tests were conducted to determine the changes in the basic performance characteristics after the boost pressure of the M-105PF engine had been increased from 1,050 mm to 1,100 mm Hg.

Initially Vladimir Ya. Klimov, Chief Designer of the engine, gave his consent to increasing the boost pressure only at the first supercharger speed. Additional tests showed that augmenting the engine power by increasing the boost pressure produced a 6-7 km/h (3.7-4.35 mph) gain in maximum speed at low altitude, reducing the climb time to 5,000 m (16,400 ft) by 0.1 minutes and affording an extra 50 m (164 ft) in altitude gain during a combat turn. It also led to a marginal improvement of field performance and entailed virtually no change in the engine's water and oil temperature.

Later theoretical calculations showed the possibility of boosting the engine at medium altitude as well. The boost pressure was also increased at the second supercharger speed. While the M-105PF engine developed 1,180 hp at 2,700 m (8,860 ft) without regard to the dynamic pressure, after boosting its output rose to 1,244 hp at 2,100 m (6,890 ft). With the manifold pressure increased to 1,100 mm Hg the engine was designated M-105PF-2, and from the spring of 1944 on it bore the designation VK-105PF-2.

As regards performance, the Yak-1M was on a par with the best fighters of the final stage of the Second World War. For example, in terms of maximum speed it surpassed production Yak-9s throughout the altitude envelope by at least 25-35 km/h (15.5-21.7 mph); it outperformed the Fw 190A-4 up to 8,300 m (27,220 ft) and the Bf 109G-2 up 5,700 m (18,700 ft), judging by the results of tests at NII VVS. The Yak-1M enjoyed the greatest advantage at low altitude, whereas at high altitudes the German fighters were faster, owing to the better altitude performance of the Daimler-Benz DB 605 and BMW 801 engines. For example, at the altitude of 7,000 m (22,960 ft) the Bf 109G-2 surpassed the Soviet fighter in speed by nearly 50 km/h (31 mph).

With regard to the rate of climb up to 5,000 m, the Yak-1M was unrivalled among the contemporary fighters of the world. All known versions of the Bf 109, which was justly considered to be one of the best in performing upward vertical manoeuvres, were somewhat inferior to the Soviet fighter. The reduction of the wing area was accompanied by a reduction of the Yak-1M's all-up weight; as a result, it did not cause a deterioration of field performance, spinning and diving characteristics. The handling of the prototype, like that of the Yak-1, Yak-7 and Yak-9, was within the capabilities of wartime pilots having an average and even below-average skill level.

The test report noted: *'As regards the effectiveness and harmonious action of controls (from the point of view of stick forces from control surfaces) the Yak-1M, along with the Spitfire Mk VB* (it was tested at NII VVS in June 1943 – *Auth.*), *is exemplary for all fighter aircraft, both indigenous and foreign.'*

A few shortcomings of the prototype, typical of the Yaks (such as overheating of oil during climb at the optimum climb rate, poor functioning of the breather, oil leakage from various sealed joints, inadequate range of air-to-air and air-to-ground radio communication and the like) could not spoil the favourable overall impression produced by the machine.

Yak-1M *dooblyor* (second prototype)

While the first prototype Yak-1M was undergoing State trials, construction of the second prototype Yak-1M, dubbed *dooblyor* in keeping with the day's terminology, was nearing completion under the direction of engineer M. Grigor'yev. It turned out to be still more refined and well thought out in every respect.

For example, the fuel tank bays were separated from the cockpit by sealed bulkheads; the fabric skinning of the fuselage was replaced by plywood; the cockpit hood was provided with an emergency jettisoning system. Some other improvements were introduced as well. The engine drove a new

VISh-105SV-01 three-blade propeller featuring a lightened hub and blades with airfoil-section root cuffs.

There were also changes in the armament. Instead of the ShVAK cannon the designers installed a prototype ShA-20M lightened cannon designed by Boris Shpital'nyy and reinstated the second UBS-12.7 heavy machine-gun *à la* Yak-7B. At the same time the avionics suite was expanded. A feature that was bound to attract attention was the radio which could be remotely controlled by a push-button on the throttle lever. This was a novelty in the Soviet aircraft construction. The Yak-1M *dooblyor* dispensed with the aerial mast, making use of a single-wire aerial.

As a result, the all-up weight remained virtually the same at 2,660 kg (5,865 lb). Construction of the aircraft was completed on 9th September 1943. At the beginning of October, after brief manufacturer's flight tests, the second Yak-1M was handed over to NII VVS for State trials, which were successfully conducted by pilot A. G. Proshakov and engineer G. A. Sedov within ten days.

The tests showed an improvement of performance: the *dooblyor* attained a speed of 570 km/h (354 mph) at sea level and 651 km/h (405 mph) at 4,300 m (14,100 ft); it gained 1,280 m (4,200 ft) of altitude in a combat turn and could perform a full-circle banking turn at low altitude within a mere 16-17 seconds. In addition, it was noted that, thanks to the installation of more effective radiators, changes in the ducting and greater maximum opening of the radiator shutter, the temperature conditions of the engine had been substantially improved. For the first time in aircraft of the Yakovlev family it was possible to perform prolonged horizontal flight at maximum speed and gain altitude in the maximum rate of climb mode with the engine at nominal revs (2,700 rpm).

Thanks to the thorough electric bonding and shielding of the basic metal elements of the structure, the ground-to-air radio reception range rose to 90 km (56 miles) and pilots could maintain stable communication between themselves at up to 20 km (12.4 miles). This was very good performance for the radio equipment of Soviet fighters at that time.

It is difficult to single out another Soviet aircraft that has received as many laudatory comments from test pilots – both company pilots and military ones. For example, A. Koobyshkin considered it to be the best among the known fighters of World War Two in overall performance. Test pilot V. Khomiakov, who had flown the prototype fighter, wrote in his report:

'The cockpit is comfortable. Forward visibility has been improved. The instruments

Yakovlev OKB

Yakovlev OKB

Top and above: The Yak-1M prototype after modifications as recommended by TsAGI. The pointed spinner and chin-mounted extra oil cooler gave the fighter a certain similarity with the Martin baker MB-5 fighter – or a Supermarine Spitfire fitted with a Vokes dust filter.

Yakovlev OKB

A front view of the modified first prototype Yak-1M.

Yakovlev OKB

Above and below: The second prototype differed from the first aircraft in having an M-105PF-2 engine. This was, in effect, the prototype of the Yak-3.

Yakovlev OKB

and control levers are well placed and their arrangement almost fully conforms to the standard cockpit. Taxying is easy. […] When airborne, the aircraft is stable and simple in handling. The machine has an excellent rate of climb and manoeuvrability, both in the vertical and in the horizontal plane. Performance has been considerably improved in comparison with its predecessor, the [first prototype] Yak-1M.'

Bearing in mind these and other comments, the Government decision was not long in coming. Already in October 1943 the Yak-1M *dooblyor* was put into production under the designation Yak-3, supplanting the 'straight' Yak-1. As noted earlier, the designation Yak-3 had already been assigned to one of Yakovlev's fighters – the would-be production version of the I-30, but this fighter failed to reach production in 1941 or at any later time. The fate of the new machine proved to be much happier.

Yakovlev OKB

One of the first Saratov-built Yak-3s ('11 White', c/n 0111) during State acceptance trials at NII VVS.

Yak-1 M-107 (Yak-1M M-107) fighter prototype

The work conducted by the Yakovlev OKB on the Yak-1M ran in parallel with the efforts, started somewhat earlier, on adapting the Yak-1 to accept the 1,600-hp Klimov M-107A engine, then under development. Quite naturally, the OKB found it wise to incorporate into the M-107-powered prototype all the aerodynamic and structural refinements that had been introduced on the Yak-1M. The resulting aircraft was variously referred to as the 'M-107-powered Yak-1', 'M-107-powered Yak-1M' or 'Moskit's *dooblyor*'. Installation of the M-107A led to a 100-mm (3$\frac{15}{16}$ in) increase in the length of the front fuselage. Aileron area was further reduced.

Outwardly the new machine differed from the original Yak-1M in having seven instead of six exhaust stubs on each side of the engine cowling. It was completed by April 1943 and, after some modifications to the cooling system, made its first flight on 5th October 1943. In the course of testing it was further modified in accordance with TsAGI recommendations, receiving an additional oil cooler in a chin position. On 18th November 1942 the M-107-powered Yak-1M attained a speed of 716 km/h (445 mph) at 6,250 m (20,500 ft). However, the M-107A engine was not yet fully developed, and the machine remained a prototype.

Production Yak-3 fighter

Preparations for series production of the Yak-3 started as far back as the summer of 1943. Initially it was envisaged that plant No.292 would start producing the new machines immediately upon completion of its restoration after the enemy bombardment. However, development of the Yak-1M prototypes dragged on well into the autumn; still, the war was at its height and it was necessary to turn out combat machines at an unabated rate while switching to the new model. Therefore, in the winter of 1944 the Saratov plant tooled up for the new machine while continuing to manufacture an average of 250 Yak-1s every month.

The work was getting under way slowly, and the command was well aware of the fact that the delivery schedule of the most modern Soviet fighters (such as the Yak-3, Yak-9U and the Lavochkin La-7) to front-line units would not be met. A. K. Repin, Chief

Engineer of the Red Army Air Force, even suggested in a letter to the Central Committee of the Communist (Bolshevik) Party that production plants be allowed to reduce the output of aircraft considered outdated by that time for the sake of speeding up the work on new types.

Although the Yak-3 had much in common with the Yak-1, the plant was making slow progress. What was the impediment, you may ask? The Yak-3 proved to be much more labour-intensive than its predecessor because it required extreme care in the surface finish: initially the number of man-hours spent on manufacturing a single Yak-3 was 2.5 times greater than in the case of the Yak-1. The first production Yak-3 assembled in Saratov was completed on 1st March 1944 and test-flown a week later. Not until the end of the following month were 22 Yak-3s officially accepted by the military.

The production version differed from the Yak-1M *dooblyor* in having numerous changes which, however, were of minor importance. For example, the armament of early production machines was absolutely identical to that of their predecessor, the Yak-1 (the ShA-20M cannon never entered production). As for the production standards, a considerable number of defects was noted on the new Yaks, such as a slightly buckled wing surface, a poor fit of the engine cowling panels, dents in inspection panels and so on.

Production Yak-3s suffered a deterioration of performance compared to the *dooblyor* on account of lower production standards. For example, the speed fell by 15-20 km/h (9.3-12.4 mph), while the time to 5,000 m (16,400 ft) rose by 30 seconds. High control forces caused a worsening of horizontal manoeuvrability. It was also noted that production machines had a shorter combat radius and radio communication range.

Service units started taking delivery of the new fighter in the late spring and early summer of 1944, when the Soviet command was making preparations for major offensive operations. Several of the first fighters were delivered to the 396th IAP, and combat evaluation of the Yak-3 was conducted in the 91st IAP (commanded by Lieutenant-Colonel A. Kovalyov) of the 2nd VA in June and July of 1944.

The regiment was assigned the task of keeping the air superiority in the course of the L'vov offensive operation, and approximately a month before the beginning of the operation's active phase the unit received 41 brand-new Yak-3s. Nearly half of the unit's flying personnel had not seen actual combat by then, but all the airmen had taken a good conversion training course before leaving for the front. Furthermore, for training purposes NII VVS arranged a session of mock combat between a Yak-3 and a captured Bf 109G-4, which demonstrated considerable superiority of the Soviet fighter at altitudes from ground level to 7,000 m (22,960 ft) and allowed the most expedient combat tactics for the Red Army Air Force pilots to be determined.

Operational evaluation confirmed the conclusions that had been made in advance. The unit flew 431 combat sorties, including interception of enemy aircraft, escorting aircraft from other units, 'free chase' missions and other duties. Twenty enemy fighters and three Ju 87s were shot down in aerial combat. Own losses included two Yak-3s shot down and three that were

Central National Economy Archive

Above: '96 White', a production Yak-3 M-105PF-2 (c/n 9626), during checkout tests at NII VVS.

Central National Economy Archive

Central National Economy Archive

Top and above: Two views of the first Yak-3 manufactured by the Tbilisi aircraft factory No.31 ('01 White'). Note the unusually small star on the fin.

Central National Economy Archive

Above: A production Yak-3 manufactured by the Saratov aircraft factory No.292 undergoing tests.

Yefim Gordon archive

Above: A production Yak-3 seen during special tests at LII.

Yefim Gordon archive

'4 White', one of the Yak-3s operated by the 'Normandie-Niémen' regiment.

damaged by flak but managed to reach the area held by Soviet troops.

Combat experience demonstrated that the new Soviet fighter caught up with the enemy fighters both in level flight and in upward vertical manoeuvres. Usually the Yak-3 could latch onto the tail of the Fw 190A at the second 360° turn, and of the Bf 109 at the third full circle. It was the Messerschmitts that remained the main opponents of the Soviet Yaks in the struggle for air superiority over the Western Ukraine.

The biggest air battle took place on 16th July 1944. Both sides called in reinforcements and, when the skirmish was in full swing, 18 Yak-3s were opposed by 24 enemy fighters. According to the claims of the Soviet airmen, they succeeded in shooting down 15 German aircraft at the cost of one Yak-3 shot down and one damaged. It is worth noting that on the following day the Luftwaffe was appreciably less active at this sector of the front.

As a result of the operational evaluation it was determined that the Yak-3 would best be used against the enemy's fighter aircraft. Its employment for providing cover to ground troops by means of patrolling, for escorting bombers or for other similar duties was less expedient because of the limited amount of fuel on board (the average duration of a sortie was about 40 minutes).

The Yak-3's special feature was its ability to accelerate very quickly – for example, in a shallow dive. In such cases the pilots had to 'hold the fighter back' because its good aerodynamics made it easy to exceed the limit of 650 km/h (404 mph) indicated airspeed, which was dangerous due to insufficient structural strength. One more peculiarity that had to be taken into account was the marked compression of the undercarriage oleos during ground running of the engine. There was a danger of nosing over during landing.

The service tests also revealed some obvious shortcomings of the early Yak-3s. There were cases of the undercarriage legs collapsing during landing runs and taxying; the main gear breaker struts and oleo leg attachment points sometimes failed. Generally, however, the maintenance of the Yak-3 was simple and the aircraft could be mastered by flying personnel and ground crews without any difficulties.

At the fronts and in the factory workshops

In the late summer of 1944 Yak-3 operations reached a fairly wide scale. The Bf 109G was the fighter's main opponent. The mass-produced G-6 version, being heavier than its predecessors, lost its superiority in speed and vertical manoeuvre over the Soviet fighter for the first time since the beginning

Yakovlev OKB

Above: The sleek lines of the Yak-3 are well illustrated by this example. Yakovlev was big on photographing his aircraft, leaving a lot of excellent pictures for posterity. This example features an extended rear canopy section for lower drag.

Yakovlev OKB

A view of the same Yak-3 in identical aspects following installation of ejector fairings on the exhaust stubs. This measure increased top speed somewhat, even though it clearly did not make the fighter more aesthetically pleasing.

of the war. This fact had to be recognised, albeit painfully, by German pilots. They noted that the Messerschmitt was 10-20 km/h (6.2-12.4 mph) slower than the Yak-3 at altitudes up to 4,000 m (13,120 ft) even when using the MW-50 methanol/water injection system for engine boost and when flying without underwing cannon pods, and that the German fighter was inferior in performing steep climb and combat turns. At the same time, in horizontal dogfighting the Yak-3's advantage proved insignificant. Still less favourable was the Yak-3's position in a dogfight against the Fw 190A.

Checkout tests conducted by NII VVS on Yak-3 c/n 11-12 (that is, 11th aircraft in Batch 12) in August 1944 showed that, with an all-up weight of 2,675 kg (5,900 lb), the fighter attained a speed of 555 km/h (345 mph) at sea level and 631 km/h at 4,200 m (13,780 ft), climbing to 5,000 m (16,400 ft) within 4.5 minutes. The military were also satisfied with other performance figures: a full-circle banking turn was performed in 21 seconds, the altitude gain in a combat turn amounted to 1,200 m (4,560 ft), the take-off run was a mere 280 m (918 ft) and the service ceiling was 10,400 m (34,110 ft).

Although the Yak-3s were not fitted with rocket launch rails or bomb racks, strafing ground targets was included into the range of their missions. Yak-3 pilots noted that the aircraft offered a marginally better rearward view compared to the 'bubbletop' Yak-1 because the latter's glazing was prone to misting up near the bulletproof glass panels. The absence of a cockpit hood emergency jettisoning feature compelled Yak-1 pilots to fly with the cockpit hoods open. On the other hand, the Yak-3's moulded cockpit visor noticeably distorted the objects observed through it, hampering the use of the gunsight.

In mid-August 1944, when the 240th IAD was waging hard battles west and north-west of Kaunas, two Yak-3 fighters landed at Orany airfield. The command and the headquarters personnel of the formation knew about the new fighter: a short while earlier the *Pravda* daily had published a photo of the machine, the caption stating that this was the best and the most lightweight fighter in the world. After landing one of the pilots who had ferried the aircraft handed his machine over to Colonel G. Zimin (later Air Marshal) with the words: 'For your personal use. To the flying division commander from Marshal A. Novikov, commander of the Red Army Air Force'.

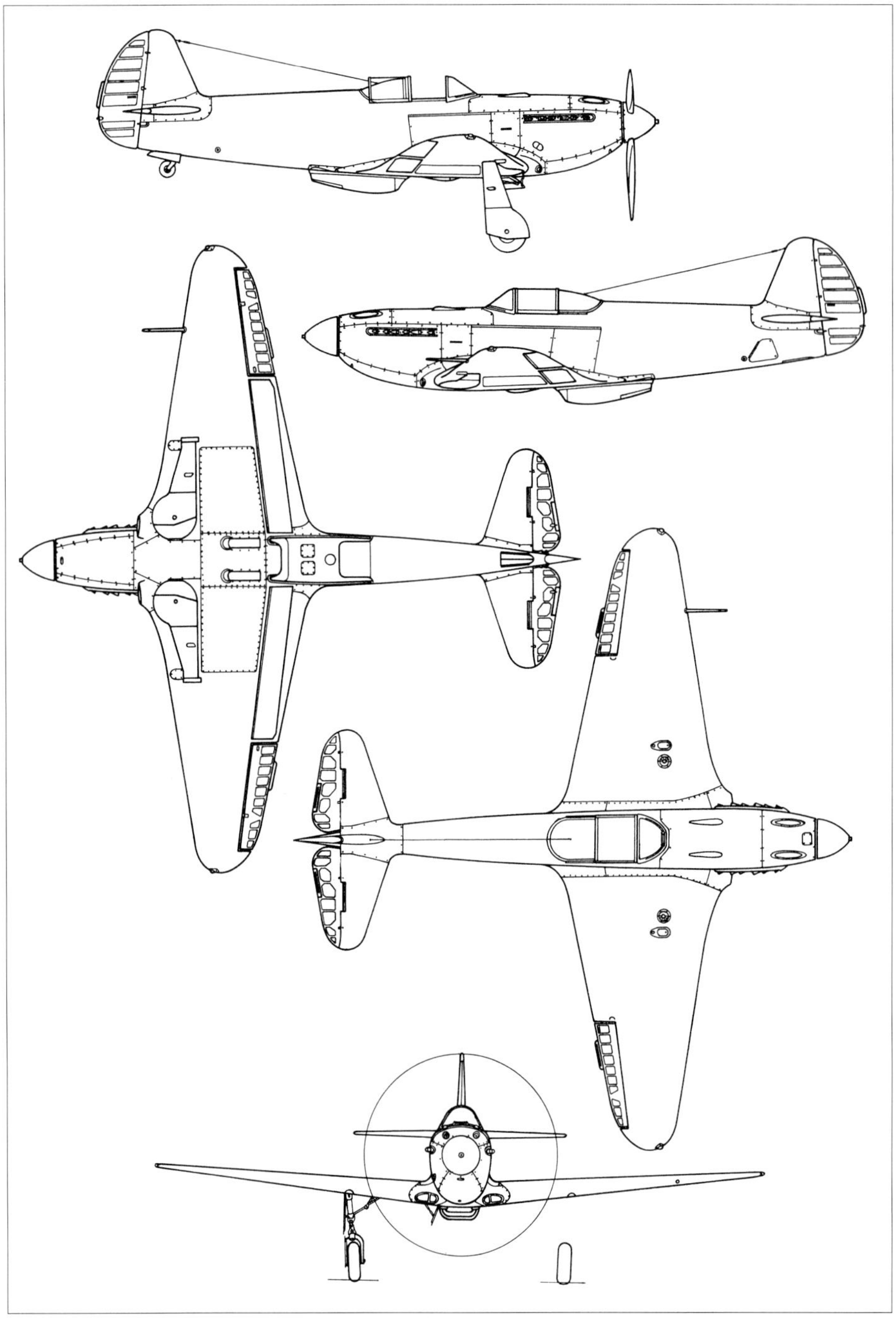

Five views of the standard production Yak-3.

Another Division CO who personally flew the Yak-3 in combat was General G. Zakharov. As soon as the new machines had been delivered to the 303rd IAD, he performed up to 40 flights in Yak-3 c/n 310147 (that is, Tbilisi aircraft factory No.31, Batch 01, 47th aircraft in the batch). His conclusion was that the fighter had no equals. The General expressed the wish that the Soviet Air Force be equipped with such fighters as quickly as possible.

The commanders of the Luftwaffe formations largely shared the high opinion of the Yak-3 that had been expressed by Soviet and British generals and senior officers. Appraising the characteristics of the new Soviet fighters, Lieutenant-General Walter Schwabedissen wrote in his book *The Russian Air Force in the Eyes of German Commanders*:

'Whereas the German Bf 109G and Fw 190 models were equal to any of above Soviet models in all respects, this does not apply to the Soviet Yak-3, which made its first appearance at the front in the late summer of 1944. This plane was faster, more manoeuvrable and had better rate of climb than the German Bf 109G and Fw 190, to which it was inferior only as regards armament.'

In the summer and autumn of 1944 a number of alterations was introduced into the design of the Yak-3 at plant No.292:

- from Batch 13 (August) onwards the second UBS-12.7 machine-gun was reinstated because this armament version suited the needs of the front better (about 500 production machines had been equipped with one cannon and one synchronised machine-gun);
- from Batch 16 (September) onwards the total fuel tankage was increased by approximately 20 litres (4.4 Imp gal);
- from Batch 17 (September) onwards the fighter was provided with modified radio equipment and dust filters which prevented excessive engine wear at the cost of a slight reduction in level speed;
- from Batch 19 (October) onwards the fabric skinning was replaced by plywood skinning, as on the second prototype.

At the same time the Saratov plant built up a good production tempo: in the autumn of 1944 the monthly output averaged approximately 250 machines. A further 80 Yak-3s per month began rolling off the assembly lines of plant No.31 in Tbilisi which gradually switched its production from the LaGG-3 to Yakovlev's fighter. From the outset the Yak-3s manufactured in Tbilisi were armed with two synchronised heavy machine-guns in addition to the engine-mounted cannon; outwardly they differed mainly in the contours of the rear part of the radiator bath. Tbilisi-produced Yak-3s proved to be heavier by an average 25 kg (55 lb), which was due partly to the additional armament and ammunition and greater tankage; their level speed was higher by some 15 km/h (9.3 mph) compared to Saratov-built machines.

The service introduction of the Yak-3 was accompanied by its share of troubles. Numerous accidents and incidents took place in many regiments. The gravest accident occurred in the 402nd IAP of the 3rd IAK, when Captain P. Tarasov, one of the most proficient aerobatics performers in the corps, was killed in a crash. Investigation showed that the wing skin had loosened in the area of the wheel wells and, when subjected to high loads, broke away from the ribs along the lines of adhesive bonding, whereupon the spars and the wing as a whole disintegrated completely.

All such incidents involved Saratov-built fighters. An investigation conducted by TsAGI confirmed that the wing strength had deteriorated due to breaches of the bonding procedures. After a sample check of the condition of wing surfaces conducted by the Chief Engineer of the Air Force and by subsequent commissions 114 Yak-3s had to be grounded; 68 of them were in the 3rd IAK.

Thanks to urgent measures and the tremendous efforts taken by ground crews of front-line units, factory repair teams and various scientific institutions, it proved possible to eliminate the defect expeditiously, making the fighter safe to fly. According to TsAGI conclusions, the fighter's structural strength came to meet the requirements of normal operation.

No less drastic measures were taken at plant No.292. Starting with Batch 28, the casein glue was supplanted by the VIAM-B3 adhesive that had proved its worth; from the following batch the area of contact between the ribs and the skin was doubled and the wheel well domes were strengthened in their uppermost part. By the beginning of the New Year of 1945 the number of unserviceable Yak-3s in active units had been reduced to 12.8%.

Acting in concert with LII, TsAGI and production plants, the Yakovlev OKB conducted serious work for the purpose of improving the fighter's performance. As a result, from October 1944 onwards the performance was, in effect, improved enough to match the characteristics of the Yak-1M *dooblyor*.

This work was conducted in parallel with building up the output of production machines. Whereas in May 1944 the Yak-3 made up only 29% of the total production volume at the Saratov plant, it accounted for 52% in June, 84% in July, and in August the share rose to 100%. Importantly, switching to the new fighter type was effected practically without detriment to the overall output of aircraft.

Yefim Gordon archive

Above: As the legend on the fuselage reveals, this Yak-3 has been presented to fighter pilot Semyon Rogovoy by the Soviet Navy personnel serving on the Amur River in the Soviet Far East.

Still more successful was the mastering of Yak-3 manufacture by plant 31 in Tbilisi. From the outset the Yak-3 was built here in a version featuring better performance and heavier armament compared to machines produced in Saratov. And, most importantly, Tbilisi-produced fighters were not afflicted with the inadequate wing strength resulting from defective bonding of the skinning to the wing framework.

By the end of 1944, Yak-3 combat losses amounted to only 56 machines (of which 42 were reported as missing, that is, fate unknown); a further 34 fighters were written off for non-combat reasons. In all, by the beginning of 1945 the fighter element of the Soviet Air Force had 5,810 machines in its inventory, including 735 Yak-3s. Of these, 44%

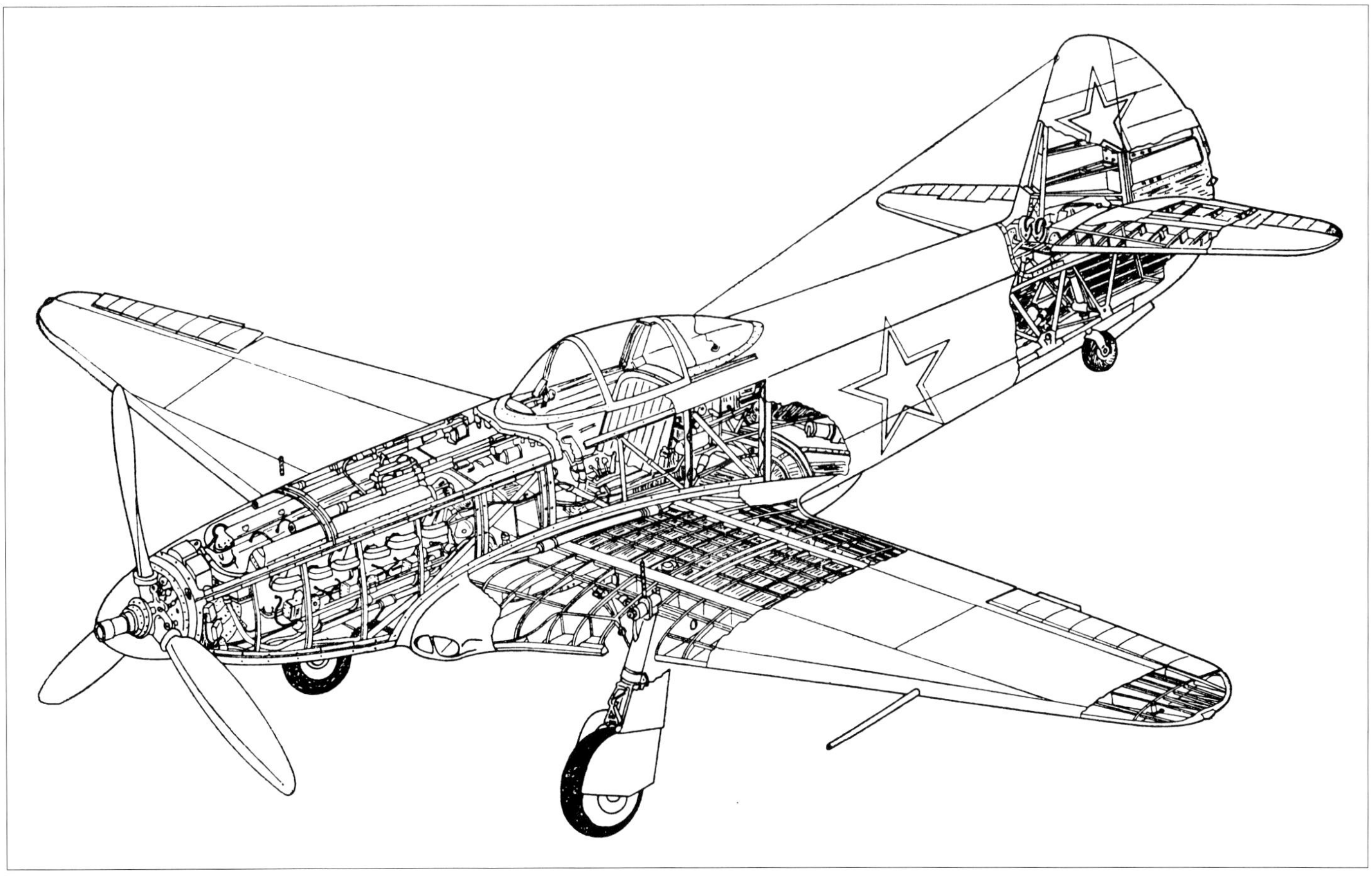

A cutaway drawing of the Yak-3.

The prototype of the Yak-3P armed with three cannons seen during State acceptance trials.

saw service with the 16th VA commanded by General S. Rudenko, which joined in providing air cover for the troops at the Berlin direction from the second half of January.

The 10th Fighter Air Corps operated in the skies of southern Poland, Austria and Czechoslovakia where the Luftwaffe was not very active. Still fewer air battles were fought by the new Yaks in the course of hostilities with Japan in August 1945. By that time the Soviet command had succeeded in redeploying 106 Yak-3s (78 of them combat-ready) to the Far East where they were integrated into the forces of the Trans-baikalian Front and the two Far Eastern Fronts.

Presumably the only Japanese fighter that could rival the performance of the Yak-3 at that time was the Nakajima Ki-84 Hayate, but Japanese airmen put up no active opposition in the air and there is no evidence of air combat between these aircraft. As for the Nakajima Ki-43 Hayabusa that was deployed in large numbers in northern China, it was more than 100 km/h (62 mph) slower than the Yak-3 and was also inferior to the Soviet fighter in vertical manoeuvrability and armament.

One Yak-3 was preserved and exhibited in the Yakovlev OKB's museum until a few years ago. This aircraft flown by Boris Yeryomin, Commander of the 31st GvIAP, had been presented to him by Ferapont Golovatyy, a collective farmer from the Saratov region. Regrettably, this legendary machine is now in thc USΛ.

A total of 4,848 Yak-3s were built, including 3,840 manufactured by the main plant in Saratov. The production run in Saratov comprised 59 batches (50 of them completed before the end of the war), while eight batches were produced in Tbilisi (the sixth batch was completed shortly after VE-Day). Of the overall number, 737 machines were built after the war.

Yak-3P production fighter

After the war the two factories built primarily the Yak-3P (***push**echnyy* – cannon-armed). This version differed mainly in having revised armament. The ShVAK cannon and the UBS machine-gun were replaced by new cannons designed by Berezin in two versions: the engine-mounted B-20M and the synchronised B-20S, the suffixes denoting ***motor**naya* (engine-mounted) and *sin**khron**naya* (synchronised) respectively. The three barrels together produced a weight of fire of 3,52 kg (7.8 lb) per second versus 2,72 kg (6.0 lb) for the 'standard' Yak-3; the new installation had virtually no effect on the weight of the airframe.

The Yak-3P showed good stability when firing the B-20 cannons within the entire speed envelope and in all attitudes. The weapons' recoil affected the aiming precision only to a negligible extent. The modified Yak-3P passed testing at NII VVS in March-April 1945 and was put into series production.

However, far more stringent demands concerning quality standards in aircraft manufacturing were introduced after the war. During tests conducted by NII VVS in October-November 1945 a production Yak-3P (c/n 02-54) demonstrated speeds superior to those of the prototype: 572 km/h (355 mph) at sea level and 646 km/h (401 mph) at 3,900 m (12,790 ft), the time to 5,000 m (16,400 ft) being 4.8 minutes. Nevertheless, the results of the tests were pronounced unsatisfactory. This was due, among other things, to persistent failures of the special brackets on which the B-20 cannon were attached.

In all, 596 Yak-3Ps were produced. None of them took part in combat.

Yak-3T fighter prototype

Built in the OKB in January 1945, the Yak-3T (*tyazhe**lovo**'orou**zhon**nyy* – heavily armed) possessed even more potent armament. It retained the pair of synchronised cannons in the front upper decking, as on the Yak-3P, while the engine-mounted B-20 cannon was replaced by a lightened 37-mm N-37 cannon designed by A. Nudel'man, with 25 rounds of ammunition.

Externally the Yak-3T differed from production machines in having a muzzle brake mounted on the barrel protruding from the propeller spinner; it absorbed up to three-quarters of the cannon's recoil energy. Installation of the heavy cannon necessitated some design changes. The cockpit was moved aft 400 mm (1 ft 3¾ in) in a fashion similar to the Yak-9T; the fuel capacity was reduced by 17 litres (3.74 Imp gal) and the self-sealing coating of the fuel tanks was deleted.

Despite the weight-saving measures, the AUW rose to 2,756 kg (6,077 lb) and the performance deteriorated marginally. Top speed was 560 km/h (348 mph) at sea level and 629 km/h (391 mph) at 4,000 m (13,120 ft), while the cruising range was a mere 559 km

The prototype of the Yak-3T armed with a 37-mm engine-mounted cannon.

Yakovlev OKB

The prototype of the high-altitude Yak-3PD seen during manufacturer's flight tests.

(347 miles). The designers succeeded in keeping the CG of the fully equipped Yak-3 at 25.7% of the mean aerodynamic chord, thanks to which good controllability was retained.

Tests showed that the fighter remained as easy in handling as ever. The armament of the Yak-3T ensured effective engagement of aerial targets and enabled attacks against railway trains, soft-skinned and armoured vehicles. There was virtually no loss of accuracy when firing a burst of four rounds from the engine-mounted cannon. This made it possible to fire long bursts while retaining accurate sighting – something that could not be done on the Yak-9T.

However, major defects were also revealed. The combination of the engine with the new cannon was not yet trouble-free, the engine being plagued by overheating of water and oil, which precluded normal operation. The engine persistently malfunctioned; this included emission of smoke, vibration, spark plug failures, drops in the fuel pressure and the like. All this prevented the Yak-3T from being ordered into production.

Yak-3PD interceptor prototypes

Using the airframe of the Yak-3 as a basis, the Yakovlev OKB created one of the most successful interceptors. In September 1944 a production example of the fighter was fitted with a VK-105PD engine – one of the several high-altitude engine versions provided with a supercharger designed by V. Dollezhal' (hence the D).

Testing of this aircraft (designated Yak-3PD, that is, *pere**khvaht**chik [s **toor**bokom**press**sorom] Dolle**zha**lya* – interceptor with a Dollezhal'-designed supercharger) at LII was conducted right before the end of the war. According to the test report, the aircraft had the following special features: the wing area was increased by 0.5 m^2 (5.4 sq ft), an experimental high-altitude propeller was installed and the engine's air inlet was moved from its original location to a place ahead of the radiator bath. The engine featured a system injecting an alcohol/water mixture between the supercharger stages in order to reduce the air temperature.

By restricting the armament to one NS-23 cannon the designers succeeded in reducing the all-up weight to 2,616 kg (5,768 lb). Pilot Sergey N. Anokhin reached an altitude of 11,500 m (37,730 ft), but calculations showed that the service ceiling could be as high as 13,000 m (42,640 ft). A speed of 692 km/h (430 mph) was attained at the altitude of 10,850 m (34,700 ft).

Certain imperfections of the engine caused the testing to be suspended. Then the aircraft was re-engined with the VK-105PV that had been verified on the Yak-9. Before resuming the tests, some development work was done by LII. At the suggestion of pilot-engineer I. Shooneyko who was conducting the flight tests, the aircraft was fitted with a valve spilling excess air from the supercharger into the atmosphere. This completely eliminated the instability in the supercharger's operation at high altitudes, enabling the machine to fly at altitudes in excess of 13,000 m (42,640 ft) with the engine running normally.

In the course of a flight performed on 26th June 1945 Shooneyko attained a speed of 710 km/h at 11,000 m (36,080 ft), and on 6th July he reached a service ceiling of 13,300 m (43,624 ft). It was noted that flights to altitudes in excess of 13,000 m with the pilot provided only with an oxygen mask required special training of pilots for high-altitude flights – for example, in an altitude chamber.

The Yak-3PD was not built in series, but the work at LII continued. The materials of the research conducted there formed the basis for a report issued in March 1947 on the functioning of engine radiators in high-altitude conditions.

Yak-3RD experimental fighter

In December 1944 an experimental fighter featuring a mixed powerplant was created on the basis of the Yak-3. In addition to the production VK-105PF-2 engine, the machine was equipped with an RD-1 liquid-propellant rocket motor designed by D. Glushko; it was intended to serve as a booster.

Yakovlev entrusted engineer B. Motorin with the design of the unorthodox machine designated Yak-3RD (*ra**ket**nyy **dvig**atel'* – rocket motor). Motorin installed the RD-1 rated at 300 kgp (661 lbst) in the aft fuselage under the suitably cropped rudder and enclosed it with an easily detachable cowling faired into the fuselage contours.

In the course of manufacturer's trials test pilot Viktor Rastorgooyev performed 21 flights in the Yak-3RD. In eight of these flights he switched on the rocket booster. On one occasion, on 11th May 1945, the aircraft attained a speed of 782km/h (486 mph) at 7,800 m (25,580 ft), which was 182 km (113 mph) higher than the speed attained at this altitude by the standard Yak-3. However, the automatic controls of the rocket motor were faulty and there were several cases of the engine cutting unexpectedly in flight. On one occasion an explosion occurred, causing severe damage to the engine nozzle.

The faults were rectified and repairs were made, whereupon the OKB began preparations for the aircraft to take part in the 1945 Aviation Day air display. However, two days before the event, on 16th August, the aircraft suddenly entered a steep dive while performing a regular flight and crashed, killing V. Rastorgooyev. The cause of the accident was not established, but it is known with certainty that there was no explosion of the rocket booster; neither did the engine disintegrate in flight. Apparently there was a failure in the control system linkage.

In connection with the emergence of turbojet engines in the post-war years the subject of fitting liquid-propellant rocket boosters to piston-engined aircraft became moot and the work was discontinued.

VK-107A-powered Yak-3 fighter prototype

An important direction of work on the Yak-3 at the Yakovlev OKB consisted in the installation of more powerful engines. This work was conducted by a group of designers under Yevgeniy G. Adler's direction. Installation of the VK-107A engine did not require any major modifications to the airframe, apart from moving the cockpit aft 400 mm (1 ft 3¾ in). The more powerful engine neces-

Yakovlev OKB

Above: The Yak-3R mixed-power fighter. The rocket booster nozzle is closed by a conical fairing. Below: The RD-1KhZ liquid-fuel rocket motor with the cowling removed.

Yakovlev OKB

Yakovlev OKB

Above and below: The prototype of the Yak-3 VK-107A seen during manufacturer's flight tests. Note the carburettor air intake immediately aft of the spinner.

Yakovlev OKB

sitated an increase of the fuel capacity to 518 litres (114 Imp gal), and the armament was changed to comprise two B-20S synchronised cannons in the front upper decking.

Two prototypes of this version were completed in early 1944. The first of them was used for manufacturer's tests which lasted until November 1944; it was flown by Pavel Ya. Fedrovi. The other prototype was handed over for State trials after a few development flights for the purpose of rectifying engine faults. The trials were started in February 1944 by G. A. Sedov, the NII VVS project engineer assigned to the Yak-3 VK-107A. Pilots Yu. A. Antipov and A. G. Proshakov had to interrupt the flights eight times; in one such case the flights were suspended for almost two months (from mid-May to mid-July) because serious defects of the powerplant had to be remedied. Up to late August 1944 the machine performed only 44 flights, logging nearly 26 hours.

On account of the heavier engine the all-up weight rose to 2,984 kg (6,580 lb). In spite of this, the VK-107A-powered Yak-3 remained a lightweight fighter with excellent specific characteristics: the wing loading was 201 kg/m^2 (41.18 lb/sq ft) and the power loading 2.0 kg/hp (4.41 lb/hp). The fighter's handling the enabled it to be flown by medium-skilled pilots, as had been the case with its predecessor, the VK-105PF-2-powered Yak-3. All this, coupled with the excellent aerodynamics, endowed the fighter with outstanding performance. The aircraft attained a maximum speed of 611 km/h (380 mph) at sea level and 720 km/h (447 mph) at 5,750 m (18,860 ft). Time to 5,000 m (16,400 ft) was 3.9 minutes and the altitude gain in a combat turn equalled 1,500 m (4,920 ft). All these characteristics constituted record-breaking performance for the final stage of the war.

However, test pilots noted that the VK-107A engines failed to log their designated service life hours because the crankshaft's main bearings failed. Also, the unreliable functioning of the throttle control system precluded flight at the specified cruising speed, and oil spill spraying the entire windshield prevented the pilot from closing the cockpit hood and making use of the gunsight. At that time Soviet specialists were not yet in possession of the experience with the VK-107A engine gained subsequently on the Yak-9U.

Work on the Yak-3 powered by the VK-107A engine was continued after the end of hostilities.

Yak-3 VK-107A – production version

In the spring of 1945 the People's Commissariat of Aircraft Industry took a decision calling for the continuation of work on the

Yakovlev OKB

Above: A production Yak-3 VK-107A manufactured by the Tbilisi aircraft factory; note the aft location of the carburettor intake. The fighter must be a test aircraft, hence the silver and red paint job.

VK-107A-powered Yak-3 in Tbilisi. As noted earlier, the Yak-3s that had been built there during the war years had an edge in production quality standard compared to fighters of the same type assembled at plant No.292; therefore, it was presumed that the VK-107A-powered machines would have reasonably high chances of being a success.

Modifications were incorporated by a team of designers under Nikolay K. Skrzhinskiy in a production machine (c/n 70-03). It was fitted with a VK-107A engine from the final batch featuring an additional oil pump; the designers replaced the plywood skin of the fuselage and the wings with duralumin, and the control surfaces received a skin of *elektron* magnesium alloy instead of the standard fabric covering. Furthermore, the new fighter had a slightly increased fuel and oil capacity which, however, was inferior to the fuel and oil tankage of the Yak-3 VK-107A prototype. As a result, the all-up weight was 2,935 kg (6,472 lb).

Manufacturer's flight tests lasted less than a month; on 11th May 1945 the all-metal Yak-3 was handed over to I. Kolosov, leading engineer of NII VVS, for testing. The institute's project test pilots noted that the aircraft had better controllability as compared to the similarly powered prototype of mixed construction, more effective wheel brakes, and its engine functioned better at medium and high altitudes.

The results obtained in 44 flights confirmed the machine's high speeds and good manoeuvrability, even though the use of the 'combat' mode of the VK-107A (3,200 rpm) had to be relinquished because of the engine's unreliability. The intention was to eliminate the aircraft's main defects prior to launching series production. However, in actual fact development of this particular machine had to be continued concurrently with the assembly of other examples of the fighter. Plant No.31 assembled 40 all-metal Yak-3s in 1945 and a further eight in the following year. Work on weeding out various defects on these aircraft continued throughout the year of 1946, whereupon the Tbilisi plant switched to mastering production of the jet-powered Yak-15.

The next stage in the work on the VK-107A-powered machines was marked by tooling up to build a batch of 30 fighters of his model at plant No.292 in the spring of 1946. They differed from the two prototypes in wing design and in having various changes to the powerplant. Whereas the prototype fighters had wings with metal spars and plywood skin, just like the production Yak-3 and Yak-9, now the government demanded that all-metal wings incorporating new technology be installed. Besides, the fighters were fitted with the improved version of the VK-107A. Based on operational experience with the Yak-9U, it featured an air cooling system for the exhaust manifolds, more capacious radiators and dust filters in the carburettor air intakes.

Changes were introduced into the radio equipment and armament. The latter was restricted to two B-20S synchronised cannons on some of the fighters, while others also had an engine-mounted B-20M. Testing of the first three Saratov-built production machines (c/ns 01-01, 03-01 and 04-01; c/n 02-01 may be a static test airframe) conducted in April and May 1946 showed that their performance, high though it was, fell

Yakovlev OKB

This silver/red Yak-3 equipped with a gun camera on top of the windscreen was used by LII as a testbed of unknown purpose.

Yakovlev OKB

Above and below: The prototype Yak-3 VK-108. Note the double rows of six exhaust stubs (dorsal and lateral), with heat-resistant steel panels aft of them, and the armoured seat back.

Yakovlev OKB

Specifications of Yak-3 fighter variants

	Yak-1M *dooblyor*	Yak-3	Yak-3
Year of manufacture	1943	1943	1944
Powerplant	M-105PF-2	M-105PF-2	VK-107A
Power at altitude, hp	1,240	1,240	1,500
Length	8.5 m (27 ft 10½ in)	8.5 m (27 ft 10½ in)	8.5 m (27 ft 10½ in)
Wing span	9.2 m (30 ft 2 in)	9.2 m (30 ft 2 in)	9.2 m (30 ft 2 in)
Wing area, m^2 (sq ft)	14.85 (159.8)	14.85 (159.8)	14.85 (159.8)
Empty weight, kg (lb)	2,105 (4,640)	2,128 (4,691)	2,346 (5,171)
All-up weight, kg (lb)	2,660 (5,860)	2,697 (5,945)	2,964 (6,578)
Speed at sea level, km/h (mph)	570 (354.1)	565 (350)	611 (379.6)
Speed, km/h (mph)	651 (404.5)	640 (397)	720 (447.3)
at altitude, m (ft)	4,300 (14,000)	4,400 (14,430)	5,750 (18,750)
Climb to 5,000 m (16,400 ft)	4.1 min	4.1 min	3.9 min
Service ceiling, m (ft)	10,800 (35,500)	10,400 (34,000)	11,800 (38,750)
Turn time, seconds	17	19	18
Operational range, km (miles)	900 (559)	850 (528)	1,060 (658)
Take-off run, m (ft)	275 (900)	290 (950)	345 (1,130)
Landing roll, m (ft)	485 (1,590)	480 (1,575)	590 (1,935)
Armament	1 x 20-mm B-20	1 x 20-mm B-20	1 x 20-mm B-20
	2 x 12.7-mm UB	2 x 12.7-mm UB	2 x 12.7-mm UB

somewhat short of the Government specifications for fighters of this type at the end of December 1944. For example, maximum speed at sea level was within the specified 600 km/h (373 mph), whereas the speed at the altitudes of 5,400-5,900 m (17,710-19,350 ft) was 695-697 km/h (431.9-433.2 mph) instead of the specified 700 km/h (435 mph). Time to 5,000 m (16,400 m) was 4.2-4.5 minutes (the requirement called for no more than 3.9 minutes).

The testing was very thorough and comprehensive. In particular, pilots A. G. Proshakov, Yu. A. Antipov and V. Khomiakov determined the new Yaks' spinning characteristics, establishing that they were virtually identical to those of the production VK-105PF-2-powered Yak-3. However, high stick forces from the ailerons and elevator made the VK-107A-powered Yak-3's handling unpleasant. This shortcoming fully came to the fore during mock combat sessions with a Spitfire Mk IX, especially when performing violent manoeuvres in the vertical plane.

Furthermore, project engineer A. T. Stepanets noted persistent powerplant troubles. The engine began vibrating when throttled back, and the oil pressure dropped below the admissible level at altitudes in excess of 4,000 m (13,120 ft). On one of the machines the engine suffered a complete breakdown; overheating of water and oil was occasionally noted on the others. The tests also revealed incomplete consumption of fuel from the tanks – a defect characteristic of many Yaks.

Modifications made to these aircraft at Yakovlev's experimental production facility and at LII failed to produce the desired result. The test report signed in June 1946 stated: *'Modified Yak-3 aircraft c/ns 01-01, 03-01 and 04-01 powered by VK-107A engines and featuring metal wings, produced at the Volga plant* (Saratov plant No.292 – *Auth.*) *have failed the State trials due to numerous serious defects.'*

Besides, the mixed construction (metal wings and fuselage with plywood skinning) did not meet the demands to post-war aviation that had been formulated in the Soviet Union. In many respects (stick forces, controllability, survivability) the mixed-construction Yak-3s proved to be a step backward as compared to the all-metal machines built in Tbilisi. Therefore, the decision was taken to limit the production run to the three machines already manufactured, and all unfinished airframes were scrapped.

The Ministry of Aircraft Industry (formerly NKAP) considered that the stronger airframe of the Yak-9 was more suitable for comprehensive development of the VK-107A engine. Also, the Yak-9 permitted the installation of potent 37-mm and 45-mm cannons between the cylinder banks and could be transformed into a long-range fighter – unlike the Yak-3. These were the reasons for the Yak-9's longevity.

VK-108-powered Yak-3 fighter prototype

Installation of the VK-108 engine in the Yak-3 was, in effect, purely experimental. The VK-107A with a nominal rating of 1,500 hp was replaced by the VK-108 delivering 1,550 hp and operating in very adverse temperature conditions because the radiators were the same as on the Yak-9U. Owing to special features of the engine design the fighter's armament had to be restricted to one engine-mounted NS-23 cannon firing through the propeller hub.

The aircraft was constructed under the direction of engineer A. Kanookov; it was rolled out at the OKB's experimental production facility on 1st October 1944. Following a brief development period test pilot Viktor Rastorgooyev performed the first flight, characterising the aircraft's performance as outstanding. On 21st December 1944 the Yak-3 VK-108 (lightened to the utmost, with the armament deleted and a reduced amount of fuel in the tanks) attained a speed of 745 km/h (463 mph) at 6,290 m (20,630 ft). That was just 10 km/h (6.2 mph) less than the world speed record established in 1939 by a specially modified Bf 109. The Yak-3 climbed to 5,000 m (16,400 ft) within just three and a half minutes, literally leaping into the air while still in tail-down position (that is, without requiring the tail to be raised).

However, once again unreliable engine operation prevented the tests from proceeding at a normal pace. The flight schedule kept slipping because of vibration, smoke emissions and numerous powerplant faults. Virtually every time the fighter covered a stretch in level flight, the oil temperature exceeded 110°C (230°F), which was the absolute limit for this engine. The work had to be discontinued on 8th March 1945.

Yak-3U fighter prototype

In the Yak-3U fighter the Yakovlev OKB made an attempt to combine its achievements in aerodynamic refinement with the merits of a Shvetsov-designed radial engine. In January 1945, when the aircraft was under construction, the ASh-82FN radial was a reliable and trouble-free engine, while the VK-107A (to say nothing of the VK-108) remained capricious and persistently malfunctioned.

The Yakovlev design team succeeded in creating a fighter that was very lightweight for a machine powered by the ASh-82FN – it had an all-up weight of only 2,792 kg (6,156 lb) versus 3,250 kg (7,166 lb) in the case of the

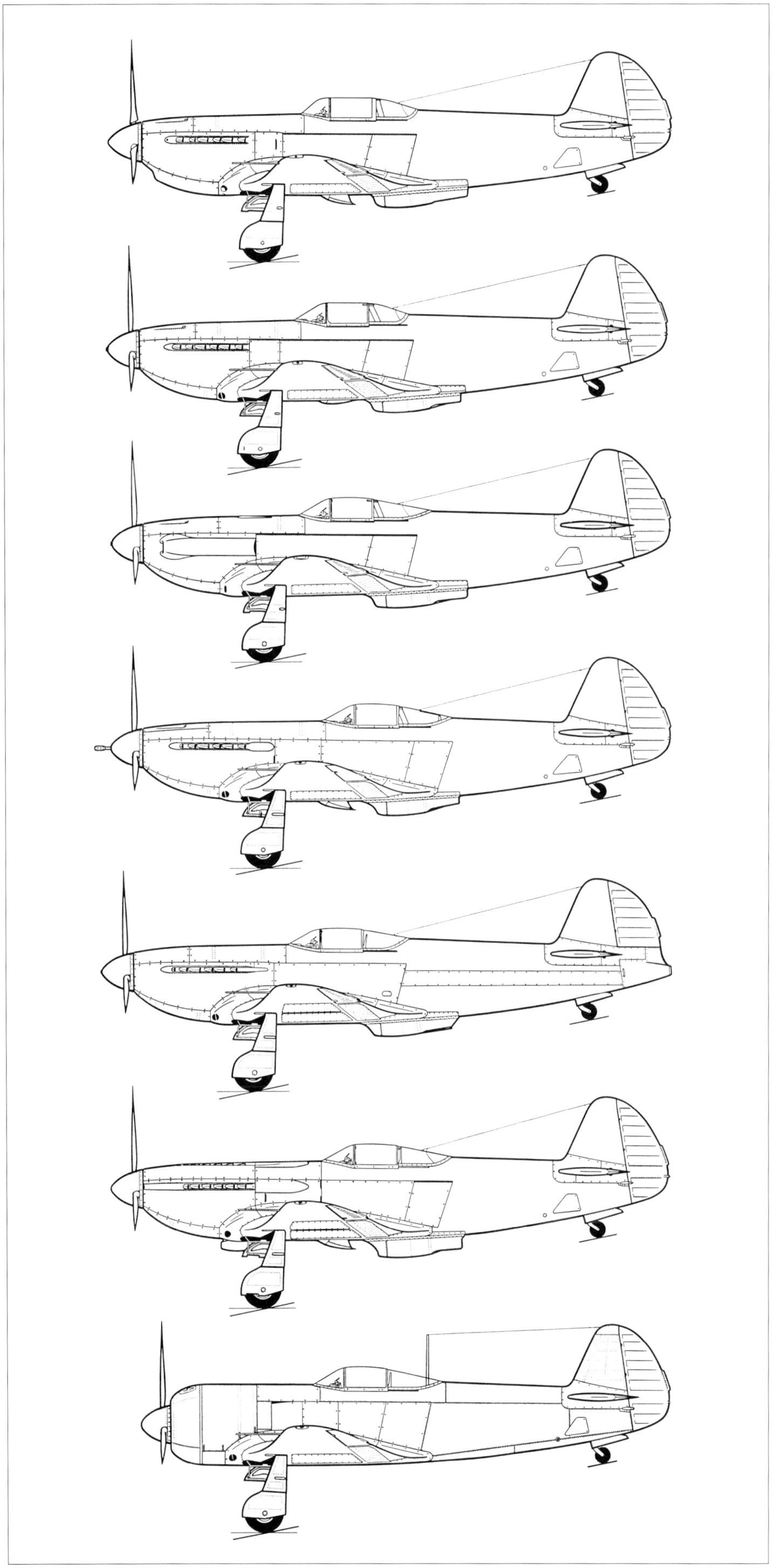

Top to bottom: The Yak-1M prototype, a standard Yak-3, a Yak-3 with ejector fairings on the exhaust stubs, the Yak-3T, the Yak-3RD, the Yak-3 VK-108 and the Yak-3U.

Yakovlev OKB

Above: The Yak-3 M-82 development aircraft during manufacturer's flight tests. Note the large oil cooler air intakes in the wing roots and the heat-resistant panels aft of the exhaust flaps.

La-7. The maximum speed of the Yak-3U – 705 km/h (438 mph) at 6,100 m (20,998 ft) – was also superior to that of the La-7. The new machine proved to have an excellent rate of climb. However, the forward shift of the wings, while improving stability, resulted in an inadmissible reduction of the aircraft's ground angle, increasing the risk of nosing-over; the aircraft became dangerous during landing and taxying.

In the course of development work the main defects of the aircraft were rectified; however, in the view of Aleksandr S. Yakovlev, this fighter was no longer needed in October 1945, and the work was discontinued.

Specifications of Yak-3 variants with VK-108 and ASh-82FN engines

	Yak-3	Yak-3U
Year of manufacture	1944	1945
Powerplant	VK-108	ASh-82FN
Power at altitude, hp	1,800	1,630
Length	8.5 m (27 ft 10½ in)	8.17 m (26 ft 9½ in)
Wing span	9.2 m (30 ft 2 in)	9.74 m (31 ft 11½ in)
Wing area, m^2 (sq ft)	14.85 (159.8)	17.15 (184.6)
Empty weight, kg (lb)	n.a.	2,273 (5,011)
All-up weight, kg (lb)	2,830 (6,238)	2,792 (6,155)
Speed at sea level, km/h (mph)	n.a.	n.a.
Speed, km/h (mph)	745 (463)	705 (438)
at altitude, m (ft)	6,000 (19,680)	6,100 (20,998)
Climb to 5,000 m (16,400 ft)	3.5 min	n.a.
Service ceiling, m (ft)	n.a.	n.a.
Turn time, seconds	n.a.	n.a.
Operational range, km (miles)	n.a.	n.a.
Take-off run, m (ft)	n.a.	n.a.
Landing roll, m (ft)	n.a.	n.a.
Armament	1 x 23-mm NS-23	2 x 20-mm B-20

Yak-3UA (Yak-3M) replica

Sadly, only a handful of the original Yak-3s has survived to this day, none of them in flying condition. However, new-build machines can to some extent compensate for this now. In 1991 Yakovlev Aircraft Corporation signed a contract with the Gunnell Museum in Santa Monica, California, under the terms of which 20 replica Yak-3 fighters with airframes cleared to the original limits would be supplied for private customers in the USA and elsewhere.

The aircraft were built mostly on the basis of original drawings, some new drawings being necessitated by the modified all-metal airframe and the totally different engine installation. The original VK-105PF2 was replaced by a reconditioned Allison V-1710 designated Allison 2L, rated at 1,240 hp and driving a Hamilton Standard propeller. External differences compared with the wartime fighter include the carburettor inlet located well aft on top of the cowling and the slightly altered lower gear door segments attached to the mainwheel legs (that is, slightly bulged instead of being flat in order to accommodate the wider modern wheels). No armament is fitted, though imitation gun barrels are provided for the sake of authenticity; civil avionics are installed to customer order.

The aircraft were built at the Strela plant at Orenburg in 1993-95. The first of the batch (c/n 0470101 – that is, plant No.47, Batch 01, 01st aircraft in the batch; the meaning of the first digit is unknown) was painted, like most others, in wartime Soviet two-tone grey camouflage and was displayed at the 1993 Paris airshow. The same aircraft was demonstrated at the MAKS-93 airshow in Zhukovskiy where it was designated Yak-3M; it was subsequently registered N854DP. Other examples include VH-YZK (c/n 0470102) in Australia, N42YK (c/n 0470103), N529SB (c/n 0470104), N74FT (c/n 0470106) and N551BH (c/n 0470107) in the USA; the latter aircraft being later sold to Germany as D-FJAK.

Data differences from Yak-3/VK-105PF2 are as follows: length 8.49 m (27 ft 10¼ in); empty weight 2,210 kg (4,872 lb); maximum loaded weight 2,697 kg (5,945 lb). Maximum speed 570 km/h (354 mph) at sea level and 648 km/h (402 mph) at 4,500 m (14,750 ft).

THE JET FIGHTERS OF 1945-1955

Yak-15 jet fighter

The Yakovlev OKB was among the first to take on the task of developing jet fighters. As related in Chapter 2, the design bureau already had some experience with jet propulsion, having built and tested a number of mixed-power aircraft based on production or experimental piston-engined fighters during the Great Patriotic War; these featured ramjet boosters under the wings or a liquid-fuel rocket motor in the aft fuselage. In all cases, however, the speed increase turned out to be rather insignificant; also, the reliability of the boosters left something to be desired. Thus, the Yakovlev OKB's practical experience with jets in the first post-war years was extremely limited.

When three Soviet aviation research establishments – GK NII VVS, LII and TsAGI – had evaluated several captured German jets, the results of these tests were distributed to the leaders of virtually all Soviet aviation design bureaux, including Aleksandr S. Yakovlev. Indeed, being Vice People's Commissar of Aircraft Industry, he was one of the first to receive these data – he was entitled to it. Yakovlev was one of the main opponents of copying the German designs completely, an idea floated by some in the commissariat. Moreover, his design team had always favoured the lightweight single-engined agile fighter concept, whereas most of the German jet aircraft tested in the USSR were twin-engined. Thus, OKB-115 decided to take a different road.

On 9th April 1945 the State Defence Committee issued a directive ordering the OKB of plant No.115 to design, build and submit for testing a single-seat fighter powered by a 'Jumo 004 engine'. In order to save time and fulfil the government orders as quickly as possible the engineers selected the proven piston-engined Yak-3 as the starting point for the development of Yakovlev's first jet fighter. Retaining the low-wing, tailwheel-gear layout, the new fighter had the 900-kgp (1,984-lbst) Junkers Jumo 004B axial-flow turbojet mounted beneath the forward fuselage so that the nozzle was located under the centre fuselage, resulting in a pod-and-boom arrangement. This approach allowed the Yak-3's good field performance and agility to be retained.

Yak-Jumo (Yak-3-Jumo) first prototype

Structurally the fighter (provisionally designated Yak-Jumo or Yak-3-Jumo at this stage) was largely identical to the Yak-3 representative of the final production batches – that is, the all-metal version powered by the Klimov VK-107A engine. The fuselage was a welded steel tube truss skinned with duralumin; its front portion incorporated attachment points for the engine. A heat shield made of heat-resistant steel was installed under the centre fuselage to protect the skin from the jet blast.

The two-spar all-metal wings (which, because of the deeper forward fuselage, were now effectively mid-set rather than low-mounted) were built as a one-piece structure; the wing area was identical to that of the Yak-3, being 14.85 m^2 (159.67 sq ft). The only major difference from the Yak-3 was that the front spar was bent to an inverted-U shape at the wing/fuselage joint where it passed over the engine. The wings featured duralumin-skinned flaps and ailerons hinged to the rear spar. The front spar carried the main landing gear pivots; the inner portions of the wings accommodated the main fuel tanks. The oil cooler air intakes built into the Yak-3's wing leading edge at the roots were eliminated and a pitot tube was installed on the port wing near the tip.

The cantilever tail surfaces were likewise of all-metal construction. The horizontal tail was taken straight from the Yak-3 but the area of the vertical tail was slightly

Yakovlev OKB

Yakovlev OKB

Top and above: Yakovlev's first jet, the Yak-Jumo prototype, on the territory of plant No.115 in Moscow.

increased. The rudder and elevators had a metal framework and a fabric skin.

The inward-retracting main gear units had oleo-pneumatic shock absorbers and 600 x 180-mm (23.62 x 7.08-in) wheels equipped with brakes. The aft-retracting tail strut with a 230 x 110-mm (9.05 x 4.33-in) wheel was unusual in having a shock absorber made up of several compact steel leaf springs.

The cockpit (including the instrument panel), electric system and many other items and assemblies of the Yak-Jumo were completely identical to those of its piston-engined forebear. The cockpit was heated but not pressurised. Only minute changes were made to the tried and tested flight control system, pneumatic system (which operated the landing gear, flaps and wheel brakes) and avionics. As on the Yak-3, the armament was housed in the forward fuselage upper decking (above the engine); it consisted of two 23-mm (.90 calibre) Nudel'man/Sooranov NS-23K cannons. The fuel system comprised four tanks in the wings, a fifth tank in the forward fuselage above the engine and a small reserve tank beneath the engine. Thus, the Yak-Jumo was in effect a straightforward conversion of the Yak-3 to take a Jumo 004B turbojet.

The State Defence Committee's directive *'On studying and mastering production of German jet aircraft designs'* issued on 20th July 1945 required NKAP in general and several design bureaux in particular, not only to launch production of the captured German turbojets but to create fighters to be powered by these engines – as soon as possible! Being one of the bosses of the Soviet aircraft industry, Yakovlev fully realised the importance of this task – and of the consequences for him and the other persons responsible should they fail. Hence the prototype of the Yak-Jumo (called 'A. S. Yakovlev's experimental fighter' in some documents) was completed on time in October 1945. However, several blunders made by the designers became apparent no sooner than the tests began. During the very first engine run the exhaust gases melted the duralumin skin of the lower fuselage beyond the heat shield (which turned out to be too short) and burned up the tailwheel; apparently nobody had realised that the tailwheel undercarriage was incompatible with the chosen engine placement and the rubber tyre stood no chances in the hot exhaust! The aircraft had to be returned to the experimental plant for repairs and modifications which were completed in late December.

Yakovlev OKB

Above and below: The second prototype designated Yak15-RD10 seen during manufacturer's flight tests. The two cannons are clearly visible.

Yak-Jumo (Yak-15, Yak15-RD10, Yak-RD) second prototype

That same month the quality control department of MMZ No.115 accepted the second prototype Yak-Jumo. It featured a new horizontal tail of increased area and an indestructible solid steel tailwheel. Five days after the roll-out Yakovlev OKB test pilot Mikhail I. Ivanov started making taxi runs and short hops in this aircraft; soon, however, the second prototype was transferred to TsAGI for the purpose of performing full-scale wind tunnel tests. The fighter made numerous 'flights' in the T-101 wind tunnel with all thinkable combinations of true airspeed and angle of attack, and with the

Yakovlev OKB

engine running at various speeds. This yielded valuable data on the change of longitudinal trim due to the thrust line being located below the aircraft's CG.

Learning from the many fatal accidents with early jets abroad (and the unfortunate statistics of the same kind accumulated in the USSR), the engineers gave a lot of thought to ensuring the structural integrity and reliability of Yakovlev's jet firstling. Sure enough, using the pod-and-boom arrangement and existing wing airfoils (which were rather thick, considering the high speeds for which the new fighter was designed) saved valuable time, but it also increased overall drag and imposed a strict limit on the top speed for structural strength reasons. During the wind tunnel tests the Yak-Jumo was examined not only by Air Force representatives and NKAP top brass but also by the heads of other Soviet aircraft design bureaux (who obviously learned a thing or two from this).

The wind tunnel research programme at TsAGI lasted until February 1946, whereupon the fighter was returned to LII's airfield to resume taxi trials. These were quite a sight; the solid steel tailwheel made a tremendous racket and struck sparks from the concrete runway, but at least it did not burn up any more.

Ground tests and minor modifications based on the results of these lasted another month; during this period the original engine, which turned out to be troublesome, was replaced with another Jumo 004B tested and approved by TsIAM. The prototype's avionics and equipment included an RSI-6MU transmitter, an RSI-3M1 receiver, a GS-15-500 DC generator, a 12-A-10 DC battery and a KP-14 breathing apparatus with a 2-litre (0.44 Imp gal) oxygen bottle. The aircraft's CG range was 21.7% to 17.7% MAC.

On 26th February 1946, when the Yak-Jumo was still undergoing ground tests, the Council of People's Commissars issued yet another directive (No.472-191ss). This document defined the main assignments for the Soviet 'fighter makers' regarding jet fighter development, set the deadlines by which the new fighters were to enter flight test and required NKAP to concentrate on the eminently important tasks of *'...dramatically improving the flight performance of development aircraft and creating new models of aircraft'*. The appropriate MAP order No.162ss appeared on 27th March (by then the Council of People's Commissars had become the Council of Ministers, and all the People's Commissariats were accordingly renamed into ministries). This document said, among other things:

'...Chief Designer and Director of plant No.115 A. S. Yakovlev [...] shall design and build [...] a single-seat fighter powered by a Jumo 004 engine, with the following performance characteristics:

- *maximum speed at sea level, 770 km/h* (478 mph);
- *maximum speed at 5,000 m (*16,400 ft), *850 km/h* (527 mph);
- *range at 0.9 V_{MAX}, 500 km* (310 miles);
- *endurance at 0.9 V_{MAX}, 40 minutes;*
- *endurance in optimum cruise mode, 60 minutes;*
- *climb time to 5,000 m, 4.5 minutes;*
- *service ceiling, 14,500 m (*47,570 ft);
- *armament, two 23-mm* (.90 calibre) *cannons.*

Two prototypes are to be built, the first of which shall be ready for flight testing on 1st September 1946.'

In the spring of 1946 the efforts of the Soviet Union's two leading fighter design bureaux finally bore fruit. The first pure jet aircraft developed by Mikoyan's OKB-155 (the I-300) and Yakovlev's OKB-115 (the Yak-Jumo) successfully made their maiden flights on the same day, 24th April. Taking off at 13:56 Moscow time – several hours after the I-300 – with Mikhail I. Ivanov at the controls, the Yak made a single circuit of the airfield and landed. Thus, seven months had elapsed from the rollout to the first real flight.

Officially the manufacturer's flight tests of Yakovlev's jet fighter, which had by then acquired a spate of official and semi-official designations (it was referred to in various documents as the Yak-15, Yak15-RD10 (*sic*) or even Yak-RD), began on 9th April, continuing until 22nd June 1946. Mikhail I. Ivanov was project test pilot, with Yevgeniy G. Adler as engineer in charge of the test programme. The protocol on the test results said that the fighter was *'a version of the Yak3-VK107A'* and was *'cleared for flight tests with a V_{NE}* (never-exceed speed) *of Mach 0.7'*. The specifications of the prototype Yak-Jumo (Yak15-RD10) as recorded during the manufacturer's flight tests are given in the table below.

RD-10-powered Yak-15 (re-engined Yak-Jumo prototypes)

Elated by this success, A. S. Yakovlev hastened to report to the government that the assignment to design a single-seat jet fighter had been successfully accomplished. Prompted by Yakovlev, on 29th April 1946 the Council of Ministers issued directive No.952-397ss followed by MAP order No.282ss to the same effect on 8th May. These documents officially tasked OKB-115 with developing and building a single-seat fighter powered by the Soviet-built version of the Jumo 004B – the RD-10. The aircraft was to have a top speed of 770 km/h at sea level and 850 km/h at 5,000 m (16,400 ft), reaching this altitude in 4.5 minutes. Range in optimum cruise mode was to be 700 km (437 miles), the service ceiling was to be 14,000 m (45,930 ft) and the take-off run no more than 600 m (1,970 ft). As before, the specified armament consisted of two 23-mm cannons.

Once again, two prototypes were to be manufactured, the first aircraft entering flight test in the third quarter of 1946. This tight schedule probably suited Yakovlev fine, since he had chosen the easy way – even though he was aware that this design approach would hardly yield good results. As a matter of fact, both prototypes were already completed, requiring only new engines to be installed; conversely, the competing design bureaux were at least several months behind in their attempts to create jet fighters (with the exception of the Mikoyan OKB, with whom it was a neck-and-neck race). One can hardly blame Yakovlev for wanting to be the first past the post; besides, the Soviet Air Force urgently needed jet fighters, and every little bit helped. Yakovlev's fighter could fill the void until something better was available.

As already noted, the manufacturer's flight tests of the Yak15-RD10 (this rather cumbersome appellation was soon changed to the definitive service designation Yak-15) ended in June 1946. Oddly, the Yak15-RD10 designation was used even before the prototypes were fitted with the actual RD-10 engines (that is, still had German engines)! Thus, despite the pressure put by the government on the aircraft industry, only two jet fighters – the I-300 (by then officially designated MiG-9) and the Yak-15 – were ready for the traditional Tushino flypast in August 1946; the Yak-15 was flown by Mikhail I. Ivanov on that occasion. After this event the fighter was allocated the NATO reporting name *Feather*.

Yak-15 (Yak15-RD10) initial production fighter (without armament or with reduced armament)

As a result of the important government assignment facing serious delays (many more jets were to have participated in this flypast), on 12th September MAP let loose with an order, demanding that small batches of production jet fighters – at least two aircraft of each model – be manufactured, test-flown and transferred to GK NII VVS for State acceptance trials in time to take part in the 7th November military parade in Moscow's Red Square.

Curiously, pursuant to the government directive the first 15 production Yak-15s earmarked for the said parade were to be manufactured without armament and cockpit armour, with a full-size forward fuselage fuel

Yakovlev OKB

Yakovlev OKB

Yakovlev OKB

Above: Three views of the 15th production Yak-15 (35 Yellow, c/n 31015) at GK NII VVS during State acceptance trials. The fighter was built with a single NS-23 cannon on the starboard side. Note the DF loop aerial. The meaning of the '3' on the rudder is obscure.

Yakovlev OKB

The forward fuselage of Yak-15 c/n 31015 with the cowling removed, showing the cannon installation and the upper fuel tank.

tank (installing the cannons would require the capacity of this tank to be reduced) and an incomplete avionics fit. The day after the MiG-9's and Yak-15's Tushino debut – almost a full month before the abovementioned directive appeared – I. V. Stalin summoned Chief Designers A. I. Mikoyan and A. S. Yakovlev to the Kremlin and personally instructed them to build fifteen copies of each fighter with a view to displaying them publicly at the 7th November parade.

The Tbilisi aircraft factory (plant No.31) was chosen to build the Yak-15; it was still turning out Yak-3s at the time, which undoubtedly was one of the reasons for the choice. The added workload of manufacturing the Yak-15s meant the plant had to work almost round the clock to cope with the task; yet cope it did, meeting the deadline – thanks not least to the considerable structural commonality between the two types. All fifteen aircraft were assembled and test-flown in time for the parade. Eventually, however, the flying display had to be cancelled due to bad weather.

The autumn of 1946 was a hectic time for plant No.31. Not only did the enterprise have to master jet aircraft production and complete the initial production batch of single-seat Yak-15s within the shortest possible time but it was also ordered to build and prepare for testing, as a matter of urgency, the prototype of the two-seat trainer version described later. Since the trainer's development schedule was slipping catastrophically, this aircraft was laid down first, receiving the c/n 31001 (that is, plant No.31, 001st Yak-15 built there).

The next aircraft in the sequence was the first production single-seater (c/n 31002) which received the serial '32 Red'; this fighter and the final aircraft of the initial production batch ('35 Red', c/n 31015) were delivered to GK NII VVS for State acceptance trials. Both fighters were armed with a single NS-23 cannon on the starboard side. The all-up weight varied from 2,638 to 2,742 kg (5,815-6,045 lb); the latter option included 90 kg (198 lb) for the pilot, 580 kg (1,278 lb) of jet fuel, 14 kg (30 lb) of petrol for the start-up system, 11 kg (24.25 lb) of engine oil, 37 kg (81.5 lb) for the cannon, 23 kg (50.7 lb) of ammunition and 25 kg (55 lb) of removable equipment. The State acceptance trials, which lasted until April 1947, were performed by Air Force test pilots Gheorgiy A. Sedov and Aleksandr G. Proshakov; the latter was also the institute's engineer in charge of the Yak-15's trials.

Predictably, the fighter's performance was inferior not only to that of the MiG-9 but also to that of the captured Messerschmitt Me 262A-1a evaluated by GK NII VVS in late 1945. The drag generated by the thick wing airfoil ate up a lot of speed, preventing the aircraft from making full use of the engine's thrust; the production Yak-15's top speed was limited to 700 km/h (434) below 3,150 m (10,330 ft) and Mach 0.68 at higher altitudes for structural strength reasons. The powerplant itself was a whole can of worms; three engine changes had to be made on the two aircraft in the course of the State acceptance trials, and frequent engine troubles (the RD-10 could not run at nominal power for more than ten minutes) prevented GK NII VVS from establishing the fighter's performance throughout the altitude envelope. Ascent in maximum rate-of-climb mode was only permitted up to 8,370 m (27,460 ft) in order to avoid damage to the engine; due to the engine's limitations the service ceiling turned out to be 10,000 m (32,810 ft) instead of the estimated 12,000 m (39,370 ft).

The military also voiced a long list of complaints regarding the design. Among other things , the metal tailwheel caused the rudder (and hence the rudder pedals) to vibrate during taxying. When the engine was ground-run it kicked up stones, damaging the elevators; the underside of the horizontal tail iced up during prolonged taxying on snow-covered airfields (quite simply, the snow melted in the jet exhaust and the water then froze on the stabiliser skin); the jet blast eroded the airfield surface and so on. The chief source of annoyance was the smoke entering the cockpit during long missions (20-25 minutes), irritating the pilot's eyes; flying goggles were of no avail. The problem was caused by fuel and oil perpetually leaking onto the engine and then decomposing when the engine ran, emitting smoke.

GK NII VVS demanded that modifications be made to the fighter. True, the military were not just complaining – they performed what modifications they could themselves (such as replacing the original single-wire aerial with a twin-wire aerial to increase radio reception range). In the State acceptance trials protocol concerning Yak-15s c/ns 31002 and 31015 the State commission suggested equipping the Yak-15 with a tricycle landing gear, modifying the horizontal tail and installing special fuel tanks to ensure stable engine operation under negative G-loads.

Concurrently, another Yak-15 (c/n 31009) underwent a special test programme in order to establish the type's aerobatic capabilities. GK NII VVS test pilot Pyotr M. Stefanovskiy performed the first aerobatics session in this aircraft on 25th February 1947. Of all Soviet jet fighters existing in the spring of 1947 the Yak-15 was easiest to fly, presenting no problems for the average fighter pilot; airmen who had flown the Yak-3 transitioned to its jet stablemate without any trouble. As a combat aircraft, however, the Yak-15 had inadequate longitudinal stability and controllability (the insufficient elevator authority was especially disappointing). Also, an unpleasant side effect of the pod-and-boom arrangement was the marked change in stick forces from the elevators when the throttle was advanced or retarded.

Despite the shortcomings described above, the fighter received a generally positive appraisal. The aforementioned protocol signed in May 1947 said:

'The Yak-15 aircraft [...] meets the Air Force's requirements for a jet fighter trainer... (sic); [...] Due to the extremely short range of

Yakovlev OKB

Yakovlev OKB

Top and above: Yak-15 '32 Yellow' (c/n 31002) seen during State acceptance trials following modifications. The aircraft has two cannons; note also the absence of the exposed DF loop aerial.

Yakovlev OKB

Yakovlev OKB

Top and above: Two views of Yak-15 '57 Yellow' during checkout trials at GK NII VVS. The yellow numeral '2' on the rudder did not mean anything at all and was intended to confuse would-be spies.

Yakovlev OKB

Above: '37 Yellow', one of the Yak-15s of the initial-production ('parade') batch. This aircraft lacks cannons altogether.

Yakovlev OKB

Above: Another view of the same aircraft, with the Major Caucasus Ridge as a backdrop.

Yakovlev OKB

'12 Yellow', an early-production Yak-15, on the factory apron in Tbilisi. The aircraft is again marked '2 Yellow' on the rudder.

only 300 km (186 miles) *in optimum cruise at 1,000 m* (3,280 ft) *[...] it is inexpedient to develop the aircraft into a fully-capable combat type, since this will impair its handling and manoeuvrability due to [...] the much higher all-up weight resulting from the additional fuel required to provide a range of 800-900 km* (497-559 miles).

Because of the simple design, low thrust of the RD-10 engine and simple piloting techniques the State commission considers it advisable [...] to develop the Yak-15 into a dual-control trainer with a tricycle landing gear...'

Thus, the decision to launch production of the Yak-15 was dictated solely by its commonality with the Yak-3. The big idea was that the type would be used as a stepping stone from piston-engined fighters to more capable ('real') jet fighters and as a proficiency trainer.

Yak-15 production fighter with full armament

On 16th December 1946, when the State acceptance trials were still in progress, the Council of Ministers issued directive No.2698-1114ss, the appropriate MAP order No.801ss following ten days later. These documents ordered the ministry's 1st Main Directorate chief Ter-Markaryan and the Tbilisi aircraft factory's director V. Saladze to build and deliver 50 Yak-15 fighters powered by RD-10 turbojets; 25 of these were to be standard single-seat aircraft and the rest were to be completed as dual-control trainers with a reduced armament fit. The production schedule was spelled out very specifically: plant No.31 was to manufacture five fighters in January 1947, 15 in February, 20 aircraft in March (including 15 two-seaters) and the final 10 (all of them trainers) in April. Thus, by the end of April 1947 MAP planned to deliver a total of 65 Yak-15s with Soviet-built engines. As of 31st December 1946 the plant had built the 15 unarmed examples mentioned earlier (the ones which were to participate in the 7th November parade), and 8 more Yak-15s were in various stages of manufacturing.

The engines powering production Yak-15s were produced by aero-engine plant No.26; the first production examples were in fact assembled from original Jumo 004B1 parts delivered from Germany by way of war reparations. The RD-10 had a ten-stage axial compressor (hence the relatively small casing diameter), six straight-through combustion chambers and a single-stage axial turbine. The nozzle area was adjusted by a translating cone (centrebody) according to engine rpm. Starting was by means of a two-stroke two-cylinder Riedel engine, which in turn was started electrically or manually (by means of a cord).

Officially (according to the engine ID cards) the service life of the first production RD-10s was only 25 hours. In reality, however, it was even shorter, being an appalling 17 hours at best. On the other hand, during this brief period the engine's reliability was reasonably high.

In January 1947, when the Yak-15 was undergoing State acceptance trials, Vice Minister of Aircraft Industry Pyotr V. Dement'yev wrote to the Air Force's Chief Engineer Markov, expressing his ideas about Yak-15 production:

'In my opinion, the 65 Yak-15s with RD-10 engines currently being manufactured by plant No.31 for the May Day parade shall be completed to exactly the same standard as the ones [...] built for the 7th November 1946 parade, that is, without armament and cockpit armour, with a full-size upper kerosene tank (the reference to kerosene is due to the fact that, as the reader remembers, there was also a petrol tank for the starter engine – *Auth.*), [...] *with a [communications] radio receiver and transmitter but no RPKO-10M direction finder...*

The NS-23K cannon and the RPKO-10M have failed to pass State acceptance trials and are not manufactured in series (at that point – *Auth.*). *On these aircraft all the defects discovered in the course of preparations for the 7th November [1946] parade, as listed in your report, will be eliminated.*

After delivering 65 aircraft [...] in the configuration stated above plant No.31 will start manufacturing aircraft equipped with armament, cockpit armour and a DF with a flat (that is, buried – *Auth.*) *loop aerial.'*

Thus by the early summer of 1947 the Soviet Air Force had more than 50 Yak-15 fighters on strength – or rather in service, since some of them were in reality not fighters at all, lacking armament or having but one cannon. Anyway, the rapid build-up of the Soviet jet fighter force was demonstrated to good effect on 1st May 1947 when 50 Yak-15s and 50 MiG-9s streaked over Moscow's Red Square. That year also saw the Soviet Union's first aerobatic displays by jet aircraft. Even before GK NII VVS had officially pronounced the Yak-15 fully aerobatic, service pilots had illegally performed aerobatics in it at the Soviet Air Force's Conversion training centre. Now, Colonel I. Poloonin was the first to perform solo aerobatics in a Yak-15 in front of the watching crowds at Tushino; this was followed by a breathtaking aerobatics display by a flight of Yak-15s led by Lieutenant-Colonel N. I. Khramov, Hero of the Soviet Union.

The 1948 Tushino air display featured aerobatics performed by assorted jets flown by I. Poloonin, Yuriy A. Antipov, A. G. Proshakov and other Air Force pilots. By then General Yevgeniy Ya. Savitskiy, twice HSU, had organised a display team flying the Yak-15; it was staffed by four well-known military pilots – P. Sereda (HSU), N. I. Khramov, V. Yefremov and P. Solov'yov. At the 1949 Tushino event group aerobatics were performed by a quintet of Yak-15s led by Colonel P. Choopikov (HSU).

Production of the type ended in 1947 after a total of 280 Yak-15s had rolled off the

Yefim Gordon archive

Above: A still from a cine film showing a flight of initial-production Yak-15s in Vee formation during a rehearsal of the 7th November 1946 military parade. The foremost aircraft has two cannons

Yakovlev OKB

Above: A famous picture of a production Yak-15 in flight. '20 Yellow' is a two-cannon example marked '1 Yellow' on the rudder.

Yakovlev OKB

Following the recommendations of GK NII VVS, the Soviet Air Force used the Yak-15 as a training aircraft to ease the transition to more capable jet fighters.

Yakovlev OKB

Yakovlev OKB

Top and above: The Yak17-RD10 development aircraft. These views illustrate well the odd main landing gear design, as well as the narrow wheel track.

Yakovlev OKB

The Yak17-RD10 runs up its engine on the ground, kicking up dust. Note that the upper half of the engine cowling has been removed.

assembly line in Tbilisi. Most of them remained in Air Force service for a few more years. The most widespread malfunctions occurring in service included seepage of oil from the main gear oleos, snapping of individual wires in the rudder control cable and breakage of the tailwheel strut's leaf springs (probably due to overheating of the wheel).

As noted earlier, the Yak-15's chief shortcoming was its extremely limited range which was less than half that of piston-engined fighters due to the RD-10's high fuel consumption. With a full fuel load of 590 kg (1,300 lb) the Yak-15 had a range of only 510 km (316 miles); in contrast, the Yak-3 whose full fuel load was only 350 kg (771 lb) had a maximum range of 1,060 km (658 miles)! This problem became especially acute when preparations for the 1947 May Day parade began. During formation flying the 50 Yak-15s drank up their fuel supply in just 30 minutes, which not only required the fighters' route to and from the Red Square to be carefully planned but also meant that large groups of aircraft had to take off at minimum intervals.

Despite all its limitations, the Yak-15 is nevertheless not to be dismissed so easily. In its day it played an important role, helping to train hundreds of jet fighter pilots who later transitioned to the much more capable MiG-15. Incidentally, the unavailability of the two-seat trainer version led some Soviet Air Force units to convert a few single-seat Yak-15s to dual-control trainers without bothering about authorisation from the OKB or the factory. The Yak-15 was the first jet aircraft officially included into the Soviet Air Force inventory and the first jet fighter on which VVS pilots performed aerobatics.

The *Feather* became the progenitor of a small family of fighters which, though none too popular with the pilots who flew them, saw service not only with the VVS but also with the air arms of some of the Soviet Union's allies.

Some Western sources claim that on 10th May 1953 a Yak-15 was shot down in Korea by a US Air Force (319th Fighter Interceptor Squadron) Lockheed F-94 Starfire piloted by Captain D. Phillips. However, this is nonsense, as Yakovlev jets never participated in the Korean War. (A possible explanation is that someone mistakenly wrote 'Yak-15' instead of 'MiG-15'.)

Currently only a single Yak-15 survives in Russia. Coded '37 Yellow', this aircraft was probably a development machine owned by the Yakovlev OKB and is now preserved in the company museum. The aircraft left the museum's premises at least twice to be displayed publicly during airshows (at Moscow-Tushino in August 1977 and at Moscow-Khodynka in August 1989).

Yak17-RD10 (Yak-RD, Yak-RD10) fighter (first use of Yak-17 designation)

Now we have to travel back in time a bit. On 29th April 1946, three days after the maiden flights of the I-300 and Yak-Jumo prototypes, the Council of Ministers issued directive No.952-397ss ordering Chief Designer and

Director of plant No.115 Aleksandr S. Yakovlev to develop and build a new single-seat fighter powered by an RD-10 turbojet and featuring improved aerodynamics. The directive specified a top speed of 850 km/h (528 mph) at 5,000 m (16,400 ft) and a range increased from the Yak-Jumo's 500 km (310.5 miles) to 700 km (437 miles) at 0.9 V_{MAX}. The appropriate MAP order No.278s *'On the manufacturer's flight tests of A. S. Yakovlev's experimental fighter with an RD-10 engine'* followed on 7th May.

The ADP of the aircraft provisionally designated Yak17-RD10 did not take long to complete (in some documents the fighter was referred to simply as the Yak-RD or Yak-RD10). Outwardly the new fighter closely resembled the original Yak15-RD10 (Yak-15) but incorporated major changes, featuring redesigned wings of a different planform utilising high-speed laminar airfoils, redesigned tail surfaces of increased area and an ejection seat. The latter had an armoured back; the pilot was also protected by a 55-mm (2⅝₂ in) bulletproof windscreen.

Because of the new airfoils the wings were too thin to accommodate the mainwheels. Hence the main gear units had to be redesigned; they were attached to one of the fuselage mainframes, retracting aft into the fuselage in a rather complex fashion, and featured a levered suspension with leading arms. The mainwheel size remained the same at 600 x 180 mm (23.62 x 7.08 in) but the metal tailwheel was widened considerably to 230 x 180 mm (9.05 x 7.08 in) and the landing gear track was narrowed to only 2.0 m (6 ft 6¾ in).

Two wing versions with an area of 15.0 m^2 (161.29 sq ft) and 13.5 m^2 (145.16 sq ft) respectively were considered for the Yak17-RD10. The avionics and equipment suite was standard for Soviet fighters of the period. It included a radio set with an RSI-6 transmitter and RSI-6M receiver, an RPKO-10M DF, an SCh-3 IFF transponder, a PBP-1A gunsight, a GS-15-500 DC generator, a 12-A-10 DC battery and a KP-14 breathing apparatus with a 2-litre (0.44 Imp gal) oxygen bottle.

Prototype construction proceeded quickly; the aircraft was rolled out on 3rd September 1946 and the first flight date was tentatively set for 20th September. However, the ground tests continued until 26th September and were terminated after OKB test pilot G. S. Klimooshkin had made a few taxi runs. Theoretically the thinner wings would have given the Yak17-RD10 a certain speed advantage over the Yak15-RD10; however, the latter had already been ordered into production and the tailwheel landing gear was outdated anyway. Therefore, the sole Yak17-RD10 was never flown and the designation Yak-17 eventually passed to a closely related but different aircraft described later.

Yak-15 IFR system testbed

A production Yak-15 serialled '47 Yellow' was used by LII as an in-flight refuelling (IFR) system testbed in the late 1940s. In 1948 a design team at LII headed by Vladimir Sergeyevich Vakhmistrov (this engineer had gained fame in the 1930s for his fighter/bomber combinations called *zveno* – flight, as in tactical unit) developed an IFR system. The tanker and the receiver aircraft deployed special cables which became intertwined due to the tanker manoeuvring in a certain fashion. Then a hose would be unwound from the tanker's wing or fuselage and the cable rewound until the hose engaged a receptacle under the receiver aircraft's wingtip. Once the hose had been locked securely into position, gravity fuel transfer could begin.

Soon afterwards, however, LII test pilots Igor' Shelest and Viktor Vasyanin developed a slightly different IFR system at their own risk. They rightly believed that the complex refuelling procedure should be simplified and automated as much as possible. The Shelest/Vasyanin system involved flying on parallel courses, which meant the receiver aircraft did not need to enter the tanker's wake turbulence which might lead to loss of control. The tanker deployed a hose stabilised by a small parachute from one wing and the receiver aircraft placed its opposite wing over the hose. After that, the hose was rewound until a fitting at the end engaged the other aircraft's receptacle. The receiver then increased speed so that the hose formed a loop and rotated the receptacle, opening a valve. Another important difference was that fuel was transferred under pressure, which speeded up the process considerably.

The Soviet Air Force top brass decided that the Shelest/Vasyanin system held promise, so two Tupolev Tu-2 bombers were set aside for conversion into IFR system testbeds with a dummy installation. Flight tests with these aircraft began in the summer of 1949; later the abovementioned Yak-15 '47 Yellow' was similarly outfitted with a dummy receptacle, making contact with the Tu-2 'tanker'.

Yak-21 (Yak-21V, Yak-Jumo *vyvoznoy*, Yak-15V, Yak-15UT) fighter trainer

As mentioned earlier, even though it had been decided to relegate the Yak-15 to the conversion trainer role, a dual-control version

Yefim Gordon archive

Above: Yak-15 '47 Yellow' was used as a testbed for the Shelest/Vasyanin wing-to-wing in-flight refuelling system. The refuelling hose can just be seen beneath the tail.

Yefim Gordon archive

The Yak-15 formates with the Tu-2 'tanker' during a simulated in-flight refuelling.

Yakovlev OKB

Above and below: '101 Yellow', the sole prototype of the Yak-21 conversion trainer derivative of the Yak-15. The Cyrillic letter V on the rudder probably stood for *vyvoznoy*, signifying the aircraft's role.

Yakovlev OKB

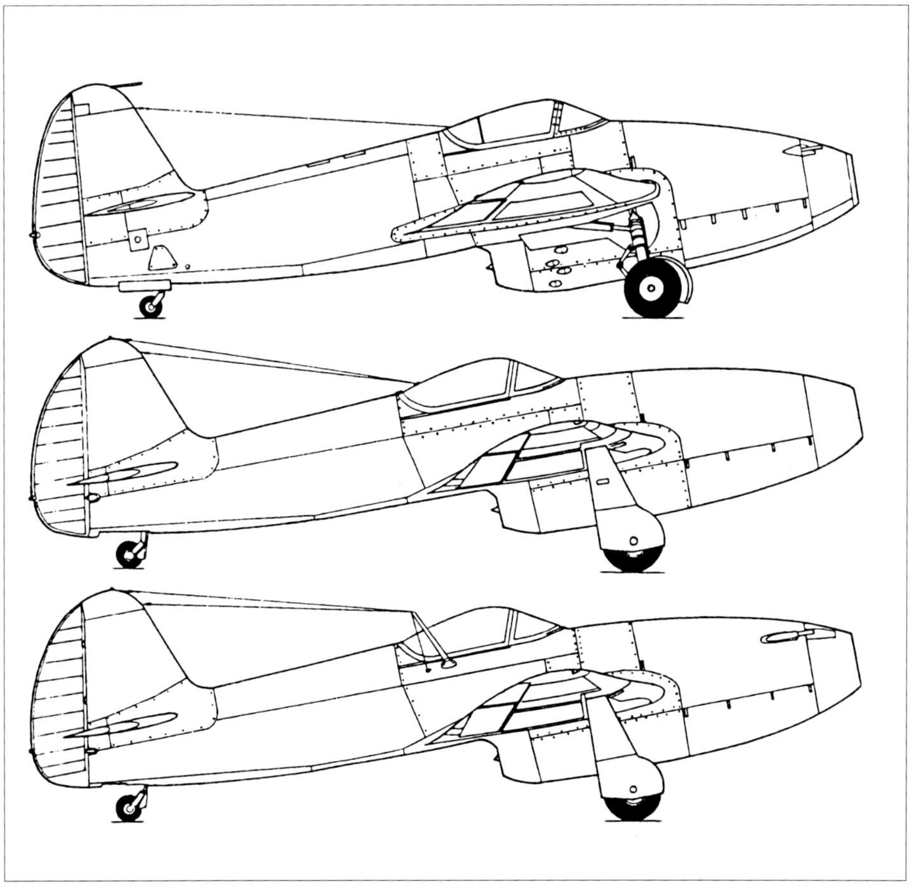

Top to bottom: The Yak17-RD10, the Yak-Jumo (Yak-3 Jumo) prototype and a standard Yak-15.

of the fighter was sorely needed. As far back as 9th July 1946, when the Yak-15 had yet to commence State acceptance trials, the Council of Ministers issued directive No.1498-665s concerning *'...development of dual-control trainers in order to facilitate the transition of fighter pilots to the new jet hardware'*. MAP order No.525s to the same effect was signed on 6th August; among other things, it said:

'1. Chief Designer A. S. Yakovlev shall [...] develop and build a single prototype of [...] a fighter with a Jumo 004 (RD) engine based on a single-seat fighter currently in production and submit it for State acceptance trials by 1st November 1946;

2. Two-seat conversion and proficiency trainers shall have the same field performance as the respective single-seat combat aircraft. Both cockpits shall be fully equipped so as to permit flying the aircraft. Trainer aircraft may have a reduced weapons fit and, if necessary, a lower top speed.'

Pursuant to this order OKB-115 started developing a two-seat trainer derivative of the Yak-15. Initially it was known as the Yak-Jumo *vyvoz***noy** (the contemporary Soviet term for familiarisation and conversion trainers); later such designations as the Yak-15V (*vyvoznoy*) and Yak-15UT (*oo***cheb***no-treni***rov***ochnyy* – training, used attributively) came into being. Eventually, however, the trainer was officially designated Yak-21 (some documents call it Yak-21V). The trainer differed from the Yak-15 primarily in having a redesigned forward fuselage. The armament was deleted and a second cockpit for the trainee was provided where the cannon(s) had been; both cockpits were enclosed by a common canopy. The instrumentation and location of controls in both cockpits was similar to that of the single-seat Yak-15.

The prototype was built by plant No.31 in Tbilisi in the autumn of 1946. Since the Council of Ministers and MAP demanded that a dual-control fighter trainer be submitted for State acceptance trials by 1st November, this machine, as noted earlier, was the first to be laid down in the initial batch of single-seat Yak-15s. Additionally, to meet the demands of the CofM and MAP, the Tbilisi plant was to complete 25 dual-control machines by the end of April 1947. For various reasons, however, the entire production and test schedule went down the drain. Serialled '101 Yellow' and bearing a yellow 'B' on the rudder (the Cyrillic letter 'V', which probably stood for *vyvoznoy*), the Yak-21 prototype commenced manufacturer's flight tests as late as 5th April 1947 instead of the autumn of 1946, as originally planned. The maiden flight was performed by test pilot P. U. Fokin.

The machine flew well; nevertheless, in the course of the tests it became clear that a more promising dual-control version of the *Feather* (the Yak-21T, described below), which had been developed by then, would have better performance. Hence all further work on the 'taildragger' trainer was suspended. It is not known whether the aircraft's performance characteristics were determined, but according to OKB-115's work plan for 1947 the Yak-21 with the tailwheel undercarriage was to have a top speed of 770 km/h (478 mph) at sea level and 800 km/h (496 mph) at altitude, a service ceiling of 10,000 m (32,810 ft), climbing to 5,000 m (16,400 ft) in 4.3 minutes. The take-off run and landing run were to be 350 m and 400 m (1,150 ft and 1,310 ft) respectively.

Yak-15U (Yak15U-RD10) (Yak-17) fighter prototype

As noted earlier, the Yak-15 had such serious shortcomings as inadequate range and an unsuitable undercarriage design. Back in May 1947 the State acceptance trials protocol stated that it was *'advisable [...] to develop the Yak-15 into a dual-control trainer with a tricycle landing gear'*. Therefore, almost concurrently with the design work on the single-seat Yak17-RD10 and the two-seat Yak-21 the Yakovlev OKB set about developing a new version of the RD-10-powered single-seat fighter and its two-seat trainer variant, taking due regard to the new, more stringent requirements concerning stability, controllability and cockpit visibility of trainer aircraft, as well as to the requirement for a tricycle undercarriage. The single-seat fighter was designated Yak-15U (*u**looch**shennyy* – improved; in documents it was referred to as the 'Yak15U-RD-10 with a tricycle undercarriage and drop tanks'), while the trainer version was designated UTI Yak17-RD10 or Yak-21T (for ***tryokh**ko**lyos**noye shas**see*** – tricycle landing gear).

Like the baseline Yak-15, these aircraft featured a pod-and-boom layout. This time, however, the designers made use of the new nosewheel undercarriage facilitating take-off and landing; this necessitated some changes to the inboard portions of the wings and the forward fuselage structure. An auxiliary spar and additional ribs were introduced in the forward parts of the wings where the mainwheel wells were located; this not only enhanced the wing skin stiffness but also made it possible to place fuel tanks in the wings. The wing tanks were separated from the mainwheel wells by a curved spar; they were accessible via a detachable stressed cover plate on the wing undersurface.

The nose gear unit was semi-retractable; it pivoted aft to lie against the fuselage

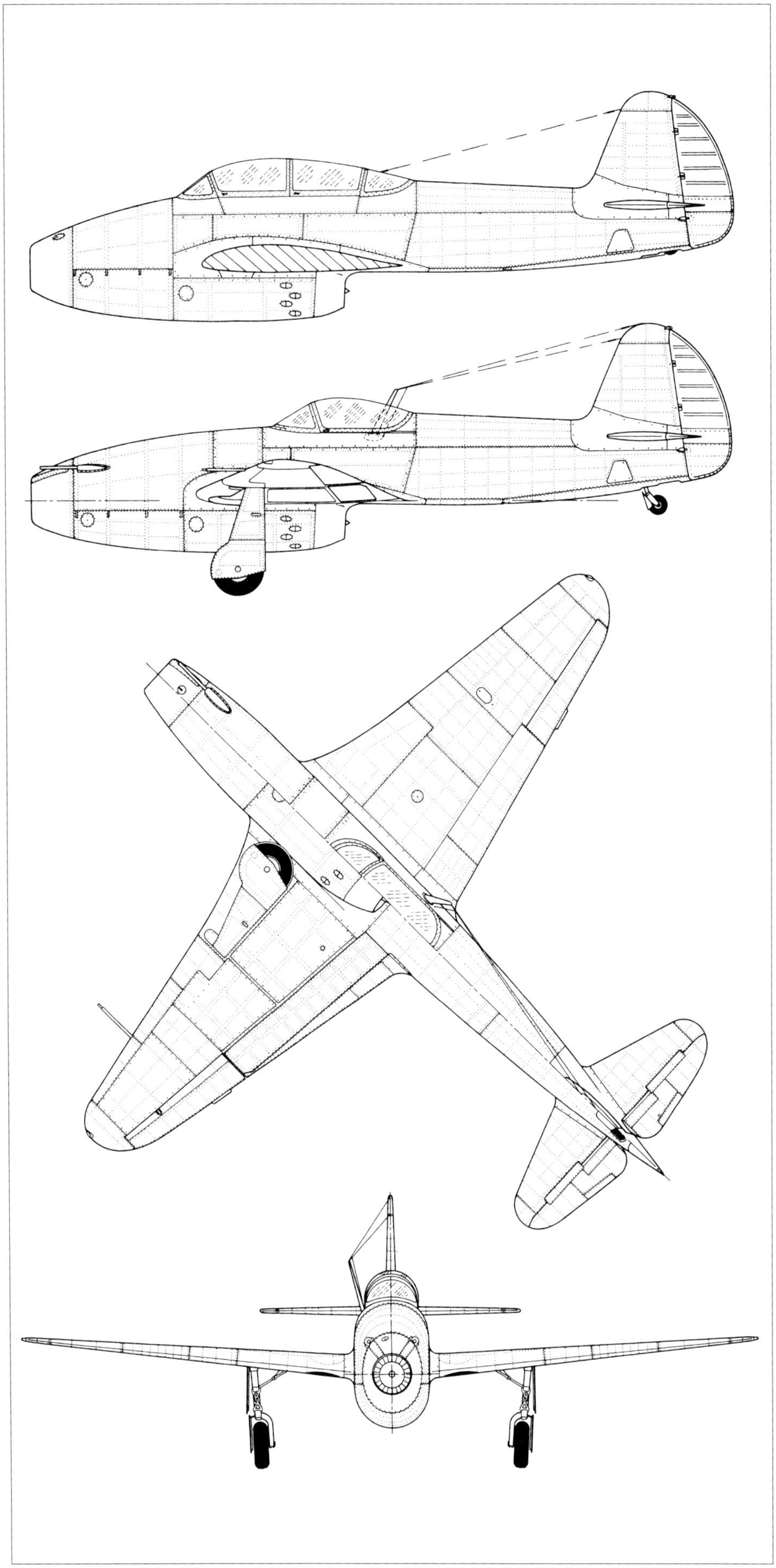

A three-view of the Yak-15, with an additional side view of the Yak-21 dual-control trainer.

Yakovlev OKB

Above and below: The tricycle-gear Yak15U-RD10 was the prototype of the Yak-17. The serial looks very dark in these views but is actually yellow, not red. Note the windshield frame and the shape of the tail.

Yakovlev OKB

Yakovlev OKB

Above: The Yak15U-RD10 prototype with drop tanks attached at GK NII VVS during State acceptance trials.

Yakovlev OKB

An air-to-air shot of the Yak15U-RD10 prototype. Note the exposed position of the retracted nose gear unit; the rear fairing visible in the upper photos has been removed.

undersurface and was covered by a teardrop fairing. The wheel size was changed, too: the mainwheels now measured 570 x 140 mm (22.23 x 5.46 in), the nose unit had a 400 x 150 mm (15.6 x 5.85 in) wheel. The cockpit of the Yak15U-RD10 was increased in length and height to improve the pilot's posture, but the teardrop-shaped canopy of the standard Yak-15 was retained.

To increase range, drop tanks were suspended under the wingtips, which had to be suitably modified. In all, the drop tanks held up to 331 litres (72.8 Imp gal) of kerosene.

To improve the new machines' stability and handling the designers made some alterations to the tail unit. The fin was increased in height and the rudder chord was reduced. The area of the horizontal tail was also increased by 0.1 m^2 (1.07 sq ft); accordingly the aerodynamic compensation of the elevators was reduced to 19%. A certain revision of the size of the tail surfaces necessitated the addition of a spring and a bellcrank in the control system, which increased the gearing ratio from the control stick to the elevators by 20%.

UTI Yak17-RD10 (Yak-21T) fighter trainer prototype

On the UTI Yak17-RD10 (Yak-21T) the common canopy enclosing the cockpits had two separate sliding hoods; it dispensed with the moulded windscreen which was believed to distort the view. The avionics and equipment suite was somewhat simplified as compared to the combat version (the trainer featured an RSI-6 radio, a RPKO-10M DF, a GSK-1500 generator, a 12-A-10 DC battery and KP-14 breathing apparatuses with two oxygen bottles holding 2 litres (0.44 Imp gal) each) One thing the single-seater did not have, however, was the SPU-2M intercom. The fuel system of the UTI Yak17-RD10 (Yak-21T) was designed so as to enable the aircraft to perform aerobatic manoeuvres in inverted flight. The rest of the design was identical to that of the single-seat Yak-15U.

The Yak-21T prototype serialled '22 Red' was manufactured at plant No.464, making use of parts and assemblies of a production Yak-15 delivered from Tbilisi. This was the first jet aircraft produced by that factory, hence the machine received c/n 01464. It took to the air in May 1947; in the subsequent two weeks it passed manufacturer's tests which were conducted by OKB-115 test pilot G. S. Klimooshkin and leading engineer N. I. Leonov. In the same month the new ***spar**ka* (the colloquial name for a dual-control trainer) was transferred to GK NII VVS for State acceptance trials which lasted from 31st May to 10th August (though the protocol on the test results was approved as early as 4th August).

It turned out that the new trainer did not meet the Air Force's requirements on many counts. The top speed fell short of the expectations, the range was little more than half of the required figure; besides, the 10-minute limit imposed on continuous engine operation at maximum thrust rating made it possible to climb at the maximum rate only up to 8,000 m (26,250 ft). Also, the possibility of flying at high altitude was restricted by the insufficient fuel amount. The instructor could not start the engine, operate the undercarriage and flaps, nor could he maintain radio communication. Contrary to official requirements, the machine could not perform flight with negative G-loads.

Still, the military testers also noted the trainer had its virtues. It could fly stably at high angles of attack, the trainer's strength margin enabled it to pull 8 Gs when performing aerobatics, and spin recovery was virtually instantaneous.

Bearing in mind the absence of a two-seat jet trainer in the Air Force, the State commission found it possible to recommend the UTI Yak17-RD10 for series production. Since the trainer was passing the tests with no armament installed (such were the initial requirements), the concluding part of the test protocol suggested that *'...one NS-23 cannon and an S-13 gun camera be installed on production aircraft'*.

Yak-17UTI production fighter trainer

The results of the State acceptance trials were approved by the Council of Ministers in its directive No.3284-1071 dated 17th September 1947. The directive required the trainer to be put into production at plant No.31 under the service designation Yak-17UTI.

Checkout tests of an early production Yak-17UTI (still lacking armament) were completed in the spring of 1948. They revealed a certain deterioration of the aircraft's performance (possibly caused by an increase in the all-up weight, including the fuel load). Nevertheless, thanks to its ease of handling and good performance the production jet trainer quickly won recognition among the pilots of the Soviet Air Force's fighter regiments. It played an important role in helping pilots master jet aircraft *en masse*. A few shortcomings were revealed in squadron service, notably the lack of an artificial horizon (which hampered poor-weather operations) and the absence of a boarding step.

Curiously, a model exists (presumably originating from the Yakovlev OKB) which depicts a Yak-17UTI (with its characteristic cockpit canopy and vertical tail) with a tail-wheel undercarriage! This may have been an intermediate project study that was abandoned at an early stage.

Yakovlev OKB

Yakovlev OKB

Yakovlev OKB

Top, centre and above: '22 Yellow', the UTI Yak17-RD10 (Yak-21T) trainer prototype. Note the tail bumper and the absence of the nosewheel's rear fairing.

Yakovlev OKB

Above: A production Yak-17UTI serialled '41 Red' seen during flight tests. The canopy design makes an interesting comparison with the Yak-21.

Yakovlev OKB

Above: A production Yak-17 serialled '20 Yellow' in flight. Note the gun camera fairing atop the windshield and the landing light buried in the port wing leading edge.

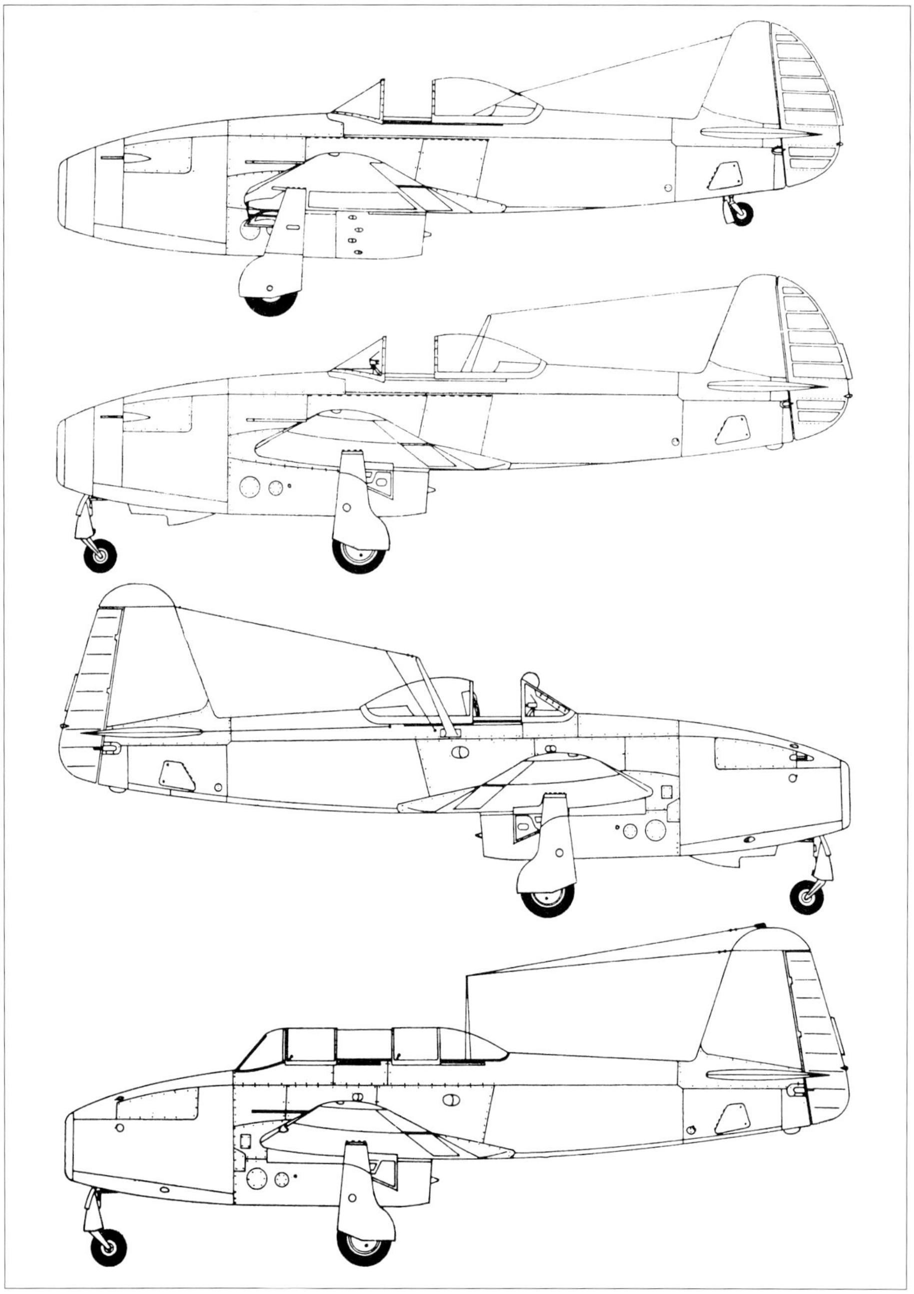

Top to bottom: The Yak-15 (given for comparison), the Yak-15U (Yak-17) prototype as originally completed, a production Yak-17 and a Yak-17UTI.

Yak-17 (Yak-15U) production fighter

The first prototype of the single-seat Yak15U-RD10 with a tricycle undercarriage was completed by plant No.464 in the summer of 1947, a while later than the Yak-21T; this was done by converting a stock production Yak-15. The first flight took place in June, and manufacturer's tests conducted by engineer V. I. Yemel'yanov and test pilot G. S. Klimooshkin were completed on 10th August (the test protocol was endorsed on 23rd August), whereupon the machine was transferred to GK NII VVS. On 3rd August 1947, when the manufacturer's tests were drawing to a close, the aircraft took part in the annual Tushino flypast with Klimooshkin at the controls.

State acceptance trials of the Yak15U-RD10 prototype began at GK NII VVS on 27th August 1947, ending on 15th November. The aircraft was tested with the armament installed, complete with an ASP-1 gunsight and a PAU-22 gun camera housed in the starboard outer wing panel. The normal all-up weight was 2,890 kg (6,370 lb), including 90 kg (198 lb) for the pilot, 555 kg (1,223 lb) of fuel, 10 kg (22 lb) of oil, 11 kg (24.25 lb) of start-up petrol, 74 kg (163 lb) for the two NS-23 cannons, 46 kg (101 lb) of ammunition and 25 kg (55 lb) of detachable equipment. With drop tanks the AUW rose to 3,240 kg (7,140 lb). The trials were conducted by GK NII VVS project engineer-cum-test pilot Gheorgiy A. Sedov and project test pilot Pyotr M. Stefanovskiy; the aircraft was also flown by A. G. Kochetkov, A. G. Proshakov, A. P. Sooproon, Yuriy A. Antipov, L. M. Koovshinov, V. G. Ivanov and V. I. Khomiakov.

In 'clean' condition the fuel amount available for level flight to a range of 395 km (245 miles) at 1,000 m (3,280 ft) and 407 km/h (252 mph) was 349 kg (769 lb), or 410 litres (90.2 Imp gal). With drop tanks, available fuel to cover a distance of 717 km (446 miles) at 8,000 m (26,250 ft) and 502 km/h (311 mph) was 693 kg (1,528 lb) or 815 litres (179.3 Imp gal). The rest of the fuel was used for a trial engine run on the ground, for gaining altitude and for descent and landing approach. Two unscheduled engine changes had to be made in the process of the State acceptance trials.

The trials protocol said that, as far as piloting techniques, stability and controllability were concerned, the Yak-15U was virtually identical to the modified Yak-15 c/n 31002 which had been tested at GK NII VVS a while earlier. The landing procedure, however, was easier. During flights in turbulent conditions with the wingtip tanks in place the aircraft tended to yaw, this tendency being especially pronounced during turns. When put into a spin, the fighter transitioned to a smooth steep spiral and, in the event of a stall, recovered from the spin without delay. True enough, the State commission made quite a few critical remarks about the fighter; nevertheless, the concluding part of the Protocol on the results of the State acceptance trials ran as follows:

'1. The aircraft has passed State acceptance trials satisfactorily.

2. In its performance and armament it is superior to the production Yak-15 and can be introduced into Air Force service.'

After some modifications effected during the winter, in March 1948 the Yak-15U passed a new round of State acceptance tests and was once more recommended for service introduction.

The new version of the single-seat fighter was cleared for series manufacture at aircraft plant No.31 in Tbilisi by Council of Ministers directive No.878-280s issued on 20th March 1948; simultaneously, the aircraft received the designation Yak-17.

In 1948-49 the plant produced a total of 430 aircraft (single-seat Yak-17s and two-seat Yak-17UTIs), of which 279 were manufactured in 1948. In 1949 an aerobatic team comprising nine Yak-17s and led by Lieutenant-Colonel N. Shool'zhenko put on an aerobatics display at the traditional Tushino air show.

Both versions were on strength with the Soviet Air Force (VVS) and the Air Defence Command (PVO); in addition, they were supplied to four 'friendly nations' – Poland, Bulgaria, Czechoslovakia and China. Like its precursor, the Yak-17 had the NATO codename *Feather*.

Production machines differed from the prototype in some respects. Thus, the horizontal tail span was increased from 3.25 m (10 ft 7 61/64 in) to 3.6 m (11 ft 9 47/64 in); accordingly, its area rose from 2.83 to 2.916 m^2 (from 30.43 to 31.35 sq ft). Conversely, the vertical tail area was reduced from 1.86 to 1.813 m^2 (from 20.0 to 19.49 sq ft). The original NS-23 cannons gave way to the NS-23K version featuring a longer barrel, and the PAU-22 gun camera was replaced with the new S-13 (it was mounted on top of the cockpit windshield).

Yak-17 c/n 311177 (that is, plant No.31, Batch 11, 77th aircraft in the batch) and examples with c/ns 311205 through 3112118 were fitted with ASP-3M optical gunsights. The USB-1 ammunition counter was introduced from c/n 311222 onwards, enabling the pilot not only to see how many rounds remained but also to fire short bursts. Late-production machines were powered by RD-10A engines having a TBO increased to 50 flying hours.

The dimensions and performance characteristics of different versions of the Yak-15 and Yak-17, based on the results of calculations, manufacturer's flight tests and State acceptance trials are compared in the table.

Specifications of the Yak-15 and its derivatives

	Yak-15 **c/ns 31002, 31005** *Trials, April 1947*	**Yak15U-RD10** **(prototype)** *State trials*	**Yak-17UTI** **(production)** *Checkout trials*
Powerplant	RD-10	RD-10	RD-10
Thrust, kgp (lbst)	900 (1,984)	900 (1,984)	900 (1,984)
Length overall	8.7 m (28 ft 6½ in)	8.78 m (28 ft 9⅝ in)	8.7 m (28 ft 6½ in)
Wing span	9.2 m (30 ft 2 13/64 in)	9.2 m (30 ft 2 13/64 in)	9.2 m (30 ft 2 13/64 in)
Wing area, m^2 (sq ft)	14.85 (159.67)	14.85 (159.67)	14.85 (159.67)
Empty weight, kg (lb)	1,852 (4,082)	2,081 (4,587)	2,148 (4,735)
Take-off weight, kg (lb):			
normal	2,638 (5,815)	2,890 (6,370)	2,906 (6,406)
maximum (with drop tanks)	2,742 (6,045)	3,240 (7,140)	n.a.
Fuel load, kg (lb):			
normal	n.a.	553 (1,219)	518 (1,142)
maximum (with drop tanks)	n.a.	884 (1,948)	n.a.
Wing loading, kg/m^2 (lb/sq ft)	197 (40.45)	189 (38.8)	n.a.
Thrust loading, kg/kgp (lb/lbst)	3.05	3.6	n.a.
Top speed, km/h (mph):			
at sea level	700 (434) *	702 (436)	702 (436)
at 5,000 m (16,400 ft)	786 (488)	748 (464)	719 (446)
Landing speed, km/h (mph)	135 (83)	142 (88)	n.a.
Rate of climb, m/sec (ft/min):			
at sea level	21.6 (4,250.8)	17.6 (3,463)	18.4 (3,261)
at 5,000 m (16,400 ft)	n.a.	11.0 (2,164)	n.a.
Time to height, minutes:			
to 5,000 m	4.8	5.8/7.0	5.8
Service ceiling, m (ft)	12,000 (39,370)	12,750 (41,830)	n.a.
Range, km (miles):			
in 'clean' condition	510 (316)	395 (245)	330 (205)
with drop tanks	n.a.	717 (445)	n.a.
Endurance:			
in 'clean' condition	1 hr 4 min	1 hr 1 min	40 min
with drop tanks	n.a.	1 hr 34 min	n.a.
Take-off run , m (ft)	600 (1,968)	635 (2,083)	640 (2,100)
Take-off distance, m (ft)	n.a.	n.a.	n.a.
Landing run, m (ft)	530 (1,740)	560/800 (1,840/2,620)	700 (2,300)
Armament	1 x NS-23	2 x NS-23	n.a.
Ammunition supply	n.a.	105 rounds	n.a.

* With drop tanks.

The Yak-17 in service

Re-equipment with the Yak-17 turned out to be a lengthy process, and the time at which Soviet fighter regiments took delivery of the new machines varied widely. Thus, for instance, the 472nd IAP based at Kursk-Vostochnyy (= Kursk-East) re-equipped with Yak-17s in the spring of 1950 immediately after being established; as early as 22nd July it was pronounced fully combat-ready and placed on combat duty in the PVO system.

The limited amount of fuel would be used up in just 25-28 minutes, which meant Yak-17 pilots could perform only short flights Besides, the engine starting procedure was rather complicated. On the other hand, the Yak-17 was easy and pleasant to fly. A low level of noise in the cockpit, good forward visibility, a short take-off run and a relatively high flight speed were the jet fighter's strong points in comparison with piston-engined machines.

The first-generation Yak fighter jets did not enjoy a long service career with the Soviet Air Force. For example, the above-mentioned 472nd IAP started converting to the MiG-15, the most advanced Soviet

Yakovlev OKB

A pilot climbs into the font cockpit of Yak-17UTI '07 Yellow'.

fighter at the time, as early as December 1950 – a mere six months after the Yak-17 achieved initial operational capability.

A single-seat Yak-17 with the tactical code '02 Red' survives in Russia; it has been on display in the Air Force Museum in Monino for many years.

As noted earlier, the Yak-17 and Yak-17UTI were supplied to Bulgaria, China, Czechoslovakia and Poland.

Bulgaria: The Bulgarian Air Force (BVVS – *Bolgarski Voyenno Vozdooshni Seeli*) operated the *Feather* in both fighter and trainer versions. Unfortunately no details are known.

Czechoslovakia: The Czechoslovak Air Force (*Československé Vojenské Létectvo*) took delivery of a small number of Yak-17s (possibly ex-VVS aircraft). One of them is now on display at Prague-Kbely.

China: The year of 1950 saw the beginning of large-scale deliveries of combat aircraft to China, together with the deployment of Soviet Air Force units to that country where a civil war had erupted. Soviet instructors trained the flight and ground crews of the nascent People's Liberation Army Air Force (PLAAF, or *Chung-kuo Shen Min Taie-Fang-Tsun Pu-tai*); this included conversion to jet hardware, for which purpose MiG-9 fighters and Yak-17UTI trainers were sent to China. In particular, the 29th GvIAP of the 324th IAD was engaged in training Chinese pilots on the Yakovlev trainers.

One of the surviving Chinese Yak-17UTIs is now on display at the PLAAF Museum at Datangshan AB.

Poland: In late 1949 and early 1950 the Soviet and Polish governments considered the possibility of building the Yak-17 in Mielec (pronounced '***Mel****ets*'), Poland under licence. Eventually, however, these plans were dropped because the Yak-17 had been overtaken by the rapid development of jet fighter technology. Instead, the Polish aircraft industry started investigating the general issue of manufacturing aircraft of all-metal semi-monocoque design which were suited for a greater variety of roles. Later, the aero-engine factory in Rzeszów (pronounced '***Zheh****-show*') manufactured a batch of 30 RD-10A turbojets under licence; the engines were intended for the *Jak-17W* trainers, as the Yak-17UTI was designated by the Polish Air Force.

In July 1950 the command of the Polish Air Force (PWL – *Polskie Wojsko Lotnicze*) sent a group of four pilots to a Soviet Air Force/4th Air Army unit stationed in Poland for conversion training so that they could master jet fighters. The first three Yak-17s for the PWL arrived in Warsaw in July 1950. On 20th August 1950 the Yak-17 was presented to the general public for the first time at Warsaw-Okęcie airfield (now the city's international airport) during the Aviation Day festivities.

By mid-1951 the Polish Air Force already had several types of jet fighters in its inventory. The Yak-17s were assigned to fighter air units where they were unofficially dubbed 'Agata'. During the Aviation Day festivities in 1951 the Yak-17Ws took part in flights together with the Yak-23s. In 1953 the Officers' Air Training School No.5 set up in Radom in 1951 received Yak-17Ws which were used for conversion training to the Yak-23. These aircraft were used by the school relatively briefly, soon yielding their place the UTI-MiG-15.

In 1957 two Yak-17Ws were transferred to the Aviation Institute (*Instytut Lotnictwa*). The first of them, serialled '1 Red', landed on the Institute's airfield on 21st January; the other one ('4 Red') followed on 4th March. The latter aircraft subsequently received the civil registration SP-GLM; it was used for training the Institute's pilots who were expected in the near future to test the prototype of the locally designed PZL TS-11 Iskra (Spark) jet-powered advanced trainer. The last flight of SP-GLM took place on 3rd February 1960. From 1963 onwards the aircraft was used primarily as a static exhibit at various airshows, and PWL insignia were again applied to it in addition to the civil registration (!). In August 1964 the same machine (this time in full military colours and with the tactical number '02 Red') was shown at an aviation hardware display at Rakowice airfield in Kraków. At present this aircraft constitutes a unique exhibit of the Aerospace Museum (Muzeum Lotnictwa i Astronautyki) in Kraków.

Yak-23 fighter

The first Soviet jet fighters (the MiG-9, Yak-15, Yak-17 and Semyon A. Lavochkin's 'aircraft 150', 'aircraft 152' and 'aircraft 156') were designed around captured German engines rated at only 800 to 900 kgp (1,764-1,984 lbst). Understandably, they were no match for American and especially British fighters powered by engines delivering nearly twice the thrust. To close this gap the Soviet leaders decided to resort temporarily to purchasing British centrifugal-flow turbojet engines – the 1,625-kgp (3,580-lbst) Rolls-Royce Nene I and the 2,270-kgp (5,000-lbst) Rolls-Royce Derwent V – until similarly rated Soviet engines entered mass production. In accordance with a Council of Ministers directive dated 11th March 1947 several design bureaux were to rapidly develop new types of aircraft around these engines. In particular, OKB-115 was tasked with creating a tactical fighter powered by a Derwent V and featuring straight wings.

The choice of unswept wings may be tentatively explained as follows. Firstly, quite probably due regard was taken to the wishes of Aleksandr S. Yakovlev whose attitude towards swept wings was still cautious. Secondly, both the ministry and the Government obviously decided to play safe, in case the future Mikoyan and Lavochkin aircraft with swept wings proved to be a failure. Yet, no one questioned the assumption that only swept-wing aircraft could reach high speeds and crack the sound barrier.

Quite possibly Aleksandr S. Yakovlev fully realised that if the Mikoyan and Lavochkin OKBs scored a success, his design bureau would lose its leading position as a 'fighter maker' and would be relegated to a marginal role. Therefore he decided to work on two fighters in parallel; one would be developed in accordance with the Government directive and the other, a more agile aircraft, as a 'private venture'. The first of the two OKB-115 projects received the designation Yak-25 (this aircraft is described below in this chapter) while the other fighter was designated Yak-23.

The Yak-23 jet fighter developed 'beyond the call of duty' employed the pod-and-boom layout already thoroughly studied by Yakovlev's design team; all the elements of its aerodynamic layout had been verified on the preceding models. The fighter was of all-metal construction. The mid-set two-spar wings fitted with TsAGI-type slotted flaps used a laminar airfoil section. The designers placed the engine in the forward fuselage, the nozzle being located under the centre fuselage. The fuselage itself was a semi-monocoque structure featuring a maintenance break (the nose section was detachable for the purpose of installing the engine).

The tail unit had a cruciform layout, the fin and the dihedral stabilisers featuring a two-spar structure. The aircraft had a fully retractable tricycle undercarriage, the main units retracting inwards so that the mainwheels were stowed partially in the wings and partially in the fuselage; the nose unit retracted forward to lie in the air intake splitter. As had been the case with the OKB's previous products (the Yak-15 and Yak-17), the Yak-23's armament consisted of two 23-mm NR-23K cannons in the forward fuselage, except that these were accommodated beneath the engine instead of above it. Drop tanks could be carried under the wingtips.

The designers embarked on projecting the machine without even having seen the

Yakovlev OKB

Above: '52 Yellow', the first prototype Yak-23, nearing completion in the assembly shop of MMZ No.115.

Yakovlev OKB

Above and below: Two views of the first prototype Yak-23 during manufacturer's flight tests. Note the nosewheel mudguard and the landing light buried in the air intake splitter.

Yakovlev OKB

Yakovlev OKB

Yakovlev OKB

Top and above: In an effort to confuse spies, the second prototype was also serialled '52 Yellow'; however, the digit on the rudder was 2 instead of 3. The triangle marking on the fuselage identifies the fuel filler.

Rolls-Royce Derwent engine in actual hardware, having at their disposal only a drawing outlining its basic dimensions. Therefore, it says a lot for the design team that when the real engine eventually reached the OKB, its installation required almost no alterations to the prototype's airframe.

Yak-23 prototypes

In accordance with the MAP order *'On the preparation of aircraft for the 18th August 1947 flypast'* Chief Designer and Director of plant No.115 A. S. Yakovlev was instructed to *'...complete construction of the Yak-23 powered by the Derwent V turbojet by 15th June and initial flight testing of the aircraft by 30th July.'* As was his wont, Yakovlev met the deadline. Serialled '52 Yellow' and bearing a yellow numeral '3' on the rudder, the first prototype (c/n 115001 – that is, plant No.115, 001st aircraft) was completed in June; a month later OKB-115 test pilot Mikhail I. Ivanov performed the fighter's maiden flight. The first prototype was powered by one of the genuine Derwent V engines bought in the UK and rated at 1,625 kgp.

The manufacturer's tests conducted by leading engineer V. Shelepchikov and pilot M. I. Ivanov proceeded quickly (they were completed on 24th September 1947) and, in the opinion of the OKB, successfully. A maximum speed of 932 km/h (578 mph) was reached at sea level, and the maximum Mach number attained at high altitude was 0.845. The fighter possessed excellent handling and easily performed all aerobatic manoeuvres. On the other hand, some performance parameters were not determined: even the airspeed indicator was not calibrated, which could lead to incorrect readings, so great was the hurry to submit the fighter for State acceptance trials.

On 3rd August, while the manufacturer's tests were still in progress, the first prototype Yak-23 was demonstrated at the Tushino air display (flown by M. I. Ivanov).

On 22nd October the second prototype (c/n 115002) – confusingly, bearing the same serial '52 Yellow' but marked with a yellow '2' on the rudder – was submitted to GK NII VVS for State acceptance trials (this elaborate deception was part of the OKB's security measures). The trials were conducted by project test pilot Gheorgiy A. Sedov, who had conducted the State acceptance trials of all Yak jets featuring the pod-and-boom layout, and pilots Pyotr M. Stefanovskiy, A. G. Kochetkov, A. G. Proshakov, V. I. Khomiakov, I. Dziuba, L. M. Koovshinov and OKB test pilot V. I. Ivanov. In the course of the State acceptance trials which began on 11th November, the Yak-23 attained a maximum speed of 925 km/h (575 mph) at sea level and displayed an excellent rate of climb: rate of climb at sea level (without drop tanks) was 41.4 m/sec (8,147 ft/min); time to 5,000 m (16,400 ft) was 2.3 minutes. Range at 5,000 m with wingtip tanks was 1,035 km (643 miles), and service ceiling was 14,800 m (48,560 ft). The majority of the fighter's performance characteristics were determined at the normal all-up weight of 2,965-3,036 kg (6,536-6,693 lb) in 'clean' configuration and 3,306-3,389 kg (7,288-7,471 lb) with drop tanks. In 'clean' condition the fighter had a short take-off run of only 440 m (1,440 ft), the landing run being 540 m (1,770 ft).

Test pilots came to the conclusion that the Yak-23 could be used quite successfully against enemy fighters. The State acceptance trials protocol said that the Yak-23 was both faster than the indigenous MiG-9 and Su-9 (the first aircraft to have this designation – a twinjet fighter of 1947) and possessed a much better rate of climb than either of these types. Being superior in basic performance to all existing Soviet jet fighters, the Yak-23 was second-best to the Su-9 only as far as the deceleration time in level flight was concerned.

Taking note of the Yak-23's undoubted merits, the pilots nevertheless placed their main emphasis the fighter's shortcomings in their reports. Regarding the slow deceleration, they rightly considered that the machine's manoeuvrability characteristics are restricted by the absence of airbrakes. Other drawbacks noted included the insufficient range of air-to-ground radio communication, excessive stick and rudder pedal forces and the unpressurised cockpit which, in the event of aerial combat at high altitude and with G-loads exceeding 5 Gs, required great physical endurance on the part of the pilot due to the insufficient amount of onboard oxygen. In the course of the State acceptance trials the Yak-23 successfully waged two sessions of mock combat against the Tu-12 ('77') – the first jet bomber produced by Andrey N. Tupolev's OKB-156. The concluding part of the State acceptance trials protocol said that the fighter *'...can be introduced into service'*.

Yak-23 production fighter

Notwithstanding the many merits of the Yak-23, its equally numerous shortcomings delayed the final decision as to whether the aircraft should be put into production. This was possibly influenced by the fact that the first prototype Yak-25 (see below) entered initial flight tests in October 1947. This fighter had been created in accordance with the

CofM directive of 11th March 1947 and met most of the Air Force's requirements (it was fitted with a pressurised cockpit, and many of its performance characteristics were superior to those of the Yak-23).

The final decision on the expediency of launching Yak-23 production was delayed by a year. The prototypes of the MiG-15 (I-310) and La-15 ('174') fighters passed their tests in 1948, demonstrating the undoubted advantages of swept-wing aircraft. Nevertheless, at the end of 1948 the Government took the decision to put the Yak-23 into production. This was due to the fact that the MiG-15 with its more advanced layout and 2,270-kgp (5,005-lbst) RD-45 turbojet (a licence-built version of the Rolls-Royce Nene I) was still suffering from a number of teething troubles. Therefore, as a kind of insurance policy, it was decided that the Yak-23, which was not expected to harbour any surprises, would be built in parallel with the introduction of the MiG-15 into mass production.

The dimensions and performance characteristics of the Yak-23 as recorded during the State acceptance trials of the second prototype (No.2, as it was referred to in GK NII VVS documents) are presented in the table on this page.

On 14th July 1948 one of the Yak-23 prototypes piloted by Mikhail I. Ivanov crashed under mysterious circumstances during the 'dress rehearsal' of the Tushino flypast. Investigation revealed that one of the fighter's wings was struck by a mass balance separating from the rudder of a Tu-14 ('78') bomber which, piloted by Fyodor F.

Specifications of the Yak-23 fighter

Length overall	8.10 m (26 ft 6 $^{57}/_{64}$ in)
Wing span	8.73 m (28 ft 7 $^{45}/_{64}$ in)
Wing area, m^2 (sq ft)	13.7 (147.31)
Empty weight, kg (lb)	1,902 (4,194)
All-up weight, kg (lb):	
normal (without wingtip tanks)	2,965-3,036 (6,536-6,693)
maximum (with wingtip tanks)	3,306-3,389 (7,288-7,471)
Fuel load, kg (lb):	
normal (without wingtip tanks)	790 (1,740)
maximum (with wingtip tanks)	1,109 (2,450)
Top speed, km/h (mph):	
at sea level	925 (574)
at 5,000 m (16,400 ft)	910 (565)
Landing speed, km/h (mph)	157 (97.5)
Time to height, minutes:	
to 5,000 m	2.3
to 10,000 m (32,810 ft)	6.2
Rate of climb, m/sec (ft/min)	
at sea level	41.4 (8,146)
at 5,000 m	30.4 (5,983)
Range in 'clean' condition/with drop tanks, km (miles):	
at 10,000 m	1,080/1,475 (671/916)
Endurance in 'clean' condition/with drop tanks:	
at 10,000 m	1 hr 44 min/2 hrs 21 min
Take-off run, m (ft):	
without drop tanks	440 (1,440)
with drop tanks	560 (1,840)
Take-off distance, m (ft):	
without drop tanks	1,075 (3,530)
with drop tanks	1,360 (4,460)
Landing run, m (ft)	540 (1,770)
Landing distance, m (ft)	1,170 (3,840)
Armament	2 x NR-23K

Central National Economy Archive

This Yak-23 serialled '01 Red' and equipped with drop tanks underwent checkout trials at GK NII VVS. Note the larger gun blast plates.

Opadchiy, was flying a considerable distance away The wing broke away and the fighter tumbled earthwards, rolling uncontrollably. The cockpit canopy was torn off and Ivanov was thrown clear of the aircraft, falling to his death; apparently he was incapacitated by this enforced 'ejection' and unable to open his parachute.

There were no doubts as to the cause of the crash. Upon landing the Tu-14 was found to lack the rudder mass balance; also, traces of the green paint with which the mass balance was coated were discovered on the wing of the Yak-23, together with traces of the impact produced by the mass balance. Still, despite all efforts, the accident investigation board could not find the answer to the main question, which boils down to the following. There were a lot of aircraft in the air. Fighters were passing one after another at low altitude along a highway past the main grandstand; the flight altitude was roughly 100 m (330 ft). Conversely, the groups of bombers and transport aircraft were passing at high altitude and far away from the route along which the fighters were flying. Assuming that every machine followed its predesignated route, it is indeed hard to understand how the truant mass balance could hit the fighter at such a long distance.

Owing to the delayed decision on the fate of the Yak-23 its production at plant No.31 did not began until 1949. Since the MiG-15 had successfully completed State acceptance trials by then, the Government decided to build this fighter at several factories at once, including the one in Tbilisi. Tooling up for MiG-15 production, which began in mid-1949, diverted a considerable number of employees from the manufacture of the Yak-17 and Yak-23 because the new machine was intended to supersede both Yaks on the production line (preparations for Yak-23 production had begun in February 1949). At the end of August new instructions were issued: the Tbilisi plant would not build the MiG-15 (eventually it *did* produce the type, but that's another story). Still, the tempo of tooling up for the Yak-23 had already been slackened, and the first machines were not completed until October.

A total of 310 production Yak-23s rolled off the assembly lines in Tbilisi before the end of 1950. Together with the three machines built in Moscow at plant No.115 (including the static test airframe), the production run totals 313. Production fighters were powered by Soviet-built RD-500 engines (a licence-built version of the RR Derwent V) delivering 1,590 kgp (3,500 lbst). Besides equipping the fighter units of the Soviet Air Force, they were supplied to some East European nations referred to in the USSR as 'countries of people's democracy'. There were plans to put the aircraft into licence production in Czechoslovakia, but they never materialised.

In the Soviet Union the Yak-23 saw service primarily with the fighter regiments of the North Caucasus and Volga Defence Districts. The fighter was intended mainly for air-to-air combat at low and medium altitudes. In the opinion of pilots who flew it, the Yak-23 was a good aircraft; it was easy to fly, displaying a high rate of climb and good manoeuvrability. Among the important merits of the aircraft were its ease of maintenance and the ability to operate from grass strips.

It was in the North Caucasus area, at an airfield near the city of Groznyy (Chechen-Ingush Autonomous SSR), that service trials of the first production machines took place. Comments coming from the regiment's pilots were mostly favourable; notably, the fighter could easily perform aerobatic manoeuvres. However, in the process of operation some production defects came to light (for example, the cockpit canopy jammed and could not be jettisoned in an emergency; fumes from the engine cooling manifold entered the cockpit and cracks were discovered in the main undercarriage struts). There were also cases of the air bleeding from the shock absorber of the nose landing gear strut.

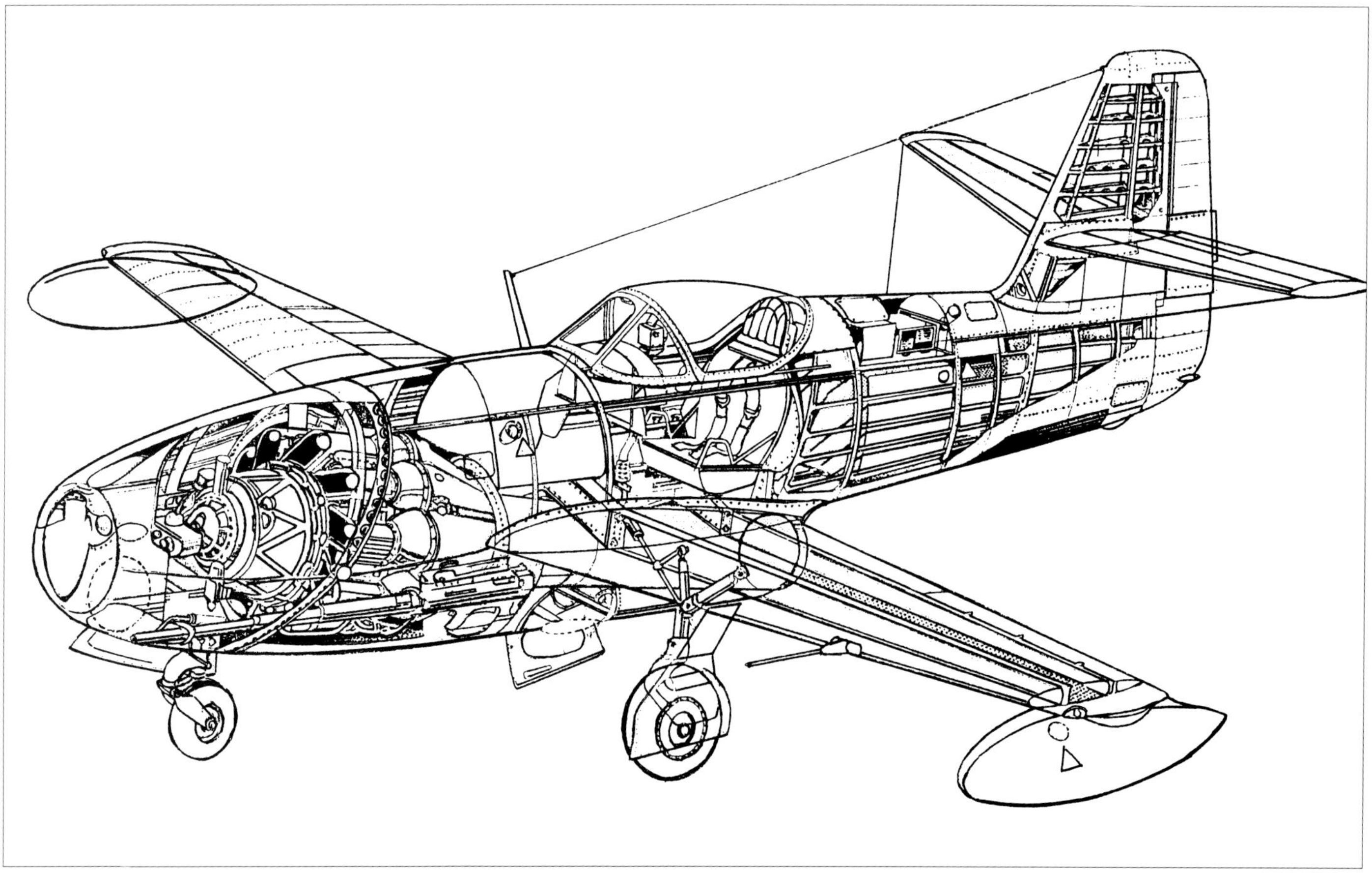

A cutaway drawing of the Yak-23. Note the sloping bulkhead at the forward fuselage break point.

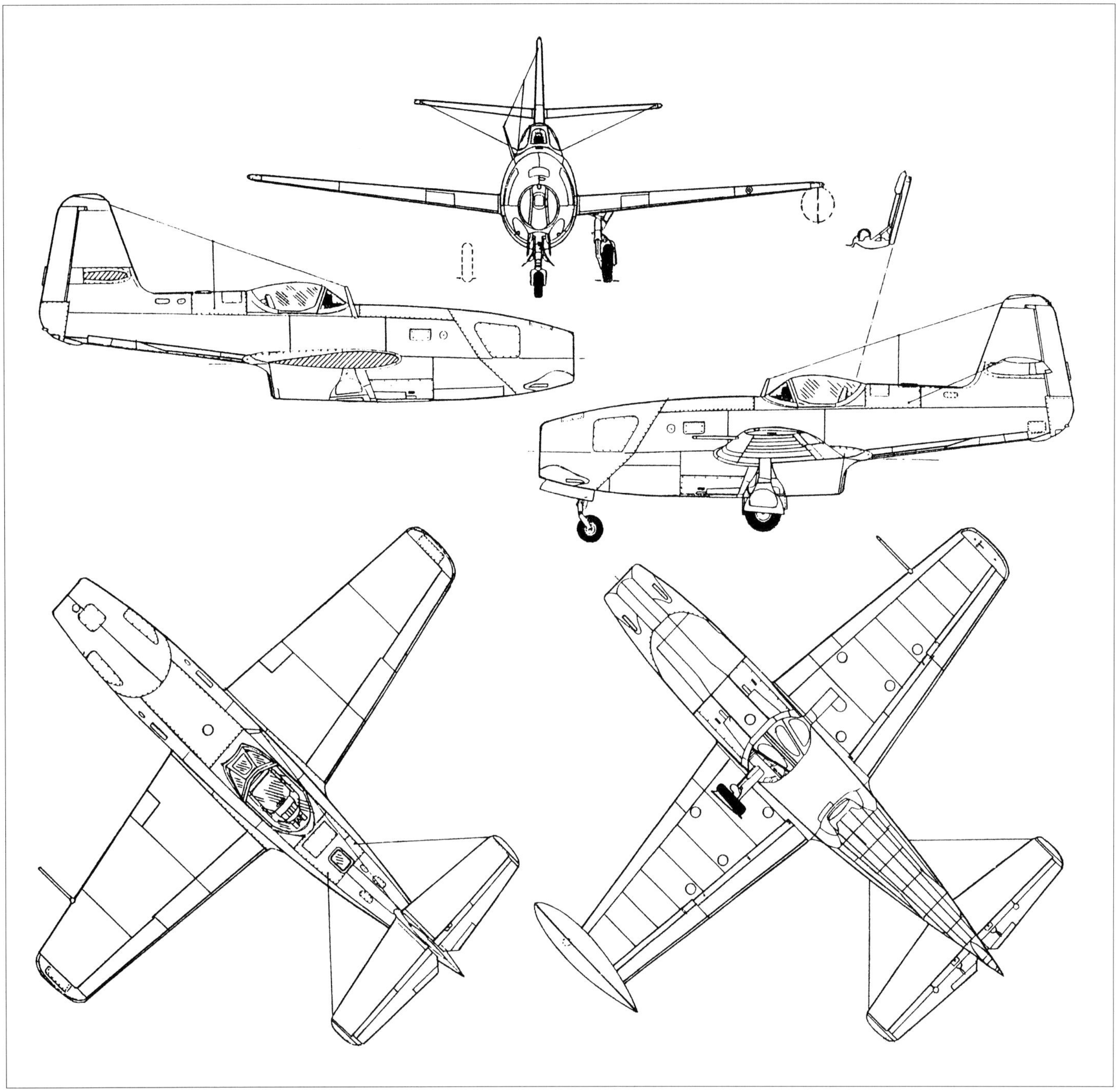

Five views of the Yak-23; the starboard main gear unit is shown extended in the lower view.

The Yak-23 also had other shortcomings which were revealed during operation of the type in other units. At Mach 0.8 the aircraft became directionally unstable. The engine was installed at an angle to the aircraft's axis, which resulted in a significant pitch-up force when the throttle was advanced rapidly; this hampered the conduct of air combat where precise manoeuvring was essential.

Such was the experience of the Soviet pilots who had volunteered for participation in the Korean War. Quite possibly all these shortcomings contributed to the decision to terminate production of the Yak-23 in favour of the MiG-15. The fighter's NATO reporting name was *Flora*.

Yak-23 experimentally towed by Tu-2

In accordance with the Council of Ministers directive No.760-288ss dated 22nd February 1949 the Yak-23 was used for conducting ground experiments associated with the development of a system for towing a jet fighter behind the Tu-2 bomber. Later the towing method that had been thus evolved was verified during flight tests of the automatic coupling on the Yak-25 jet fighter to the B-25 bomber.

Yak-23UTI (UTI Yak-23) prototype, initial configuration

Shortly after submitting the single-seat fighter for testing, the staff of OKB-115 developed a two-seat trainer version designated UTI Yak-23. In its first version, in an effort to avoid major structural changes to the single-seat machine, a second cockpit was placed ahead of the existing one.

That was the configuration of the prototype of the new dual-control aircraft built by converting the first prototype of the Yak-23 (the UTI Yak-23 retained its original c/n 115001, but the serial was changed to '50 Yellow' and the digit '3' on the rudder was replaced by a yellow '2'). The equipment fit was reduced, as well as the armament complement; the latter included one 12.7-mm Berezin UB machine-gun. However, visibility from the rear cockpit proved to be rather

Yakovlev OKB

Yakovlev OKB

Top and above: The prototype of the UTI Yak-23 trainer as originally flown.

limited. To remedy this, the machine had to be modified twice.

Yak-23UTI (UTI Yak-23) prototype, second version

In response to demands from test pilots, the modified UTI Yak-23 received a completely new layout. The engineers moved both cockpits forward, increasing the fuselage length in so doing. The cockpits canopies were no longer flush with the rear fuselage contour but protruded above it. Initially, the windshield and the front canopy were frameless. The rear seat canopy was level with the forward one, forming a smooth contour.

In further modified form, this version featured additional framing in the cockpit visor and the front canopy; the rear canopy remained unchanged. However, the instructor's visibility left something to be desired, which prompted the next step in modification.

Yak-23UTI (UTI Yak-23) prototype, third version

In the UTI Yak-23's ultimate form the instructor's seat was raised, resulting in a curious canopy shape with blister-like bulge above the rear seat. The front seat's canopy was also redesigned, dispensing with the intermediate frame at the rear and featuring a new, deeper windshield. Placed on top of it was a gun camera.

The third version of the UTI Yak-23 finally found approval on the part of test pilots. At an all-up weight of 3,410 kg (7,520 lb) with drop tanks the aircraft had an endurance of 1 hour 50 minutes. The top speed reached 816 km/h (506 mph) at sea level and 850 km (528 mph) at 5,000 m (16,400 ft). The rate of climb came very close to that of the combat version; the trainer climbed to 5,000 m within 2.5 minutes.

Like most of the trainer machines, the UTI Yak-23 had a reduced armament fit – one 12.7-mm (.50 calibre) Berezin UB machine-gun – and equipment complement (on the other hand, a gun camera was invariably installed on the dual-control machines). The protracted testing, modification and development of the UTI Yak-23 lasted until the end of 1950 when the single-seat Yak-23 was already being phased out of production; besides, the Air Force units had already taken delivery of Yak-17UTI trainers and there was no longer any need for a trainer version of the Yak-23. Finally, the UTI Yak-23 had cramped cockpits (not surprisingly, given the aircraft's small dimensions). The trainer's insufficient range (the new cockpit 'ate up' much of the fuel tankage) also reduced the usefulness of the aircraft.

The prototype UTI Yak-23 was subsequently used by MMZ No.115 for submitting an ejection seat developed in-house for State acceptance trials at GK NII VVS. This was accompanied by one interesting episode. When the two-seat trainer fitted with the ejection seat was delivered to the military institute, it was placed in a hangar. Representatives of the plant warned the military that they had better avoid approaching the ejection mechanism because this was dangerous. Work on it could be performed only by personnel who had received special instructions.

Before starting the briefing one of the OKB-115 specialists placed his foot on the seat, sat down astride the cockpit sill (facing the tail), apparently wishing to demonstrate something. At that moment the ejection mechanism fired! The luckless man was ejected from the cockpit, sustaining a heavy injury (a fractured thigh). The ejection seat was thrown up nearly to the roof of the hangar, falling squarely on the aircraft – fortunately, without damaging it. Those with a penchant for humorous comparisons immediately recalled a similar episode that had happened in Britain during the testing of the Gloster Meteor. In that case, too, a man who was supposed to give instruction to the personnel placed himself into the seat, and it fired accidentally. The seat hit the hangar roof, but the engineer escaped nearly unscathed.

This is how the first ejection of a 'Yakovlev seat' was effected at GK NII VVS. Moral: doctor, heal thyself!

Yak-23K liaison aircraft

Some sources assert that the Yak-23UTI prototypes, upon completion of testing, were not used as trainers; they allegedly were converted into liaison aircraft with suitable modifications to the rear cockpit intended for a passenger and were used in this role under the designation Yak-23K (*koor'**yer**skiy* – courier, used attributively)

Two examples of the Yak-23 have been preserved in Russia. A standard single-seat Yak-23 with the tactical code '15 Red' is part of the Monino museum's collection, while the Yakovlev OKB has managed to preserve the sole example of the dual-control machine (painted green, with no insignia whatever and with a new rear cockpit canopy resembling the original second version; apparently the bulged canopy was lost somehow). The trainer was exhibited at Moscow-Khodynka airfield in August-September 1993 during the MAKS-93 airshow.

Yak-23 Bulgarian trainer conversion?

There are somewhat vague reports asserting that one of the Yak-23 fighters serving with the Bulgarian Air Force was converted locally into a two-seat dual-control trainer. No details are available, and the story requires confirmation.

Yak-23DC (Romanian trainer conversion)

In 1956 one of the Yak-23 fighters delivered to Romania was converted into a two-seat training version. The project was evolved by CTIA, the military aeronautical institute situated in the town of Piperz. The aircraft received the designation Jak-23DC (*dubla comanda* – dual-control). The second cockpit was inserted in the aft fuselage, taking up the place previously occupied by the rear fuel tank; the front cockpit remained in its normal position. The fuselage section between the glazing of the new cockpit and the fin was suitably modified. The front cockpit was occupied by the trainee, the instructor sitting in the back seat. The aircraft had no armament.

Yak-23 in foreign service

The Yak-23 was in the inventory of the air forces of a number of 'friendly nations'.

Czechoslovakia: The Czechoslovak Air Force was the first foreign operator of the Yak-23. Czechoslovakia even intended to start licence production of the type, but these plans never materialised. The first 12 aircraft were delivered to the CzAF in late 1949, almost concurrently with the Yak-23's service entry with the Soviet Air Force; later Czechoslovakia received a further nine examples. Until the mid-1950s the CzAF had a habit of allocating its own designations to foreign-designed and/or -built aircraft in Czech service; hence the Yak-23 was known locally as the S-101, the S denoting *stíhaci [letoun]* – fighter; its RD-500 engine was also redesignated, becoming the M-02.

Bulgaria: The Bulgarian Air Force also introduced the Yak-23 in 1949. Initial deliveries are known to have been 12 aircraft but the total number of Yak-23s supplied to Bulgaria is unknown. There have been reports in the press that Bulgaria attempted to develop a two-seat trainer version, but so far no confirmation has been found. Two Yak-23s are preserved in the BVVS museum at Graf Ignatiev AB in Plovdiv, home of the 19th IAP (*Iztrebitelen Aviopolk*).

Romania: The first 12 Yak-23 fighters were delivered to Romania in 1951. In all, the Romanian Air Force (*Fortele Aeriene ale Republicii Socialiste Române*) received 40 machines of this type. As noted above, at least one was converted into the Jak-23DC trainer.

Albania: It is known that the first Yak-23s were delivered to the Albanian People's Republic Air Force (*Forcat Ushtarake Ajore Shquipëtare*) in 1951. It is also known that a certain number of these aircraft, still operational, and a large amount of spares for them were transferred to Albania from Poland.

North Korea: Some reports made by American pilots during the Korean War mention air encounters with Yak-23s.

Poland: The first Yak-23s reached Poland in December 1950. Later the Polish Air force received nine aircraft from Czechoslovakia and a certain number of Yak-23s from units of the Soviet Air Force. The total number of aircraft of this type operated by the PWL exceeds 100.

The aircraft delivered in the first batch were assembled by Soviet specialists who concurrently trained the Polish ground crews. The reassembled machines were test-flown by Soviet instructors who had been sent to undertake conversion training of Polish pilots *in situ*. Starting in April 1951, the newly assembled machines were reflown by Polish pilots from the first group of airmen who had received training on this

Yakovlev OKB

Yakovlev OKB

Top and above; Two views of the UTI Yak-23 following its second update. Note the dipole aerials of the radio altimeter under the port wing and port stabiliser.

Yakovlev OKB

A view of the UTI Yak-23's canopy in its ultimate configuration. The aft portion of the sliding rear canopy lifts up on a special rocker lever. The canopy of the forward cockpit hinges open to starboard.

aircraft type. They also became the first instructors who were granted the right to undertake training on jet aircraft. In January 1951 these pilots were sent on a mission to fighter air regiments of the Soviet Air Force.

Concurrently with new deliveries of new hardware most of the pilots of the 1st and 2nd Fighter Air Regiments of the Polish Air Force (1. PLM 'Warszawa' and 2. PLM) underwent conversion training. As early as September 1951 squadrons commanded by Maj. Ulanowski and Capt. Rybacki were granted the right to be on combat duty. For the first time mock combats between jet aircraft were staged for training purposes in the skies of Poland. A while earlier, on 26th August 1951, the Yak-23 was presented to the public for the first time at Warsaw-Okęcie on occasion of the Aviation Day.

The Yaks were also demonstrated on 22nd July 1952. On that day inauguration of the Dzierżinski Square took place in Warsaw; on that occasion a flypast was staged and aircraft from three regiments were allotted for participation in it.

The Yak-23 aircraft remained in the first line of Poland's Air Defence until the end of 1955 and the beginning of 1956, whereupon it was gradually supplanted by more modern aircraft. The Yak-23s were also placed at the disposal of Officers' Flying School No.5 in Radom, which specialised in training pilots for fighter units. The trainees performed their first flights on the Jak-17W (Yak-17UTI), after which they transitioned to the Yak-23.

In the early 1950s the Polish aircraft industry began preparations for manufacturing the Yak-23 under licence. The production was planned to be undertaken at WSK plants in Mielec and Świdnik. The fighter was expected to be allocated the Polish designation G-3. Tooling up for production was already under way when these plans were cancelled in favour of the license manufacture of the MiG-15.

In 1956 the command of the Polish Air Force handed two Yak-23 over to the Aeronautical Institute in Warsaw. The aircraft received civil registrations SP-GLK and SP-GLL. The latter aircraft, however, did not fly, becoming a source of spares for SP-GLK. The many experiments conducted on the Yak-23 included evolving the methods of flutter tests. The aircraft was also intended for use as a testbed for ramjet engines to be placed under the wings close to the fuselage; eventually, however, these tests were never started.

On the other hand, it is worth noting that on 21st September 1957 test pilot Andrzej Abłamowicz performed a record-breaking flight Yak-23 SP-GLK, setting two international time-to-height records: he climbed to 3,000 m (9,842 ft) within 119 seconds and to 6,000 m (19,685 ft) within 197 seconds.

One of the ex-Polish Air Force Yak-23s aptly serialled '23 Red' (the c/n has been reported as 1017 but the true c/n is presumably 3123017 – that is, plant No.31, Yak-23, 017th production example) was handed over to the Polish Armed Forces Museum (*Muzeum Wojska Polskiego*) in Warsaw in 1957. One more machine serialled 16 Red was handed over to the Aerospace Museum in Kraków in 1963.

USA: One Yak-23 mysteriously found its way to the United States and was evaluated at Wright-Patterson AFB.

Wojskowas Agencja Fotograficzna

Above: A Polish Air Force Yak-23 serialled '09 Red' sits parked on a snow-covered airfield with a battery cart hooked up for engine starting. Note the open ground power connector panel cover.

Letectví+Kosmonautika

This Czechoslovak Air Force Yak-23 wearing the pre-1957 alphanumeric serial HX-51 is currently on display at the Military Museum at Prague-Kbely.

Yak-19 fighter prototype

Although being one of the lesser-known designs of the Yakovlev OKB, the Yak-19 occupies, in a way, a special place among Yakovlev's early jet fighters. On the one hand, this fighter, designed and built in 1946, is his last fighter to be powered by the clearly inadequate RD-10 engine before switching over to the more powerful RD-500 and RD-45 turbojets. On the other hand, this was his first fighter to mark a departure from the pod-and-boom configuration in favour of the more advantageous layout in which the engine was placed amidships in the fuselage, with a nose air intake and the jet nozzle at the end of the fuselage.

This was also the first all-metal aircraft to originate from the Yakovlev OKB. Until then, all Yakovlev designs had been of mixed construction, utilising a welded steel-tube fuselage truss with plywood, metal or fabric skin, wings with wooden or metal spars and plywood/fabric skin, and fabric-covered control surfaces. Finally, the Yak-19 was the first Soviet aircraft to be powered by an afterburning turbojet; the RD-10F was a derivative of the Jumo 004 equipped with an indigenous afterburner designed jointly by OKB-115 and TsIAM.

Despite the extremely limited time available, the engineers did an excellent job, meeting all design specifications except empty weight. The aircraft was 225 kg (496 lb) overweight, much to the annoyance of Aleksandr S. Yakovlev who always prided himself on the high weight efficiency of his designs.

Initially Leon M. Schekhter was the Yak-19's chief project engineer. Later he was succeeded in this capacity by Nikolay K. Skrzhinskiy, who in turn was succeeded at the flight test stage by Leonid L. Selyakov who had joined the Yakovlev OKB a short while earlier after the closure of Vladimir M. Myasishchev's OKB-482. It was under Selyakov's guidance that a full-scale wooden mock-up showing the layout of all control runs, piping, wiring and so on was built for the first time in OKB-115's practice.

Outwardly the Yak-19 was very different from the rather ungainly Yak-15 and Yak-17 fighters; in fact, the chosen general arrangement became a sort of standard for many fighters developed in the USSR during the 1950s and 1960s. The design incorporated a number of revolutionary features which found active support not only from Yakovlev himself but from his closest aides – K. A. Vigant, S. Ya. Makarov and the OKB's chief technologist S. S. Bekin. The pod-and-boom layout with the engine(s) in the forward fuselage shared by the first Soviet jet fighters gave way to a more aerodynamically efficient arrangement with the engine buried in the aft fuselage, exhausting via a long extension jetpipe – for the first time in Soviet design practice. The monocoque fuselage had a break point aft of the wings which allowed the entire aft fuselage to be removed, exposing the engine. The unpressurised cockpit was located well forward, which significantly improved cockpit visibility; the pilot was provided with an ejection seat and protected by a 57-mm (2¼ in) bulletproof windscreen, a 6-mm (0$\frac{15}{64}$ in) forward armour plate and an 8-mm (0$\frac{5}{16}$ in) armoured seat back.

The Yak-19 had mid-set two-spar wings of trapezoidal planform, with a span of 8.72 m (28 ft 7$\frac{5}{16}$ in) and an area of 13.56 m^2 (145.8 sq ft). The relatively thin wings with a 12% thickness/chord ratio at the root employed the TsAGI S-1-12 laminar airfoil verified on the Yak-RD-10 (Yak-RD) experimental fighter in the summer of 1946; they featured TsAGI flaps (modified Fowler flaps) and Frise ailerons with a TsAGI KV-3-12 laminar airfoil. The cruciform tail unit had a stabiliser span of 3.5 m (11 ft 5$\frac{51}{64}$ in); vertical tail area and horizontal tail area was 2.025 m^2 (21.77 sq ft) and 3.0 m^2 (32.25 sq ft) respectively.

The tricycle landing gear comprised a forward-retracting nose unit with a 400 x 150 mm (15.7 x 5.9 in) wheel and inward-retracting main units, the 570 x 140 mm (22.44 x 5.51 in) mainwheels stowing in the centre fuselage.

The Yak-19 was armed with two 23-mm (.90 calibre) Shpital'nyy Sh-3 cannons with 75 rounds each. Avionics and equipment included an RSI-6 radio, an RPKO-10M direction finder, an SCh-3 IFF transponder (SCh = *svoy/choo**zhoy*** – friend/foe), a GSK-1200 generator, a 12-A-10 DC battery, a PAU-22 gun camera, a KP-14 breathing

Yakovlev OKB

Above: '24 Yellow', the first prototype Yak-19 (c/n 01115001), during a test flight. The rounded tips of the wings and tail surfaces are clearly visible, as are the zero-dihedral tailplanes.

Yakovlev OKB

Yakovlev OKB

Yakovlev OKB

Three views of the first prototype Yak-19 during manufacturer's flight tests. Quite possibly the Yak-19 was not the most elegant fighter jet, but it sure was an improvement on the preceding pod-and-boom designs! Note the yellow '5' on the lower rudder segment.

Yakovlev OKB

Yakovlev OKB

Yakovlev OKB

Top, centre and above: Appropriately enough, '25 Yellow', the second prototype, was marked '2 Yellow' on the rudder. Note the drop tanks carried in the same fashion as on previous Yakovlev jets, the stabiliser dihedral and the slightly different shape of the vertical tail; the angular stabiliser tips are not so obvious.

apparatus with a 2-litre (0.44 Imp gal) oxygen bottle and other items.

Serialled '24 Yellow' and bearing the yellow numeral '5' on the rudder, the first prototype Yak-19 (c/n 01115001 – that is, Batch 01, Moscow Machinery Plant No.115, 001st aircraft in the batch) was cleared for manufacturer's flight tests on 29th November 1946. Next day the aircraft was trucked to the nearby Central airfield (Khodynka) to begin taxying trials. On 12th December the Yak-19 suffered a fire; when a technician started the engine as part of the ground tests, a defective gasket in a fuel line gave way, causing the fuel to leak out and ignite. After repairs, which took two weeks, the prototype was trucked to LII's airfield in Zhukovskiy on 26th December and taxying trials resumed there the following day. Finally, on 8th January 1947 the Yak-19 performed its maiden flight with Colonel Mikhail I. Ivanov (HSU) at the controls. Sergey N. Anokhin was the Yak-19's project test pilot for the rest of the manufacturer's flight test programme.

The aircraft had to be grounded from 2nd March to 18th May for an engine change; the long lay-up was caused by the late delivery of the replacement engine. As already mentioned, the Yak-19 was the first Soviet aircraft to be powered by an afterburning turbojet – but not the first to actually *use* afterburning; the afterburner was not ignited in flight until 21st May 1947, more than a month later than on Lavochkin's rival '156' (La-156) fighter. On that day the afterburner was engaged at 3,000 m (9,840 ft), increasing the fighter's indicated airspeed from 580 km/h (360 mph) to 670 km/h (416 mph). A second attempt was made on 22nd May and the Yak-19 accelerated from 600 km/h (372 mph) to 700 km/h (434 mph) when the afterburner kicked in.

The second prototype (c/n 115002) joined the test programme in the summer of 1947; it was referred to in official documents as the Yak-19 *dooblyor* and differed in having fittings for drop tanks under the wingtips. Serialled '25 Yellow' and marked '2 Yellow' on the rudder, this aircraft first flew on 6th June at the hands of Sergey N. Anokhin.

All in all, the manufacturer's flight tests included 64 flights; the first prototype made 27 flights and the second prototype 37, logging 23 hours 30 minutes between them. The Yak-19 became the first Soviet aircraft to crack 900 km/h (559 mph); with a 3,000-kg (6,610-lb) TOW the fighter had a top speed of 904 km/h (561 mph) at 5,000 m (16,400 ft), reaching this altitude in 3.9 minutes with the afterburner engaged. Anokhin reported that the Yak-19 presented no problems for the average pilot; the fighter easily performed the entire range of aerobatics, including inverted flight.

The manufacturer's flight test report was signed on 27th September 1947. It said: *'The Yakovlev-19 aircraft powered by an afterburning RD-10 turbojet has duly completed manufacturer's flight tests and can be submitted for State acceptance trials at GK NII VVS.'*

The test programme included canopy jettison trials; the process was filmed by a movie camera from a chase plane to determine the canopy's trajectory after separation. When the canopy locks were released the sliding portion popped up about an inch, then rotated rear end uppermost and slid aft along the fuselage in this nearly vertical position, gouging out deep scratches in the fuselage skin as it did. Moving to one side, the canopy finally fell clear of the aircraft, but the fuselage was severely damaged.

Back in the spring of 1947 Minister of Aircraft Industry Aleksey I. Shakhoorin had issued an order requiring General Designer and Director of MMZ No.115 Aleksandr S. Yakovlev to 'speed up flight tests of two Yak-19 fighters with afterburning RD-10 gas turbine engines so as to submit them for State acceptance trials in May'. However, the protracted construction of the second prototype caused the manufacturer's flight tests to last longer than anticipated, which meant the deadline stated in this document could not be met.

On 3rd August 1947 the second prototype Yak-19 piloted by Sergey N. Anokhin led a group of jet aircraft from various bureaux during the traditional flypast at Moscow-Tushino. Exactly two months later, on 3rd October, this aircraft was turned over to GK NII VVS for State acceptance trials which took place at Chkalovskaya AB between 17th October 1947 and 30th January 1948. Lieutenant-Colonel A. G. Proshakov was assigned project test pilot and Captain V. P. Belodedenko was the engineer in charge of the tests. At this stage the Yak-19 showed a top speed of 907 km/h (563 mph) in afterburner mode at 5,000 m with a 3,050-kg (6,724-lb) TOW, reaching this altitude in four minutes. The service ceiling was 12,100 m (39,700 ft). At full military power the fighter reached a top speed of 782 km/h (485 mph), climbing to 5,000 m in 7.7 minutes. Range in 'clean' configuration was 550 km (341 miles), increasing to 895 km (556 miles) with drop tanks.

Using the afterburner reduced the take-off run by some 24% and the balanced field length by 20%. Maximum speed at high altitude increased by 125 km/h (77 mph), or by 102 km/h (63 mph) as compared to the Yak-15's top speed. The State acceptance trials corroborated all the performance figures stated by the manufacturer. The trials report released by GK NII VVS (curiously, this document was not signed until 8th May 1948, more than three months after the completion of the trials!) said: '*...Uprating turbojet engines by combusting additional fuel in the jet nozzle is an effective way of increasing engine thrust during take-off and climb, as well as in level flight for a period of 2-4 minutes.*'

However, the report also mentioned numerous problems and shortcomings. For instance, Item 3 of the report's concluding part said:

'3. The State acceptance trials of the Yak-19 showed that:

- afterburner operation is unreliable and the afterburner cannot be used in air-to-air combat due to the problems associated with controlling it;

- the aircraft has numerous deficiencies, namely high stick forces from the ailerons, the lack of heating and ventilation systems in the cockpit, inadequate armour protection and so on.'

In Item 4 of the Conclusion the military stated their verdict unambiguously: '*...the Yak-19 fighter powered by the a/b RD-10 engine cannot be recommended for Air Force service.*'

Instead of building a third prototype with appropriate modifications and refining the RD-10F to a degree when it could be put into production (as was normal practice at OKB-115), Yakovlev chose to pull the plug on the Yak-19. The reason was that in early 1948 the Soviet aircraft industry had launched production of the new RD-500 non-afterburning centrifugal-flow turbojet – a licence-built Rolls-Royce Derwent V delivering 2,270 kgp (5,000 lbst) for take-off. The PD section of OKB-115 did begin preliminary calculations for a Derwent-powered version of the Yak-19 but this aircraft was

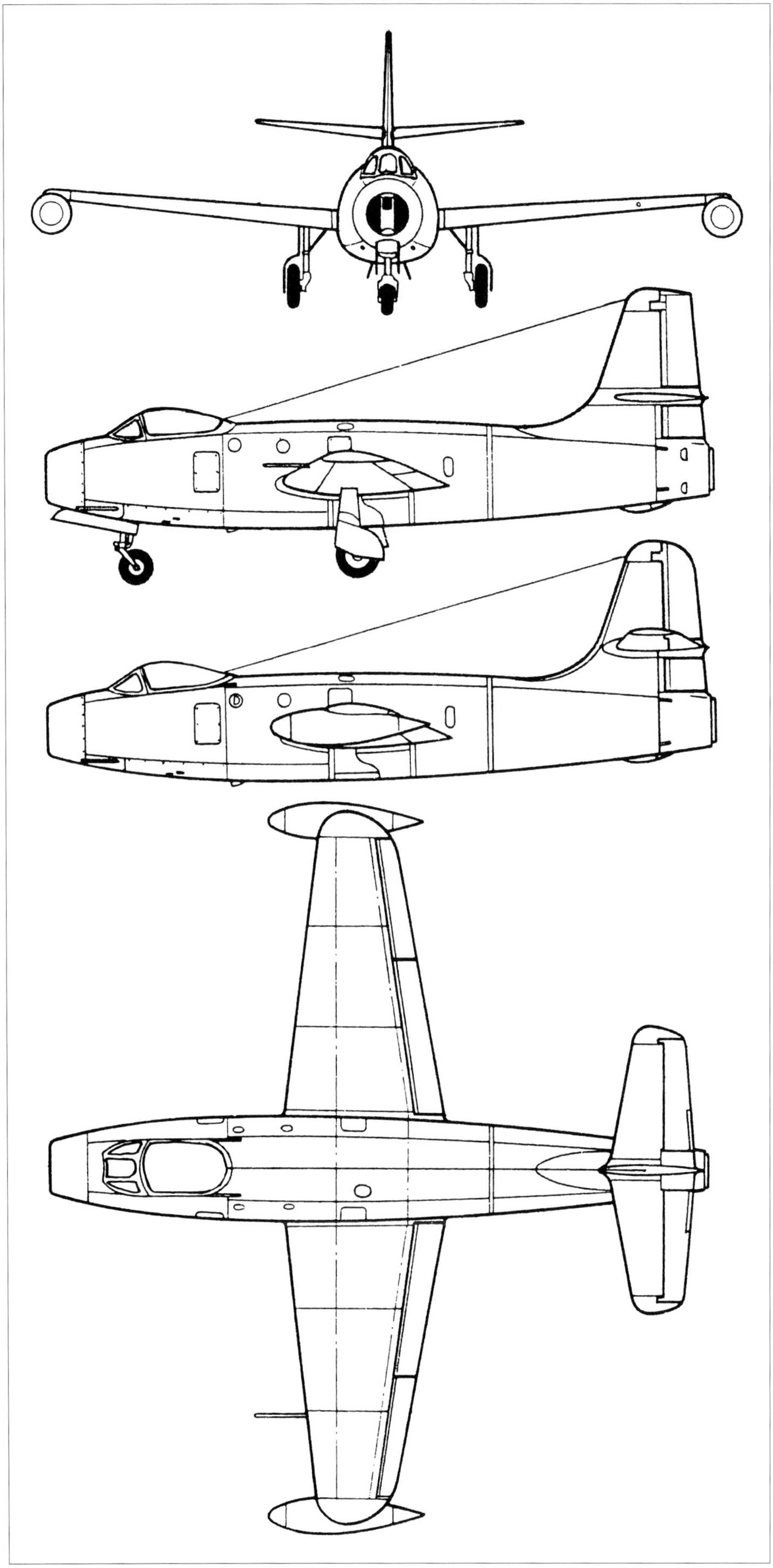

A three-view of the second prototype Yak-19, with an additional side view (with the undercarriage extended) of the first prototype.

Specifications of the Yak-19

	Manufacturer's flight tests	State acceptance trials
Length overall	8.36 m (27 ft 5 9/64 in)	8.357 m (27 ft 5 1/64 in)
Wing span (less drop tanks)	8.70 m (28 ft 6 33/64 in)	8.72 m (28 ft 7 5/16 in)
Wing area, m^2 (sq ft)	13.50 (145.16)	13.56 (145.80)
Operating empty weight without drop tanks, kg (lb)	2,192 (4,832)	2,151 (4,742)
Payload, kg (lb):		
normal	808 (1,781)	899 (1,982)
in overload configuration	1,138 (2,508)	1,249 (2,753)
Fuel load, kg (lb):		
without drop tanks	650 (1,430)	650 (1,430)
with drop tanks	980 (2,160)	973 (2,145)
All-up weight, kg (lb):		
normal (in 'clean' condition)	3,000 (6,613)	3,050 (6,724)
maximum (with drop tanks)	3,350 (7,385)	3,400 (7,495)
Wing loading, kg/m^2 (lb/sq ft)	222 (45.51)	226 (46.33)
Power loading, kg/kgp (lb/lbst)	2.91	2.74
Top speed at full military power/ with afterburner, km/h (mph):		
at sea level	760/875 (472/543)	714 (443)/n.a.
at 10,000 m (32,810 ft)	788/875 (489/543)	747 (463)/n.a.
Landing speed, km/h (mph)	180 (111)	155 (96)
Time to height, minutes:		
to 5,000 m	n.a./3.9	7.7/4.0
to 10,000 m	n.a./10.5	24.0/n.a.
Rate of climb, m/sec (ft/min):		
at sea level	16.4/25.8	13.8/23.0 (2,715/4,526)
at 5,000 m (16,400 ft)	n.a.	8.3/17.8 (1,633/3,503)
Service ceiling, m (ft)	n.a.	12,100 (39,700)
Range at 0.9 V_{MAX} at 8,000 m (26,250 ft), 'clean'/with drop tanks, km (miles)	n.a.	545/900 (338.5/559)
Endurance without/with drop tanks at 8,000 m	n.a.	1 hr 10 min / 1 hr 44 min
Take-off run in 'clean' condition (full military power/afterburner), m (ft)	n.a./550 (1,800)	685/550 (2,250/1,800)
Take-off run with drop tanks (full military power/afterburner), m (ft)	n.a.	925/675 (3,304/2,214)
Landing run, m (ft)	520 (1,706)	550 (1,804)

never built because the OKB had already developed two new fighters designed around this engine – the Yak-23 and the Yak-25 (see next entry). Nevertheless, afterburners were quickly brought up to an adequate reliability level, becoming standard equipment on all jet fighters designed from the mid-1950s onwards.

Yak-25 single-seat fighter (first use of designation)

OKB-115's next step in jet fighter development was the Yak-25 of 1947 – yet another straight-wing aircraft (the first to bear this designation). As already mentioned, the Council of Ministers directive issued on 11th March 1947 tasked several design bureaux with developing new tactical fighters; specifically, the Yakovlev OKB was to create a straight-wing fighter powered by a single Rolls-Royce Derwent V. The straight-wing clause was very probably included at the request of Aleksandr S. Yakovlev himself who was still wary of swept wings.

In designing the Yak-25 the engineers had, first of all, to eliminate the Yak-23's main deficiencies: the pod-and-boom arrangement (which was detrimental to flight performance) and the poor cockpit visibility caused by the aft position of the cockpit. Hence the Rolls-Royce V turbojet was relocated to the rear fuselage, while the pressurised cockpit equipped with a bulletproof windscreen and an ejection seat was placed in the forward fuselage between the engine's inlet ducts. This arrangement had been first used by OKB-115 on the Yak-19. As on the latter type, the Yak-25's fuselage had a break point aft of the wings allowing the aft fuselage to be detached for engine maintenance or removal; the fuselage break was sealed by a wraparound duralumin strip. The fuselage was 8.66 m (28 ft 4 15/16 in) long (the test reports quote this figure as the overall length), with an almost circular section – 1.36 m (4 ft 5 35/64 in) wide and 1.34 m (4 ft 4¾ in) high; cross-section area was 1.4 m^2 (15.05 sq ft).

For the first time in Soviet design practice, airbrakes were installed on the aft fuselage sides beneath the horizontal tail, deploying automatically when the aircraft approached V_{NE}. Also, the Yak-25 was the first Soviet aircraft to feature an air conditioning system using engine bleed air; the system was designed by Yakovlev OKB engineer G. D. Protasov.

The Yak-25 had two-spar wings with a span of 8.88 m (29 ft 1 19/32 in), an area of 14.0 m^2 (150.53 sq ft) and a thickness/chord ratio of 9% throughout. Sweepback at quarter-chord was 2°45' but for all practical purposes the wings can be considered unswept; the aspect ratio was 5.64 and wing taper was 2.5. The straight-through wing spars were bent to a U-shape where they passed through the fuselage to avoid infringing on the inlet ducts; the centre section incorporated an abbreviated auxiliary spar (also U-shaped) to which the main landing gear units were attached. The spar flanges were made of steel T-beams and the spar webs of V-95 grade duralumin sheet. The Yak-25 utilised TsAGI laminar airfoils developed by G. P. Svischchev, ranging from S-9S-9 at the root to KV-4-9 at the tip. The wings featured TsAGI slotted flaps deflected 25° for take-off and 55° for landing.

The swept tail surfaces were of cruciform layout, as was customary for Soviet fighters at the time. The stabilisers were swept back 45° (though it is not clear whether this was leading-edge sweep or quarter-chord sweep). Once again the vertical and horizontal tail were of two-spar construction and utilised TsAGI laminar airfoils; stabiliser span was 3.208 m (10 ft 6 19/64 in), vertical tail area 2.125 m^2 (22.85 sq ft) and horizontal tail area 3.0 m^2 (32.25 sq ft).

The tricycle landing gear was similar to that of the Yak-19, with a 400 x 150 mm (15.7 x 5.9 in) nosewheel and 570 x 140 mm (22.44 x 5.51 in) mainwheels. The armament was different, comprising three 23-mm (.90 calibre) Nudel'man/Rikhter NR-23 cannons with 75 rpg housed in the forward fuselage beneath the cockpit. To extend range the Yak-25 could be fitted with 380-litre (83.6 Imp gal) teardrop-shaped drop tanks carried under the wingtips, as on the Yak-19.

The avionics and equipment fit was updated to include an RSI-6M radio, an

RPKO-10M direction finder, an SCh-3 IFF transponder, a more powerful GSK-1500 generator, a 12-A-10 DC battery, an S-13 gun camera and a KP-14 breathing apparatus with a 2-litre (0.44 Imp gal) oxygen bottle and more. As on the Yak-19, the pilot was protected by a 57-mm bulletproof windscreen, a 6-mm forward armour plate and an 8-mm armoured seat back; the GK NII VVS test report, however, states the bulletproof windscreen was 60 mm ($2\frac{23}{64}$ in) thick.

As already mentioned, the Yak-25 had a pressurised cockpit featuring a canopy jettison system. The original version was designed by Yakovlev's aide N. K. Skrzhinskiy and was a decidedly complex affair that could prove troublesome if fitted to a combat aircraft. Hence, after discussing this cockpit (which had by then been almost completed at OKB-115's experimental shop) with his colleagues, Yak-25 project chief Leonid L. Selyakov proposed a simpler version, which was eventually installed in the aircraft.

The 'step by step' approach taken by OKB-115 had its advantages; the trials of a new aircraft (with any changes that needed to be made in the course of the trials) and its eventual entry into full-scale production usually proceeded quickly and without a hitch. Yet there is always an exception to the rule, and the Yak-25 was one.

The design stage gave no reason for gloomy forebodings. The fighter was designed and built in an amazingly short time; A. S. Yakovlev approved the ADP project on 1st February 1947 and the first prototype ('55 Yellow', c/n 115001) was completed in October. As per OKB-115 custom, a yellow numeral (in this case, 2) was painted on the rudder; this numeral meant nothing at all and was intended to confuse would-be spies by falsely suggesting how many development aircraft had been built and in what order. In internal documents the aircraft was referred to as 'Yak-25 No.01' for the sake of clarity.

On 29th October '55 Yellow' was cleared for manufacturer's flight tests which officially began two days later. The ground test stage was extremely brief, and on 2nd November 1947 test pilot Sergey N. Anokhin took the Yak-25 into the air for the first time.

The tests showed that generally the Yak-25 outperformed both the experimental Yak-19 and the production Yak-23. The fighter was easy to fly and handled well, performing the full range of aerobatics, and, importantly, it was absolutely spin-proof. However, the first test flights brought bad news as well. The laminar airfoils which worked well in the straight wings turned out to be absolutely unsuitable for the swept tail surfaces (or, to be precise, the elevators). Anokhin discovered that severe buffeting set

Yakovlev OKB

Yakovlev OKB

Top and above: Two views of '55 Yellow', the first prototype Yak-25, during manufacturer's flight tests, showing to advantage the swept tail. Note the kink in the wing leading edge at the roots.

Yakovlev OKB

Above: '15 Yellow', the second prototype Yak-25 (c/n 115002). Note that the numeral on the rudder is the same but the fin cap is more angular.

Yakovlev OKB

An air-to-air shot of the second prototype Yak-25 with drop tanks during a test flight.

Yakovlev OKB

Yakovlev OKB

Top and above: No, this is not the first Yak-25 prototype, it is the third one (c/n 115003)! Compare the fin cap shape to the aircraft on the preceding page.

in at 500 km/h (310 mph), increasing dramatically as the speed grew; at 550 km/h (341 mph) the pilot could no longer hold the control stick. According to L. L. Selyakov, the buffeting was so bad that 'the pilot was thrown about, banging his head on the canopy, and all the needles of the flight instruments came off'.

In an effort to cure the problem a team of Yakovlev engineers supervised by M. A. Teitz studied piles of British and captured German materials on aerodynamics dealing with high-speed buffet. It transpired that the laminar airfoil with the maximum thickness well aft resulted in large upwash and downwash angles, generating vortices which rocked the elevators up and down. The problem became especially acute when the elevators were deflected, as the vortices grew stronger and the oscillation frequency increased. The Germans and the British had tried to fix the problem (on the Messerschmitt Me 262 and the Gloster Meteor respectively) by riveting T-profiles or attaching heavy-gauge round wire along the trailing edges of the elevators and ailerons. This postponed the onset of buffeting but did not eliminate it completely; also, while the speed envelope was slightly expanded, elevator forces became excessively high.

As a result, the first prototype Yak-25 was provisionally fitted with T-profiles 20 mm ($0\frac{25}{32}$ in) wide along the elevator trailing edges. (Yakovlev had had to resort to this earlier with the Yak-19, installing a T-profile on the rudder trailing edge.) The problem was ultimately solved by fitting the second and third prototypes with a redesigned horizontal tail. The new stabilisers featured the specially developed NACA 004 airfoil with the maximum thickness at the quarter-chord line (the thickness/chord ratio was 20%); this made for very small upwash and downwash angles. Changes were also made to the vertical tail; outwardly this was revealed by the more streamlined fin cap.

The redesigned tail unit eliminated the buffeting problem completely, enabling Anokhin to reach an unprecedented top speed of 982 km/h (610 mph), according to the manufacturer's test report. Some documents state a more modest but still impressive achievement of more than 970 km/h (602 mph). During manufacturer's flight tests the empty weight was recorded as 2,285 kg (5,037 lb); take-off weight in 'clean' configuration was 3,235 kg (7,130 lb), increasing to 3,580 kg (7,890 lb) with drop tanks.

The story of the first fighter to bear the Yak-25 designation would be incomplete if details of individual aircraft identities and the Yakovlev OKB's counter-intelligence tricks were left out. Ironically, these tricks aimed at the 'potential adversary' proved so effective that they fooled many Soviet/Russian aviation researchers, including the late Vadim B. Shavrov whose authority as an aviation historian is universally acknowledged. For many years it was erroneously believed that only two 'first-generation' Yak-25s ('55 Yellow' and '15 Yellow') were ever built; in fact, there were *three* flying prototypes. The key to this puzzle is that the first prototype (c/n 115001, referred to in paperwork as 'aircraft No.01') was serialled '55 Yellow' and marked '2 Yellow' on the rudder. The second aircraft (c/n 115002 or 'aircraft No.02') was indeed serialled '15 Yellow' and likewise had a yellow numeral '2' on the rudder, while the third Yak-25 (c/n 115003 or 'aircraft No.03') was again serialled '55 Yellow' and marked '2 Yellow'! The trick was disclosed when high-quality pictures of the three prototypes showing the c/n clearly stencilled on the fin became available.

The attempts to cure the fighter's pitch control problem coincided with the preparations for the annual Tushino flypast. In August 1948 Sergey N. Anokhin was to lead the parade in the Yak-25 (he was even allocated the callsign ***Pa**roos*, 'Sail', in advance). The aircraft was to pass in front of the grandstand with the Soviet government and Communist Party leaders at 100 m (330 ft), doing 500 km/h (310 mph). Leonid L. Selyakov recalls that he was summoned to the Ministry of State Security (MGB, the forerunner of the KGB) shortly before the show; after asking a few questions about the flypast the security officer demanded a written statement that the demonstration would be perfectly safe. Selyakov wrote and signed the required statement without hesitation and the Yak-25 put on a spectacular performance at Tushino – just like the Yak-19 had done a year earlier. Curiously, Aleksandr S. Yakovlev chose not to intervene before the show; afterwards, however, he summoned Selyakov and inquired on what grounds the latter had ventured to give such guarantees. It is easy to guess the consequences, had the aircraft crashed, especially with fatalities on the ground! Selyakov replied earnestly that technically there was no risk – and should anything go wrong, he had counted on Yakovlev (who was still influential then) to step in and save him. This last bit, Selyakov recalled, 'amused Yakovlev immensely'. It is easy to see why. In those days retribution came swiftly, starting with those at the top of the pile; Yakovlev would hardly have been able to save himself, much less save others!

Because of the abovementioned problems and modifications the manufacturer's flight test programme dragged on until mid-1948; the test report was finally signed on 3rd July. This in turn caused the State acceptance trials to be postponed; besides, GK NII VVS encountered difficulties in filling out the associated paperwork because the Yak-25 was the first Soviet aircraft to feature a fully capable pressurised cockpit.

The cockpit was pressurised at 0.3 kg/cm² (4.28 psi). The engineers had to prove that if the canopy jettison system failed the pilot would manage to jettison it manually in order to eject. OKB-115 had already had some negative experience with the Yak-19; as recounted earlier, trials of canopy jettison system were dangerous for aircraft and pilot alike.

Selyakov suggested a different *modus operandi* – the system was to be tested on the ground. One Sunday, when the territory of MMZ No.115 was almost empty, he ordered the Yak-25 prototype wheeled into the assembly shop, then climbed into the cockpit, strapped himself in and closed the canopy. When the nominal pressure differential of 0.3 kg/cm² had been attained, using a ground compressor, Selyakov jettisoned the canopy as an Air Force commission watched. He did not feel too well immediately afterwards, but it was worth it: the military were convinced by this demonstration and signed an act to the effect that the cockpit and canopy jettison system were OK. In his memoirs, Selyakov later wrote he was confident that everything would work; after all, he had designed the system himself. Besides, the canopy and the jettison system had been manufactured and tested under his personal supervision.

Another unpleasant surprise (fortunately, again with no disastrous consequences) occurred during the manufacturer's tests. Anokhin had test flown the MiG-9 before transitioning to the Yak-25, and it took a while for him to get used to the latter's slightly different cockpit layout. Standardised cockpits were still unheard-of in those days, and the different location of certain control levers nearly resulted in a major accident. The port cockpit console of early Soviet fighters featured the throttle, the engine shutdown lever and the airbrake control lever. These were located differently on the MiG-9 and the Yak-25 (left to right: throttle, shutdown lever and airbrake control on one aircraft and throttle, airbrake control and shutdown lever on the other). Attempting to deploy the airbrakes in one of the flights, Anokhin grabbed the wrong lever and shut down the engine instead. The aircraft was at low altitude, which left no time to try and restart the engine, and the pilot did the only possible thing, landing *across* the runway. Luckily, both he and the aircraft stayed in one piece.

On 12th June 1948 the Council of Ministers issued directive No.2052-804ss which, among other things, gave a more specific operational requirement concerning the Yak-25 and demanded that the fighter be submitted for State acceptance trials. On 5th July Vice-Minister of Aircraft Industry Shishkin signed the appropriate papers, authorising the aircraft to be transferred to GK NII VVS. A day later a LII test pilot ferried the third prototype to Chkalovskaya AB, a complete set of documents following three days later.

The Act of acceptance (stating that the aircraft had begun State acceptance trials) said that the fighter had been delivered with

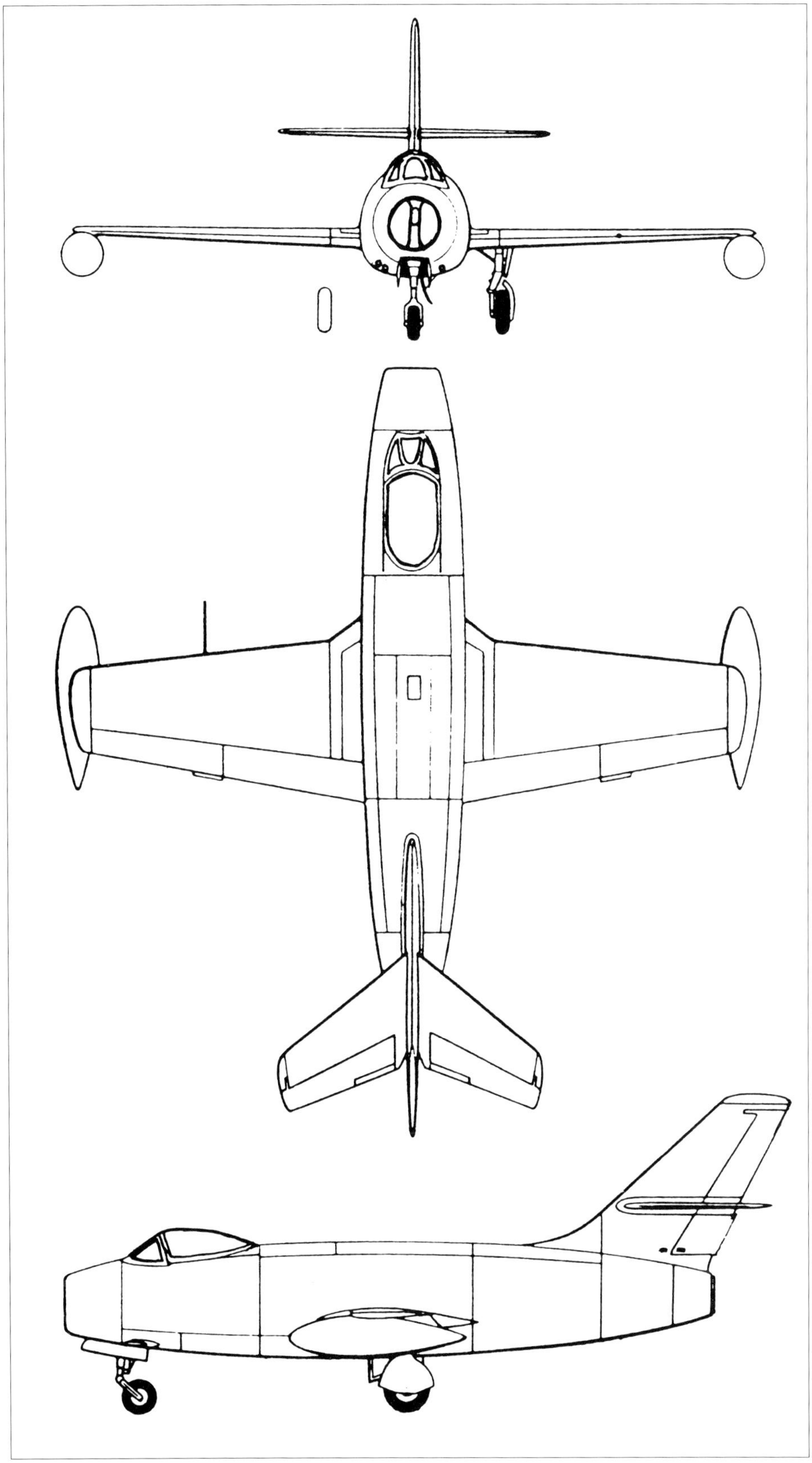

A three-view of the first prototype Yak-25.

TsAGI

The third prototype Yak-25 ('aircraft 03') with drop tanks sits on a rain-soaked hardstand at GK NII VVS during State acceptance trials.

an incomplete avionics and equipment fit. Missing items included the RSI-6 communications radio, the *Bariy-1* (Barium) IFF transponder, the NI-46 navigation display (*navigatsi**on**nyy indi**kah**tor*), the gun camera and the fire extinguishing system; additionally, a 2-litre (0.44 Imp gal) oxygen bottle was fitted instead of the specified 4-litre (0.88 Imp gal) bottle. Moreover, the manufacturer's flight tests of the Yak-25 had not included several important aspects. Among other things, the fighter's behaviour at top speed below 2,000 m (6,560 ft) had not been studied; nobody had tried restarting the engine after an in-flight shutdown or studied the effect of firing the cannons on engine operation above 3,000 m (9,840 ft); the fighter's ability to withstand the specified maximum operational G-loads had not been verified, and so on. Finally, some of the features requested by the military had not been incorporated; for instance, the ammunition supply was only 75 rpg instead of the required 100 rpg, there was no wire mesh screen in the inlet duct to protect the engine from foreign object damage (FOD) and the mechanical canopy jettison system was inoperative, which is why it could not be tested in flight.

Still, the VVS agreed to hold State acceptance trials of the Yak-25; the formal Act of acceptance was signed on 29th July. For the reasons stated above it was decided to abbreviate the trials programme, limiting it to the bare necessities. Still, there was really no point in holding these trials because other fighter types had entered production in the meantime.

Officially State acceptance trials of Yak-25 No.03 lasted from 23rd July to 9th September 1948. As with the Yak-19, Lieutenant-Colonel A. G. Proshakov was project test pilot and Captain V. P. Belodedenko was the engineer in charge of the tests. The fighter fully met the specifications to which it was designed, and despite a 270 kg (595 lb) higher all-up weight than the Yak-23 it was markedly superior to the latter aircraft in speed and service ceiling. The trials showed that the new general arrangement increased the speed by 25-50 km/h (15.5-31 mph) as compared to the Yak-23, depending on the altitude; the Yak-25 had a top speed of 950 km/h (590 mph) at sea level and 972 km/h (603 mph) at 3,000 m (9,840 ft). The fighter climbed to 5,000 m (16,400 ft) in 2.6 minutes versus 2.3 minutes for the lighter Yak-23; on the other hand, the service ceiling was 15,200 m (49,870 ft), or 400 m (1,312 ft) higher than the Yak-23's. At 5,000 m the Yak-25 could make a 360° turn in 29 seconds. The take-off run was only 540 m (1,770 ft) and the landing run 500 m (1,640 ft). Maximum range with drop tanks at 10,000 m (32,810 ft) was 1,445 km (897.5 miles) – 140 km (87 miles) shorter than the Yak-23's, while endurance (as per manufacturer's flight test report) was 2 hours 16 minutes.

As already mentioned, the canopy jettison system could not be tested during the State acceptance trials, so a separate test programme was held afterwards (the test report was signed on 29th September 1948). The system worked as it should; the problem that had affected the Yak-19 was not repeated.

Specifications of the Yak-25 fighter prototypes

	Yak-25 c/n 115001	Yak-25 c/n 115003 State acceptance trials
Length	8.66 m (28 ft 5 in)	n.a.
Wing span	8.88 m (29 ft 1⅝ in)	n.a.
Wing area, m² (sq ft)	14.0 (150.7)	n.a.
Empty weight, kg (lb)	2,285 (5,037)	n.a.
All-up weight, kg (lb):		
'clean'	3,185 (7,022)	3,235 (7,132)
with tip tanks	3,535 (7,793)	3,580 (7,892)
Speed at sea level, km/h (mph)	982 (610)	950 (590)
Speed at 3,000 m (9,840 ft), km/h (mph)	953 (592)	972 (504)
Landing speed, km/h (mph)	172 (107)	160 (99.4)
Climb time, minutes:		
to 5,000 m (16,400 ft)	2.5	2.6
to 10,000 m (32,810 ft)	6.3	n.a.
Service ceiling, m (ft)	14,000 (45,930)	12,500 (41,010)
Operational range, km (miles):		
'clean'	1,100 (683.5)	n.a.
with tip tanks	1,600 (994)	1,445 (898)
Take-off run, m (ft)	510 (1,670)	540 (1,770)
Landing roll, m (ft)	825 (2,710)	500 (1,640)

The tests of the straight-wing Yak-25 held in 1948 were no more than a formality; the 'step by step' development policy had backfired on the Yakovlev OKB in this instance. While the design staff of OKB-115 methodically refined straight-wing fighters from the Yak-15 to the Yak-25, the rival OKB-155 headed by Artyom I. Mikoyan brought out and put into production the highly successful swept-wing MiG-15 fighter which outclassed the Yak-25 in many respects. (Interestingly, the Yak-25 reached a top speed of Mach 0.85 at 1,000 m (3,280 ft), which was just a trifle slower than the MiG-15's Mach 0.902.) Thus Yakovlev finally conceded it was time to switch to swept-wing designs; therefore, even as the State acceptance trials proceeded, all further attempts to improve the Yak-25 were abandoned.

Two of the prototypes were later used for test and development purposes. The second aircraft was modified under the Burlaki (pronounced *boorla***kee**) captive escort fighter programme. A pneumatically-operated telescopic probe (dubbed 'harpoon') was installed atop the nose; its barbed tip locked into a drogue at the end of a steel cable paid out by the bomber to be escorted. (In 19th-century Russia, the *burlaki* were teams of strongmen whose job was to haul barges up rivers by means of ropes; the analogy with the towed fighter concept was obvious.) The third (or first?) prototype was fitted experimentally with a non-retractable bicycle landing gear.

Yak-27 single-seat fighter (project, first use of designation)

In 1947, concurrently with the development of the Yak-25 fighter, the OKB undertook design work on a similar fighter project designated Yak-27 (the first to bear this designation). Like its stablemate, it featured a combination of straight wings and swept-back tail surfaces. The aircraft was to be powered by a single Rolls-Royce Nene turbojet installed in the rear fuselage. Initial project studies conducted in February 1947 envisaged a circular nose air intake; in May 1947 a version was under consideration featuring wing root air intakes, like those of the future Yak-30 trainer. The project was not proceeded with.

Yak-29 single-seat fighter (project)

Between July and October 1947 the Yakovlev OKB conducted design work on yet another fighter. Designated Yak-29, it was a machine of very modest dimensions, having a wing span of a mere 6.3 m (20 ft 8 in), a length of 8.64 m (28 ft $4\frac{5}{32}$ in) and a wing area of 8.0 m^2 (86.12 sq ft). The all-up weight was 2,300 kg (5,070 lb), resulting in a wing

Yakovlev OKB

Yakovlev OKB

Yakovlev OKB

Top, centre and above: The Yak-30 fighter combined the fuselage, tail surfaces and powerplant of the Yak-25 with new swept wings. This is the first prototype, '42 Yellow'. Note the wing fences.

Yefim Gordon archive

The first prototype Yak-30 at the 1948 Aviation Day flypast at Moscow-Tushino.

Yakovlev OKB

Yakovlev OKB

Top and above: Two aspects of the second prototype ('54 Yellow'), sometimes called Yak-30D. A close look at the lower photo reveals that the aircraft had a slightly longer rear fuselage to create room for an extra fuel tank (see previous page).

Specifications of the Yak-30 fighter prototypes

Type	Yak-30	Yak-30D
Length	8.58 m (28 ft 1¼ in)	8.96 m (29 ft 5 in)
Wing span,	8.65 m (28 ft 4½ in)	8.65 m (28 ft 4½ in)
Wing area, m² (sq ft)	15.1 (162.5)	15.1 (162.5)
Empty weight, kg (lb)	2,415 (5,324)	n.a.
All-up weight, kg (lb)		
'clean'	3,305 (7,286)	3,440 (7,584)
with drop tanks	3,630 (8,003)	4,015 (8,851)
Speed at sea level, km/h (mph)	930 (578)	1,025 (637)
Speed at altitude, km/h (mph):		
at 5,500 m (18,050 ft)	1,010 (627.7)	n.a.
at 10,000 m (32,810 ft)	970 (602.8)	n.a.
Landing speed, km/h (mph)	166 (103)	n.a.
Climb time, minutes:		
to 5,000 m (16,400 ft)	2.6	n.a.
to 10,000 m	6.6	n.a.
Service ceiling, m (ft)	15,000 (49,210)	n.a.
Operational range, km (miles):		
on internal fuel	1,000 (621)	n.a.
with drop tanks	1,500 (932)	1,950 (1,212)
Take-off run, m (ft)	510 (1,670)	n.a.
Landing roll, m (ft)	610 (2,000)	n.a.

loading close to 300 kg/m² (61.45 lb/sq ft), which was a fairly high figure. A twin-cannon version and a version without armament were studied.

Yak-30 single-seat fighter prototype (first use of designation)

In 1948 the Yakovlev OKB produced one more fighter prototype which, significantly, was Yakovlev's first swept-wing aircraft. Apart from the wings, the new fighter bore an obvious relationship to the preceding Yak-25. For some reason, the OKB chose to allocate to it the designation Yak-30, which was out of sequence; moreover, the even number ran counter to the established practice of assigning odd numbers to fighters. Actually, this was the first of a series of Yakovlev's fighter projects whose designations used 'round' numbers ending in a zero, possibly in order to sound more impressive.

The new wings had 35° leading-edge sweep, 2° anhedral from the roots and a taper ratio of 1.5. The wing root sections had a straight trailing edge, thus increasing the root chord. Mounted on the upper surface were two pairs of shallow full-chord boundary layer fences. Inboard of the hydraulically boosted ailerons, the wing trailing edge was occupied by slotted flaps.

The circular-section fuselage accommodated an RD-500 turbojet accessible by detaching the entire rear fuselage and tail unit. The tail was almost identical to that of the Yak-25, with some minor changes. The undercarriage, again, repeated the pattern adopted for the Yak-25, with a forward-retracting nose unit and main units retracting inward so that the mainwheels stowed in the fuselage. The cockpit, armament and fuel systems were virtually unchanged, but the drop tanks fitted flush under the wings at about mid-span rather than at the tips.

The first prototype, serialled '42 Yellow' and carrying a yellow '5' on the rudder, passed its manufacturer's tests between 4th September and 16th December 1948. The Yak-30 displayed good handling, demonstrating the advantage of the swept wings by attaining speeds in excess of the Yak-25's (at altitude) and a better rate of climb. However, the swept wings entailed some increase of the landing speed and landing run.

In early 1949 the Yakovlev OKB built the second prototype which was known as the Yak-30D (presumably denoting *doob**lyor*** – second prototype). It had a slightly longer fuselage providing room for an additional tank. The wing root structure was modified; incorporating the so-called 'internal bracing strut'. Door-type airbrakes were added on the rear fuselage. The main undercarriage was redesigned, the slotted flaps gave place to extension flaps, changes were introduced into the communications and oxygen systems. The fuel load and ammunition supply were increased. All this increased the all-up weight slightly and resulted in marginally better performance, the maximum speed at sea level reaching 1,025 km/h (627 mph) versus the first prototype's 930 km/h (578 mph); the range increased to 1,950 km (1,212 miles).

However, by the standards of 1949 these performance figures were already considered insufficient; the Yak-30 and Yak-30D did not proceed beyond the prototype stage. The Yak-30 designation was later re-used twice for quite different aircraft.

Yak-40 fighter (project, first use of designation)

Between January and June 1948 the Yakovlev OKB conducted project studies of a single-seat fighter powered by two ramjets. The engines were mounted at the tips of the wings featuring 45° sweepback. The ramjets had a thrust rating of 850 kgp (1,874 lbst) apiece and were housed in nacelles measuring 2.50 m (8 ft 2$\frac{27}{64}$ in) in length and 0.55 m (1 ft 9$\frac{21}{32}$ in) in diameter. The aircraft had a length of 7.50 m (24 ft 7$\frac{9}{32}$ in) and an all-up weight of 1,800 kg (3,970 lb). The take-off was to be assisted by solid-fuel rocket boosters mounted on a wheeled launch trolley. A modified version of the fighter was designated Yak-40A (see below). It did not reach the hardware stage.

Interestingly, the Yak-40 was to be used in combination with a carrier aircraft, the Tupolev Tu-4 bomber being envisaged as the 'mother ship'. Up to six Yak-40 fighters (!) were to be suspended under the bomber's outer wing panels, the wings of the fighters overlapping to enable the stowage of three diminutive machines within a restricted space outside the discs of the outer engines' propellers.

Subsequently the Yak-40 designation was re-used for a short-haul airliner.

Yak-40A fighter (project)

This version of the ramjet-powered fighter differed from the original project variant in operational procedures. The Yak-40A was to take off with the assistance of two U-93-1 rocket boosters delivering a thrust of 750 kgp (1,650 lbst) apiece and mounted under the wings of the fighter, not on a trolley. This version did not reach the prototype construction stage either.

Yak-41 fighter (project, first use of designation)

Concurrently with the work on the Yak-40 and Yak-40A ramjet-powered fighters the Yakovlev Design Bureau conducted studies on yet another project of a similar kind

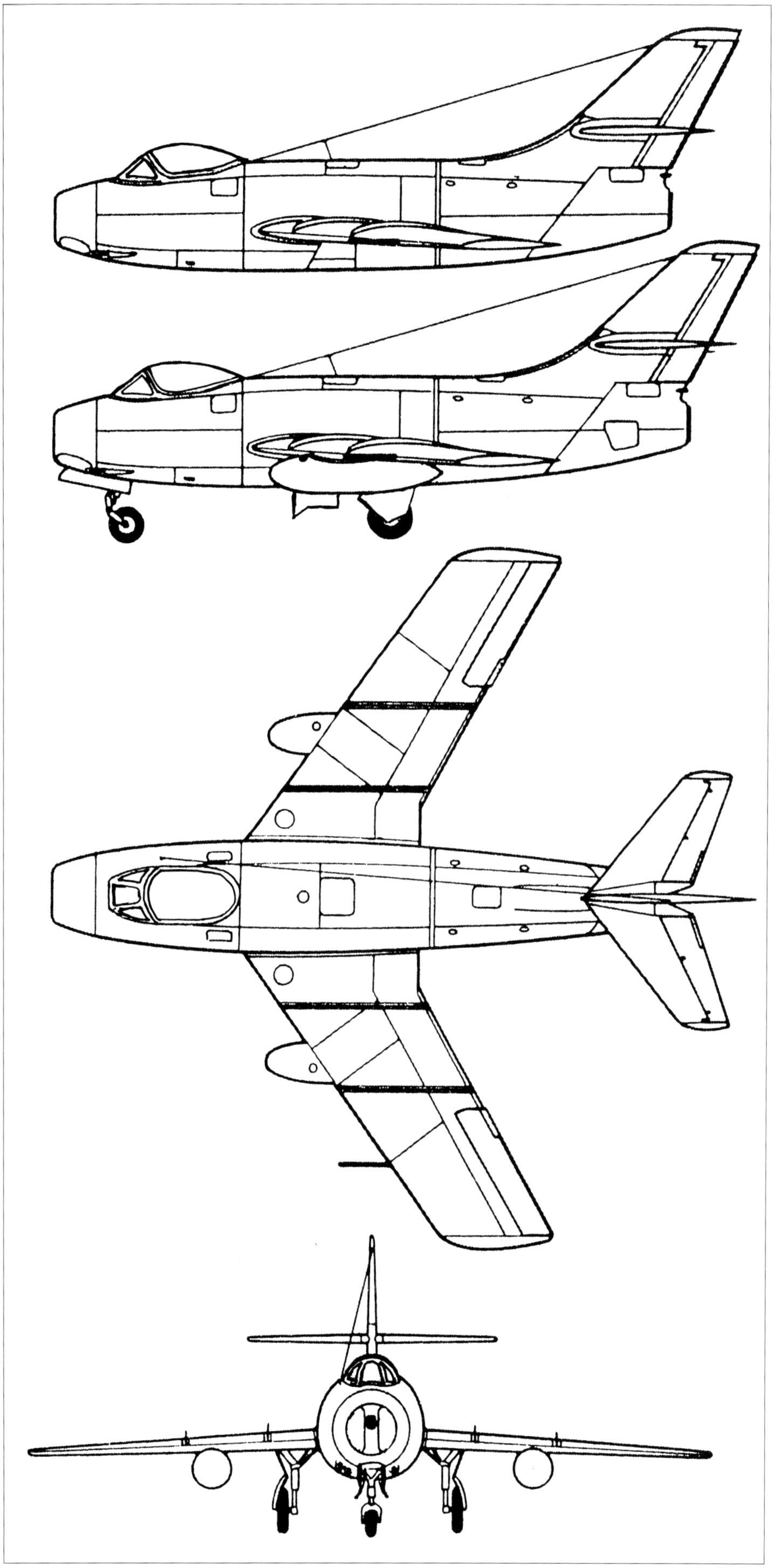

A three-view of the Yak-30D, with an additional side view (top) of the Yak-30 *sans suffixe*.

Yakovlev OKB

Above, below and bottom: The first prototype Yak-50 all-weather fighter prototype during manufacturer's tests. Note that all three boundary layer fences on each wing are of different height. Note also the shallow twin ventral strakes and the bullet fairing at the fin/stabiliser joint.

Yakovlev OKB

Yakovlev OKB

designated Yak-41. Featuring the same basic layout, this aircraft had a wing span of 4.5 m (14 ft $9^{11}/_{64}$ in), a length of 7.9 m (25 ft 11 in) and a take-off weight of 2,300 kg (5,070 lb). The take-off was to be effected with the help of a 300-kgp (660-lbst) liquid-fuel rocket booster installed in the rear fuselage. The aircraft was not built.

The Yak-41 designation was later re-used for a V/STOL jet fighter better known as the Yak-141.

Yak-50 all-weather fighter (first use of designation)

The Yak-50 fighter was designed to meet a specification for an all-weather/night interceptor equipped with a gun-ranging radar. The design philosophy adopted by the OKB envisaged a lightweight machine based on the use of the wings with increased sweepback and powered by a single Klimov VK-1 centrifugal-flow turbojet – an uprated version of the RD-45F delivering 2,700 kgp (5,590 lbst) for take-off.

The design performance stipulated by a government directive issued on 21st February 1949 included a speed of 1,135 km/h (705.4 mph) at 4,200 m (13,780 ft). The aircraft was to climb to 10,000 m (32,810 ft) within 3-3.5 minutes, have a service ceiling of 15,000-16,000 m (49,210-52,500 ft) and an endurance of 40-45 minutes.

The Yak-50 was a single-seat monoplane with mid-set wings having a sweepback of 45° at quarter-chord and 5° anhedral. Each wing had three shallow fences, the inboard pair being full-chord and the outer fences terminating ahead of the ailerons. The wings were provided with hydraulically actuated ailerons, slotted track-mounted flaps occupying the trailing edge inboard of these.

The circular-section semi-monocoque fuselage featured a removable tail section for engine maintenance and change. The fuselage nose accommodated the ***Korshun*** (kite, the bird) radar developed by NII-17, its scanner being housed in a prominent bullet-shaped radome above the engine air intake which was split into inlet ducts flanking the cockpit. Door-type airbrakes were hinged on each side of the rear fuselage. The vertical tail had a quarter-chord sweep of 54°30', the fixed-incidence tailplane had the same sweepback as the wings.

A novel feature of the Yak-50 was its bicycle undercarriage. The forward-retracting main unit with twin drum-braked wheels supported 85% of the weight; the rest was supported by a steerable aft-retracting single-wheel nose unit. Outrigger struts with small solid-tyre wheels were mounted under the wingtips, retracting aft into cigar-shaped fairings.

The pressurised cockpit enclosed by an aft-sliding bubble canopy was fitted with an ejection seat incorporating an 8-mm ($0\frac{5}{16}$ in) armoured back. The armament comprised two NR-23 cannons, each with 80 rounds, mounted low on the centre fuselage sides.

Serialled '20 Yellow' and bearing a yellow figure '3' painted on the lower part of the fin, the first of the two prototypes entered flight test in July 1949, making its first flight with test pilot Sergey N. Anokhin at the controls. This machine, known as the Yak-50-I, had two parallel ventral strakes set at 90° to the skin under the rear fuselage. Later the second prototype (Yak-50-II) joined the programme. This aircraft, which was serialled '35 Yellow' and marked '2' on the fin, differed in having a single deeper ventral strake on the centreline and a dielectric fin cap incorporating a wire mesh antenna instead of the first prototype's exposed wire aerial and associated mast.

The factory testing lasted for almost a year and was completed on 30th May 1950. The aircraft attained a maximum speed of 1,065 km/h (661 mph) at 10,000 m; in a shallow dive it could exceed the speed of sound at an altitude of 10-11 km (32,810-36,090 ft), reaching Mach 1.03.

In June 1950 the Yak-50 was submitted for State acceptance trials which lasted for more

Yakovlev OKB

Yakovlev OKB

Top and above: The second prototype (Yak-50-II) was outwardly identifiable by the lack of the aerial mast aft of the cockpit and the single but deep ventral strake on the centreline.

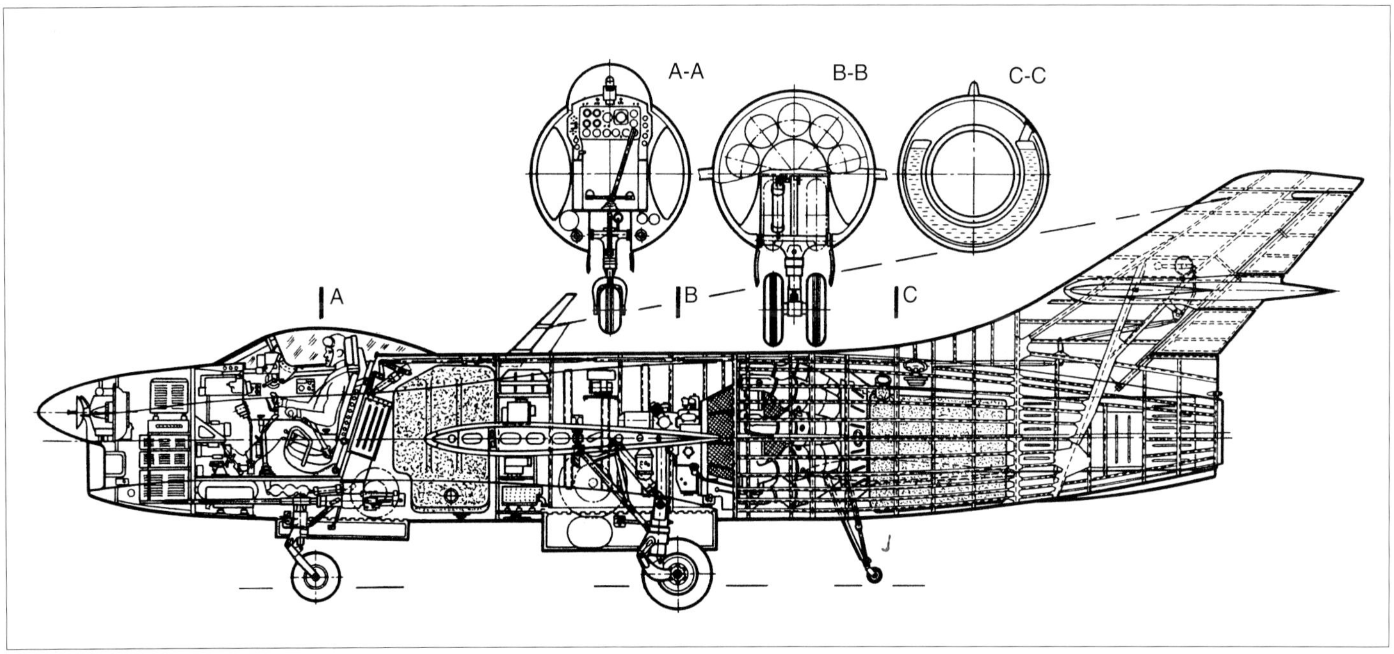

A cutaway drawing of the Yak-50-I.

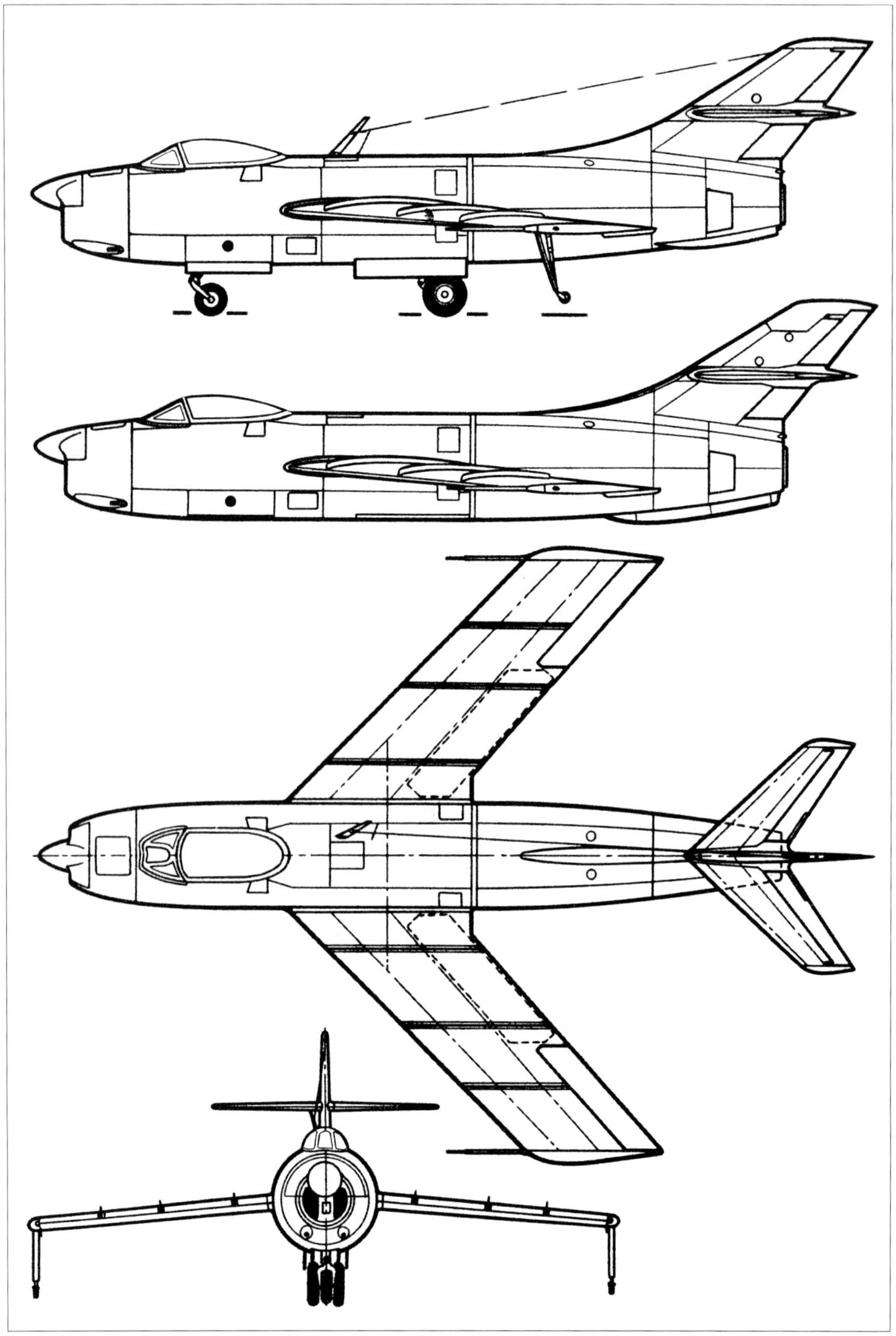

A three-view of the first prototype Yak-50 fighter, with an additional side view (second from top) of the second prototype.

than a year but were never completed. In their preliminary conclusions the military test pilots stated that the Yak-50 had an advantage over other fighter types as far as maximum level speed, service ceiling and especially rate of climb were concerned. However, the Yak-50 also had some deficiencies which adversely affected its operational capabilities. It was noted that the airbrakes were insufficiently effective; the aircraft suffered longitudinal oscillations in horizontal flight at speeds within the range of Mach 0.92-0.97, which made accurate gun aiming very difficult. Lastly, the aircraft had difficulty maintaining a straight course during a landing run in a crosswind of more than 4 m/sec (8 kts). This was compounded by the difficulties associated with the rather imperfect radar, which made the presence of a second crew member on board essential. All this prevented the Yak-50 from entering production and service.

Yak-M single-seat light fighter (project)

A project of a single seat-fighter powered by one Mikulin AM-5 turbojet and provisionally designated Yak-M was under development between November 1950 and February 1951. It was based on the Yak-50 and differed from it in having small lateral air intakes flanking the cockpit in the manner of the Lockheed P-80 Shooting Star or the SAAB 32 Lansen.

The Yak-M was to be 10.25 m (33 ft 7½ in) long, with a wing span of 7.7 m (25 ft 3$\frac{3}{32}$ in); the normal all-up weight was 3,000 kg (6,610 lb), increasing to 3,450 kg with an extra fuel tank. Design performance included a maximum speed of 1,158 km/h (720 mph) at 1,000 m (3,280 ft) and 1,068 km/h (664 mph) at 10,000 m (32,810 ft), the service ceiling being 17,500 m (57,400 ft). The armament comprised one 37-mm NS-37 cannon and two 23-mm NR-23 cannons.

Yak-U single-seat twin-engined fighter (project)

This project of a single-seat fighter powered by two AM-5 turbojets dates back to 1951. The engines were arranged side by side in the rear fuselage having an oval cross-section, breathing through a nose air intake (an arrangement similar to the future MiG-19). The aircraft had a bicycle landing gear. The fuel was housed in three fuselage fuel tanks

Yak-U (L) single-engined fighter (project)

This project studied in April-May 1951 had no relationship with the previous Yak-U, being a variant of the Yak-M aircraft (see above). No further details are available.

Yak-1000 single-seat fighter prototype

The unusual type number seems to accentuate the unorthodox – indeed, unique features of this aircraft, which deserves attention despite the fact that it never got off the ground. Some interesting facts concerning this fighter were revealed in 2000 by Yevgeniy G. Adler, one of Yakovlev's close associates, who had first-hand knowledge of the subject.

The concept of the Yak-1000 was born under the influence of several circumstances. One of these was Yakovlev's wish to produce a tactical fighter that would become a worthy competitor to the successful MiG-15 and other fighters from the rival Mikoyan OKB. This was backed up by a Council of Ministers directive issued in June 1950 which tasked the Yakovlev OKB with building a fighter powered by the new 5,000-kgp (11,020-lbst) Lyul'ka TR-3 (AL-5) turbofan, then under development. Finally, the layout of the future Yak-1000 reflected insistent recommendations from Pyotr Krasil'shchikov, one of TsAGI's aerodynamics experts, to make use of the captured German research materials on the so-called rhomboid-shaped wings. (In fact, this name is a misnomer, the planform of the wings in question being close to a cropped delta with a sharply swept leading edge and moderate forward sweep on the trailing edge.) Later, wings of this planform found wide use on guided missiles. Indeed, the German research envisioned exactly this application for them, but this important detail was overlooked by Krasil'shchikov in his recommendations…

The specifications laid down in the CofM directive stipulated performance characteristics that were extremely high by the day's standards. The aircraft was to attain a speed of Mach 1.7. When the Yakovlev OKB embarked on the project, it was faced with the fact that the AL-5 engine was still unavailable (it was all too common for the engine's development to lag behind that of the airframe which it was to power). This prompted a decision to build a kind of a technology demonstrator fitted with the rhomboid wings and powered by an available production engine. The RD-500 was chosen for this purpose. The aircraft was intended to test the aerodynamic layout and field performance. Accordingly, the design performance figures were much more modest and included a maximum speed of 1,100 km/h (683 mph) at a take-off weight of 2,407 kg (5,307 lb). The actual design work was handled by a team of engineers headed by Leonid L. Selyakov (who later worked in the design bureaux of Myasishchev and, finally, Tupolev).

The Yak-1000 prototype was completed on 27th February 1951. The aircraft featured a circular-section stressed-skin semi-monocoque fuselage with a nose air intake; the engine was installed amidships behind the pressurised cockpit, exhausting via a long extension jetpipe terminating at the rear extremity. The ducts from the air intake flanked the cockpit. The rear fuselage incorporated two flush-fitting airbrakes. There were two fuselage fuel tanks holding 430 and 167 litres (94.6 and 36.74 Imp gal), accommodated in the front and rear fuselage

Specifications of the Yak-50 fighter prototypes

	Yak-50-I	Yak-50-II
Length	11.185 m (36 ft 8⅜ in)	11.185 m (36 ft 8⅜ in)
Wing span	8.01 m (26 ft 3⅛ in)	8.01 m (26 ft 3⅛ in)
Wing area, m^2 (sq ft)	16.0 (172)	16.0 (172)
Empty weight, kg (lb)	3,085 (6,901)	3,125 (6,890)
All-up weight, kg (lb)	4,100 (9,040)	4,155 (9,160)
Speed at sea level, km/h (mph)	1,170 (727) estimated	n.a.
Speed at altitude, km/h (mph):		
at 5,000 m (16,400 ft)	1,135 (705)	1,120 (696)
at 10,000 m (32,810 ft)	1,065 (662)	1,057 (657)
Landing speed, km/h (mph)	200 (124)	196 (121.8)
Climb time, minutes:		
to 5,000 m	1.5	n.a.
to 10,000 m	3.5	3.6
Service ceiling, m (ft)	16,600 (54,460)	16,500 (54,120)
Operational range, km (miles)	1,100 (864)	850 (528)
Take-off run, m (ft)	587 (1,926)	850 (2,790)
Landing roll, m (ft)	965 (3,166)	765 (2,592)

Yakovlev OKB

Yakovlev OKB

Top and above: A subscale wind tunnel model of the Yak-1000 with a parabolic nose fairing instead of the nose air intake.

Yakovlev OKB

Yakovlev OKB

Above left: The Yak-1000 prototype, '34 Yellow'. Note the air data boom mounted on the air intake splitter and the extremely narrow landing gear track.

Above right: This view shows the ailerons inset into the flaps and accentuates the small area of the wings and horizontal tail.

Yakovlev OKB

Another aspect of the Yak-1000. Note the wide and shallow cockpit canopy.

respectively. The length of the aircraft, less nose probe, was 11.69 m (38 ft 4¼ in).

The mid-set wings of rhomboid planform had a very small span of only 4.59 m (15 ft 0¾ in) and an aspect ratio of only 1.46; it featured 60° leading-edge sweepback, 51°11' sweepback at quarter-chord and 11° forward sweep on the trailing edge. The latter was occupied by full-span flaps which incorporated inset ailerons with a separate axis (!). Each wing panel had three spars and ten ribs.

The bicycle undercarriage comprised two single-wheel fuselage units; both were fitted with wheels measuring 570 x 140 mm (22.44 x 5.5 in) and retracted aft. They were supplemented by two outrigger struts which retracted into underwing fairings located at mid-span.

Taxi runs were conducted on 2nd and 3rd March 1951, with speeds increasing gradually from 50 km/h (31 mph) to a maximum of 250 km/h (155 mph). The aircraft proved vulnerable to the influence of a crosswind which caused it to bank, compressing the opposite underwing strut. Application of ailerons in an attempt to counteract the bank was of no avail. The aircraft veered and ran off the side of the runway. The tests had to be suspended.

For starters, some work was done in an effort to remedy the faults; planned modifications to the airframe included alterations in the rear fuselage, control system, undercarriage and so on (see Yak-1000M version below). The suitably altered mock-up was subjected to wind tunnel tests at TsAGI. One of the proposals coming from Yevgeniy Adler envisaged, in particular, a radical redesign of the vertical tail which was to be replaced by two endplate fins at the tips of the horizontal tail.

However, Yakovlev considered further work to be pointless, having been sceptical about the rhomboid-shaped wings from the outset. The flight testing of the Yak-1000 never resumed, and in October 1951 all work on the project was terminated. Yevgeniy Adler regrets Yakovlev's decision to this day, maintaining that the aircraft could be developed into a flyable machine. Other specialists disagree with him and find Yakovlev's step to be a wise one.

As a matter of fact, problems with the undercarriage during the take-off were only part of the trouble. In addition to that, the new wing layout posed serious flight stability problems which could not be readily tackled at that stage. Delta wings were successfully tested in the Soviet Union only four to five years later when Artyom I. Mikoyan's Ye-4 and Pavel O. Sukhoi's T-3 fighters were first flown in June 1955 and May 1956 respectively. By that time special oscillation dampers in the control system

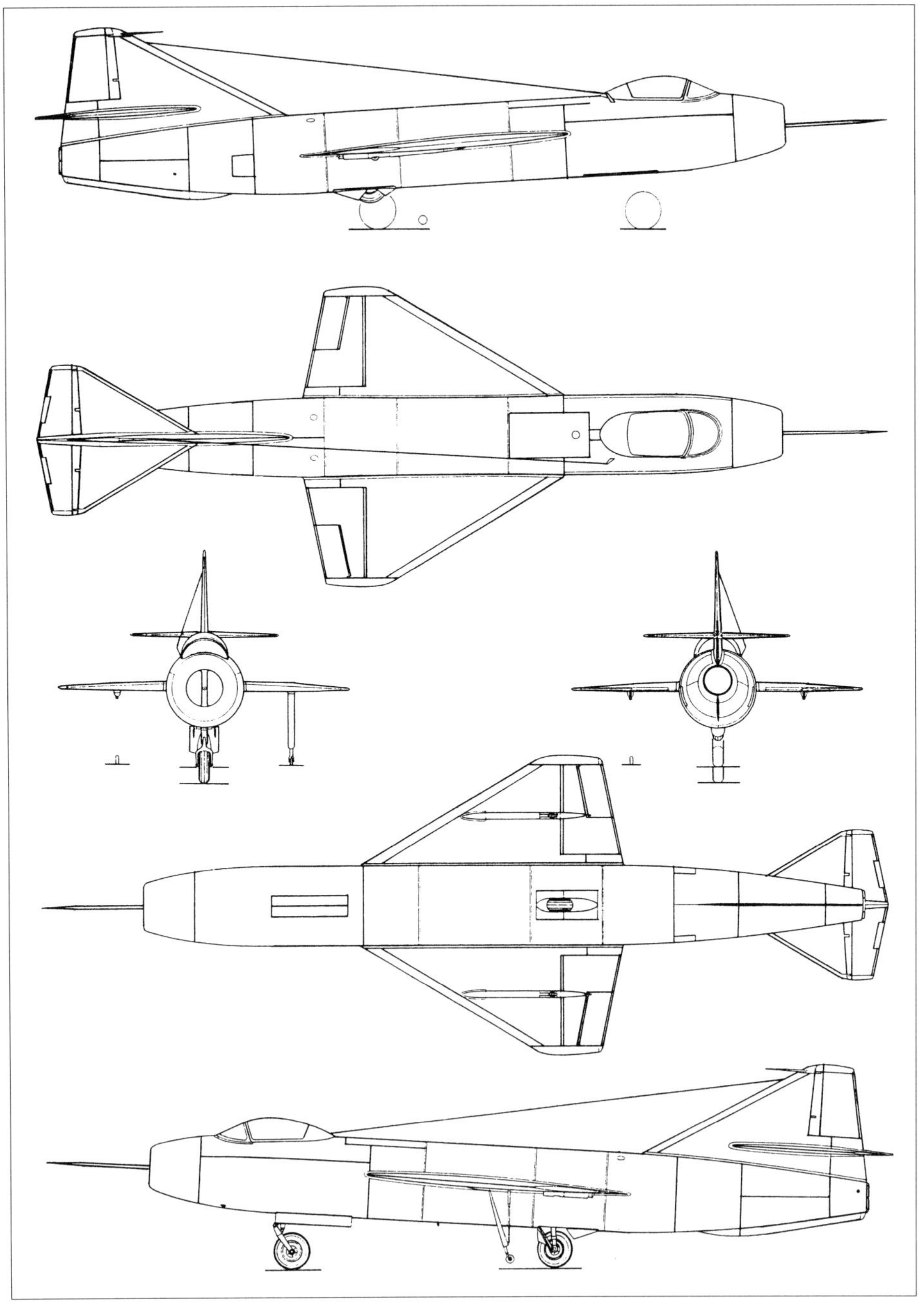

Six views of the Yak-1000. Note the semi-exposed position of the mainwheel when retracted.

ensured safe piloting of such unorthodox aircraft, and the delta wing layout itself had undergone a thorough re-thinking at TsAGI – prompted, among other things, by the negative experience with the Yak-1000.

The termination of work on the Yak-1000 sealed the fate of the original fighter project based on the AL-5 engine and developed in parallel; it was also abandoned in October 1951.

Yak-1000M (M-33) experimental fighter (project)

Albeit curing the illnesses of the Yak-1000 was brought to a halt before it could bear fruit, it is interesting to view one of the solutions proposed. It was a project dubbed Yak-1000M. A three-view drawing sketching the general layout of the new version was signed by L. Selyakov on 17th March 1951. The following changes were introduced: the wings were set lower on the fuselage and given marked anhedral; the wing incidence was changed; the vertical tail was redesigned, acquiring greater area and sweepback; the tailplane was relocated from the fin to the rear fuselage.

Apparently the project did not receive the go-ahead at the Yakovlev OKB. In the meantime, in March 1951 the government issued a directive establishing a special design bureau under the direction of Vladimir M. Myasishchev for the purpose of developing a strategic bomber (the future M-4). On 31st March Selyakov, like many other engineers, was transferred from the

Yakovlev OKB

Above: The Yak-140 prototype prior to the commencement of the tests. Note the metal air intake centrebody and the metal plugs closing the apertures for the cannon muzzles.

Yakovlev OKB

Yakovlev OKB

Above and left: Two more aspects of the ill-starred Yak-140. Note the airbrakes aft of the wings. The nose gear doors open only when the gear is in transit; the main gear doors open through 180°.

Yakovlev OKB to the newly-formed OKB-23. He took the project of the Yak-1000M to the new design bureau when he left. In accordance with the system of designations adopted at OKB-23 this project was renamed M-33 (M for Myasishchev). There were plans to use this aircraft for experiments aimed at researching the behaviour of low-aspect-ratio delta wings at transonic speeds. However, the project did not reach the hardware stage at OKB-23 either.

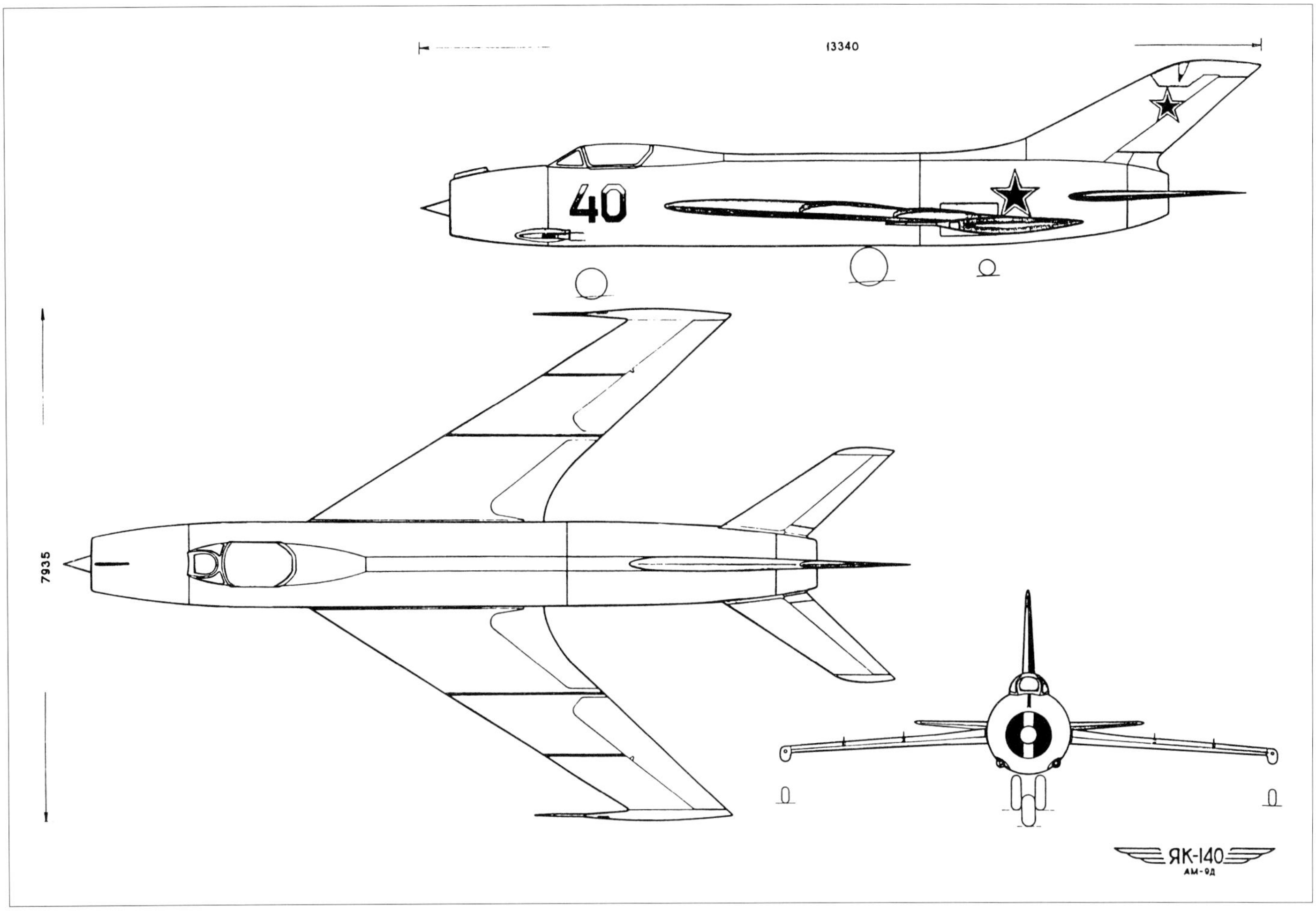

A three-view of the Yak-140 from the project documents.

Yak-60 single-seat fighter (project)

This single-seat fighter was very similar in layout to the Yak-50 interceptor described above and was also powered by the VK-1 turbojet rated at 2,700 kgp (5,950 lbst). It featured wings with 45° sweepback, an aspect ratio 4.0 and a taper of 1.2. The wing area was increased from 16 to 23 m² (from 172.2 to 247.6 sq ft). The fuselage length was increased from 9.465 m (31 ft 0$\frac{41}{64}$ in) to 11.0 m (36 ft 1 in). The aircraft had a tricycle undercarriage and was expected to have an all-up weight of 4,000 kg (8,820 lb). Provision was made for capacious underwing drop tanks with a diameter of 0.66 m (2 ft 1$\frac{63}{64}$ in) to increase range and endurance.

The general aerodynamic configuration of the Yak-60 was endorsed by Yakovlev on 20th November 1948. The results of wind tunnel tests suggested that the aircraft would be able to attain a speed in excess of Mach 0.9.

The second version of the Yak-60 project differed from the one described above in having a bicycle undercarriage and a radar in a bullet fairing mounted above the nose air intake.

The Yak-60 project remained on the drawing board, but aerodynamically it formed the basis for the later Yak-25 twin-engined interceptor (described in the following chapter). The main difference was the substitution of two engines for the under-

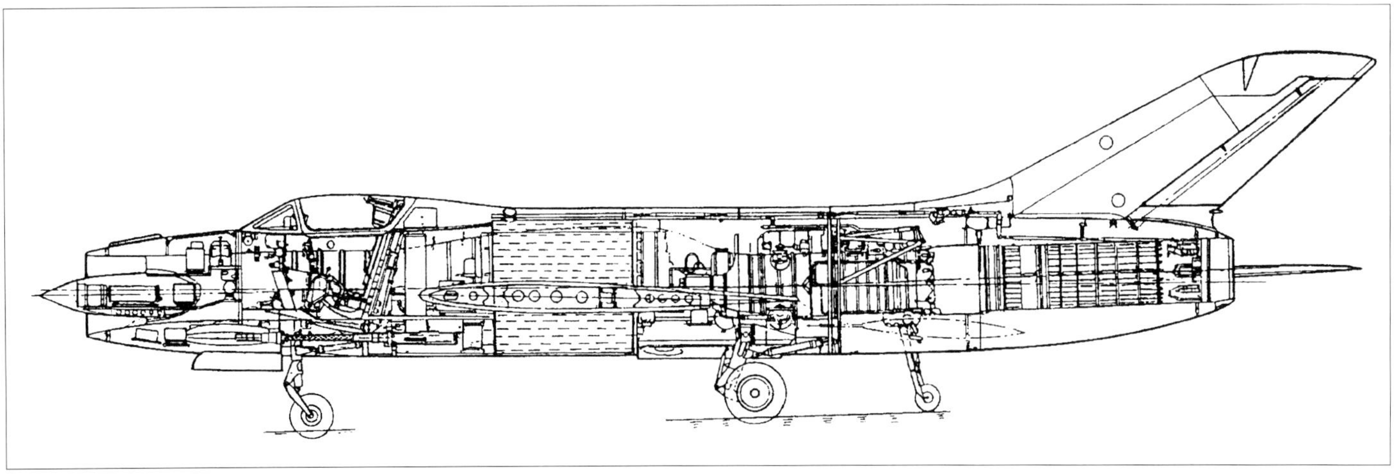

A cutaway drawing of the Yak-140.

Specifications of the Yak-140 fighter, original project and prototype (design performance in both cases)

	Yak-140 project	Yak-140 prototype
Engine type	AM-11	AM-9D
Engine thrust, kgp (lbst)	4,000 (8,820) dry/ 5,000 (11,025) reheat	2,600 (5,730) dry/ 3,250 (7,170) reheat
Length	12.95 m (42 ft 5⅞ in)	13.34 m (43 ft 9¼ in)
Wing span	8.0 m (26 ft 3 in)	7.395 m (26 ft 0½ in)
Wing area, m^2 (sq ft)	20.0 (215.3)	19.0 (205)
Wing loading, kg/m^2 (lb/sq ft)	242.5 (49.67)	n.a.
Empty weight, kg (lb)	3,324 (7,329)	3,315 (7,308)
All-up weight, kg (lb)	4,850 (10,692)	4,500 (9,920) 4,850 (10,692)
Max speed, km/h (mph)	1,650-1750 (1,025-1,088) 1,715 (1,066) at 12,000 m*	1,275 (792) at sea level 1,250 (777) at altitude
Landing speed, km/h (mph)	n.a.	189 (117)
Rate of climb, m/sec (ft/min):		
at sea level	200 (39,370)	n.a.
at 15,000 m (49,210 ft)	30 (5,900)	n.a.
Climb time, minutes:		
to 10,000 m (33,000 ft)	1.15	2.1
to 15,000 m (49,210 ft)	2.6.	4.4
Service ceiling, m (ft)	18,000 (59,060) 18,300 (60,040)*	18,000 (59,055)
Range, km (miles)	1,800 (1,118) 1,900 (1,181)*	1,900 (1,181)
Take-off run, m (ft)	400 (1,312)	400 (1,312)
Landing roll, m (ft)	600 (1,970)	600 (1,970)
Armament, mm	3 x 30	2 x 23

*Alternative figures from a different source

wing tanks and deletion of the engine in the fuselage, coupled with an increase in the wing area to 28 m^2 (301.4 sq ft).

Yak-70 single-seat fighter (project)

In April-May 1950 the Yakovlev OKB worked on a project of a single-seat fighter powered by one 5,000-kgp (11,020-lbst) Lyul'ka TR-3 turbojet. It had a fuselage with a length of 15.24 m (50 ft 0 in) and a diameter of 1.6 m (5 ft 3 in) which featured a circular air intake with a dielectric centrebody housing a radar antenna. The aircraft was to be fitted with a bicycle undercarriage.

Yak-135 light tactical fighter (project)

Little is known about this project, the very existence of which came to light by sheer accident. In 1999 a Russian aeronautical magazine published a picture showing Yakovlev OKB designers discussing some subject. Behind their backs, on the wall, a large poster could be seen depicting an aircraft and inscribed 'Yak-135 light tactical fighter'. The aircraft shown on the drawing had a pointed nose, lateral air intakes, wings with moderate leading-edge sweepback and an unswept trailing edge, a horizontal tail of similar shape and a sweptback vertical tail. The jet nozzle could not be clearly seen, but the machine left the impression of featuring a pod-and-boom configuration. No further information is available and it is not even clear when this project was conceived (it is mentioned in this chapter on the assumption – possibly erroneous – that it preceded the Yak-140, which is described below.)

Yak-140 supersonic single-seat tactical fighter prototype

Advanced development project of this lightweight single-seat single-engined supersonic fighter was approved by Aleksandr S. Yakovlev on 10th July 1953. In accordance with a government directive a prototype of this fighter was to be submitted for State acceptance trials in March 1955. Performance characteristics stipulated by the government included a maximum speed of 1,650-1,750 km/h (1,025-1,088 mph), a service ceiling of 18,000 m (59,040 ft), a range of 1,800 km (1,118 miles), a take-off run of 400 m (1,310 ft) and a landing run of 600 m (1,970 ft). These figures were duly reflected in the design performance of the aircraft. Attaining this performance presupposed the use of the new Mikulin AM-11 lightweight turbojet rated at 4,000 kgp (8,820 lbst) dry and 5,000 kgp (11,025 lbst) in full afterburner. The aircraft was to have an AUW of 4,850 kg (10,690 lb). Its armament comprised three 30-mm (1.18 calibre) cannons with 50 rpg, supplemented by several options of unguided rockets (from two to 16, depending on the calibre) or by a 200-kg (440-lb) bomb load.

The Yak-140 was a mid-wing monoplane with sharply swept wings and tail surfaces. The wings, which had 55° sweep at quarter-chord, were fitted with aerodynamically balanced ailerons and TsAGI-type extension flaps. The circular-section monocoque fuselage featured a thick skin. The circular nose air intake was provided with a conical fixed centrebody housing a radar ranging device. The tailplane with separate elevators was set in a mid position on the rear fuselage. The aircraft featured a bicycle undercarriage which comprised a single-wheel nose unit and a twin-wheel main unit, supplemented by outrigger wheels retracting into wingtip fairings.

The AM-11 engine's development schedule was slipping, and presentation of the Yak-140 for the State acceptance trials had to be postponed until the first quarter of 1956. However, this proved of little help, and the aircraft had to be fitted experimentally with the AM-9D afterburning turbojet delivering a thrust of 2,600 kgp (5,730 lbst) dry and rating of 3,250 kg (7,170 lbst) reheat. The armament was reduced to two 23-mm NR-23 cannon, each with 75 rpg. The installation of a less powerful engine entailed a marked deterioration of the design performance, which now included a maximum speed of only 1,275 km/h (792 mph) at sea level and 1,250 km/h (777 mph) at 13,500 m (44,280 ft). But the AM-9D was a strictly temporary solution pending availability of the AM-11.

The Yak-140 prototype serialled '40 Yellow' was completed in December 1954. It successfully passed all necessary ground tests, including taxying and high-speed runs up to the attainment of the unstick speed. All systems functioned faultlessly, and on 10th February 1955 the Yak-140 was cleared for manufacturer's flight testing with no limits imposed on the maximum speeds. However, the aircraft was barred from entering the flight test stage because the Ministry of Aircraft Industry was determined to support a fighter from another Design Bureau (as is well known, Mikoyan and Sukhoi were actively pursuing their own programmes of supersonic fighter development at that time). On 28th March 1956 the Council of Ministers issued a directive closing down the Yak-140 programme, the corresponding MAP order following on 6th April. The Yak-140 was the last lightweight single-seat tactical fighter built by the Yakovlev OKB.

THE TACTICAL TWINJET FAMILY

This chapter deals with a large family of combat aircraft created by the Yakovlev OKB between the early 1950s and the early 1970s which remained in service until the mid-1980s. Starting with the Yak-25 twin-engined interceptor, it comprised a great number of major and minor variants and derivatives spanning over a wide variety of roles, the most important being interceptors, tactical bombers and tactical reconnaissance aircraft. While the Yak-26 and Yak-27 remained a transitional stage in the family's development, the Yak-28 with its numerous versions came to occupy a notable place in the nation's air defence and in the front-line element of the Soviet Air Force. The story of this aircraft family brings to the mind the OKB's wartime experience in developing dozens of fighter aircraft versions stemming basically from the initial Yak-1.

Yakovlev OKB

Yakovlev OKB

Top and above: The uncoded first prototype of the Yak-120 interceptor. The fuselage nose and tailcone, fin (but not rudder), airbrakes and front portions of the engine nacelles are painted light grey.

Yak-25 (Yak-120) interceptor
Yak-13 twinjet interceptor (project, second use of designation)

This fighter project dates back to August 1951 (the Yak-13 designation previously applied to a light touring aircraft of 1946) and can be regarded as an immediate predecessor of the Yak-25 (Yak-120). It was to be powered by two turbojets mounted under the wings. Little is known about the particulars of the project, apart from its dimensions: a wing span of 10.6 m (34 ft 9$\frac{5}{16}$ in), a length of 15.4 m (50 ft 6$\frac{19}{64}$ in) and a wing area of 28 m^2 (301.4 sq ft).

Yak-120 patrol interceptor prototype

On 10th August 1951 the government formalised its interest in a previously made Yakovlev proposal when the Soviet Council of Ministers issued directive No.2929-1379. This document tasked OKB-115 with developing a two-seat twin-engined patrol interceptor bearing the in-house designation Yak-120 and the Yak-125 tactical photo reconnaissance aircraft based on same; the two types were to commence trials in August and October 1952 respectively.

The Yak-120/Yak-125 design effort was largely completed and full-scale mock-ups of both aircraft were constructed in the autumn of 1951; the mock-up review commission signed its Protocol of Acceptance on 12th December.

The Yak-120 was very different from anything the Yakovlev OKB had brought out before. The thin mid-set wings were swept back 45°, with parallel leading and trailing edges and a single boundary layer fence on each wing in line with the inboard end of the aileron. The wings were two-spar structures built in two sections and featuring large two-section flaps. The cruciform tail surfaces were likewise sharply swept; there was a shallow ventral fin for greater directional stability. Two rectangular airbrakes were provided on the aft fuselage sides.

Two Mikulin AM-5 axial-flow turbojets rated at 2,000 kgp (4,410 lbst) were mounted in slender nacelles adhering directly to the wing undersurface. The inner/outer wing joints were just outboard of the nacelles; the wing trailing edge was kinked about halfway between the nacelles and fuselage, running at right angles to the fuselage on the innermost portions. The engines were placed fairly low, so special grilles or screens were incorporated in the inlets as a protection against foreign object damage (FOD). Placing the engines under the wings allowed a sizeable fuel load to be carried in the fuselage, despite the two-crew cockpit. For added endurance a conformal drop tank could be carried on the fuselage centreline.

However, the chosen layout precluded the use of a tricycle undercarriage; having the mainwheels retract into the fuselage would encroach on fuel capacity, and having them retract into the wings or engine nacelles would pose problems from an engineering standpoint. Hence the Yak-120 received a bicycle undercarriage with a single-wheel nose unit, a twin-wheel main unit located aft of the CG and outrigger struts mounted under the wingtips. All units retracted aft, the outrigger struts stowing in fairly large cigar-shaped fairings which also carried the pitot booms.

The huge dish of the RP-6 ***Sok**ol* (Falcon) gun-ranging radar (RP = ***rah**dioprit**sel*** – radio sight) developed by OKB-339 dictated a basically circular fuselage cross-section and was enclosed by a large glassfibre radome. The latter had an almost hemispherical front end. The crew sat in tandem under a common aft-sliding canopy. The fixed windshield incorporated a bulletproof glass panel no less than 105 mm ($4\frac{9}{64}$ in) thick; protection was also provided by a forward armour plate, armoured seat backs and headrests, and side armour, all of them 10 mm ($0\frac{13}{32}$ in) thick.

The armament comprised two 37-mm (1.45 calibre) Nudel'man N-37L cannons mounted low on the centre fuselage sides. Normal ammunition supply was 50 rpg. There was also provision for two 212-mm (8.35-in) ARS-212 (aka S-21) unguided rockets on underwing pylons (ARS = *aviatsi**on**nyy reak**tiv**nyy sna**ryad*** – [high-velocity] aircraft rocket, HVAR).

The interceptor was carefully designed, which ensured its operational success and allowed the aircraft to be developed into a range of specialised versions. The avionics suite enabled the Yak-120 to navigate and intercept its targets in any weather/visibility conditions right up to the aircraft's service ceiling. Apart from the radar, it included an SRO identification friend-or-foe (IFF) transponder, an RSIU-3 Klyon (Maple) VHF radio and an AP-28 autopilot. For automatic landing approach in poor weather the interceptor was equipped with a *Mate**rik*** (Continent) instrument landing system (ILS) comprising an ARK-5 Amur (a river in the Soviet Far East; pronounced like the French word *amour*) automatic direction finder, an RV-3 low-altitude radio altimeter and an MRP-48 ***Dya**tel* (Woodpecker) marker beacon receiver. Other avionics items included a Pozitron-1 system (probably a command link system), an SPU-2R intercom and an S-13 gun camera. The MRP-48 is called *Khrizan**tem**a* (Chrysanthemum) in some sources.

The crew was equipped with KP-18 oxygen masks and the life support system featured 16 spherical bottles, each holding 2 litres (0.44 Imp gal) of liquid oxygen. Electric power was provided by two 6-kilowatt KSTGS-6000 DC generators, one SGS-7,5B alternator and two 12A-30 DC batteries.

The wings, tail unit and air intakes were de-iced by engine bleed air, while the engine FOD protection screens and intake centrebodies were electrically de-iced. This enabled the Yak-120 to loiter for an extended time at high altitudes with low ambient temperatures and operate in cold climatic regions.

After reviewing the advanced design project the VVS top brass and the mock-up review commission were well pleased with the interceptor's design and specified performance figures, giving the go-ahead to complete a set of manufacturing drawings and begin prototype construction. M. I. Leonov was appointed engineer in charge of the flight tests. At this stage vertical tail area was increased from 5.1 m^2 (54.84 sq ft) to 5.329 m^2 (57.3 sq ft).

Two prototypes and a static test airframe were built at the OKB's experimental shop, MMZ No.115 ***Sko**rost'* (Speed). The unserialled first prototype was rolled out in the summer of 1952. However, development of the Sokol radar was taking longer than predicted, and as the radar was unavailable it was substituted with ballast. Also, an RV-2 *Kris**tall*** (Crystal) radio altimeter was fitted instead of the planned RV-3; thus the aircraft had the earlier OSP-48 ILS instead of the Materik system.

On 19th June 1952 the Yak-120 took to the air for the first time with OKB test pilot V. M. Volkov at the controls. The manufacturer's flight tests continued until November.

Despite the complex and heavy avionics, the Yak-120 had a lightweight airframe for a twin-engined fighter. Apart from the bicycle undercarriage, this was due to a new internally braced wing design. The engineers had taken great care to minimise structural weight; as a result, airframe weight made up only 29% of the MTOW – the absolute minimum possible for an all-metal swept-wing combat aircraft stressed for 7 Gs.

Even though the Yak-120's normal all-up weight of 8,530 kg (18,800 lb) was higher than that of traditional single-engined air superiority fighters of the time, the aircraft was smaller and lighter than the La-200B interceptor prototype. The compact engines and generally clean lines enabled the Yak-120 to outperform its bulkier Mikoyan and Lavochkin competitors; moreover, the aircraft surpassed the Air Force's SOR in all respects except range and endurance. With a 7,650-kg (16,865-lb) AUW the Yak-120 had a top speed of 1,140 km/h (708 mph) at 4,000 m (13,120 ft) and 1,075 km/h (667 mph) at 10,000 m (32,810 ft). The latter altitude was reached in 4.3 minutes and the service ceiling was 300 m (980 ft) better than specified.

Range fell a little short of the requirements; at 12,000 m (39,370 ft) it was 2,800 km (1,739 miles) on internal fuel only and 3,250 km (2,018 miles) with a drop tank. Still, this did enable the Yak-120 to patrol an assigned area at a considerable distance from its home base.

The aircraft responded quickly to control inputs and was quite agile, if not fully aerobatic. Having a two-man crew increased combat efficiency, as in adverse weather the weapons systems operator (WSO) would

Yakovlev OKB

Yakovlev OKB

Top and above: These views of a brand-new Yak-25M emphasise the large fuselage diameter and low-slung engines. Note the trapezoidal blast panels of the guns flanking the lower centre fuselage.

handle target search and help the pilot (flying the aircraft in instrument mode) guide the aircraft towards the target. Additionally, the Yak-120 had dual controls, which meant the WSO could take over if necessary.

The results of the manufacturer's flight tests convinced the military leaders and the Ministry of Aircraft Industry that the Yakovlev contender was superior to the La-200B. However, the continued unavailability of the intended RP-6 Sokol radar meant that the Yak-120 could not be submitted for State acceptance trials.

Hence in early December 1952 MAP issued an order to the effect that the prototype be urgently fitted with the RP-1 *Izum**rood*** (Emerald) radar and transferred to GK NII VVS for State acceptance trials. The RP-1 had been developed by NII-17 under Viktor V. Tikhomirov; this became the Moscow Research Institute of Instrument Engineering (MNIIP – *Mos**kov**skiy na**ooch**no-is**sle**dovatel'skiy insti**toot** pri**bor**ostro**yen**iya*), aka NPO Vega-M, in 1967. It had automatic target-tracking capability and had been tested successfully in January-March same year on the MiG-15*bis* (*izdeliye* SP-5) development aircraft; later it was fitted to the production MiG-17P and MiG-17PF interceptors. Unlike the RP-6, the RP-1 had separate search and tracking antennas which were very compact; as a result, they looked lost inside the Yak-120's huge radome.

From March to June 1953 the Yak-120 was tested at NII VVS with the provisional radar installation. The general opinion was favourable and the aircraft was recommended for production under the service designation Yak-25 (which, as the reader remembers, had been used for a fighter prototype in 1947) – providing some shortcomings were corrected. The prototype's test results obtained at NII VVS were approved on 8th September 1953 by Council of Ministers directive No.2359-965 as a specification for production Yak-25s.

In accordance with the commission's findings some changes were made to the Yak-120 prototypes (among other things, a second pair of boundary layer fences was added on the outer wings at about mid-span of each aileron). These changes added 25 kg (55 lb) to the aircraft's empty weight, increasing the AUW to 8,700 kg (19,180 lb). On 1st-26th October 1953 the modified interceptor underwent renewed manufacturer's flight tests at the hands of V. M. Volkov. Subsequent checkout trials at NII VVS confirmed the Yak-120's high performance, clearing the way for production.

Yefim Gordon archive

Five early-production Yak-25 interceptors ('129 Red', '089 Red', '069 Red', '099 Red' and '209 Red') make a flypast at the 1955 May Day parade.

Yak-25 patrol interceptor

As noted earlier, debugging and testing of the Sokol radar was taking longer than anticipated. Therefore, the Powers That Be decided to launch Yak-25 production with the interim RP-1 Izumrood radar. The version for the Yak-25 incorporated minor changes from the basic model fitted to the MiG-17P/PF and thus was designated RP-1D (*dora**bot**annyy* – modified or improved).

Yak-25 production was assigned to factory No.292 in Saratov. The first production interceptors were completed in September 1954. Very few were built with the Izumrood radar, as at the end of the year the initial version was superseded by the Yak-25M. Outwardly production Yak-25s differed from the prototypes in having a dielectric fin cap housing navigation antennas. The Yak-25 was allocated the NATO reporting name *Flashlight*; later, when supersonic derivatives of the aircraft (the Yak-26-1 modified and Yak-27) became known, this was changed to *Flashlight-A*.

The Yak-25 proved indispensable in the remote northern and eastern regions of the USSR. Having a second crewmember who could take over the controls gave the pilot extra confidence on long overwater missions – a thing not to be treated lightly. Besides, having two engines spaced far apart was an important safety factor increasing the chances of coming home in the event of an engine failure.

Yak-25M patrol interceptor (*izdeliye* 12/15)

By the end of 1953 the Sokol (RP-6) radar had been brought up to scratch on the La-200B testbed. This radar proved to be much more capable than the RP-1D; it had not only greater detection range (30 km/18.6 miles versus 12 km/7.45 miles) but also limited 'look-down' capability, detecting targets flying as low as 300 m (980 ft). Hence in November 1953 MAP issued an order to the effect that the RP-6 radar be installed on the Yak-120 interceptor.

In April 1954 the upgraded Yak-120 prototype with the intended RP-6 radar successfully completed the final stage of the State acceptance trials and the radar was cleared for production. On 13th May 1954 the Council of Ministers issued directive No.899-385 determining the main specifications of the production interceptor with the Sokol radar as detailed below. The directive ordered the production of 30 aircraft to this standard in 1954.

The version with the RP-6 was designated Yak-25M (*modifi**tsee**rovannyy* – modified). In service, however, this designation did not find wide use; Air Force documents usually refer to production aircraft simply as 'Yak-25', regardless of radar type. The in-house product code at the Saratov aircraft factory was '*izdeliye* 12/15'.

Apart from the radar, the Yak-25M had a number of detail changes. The AM-5A Srs 1 turbojets of the original Yak-25 were replaced with identically rated RD-5A (AM-5A) Srs 2 engines with design improvements. Aleksandr A. Mikulin's innovations and his enchantment with fighter engines made him some bitter enemies among his colleagues who sent an anonymous letter to the Soviet government, alleging that Mikulin was 'disorganising the work of OKB-300'. On 20th January 1955 Mikulin was removed from office and succeeded by Prokofiy F. Zoobets who had his own ambitious designs of large turbojets. Since Mikulin had fallen from grace, the initials 'AM' in the designations of his engines were replaced by the anonymous-sounding acronym RD (*reak**tiv**nyy **dvig**atel'* – jet engine).

In order to improve directional stability during take-off and landing the Yak-25M's wheelbase was increased by moving the

Sergey and Dmitriy Komissarov archive

Above: '27 Red', a Yak-25 converted for ejection seat trials. The seat is accommodated in the former WSO's cockpit, requiring the canopy to be modified (the remaining portion opens to starboard).

Yefim Gordon archive

Above: This *Flashlight-A* coded '18 Red' was used for canopy jettison system trials. The photo was taken after 1955, as indicated by the absence of the star insignia on the fuselage.

Yefim Gordon archive

'21 Blue', the much-modified Yak-25M used for testing the Yak-28's engine intake de-icing system. The grid with water spray nozzles is clearly visible.

nose gear unit forward 33 cm (1 ft 1 in). The cannons were fitted with muzzle brakes and some other changes were introduced. Four bladder tanks in the fuselage held 3,445 litres (757.9 Imp gal) of fuel, and a 670-litre (147.4 Imp gal) drop tank could be carried.

A total of 406 *Flashlight-As* were built to Yak-25M standard. Curiously, the aircraft was never officially on strength with the PVO (it should be noted that in the Soviet Union, combat aircraft were often not added to the inventory until after they had been in service for years!). Still, this did not prevent the Yak-25 and Yak-25M from soldiering on with the Air Defence Force until the early 1960s.

Yak-25M (SM-6) testbeds

Two production Yak-25Ms, redesignated SM-6, were intended to be used for testing K-6 air-to-air missiles. However, the development of these missiles was halted, and the two aircraft were modified for other uses.

Yak-25MG patrol interceptor

In the course of their service career some Yak-25Ms were retrofitted with the new *Gori**zont**-1* (Horizon-1) command link system for working with ground controlled intercept (GCI) stations using target information from air defence radars. These upgraded aircraft were sometimes referred to as Yak-25MG.

Yak-25L ejection seat testbed

The Yak-25 also spawned several specialised versions. One of these was the Yak-25L, a testbed for new ejection seats, ejection guns and pressure suits. The L suffix denoted (*le**ta**yuschchaya*) *labora**tor**iya* – 'flying laboratory'; this Russian term is used indiscriminately for any kind of testbed or research/survey aircraft.

Coded '01 Red', the Yak-25L was converted from an early-production *Flashlight-A*. The cockpit section was extensively modified, featuring two separate pressurised cockpits with individual canopies. The experimental seat was installed in the rear cockpit. For safety reasons the rear canopy was often removed on test missions and a fairing was fitted over the rear cockpit to shield it from the slipstream. The cannons and some equipment items were deleted to save weight, and high-speed cine cameras were installed at the wingtips for capturing the ejection sequence. Numerous test ejections were performed with this aircraft; the ejection seat was invariably occupied by a dummy.

Another Yak-25M coded '27 Red' was also converted into an ejection seat testbed with a different cockpit design. The front cockpit was enclosed by an abbreviated version of the standard canopy, apparently opening sideways, while the rear cockpit was completely open.

Yak-25 flying testbeds for LII

Several Yak-25s served with the Flight Research Institute as testbeds of various kinds. One of these was used in 1956-57 to develop a remote control system for target drones; the results obtained were incorporated into the design of a target drone version of the Yak-25. Another Yak-25 testbed was used in 1957-58 for developing remote control systems for cruise missiles. In yet another case the Yak-25 was used for the testing and flight development of the Sokol-2K radar; in 1956-57 a Yak-25 served to verify the navigation system developed for the Lavochkin OKB's ***Boo**rya* (Storm) intercontinental cruise missile. This list is by no means exhaustive.

Yak-25 cruise missile simulator

At least two Yak-25s were intended to be used for 'simulating in flight the P-6 and P-35 sea-launched cruise missiles'. Apparently this means they were to be used as guidance system testbeds.

Yak-25M testbeds

Several Yak-25Ms were also modified for various test and development programmes. For example, one aircraft coded '18 Red' was used to test the canopy jettison system.

Another example, '21 Blue', was a de-icing system testbed for the Yak-28I and Yak-28L tactical bombers described later in this chapter. The front portions of the engine nacelles were redesigned to replicate those of the Yak-28I/Yak-28L powered by Tumanskiy R11AF2-300 afterburning turbojets, featuring extended sharp-lipped air intakes with movable shock cones. A circular grille with water sprinkler nozzles was installed on struts in front of the starboard engine. Other non-standard features were an additional aerial on the fuselage spine just ahead of the fin fillet and a 'towel rail' aerial (possibly for telemetry downloading) on the fin leading edge above the horizontal tail. This aircraft has likewise been referred to as Yak-25L.

Other Yak-25Ms were converted into engine testbeds. Little is known about these aircraft, except that the development engines were mostly those envisaged for future versions of the *Flashlight*.

Yakovlev OKB

'15 Yellow', the Yak-120M prototype. The tail code '3 Yellow' means nothing, as usual.

Yak-120M patrol interceptor prototype

The abovementioned Council of Ministers directive No.2359-965 of 8th September 1953 issued shortly after Stage 1 of the Yak-120's State acceptance trials contained several items. One of them ordered the Yakovlev OKB to re-engine the second prototype with the new Mikulin AM-9A afterburning turbojets (a version of the engine powering the future MiG-19). Rated at 2,650 kgp (5,840 lbst) dry and 3,250 kgp (7,160 lbst) reheat, the AM-9A was a straightforward development of the AM-5 featuring an extra compressor stage, a new can-annular combustion chamber and an afterburner with a three-position convergent-divergent nozzle.

Another change concerned the armament. The cannons were to be supplemented with rocket pods for 57-mm (2.24-in) ARS-57 Skvorets (Starling) folding-fin aircraft rockets (FFARs; ARS = *aviatsi**on**nyy reak**tiv**nyy sna**ryad*** – aircraft rocket) and appropriate modifications were to be made to the RP-6 Sokol radar. The FFARs were to be used against enemy bomber formations. Designated Yak-120M, the modified interceptor was to commence State acceptance trials in September 1954 pursuant to CofM directive No.49-34 of 13th January 1954 and MAP order No.26 of 15th January.

Actually work on an improved Yak-120 with uprated engines and a Sokol-M radar began before all these government orders. However, conversion of the second prototype was constantly delayed – first by late delivery of the engines and then by the unavailability of the radar; eventually a standard RP-6 (which had by then entered production) had to be installed as a stopgap measure. Serialled '15 Yellow' and carrying the tail code '3', the aircraft was completed at the end of 1954.

Of course the intended test schedule had gone to the dogs. Fully realising that the Yakovlev OKB was not to blame for this, on 20th November 1954 the Council of Ministers issued a new directive (No.2337-1111) which postponed Stage 1 of the State acceptance trials – that is, with the standard Sokol radar and without rocket armament – until January 1955. By March 1955, however, the Yak-120M was to be ready for Stage 2 with the Sokol-M radar and FFARs. Interestingly, this document also specified 190-mm (7.48-in) TRS-190 rockets (***toor**boreak**tiv**nyy sna**ryad*** = spin-stabilised rocket) as alternative armament.

As noted earlier, the Yak-120 retrofitted with the Sokol radar had successfully passed its State acceptance trials earlier that year. Thus the Yak-120M was, in effect, a derivative of the future production Yak-25M, differing from it in the following respects. The engine nacelles were redesigned to accommodate the afterburning AM-9A engines, being both longer and fatter, and the nacelle attachment fittings were beefed up. The increased nacelle diameter necessitated a slight reduction in flap span on both inner and outer wings, and the take-off flap setting was reduced by 10°. The boundary layer fences were made 50 mm ($1\frac{31}{32}$ in) taller and moved outward 250 mm ($9\frac{27}{32}$ in). An 11-kg (24.25-lb) anti-flutter weight was fitted at each wingtip near the pitot boom attachment.

Two recesses accommodating launchers for ARS-57 or TRS-190 FFARs were provided in the lower centre fuselage sides; when no rocket launchers were carried the recesses were closed by detachable covers to reduce drag.

The Sokol-M radar replacing the standard RP-6 unit enabled targeting and accurate fire with both cannons and rockets in all weather conditions, day and night. Speaking of cannons, the Yak-25's standard N-37Ls were substituted with lighter 23-mm (.90 calibre) Nudel'man/Rikhter NR-23s with 50 rpg to offset the extra weight of the rocket armament.

The Yak-120M's normal fuel load was 2,700 kg (5,950 lb) of internal fuel, increasing to 3,400 kg (7,495 lb) with a drop tank; the latter figure includes 540 kg (1,190 lb), which means that with the normal fuel load the internal fuel cells were not quite full. The wheelbase was identical to that of initial production Yak-25s (that is, short).

Stage 1 of the trials began as planned in January 1955 and was completed same month; however, it was performed solely by OKB-115 (these were manufacturer's flight tests). By then OKB-339 had still failed to deliver the radar; therefore the Yak-120M was turned over to NII VVS with the old RP-6 radar and minus rocket armament. Stage 2 – or, officially, Stage 1 of the State acceptance trials – also began on schedule (in March 1955) and was duly completed on 26th April. In the course of the trials a long pitot was added to the fin leading edge and a forked ILS aerial mounted atop the fin was replaced by a strake aerial.

Since the afterburners could not be engaged below 5,000 m (16,400 ft), the aircraft's performance fell a little short of the requirements. Top speed at 5,000 m was 1,122 km/h (696 mph) instead of the specified 1,150 km/h (714 mph); the service ceiling was 16,300 m (53,480 ft) instead of 16,500-17,000 m (54,130-55,770 ft) and the time required to reach 10,000 m (32,810 ft) exceeded the specified 2 minutes by far.

Eventually the Sokol-M radar never performed as it should. Besides, in the meantime the military had drawn up new and more stringent operational requirements for interceptors, demanding first and foremost higher speed. To meet these the Yakovlev OKB began development of a supersonic derivative of the *Flashlight-A* and thus the Yak-120M was abandoned.

Yak-120MF engine testbed

On 28th March 1956 the Council of Ministers issued directive No.424-261 followed on 6th April by MAP order No.194 to the same effect. These documents relegated the Yak-120M prototype to a role as a testbed for

Yakovlev OKB

The Yak-125 reconnaissance aircraft at a late stage of the trials with two boundary layer fences each side.

the RD-9F turbojets. At this stage the aircraft was known as the Yak-120MF.

Yak-125 tactical photo reconnaissance aircraft prototype

As noted earlier, Council of Ministers directive No.2929-1379 tasked the Yakovlev OKB

Yak-125 specifications

Wing area, m^2 (sq ft)	28.98 (311.61)
Empty weight	
without drop tank, kg (lb)	n.a.
Normal AUW, kg (lb)	9,177 (20,231)
Maximum AUW with drop tank, kg (lb)	9,785 (21,570)
Fuel load, kg (lb):	
internal fuel (normal)	3,250 (7,165)
with drop tank	3,800 (8,380)
Wing loading, kg/m^2 (lb/sq ft)	316.6 (64.91)
Power loading, kg/kgp (lb/lbst)	2.3
Top speed, km/h (mph):	
at sea level	910 (565) *
at 5,000 m (16,400 ft)	1,100 (683)
at 10,000 m (32,810 ft)	1,052 (653)
Landing speed, km/h (kts)	203 (109.73)
Climb time, min:	
to 5,000 m	2.15
to 10,000 m	5.65
Rate of climb, m/sec (ft/min):	
at sea level	n.a.
at 10,000 m	n.a.
Service ceiling, m (ft)	14,900/14,140 (48,884/46,390) †
Range at 12,000 m (39,370 ft), km (nm):	
on internal fuel only	2,650 (1,432)
with drop tank	3,040 (1,643)
Endurance at 12,000 m:	
on internal fuel only	3 hrs 37 min
with drop tank	4 hrs 6 min
Take-off run, m (ft)	850 (2,778)
Landing run, m (ft)	975 (3,199)

Notes:
* Limited by dynamic pressure.
† In the event of take-off with full fuel.

with developing the Yak-125 two-seat tactical reconnaissance aircraft based on the Yak-120 interceptor and submitting it for State acceptance trials in October 1952. Development of the two versions proceeded in parallel, and thus structurally and outwardly the two aircraft were almost identical.

The fuselage nose was all-metal, albeit with nearly identical contours, and the ILS glideslope antenna was mounted under the nose (on the interceptor it was hidden inside the radome). The radar was deleted to make room for two AKAFU automatic tilting camera mounts (*avtoma**ticheskaya** ka**cha**yushchayasya **a**ero**foto**'usta**nov**ka*) installed side by side. These mounts could carry AFA-33/100M, AFA-33/75M or AFA-33/50M cameras for two-strip, three-strip or four-strip vertical photography which covered a strip 3-8 km (1.86-4.97 miles) wide when the aircraft was flying at 2,800-15,000 m (9,190-49,210 ft). Ahead of these was a flexible oblique camera mount with an AFA-33/75M or AFA-22/50M camera shooting to port or to starboard. The camera ports were protected by hinged doors on take-off and landing. To accommodate the bulky camera equipment the nose had to be extended slightly.

The camera and camera port control panels were located in the rear cockpit occupied by the reconnaissance systems operator (RSO). There was a viewing window in the rear cockpit floor; additionally, the RSO could find his targets by means of an OBP-1R optical sight. An AK-2 automatic heading control system (*avto**maht** **koor**sa*) ensured the aircraft stayed on its planned course in the event of an engine failure, compensating for the resulting asymmetric thrust. The principal avionics were identical to those of the interceptor version.

The armament was reduced to a single NR-23 cannon with 80 rounds on the starboard side. Internal fuel capacity in four bladder tanks was increased to 3,925 litres (863.5 Imp gal); as with the interceptor, a 670-litre (147.4 Imp gal) centreline drop tank could be carried.

Powered by AM-5 turbojets, the unseriailled Yak-125 prototype was completed in August 1952. On 26th August the aircraft made its maiden flight, piloted by V. M. Volkov. The manufacturer's flight test programme was completed just over a year later, on 3rd October 1953. According to the OKB's test report, by then the Yak-125 had logged 83 hrs 25 min in 79 flights; the GK NII VVS report, however, gives different figures for this period – 82 flights totalling 82 hrs 18 min. As originally flown the aircraft had a single boundary layer fence on each wing. Later, a second pair of fences was added and the ILS aerial under the nose was reversed, the fork pointing forward instead of aft.

Apart from Volkov, the aircraft was flown by LII test pilots Sergey N. Anokhin and Yakov I. Vernikov. Normal AUW with drop tank was 9,630 kg (21,230 lb). The aircraft reached a top speed of 1,120 km/h (695 mph) at 3,000 m (9,840 ft) and 1,075 km/h (667 mph) at 10,000 m (32,810 ft). The service ceiling was 15,000 m (49,210 ft); with the drop tank fitted, range and endurance were 3,600 km (2,236 miles) and 4 hrs 25 min respectively.

The Yak-125's State acceptance trials at NII VVS lasted a mere two months, commencing on 25th December 1953 and ending on 26th February 1954. The reconnaissance version was almost equal in performance to the baseline interceptor. During the State acceptance trials the normal AUW was reduced to 9,177 kg (20,231 lb).

The State commission ruled that the Yak-125 was superior to production tactical reconnaissance aircraft then in service. However, the aircraft had a serious shortcoming – the longer nose impaired forward visibility for the pilot during take-off and landing, to say nothing of the RSO who had an extremely limited view through the ventral window and bombsight. Hence the commission recommended that modifications be made and a small pre-production batch be built for service trials.

Yak-125B (Yak-25B) tactical nuclear strike aircraft prototype

A special joint directive of the Communist Party Central Committee and the Council of Ministers tasked the Yakovlev OKB with developing a 'high-speed special-mission bomber' based on the Yak-125 and designated Yak-125B (*bombardi**rov**shchik*). The quaint term 'special-mission' meant that the aircraft would carry a small nuclear bomb intended for a nuclear strike against targets of importance behind enemy lines.

Since the aircraft would have a crew of two, choosing the correct location for the navigator/bomb-aimer's station was essential. Yakovlev engineers decided on the following arrangement. The navigator sat in the extreme nose; his compartment was accessed via a dorsal hatch. It had a large hemispherical glass nose dome and four curved Perspex side windows on each side; an optically flat silicate glass panel of elliptical shape was provided on the underside of the nose for the optical bomb sight. The pilot's cockpit was located between frames 6 and 10.

The nuclear bomb (euphemistically called ***spets**pod**ves**ka* – lit. 'special slung load' or special store) was housed internally in a bay located near the aircraft's CG. This and the much bigger payload necessitated a major redesign of the undercarriage. The nose unit was beefed up, with twin wheels of about the same size as the main unit, and retracted forward, not aft. It was also rather longer than the interceptor's, giving the aircraft a pronounced nose-up attitude on the ground to facilitate bomb loading. The main unit was moved aft to frame 29 (immediately aft of the bomb bay). This arrangement was retained on all subsequent Yakovlev tactical bombers up to and including the Yak-28.

The bomb loading procedure was a rather complicated one. Because of the small ground clearance the tail of the aircraft had to be jacked up; then the bomb would be wheeled in on a special dolly and raised into position by means of hand-driven hoists, their cables passing through special apertures in the fuselage skin.

The Yak-125 was the first Yakovlev aircraft to feature a 360° ground mapping/bomb-aiming radar – the RMM-2 *Ru**bid**iy* (Rubidium). The radar set and the revolving antenna enclosed by a rather large teardrop radome were located directly below the cockpit. The radarscope and the OPB-P5 optical sight were located in the navigator/bomb-aimer's compartment. The RMM-2 was to be replaced by the more capable RBP-3 radar (RBP = ***rah**diolokatsi**on**nyy bombardi**rov**ochnyy pri**tsel*** – radar bomb sight) later on. Another distinguishing feature was the long strake aerial (probably for the ADF) running along the starboard side of the aft fuselage from wing to tailcone.

Serialled '25 Yellow', the Yak-125B entered flight test in 1955. In the course of the trials the optically flat silicate glass panel was enlarged considerably to improve the bomb sight's field of view.

The production version was to have the service designation Yak-25B. Yet the aircraft never entered production, as it was already obsolete; the Yakovlev OKB was already testing a supersonic stablemate to the Yak-125B with more powerful engines and improved aerodynamics.

Above: The first Yakovlev jet bomber – the Yak-125B – in ultimate form, showing the radically redesigned nose, the taller landing gear with a forward-retracting twin-wheel nose unit and the ventral radome.

Close-up of the Yak-125B's nose. Note the dummy nuclear bomb on a dolly beside the aircraft.

Yak-25R tactical photo reconnaissance aircraft

As noted earlier, the chief shortcoming of the Yak-125 tactical reconnaissance aircraft was the RSO's extremely limited field of view. The need to correct this led to a massive redesign of the entire forward fuselage; the result was very similar to the Yak-125B bomber. The RSO sat in the nose, which featured a dorsal entry hatch, a hemispherical glass dome and five Perspex windows on each side (these were located differently from the Yak-125B). Oblique and vertical cameras were installed in the nose (ahead of the RSO's seat) and the centre fuselage.

The rest of the airframe, including the landing gear, and the powerplant (two AM-5As) remained unchanged. So did the armament consisting of a single NR-23 on the starboard side; a long strake aerial ran from the cannon fairing to the tailcone, just like on the Yak-125B. The resulting aircraft was designated Yak-25R ([*samo**lyot***-] *raz**ved**chik*, reconnaissance aircraft).

On 10th March 1955 the Council of Ministers issued directive No.448-271 followed next day by MAP order No.160. These documents tasked OKB-115 and the Saratov aircraft factory No.292 with building a pre-production batch of ten Yak-25Rs. During State acceptance trials the aircraft had a 9,840-kg (21,690-lb) take-off weight and a top speed of 1,080 km/h (670 mph).

Still, the Yak-25R shared the fate of its predecessor – for much the same reasons as the Yak-125B: it no longer met the growing requirements and Yakovlev had a supersonic successor ready for testing. Besides, by then the Il'yushin IL-28R tactical reconnaissance aircraft had been built in substantial numbers. Even though it, too, was obsolete, the IL-28R still suited the Soviet Air Force's tactical arm just fine.

Yak-25MR maritime reconnaissance aircraft prototype

With the onset of the Cold War the Soviet leaders grew increasingly apprehensive of

Yakovlev OKB

Above: One of the pre-production Yak-25R reconnaissance aircraft which were a cross-breed between the Yak-25 and the Yak-125B. The glazed nose, the cockpit design and the short wheelbase are clearly visible.

the ever-growing military might of the 'potential adversaries' in the West – including the navies of the NATO countries. Keeping an eye on the movements of Western navies became a priority task.

The Soviet government tasked Yakovlev with developing a fast maritime reconnaissance version of the Yak-25 which had just entered production; the aircraft would be designated Yak-25MR (*mor**skoy** raz**ved**chik* – maritime reconnaissance aircraft). The appropriate CofM directive No.1968-911 and MAP order No.597 were issued on 18th and 21st September 1954 respectively. The documents required the Yak-25MR to match – or exceed – the performance of the basic *Flashlight-A*. The aircraft was to commence State acceptance trials in February 1955.

Development dragged on rather longer than intended; the ADP was approved on 4th March 1955 (in other words, a month after the intended beginning of the certification trials the Yak-25MR was still a 'paper aeroplane'). Prototype construction (or rather conversion), however, was completed in a remarkably short time. Serialled '43 Yellow', the aircraft differed from the standard Yak-25 in having a forward fuselage extended by 0.5 m (1 ft 7$\frac{11}{16}$ in) to house an AFA-33/75M reconnaissance camera for vertical and oblique photography. A second oblique camera (AFA-33/50) was optional.

The huge dielectric radome was replaced by a more elongated metal structure with a smaller radome for the SPRS-1 Koors (Heading) radar. This would be used for detecting NATO carrier groups or single warships.

The SRO IFF transponder gave way to a more modern SRZO-1 (*samo**lyot**nyy **rah**di-olokatsi**on**nyy za**pros**chik-ot**vet**chik* – aircraft-mounted radar interrogator/responder). The SPU-2R intercom was replaced by an SPU-5 and the RSIU-3M communications radio by an RSIU-3. Additionally, a 1RSB-70M radio and a Sirena-2S radar homing and warning system (RHAWS) were installed.

The port N-37L cannon and its ammunition box were deleted to offset the extra weight of the camera. So were the airbrakes; the engineers used the recess remaining from the port airbrake to accommodate an inflatable rescue dinghy

Manufacturer's flight tests were completed on 25th July 1955. Though having marginally shorter range than the Il-28R, the Yak-25MR was 170-200 km/h (105-124 mph) faster and had a 2,400-m (7,870-ft) advantage in service ceiling. Nevertheless, the aircraft suffered the same ignominious fate as the 'landlubber' Yak-25R. The naval Air Arm (AVMF) already operated a substantial number of IL-28Rs (which, of course, had a common type rating with the Navy's IL-28T torpedo-bomber) and didn't want to go through all the trouble with converting to a new type. Thus the Yak-25MR never progressed beyond the prototype stage.

The Yak-25MR's range and endurance were not measured during manufacturer's flight tests and can be assumed about equal to those of the Yak-25M interceptor.

Yakovlev OKB

'43 Yellow', the Yak-25MR maritime photo reconnaissance aircraft prototype. Note the elongated nose.

Specifications of the Yak-25MR (as per manufacturer's flight tests)

Length overall	16.165 m (53 ft 0 13/32 in)
Wing span	10.964 m (35 ft 11 21/32 in)
Wing area, m^2 (sq ft)	28.94 (311.18)
Empty weight without drop tank, kg (lb)	5,739 (12,652)
Normal AUW, kg (lb)	8,675 (19,120)
Maximum AUW with drop tank, kg (lb)	9,450 (20,830)
Fuel load, kg (lb):	
internal fuel (normal)	2,650 (5,842)
with drop tank(s)	3,385 (7,462)
Top speed, km/h (mph):	
at 5,000 m (16,400 ft)	1,100 (683)
at 10,000 m (32,810 ft)	1,040 (646)
Service ceiling, m (ft)	14,700 (48,230)

By the mid-1950s both East and West were facing a requirement for ever more capable military aircraft of all types. The Soviet leaders closely followed the development of Western military hardware and promptly took steps to offset any advantage of the 'potential adversary'.

The development of new strategic bombers in the USA and Great Britain, along with the establishment of a wide network of NATO bases along the borders of the Soviet Union and its allies, necessitated the development of new air defence systems – both surface-to-air missiles and interceptor aircraft. The latter were required to have a much higher speed, rate of climb and service ceiling than hitherto. On the other hand, the immediate retaliation doctrine pursued by the Soviet Union involved a massive counter-offensive after a hypothetical NATO 'first strike'; therefore, considerable importance was attached to tactical strike systems – including aircraft.

The availability of new powerful jet engines enabled OKB-115 to develop a whole family of supersonic aircraft based on the original Yak-25 design. In so doing Yakovlev engineers made good use of their experience with the experimental Yak-125B tactical bomber and Yak-25R reconnaissance aircraft featuring a new crew section design. Starting in 1954, the Yakovlev OKB brought out three twin-engined two-seat supersonic aircraft to fill three different roles – the Yak-121 interceptor, the Yak-122 tactical photo reconnaissance aircraft and the Yak-123 tactical bomber.

Yak-2AM-11 interceptor and tactical reconnaissance aircraft projects

On 10th June 1954 the Council of Ministers issued directive No.1144-504 ordering the Yakovlev OKB to develop two derivatives of the Yak-25 (an interceptor and a tactical reconnaissance version) powered by the new Mikulin AM-11 afterburning turbojet rated at 4,000 kgp (8,820 lbst) dry and 5,000 kgp (11,020 lbst) reheat. The two versions were to commence State acceptance trials in the fourth quarter of 1955 and the second quarter of 1956 respectively. The aircraft was tentatively designated Yak-2AM-11 – that is, Yakovlev aircraft (unnumbered) with two AM-11s. The corresponding MAP order No.368 was issued four days later.

In overload configuration the fighter could also carry various types of missiles. The reconnaissance version would feature a battery of four cameras: two AFA-30/50s, AFA-30/75s or AFA-30/100s (depending on the mission) on an AKAFU automatic tilting mount, one AFA-33/20 vertical camera and one AFA-33/50 or AFA-33/75 oblique camera. The interceptor was to be submitted for State acceptance trials in the fourth quarter of 1955, the reconnaissance version following in the first quarter of 1956.

However, development and testing of the AM-11 turbojet was taking longer than expected, and as the new engine (known in production form as the Tumanskiy R11-300) was intended mainly for the prototypes of a new lightweight tactical fighter then under development at Mikoyan's OKB-155 (which ultimately entered service as the MiG-21F), not enough prototype engines could be produced to power the Yakovlev twinjets. Hence on 30th March 1955 the Council of Ministers issued directive No.616-381, cancelling the Yak-2AM-11 project; pursuant to this document the Yakovlev OKB was to design new supersonic tactical aircraft around the proven Mikulin RD-9 (AM-9) afterburning turbojet which was more mature and more readily available.

Yak-25K interceptor

In the early 1950s the Soviet government issued a number of directives concerning the development of missile-armed interceptors. Several design bureaux were tasked with the development of AAMs, among them KB-1, a division of the Ministry of Defence Industry, which began development of the K-5 missile (K = ***kom**pleks [vo'orou**zhen**iya]* – weapons system). KB-1 is now called NPO Almaz ('Diamond' Research & Production Association; NPO = *na**ooch**no-proiz**vod**stvennoye obyedi**nen**iye*).

One of the said directives issued on 26th November 1953 marked the birth of OKB-2 within the MAP framework; one of its principal tasks was AAM development. The bureau was led by Pyotr D. Grooshin; by then he already had some experience in rocket weapons design.

The newly established bureau took up residence at plant No.293 previously allocated to the defunct Bolkhovitinov OKB. Development of the K-5 was transferred to OKB-2 (except for the missile's radio command guidance system which remained the responsibility of KB-1). At the same time the Grooshin OKB worked on the K-6 and K-51 missiles. Two other bureaux, OKB-134 under I. I. Toropov and OKB-4 under Matus R. Bisnovat, were also working in the same area. The former company developed the K-75 and K-7 AAMs (in two versions, the semi-active radar homing K-7L and the active K-7S), while the Bisnovat OKB was busy with the K-8 missile which again had SARH and active versions.

On 30th December 1954 the Council of Ministers issued directive No.2543-1224 ordering these missile systems to be tested on interceptors. The corresponding MAP

The Specific Operational Requirement data for the Yak-2AM-11

	Interceptor version		Reconnaissance version	
Engine operating mode	full mil power	full afterburner	full mil power	full afterburner
Top speed at 10,000 m (32,810 ft), km/h (mph)	1,200 (745)	1,350-1,400 (838-869)	1,300 (807)	1,400-1,500 (869-931)
Climb time to 10,000 m	3.5 min	1.8 min	3.5 min	2.0 min
Service ceiling, m (ft)	15,000-16,000 (49,210-52,490)	17,000-18,000 (55,770-59,055)	15,500 (50,850)	16,000-17,000 (52,490-55,770)
Range at 12,000-14,000 m (39,370-45,930 ft), km (miles):				
on internal fuel only	2,500 (1,552)	2,000 (1,240)	3,500 (2,179)	3,000 (1,863)
with drop tank	3,000 (1,863)	2,500 (1,552)	4,000 (2,474)	3,500 (2,179)
Endurance at 12,000-14,000 m:				
on internal fuel only	3 hrs 00 min	2 hrs 25 min	–	–
with drop tank	3 hrs 30 min	2 hrs 55 min	–	–
Armament	Three 30-mm (1.18 calibre) NR-30 cannons with 100 rpg		One 30-mm (1.18 calibre) NR-30 cannon with 50 rounds	
Targeting equipment	RP-6 Sokol radar		Computing optical sight	

Yakovlev OKB

The Yak-25K-75 (c/n 1608) armed with four Toropov K-75 missiles. Note the non-standard pitot on the fin leading edge.

order No.704 did not appear until 5th November 1955.

The Soviet Air Force and MAP had a hard time choosing a good missile platform. Hence all four Soviet 'fighter makers' were ordered to adapt their production and/or experimental fighters and present them for evaluation in this capacity. The Yakovlev OKB was to retrofit a Yak-25M with the K-5 weapons system and test it jointly with OKB-2.

In 1955 a production Yak-25M was converted to take the K-5 system. Two RS-1-U missiles, as the K-5 AAM was known in production form (NATO codename AA-1 *Alkali*), were carried under each inner wing on pylons equipped with APU-3 launch rails (*aviatsi**on**naya pooskov**a**ya oostan**o**vka* – aircraft-mounted launcher). The RP-6 Sokol radar was replaced by the RP-1-U – a modified version of the MiG-17PF's Izumrood radar, the suffix letter standing for *oopravl**en**iye* [*snar**ya**dami*] – missile control or guidance. The cannons were deleted but the lateral cannon fairings were retained (only the muzzle openings were faired over). Of course, appropriate changes were made to the electrics and data recording equipment was fitted.

The converted aircraft received the designation Yak-25K and the weapons system built around it was designated Yak-25K-5. The trials programme involved flights with both inert and live missiles; several live launches were made, including at least one against a full-size aircraft converted into a target drone. Concurrently the K-5 AAM was tested on five suitably converted prototypes of the MiG-17P interceptor (*izdeliye* SP-6); these were effectively prototypes of the MiG-17PFU missile-armed interceptor.

The trials proved successful; the K-5 weapons system was recommended for production and included into the PVO inventory, as were the Yak-25K and MiG-17PFU interceptors. The RS-1-U AAM was manufactured under the product code *izdeliye* M (later changed to *izdeliye* I). The Yak-25K was built on a very small scale. In fact, most Yak-25Ks eventually became weapons testbeds as described below.

Specifications of the Yak-25 interceptor variants

	Yak-120	Yak-25M	Yak-25K-75
Length overall	15.665 m (51 ft 4¾ in)	15.665 m (51 ft 4¾ in)	15.665 m (51 ft 4¾ in)
Wing span	10.964 m (35 ft 11½ in)	10.964 m (35 ft 11½ in)	10.964 m (35 ft 11½ in)
Wing area, m² (sq ft)	28.94 (311.18)	28.94 (311.18)	28.94 (311.18)
Powerplant	2 x AM-5A	2 x RD-5A	2 x RD-5A
Rating, kgp (lbst)	2 x 2,000 (2 x 4,410)	2 x 2,000 (2 x 4,410)	2 x 2,000 (2 x 4,410)
Empty weight, kg (lb)			
in clean condition	5,675 (12,510)	n.a.	n.a.
with drop tanks	5,720 (12,610)	n.a.	n.a.
TOW, kg (lb):			
in clean condition	8,675 (19,124)	9,220 (20,330)	8,830
with drop tanks	9,450 (20,830)	10,045 (22,145)	n.a.
Top speed, km/h (mph):			
at 5,000 m (16,400 ft)	1,090 (677.4)	n.a.	1,000 (621)
Rate of climb at S/L, m/sec (ft/min)			
in 'clean' condition	50 (9,840)	44 (8,659)	n.a.
with drop tanks	n.a.	37 (7,281)	n.a.
Service ceiling , m (ft)	14,500 (47,572)	n.a.	13,600 (44,619)
Range, km (miles)			
on internal fuel	2,100 (1,305)	2,010 (1,249)	n.a.
with drop tanks	2,700 (1,680)	2,560 (1,590)	n.a.
Take-off run, m (ft)	735 (2,410)	800 (2,620)	n.a.
Landing run, m (ft)	800 (2,620)	850 (2,790)	n.a.
Armament	Two N-37L cannons	Two N-37L cannons	Four K-75 AAMs

Yak-25K-75 interceptor prototype/weapons testbed

Pursuant to the abovementioned Council of Ministers directive No.2543-1224 and MAP order No.704 one of the Yak-25Ks (c/n 1608) was modified in early 1956 for testing the K-75 missile system developed by the Toropov OKB. The conversion was performed by the OKB's experimental shop (plant No.134).

The aircraft carried four K-75 (*izdeliye* 129) AAMs on the existing wing pylons; appropriate changes were made to the Izumrood radar, and an ASP-3NM computing gunsight (*avtomat**i**cheskiy strel**ko**vyy prit**sel***) was installed. AUPT fuel flow limiters (*avto**maht** oopravl**en**iya po**dah**chey **top**liva* – automatic fuel feed controller) were incorporated into the fuel system; they throttled the engines back automatically when the missiles were fired to preclude surging caused by missile exhaust gas ingestion. Camera pods were fitted to record missile launches.

Appropriately designated Yak-25K-75, the uncoded aircraft underwent tests between 5th March and 10th July 1956. The all-up weight was 8,830 kg (19,470 lb), including 324 kg (714 lb) for the four missiles and 2,650 kg (5,840 lb) of fuel. At 11,000 m (36,090 ft) the Yak-25K-75 cracked the sound barrier, attaining Mach 1.07; top speed was 1,000 km/h (621 mph) at 5,000 m (16,400 ft) and 950 km/h (590 mph) at 10,000 m (32,810 ft). As compared to the standard Yak-25M the service ceiling decreased to 13,600 m (44,620 ft) and the time required to reach 5,000 m and 10,000 m increased to 2.5 and 6.0 minutes respectively, but manoeuvrability and handling remained unaffected.

Most flights were made with dummy missiles. The radar could detect a medium bomber such as the IL-28 (or English Electric Canberra) flying at 5,000-8,000 m (16,400-26,250 ft) at a range of 7-7.5 km (4.34-4.65 miles), getting a target lock-on at 4-4.5 km (2.48-2.79 miles). Generally the test results were satisfactory; still, the K-75 weapons system did not enter production.

Yak-25K-7L interceptor prototype/weapons testbed

Another uncoded Yak-25K (c/n 0109) became a testbed for the Toropov K-7L AAM. The new missile was larger and heavier than the K-5 (RS-1-U) and K-75 (*izdeliye* 129), so the complement had to be limited to two missiles. This aircraft also had the same camera fairings on the aft fuselage sides and under the wings. Designated Yak-25K-7L, the aircraft was tested but once again the K-7L weapons system did not proceed to production.

Yak-25K-8 (Yak-25S K-8) interceptor prototypes/weapons testbeds

By the end of 1956 the Bisnovat OKB was ready to test the K-8 missile system. Consequently four interceptors, including two Yak-25Ks, were converted into weapons testbeds for this AAM. These Yak-25s were equipped with the Sokol-2K radar, a version of the RP-6 specially modified to work with the Bisnovat missile.

The K-8 was a big and heavy weapon, which again meant only two of these missiles could be carried. The long trapezoidal pylons were located under the inner wings and were different on the two testbeds. The first aircraft, serialled '110 Red' (c/n 1110) had pylons with a forward-swept leading edge, while on the other aircraft, which wore no tactical code (c/n 0119), the pylons had a sweptback leading edge. The reason for this was the need to determine the best location of the missiles' attachment points with respect to the wings.

The aircraft were designated Yak-25K-8. Some documents, however, refer to '110 Red' as 'Yak-25S K-8', the S standing for *se**ree**ynyy* (production, used attributively); thus the aircraft was meant to be the standard-setter for production Yak-25Ks equipped with the K-8 weapons system. Yet this never happened, as we will see.

Between them the two Yak-25K-8s made 111 test flights, and performed 42 launches, including two live ones. The K-8 had passive infra-red homing with an S1-U seeker head (not to be confused with the similarly designated Grooshin weapons system!) and a high-explosive/fragmentation warhead.

The K-8 (NATO codename AA-3 *Anab*) was intended for several Sukhoi interceptors, the Mikoyan I-75 interceptor – and the Yak-27 described below. Since the latter aircraft was already in the middle of its flight test programme, there was no point in fielding the Yak-25K-8 weapons system based on an outdated missile platform. The missile itself, however, did enter production after passing its State acceptance trials; it was later developed into several improved versions.

Yakovlev OKB

Above: The Yak-25K-7L (c/n 0109) armed with four Toropov K-75 missiles. Again a pitot is installed on the fin leading edge. Note the camera pod under the port outer wing.

Yakovlev OKB

Above: One of the two Yak-25K-8 testbeds (c/n 0119). This views illustrates the size of Bisnovat's K-8 missile. Note the shape of the pylons making an interesting comparison with the preceding version.

Yefim Gordon archive

Above: '14 Red', one of the Yak-25MSh target drone prototypes. Note the whip aerials above and below the cockpit, the nose probe aerial and the wingtip antenna blisters.

Yefim Gordon archive

'10 Blue', another Yak-25MSh, in flight with clearly no-one in the cockpit.

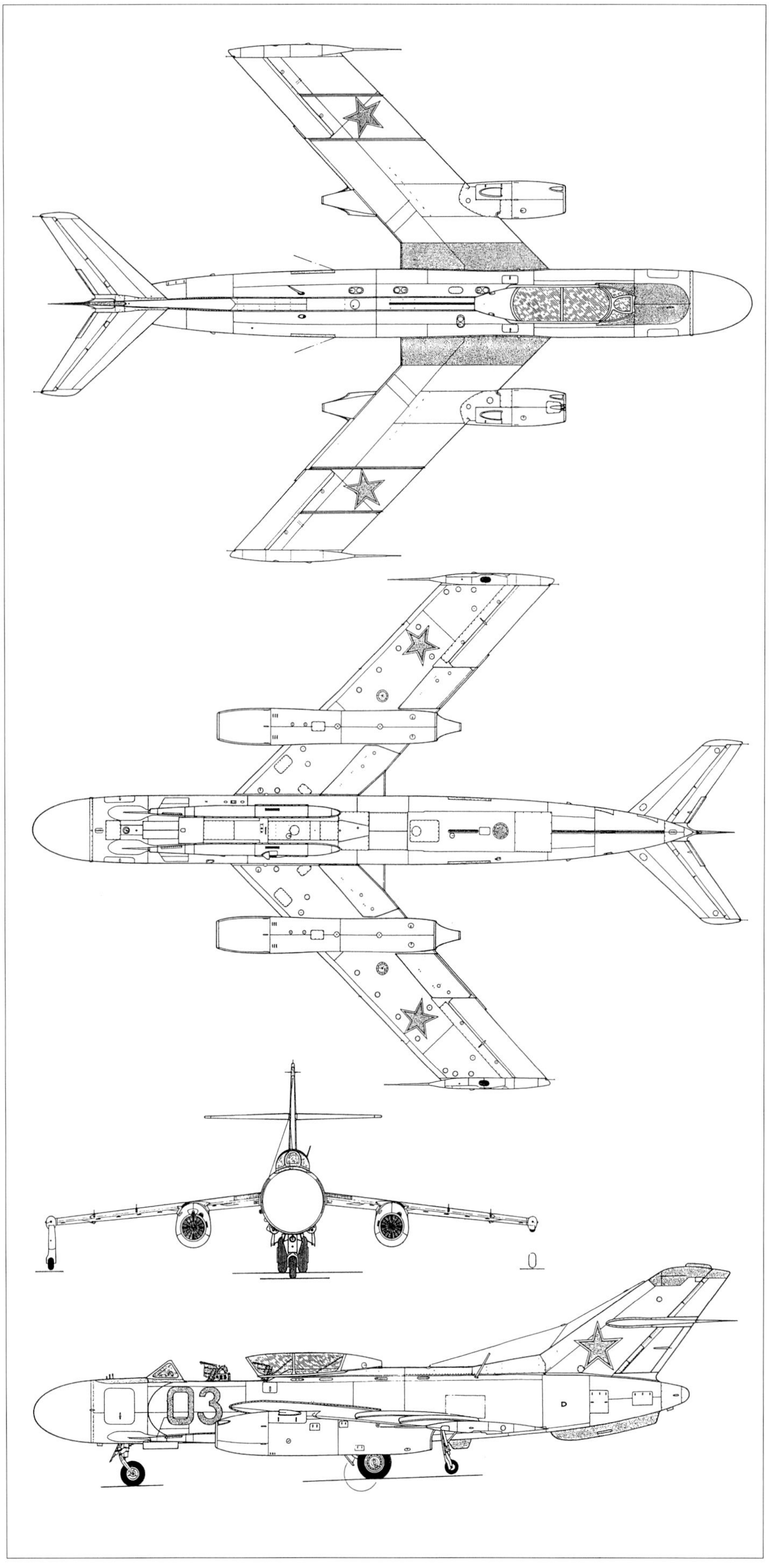

Five views of the Yak-25M.

Yak-25MSh target drone

As the *Flashlight-A* was displaced from first-line service by more capable interceptors, many Yak-25s would become surplus. Realising this, in 1957 the Yakovlev OKB developed a target drone version of the Yak-25M for training surface-to-air missile (SAM) crews designated Yak-25MSh (*mi**shen'*** – target). The conversion involved removal of the armament, radar and much of the avionics, whereupon a remote control system was installed. The aircraft was controlled by an operator sitting in a drone director aircraft. Outwardly the drone could be identified by two whip aerials ahead of the windscreen and under the front cockpit, plus two angular antenna blisters on the wingtips.

Coded '10 Blue', the Yak-25MSh prototype was successfully tested at LII; a suitably modified UTI-MiG-15 trainer acted as the drone director aircraft. Another prototype coded '14 Red' differed in having an additional probe aerial on the radome. The Yak-25MSh was not built as such, but a substantial number of Yak-25s and Yak-25Ms were converted into target drones after the type had been phased out.

The Yak-25 and Yak-25M were in production at the Saratov aircraft factory No.292 in 1954-57. A total of 480 was built, including seven aircraft (six flying prototypes and a static test airframe) manufactured by the Yakovlev experimental shop in Moscow, 65 initial-production Yak-25s and 409 (some sources say 406) Yak-25Ms; the maximum number of aircraft per batch was probably 20.

Yak-25RV high-altitude reconnaissance aircraft

Three years after its service entry the Yak-25 evolved into a very different aircraft – the Soviet Union's answer to the Lockheed U-2. In 1957-58 the Yakovlev OKB developed the Yak-25RV strategic reconnaissance aircraft (RV = *raz**ved**chik vy**sot**nyy* – reconnaissance aircraft, high-altitude). Some early documents referred to the aircraft simply as the 'Yak-RV'.

Structurally the Yak-25RV had virtually nothing in common with the original interceptor. Firstly, wing loading had to be reduced considerably to ensure good high-altitude performance. Hence the wings were all-new; they were straight, with more than double the span – 23.5 m (77 ft $1\frac{13}{64}$ in) versus 11.0 m (36 ft $1\frac{5}{64}$ in), an area of 55 m^2 (591.39 sq ft) versus 28.96 m^2 (311.39 sq ft) and an aspect ratio of about 10. As a result, wing loading decreased from 310 kg/m^2 (63.55 lb/sq ft) to a mere 175-178 kg/m^2 (35.87-36.49 lb/sq ft) – less than that of the World War Two-vintage Yak-3 fighter!

Secondly, the thrust/weight ratio was improved dramatically by installing two Tumanskiy R11V-300 turbojets – a special non-afterburning high-altitude version of the engine powering the MiG-21F (V = *vysotnyy*). Hence the nacelles were new, too. The engines delivered 3,900-4,000 kgp (8,600-8,820 lbst) at full military power and 3,250 kgp (7,165 lbst) at cruise power.

The forward fuselage was redesigned, featuring a single-seat cockpit and an all-metal recontoured nosecone housing reconnaissance cameras; more cameras were located in the centre fuselage. The vertical tail was also redesigned, with slightly increased leading-edge sweep and a larger fin fillet. The outrigger struts now retracted forward instead of aft, and the wingtip fairings looked disproportionately large because of the narrow tip chord. The twin wingtip pitot booms were replaced by a single nose pitot. Finally, the Yak-25RV was unarmed.

Prototype construction was completed in early 1959. Coded '75 Yellow', the aircraft entered flight test on 1st March 1959 with project test pilot V. P. Smirnov at the controls; the test programme was completed successfully on 29th May 1959. The prototype's empty weight was 6,175 kg (13,610 lb) and all-up weight was 9,800 kg (21,600 lb). During trials the Yak-25RV attained a maximum altitude of 21,000 m (68,900 ft) and a top speed of Mach 0.82, which was considered adequate for a high-altitude reconnaissance aircraft.

Smirnov reported that climbing to 18,500-19,500 m (60,700-63,980 ft) presented no problems. Above that, however, the pilot had to exercise special care because the permissible speed range was narrowed (the never-exceed speed was uncomfortably close to the stalling speed because of the rarefied air). The service ceiling could not be determined because the engines tended to flame out at 19,600-20,100 m (64,300-65,940 ft). The pilot had to wear a special SI-3M pressure suit which also drew some criticism.

A curious feature of the Yak-25RV was its reluctance to descend from high altitude. In one of the test flights, when Smirnov had reached maximum altitude, he discovered that the aircraft was firmly intent on staying up there. Only when the landing gear was extended did the prototype start descending slowly.

Upon completion of the manufacturer's flight tests the Yak-25RV was prepared for an attempt on the world altitude record. On 13th July 1954 V. P. Smirnov reached 20,456 m (67,112 ft) with a 1,000-kg (2,204-lb) payload; sixteen days later he set a second record by reaching 20,174 m (66,187 ft) with a 2,000-kg (4,410-lb) payload. In the

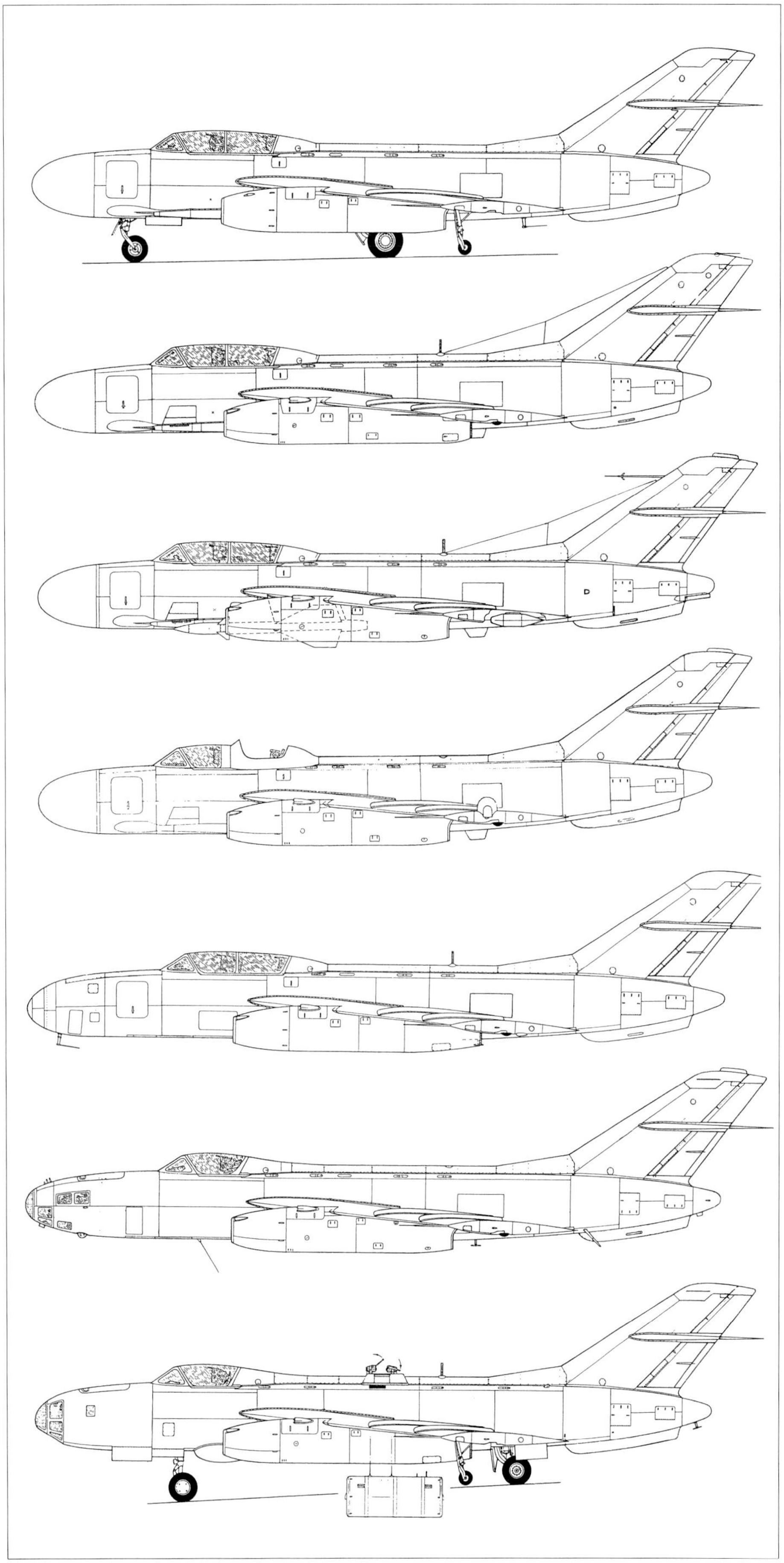

Top to bottom: The yak-120 prototype; the production Yak-25; the Yak-25K-8; the Yak-25L ejection seat testbed; the Yak-125 reconnaissance aircraft; the Yak-25R reconnaissance aircraft; and the Yak-125B (Yak-25B) bomber shown with an extra fuel tank.

Yakovlev OKB

Above: An air-to-air study of the first prototype Yak-25RV ('75 Yellow'). This view clearly shows how little the *Mandrake* had in common with the *Flashlight*.

Yakovlev OKB

Above: This forward view illustrates the huge wingspan of the 'Soviet U-2', which is not surprising since both aircraft were designed to fill the same role.

Fédération Aéronautique Internationale (FAI) papers acknowledging the records the aircraft type is stated simply as 'RV', which some Western analysts deciphered erroneously as *re**kord** vyso**ty*** (altitude record).

The Air Force was not completely satisfied with the aircraft's performance; still, there was no alternative design. As the Yak-25RV did have a high service ceiling and a good endurance, the VVS reluctantly gave the go-ahead to launch production.

After a few detail changes the Yak-25RV passed its State acceptance trials, entering production at the Ulan-Ude aircraft factory No.99 under the in-house product code *izdeliye* 25RV. Production aircraft differed in having two additional pitots flanking the nose air data boom; late-production Yak-25RVs also had a flat-bottomed dielectric fairing aft of the main gear unit. The avionics suite included an RSIU-5 UHF radio, an ARK-9 ADF, an MRP-56P marker beacon receiver, an SOD-57M radio rangefinder, an SRO-2M *Khrom-Nikel'* (Chromium-Nickel; NATO *Odd Rods*) IFF transponder and so on. The Yak-25RV's NATO reporting name was *Mandrake*.

On 11th August 1965 test pilot Marina L. Popovich set a world speed record on a modified Yak-25RV, averaging 753.048 km/h (467.73 mph) over a 2,000-km (1,242-mile) closed circuit. On 18th September 1967 she set another official world record, covering a distance of 2,497.009 km (1,550.93 miles) on a closed circuit.

Specifications of the Yak-25RV high-altitude reconnaissance aircraft

Length overall	15.93 m (52 ft 3 in)
Wing span	23.5 m (77 ft 1¼ in)
Wing area, m^2 (sq ft)	55 (591.39)
Powerplant	2 x R11V-300
Rating, kgp (lbst)	3,900 (8,600)
Empty weight, kg (lb)	6,175 (13,610)
TOW, kg (lb)	9,800 (21,605)
Top speed	Mach 0.82
Service ceiling , m (ft)	21,000 (68,900)
Armament	None

Yak-25RV-I target aircraft

In 1958, even before the Yak-25RV had completed its test programme, the Yakovlev OKB began development of a manned target aircraft version of the *Mandrake*. The mission was to emulate high-flying spyplanes which PVO fighters would intercept, making simulated attacks; only the gun cameras would be used to record a 'kill'. The interception would take place outside the usual shooting ranges, thus providing for more realistic training.

Designated Yak-25RV-I (the last symbol is a Roman numeral), the aircraft differed from the basic *Mandrake* mainly in lacking reconnaissance cameras. According to the Council of Ministers directive ordering its development, the target aircraft was to have a top speed of 900 km/h (559 mph) at 15,000 m (49,020 ft), a service ceiling of 20,000-21,000 m (65,620-68,900 ft) and a range of 2,500 km (1,552 miles) at 20,000 m.

After going through manufacturer's flight tests the Yak-25RV-I prototype was handed over to NII VVS for State acceptance trials which were completed in 1961. During State acceptance trials the aircraft had an empty weight of 6,285 kg (13,855 lb) and a take-off weight of 9,935 kg (21,900 lb); otherwise, performance was extremely similar to that of the reconnaissance version prototype.

A small production run of the Yak-25RV-I was built at the Ulan-Ude factory. In case of need the aircraft could be fitted with a remote control system and used as an expendable target drone.

In August-September 1962 NII VVS held checkout trials of a production Yak-25RV-I (c/n 25990302 – that is, *izdeliye* 25RV, plant No.99, batch 03, 02nd aircraft in the batch). This aircraft turned out to be almost 240 kg (530 lb) lighter than the prototype, with an

Yakovlev OKB

'20 Yellow', a production Yak-25RV.

empty weight of 6,043 kg (13,322 lb) and a take-off weight of 9,693 kg (21,369 lb); this was important, as each kilogram of weight saved had a positive effect on the aircraft's service ceiling.

Yak-25RV-II target drone

A second Yak-25 version was built 'for the killing' as the Yak-25RV-II. Unlike the version described above, this was designed from the outset as a remote-controlled target drone for live weapons training. The drone differed from the Yak-25RV-I in having an AP-28 autopilot, a control signals receiver and other equipment; external identification features were a small boxy fairing at the base of the rudder and an L-shaped aerial under the cockpit.

Coded '76 Yellow', the Yak-25RV-II prototype was probably converted from the original second prototype of the *Mandrake*. A modified Yak-30 jet trainer fitted with guidance equipment acted as the drone director aircraft.

Yak-25RV testbeds

An example of the Yak-25RV was used by LII for developing a remote flight control system. The results of its testing were used in the radio-controlled Yak-25RV target drone (see Yak-25RV-II above).

Another Yak-25RV was used for exploring critical phenomena associated with the flight dynamics and controllability of an aircraft and caused by imperfections of the surface finish and by failures of onboard flight systems.

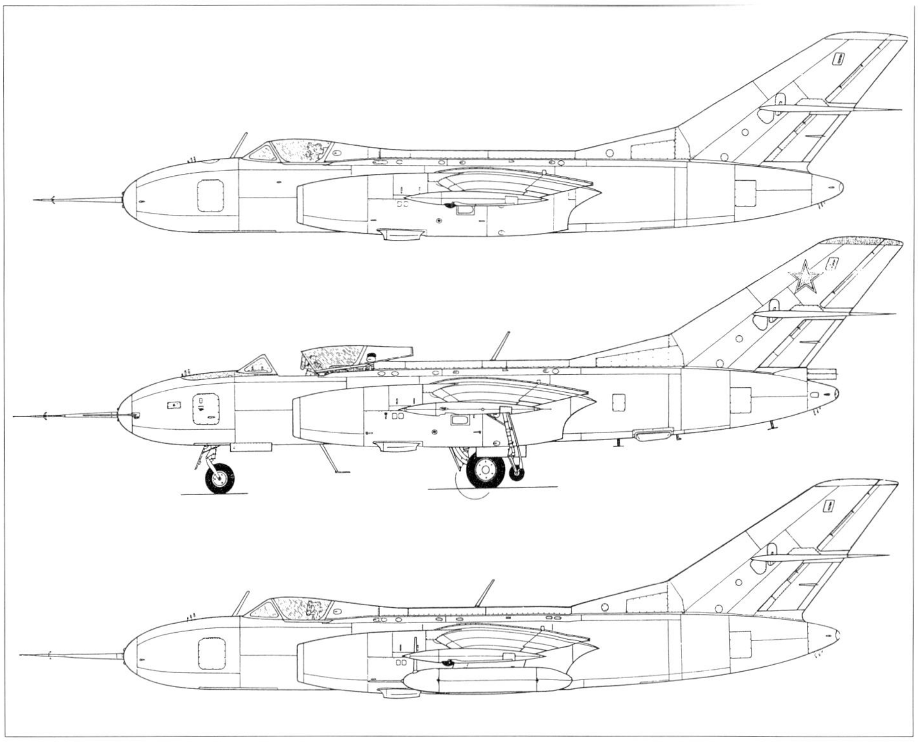

Top to bottom: The Yak-25RV high-altitude photo reconnaissance aircraft, the Yak-25RV-II high-altitude target drone and the Yak-25RRV high-altitude ELINT aircraft with the IRIS pods.

Yefim Gordon archive

A line-up of production Yak-25RV-IIs on a snow-covered hardstand. The type stayed in service until the early 1980s.

Yak-25RR radiation intelligence aircraft

In the late 1960s a handful of Yak-25RVs was converted for radiation intelligence (RINT) duties. Designated Yak-25RR (*radiatsi**on**nyy razv**ed**chik* – RINT aircraft), they had two underwing pylons (strut-braced on the inboard side) for carrying standard RR8311-100 air sampling pods. These pods, originally developed in 1964 for the Yak-28RR (described later), had a nose intake closed by a movable cone and a paper filter which arrested dust particles, enabling their radiation level to be measured. They were also carried by the Antonov An-12RR, An-24RR and An-30R RINT aircraft, one of several Tupolev Tu-16R reconnaissance versions (*Badger-F*) and occasionally the Tu-95K-22 naval missile strike aircraft.

Yak-25RRV ELINT aircraft

On 9th-26th October 1971 a modified Yak-25RR (c/n 25991201) underwent ground and flight tests at the Yakovlev flight test facility at LII. Designated Yak-25RRV (***rah**diotekhn**ich**eskiy razv**ed**chik vys**ot**nyy* – high-altitude electronic intelligence aircraft), the aircraft was developed pursuant to ruling No.111 of the CofM Presidium's Commission on Defence Industry Matters (VPK – *Vo**yen**no-pro**mysh**lennaya ko**mis**siya*) dated 7th May 1968 to suit a Ministry of Defence requirement.

The standard underwing air sampling pods were replaced with special pods housing the IRIS signals intelligence system (*izme**rit**el'-reghi**strah**tor **im**pool'snykh sig**nah**lov* – [electromagnetic] pulse signal measurement and recording device), aka *Vol**na**-S* (Wave-S), designed to detect and record electromagnetic pulses emitted by radars and the like. The cylindrical pods had dielectric front and rear ends, with two probe aerials pointing forward and up at the front; they were pressurised and heated by engine bleed air to make sure the equipment would operate normally. The cockpit featured a new control panel for the SIGINT suite; some minor changes were made to other systems.

Tests showed that the new pods and their air conditioning ducts had virtually no negative effect on the aircraft's performance. The Yak-25RRV was built in very small numbers (the exact quantity is unknown).

Yak-25PA 'balloon killer' aircraft project

The final spinoff of the Yak-25RV, which never got off the drawing board, was a highly specialised version designed to intercept and destroy balloons carrying reconnaissance equipment or printed matter. Launched in large numbers from Western Europe, these balloons were a major nuisance for the Soviet air defences in the 1950s and 1960s. Shooting them down was considered imperative – for two reasons. Firstly, they could survey secrets or deliver subversive Western literature to gullible Socialist readers. Secondly (perhaps even more importantly), they presented a serious danger for civil and military aircraft, especially because the balloon's equipment container was painted blue for low visibility; several aircraft crashed after colliding with such 'visitors from across the wall'.

The aircraft was designated Yak-25PA (*pere***khvaht***chik* **a***ero***stah***tov* – balloon interceptor). It differed from the Yak-25RV mainly in having a wing area enlarged from 55 m² (591.39 sq ft) to 58 m² (623.65 sq ft), which would enable the aircraft to get at balloons flying as high as 19,400 m (63,648 ft) – the Yak-25PA's service ceiling. Finally, the aircraft was armed with cannons.

Development began in August 1971 but was discontinued because of the lack of a suitable powerplant. With the only available high-altitude turbojet – the R11V-300 – the aircraft would be underpowered.

A source mentions the aircraft described above with the designation Yak-25ASh; it also mentions the Yak-25PA, but infers that this was a straightforward adaptation of the standard (swept wing) interceptor.

Yak-26 tactical bomber

Yak-123 (Yak-26) experimental tactical bomber

Encouraged by the success of the Yak-25, General Designer Aleksandr S. Yakovlev continued his attempts to develop the *Flashlight-A* into fully-fledged tactical bomber and reconnaissance versions. As it became clear that the subsonic Yak-125B and Yak-25R could not meet the Air Force's growing demands, Yakovlev decided to increase the speed of these types by installing more powerful afterburning engines – again from the Mikulin house.

Yakovlev soon managed to secure government support for the two projects. On 30th March 1955 the Council of Ministers issued directive No.616-381. This document and MAP order No.240 dated 5th April tasked the Yakovlev OKB with building two prototypes of a supersonic light tactical bomber designated Yak-123 with a top speed of at least 1,350-1,400 km/h (838-869 mph). The first prototype powered by RD-9AK engines was to be handed over to NII VVS for State acceptance trials in September 1955, followed by the second aircraft with uprated RD-9Fs in the first quarter of 1956.

The normal bomb load was specified as 1,200 kg (2,645 lb), increasing to 2,000 kg (4,410 lb) for short-range missions. The aircraft was primarily intended to carry one RDS-4 'special store'. Apart from the powerplant, the two prototypes would differ in avionics and armament. The first prototype would have a PSBN-M 360° ground mapping/bomb-aiming radar, while the second aircraft would have a Rym-S system – probably a data link system guiding the aircraft to its target, using information downloaded from command, control, communications and intelligence (C³I) centres ('rym' is a nautical term meaning 'lifting lug'). Also, the second prototype would have provisions for pods with 24 ARS-57 or ARS-70 FFARs.

The abovementioned RDS-4 was, in fact, not a bomb but a 30-kiloton nuclear warhead – the first Soviet production nuclear warhead developed by OKB-11 under Yuriy B. Khariton. It was designed to be fitted to both ballistic missiles and free-fall bombs, including the 8U-49 which was codenamed 'Natasha' in MoD documents. An improved version, the RDS-4T, was installed in the 'Tat'yana' bomb which was taken on strength by the Soviet Air Force's tactical arm (FA – *Fronto***va***ya avi***ah***tsiya*) and strategic bomber arm (DA – ***Dahl'****nyaya avi***ah***tsiya*) in the mid-1950s. Both munitions weighed 1,200 kg.

The RD-9AK afterburning turbojet specified for the first prototype was a version of the RD-9B powering most versions of the MiG-19. It had an identical thrust rating – 2,600 kgp (5,730 lbst) dry and 3,250 kgp (7,160 lbst) reheat, differing only in accessory gearbox location so as to permit installation in an underwing nacelle, hence the K for *kryl'ye***voy** – wing-mounted.

Development of the Yak-123 light tactical bomber was completed in 1955; at the ADP stage the aircraft received the service designation Yak-26. Outwardly it was very similar to the Yak-125B. The main external differences were the new engine nacelles, which were rather more streamlined, with sharp-lipped air intakes, and the crew section.

Designing the forward fuselage of a supersonic bomber turned out to be a major task. The part of the SOR which concerned the optical bombsight specified a 90° forward and downward field of view from the bomb aimer's station (the upper limit of the field of view was at just 5° to the fuselage waterline). However, providing a fully glazed nose, as had been the case with the predecessor, was not possible; a hemispherical nosedome *à la* Yak-125B would reduce the aircraft's top speed unacceptably, and making a conical glass transparency that was strong enough was technically impossible at the time.

Hence the Yak-123 had an ogival metal nosecone incorporating three large windows and one small window on each side, plus a much larger lower optically flat glass panel of elliptical shape. The mock-up review commission was quite pleased with the new design. The pilot's canopy was also

Yakovlev OKB

Above: The Yak-123-1, the first supersonic Yakovlev bomber, as originally flown (with a largely metal nosecone). Note the flattened radome of the PSBN-MA radar.

Yakovlev OKB

The same aircraft after the first round of updates, with a redesigned nose featuring a greater glazing area; even though the tactical code appears darker in this view, it is still yellow.

redesigned: the windshield was more sharply raked to reduce drag and the bullet-proof windscreen was now purely elliptical (not cut away at the bottom).

In order to increase speed it was also necessary to reduce the drag generated by the wings and empennage. There were two possible ways of doing this: by increasing their sweepback or by using thinner airfoils.

Using a new airfoil was less time-consuming; hence, Aleksandr S. Yakovlev made the decision to reduce the thickness/chord ratio from 12% to 6% on the Yak-123. Wing span and wing sweep remained unchanged, but wing area was increased slightly by adding small leading edge root extensions (LERXes) to improve the wings' performance at high speeds. The shape, sweepback and area of the tail unit also remained unchanged.

According to Yevgeniy G. Adler, Yakovlev did not consult the OKB's structural strength department before making his decision. This was a mistake. No-one had considered the torsional stiffness problem. As it turned out, this light-hearted approach crippled the aircraft's chances.

The fuselage accommodated six fuel cells and a large bomb bay. Possible bomb loads included eight 100-kg (220-lb) or 250-kg (551-lb) HE bombs, a single 1,500-kg (3,306-lb) HE bomb or an RDS-4 (RDS-4A) nuclear bomb.

The avionics suite included RSIU-4 and 1RSB-70M radios, an AP-40 autopilot, an OSP-48 ILS and the like. As on the Yak-125B and Yak-25R, a long strake aerial ran from the starboard wing to the tailcone. Bomb-aiming was done by means of the PSBN-MA radar whose radar set and revolving antenna enclosed by a flattened teardrop radome were located beneath the cockpit. The radar was linked to the OPB-11P optical sight, the RV-17 radio altimeter and the autopilot.

The engineers also paid attention to providing a defensive capability against enemy fighters. The Yak-123 featured a remote-controlled tail gun barbette which could move 15° up, down and to either side; it mounted a single 23-mm (.90 calibre) Afanas'yev/Makarov AM-23 cannon with 100 rounds. However, as the first prototype was nearing completion the barbette was still undergoing tests and the aircraft had to be rolled out *sans* defensive armament.

Aleksandr S. Yakovlev managed to get the mock-up approved and receive the go-ahead to begin prototype construction. Coded '50 Yellow', the first prototype (c/n 01), sometimes referred to as Yak-123-1, was rolled out at MMZ No.115 in 1955. After a brief manufacturer's flight test programme the aircraft was turned over to NII VVS for State acceptance trials; these lagged

Yefim Gordon archive

This poor but interesting shot depicts the Yak-123-1 during its public debut at the 1956 Tushino flypast.

behind schedule, the Stage 1 completion report being signed on 25th June 1956.

At 10,000 m (32,810 ft) the RD-9AK-powered aircraft was required to have a top speed of 1,225-1,250 km/h (760-776 mph) in full afterburner and 1,100 km/h (683 mph) at full military power. In reality, however, the highest speed attained during tests was 1,230 km/h (764 mph) at 10,600 m (34,780 ft). The Yak-123-1 had a service ceiling of 16,000 m (52,490 ft) whereas the requirement was for 16,000-17,000 m (52,490-55,770 ft).

Even with the interim RD-9AK engines the Yak-123-1 was markedly superior in speed and service ceiling to the subsonic IL-28; in fact, it was the Soviet Union's first supersonic tactical bomber. However, tests revealed a spate of major deficiencies, including instability at high angles of attack, poor roll control, and aileron reversal at 4,000-6,000 m (13,120-19,685 ft) which put a limit on the aircraft's top speed.

The biggest bad news was the aileron reversal caused by the wings' inadequate torsional stiffness. When the ailerons were deflected at high speed the outer wings would warp, causing the aircraft to roll in the opposite direction to the one the pilot wanted it to. This phenomenon manifested itself during the very first attempts to reach the limits of the aircraft's speed envelope; the Yak-123-1 could go supersonic at 8,000-12,000 m (26,250-39,370 ft) but could not do so at lower altitudes. Eventually, when the aircraft had completed its State acceptance trials, the aileron reversal was found to be 'incurable' (*sic*), quite apart from the fact that the aircraft did not meet some items of the SOR and some of the mock-up review commission's findings had been ignored.

Nevertheless, Yakovlev persuaded the VPK to order the construction of a pre-production batch of Yak-26s for service trials. On 5th January 1956 – even before the aircraft commenced State acceptance trials – the Council of Ministers issued directive No.16-10 followed a week later by MAP order No.20. These documents tasked MMZ No.30, a production factory located at Moscow-Khodynka airfield, with building an experimental batch of ten Yak-26 bombers.

The engineers tried to eliminate the deficiencies discovered during trials by redesigning the wings, tail unit and forward fuselage of the Yak-123-1 and subsequent prototypes.

Specifications of the Yak-123-1 experimental tactical bomber

Length overall	17.16 m (56 ft 3½ in)
Wing span	10.964 m (35 ft 11½ in)
Wing area, m² (sq ft)	28.94 (311.18)
Powerplant	2 x RD-9AK
Engine thrust, dry/reheat, kgp (lbst)	2 x 2,600/3,250 (2 x 5,730/7,160)
TOW (with bombs), kg (lb)	11,220 (24,735)
Top speed, km/h (mph)	1,230 (764)
Service ceiling , m (ft)	16,000 (52,490)
Armament	One AM-23 cannon; 1,200-3,000 kg (2,645-6,613 lb) of bombs

Yak-26-3 tactical bomber prototype (and Yak-123-1 modified)

The Soviet Air Force top brass kept urging the Yakovlev OKB to build and test a second Yak-26 prototype as soon as possible. On 28th March 1956 the Communist Party Central Committee and the Council of Ministers issued a joint directive (No.424-261) ordering the OKB to complete the RD-9F-powered version and submit it for State acceptance trials in the third quarter of the year.

The bomber was required to have a top speed of 1,400 km/h (869 mph), a service ceiling of 16,000-17,000 m and climb to 10,000 m in three minutes flat. Maximum range at 12,000-14,000 m with a 1,200-kg bomb load – one nuke, that is – was specified as 2,200-2,400 km (1,366-1,490 miles), given a cruising speed of 800-900 km/h

Yakovlev OKB

Above: The second prototype of the Yak-26 (Yak-26-3) as originally flown. Note the unbroken wing leading edge and the change in the leading-edge sweep inboard and outboard of the engines.

Yakovlev OKB

Above: The same aircraft following modifications, with a dogtooth leading edge and no wing fences; note the pilot's rear view mirror. The lack of a ground mapping radar is evident in this view.

Yakovlev OKB

An OFAB-1500M-54 HE bomb is loaded into the bomb bay of the Yak-26-3. Note how the bomb bay doors swing open 180° to provide clearance.

(496-559 mph). The take-off and landing run were to be no greater than 1,100 m (3,610 ft).

The second prototype had the in-house designation Yak-26-3, being the third example built (c/n 03); the second example was apparently a static test airframe. Coded '54 Yellow', it differed outwardly from the original Yak-123-1 in lacking the ventral radome (the PSBN-MA radar was substituted with the Rym-S system) and having the intended tail cannon barbette. The Yak-26-3 was powered by uprated RD-9F engines delivering 2,750 kgp (6,060 lbst) at full military power and 3,800 kgp (8,380 lbst) in full afterburner.

Finally, the aircraft could be used in the strike/close air support (CAS) role, carrying assorted unguided rockets on special launchers in the bomb bay. On short-range sorties in overload configuration additional rockets could be carried on underwing pylons.

As an alternative, the bomb bay could accommodate a battery of four downward-angled rapid-firing 23-mm (.90 calibre) cannons with 300 rpg and a 12,000-rpm aggregate rate of fire. A third option was to fit the Grad (hail; pronounced *grahd*) 'aerial minelaying' system specially designed by OKB-115 for killing off bombers in large amounts. The system consisted of a dispenser for 1,000 parachute-retarded mines to be dropped in the way of enemy bomber formations.

In the course of its flight test programme the Yak-26-3 underwent a major redesign. The wing leading edge was extended and drooped some way outboard of the engine nacelles, producing a prominent dogtooth which energised the airflow over the outer wings, preventing tip stall. The boundary layer fences were deleted at the same time. Variable-incidence tailplanes were introduced for pitch trimming at varying speeds and altitudes, and stabiliser sweep at quarter-chord was increased to 55°. The cockpit canopy was equipped with a rear view periscope.

The redesigned wings and uprated engines improved the aircraft's performance and handling. In 1956 the Yakovlev OKB and LII held a joint test programme aimed at determining the Yak-26-3's top speed and range. Generally the aircraft received a thumbs-up, but still the type's flaws were not completely eliminated.

In 1956 the first prototype (Yak-123-1) was also modified. Retaining the original wings and powerplant, it received variable-incidence stabilisers, a tail gun barbette and a pilot's rear view periscope *à la* Yak-26-3; additionally, the navigator's station was redesigned in an effort to improve visibility, resulting in an almost completely glazed nose. In this guise the aircraft participated in the 1956 Tushino flypast together with the Yak-121 interceptor prototype.

By the end of 1956 MMZ No.30 had completed the entire pre-production batch of Yak-26s. However, the military were turned off by the test results and refused to accept the bombers. Aleksandr S. Yakovlev managed to reach a compromise; the VVS formally accepted the entire production run but from then on it was up to the OKB how to use the aircraft. Eventually three of the pre-production Yak-26s were retained by the OKB, ostensibly for development work in order to eliminate the shortcomings discovered during State acceptance trials. Two aircraft were transferred to LII and one more went to the

Yakovlev OKB

Above: The former Yak-123-1 following conversion as the Yak-26-1. The nose glazing and the wing dogtooth have been redesigned again. Note the black anti-glare panels on the engine nacelles and the striped pitot boom.

Moscow Aviation Institute as a ground instructional airframe. The fate of the remaining four bombers is not known.

Yak-26-1 (Yak-123-1 further modified) tactical bomber prototype

Changes progressively incorporated on the first and second prototypes gradually brought the Yak-26 closer to the Air Force's demands. In early 1957 the first prototype (Yak-123-1, '50 Yellow') underwent checkout tests at NII VVS. By then the aircraft had undergone further modifications and had been redesignated Yak-26-1. Firstly, it was re-engined with RD-9F turbojets.

Secondly, the wings were modified – albeit differently than on the second prototype. The outer wing leading edge was extended 250 mm ($9\frac{27}{32}$ in) and drooped 100 mm ($3\frac{15}{16}$ in). This prevented tip stall and reduced drag at high alpha but also improved cruise performance and aerodynamics in maximum altitude/minimum speed mode. Unlike the Yak-26-3, however, the boundary layer fences were retained (the dogtooth was in line with the inboard fence). Both wings featured bendable trailing edge trim tabs which were adjusted individually to neutralise the aircraft's tendency to drop a wing caused by the aforementioned torsional stiffness problem). The ailerons were moved inboard 1.3 m (4 ft $3\frac{3}{16}$ in) from their original location at the wingtips; roll control was assisted by spoilers on the wing undersurface.

Thirdly, the navigator/bomb-aimer's station was totally redesigned once more. The glazing framework was different and the top of the nose was also glazed – right up to the entry hatch; the nose terminated in a long pointed air data boom. The new design turned out to be successful and was later used on production Yak-27R reconnaissance aircraft (in slightly modified form) and initial-production Yak-28 bombers. The PSBN-MA radar was retained.

The tail gun barbette was deleted and replaced by a vertically split tailcone whose clamshell doors enclosed a brake parachute to shorten the landing run. A new wheel braking system and an alpha warning system were also provided.

The modifications increased the AUW to 11,500 kg (25,350 lb). The aerodynamic refinements and uprated engines improved the aircraft's performance considerably: during trials the Yak-26-1 attained a top speed of 1,400 km/h (869 mph) at 10,000-12,000 m (32,810-39,370 ft) and a range of 2,050 km (1,273 miles). Handling was also improved.

The Yak-26-3 also participated in the checkout trials. Between them the two prototypes made 33 flights. The modified version had the NATO codename *Flashlight-B*.

Given the satisfactory results of the trials, the next logical thing to do would be to launch production of the upgraded Yak-26. By then, however, the Yakovlev OKB was already working on a much more capable tactical bomber. Designated Yak-129, this aircraft would be powered by more powerful Tumanskiy R11-300 afterburning turbojets and have a maximum bomb load of 3,000 kg (6,610 lb). Thus the VVS preferred the as-yet unflown Yak-129 (the future Yak-28) – a decision later proven wise.

Yak-122 tactical reconnaissance aircraft prototype

The abovementioned Council of Ministers directive No.616-381 of 30th March 1955 and MAP order No.240 of 5th April cancelling the Yak-2AM-11 project tasked the Yakovlev OKB with developing an improved tactical photo reconnaissance aircraft powered by two RD-9F afterburning turbojets. The aircraft was to commence State acceptance trials in the second quarter of 1956.

The operational requirements were basically the same as for the Yak-123 bomber developed pursuant to these documents. Top speed was specified as 1,400 km/h (869 mph) in full afterburner and 1,250 km/h (776 mph) full military power, the service ceiling as 16,000-17,000 ft (52,490-55,770 ft), climb time to 10,000 m (32,810 ft) as three minutes and maximum range as 3,500 km (2,179 miles) or 3,000 km (1,863 miles) with 7% fuel reserves. The camera suite comprised two AFA-33/50s, AFA-33/75s or AFA-33/100s (depending on the mission) on AKAFU automatic tilting mounts, one AFA-33/20 oblique camera in the nose and one AFA-33/50 or AFA-33/75 vertical camera.

Designated Yak-122, the first Soviet supersonic reconnaissance aircraft evolved from the subsonic Yak-25R in the same way that the Yak-123 was derived from the

Yakovlev OKB

The Yak-122 was the Soviet Union's first supersonic reconnaissance aircraft. It retained the short wheelbase of the Yak-25R from which it was derived.

Yakovlev OKB

Above: '55 Yellow', the Yak-121 supersonic patrol interceptor prototype, during manufacturer's flight tests. Note the large ogival radome.

Yakovlev OKB

This uncoded aircraft is probably the first production cannon-armed Yak-27 interceptor (c/n 0201); note the modified wings with dogtooth.

Yak-125B. In fact, the Yak-122 prototype ('45 Yellow') was converted from one of the pre-production Yak-25Rs. Thus, despite having a similar forward fuselage design to the Yak-123, the Yak-122 retained the short wheelbase and single-wheel aft-retracting nose gear unit of the *Flashlight*.

As the intended RD-9F engines were unavailable at the time of completion, the aircraft had to be fitted with RD-9AK engines. The wings and tail unit were identical to those of the Yak-123 and the Yak-121 interceptor described later – that is, the Yak-122 had an unbroken wing leading edge, dual boundary layer fences and fixed-incidence stabilisers. The armament consisted of one NR-23 cannon on the starboard side with 50 rounds.

The ogival nose was more extensively glazed than on the Yak-123-1 as originally built, though to a lesser extent than on the Yak-26-1 and Yak-26-3. Instead of an elliptical optically flat lower glazing panel the Yak-122 had two curved trapezoidal panels, with a small Perspex blister for an optical sight immediately aft of them. The oblique camera was mounted in the nose immediately aft of the RSO's station, with a dorsal access hatch and a camera port protected by a door on either side, while the vertical camera and tilting camera installation were accommodated in the centre fuselage.

The Yak-122 shared the fate of the Yak-123 (Yak-26) bomber; being a parallel development, it shared the bomber's design flaws and was surpassed by more capable types before it had a chance to mature. Unlike the Yak-26, however, no pre-production batch was built.

The Yak-122 was subsequently converted into the first prototype of the Yak-27R tactical reconnaissance aircraft which did enter production as described below.

Yak-27 family

Yak-121 (Yak-27) interceptor prototype

The production Yak-25M could not meet the growing demands of the Soviet Air Force. Hence the Yakovlev OKB kept working on a more capable interceptor. The abovementioned CofM directive No.616-381 of 30th March 1955 and MAP order No.240 of 5th April also envisaged the development of a supersonic interceptor derivative of the Yak-25 and the construction of two prototypes. The first prototype would be powered by RD-9AK afterburning turbojets, equipped with the Sokol-M radar and armed with two 37-mm N-37L cannons with 50 rpg (plus 24 ARS-57 FFARs in overload configuration). In contrast, the second prototype would have uprated RD-9F engines, a Sokol-2 radar and an armament consisting of beam-riding or passive IR homing K-8 air-to-air missiles. The two aircraft were to commence State acceptance trials in December 1955 and the second quarter of 1956 respectively.

The new two-seat patrol interceptor received the manufacturer's designation Yak-121. Coded '55 Yellow', the first prototype was completed at MMZ No.115 in 1955. The aircraft's main distinguishing feature was the long ogival radome. The sliding cockpit canopy was identical to that of the Yak-25M but the fixed windshield was more sharply raked, with an elliptical bulletproof windscreen. The wingspan was also identical but the wing planform was slightly different and wing area was slightly enlarged due to the addition of LERXes. The landing gear remained unchanged. The nacelles were redesigned to accommodate the new engines. The first prototype completed its manufacturer's flight test programme in the spring of 1956.

As with the Yak-26 tactical bomber, the Air Force kept pressuring the Yakovlev OKB to commence State acceptance trials of the definitive RD-9F-powered version. However, construction of the second prototype and hence the test schedule were delayed by late delivery of the engines. The second prototype's certification trials had to be postponed until the third quarter of the year.

Yakovlev OKB

Production cannon-armed Yak-27s differed from the prototypes in having a double-curvature radome tipped by a long pitot boom. Note the paint job on the nose similar to that of the Yak-25.

Yakovlev OKB

Above: '58 Yellow', the prototype of the missile-armed Yak-27K-8.
Below: Close-up of the Bisnovat K-8 missile under the wing of '58 Yellow'.

Yakovlev OKB

In the summer of 1956 the Yak-121 took part in the traditional Aviation Day flypast at Tushino together with the modified Yak-123-1 bomber prototype. In the West the aircraft was misidentified first as the Yak-25K (which, as the reader knows, was quite a different aircraft) and then as the Yak-27 (almost correct but still not quite). The NATO reporting name was *Flashlight-C*.

The Yak-121 was effectively the first prototype of a new interceptor which received the service designation Yak-27. Due to the installation of the interim RD-9AK engines it was underpowered. Hence even before the State acceptance trials began the Saratov aircraft factory No.292 was tasked with building a pre-production batch of Yak-27s which would incorporate all the changes recommended by the State commission. The aircraft would be powered by RD-9F engines.

Yak-27 production interceptor (*izdeliye* 21)

In Saratov the new interceptor received the in-house product code *izdeliye* 21. Production began in 1956; incorporating various changes based on the Yak-121's test results and the Air Force's revised SOR issued in 1955.

The Yak-27 was built in two versions differing in armament (cannons or AAMs). Individual Yak-27s differed in detail. The very first production examples had an identical wing design to the Yak-121 prototype from which they differed in having RD-9F engines, a slightly recontoured radome terminating in a long pitot angled slightly downward, and 30-mm (1.18 calibre) Nudel'man/Rikhter NR-30 cannons.

The first production aircraft (c/n 0201) was test flown in Saratov by factory test pilot Choovin in July 1957. As production (and tests) progressed the increased-chord outer wings with a dogtooth were introduced; the first Yak-27 to have the redesigned wings was c/n 0204.

The aircraft passed its State acceptance trials with good results, but even as it did it was clear that the cannon-armed version was doomed. Firstly, in the late 1950s the Soviet military were into missile systems, and any viable new interceptor was automatically required to have missile armament. Secondly, the rival Mikoyan and Sukhoi bureaux had by then produced several interceptor prototypes which outperformed the Yak-27 by far. Thus, logically enough, the 'pure' (cannon-armed) *Flashlight-C* was not proceeded with.

Yak-27K (Yak-27K-8) interceptor

Development of the cannon-armed and missile-armed versions of the Yak-27 progressed in parallel. The chosen missile was the Bisnovat K-8; consequently, the 'missilised' version received the designation Yak-27K. Coded '58 Yellow', the Yak-27K prototype was converted from a pre-production *Flashlight-C* at the Yakovlev OKB's experimental shop.

Structurally the Yak-27K was very similar to the Yak-121. The ogival radome was slightly shortened and the aircraft had variable-incidence stabilisers but as yet no dogtooth. The missile pylon design was borrowed from Yak-25K c/n 0119 (that is, the pylons had a sweptback leading edge).

In this guise the Yak-27K underwent manufacturer's flight tests in September-November 1956. Later the wings were modified *à la* Yak-26-1 (that is, with both dogtooth and boundary layer fences) and the aircraft was tested again at LII in May-June 1957.

Yakovlev OKB

Yak-27K c/n 0201 at the factory airfield in Saratov. The nose of the port missile peeks out from behind the engine air intake.

Yakovlev OKB

Above: In 1957 the original Yak-121 prototype was converted into the Yak-27V high-altitude interceptor.

K-8 missiles were carried throughout the test programme.

Production aircraft built in Saratov differed from the prototype in having a recontoured radome tipped with a pitot. They were equipped with the Sokol-2K radar, a version of the RP-6 optimised for the K-8 AAM, and the ***Voz**dookh-1* (Air-1) ground controlled intercept (GCI) system. A brake parachute was provided to shorten the landing run.

Production Yak-27Ks underwent manufacturer's flight tests in Saratov throughout 1958. The test programme included speed envelope exploration, aileron efficiency checks and trim/handling checks with the elevator and aileron hydraulic actuators switched off. Once again, K-8 missiles were always carried.

In addition to three Yak-27Ks built as such, the first production 'pure' Yak-27 (c/n 0201) was converted to Yak-27K standard and rewinged in 1957 (it had new outer wings with dogtooth, boundary layer fences and relocated ailerons).

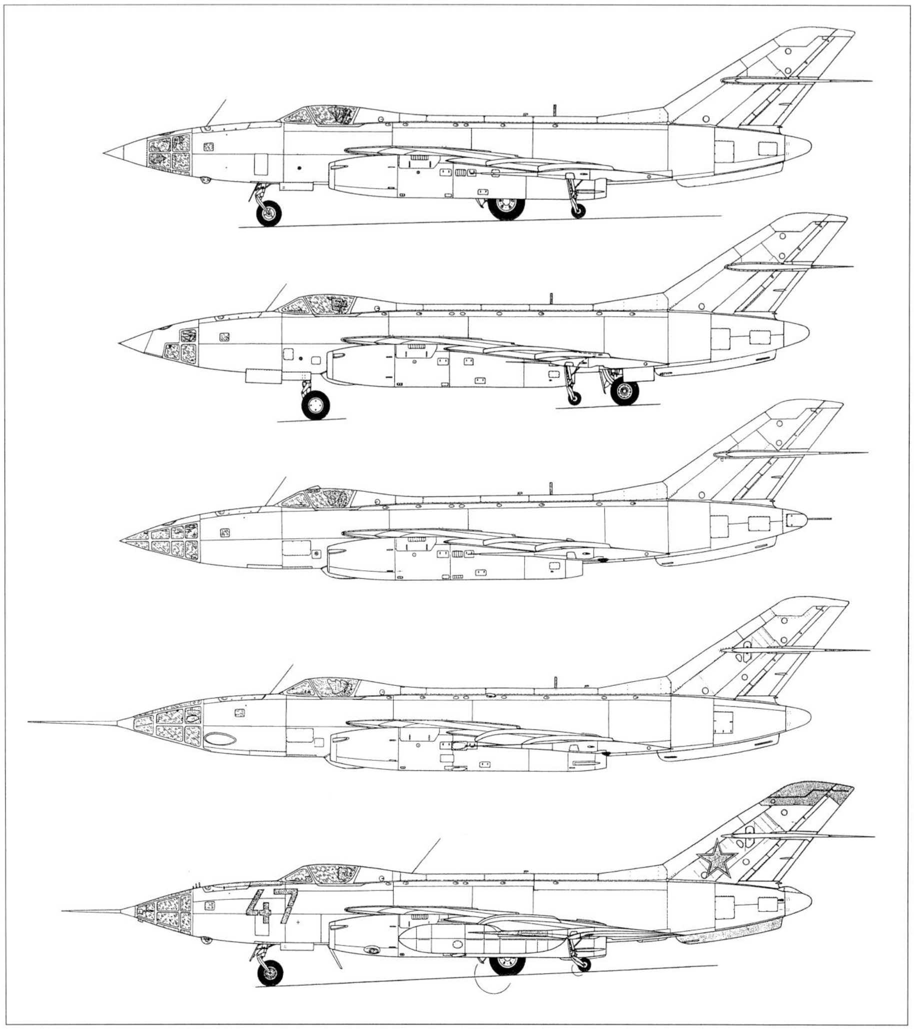

Top to bottom: The Yak-122, the Yak-123-1 as originally flown, the Yak-26-3, the Yak-26-1 in ultimate configuration and the Yak-27R.

Between 29th December 1956 and 29th May 1958 the K-8 missile system underwent manufacturer's tests, including live firing trials, at the NII VVS facilities in Akhtoobinsk in southern Russia. Together with the two Yak-25Ks, two Yak-27Ks – the prototype ('58 Yellow') and the Yak-27 *sans suffixe* conversion (c/n 0201) – took part in these trials.

All in all, the Yak-27Ks made 74 test flights between them during the various trials programmes, including 37 with K-8 missiles. Seven launches, including three live ones, were made against PM-1 parachute-retarded IR/radar signature targets and M-28 remote-controlled target drones (converted IL-28 bombers). Maximum target detection range and target lock-on range were 33-35 km (20.5-21.7 miles) and 23-30 km (14.3-18.6 miles) respectively.

The Yak-27's top speed was recorded as 1,270 km/h (788 mph) in full afterburner and 1,110 km/h (689 mph) at full military power. With a 10,680-kg (23,545-lb) take-off weight and a 9,400-kg (20,720-lb) AUW the aircraft could reach 16,300 m (53,480 ft) in full afterburner. The maximum Mach number attained during trials was 1.25.

Generally the Yak-27K earned positive comments from the pilots who flew it. Yet it was destined never to enter service because a competitor had it outclassed. In 1958 the PVO adopted the new Su-9-51 aerial intercept weapons system built around the Sukhoi Su-9 interceptor; initially the aircraft was armed with K-51 AAMs but these were to be substituted by K-8s later on. Since the Su-9 outperformed the Yak-27K by a considerable margin, the choice was obvious.

Yak-27V experimental high-altitude interceptor

In 1955 the Soviet government ordered the development of a single-seat high-altitude interceptor with a mixed powerplant (a turbojet engine and a liquid-fuel rocket booster) enabling the aircraft to reach altitudes in excess of 20,000 m (65,620 ft). The Mikoyan OKB evolved the Ye-2A experimental swept-wing fighter (a precursor of the MiG-21F) into the mixed-power Ye-50 on which the original Mikulin AM-9B was replaced with an AM-9Ye and augmented with a 1,300-kgp (2,866-lbst) S-155 rocket motor developed by L. S. Dooshkin at the base of the fin.

On 25th August 1956 the Council of Ministers issued another directive (No.1195-613) ordering the development of a single-seat mixed-power interceptor. Together with the other Soviet 'fighter makers' the Yakovlev OKB was tasked with developing such an aircraft.

The SOR for the new Yakovlev interceptor designated Yak-27V (*vysotnyy* – high-altitude) was signed on 16th November 1956. This document stated that the fighter was to have a new Almaz (Diamond) radar linked to an ASP-5NM automatic gunsight, a Gorizont-1 or Vozdookh-1 GCI data link system, an AP-28 autopilot and other special equipment. The armament was to consist of two NR-30 cannons or other weapons with a higher rate of fire (13,000-15,000 rpm). Alternatively, the fighter would be armed with unguided rockets.

The Yak-27V prototype entered flight test in late April 1957. the aircraft was converted from the original Yak-121 interceptor prototype, retaining the tactical code '55 Yellow'. The extensive conversion involved installation of RD-9AKYe afterburning turbojets (a wing-mounted version of the Ye-50's main engine; and an S-155 liquid-fuel rocket booster supplanting the tailcone.

Changes associated with the rocket booster were made to the fuel system and pneumatic system. Airframe parts and various equipment items located near the booster and its oxidiser filler caps were protected against the corrosive chemicals and their vapours.

Outwardly the Yak-27V looked like a two-seater. In reality, however, the rear cockpit had been eliminated to save weight, as was part of the equipment and the brake parachute. The radome was replaced by an ogival metal nosecone terminating in a long pitot. Finally, the aircraft was retrofitted with increased-chord outer wings featuring a dogtooth and with variable-incidence stabilisers. The armament remained unchanged, comprising two NR-30 cannons. The aircraft made its maiden flight on 26th April 1957 and the manufacturer's flight tests lasted until 3rd July.

The flight tests of the S-155 rocket booster were preceded by a series of ground runs and went according to a special programme developed by Dooshkin's OKB-1 and approved by the Yakovlev OKB.

The Yak-27V's MTOW was 12,050 kg (26,565 lb). The mixed powerplant necessitated a special flight profile. First, the aircraft climbed to 9,000 m (29,530 ft), travelling at 870 km/h (540 mph) with the engines at full military power. Then the afterburners were engaged and the aircraft climbed to 14,000 m (45,930 ft), accelerating to 940-960 km/h (583-596 mph). Finally, the rocket booster was switched on at 14,000 m, operating for 264 seconds at minimum thrust or 160 seconds at full thrust.

During trials the Yak-27V reached a maximum altitude of 23,500 m (77,100 ft) and a top speed of Mach 1.8; it could cruise at 20,000 m (65,620 ft) for up to three minutes at Mach 1.5-1.6 with the booster at minimum thrust. Unfortunately, the time required to reach various altitudes with the booster operating was not recorded during trials.

The Yak-27V was under test for two years. In spite of the promise it held, the aircraft remained a one-off. There was little hope of an aircraft with a mixed powerplant entering production because the rocket booster was rather troublesome in day-to-day operation.

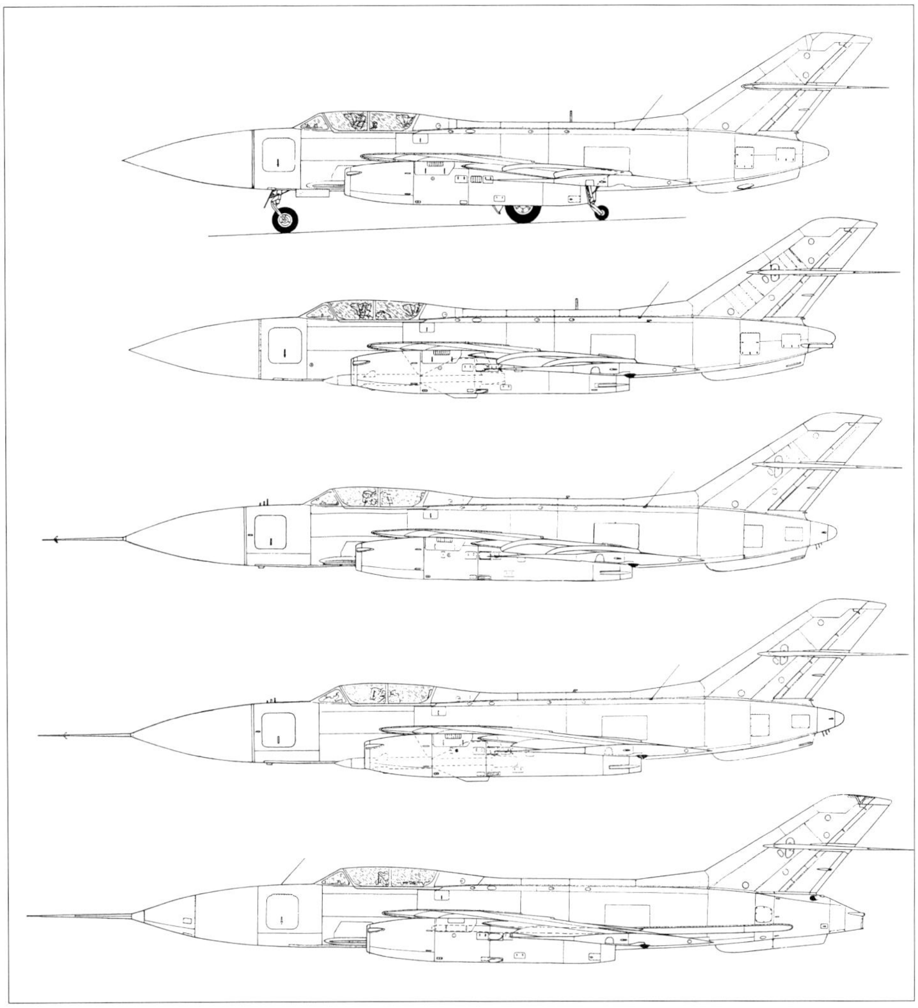

Top to bottom: The Yak-121, the Yak-27K-8, the cannon-armed Yak-27, a production Yak-27K and the Yak-27V.

Yak-27R tactical photo reconnaissance aircraft prototypes

The OKB's efforts with the Yak-27 were not in vain, as one version did enter production, after all. In 1958 the Yak-122 prototype underwent a major conversion, becoming the prototype of the Yak-27R tactical reconnaissance aircraft ('45 Yellow').

The navigator/RSO's station was reworked and the result was a cross-breed between the Yak-26-1 bomber prototype and the original Yak-122. The entire nose was glazed and terminated in an air data boom *à la* Yak-26-1, but the curved lower glazing panels and ventral sighting blister were retained. The wings and tail unit were modified to match those of the Yak-26-1 and the pre-production Yak-27/Yak-27K interceptors

Yakovlev OKB

'45 Yellow', the Yak-27R reconnaissance aircraft prototype.

Yakovlev OKB

Above: A still-uncoded production Yak-27R pictured during a pre-delivery test flight. Note the extended wingtips and the PV-2R optical sight in the ventral blister.

(that is, the Yak-27R had a dogtooth and two boundary layer fences on each outer wing, plus variable-incidence stabilisers). The aircraft was powered by RD-9Fs. The landing gear was identical to that of the interceptor and previous reconnaissance versions.

The second prototype ('46 Yellow') incorporated a few changes. The framework of the RSO's station glazing was altered, as was the dielectric fin cap. The outer boundary layer fences were deleted and the inner fences made taller and reshaped to compensate for this. Wing span was increased slightly by extending the wingtips beyond the outrigger strut fairings, and the under-nose sighting blister was enlarged to give the optical tracker head more room.

In this guise the aircraft began State acceptance trials, showing a top speed of 1,285 km/h (798 mph) at an AUW of 12,500 kg (27,550 lb). The results were good and the aircraft was cleared for production.

Yakovlev OKB

Above: A pre-production Yak-27R coded '01 Red' on test at LII. Note the highly unusual grey nose colouring *à la* Yak-25.

Yakovlev OKB

This view illustrates the difference in the shape of the dielectric fin cap between prototypes and production Yak-27Rs. This example has a small skid on the main gear unit between the wheels.

Yak-27R production model

The Saratov aircraft factory produced 180 Yak-27Rs which saw service with the tactical arm of the VVS for the next 20 years. In keeping with the day's practice the type was allocated the NATO reporting name *Mangrove* (in the 'miscellaneous' category).

Production Yak-27Rs were basically identical to the second prototype. As compared to the Yak-25R and Yak-122, the production Yak-27R could carry a much wider range of equipment. The four cameras permitted photography at subsonic and supersonic speeds in daytime VMC conditions.

A fixed AFA-41/10 vertical camera was used for small-scale strip photography at altitudes of 300-400 m (980-1,310 ft) and for taking overall views of the target area at 1,000-16,000 m (3,280-52,490 ft) for the purpose of identifying the relative position of objects photographed to a larger scale by the other cameras. Two AFA-42/75 or AFA-42/100 cameras on AKAFU-33M automatic tilting mounts permitted one-, two-, three- or four-strip photography at 1,000-16,000 m in subsonic flight or 12,000-14,500 m (39,370-47,572 ft) in supersonic mode.

The camera suite was changed in the course of production. For instance, aircraft with odd c/ns in each batch had AFA-42/100 vertical cameras, while those with even numbers were fitted with AFA-42/75s. Moreover, even-numbered aircraft in the first twelve batches were equipped with the earlier AFA-37 instead of AFA-42/75s.

Initially production Yak-27Rs had RD-9F Srs 1 engines. From c/n 1010 onwards these were replaced with Srs 2 engines which could be restarted more easily above 5,500 m (18,040 ft) in the event of a flame-out but had a higher fuel consumption. Range was 1,770 km (1,099 miles) at 10,000 m (32,810 ft) and 1,807 km (1,122 miles) at 12,000 m (39,370 ft).

Yak-27R with drop tanks

In the summer of 1962 NII VVS tested a Yak-27R (c/n 0710) equipped with two 'wet' hardpoints under the outer wings for carrying two 1,050-litre (231 Imp gal) drop tanks. These were so-called slipper tanks adhering directly to the wing undersurface.

The 2,100 litres (462 Imp gal) of extra fuel extended the *Mangrove's* range to 2,380 km (1,478 miles), providing the tanks were jettisoned when they ran dry, and increased the take-off weight from 11,970 kg (26,390 lb) to 13,633 kg (30,055 lb). On the minus side, top speed with the drop tanks in place was limited to Mach 0.9 and the service ceiling was reduced from 16,550 m (54,300 ft) to 13,450 m (44,130 ft).

Production Yak-27Rs were armed with a single NR-23 cannon with 50 rounds.

Yak-27 reconnaissance equipment testbed

In 1958 a production example of the Yak-27R was used for the development of reconnaissance equipment intended for the Yak-28R. The aircraft was fitted with the RBP-3 bombing sight, AP-28 autopilot and DAK-I remote-controlled astrocompass.

Yak-27RN reconnaissance aircraft (prototype?)

No information is available on this version of the Yak-27, apart from a brief statement that this reconnaissance aircraft underwent flight testing.

Yak-27R (Yak-27LSh) wheel/ski landing gear testbed

A production Yak-27R was used to test an experimental heavy-duty landing gear designed for operation from semi-prepared snow-covered airstrips. The twin mainwheels were augmented with a large skid which could be raised by a hydraulic ram when not in use, while the nose unit was reinforced and equipped with twin K-262 wheels measuring 600 x 125 mm (23.62 x 4.92 in). The landing gear was non-retractable because the twin nosewheels would not fit into the standard wheel well. The aircraft was appropriately designated Yak-27LSh (***lyzh**noye **sh**as**see*** – ski undercarriage). Trials began in late 1963 and continued until the spring of 1964; at one time the aircraft was also flown with a standard single-wheel nose gear unit.

Yak-27F development aircraft

This was the final spin-off of the Yak-27; the meaning of the F suffix and the aircraft's role are not known. Trials began in the summer of 1960 It is possible that the Yak-27F was a one-of-a-kind equipment testbed with two downward-looking TV cameras in a large boxy fairing under the aft fuselage and a small radome aft of the nose gear unit; this aircraft ('08 Red', c/n 0507) until recently resided in the open-air museum at Moscow-Khodynka. This may well have been an example of a version fitted with TV cameras on a flexible mount which was mentioned in the Russian aeronautical press

Yak-28

Yak-129 (Yak-28-1) tactical bomber prototype

On 28th March 1956, when the Yak-26 tactical bomber was in the middle of its flight test programme, the Communist Party Central Committee and the Council of Ministers issued directive No.424-261 tasking the Yakovlev OKB (OKB-115) with developing a new high-altitude supersonic light tactical bomber based on the Yak-26.

Specifications of the Yak-27 variants

	Yak-121	Yak-27K	Yak-27R
Length overall	17.335 m (56 ft 10½ in)	17.335 m (56 ft 10½ in)	19.0/18.52 m (62 ft 4 in/60 ft 9 in)
Wing span	10.964 m (35 ft 11½ in)	10.964 m (35 ft 11½ in)	11.82 m (38 ft 9¾ in)
Powerplant	2 x RD-9AK	2 x RD-9F	2 x RD-9F
Engine thrust, dry/reheat, kgp (lbst)	2 x 2,600/3,250 (2 x 5,730/7,160)	2 x 2,750/3,800 (2 x 6,060/8,380)	2 x 2,750/3,800 (2 x 6,060/8,380)
Empty weight, kg (lb)	7,005 (15,443)	n.a.	n.a.
TOW, kg (lb)	9,740 (21,470)	10,680 (23,545)	11,970 (26,390)
Top speed, km/h (mph)	1,235 (767)	1,270 (789)	1,285 (798)
Service ceiling , m (ft)	15,800 (51,840)	16,300 (53,480)	16,550 (54,300)
Range, km (miles)	n.a.	n.a.	2,380 (1,480)
Armament	Two N-37L cannons	Two K-8 AAMs	One N-37L cannon

The aircraft was to have a crew of two and be powered by two Tumanskiy R11-300 afterburning turbojets delivering a thrust of 3,900 kgp (8,600 lbst) dry and 5,300-5,500 kgp (11,680-12,125 lbst) reheat. Top speed at 10,000 m (32,810 ft) was specified as 1,500-1,600 km/h (931-993 mph) in full afterburner and 1,200-1,300 km/h (745-807 mph) at full military power. The bomber was required to have a service ceiling of 16,000-17,000 m (52,490-55,770 ft). The required take-off and landing run were 1,000 m (3,280 ft) and 1,100 m (3,610 ft) respectively. The normal bomb load was 1,200 kg (2,645 lb, increasing to 3,000 kg (6,610 lb) in overload configuration, and take-off weight was 12,000-13,000 kg (26,455-28,660 lb). For self-defence the bomber was to have a tail cannon barbette.

Development of the new bomber, which received the manufacturer's designation Yak-129, took 18 months. The work had scarcely begun, however, when on 15th August 1956 the Council of Ministers confused matters considerably by issuing directive No.1115-578 which ordered the Yakovlev OKB to develop in parallel a version of the Yak-129 powered by Klimov VK-11 turbojets rated at 6,100 kgp (13,450 lbst) dry and 9,000 kgp (19,840 lbst) reheat. Hence the required performance figures were now much higher. In particular, top speed rose to 2,500 km/h (1,552 mph) in full afterburner and 1,300 km/h (807 mph) at full military power, while service ceiling was increased to 20,000-21,000 m (65,620-68,900 ft).

The bomb load remained the same (normal, 1,200 kg; maximum, 3,000 kg). The first VK-11-powered prototype was to begin manufacturer's flight tests in the first quarter of 1958, commencing State acceptance trials in the fourth quarter of the year.

The VK-11 had been under development since 1954 as a joint effort of two engine design bureaux led by Vladimir Ya. Klimov and Nikolay G. Metskhvarishvili. Prototypes of the VK-11 commenced bench running in 1956. Apart from the Yakovlev bomber, the engine was to power two high-speed heavy interceptors – the Mikoyan/Gurevich Ye-150 and the rival Sukhoi T-37.

Eventually, however, neither aircraft received the VK-11, presumably due to development problems with the engine. The Ye-150 flew with the even more powerful Tumanskiy R15-300 turbojet while the T-37 built around the same engine was scrapped without ever being flown. Since the Ye-150/ Ye-152 series did not progress beyond the

Yakovlev OKB

'57 Yellow', the first prototype Yak-28 (Yak-129) converted from a pre-production Yak-26. This view shows the high-wing layout, the lack of bomb-aiming radar and the circular intakes of the R11A-300 engines.

Yakovlev OKB

The uncoded second prototype (Yak-28-2) differed in having R-11AF-300 engines with elliptical intakes and fixed shock cones. Note also the forked pitot.

prototype stage, no aircraft existed for the VK-11 and no decision to launch production was made. Hence the Soviet government had to settle for the much smaller R11-300 as a powerplant for the Yak-129, of which the Yakovlev OKB was duly notified.

General Designer Aleksandr S. Yakovlev was openly sceptical about the new assignment, believing the Yak-129 could not be developed into a viable supersonic bomber. However, some Yakovlev OKB employees, including Yevgeniy G. Adler, thought differently. After analysing the Air Force's critical comments on the Yak-26 they decided that the basic design should be retained rather than start from scratch. In order to correct the bomber's deficiencies the wings should be stiffened and the ailerons moved inboard to where torsional stiffness was greater; this would eliminate aileron reversal, or at least delay its onset until higher speeds.

The wing planform was to be changed by increasing leading-edge sweep and eliminating trailing-edge sweep between the engine nacelles and the fuselage. This allowed the wing spars to be made taller and stronger without affecting the thickness/chord ratio on the inner wings or sweep on the outer wings. Also, the inefficient Schrenck flaps on the inner wings were to be replaced by more effective Fowler flaps.

The R11-300 turbojet had bigger dimensions than the RD-9F. In order to provide adequate ground clearance for the larger nacelles the wings had to be moved up; thus the Yak-129 would have shoulder-mounted wings as opposed to the Yak-26, which was a mid-wing monoplane. The new layout gave the added bonus of increasing the bomb bay volume. Also, the air intake design needed to be optimised for cruising at high Mach numbers.

In order to speed up prototype construction one of the three pre-production Yak-26 bombers owned by the Yakovlev OKB was set aside for conversion. Coded '57 Yellow', the rebuilt aircraft was rolled out in early 1958. Despite its apparent similarity to the upgraded Yak-26-1 of 1957, it was, in effect, a completely new bomber.

The Yak-129 combined the fuselage and tail unit of the Yak-26-1 and Yak-27 with some new features as described above. The shoulder-mounted wings were likewise of basically trapezoidal shape and swept back 45° at quarter-chord, and the outer wing leading edge was extended and drooped to create a dogtooth, delaying tip stall at high angles of attack. However, wing area was increased to cater for the higher gross weight by increasing leading-edge sweep and eliminating trailing-edge sweep on the inner wings. Also, wing anhedral was perceptibly increased. The Schrenck flaps with a skewed rotation axis on the inner wings were replaced by single-slotted Fowler flaps providing a much bigger lift increase. The ailerons occupying 29.8% of the wing span and about 8% of the wing area were located immediately outboard of the engine nacelles, just like on the Yak-26/-27.

The Yak-129 was powered by R11A-300 afterburning turbojets rated at 3,900 kgp (8,600st) dry and 4,850 kgp (10,690lbst) reheat. The nacelles were totally new, having a larger diameter and a longer front portion of basically conical shape. The circular supersonic air intakes had a sharp lip and a small two-position centrebody (shock cone). The fuel system comprised seven bladder tanks holding a total of 4,025 litres (885.5 Imp gal).

Minor changes were made to the landing gear: the levered-suspension main unit incorporated an automatic 'kneeling' feature to increase the angle of attack on take-off. A brake parachute was provided to shorten the landing run. Finally, the long conical air data boom on the nose was now tipped with a characteristic forked pitot.

The Yak-129 took to the air on 5th March 1958 with project test pilot V. M. Volkov at the controls. The manufacturer's flight test programme was completed on 4th October, by which time the new bomber had been allocated the service designation Yak-28 (hence the first prototype has also been referred to as Yak-28-1).

With a 1,200-kg (2,645-lb) normal bomb load and 3,200 kg (7,050 lb) of fuel the Yak-129 grossed at 12,885 kg (28,410 lb) – a sizeable increase over the Yak-26's 11,500 kg (25,350 lb). Top speed in full afterburner was recorded as 1,500 km/h (931 mph) at 10,000 m (32,810 ft). The specified service ceiling rose from the Yak-26's 16,000-17,000 m (52,490-55,770 ft) to 17,800 m (58,400 ft). In reality the Yak-129 reached only 16,300 m (53,480 ft) during manufacturer's flight tests because the afterburners flamed out at high altitude, but even that was a modest improvement over the predecessor's 16,000 m (52,490 ft).

Pilots flying the prototype reported that the aircraft handled satisfactorily and had good field performance; the take-off and landing run was 850-950 m (2,880-3,120 ft). Also, the redesigned navigator/bomb-aimer's station with the 'fixed' ejection seat earned unanimous praise from the Yakovlev and LII navigators alike.

On the minus side, directional stability was poor, so two additional boundary layer fences on the inner wings and two shallow splayed ventral fins were fitted in the course of the trials to correct this. Also, the wingtips were extended beyond the outrigger strut fairings/anti-flutter weights by analogy with the second prototype Yak-27R.

When the Yak-129 commenced State acceptance trials, however, the Air Force voiced quite a few complaints. The engineers tried to correct the aircraft's flaws on the second prototype as described below.

Yak-28-2 and Yak-28-3 tactical bomber prototypes

The second prototype, which had the in-house designation Yak-28-2, was likewise converted from one of the three pre-production Yak-26s retained by the OKB. It differed from the first prototype mainly in being powered by R11AF-300 engines delivering 5,750 kgp (12,680 lbst) in full afterburner, the F standing for *for**see**rovannyy* – uprated. This engine was derived from the R11F-300 in the same way as the original R11A-300 had been derived from the R11-300, differing only in accessory gearbox location.

Because of the higher thrust and increased mass flow the air intakes had to be redesigned: they were cut back and had an elliptical section with the larger axis vertical; the movable shock cone was replaced by a fixed spike and a translating intake leading edge. The spike had a rather unusual shape with a conical tip followed by a perfectly cylindrical section. The nacelles featured

Laval nozzles which were expected to give an additional thrust increase.

One of the Yak-28-2's major deficiencies was poor acceleration in the Mach 1.3-1.6 range. Hence at high ambient temperatures the fuel consumption was so high that, if the aircraft was flown at top speed for an extended time, there might not be enough fuel left to return home. To reduce power losses the intakes were redesigned – the engineers reverted to the two-position shock cones which were moved forward automatically or manually when the aircraft reached Mach 1.45. Still, between Mach 1.3 and Mach 1.45 the Yak-28 was reluctant to accelerate even in full afterburner. Also, the aluminium skinning on the inner wings was replaced by steel on the Yak-28-2.

In the course of the manufacturer's flight tests the Yak-28-2 flown by OKB test pilot V. P. Smirnov made a landing and take-off at the Tret'yakovka dirt strip near Moscow with a 1,500-kg (3,310-lb) dummy bomb. Tests showed that the aircraft's top speed was 950 km/h (565 mph) at sea level and 1,250 km/h (776 mph) at 12,000 m (39,370 ft). Later, test pilot S. G. Petukhov and navigator N. M. Shipovskiy ferried the aircraft to the NII VVS test facility in Akhtoobinsk where the Yakovlev OKB and the Air Force began preparations for supersonic bombing trials. These were eventually conducted with some satisfactory results.

After this first successful supersonic drop Aleksandr S. Yakovlev ordered that bombing accuracy be increased to meet the existing standards for subsonic bombers. When this had been accomplished, Yakovlev decided to transfer the aircraft to NII VVS for State acceptance trials.

The trials went fairly quickly, since the Yak-28 was the only Soviet bomber capable of cruising at about 1,400 km/h (869 mph) at altitudes in excess of 10,000 m (32,810 ft) with a normal bomb load of 1,200 kg (2,645 lb). The bomber was recommended for production, even though it had only an optical bombsight because the intended navigation/ground mapping radar had not yet entered production. MAP aircraft factory No.39 in Irkutsk was chosen to build the Yak-28.

Coded '56 Yellow' and designated Yak-28-3, the third prototype was probably also converted from a Yak-26, retaining the latter's tactical code. It was largely identical to the second prototype and also participated in the State acceptance trials.

Yak-28 tactical bomber (*izdeliye* B)

Known at the factory by the in-house product code *izdeliye* B, the radarless initial production version was virtually identical to the second prototype and was built in extremely limited numbers. The Yak-28's avionics fit (apart from the mission avionics) remained constant in all subsequent versions, comprising the RSIU-5V command radio, the ARK-10 ADF, the MRP-45P marker beacon receiver, the SOD-57 distance measuring equipment (***stahn**tsiya opredel**en**iya **dahl'**-nosti*), the RSBN-2S Svod (Dome) short-range radio navigation system (***rah**diotekh-**ni**cheskaya sis**tem**a **blizh**ney navigahtsii* – SHORAN) and the SRZO-2M Khrom-Nikel' IFF transponder.

The Yak-28 soon achieved initial operational capability (IOC), replacing the obsolete IL-28 bomber. Probably due to the scarcity of this initial production version it apparently remained unknown to the West, and no NATO reporting name was allocated – as yet.

Yak-28B tactical bomber (*izdeliye* 28B)

Production of the radarless Yak-28 *sans suffixe* was a stopgap measure. Predictably, an extremely small number was ordered due to the aircraft's inadequate targeting equipment, since the Yakovlev OKB had two new versions coming up, equipped with the *Init-siativa-2* (Initiative-2) radar and the Lotos (Lotus) data link system respectively. However, production and service entry of these sub-types was a matter of a few more years and the VVS could not wait that long. Therefore the OKB undertook a crash programme to equip the Yak-28 with the production RBP-3 ground mapping/bomb-aiming radar linked to an OPB-115 optical sight and launch production of this interim version.

The RBP-3 had a 360° field of view and was installed beneath the cockpit, its antenna enclosed by a teardrop radome, just like on the Yak-125B and Yak-26. It enabled bomb delivery in all weather conditions, day and night, and was linked to the autopilot for automatic target approach. To increase range the bomber had two 'wet' hardpoints for carrying 1,000-litre (220 Imp gal) drop tanks, just like the Yak-27R; these 'slipper tanks' were of basically cylindrical shape with a rounded front end and were angled slightly outward to ensure safe separation from the aircraft. The bomber could carry 1,000-3,000 kg (2,204-6,610 lb) of bombs; it was also armed with one forward-firing Nudel'man/Rikhter NR-23 cannon on the starboard side of the forward fuselage.

Designated Yak-28B, the new version entered flight test in 1959. It is not certain what the B suffix stands for; it surely does not denote *bombardi**rov**shchik* (bomber), since the previous versions of the type were bombers anyway. A version designator in alphabetical order appears more logical – but then, there never was a Yak-28A!

After successfully completing its State acceptance trials in 1960 the type was cleared for production. Top speed was 1,900 km/h (1,180 mph) at 12,000-13,000 m (39,370-42,650 ft); below 8,000 m (26,250 ft)

Yakovlev OKB

Yakovlev OKB

Top and above: '74 Yellow', the Yak-28B prototype. Note the R11AF-300 engines and the radome of the RBP-3 bomb-aiming radar aft of the nose gear.

Above: The Yak-28L bomber prototype converted from the original third prototype Yak-28 *sans suffixe* ('56 Yellow'), still with R11AF-300 engines. Note the lack of the radar which is substituted by a DBS-2 Lotos radio command guidance system. The aircraft carries early-model drop tanks with blunt noses.

the bomber could not go supersonic because of dynamic pressure limits. The aircraft had a service ceiling of 16,200 m (53,150 ft) and a range of 1,950 km (1,211 miles) on internal fuel only or 2,630 km (1,633 miles) with drop tanks.

The Yak-28B succeeded the initial production version on the Irkutsk assembly line as *izdeliye* 28B. In the course of production the R11AF-300's afterburner rating was increased from the original 5,750 kgp (12,680 lbst) to 5,900-6,050 kgp (13,010-13,340 lbst) with no change in the engine's designation.

The type had its public debut in 1961 at the traditional Aviation Day flypast at Tushino. It was then that the bomber received its NATO codename, *Brewer*; this was later changed to *Brewer-A* when other versions became known. *'The USA do not have an equivalent to this airplane'*, the *New York Times* wrote.

Meanwhile, the Yakovlev OKB sought to improve the *Brewer*. Between 20th March and 4th April 1962 NII VVS tested a 'production Yak-28 bomber (c/n 1900304) with an RBP-3 radar bomb sight' (that is, a Yak-28B). The aircraft had been specially modified in order to improve field performance. Fittings for two SPRD-118 jet-assisted take-off (JATO) bottles (***star**tovyy porokho**voy** ra**ket**nyy **dvi**-**gat**el'* – solid-fuel rocket booster) were installed on the aft fuselage sides. The nose gear unit featured new KT-82 brake wheels (fitted to all Yak-28s from this aircraft onwards) and an automatic brake parachute deployment feature was added. As compared to the Yak-28 *sans suffixe* the fuel load was reduced by 755 kg (1,665 lb) at normal TOW and by 995 kg (2,190 lb) at maximum TOW. The modified aircraft was flown by NII VVS test pilots Yuriy M. Sookhov and Vladimir V. Dobrovol'skiy; upon completion of the test programme the changes were introduced on the production line in Irkutsk.

In 1963 another Yak-28 was used to test an airbrake but this remained merely an experimental modification.

Yak-28L tactical bomber (*izdeliye* 28L)

1960 also saw the first flight of a new version designated Yak-28L. The suffix letter referred to the aircraft's totally different targeting system – the DBS-2S Lotos radio command guidance (data link) system. The system received inputs from two ground control/target ranging stations located in territory held by friendly forces, computing the target coordinates by triangulation.

The Yak-28L prototype was converted from the original third prototype of the *Brewer*, retaining the latter's tactical code '56 Yellow'. State acceptance trials commenced in 1962, but a lengthy improvement programme was needed before the aircraft was satisfactory.

Among other things, the original R11AF-300 engines were substituted by R11AF2-300s rated at 6,200 kgp (13,670 lbst) in full afterburner. Hence the engine nacelles were redesigned once again, reverting to extended circular intakes of increased diameter to cope with the bigger mass flow. Finally, the glazing of the navigator/bomb-aimer's station had a simplified frame to improve visibility.

In this guise the Yak-28L entered production in Irkutsk with the product code '*izdeliye* 28L', replacing the Yak-28B. Unlike the latter model, however, the Yak-28L was never officially included in the VVS inventory, which is why the government order for the type was limited (only 111 aircraft were built). In the course of production the single-barrel NR-23 cannon was replaced by a Gryazev/Shipoonov GSh-23Ya double-barrel cannon. The capacity of the drop tanks was increased and the tanks were reshaped, receiving pointed noses. The NATO reporting name was *Brewer-B*.

The Yak-28L's maximum bombing altitude depended on the system's reception range, and this was very limited because the Lotos used the UHF waveband; sometimes range was literally confined to direct visibil-

A production Yak-28L with R11AF2-300 engines shows clearly the 'solid' underside of the nose, the increased area of the lateral glazing and the blade aerials of the Lotos system fore and aft of the nosewheel well.

Yakovlev OKB

'24 Yellow', the Yak-28I prototype, during manufacturer's flight tests with the original R11AF-300 engines (note th small intake centrebodies). This view shows clearly the enlarged radome of the Initsiativa-2 radar.

ity limits! If radio contact with the ground control stations was lost the system would compute the target coordinates, using the latest available radio data or landmarks. Working the ground part of the system was an immensely complicated task (which, of course, did not add to the system's popularity either).

The trials showed that the engines flamed out at 210-270 km/h (130-167 mph) IAS during deceleration at high alpha, just as they would in the event of a stall. Some other systems became inoperative, too, including the de-icing and cockpit heating systems.

The trials showed that the Yak-28 recovered from a spin with a delay of two or three turns after recovery was initiated. This ran contrary to the then-current OTTT VVS-58 (***Ob**shchiye **tak**tiko-tekh**nich**eskiye **trebo**vaniya* – Air Force General Operational Requirement of 1958) which required the aircraft to recover with a maximum delay of one full turn. Four spin recovery techniques were developed for the Yak-28; yet the trials report said that *'spin recovery training is inadvisable on all versions of Type 129* (*sic* – that is, Yak-129) *due to the high flying skills required for this'*.

A late-production Yak-28L ('44 Red', c/n 2920903) is on display at the Central Russian Air Force Museum in Monino.

Yak-28I tactical bomber (*izdeliye* 28I)

Concurrently with the Yak-28L the Yakovlev OKB developed and tested another bomber version with a different targeting system designated Yak-28I.

The Yak-28I was equipped with the *Initsiativa-2* (Initiative) 360° ground mapping/bomb-aiming radar, hence the I suffix. The new radar, identifiable by a slightly larger and recontoured radome, had longer detection range and higher resolution by comparison with the Yak-28B's RBP-3. The weapons control system also included an OPB-16 optical bomb sight and an AP-28K autopilot.

The new radar enabled the bomber to seek and destroy small moving targets (such as tanks) in any weather, day and night. It was linked to the optical sight and the autopilot; the latter provided for automatic guidance in the heading channel. The OPB-16 permitted bombing runs at 2,000-20,000 m (6,560-65,620 ft); speed, altitude and ballistic data were fed into the sight automatically above 3,500 m (11,480 ft) or manually at 2,000-3,500 m. Bombing runs using the optical sight could be made at 800-1,700 km/h (496-1,055 mph) true airspeed with headwinds or tailwinds of up to 350 km/h (217 mph).

In order to accommodate the bulky radar set of the Initsiativa-2 radar a 780-mm (2 ft 6$\frac{45}{64}$ in) plug had to be inserted in the fuselage between the cockpit and the wings, increasing the landing gear wheelbase. To compensate for the extra area ahead of the aircraft's centre of gravity the boundary layer fences on the inner wings were made taller. As with the Yak-28L, the glazed nose had a simplified frame to improve visibility.

Specifications of the Yak-28 bomber variants

	Yak-28B	Yak-28I	Yak-28L
Length overall	n.a.	n.a.	20.2 m (66 ft 3¼ in)
Wing span	11.64 m (38 ft 2¼ in)	11.64 m (38 ft 2¼ in)	11.64 m (38 ft 2¼ in)
Wing area, m² (sq ft)	35.25 (379.03)	35.25 (379.03)	35.25 (379.03)
Powerplant	2 x R11AF-300	2 x R11AF2-300	2 x R11AF2-300
Engine thrust (reheat), kgp (lbst)	2 x 5,750 (2 x 12,680)	2 x 6,100 (2 x 13,450)	2 x 6,100 (2 x 13,450)
Normal TOW, kg (lb)	13,630 (30,050)	16,160 (35,630)	15,545 (24,270)
MTOW, kg (lb)	n.a.	18,080 (39,860)	17,465 (38,500)
Top speed, km/h (mph):			
at sea level	n.a.	n.a.	1,070 (665)
at 12,000-13,000 m (39,370-42, 650 ft)	1,900 (1,180)	1,850 (1,150)	1,945 (1,208)
Service ceiling , m (ft)	16,200 (53,150)	14,500 (47,570)	16,250 (53,310)
Time to 10,000 m (32,800 ft)	n.a.	n.a.	6 min
Effective range, km (miles)	1,950 (1,212)	2,070 (1,286)	2,420 (1,504)
Take-off run, m (ft)			
with normal TOW	n.a.	1,150 (3,773)	1,200 (3,937)
with MTOW	n.a.	1,803 (6,004)	1,550 (5,085)
Landing run, m (ft)			
without brake parachute	n.a.	1,160 (3,800)	1,290 (4,230)
with brake parachute	n.a.	700 (2,300)	1,000 (3,280)
Armament	1 x NR-23 cannon; 1,200-1,300 kg (2,645-6,613 lb) of bombs	1 x NR-23 or GSh-23Ya; 1,200-1,300 kg (2,645-6,613 lb) of bombs	1 x NR-23 or GSh-23Ya; 1,200-1,300 kg (2,645-6,613 lb) of bombs

Yakovlev OKB

Above: The R-11AF-300's propensity to surging led to the Yak-28 being re-engined. Here, the Yak-28I prototype is seen at a later test stage with the new R11AF2-300 engines with circular intakes.

Yefim Gordon archive

A production Yak-28I coded '58' at an airbase.

Coded '24 Yellow', the Yak-28I prototype originally flew with R11AF-300 engines having cut-back elliptical air intakes, just like the Yak-28L prototype. In the middle of the test programme, however, it was re-engined with R11AF2-300s in redesigned nacelles with extended circular intakes.

The Yak-28I entered production in Irkutsk with the new powerplant; it was produced in parallel with the Yak-28L as *izdeliye* 28I. A total of 223 were built. Unlike the Yak-28L, this version was officially included into the VVS inventory. The NATO reporting name was *Brewer-C*.

The new radar and associated structural changes increased the normal AUW of the production Yak-28I to 16,200 kg (35,710 lb) and the maximum AUW to 18,100 kg (39,900 lb), even though the bomb load remained the same. Top speed was recorded as 1,800 km/h (1,118 mph) at high altitude and 900 km/h (559 mph) at sea level. Service ceiling was 14,500 m (47,570 ft) and range with drop tanks was 2,290 km (1,422 miles) at high altitude or 875 km (543 miles) at sea level. In the course of production the original NR-30 cannon was replaced by a twin-barrel GSh-23Ya.

The Yak-28I was rushed into production before the Initsiativa-2 radar had been thoroughly tested. As a result, it was soon discovered that the radar's performance fell short of the specifications by a considerable margin. In practice this meant that bombing accuracy during checkout trials was unacceptably low.

Accordingly the VVS inspectors stopped accepting the bombers. A large team of OKB-115 engineers was dispatched to the airbase where the radar was being tested and refined. The team included chief project engineers and other specialists for all of the bomber's systems, as well as specialists of various rank from MAP and Ministry of Electronics Industry (MRP) research establishments involved in the Yak-28I programme.

The engineers at the Aircraft Systems Research Institute (NIIAS – *Na**ooch**no-is**sle**dovatel'skiy insti**toot** aviatsi**on**nykh sis**tem***) contributed immensely to perfecting the Yak-28I's targeting system. It was they who discovered that bombing accuracy was significantly affected by errors in determining the bombs' ballistic parameters (the airflow conditions during separation from the aircraft were not taken into account), errors in air data adjustment, and errors in flight instrumentation readings.

The R&D programme aimed at perfecting the Yak-28I's targeting system took almost a full year to complete. This was the first serious research programme concerned with the targeting system of a supersonic bomber.

A specially-equipped *Brewer-C* flown by test pilot Vladimir V. Dobrovol'skiy and navigator Boris I. Gherasimov was used by GK NII VVS in a flutter test programme which resulted in the wingtip anti-flutter weights being enlarged and mounted on longer booms. The aircraft was flown with full drop tanks and special electric vibrators in the wings to provoke flutter.

On 15th June – 6th July 1965 and 21st-28th January 1966 three Yak-28Is and two Yak-28Ls were used in a two-stage test programme to check the *Brewer*'s suitability for operation from dirt and snow strips respectively. The tests showed that with some minor modifications to the landing gear the Yak-28 could operate successfully from semi-prepared runways.

Work on refining the bomber and increasing its combat potential proceeded jointly with the VVS for almost ten years. In particular, between 31st March 1968 and 30th January 1969 a production Yak-28I participated in a programme to investigate the possibility of dropping bombs in a climb (after a low-level stealthy approach to the target) without making changes to the Initsiativa-2 radar or the OPB-16 bomb sight. On 6th May – 18th June 1970 a Yak-28L (c/n 3921204) was used to test a modified NR-23 cannon which could fire bursts of up to 20 rounds at altitudes in excess of 8,000 m (26,250 ft) without affecting engine operation.

Yak-28UVP tactical bomber prototype

A single Yak-28 bomber (subtype not specified, c/n ending 0304) was fitted with two 28SPRD-99 JATO bottles and twin PT-25 brake parachutes; the testing took place in the early 1960s. The aircraft has been referred to as the Yak-28UVP (*ooko**roch**ennyy vzlyot i po**sahd**ka* – short take-off and landing).

Yak-28R tactical reconnaissance aircraft (*izdeliye* 28R)

Back in 1956, government documents contained a clause ordering the Yakovlev OKB to develop the new Yak-129 (Yak-28) bomber into a high-altitude light reconnaissance aircraft, submitting it for State acceptance trials in the third quarter of 1958. Hence development of the reconnaissance version proceeded in parallel with the baseline bomber.

Development was completed in 1962; coded '45 Yellow', the Yak-28R prototype was rolled out at MMZ No.115 in 1963. The aircraft was based on the Yak-28I but incorporated major changes to the forward fuselage. The cockpit canopy was new: the sliding portion was more convex to give the pilot more headroom, while the fixed windscreen consisted of two elliptical panels joined in V fashion and two small sidelights.

Yakovlev OKB

Above: '45 Yellow', the prototype of the Yak-28R reconnaissance version. The redesigned nose glazing with a slanting rear edge is obvious but the V-shaped cockpit windshield is not. Note the 'towel rail' telemetry aerial under the fuselage and the drop tank attachment fittings under the wings.

Yefim Gordon archive

Above: Production Yak-28Rs in their natural metal factory finish on the flight line. Note the production-standard drop tanks and the ladders for access to the cockpit and the navigator's station.

Yefim Gordon archive

A Yak-28R in action with the camera ports open. The *Brewer* was nicknamed *Raschoska* (Comb) in Soviet Air Force service; the reason is obvious in this view.

The nose, which was now the RSO's station, was even more extensively glazed to improve visibility and incorporated an optically flat panel of high-quality glass. The glazing was attached to a sloping fuselage frame, that is, its rear edge was raked aft instead of being at right angles to the fuselage waterline.

(Note: Some sources say the Yak-28R prototype was not converted from a production Yak-28I but was actually rebuilt from the first prototype Yak-27R (itself a conversion of the sole Yak-122) which had been withdrawn from use by then.)

The internal layout aft of the cockpit was completely new. Internal fuel capacity was increased by 550 litres (121 Imp gal). The bomb bay was supplanted by three equipment bays accommodating five interchangeable sets of reconnaissance equipment, allowing the aircraft to be configured for a specific mission (this equipment was not necessarily cameras – one of the sets included SIGINT equipment). Accordingly, camera ports were provided in the centre fuselage underside. Finally, the Yak-28I's radar gave way to a modified Iniatsiativa-2R unit.

The Yak-28R prototype had the new R11AF2-300 engines and definitive nacelles from the outset. On 7th March 1963 the Yak-28R was transferred to NII VVS for State acceptance trials. For reasons unknown these did not commence until exactly four months later (7th July), ending successfully on 22nd August.

The Irkutsk aircraft factory launched production of the Yak-28R in an amazingly short time. Known in-house as *izdeliye* 28R, production aircraft began rolling off the line in January 1966 and the version achieved IOC that year. The Yak-28R was codenamed *Brewer-D* by NATO.

On 1st December 1966 the first production Yak-28R (c/n 6960101) embarked on an 11-month checkout test programme which was completed on 30th October 1967. Pictures were taken at 3,000-14,000 m (9,840-45,930 ft) and 600-1,500 km/h (372-931 mph), and all five reconnaissance equipment configurations were tested and received a thumbs-up. The following year the same aircraft was tested again after some changes had been incorporated – again with good results. In due course the Initsiativa-2R radar fitted to early-production Yak-28Rs was replaced with a specially-developed version, the Initsiativa-3.

The reconnaissance version had a 15,500-kg (34,170-lb) TOW and a service ceiling of 15,600 m (51,180 ft); top speed was 1,800 km/h (1,118 mph) at high altitude and 900 km/h (559 mph) at sea level. The *Brewer-D*'s effective range with drop tanks was 2,680 km (1,664 miles) at high altitude or 950 km (590 miles) at sea level.

More trials came in 1969 when NII VVS tested a modified Yak-28R (c/n 8961302) manufactured in November 1968. The aircraft had pneumatically-actuated intake shock cones, a cone position indicator system and other detail improvements.

Production of the Yak-28R ended in 1970, by which time only 183 examples had been built (this figure includes the Yak-28RL and Yak-28SR sub-types described below).

Yefim Gordon archive

Above: The Yak-28SR chemical warfare aircraft prototype. Note the chemical dispensers under the wings; unlike the standard drop tanks, they were not conformal.

Still, this was by far the most successful version; along with the Yak-28PP ECM aircraft, it outlasted the other versions of the *Brewer* family in Soviet Air Force service and was popular with its crews. The Yak-28R was also the only version to see action in a real war – the Afghan War.

Yak-28SR chemical warfare aircraft (first use of designation)

In 1962 the Yakovlev OKB started work on a chemical warfare variant of the Yak-28 intended for dispensing liquid or powdered chemical agents from low altitudes by spraying or dusting. Designated Yak-28SR (*samo**lyot**-raspy**li**tel'* – spraying/dusting aircraft), the prototype of this version was a converted Yak-28L (c/n 3920903); it underwent testing at NII VVS between 16th September and 5th November 1963. On this machine the drop tank hardpoints could be used for carrying two types of external containers equipped with devices for dispensing chemical substances. The Yak-28SR was recommended for service introduction but none were actually delivered to the VVS.

Yak-28SR tactical reconnaissance aircraft (*izdeliye* 28SR; second use of designation)

In 1970 the Yak-28SR designation was reused for a modified Yak-28R (c/n 9961309) which underwent State acceptance trials at NII VVS that year; the meaning of the SR suffix in this case is unknown. The aircraft was retrofitted with an SPS-141 active jammer (***stahn**tsiya po**mekh**ovykh sig**nah**lov*) for self-defence to disrupt the operation of enemy missile guidance radars.

Later (5th July to 23rd August 1971) the same aircraft underwent further trials at NII VVS in order to determine the Yak-28SR's range and measure the cockpit temperature. The latter was necessary because the SPS-141 jammer generated a lot of heat and there were fears that it would turn the cockpit into a steam bath. The results were good; range was identical to that of the standard *Brewer-D* (c/n 8961302) tested earlier and the cockpit temperature was bearable.

The Yak-28SR entered production but was only built on a very small scale. Some examples were equipped with an SPS-143 jammer instead of the SPS-141.

Yak-28R with TV reconnaissance system (Yak-28TARK)

Between 28th December 1966 and 6th February 1967 the Yakovlev OKB tested a converted early-production Yak-28R (c/n 6960305) equipped with the TARK-1 aerial TV reconnaissance system (*televizi**on**nyy aviatsi**on**nyy raz**ved**yvatel'nyy **kom**pleks*). The equipment bays housed TV cameras and a data link transmitter for sending the TV picture to C³I centres in real time. The Initsiativa-2R radar was deleted but the radome was retained, housing the system's data link antenna. As a backup, the aircraft was equipped with a still camera using 190-mm film.

On 6th April the aircraft was handed over to the VVS for service trials which lasted from 1st September to 15th December. The results were encouraging and the TARK-1 system was adopted by the Air Force, several *Brewer-Ds* being converted to this standard.

Yak-28RR radiation intelligence aircraft (*izdeliye* 28RR)

A number of Yak-28Rs were fitted out for RINT duties in the late 1960s, receiving the designation Yak-28RR. This version was put through its paces by OKB test pilot A. P. Kolosov and navigator N. M. Shipovskiy. The RR8311-100 air sampling pods, which had been specially developed for the Yak-28RR in 1964, were later standardised and used on such varied types as the Yak-25RR (described earlier), An-12RR, An-24RR, An-30R, Tu-16R *Badger-F* and Tu-95K-22. The pods were probably carried on the standard drop tank hardpoints.

Yak-28RL reconnaissance aircraft

This version was probably a standard Yak-28L bomber adapted for RINT duties simply by fitting RR8311-100 air sampling pods to the drop tank hardpoints. One such aircraft, coded '06 Blue', has been seen to date.

Yak-28PP ECM aircraft (*izdeliye* 28PP)

The VVS urgently needed a lightweight supersonic electronic countermeasures (ECM) aircraft which would jam enemy air defence radars and communications, operating in support of a bomber strike group. Existing ECM aircraft, such as the Tu-16PP, were ill-suited to operating over the battlefield.

Outwardly the Yak-28PP, as the new version was designated, was a cross-breed between the Yak-28R and the bomber versions. It had the *Brewer-D*'s bulged canopy and V-shaped windshield made up of two elliptical panels, but the nose glazing reverted to that of the Yak-28I and Yak-28L, with the rear at right angles to the fuselage waterline. All armament was deleted to make room for a comprehensive active ECM suite including the *Boo**ket*** (Bouquet), *Stre**la*** (Arrow), *Fa**sol'**-1* (String Bean-1) and SPS-153 *See**ren'*** (Lilac) jammers. Hence the Yak-28PP sported tell-tale 'bumps and bulges': small dielectric blisters on the forward fuselage sides aft of the cockpit and immediately aft of the nose pitot boom, a

via Polygon-Press

This pristine-looking Yak-28PP coded '45 Blue' shows off its nacelle- and fuselage-mounted ECM aerials and the underwing pods loaded with chaff-filled rockets. Note the three-tone tactical camouflage.

large semi-cylindrical fairing on the lower fuselage ahead of the main gear, a bulged equipment access panel with a cooling air intake low on the starboard side in line with the nose gear and downward-angled blade aerials on the outer faces of the engine nacelles. The ECM equipment generated a lot of heat, so two faired heat exchangers were installed under the centre fuselage.

The aircraft also had powerful passive ECM equipment. Firstly, the outer wings featured pylons for two UB-16-57UM sixteen-round FFAR pods loaded with S-5P (aka PARS-5) 57-mm (2.24-in) rockets; these were filled with chaff and exploded some way ahead of the aircraft when fired to create a cloud of chaff. Secondly, twin KDS-19 *Avtomaht-2I* (in this case, 'automatic device') chaff dispensers were built symmetrically into the lower parts of the engine nacelles, firing bundles of metal-coated fibreglass strips. Finally, the aircraft had ASO-2I flare packs firing 26-mm (1-in) magnesium flares for protection against heat-seeking missiles.

The avionics fit included the DISS-3A Doppler speed and drift sensor (***dop**lerovskiy izme**rit**el' **sko**rosti i **sno**sa*) and a modified NVU-VI-2 navigation computer (*navigatsi**on**noye vychis**lit**el'noye oo**stroy**stvo*) receiving inputs from it. The rest of the avionics and equipment matched that of the bomber versions. The internal fuel cells held 5,480 litres (1,205.6 Imp gal) of fuel and two standard 1,050-litre (231 Imp gal) slipper tanks could be carried.

After successfully passing its State acceptance trials in 1970 the ECM version entered production in Irkutsk as *izdeliye* 28PP, receiving the NATO reporting name *Brewer-E* in due course. The Yak-28PP was almost equal in performance to the bomber versions.

The Yak-28PP proved indispensable to the Soviet Air Force and stayed in service until replaced by the Su-24MP in the early 1990s, long after other versions of the *Brewer* had been retired. By then, however, podded active jammers had been developed and could be fitted to any tactical aircraft, rendering specialised tactical ECM aircraft all but obsolete. No fewer than three *Brewer-Es* (coded '43 Blue', '45 Blue' and '53 Blue', c/ns 1971002, 0970603 and 1971105) reside at the Russian Air Force Museum in Monino – unfortunately, all in varying stages of disrepair.

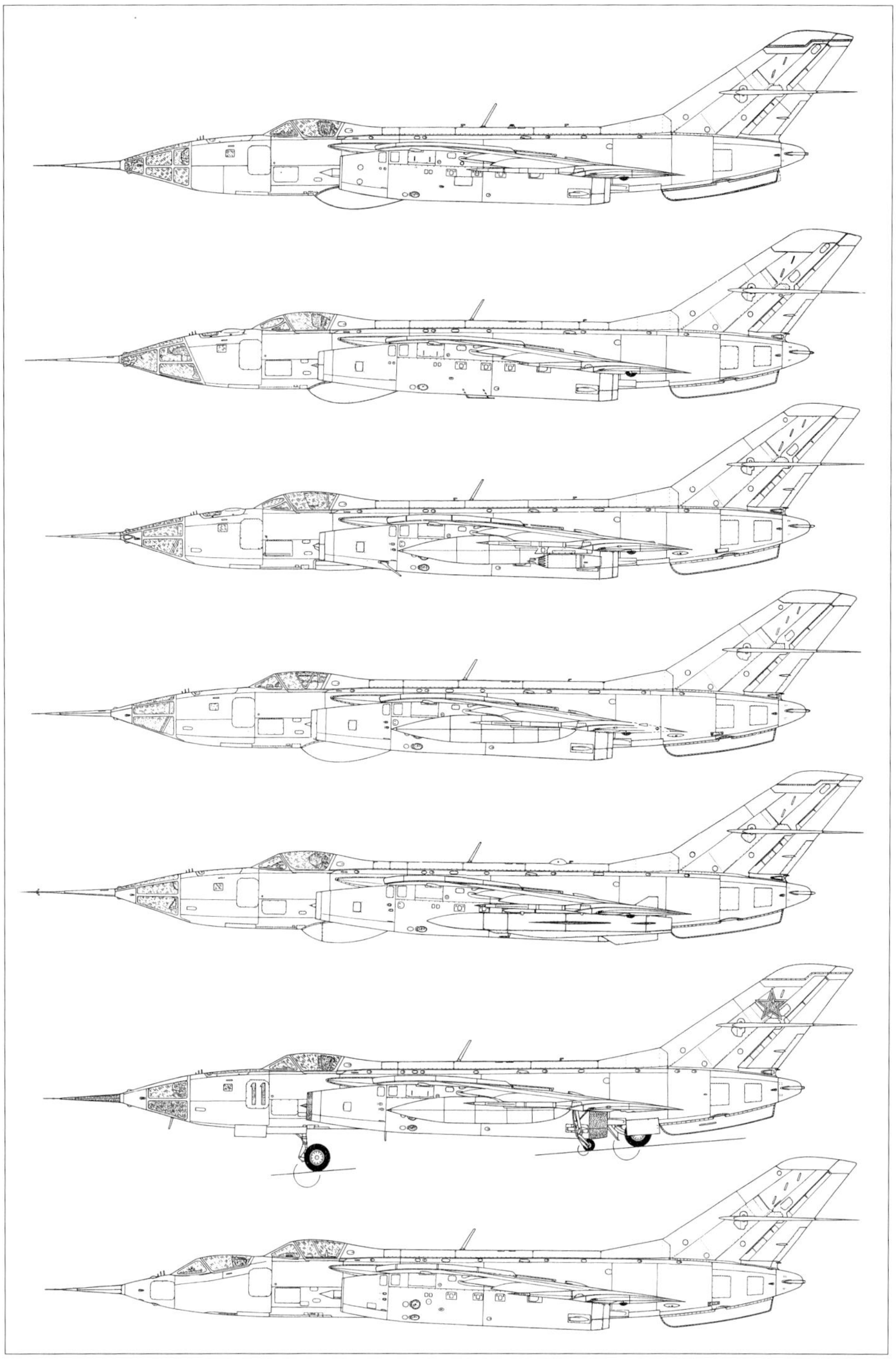

Top to bottom: The Yak-129, the Yak-28R, the Yak-28PP, the Yak-28N, the Yak-28L and the Yak-28U.

Yak-28U trainer (Yak-129U, *izdeliye* 28U)

A high-performance aircraft like the Yak-28 called for a dual-control version intended for conversion training. Development of the trainer version, initially designated Yak-129U (*oo**cheb**nyy* – training, used attributively) began straight away. Based on the second and third prototypes with R11AF-300 engines, the Yak-129U differed mainly in having a new forward fuselage. The navigator/bomb-aimer's station was replaced by a pointed metal nose incorporating the instructor's cockpit. This had a teardrop canopy opening to starboard. Thus the trainer had a stepped-tandem seating arrangement also used on the IL-28U and later on the Tu-128UT and MiG-25PU/MiG-25RU trainers. The bomb bay was retained but was normally occupied by an extra fuel tank.

The prototype (c/n 0001) was rolled out in 1962. Manufacturer's flight tests took place in July-September. Between 23rd November 1962 and 15th March 1963 the trainer (by then allocated the service designation Yak-28U) underwent State acceptance trials at NII VVS. The teething troubles discovered at this stage took some time to eliminate, and in reality the trials were not completed until 3rd September 1964.

Typical of Soviet military aircraft, the Yak-28U entered production long before trials had been completed. From 1962 on it was built in Irkutsk as *izdeliye* 28U in parallel with the Yak-28I and Yak-28L, and produc-

Specifications of the Yak-28R reconnaissance aircraft and Yak-28PP ECM aircraft

	Yak-28R	Yak-28PP
First flight	1963	1965
Length overall	20.34 m (66 ft 8¾ in)	20.34 m (66 ft 8¾ in)
Wing span	11.64 m (38 ft 2¼ in)	11.64 m (38 ft 2¼ in)
Wing area, m^2 (sq ft)	35.25 (379.03)	35.25 (379.03)
Powerplant	2 x R11AF2-300	2 x R11AF2-300
Engine thrust (reheat), kgp (lbst)	2 x 6,100 (2 x13,450)	2 x 6,100 (2 x13,450)
Normal TOW, kg (lb)	15,725 (34,670)	15,650 (34,500)
MTOW, kg (lb)	17,645 (38,900)	17,470 (38,510)
Top speed at 12,000-13,000 m		
(39,370-42, 650 ft), km/h (mph)	1,807 (1,123)	1,720 (1,069)
Service ceiling , m (ft)	15,650 (51,345)	14,600 (47,900)
Effective range, km (miles)	2,680 (1,665)	1,900 (1,180)
Take-off run, m (ft)		
with normal TOW	1,230 (4,035)	950 (3,120)
with MTOW	n.a.	1,600 (5,250)
Landing run, m (ft)		
without brake parachute	n.a.	1,160 (3,770)
with brake parachute	850 (2,790)	700 (2,300)
Armament	none	none

Yakovlev OKB

'59 Red', the first production Yak-28U trainer, seen during a test flight. Note the phototheodolite calibration markings on the fuselage and tail.

Specifications of the Yak-28U trainer

	Yak-28U
Length overall	20.2 m (66 ft 3¼ in)
Wing span	11.64 m (38 ft 2¼ in)
Wing area, m^2 (sq ft)	35.25 (379.03)
Powerplant	2 x R11AF-300
Rating, kgp (lbst)	2 x 5,750 (2 x 12,680)
Normal TOW, kg (lb)	14,680 (32,360)
Top speed, km/h (mph):	
at 12,000-13,000 m	
(39,370-42, 650 ft)	1,830 (1,137)
Service ceiling , m (ft)	15,400 (50,525)
Effective range, km (miles)	1,781 (1,107)
Take-off run	
with normal TOW, m (ft)	700 (2,300)
Landing run, m (ft)	
without brake parachute	1,050 (3,440)
with brake parachute	780 (2,560)
Armament	none

tion totalled 183. In those days NATO still assigned codenames in the 'miscellaneous' category to trainer versions of Soviet fighters, and the Yak-28U's codename was *Maestro*.

All Yak-28Us were powered by R11AF-300 engines with elliptical intakes. The take-off weight was 15,000 kg (33,070 lb); the Maestro had a top speed of 1,850 km/h (1,149 mph) and a range of 2,200 km (1,366 miles).

The fifth production Yak-28U ('63 Blue', c/n 2930105) is preserved at the Central Russian Air Force Museum in Monino.

Yak-28N experimental air defence suppression aircraft (Yak-28-64, first use of designation)

In 1964-65 the Yakovlev OKB adapted the Yak-28 for the air defence suppression role (that is, taking out enemy AD radars). The aircraft was armed with Kh-28 anti-radiation missiles (ARMs) and a weapons control system for same. In 1965 a production Yak-28I was set aside for conversion.

Coded '12 Yellow' and designated Yak-28N (*nositel'* – 'carrier', in this case, missile carrier), the 'Wild Weasel *Brewer*' commenced manufacturer's flight tests same year (it was also known as Yak-28-64, a designation that was re-used later for a totally different machine). The large Kh-28 missiles with delta wings and cruciform tail surfaces were carried on short pylons under the outer wings. The Initsiativa-2 radar was replaced by a radar pulse detection system, its antennas being housed in the standard ventral radome. Two L-shaped aerials serving the targeting station were located on the starboard engine's air intake.

The Yak-28N remained a one-off, because by the early 1970s the Yak-28 was already approaching obsolescence. The 'Wild Weasel' role was later filled by the MiG-25BM, Su-17 and Su-24. The Kh-28 missile was more lucky, entering production and being used by the VVS and some foreign air arms.

Yak-28IM development aircraft

In September-October 1969 the Yakovlev OKB tested a production *Brewer-C* (c/n 4940503) converted into the Yak-28IM development aircraft; the M suffix denoted *modifitseerovannyy* (modified) or *modernizeerovannyy* (upgraded). The bomber's punch was enhanced by fitting four underwing pylons for additional bombs or rocket pods.

The Yak-28IM commenced State acceptance trials in 1973 – too late. By then the Soviet aircraft industry had begun large-scale production of the Sukhoi Su-17 fighter-bomber and Su-24 tactical bomber, both of which could carry a wide nomenclature of weapons. These aircraft were already displacing the *Brewer* in FA service, and there was no point in upgrading a generally outmoded design.

Yak-28BI ELINT aircraft prototype

On 27th March 1965 the Council of Ministers issued a directive ordering the development of a electronic intelligence derivative of the Yak-28I. Designated Yak-28BI, this was to have a Bulat (Damask Steel; pronounced *boolaht*) synthetic-aperture side-looking aircraft radar (SLAR) while retaining the standard Initsiativa-2 search radar, hence the BI suffix.

The high-resolution SLAR was designed for ground mapping in any weather, day or night, generating a radar image comparable in detail to a regular photograph. The Bulat radar set and antenna array were housed in the bomb bay; the podded antenna array would be extended into the slipstream as the aircraft approached the target area.

The Yak-28BI prototype completed its State acceptance trials in 1972. The trials programme included flights in a straight line at 200-2,000 m (660-6,560 ft) and 600-1,200 km/h (372-745 mph); the SLAR mapped a strip 15 km (9.3 miles) wide to port or starboard of the aircraft's course at the navigator's discretion. No decision to field the Bulat system was taken and the Yak-28BI remained a one-off.

Yefim Gordon archive

Above: A production Yak-28U coded '51 Blue'. Note the sideways-opening canopy of the front cockpit.

Yak-28VV VTOL tactical bomber (project)

In the early 1960s the Yakovlev OKB developed a project of a VTOL version of the Yak-28 tactical bomber which was designated Yak-28VV (*verti**kahl'**nyy vzlyot* – vertical take-off). It was believed that VTOL capability would considerably increase the possibilities of deploying tactical bombers and their combat potential.

According to the project, the Yak-28VV was to be powered by two R27AF-300 lift/cruise engines and four R39P-300 lift engines. Estimated performance included a speed of no less than 1,700 km/h (1,056 mph), and a ceiling of 18,000 m (59,060 ft); the bomber was expected to deliver a bomb load of 600-1,200 kg (1,320-2,650 lb) at an AUW of 18,000 kg (39,690 lb) to which, if used from normal-length runways, would have roughly the same performance as the Yak-28B at a similar AUW. However, the lift engines could be installed only at the expense of fuel load with a resulting unacceptable reduction of range.

Yak-LSh light attack aircraft (project)

This project (LSh stands for ***lyokh**kiy shtoor**movik***, light attack aircraft) was submitted by the Yakovlev OKB to a competition of strike aircraft projects held in 1969, the other contenders being the design bureaux led by Sukhoi, Mikoyan and Il'yushin. Even now, the details of this project are still classified, but information published in the popular press suggests that it was based on a highly-modified version of the Yak-28 light supersonic tactical bomber. The project featured armour protection for the crew.

The Yak-LSh was eliminated from the competition at an early stage, and all further work on this particular design was terminated. It was rejected because of deficiencies in mission equipment and armament.

Yak-28 bombers as LII testbeds

Several Yak-28s were used by the Flight Research Institute (LII) as flying testbeds in various research programmes. The examples listed below are presumed to be based on bomber versions of the Yak-28.

A Yak-28 testbed was used for measuring the distribution and fluctuations of pressure on the wings and control surfaces in various flight modes. Various means of airflow visualisation were used during these experiments.

A Yak-28 was used for studying the shock waves on the wing under natural conditions.

One Yak-28 was used for measuring the strength of structural members. The aircraft was modified by installing aerodynamic force generators at the wingtips, supplemented by a set of measuring and data recording equipment.

Another Yak-28 served as an avionics testbed for antenna and feeder systems and communication radios.

An example referred to as 'Yak-28 No.720' (presumably the last digits of the c/n) was used by LII between 1967 and 1976 for studying the airflow around the fuselage at subsonic and supersonic speeds. The results of these tests were used in design and development work on the Tu-128 interceptor.

Yak-28P interceptor (*izdeliye* 40)

As described earlier in the book, the subsonic Yak-25 *Flashlight-A* evolved into the supersonic Yak-27 *Flashlight-C* interceptor and then its missile-toting version, the Yak-27K. Design flaws prevented the type from entering mass production and service with the PVO. Now, given the availability of the more capable Yak-28, it was only logical that the interceptor should re-emerge at a new performance level.

Thus in 1960 the Yakovlev OKB brought out an interceptor derivative of the *Brewer* designated Yak-28P (*pere**khvaht**chik* – interceptor). It was designed to seek and destroy enemy aircraft at low and medium altitude in any weather, day or night, within a wide speed envelope. The Yak-28P featured the K-8M-1 weapons system comprising two R-8M-1 (AA-3 Anab) air-to-air missiles, missile launch rails and an ***Oryol**-D* (Eagle-D; NATO Skip Spin) fire control radar. The R-8M-1 was an improved version of the K-8 AAM used on the Yak-27K. It came in two versions with passive IR homing and semi-active radar homing seeker heads; the idea was to carry one missile of each version and fire them in a salvo for maximum 'kill' probability, should the target use ECM. No cannons were fitted.

As compared to the Su-9 interceptor in service with the PVO since 1958, the Yak-28P had a more sophisticated weapons system. Moreover, later the aircraft was to receive the K-98 weapons system which was an advanced version of the K-8M-1.

The Yak-28P's cockpit was almost a direct copy of the Yak-27K's; the pilot and WSO sat in tandem under a common sliding

Yakovlev OKB

'12 Yellow', the experimental Yak-28N air defence suppression version. The Kh-28 anti-radiation missiles under the wings are clearly visible.

Yakovlev OKB

Above: '15 Yellow', the prototype of the Yak-28P interceptor, as originally flown (with R11AF-300 engines). Note the ventral 'towel rail' telemetry aerial and the shape of the air intake shock cones.

Yakovlev OKB

Above: The prototype's configuration changed several times during tests. Here, the dorsal blade aerial and ventral telemetry aerial have been removed and the aircraft carries evenly spaced markings on the fin leading edge (possibly for icing tests).

canopy, and full dual controls were provided. Unlike the *Flashlight-C*, however, the new interceptor carried the missiles under the outer wings rather than between the engines and the fuselage.

Coded '15 Yellow', the first prototype initially flew with R11AF-300 engines in old-model nacelles *à la* Yak-28-2 and Yak-28U. However, compressor stall was experienced during tests; the engineers tried to cure the problem by redesigning the intake shock cones, using simple cones instead of that curious cone/cylinder design without changing the elliptical shape of the intakes.

With the modified intakes and some equipment changes the aircraft was redesignated Yak-28PD (*dora**bot**annyy* – modified or improved). Still, the Air Force was not satisfied and the prototypes had to go through a lengthy modification and trials programme.

Meanwhile, preparations for series production got under way. Since the Irkutsk aircraft factory had no spare capacity, MAP assigned Yak-28P production to the Novosibirsk aircraft factory No.153.

Eventually the R11AF-300 engines gave place to R11AF2-300s in extended nacelles with enlarged circular intakes, just as had been the case with the bombers. Since the missile pylons precluded use of drop tanks, all fuel had to be carried internally; on the other hand, there was no bomb bay any longer. Hence the Yak-28P's internal fuel capacity was increased by nearly 2,000 litres (440 Imp gal), totalling 6,570 litres (1,445.4 Imp gal).

Yakovlev OKB

The same aircraft at a later date, fitted with R11AF2-300 engines and sporting four 'kill' stars under the canopy to mark successful test launches. The red-painted K-8M-1 AAMs are obviously dummies.

After completing its State acceptance trials in 1965 with the new engines and K-8M-1 or K-98 weapons system the Yak-28P entered production in Novosibirsk under the in-house product code '*izdeliye* 40'. This was by far the most prolific version of the Yak-28 family: 435 examples were built between 1962 and 1967. Still, the aircraft was never officially included into the Air Force inventory, despite serving with the PVO for many years. On 9th July 1967 the Yak-28P made its first public appearance at a big airshow at Moscow-Domodedovo airport; after this the fighter received the NATO codename *Firebar*.

At 12,500 m (41,010 ft) the Yak-28P had a top speed of 1,840 km/h (1,142 mph) with two AAMs or 2,060 km/h (1,279 mph) in 'clean' condition. With a normal TOW of 16,065 kg (35,420 lb), the service ceiling was 16,000 m (52,490 ft). With a 13,230-kg (29,170-lb) AUW, including 3,710 kg (8,180 lb) of fuel, the *Firebar* had a range of 2,050 km (1,273 miles). At 7,000 m (22,965 ft) the range decreased to approximately 1,550 km (962 miles).

As an interceptor, the Yak-28P had an important asset in the form of provision for JATO bottles permitting short take-offs. In full afterburner and with JATO bottles the aircraft would become airborne in about 400 m (1,310 ft); this and the *Firebar*'s high rate of climb enabled the fighter to close in on the target rapidly.

On 7th July – 15th September 1965 the PVO's 148th Combat & Conversion Training Centre (TsBP i PLS – *Tsentr boye***voy** *podgo*-**tov***ki i pere***oo***chivaniya* **lyot***novo sos***tah***va*) at Savasleyka AB near Gor'kiy tested a Yak-28P (c/n 0615318) modified for operation from semi-prepared tactical airstrips with a bearing strength of no more than 9 kg/cm^2 (128.5 lb/sq in). The nosewheels were fitted with mudguards, and the nose gear oleo's stroke was increased by 75 mm (2$^{61}/_{64}$ in). The main gear unit's main pivot and the wingtip outrigger struts were beefed up. Finally, the intake lips were extended by 760 mm (2 ft 5$^{59}/_{64}$ in) to stop the engines from 'eating dirt'. The results were generally good (the test report said the modified interceptor would be suited for operation from unpaved runways).

On 28th March – 6th May 1966 another modified production Yak-28P (c/n 0915310) was used to test new take-off techniques (with afterburner operation) and a new twin-canopy brake parachute. Flown by V. Baranov and S. Peterin, the aircraft made 14 sorties. The results were excellent: the take-off run was reduced to 950-1,075 m (3,120-3,530 ft), the landing run to 650-750 m (2,130-2,460 ft) and the required runway length was 1,280 m (4,200 ft). This enabled new bases in the extreme north of the Soviet Union, where construction work was difficult, to be built with short strips, saving time and effort.

Production Yak-28Ps during final assembly at the Novosibirsk aircraft factory No.153.

One more Yak-28P served as a testbed for the improved Oryol-D 66-28 radar (the numbers probably stand for '1966 model, adapted for Yak-28'). The objective of the trials was to test the radar's resistance to active ECM but the results were to be inconclusive.

Yak-28P development aircraft with blown flaps

A production Yak-28P was experimentally fitted with flaps blown by engine bleed air to give more lift. The modification was tested but did not find its way to the production line. It's hard to say if this was because the system worked unsatisfactorily or because the *Firebar* was about to be superseded by a brand-new interceptor, the Su-15.

Yak-28PM development aircraft

As the Yak-28P entered production, the Yakovlev OKB and the Novosibirsk aircraft factory kept working on improving the type. A new version of the *Firebar*, the Yak-28PM (*modifi***tsee***rovannyy* – modified or *moderni*-**zee***rovannyy* – upgraded), appeared in 1962. The aircraft was powered by experimental R11AF3-300 engines; this further uprated version was comparable in performance to the R13-300 afterburning turbojet then under development.

Manufacturer's flight tests of the Yak-28PM began in 1963; test pilots V. M. Volkov (Yakovlev OKB) and Yuriy V. Petrov attained a top speed of 2,400 km/h (1,490 mph), which was a sizeable improvement over the standard aircraft. However, the R11AF3-300 (and hence the Yak-28PM) did not enter production because the brand-new R13-300 held more promise. Besides, production of the Yak-28 family was winding down and there was no point in developing a wing-mounted version of the R13-300 for an aircraft that would soon be gone.

Specifications of the Yak-28P interceptor

Length overall	n.a.
Wing span	11.64 m (38 ft 2¼ in)
Wing area, m^2 (sq ft)	35.25 (379.03)
Powerplant	2 x R11AF2-300
Rating, kgp (lbst)	2 x 6,100 (2 x13,450)
Normal TOW, kg (lb)	16,400 (36,155)
Top speed at 12,000-13,000 m (39,370-42, 650 ft), km/h (mph)	1,840 (1,143.6)
Service ceiling, m (ft)	16,000 (52,490)
Effective range, km (miles)	2,150 (1,336)
Take-off run with normal TOW, m (ft)	1,430 (4,690)
Landing run, m (ft):	
without brake parachute	1,250 (4,100)
with brake parachute	710 (2,330)
Armament	2 x R-8M-1 AAMs (2 x K-98 + 2 x R-3S AAMs on late production aircraft)

Upgraded Yak-28P interceptor (*izdeliye* 40)

Yakovlev engineers tried to refine the aerodynamics of the *Firebar* in order to improve acceleration above Mach 1 throughout the altitude envelope. To this end the standard ogival radome tipped by a long pitot (similar to that of the Yak-27K) was replaced with a new, longer radome of simple conical shape terminating in a short pitot.

ITAR-TSS

Above: A typical 1970s publicity shot. The crew is not wearing pressure suits because the Yak-28P did not operate at high altitudes. The guy on the right must be the navigator/WSO, as his pockets are full of maps.

Yakovlev OKB

Above: '01 Red', the second prototype of the late-production Yak-28P featuring a long conical radome and armed with two R-98s for beyond visual range engagements and two R-3Ss for close combat.

Yakovlev OKB

'51 Yellow', the original prototype of the long-nosed version, was later used for development work. Here it appears to have a metal nosecone; note also the *Brewer-D/E* style V-shaped windshield.

The new radome was tested on a Yak-28P coded '51 Yellow'. The aircraft also incorporated several other changes. Firstly, it had a V-shaped windshield borrowed from the Yak-28R and Yak-28PP because the engineers feared that the longer nose might impair cockpit visibility. However, flight tests showed that the new arrangement only made things worse, distorting the view, and the standard windshield was reinstated.

Secondly, the number of missiles was doubled by adding two R-3S (K-13M1; NATO codename AA-2-2 *Advanced Atoll*) IR-homing short-range AAMs. On the prototypes ('51 Yellow' and '01 Red') these were initially carried in Yak-27K fashion on pylons under the inner wings. However, tests showed that firing these missiles caused the engines to ingest missile exhaust gases and flame out, and the extra pylons were relocated to the outer wings outboard of the standard pylons for the R-8M-1s.

After passing State acceptance trials in 1966 the four-missile version of the Yak-28P entered production, replacing the original short-nosed version armed with only two *Anabs*. Fully loaded, the aircraft had a 16,900-kg (37,230-lb) TOW, a top speed of 1,860 km/h (1,155 mph) and a service ceiling of 15,000 m (49,210 ft). It could destroy targets flying at 500-19,000 m (1,640-62,340 ft) and speeds up to 1,400 km/h (869 mph) at a distance of up to 600 km (372 miles).

On 9th July 1967 a long-nosed late-production Yak-28P with four missile pylons was displayed statically at Moscow-Domodedovo along with other aircraft during the abovementioned airshow.

Along with improving the fighter's performance, OKB-115 worked in close co-operation with the avionics designers to improve the *Firebar*'s avionics suite. For example, on 15th-26th August 1967 a Yak-28P (c/n 0415303) was used to test a BKP-1 bank compensation module (*blok* ***kren****ovykh po****pra****vok*) linked to the ARK-10 ADF. Various changes introduced over the years (such as a new autopilot) improved reliability and flight safety significantly; steps were also taken to reduce the crew workload and improve handling, especially at high speeds and G-loads and in the event of various failures.

The long-nosed version saw service mainly in the northern regions of the Soviet Union.

Yak-28P development aircraft with additional ailerons

For various reasons the Yak-28P was subject to strict speed limits. Above 7,000 m (22,965 ft) this was due to flutter and structural strength considerations; below 7,000 m it was due to aileron reversal which, as mentioned earlier, was never completely eliminated. Yakovlev engineers offered an ingenious way of correcting the latter deficiency – by installing small triangular additional ailerons at the extreme wingtips (outboard of the outrigger gear fairings).

The prototype of the 'long-nose' version ('51 Yellow') was converted for testing the idea; hence the aircraft could be described as an aerodynamics research aircraft or as a control-configured vehicle (CCV). Tests held in 1966 showed that the arrangement worked as intended, but no decision to update new-build or in-service Yak-28Ps was taken.

Yak-28URP development aircraft

In order to investigate the possibility of radically increasing the *Firebar*'s top speed and service ceiling for a brief period (when chasing a high-flying intruder) a production Yak-28P was fitted experimentally with a liquid-propellant rocket booster (housed in the rear fuselage in similar fashion to the Yak-27V described above. Designated Yak-28URP, this aircraft did not progress beyond the prototype stage.

Yak-28P testbeds

A late Yak-28P coded '57 Red' was modified by LII for some kind of test work, with two small projections (possibly cameras or aerials) on top of each wingtip fairing.

Another Yak-28P coded '20 Red' (?) was converted into a testbed for an unknown purpose (probably an avionics testbed) by LII. The radome was replaced by a conical metal nosecone terminating in a short fat probe, and a small cylindrical object was carried under the nose on a short pylon.

Yak-28-64 experimental interceptor (second use of designation)

In 1962 OKB-51 led by Pavel O. Sukhoi designed the T-58 twin-engined supersonic interceptor which, after duly passing all required trials, entered production and service with the IA PVO as the Su-15 *Flagon-A* (the second aircraft to be thus designated). Ironically, Su-15 production was allocated to the Novosibirsk aircraft factory No.153. This resulted in a unique situation when the same plant manufactured two types developed by competing design bureaux but filling the same role and sharing the same powerplant and radar (the Su-15 was powered by Tumanskiy R11F2-300s and equipped with the Oryol-D radar).

Once again, however, the choice of the plant to build the Su-15 was not a matter of chance. The Novosibirsk aircraft factory was well familiar with the design features of Sukhoi fighters, having built the Su-9 and Su-11 interceptors since 1958, and it was only natural that the *Flagon*, which drew heavily on these two types, should be produced at the same location. Also, engine and avionics commonality between the Su-15 and the Yak-28P certainly saved the plant from a lot of logistical problems.

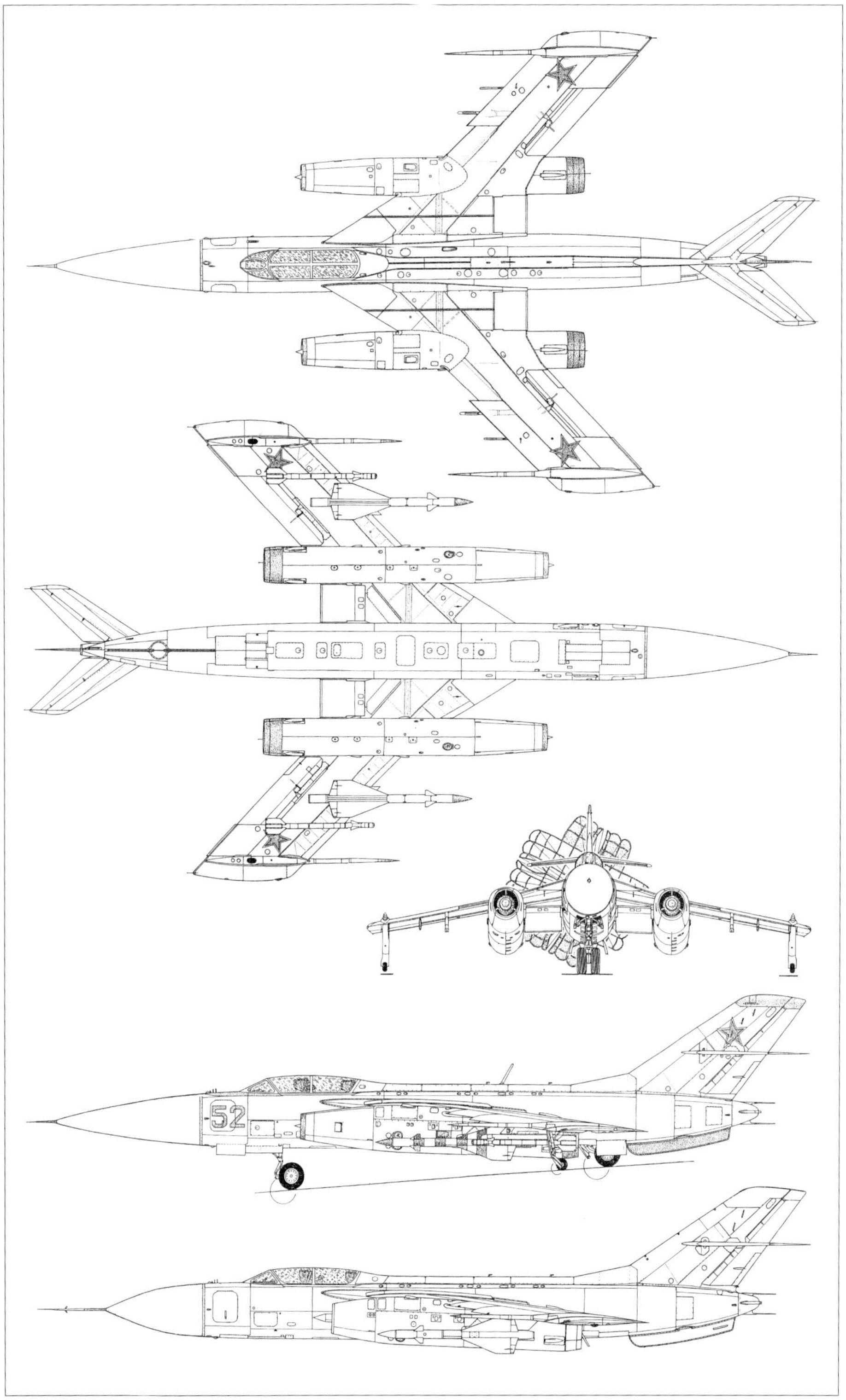

Four views of a late-production long-nosed Yak-28P with an additional side view (bottom) of the first prototype as originally flown (with R11AF-300 engines).

Yet Sukhoi's new interceptor outperformed the *Firebar* in many respects, and the advisability of building a less capable fighter seemed questionable. Realising the threat posed by the Su-15, Aleksandr S. Yakovlev paid a visit to the plant – and ordered the competitor be carefully studied.

Having obtained the information he wanted, Yakovlev tried to incorporate some of the Su-15's features into the Yak-28P in order to increase the latter's speed and service ceiling. The objective was to excel not only the production *Firebar* but also its Sukhoi competitor! Development began in

Yakovlev OKB

Above: '64 Yellow', the sole prototype of the radically redesigned (and unsuccessful) Yak-28-64. The inboard pylons carry dummy R-98s, while the dummy missiles on the outer pylons are a mystery.

1964, hence the improved interceptor received the provisional designation Yak-28-64. Actually the word 'improved' is not applicable; 'redesigned' is more correct, since the finished result was a completely different aircraft.

To reduce drag the general arrangement was altered considerably, with the engines – R11F2-300s in this case – buried in the fuselage and breathing through lateral air intakes; this resulted in a much wider centre and rear fuselage. A forked 'pen nib' fairing reminiscent of the MiG-19 was located between the engine nozzles, with an elongated brake parachute canister above it.

The large boxy intakes with vertical movable ramps and boundary layer splitter plates were copied directly from the Su-15. This was a forced measure, since the Yakovlev OKB had no prior experience with lateral intakes.

The basic wing design was identical to that of the *Firebar* but the engine nacelles were deleted, of course. This allowed the ailerons to be extended inboard and split into two sections in an attempt to finally cure the aileron reversal problem. Each wing had two pylons; the inner ones were occupied by R-8M-1 AAMs while the outer pylons carried dummy missiles and were intended for a new 'dogfight AAM' then under development.

The forward fuselage and cockpit section, tail unit, landing gear and control system remained unchanged but the Yak-28's single ventral fin was replaced by outward-canted twin strakes.

The wider fuselage of the Yak-28-64 allowed the engineers to re-introduce drop tanks, increasing the fighter's endurance. The tanks were carried on special pylons under the air intake ducts.

Aptly coded '64 Yellow', the Yak-28-64 prototype was rolled out at MMZ No.115 in 1966. Compared to the Yak-28P, which would win no prizes at a beauty contest but was still a sleek machine with an air of speed and brute force about it, the Yak-28-64 was downright ungainly, the bulky fuselage making the aircraft look definitely overweight.

Early test flights revealed appallingly poor performance (the aircraft could not even surpass its production stablemate). Moreover, the handling characteristics were unsatisfactory (the aileron reversal problem persisted, despite the redesigned ailerons). Hence the Yak-28-64 was soon abandoned.

Yakovlev OKB

The Yak-28-64 looked as if it was weighed down by an excessively heavy rear end. This side view shows the brake parachute container and the large outward-canted ventral fins.

5

COMBAT AIRCRAFT PROJECTS SINCE 1955

Yak-2VK-11 high-altitude tactical bomber and reconnaissance aircraft

On 15th August 1956 the Council of Ministers issued a directive ordering the Yakovlev OKB to develop a two-seat high-altitude tactical bomber and its reconnaissance derivative. The aircraft was to be powered by two Klimov VK-11 afterburning turbojets delivering 4,500 kgp (9,920 lbst) at full military power and 9,000 kgp (19,845 lbst) in full afterburner (some sources state this engine's dry thrust as 6,100 kgp/13,450 lbst or 5,370 kgp/11,840 lbst). The first of the five bomber prototypes ordered was to enter flight test in the first quarter of 1958 and be submitted for State acceptance trials in the fourth quarter of the same year. The reconnaissance version was to be presented for Air Force check-up trials in the third quarter of 1958.

The bomber's stipulated performance included a maximum speed of 1,300 km/h (808 mph) at full military power; top speed with reheat was to be 2,500 km/h (1,552 mph). The range with 5% fuel reserves was to be 2,500 km (1,552 miles); a service ceiling of 20,000-21,000 m (65,620-68,880 ft) was anticipated. The aircraft was to have a take-off and landing run of 1,300 m (4,265 ft). It was intended to carry a normal bomb load of 1,200 kg (2,650 lb), the maximum bomb load being 3,000 kg (6,615 lb).

The reconnaissance version was to have a range of 2,500 km (1,554 miles) and a service ceiling of 22,000 m (72,160 ft). The design performance figures quoted above were considerably superior to those stipulated for the Yak-28. The project did not reach the hardware stage (according to one account, the work was discontinued in favour of adapting the Yak-32 described below in this chapter).

Yak-30, Yak-32 and Yak-34 reconnaissance aircraft family

These three projects of supersonic photo reconnaissance aircraft had much in common. Their powerplant comprised two Klimov VK-13 turbojets rated at 7,100 kgp (15,655 lbst) dry and 10,000 kgp (22,040 lbst) with reheat. In their aerodynamic layout they bore a close similarity to the Yak-28, featuring wings with 45° sweep. They differed from the Yak-28 in having greater wing area (up to 70 m^2/753.5 sq ft) and being refined aerodynamically. The three versions are described separately below.

Yak-30 reconnaissance aircraft (first use of designation)

The Yak-30 was the initial project which underwent wind tunnel tests from September 1957 onwards. The aircraft's design performance was as follows: normal all-up weight 29,600 kg (65,268 lb); AUW with drop tanks 34,900 kg (76,955 lb); fuel load 11,000 and 16,000 kg (24,255 and 35,280 lb) respectively. The Yak-30 was to have a top speed of 2,450 km/h (1,523 mph), a service ceiling or 17,900 m (58,710 ft), a range of 3,700 km (2,300 miles) in 'clean' configuration and 5,300 km (3,294 miles) with drop tanks.

The Yak-30 designation had previously been used for a fighter developed in 1948 (see Chapter 3) and was subsequently re-used for a jet trainer of 1959 (see Chapter 7).

Yak-32 reconnaissance aircraft and bomber (first use of designation)

The Yak-32 two-seat tactical reconnaissance aircraft was developed in response to a directive issued by the Council of Ministers on 31st July 1958. The machine was to enter State acceptance trials in the first quarter of 1960. Its intended mission was reconnaissance in daytime and at night and in adverse weather conditions; the Yak-32 was expected to operate within a broad range of speeds and altitudes. In common with the Yak-30, the powerplant consisted of two Klimov VK-13 afterburning turbojets (an alternative version featured two Lyul'ka AL-7F-1 afterburning turbojets). This project also served as a basis for developing a tactical bomber to the same specification as the abandoned Yak-2VK-11 project (see above). The work on the Yak-32 proceeded as far as the advanced development project stage; the ADP was endorsed by Aleksandr S. Yakovlev on 25th May 1959.

The Yak-32 possessed the following design characteristics: wing span 14.52 m (47 ft $7\frac{21}{32}$ in), length 24.60 m (80 ft 8½ in) (dimensions of the reconnaissance version); wing area approximately 70 m^2 (753.5 sq ft), reduced to 60 m^2 (645.9 sq ft) in one of the versions studied. The normal take-off weight was to be 26,000 kg (57,330 lb) for the bomber and 23,500 kg (51,810 lb) for the reconnaissance version; maximum TOW was 27,000 kg (59,535 lb) in both cases, with a maximum ordnance load of 3,000 kg (6,610 lb) in both cases. Maximum speed was 2,700 km/h (1,678 mph) for the bomber and 2,500 km/h (1,552 mph) for the reconnaissance version. The bomber's range was 3,000 km (1,865 miles), reduced to 2,600 km (1,616 miles) in the reconnaissance version; drop tanks increased the range to 3,600 km (2,237 miles). The service ceiling was 20,000 m (65,620 ft) for the bomber and 21,000 m (68,880 ft) for the reconnaissance version.

The Yak-32's design incorporated such features as an extended and drooped leading edge on the outer wing panels to create a dogtooth which served to enhance the aircraft's manoeuvrability.

The Yak-32 designation was also used for the single-seat sports version of the Yak-30 two-seat jet trainer; there was also a proposal for a heavyweight helicopter bearing the same designation.

Yak-34 supersonic reconnaissance aircraft

This project, the most advanced of the trio, envisaged exceptionally high performance. The Yak-34 was to have a maximum speed of 3,000 km/h (1,865 mph), a service ceiling of 21,000-22,000 m (68,880-72,160 ft) and a range of 3,400 km (2,113 miles) in optimum cruise mode or 2,200 km (1,367 miles) at a speed of 2,500 km/h (1,554 mph).

In March 1962 the OKB proposed a version designated Yak-34R and powered by Tumanskiy R21-300 (or R21A-300) engines. Under the terms of the proposal, the aircraft was to be submitted for flight tests in the fourth quarter of 1965. Design work on this project was discontinued in 1963.

Yak-33 multi-purpose supersonic VTOL aircraft family

In early 1960s the Yakovlev OKB embarked on projecting a family of aircraft collectively referred to as the Yak-33. Sharing a common airframe and differing in mission equipment,

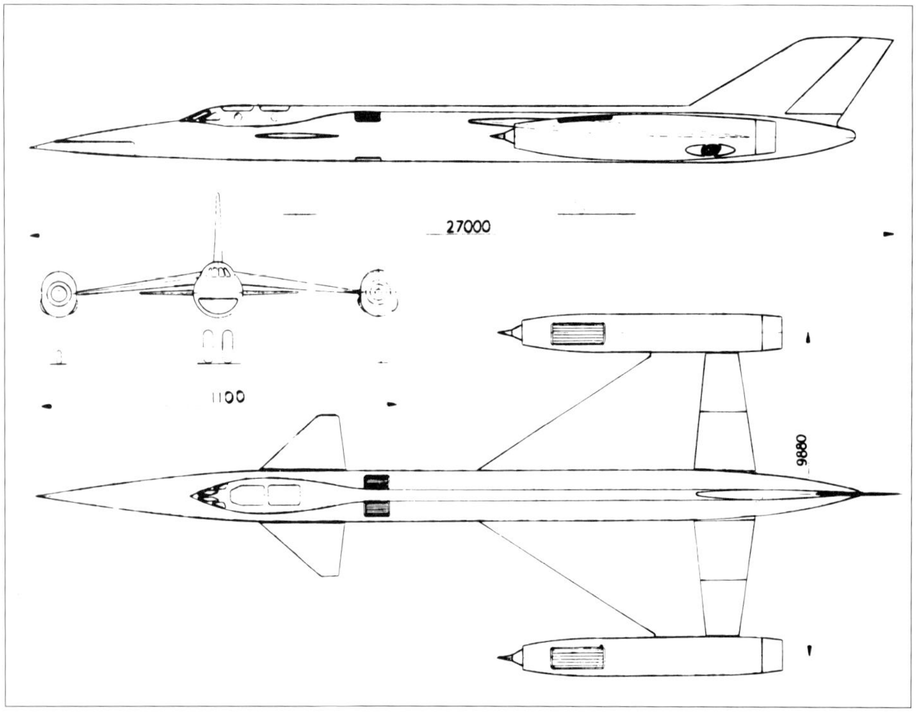

One of the Yak-33's project configurations featuring a canard layout. Note the downward-directed nozzles on the outer faces of the engine nacelles and the lift-jets buried in the forward fuselage.

the family included a bomber, an interceptor and a reconnaissance aircraft. These were supersonic VTOL aircraft (a concept that, with the benefit of hindsight, may well be regarded as being ahead of its time). The aircraft were configured for high-speed low-altitude missions. One of the projects of this family, namely the bomber, was submitted as a contender in a competition for a supersonic strike bomber in 1962.

The Yak-33 was designed in several alternative basic layouts, some of which were rather unorthodox. These included a machine with high-set delta wings and a traditional tail unit, a tailless delta-wing aircraft with high-set wings, and a version with a canard layout.

The tailless-delta version was powered by two cruise engines placed side-by-side in the rear fuselage. These were presumably Kolesov RD36-41 afterburning turbofans with a nominal rating of 7,000 kgp (15,435 lbst) dry and a reheat (take-off) rating of 10,000-16,000 kgp (22,050-35,280 lbst). On take-off the jetpipes would be diverted upstream of the afterburner to direct the jets downwards, thus helping to lift the aircraft. The cruise engines breathed through lateral air intakes (fashioned like those of the later Tu-22M3). They were to be supplemented by six to eight lift engines placed between the air intake ducts and delivering a thrust of 3,000 kgp (6,610 lbst) apiece. The bomber's armament comprised the Kh-45 anti-shipping missile with a range of 1,500 km (932 miles), which was under development at MKB ***Ra****duga* ('Rainbow' Machinery Design Bureau), and various free-fall bombs.

The alternative version featuring a canard layout was rather unorthodox in its powerplant arrangement. Two RD36-41 cruise engines (with thrust-vectoring nozzles, as above) were accommodated in nacelles attached to the tips of a delta wing; the same nacelles, measuring 9.88 m (32 ft 5 in) in length, also housed two lift engines each. In addition, a pair of lift engines was accommodated in the fuselage amidships, behind the canard foreplanes. Both this and the preceding version featured a bicycle undercarriage with outriggers placed at the wingtips and retracting into special fairings (in the case of the tailless-delta layout) or into the engine nacelles (in the case of the canard layout). In all versions the crew comprised a pilot and a navigator seated in tandem in a pressurised cockpit.

The dimensions and some design performance figures for the bomber version of the Yak-33 in the tailless delta configuration are as follows (figures for the canard layout version are given in the brackets): Length 26.375 m/86 ft $6\frac{25}{64}$ in (27.0 m/88 ft 7 in); wing span 10.25 m/33 ft 7½ in (11.1 m/36 ft 5 in). Normal take-off weight in both cases was 32,000 kg (70,560 lb) and maximum take-off weight up to 40,000 kg (88,200 lb); the lift engines in both cases were eight 3,000-kgp turbojets. Maximum and cruising speed, as stipulated by the operational requirements, were Mach 3 and Mach 2 respectively in both cases; specified range in both cases was 4,000 km (2,486 miles).

The bomber project described above was rejected because it failed to meet some of the stipulated performance figures. The lack of production engines with the required performance must have been a contributing factor. It should be noted that aircraft companies in other major aircraft-manufacturing countries gave up the idea of developing a heavy supersonic VTOL aircraft due to the excessive complexity of the task.

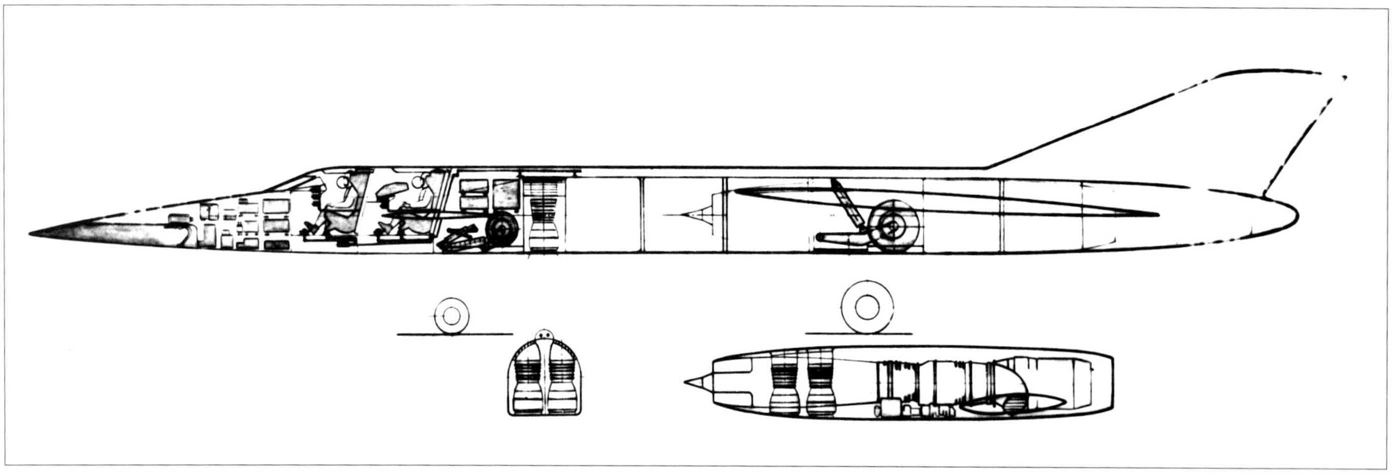

This cutaway drawing of the Yak-33 in the canard configuration depicted above shows the tandem cockpits, the lift engines side by side aft of them and the nacelle design with pairs of lift engines in the air intake centrebodies and the blocker door ahead of the afterburner directing the cruise engine exhaust downward.

Yak-35MV fighter and fighter-bomber

In the late 1950s OKB-115 developed a project for a low-altitude tactical fighter designated Yak-35MV (MV obviously stands for ***ma**lovysotnyy* – low-altitude). It was an interceptor powered by a single RD11-300 afterburning turbojet (an early designation for the R11-300). The machine was intended for destroying airborne targets at altitudes between 200 m (660 ft) and 10,000 m (32,810 ft), for which purpose it was to be fitted with the Vozdookh-1 GCI system.

The specifications for the project were formalised by a Council of Ministers' directive dated 4th June 1958. It stipulated that the aircraft was to commence State acceptance trials in the fourth quarter of 1961.

The baseline fighter version served as a basis for developing a project for a fighter-bomber (tactical bomber). The latter was expected to incorporate a number of features designed to enhance survivability under conditions of heavy opposition from enemy anti-aircraft artillery and small arms fire. This was a single-seat aircraft with a take-off weight of 15,000 kg (33,075 lb), capable of carrying a 2,000-kg (4,410-lb) weapon load.

Design performance included a maximum speed of 1,500-1,600 km/h (932-994 mph) at 1,000 m (3,280 ft), reduced to 1,300-1,350 km/h (807-839 mph) at ground level with bombs carried externally, and a service ceiling of 10,000 m. The aircraft was to be powered by two Tumanskiy R11A-300 afterburning turbojets (a source gives the engine designation as RNA-300, but this must be a misprint).

The work on this project was severely handicapped by a temporary loss of interest on the part of the nation's leaders towards the development of new manned combat aircraft types (Khrushchov's notorious missile euphoria) with the resultant lack of funding. This was compounded by delays in the development of the engines. Thus, the project failed to reach the hardware stage.

Yak-45 fighter

In 1971-72 the Ministry of Aircraft Industry held a design contest for a new fighter. In this connection the Yakovlev OKB undertook a series of studies for a single-seat fighter equipped with radar and missiles. These projects bore a common designation Yak-45. The fighter was to be powered by two Favorskiy Type 69 non-vectoring augmented turbofans, derived from the R-28, each rated at 8,000 kgp (17,635 lbst). The engines were underslung ahead of broad delta wings supplemented by large canard foreplanes. The aircraft was expected to possess outstanding manoeuvrability. Together with the Yak-47 (described below) the Yak-45 was submitted for the contest.

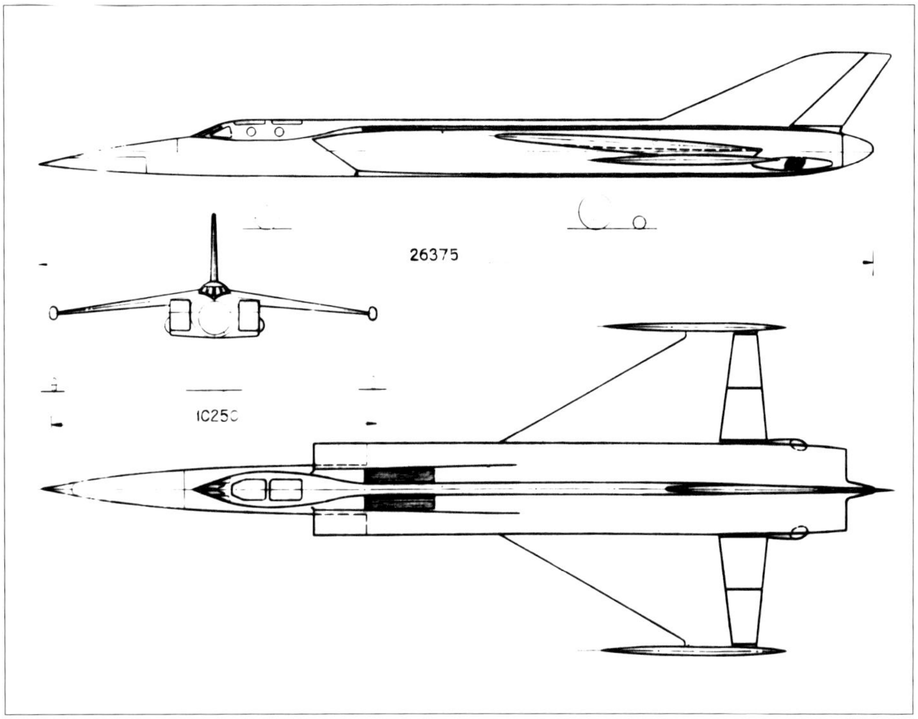

This drawing depicts an alternative tailless-delta version of the Yak-33. Note the two-dimensional air intake design, the position of the lift-jets and the main engines' lift nozzles and the 'pen nib' fairing between the rear (afterburner) nozzles.

The other contenders were the Mikoyan OKB with the MiG-29 and the Sukhoi OKB with the T-10 (the future Su-27). By the summer of 1972 the contest was concluded, MAP declaring the Sukhoi project as the winner (the MiG-29 also received a go-ahead due to the original requirement being split into heavy fighter and light fighter categories). The Yakovlev project was not proceeded with, having lost to these contenders.

Yak-45 V/STOL version

In 1978-79 Yakovlev developed a V/STOL version of the Yak-45 with rectangular vectoring nozzles and two RD-38 lift jets in the fuselage. This project fared no better than the others. It posed a major problem: how to survive a failure of any engine in the hover mode.

Yak-45I fighter

One of the Yak-45 project studies submitted for the contest in the early 1970s featured a layout differing from the one described above. Dubbed Yak-45I (some sources call it Yak-45M, but this must be an error), it had no canard foreplanes and was fitted with a horizontal tail of delta planform. A published drawing shows it to have broad delta wings similar to those of the initial project, but the engines were mounted on the wings rather than ahead of them, the nozzles protruding aft of the wing trailing edge. The engines' air intakes featured protruding shock cones. The wings had compound sweep on the leading edges (approximately 40° outboard

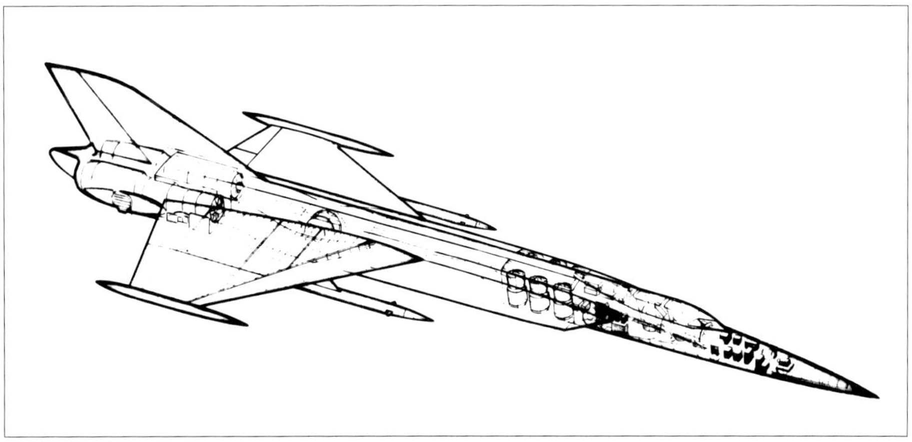

A cutaway drawing of the tailless-delta Yak-33. Note the missile pylons under the wings, the only possible position.

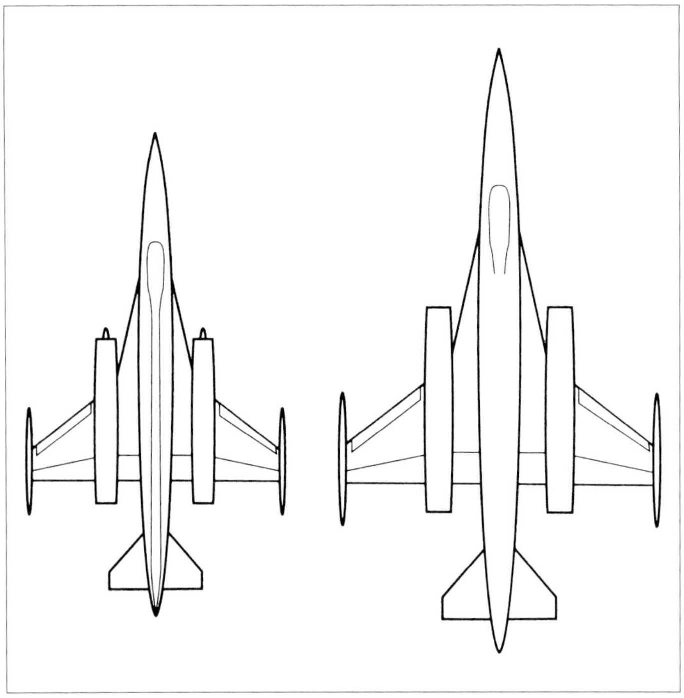

Above: These drawings to the same scale illustrate the similarity of the projected Yak-45I (left) and Yak-47 fighters developed to compete with the future MiG-29 and Su-27 respectively.

Yakovlev OKB

Yakovlev OKB

Right and above right: A wind tunnel model of Yakovlev's contender for the LFI fifth-generation fighter programme.

of the engines and 75° between the engines and the fuselage). The wingtips carried slender fairings (possibly for outrigger wheels, indicating the use of a bicycle undercarriage).

Yak-47 fighter

This fighter project, also submitted for the contest, closely resembled the Yak-45M described above, differing in having a slightly longer forward fuselage and greater wingspan. There were no protruding shock cones in the engine air intakes.

LFI light tactical fighter and other similar projects

In the 1980s the Yakovlev OKB was involved in the work on a fifth-generation fighter to succeed the Su-27 and the MiG-29, taking part in a contest under the I-90 programme (*istre**bi**tel'* – fighter; that is, the name of the programme denoted 'fighter for the 1990s'). The Soviet fifth-generation fighter concept envisaged parallel development of two basic types. One was a twin-engined long-range heavy fighter with a powerful long-range radar and a large complement of missiles, the other was a single-engined light tactical fighter having the maximum possible degree of commonality with the 'heavy' type. Yakovlev proposed a single-engined aircraft with a canard layout, delta wings and twin tails. The airframe incorporated a number of stealth features. Unusually, the wings and canards alike had a double-kinked trailing edge. The forward fuselage with sharp chines was reminiscent of the Lockheed Martin F-22 Raptor. The engine was to have a vectoring nozzle which moved only in the vertical plane.

However, the single-engined layout did not find favour with the Soviet military, and the Yakovlev fighter was rejected in favour of the multi-role tactical fighter developed by the Mikoyan OKB (eventually to be known as the 1.42 and 1.44 projects).

In early 2000 the Yakovlev OKB submitted to the Russian Ministry of Defence a project dubbed LFI (***lyoh**kiy fronto**voy** istre**bi**tel'* – light tactical fighter). It was a fifth-generation aircraft featuring VTOL capability. However, from the outset the OKB's leader Aleksandr N. Dondukov was sceptical about the possibility of obtaining state funding for the project.

As noted in the Introduction, eventually the leading role in the development of a fifth-generation fighter was allocated to the Sukhoi Corporation, with Yakovlev OKB's participation as a rather vague possibility.

Performance comparison of the Yak-45I and Yak-47 (project data)

	Yak-45I	Yak-47
Powerplant	2 x R53F-300	2 x R59F-300
Thrust in full afterburner, kgp (lbst)	2 x 8,200 (2 x 18,080)	2 x 12,500 (2 x 27,560)
Wing area, m² (sq ft)	40 (430.1)	65 (698.9)
Normal TOW, kg (lb)	13,900 (30,640)	22,800 (50,260)
Take-off thrust/weight ratio	1.18	1.1
Wing loading, kg/m² (lb/sq ft)	348 (71.34)	350 (71.76)
Maximum speed, km/h (mph):		
at sea level	2,500 (1,552)	2,500 (1,552)
at high altitude	1,500 (931)	1,500 (931)
Service ceiling, m (ft)	21,500 (70,540)	20,000 (65,620)
Maximum rate of climb at sea level, m/sec (ft/min)	340 (66,912)	275 (54,120)
Acceleration, seconds:		
600 to 1,100 km/h (372 to 683 mph)	14.5	16.0
1,100 to 1,300 km/h (683 to 807 mph)	6	7.7
Range without drop tanks, km (miles):		
at sea level	1,000 (621)	1,000 (621)
at high altitude	2,500 (1,552)	2,500 (1,552)

UTILITY AIRCRAFT AND TRAINERS OF 1941-1953

6

Yak-6 (NBB) light bomber and utility transport

In 1941-42 the harsh realities of the war revealed the need for a twin-engined light utility aircraft, the single-engined Polikarpov U-2, Yakovlev UT-2 and Polikarpov R-5 being clearly inadequate in terms of payload, range and operational safety. In April 1942 the Yakovlev OKB was entrusted with the development of such an aircraft which was designated Yak-6 (according to Yakovlev, the task came from Stalin personally). The aircraft's main role was that of liaison and utility transport, but the designers aspired from the outset to make it as versatile as possible. It was considered suitable for adaptation to the role of a light night bomber; this version was designated NBB (*noch**noy** **blizh**niy bombardi**rov**shchik*, short-range night bomber, described below).

The Yak-6 was a twin-engined cantilever low-wing monoplane with a conventional tail unit and a retractable undercarriage. Many of its sub-assemblies and systems were taken from existing aircraft; for example, the Shvetsov M-11F engines driving two-blade wooden propellers were taken from the UT-2, and electric and pneumatic system components, as well as wheels and other parts, were borrowed from fighters which were in large-scale production at that time. The airframe was made of readily available materials: spruce, birch plywood, fabric.

Yakovlev OKB

Above: The first prototype of the Yak-6 light utility aircraft. Note the exposed engine cylinder heads and the (apparently) fixed undercarriage.

Yakovlev OKB

Above: This silver-doped Yak-6 prototype features completely cowled engines, propeller spinners and obviously retractable main gear units. The strong wing leading-edge sweep is readily apparent.

Yakovlev OKB

A batch of camouflaged 'Baby Douglases' awaits delivery at one of the plants that built the type. Again, these aircraft represent the fixed-gear variety and appear to be in unarmed utility configuration.

Yakovlev OKB

Above: The NBB night bomber version of the Yak-6 undergoing State acceptance trials. Note the five 100-kg bombs under the wing centre section, the dorsal ShKAS machine-gun, the retractable landing gear and the DF loop aerial under the nose. The engines are fitted with exhaust collector rings.

Yakovlev OKB

Another Yak-6 (NBB) pictured during trials.

Specifications of the Yak-6 utility aircraft and light bomber

	Yak-6 prototype	Yak-6 production	Yak-6 modified
Engine type	M-11F	M-11F	M-11FM
Engine power, hp	140	140	145
Length	10.35 m (33 ft 11½ in)	10.35 m (33 ft 11½ in)	n.a.
Wing span	14.0 m (45 ft 11¼ in)	14.0 m (45 ft 11¼ in)	n.a.
Wing area, m² (sq ft)	29.6 (319)	29.6 (319)	n.a.
Empty weight, kg (lb)	1,368 (3,016)	1,415 (3,120)	n.a.
All-up weight, kg (lb)	2,350 (5,180)	2,300 (5,070)	2,500 (5,510)
Maximum speed, km/h (mph)	183 (114)	187 (116)	230 (143)
Landing speed, km/h (mph)	93 (58)	93 (58)	n.a.
Service ceiling, m (ft)	n.a.	3,380 (11,090)	n.a.
Range, km (miles)	n.a.	900 (559)	1,040 (646)
Take-off run, m (ft)	285 (935)	280 (920)	255 (840)
Landing run, m (ft)	265 (870)	220 (720)	115 (380)

Even the propellers were made of spruce instead of the usual hardwood. Impregnated thin plywood was used for the manufacture of the fuel tanks, normally made of metal sheet or rubber. The few metal parts that were used in the airframe were made of low-grade steel.

The fuselage comprised three sections joined together by steel fittings. The forward and centre sections had a semi-monocoque structure, while the rear fuselage was a truss with fabric covering. The two-spar wings featuring a Clark YH airfoil were a one-piece structure with detachable wingtips to facilitate transportation. The slotted ailerons were fabric-covered. The space between the spars in the centre section accommodated fuel tanks, and the aft part of the centre section incorporated a landing flap. The two-spar tail surfaces were mainly wood, with fabric covering aft of each front spar. Each control surface had a hinged trim tab. The main undercarriage units had single oleo legs with lateral jury struts, with outboard-mounted single wheels carried on steel axles and equipped with pneumatically operated brakes. They retracted aft by means of crank arms pulled by cables from a hand-driven windlass. Later many series aircraft had a non-retractable undercarriage with a rigid strut at the rear instead of the breaker strut. The non-retractable tailwheel was free to castor. The aircraft had dual controls with cable linkages, the two pilots sitting side by side.

In the utility transport and liaison version the Yak-6 could carry four passengers or 500 kg (1,102 lb) of cargo in addition to the two crewmen.

Design work and prototype construction were conducted at plant No.47 in Chkalov (later renamed Orenburg) with Mikhail Leonov as project designer. The work proceeded at a breakneck tempo (Yakovlev gave his staff a mere two months for accomplishing this task), and the prototype made its maiden flight as early as the beginning of June 1942 (according to Yevgeniy Adler).

NBB light bomber prototypes

Actually, two prototypes were built simultaneously, both of them in the NBB version. In the bomber variant the aircraft could carry five 100-kg (220-lb) bombs or two 250-kg (551-lb) bombs suspended under the wing centre section. The NBB was equipped with a radio and a direction finder; the exhaust stubs were fitted with flame suppressors. To provide protection from the rear, the NBB was provided with a 7.62-mm ShKAS machine-gun placed atop the fuselage above the passenger/cargo cabin and accessed through a hatch. The crew comprised a pilot and a navigator, the latter doubling as gunner when necessary.

The first prototype was built with a fixed undercarriage. On 20th June, after the initial flight testing at plant No.47, it was flown to Moscow where it was subjected to testing jointly by the OKB and LII with Fyodor Bolotov and Yuriy Shiyanov as test pilots respectively. The machine created a very good impression; soon the second prototype, fitted with a retractable undercarriage, was test flown and ferried to Moscow.

Production Yak-6 light bomber (NBB) and utility transport aircraft

The Yak-6 passed its State Acceptance trials in September 1942 and was put into series production at three plants: No.47 in Chkalov, No.471 in Shumerlya and No.464 in Tushino, then a suburb of Moscow (but now part of the city). Plant No.47, evacuated from Leningrad, was responsible for the bulk of Yak-6 production; it started turning out Yak-6s on 1st November 1942, manufacturing 25 machines before the end of 1942. In 1943 the plant produced 224 (according to some sources, 217) Yak-6s. Production of the other two plants did not get under way until 1943; that year they produced 50 machines (plant No.464) and 82 aircraft (plant No.471). A total of 381 production Yak-6/NBB aircraft had been completed by the time production was terminated in 1943.

The Yak-6 was produced in both armed and unarmed versions, finding use in a wide variety of roles and earning the affection of its crews and all those who benefited by its services. It was a simple, rugged and dependable aircraft, very easy to pilot and maintain. Resembling a scaled-down Douglas DC-3 (produced in the USSR as the Lisunov Li-2), the Yak-6 was affectionately dubbed *Dougla**syo**nok* (Baby Douglas). This very useful aircraft seemed to have all the prerequisites for becoming as ubiquitous as the famous U-2. Yet, its manufacture was halted after a relatively small production run. The decision was taken after several crashes which revealed some problems with the aircraft's handling that were overlooked

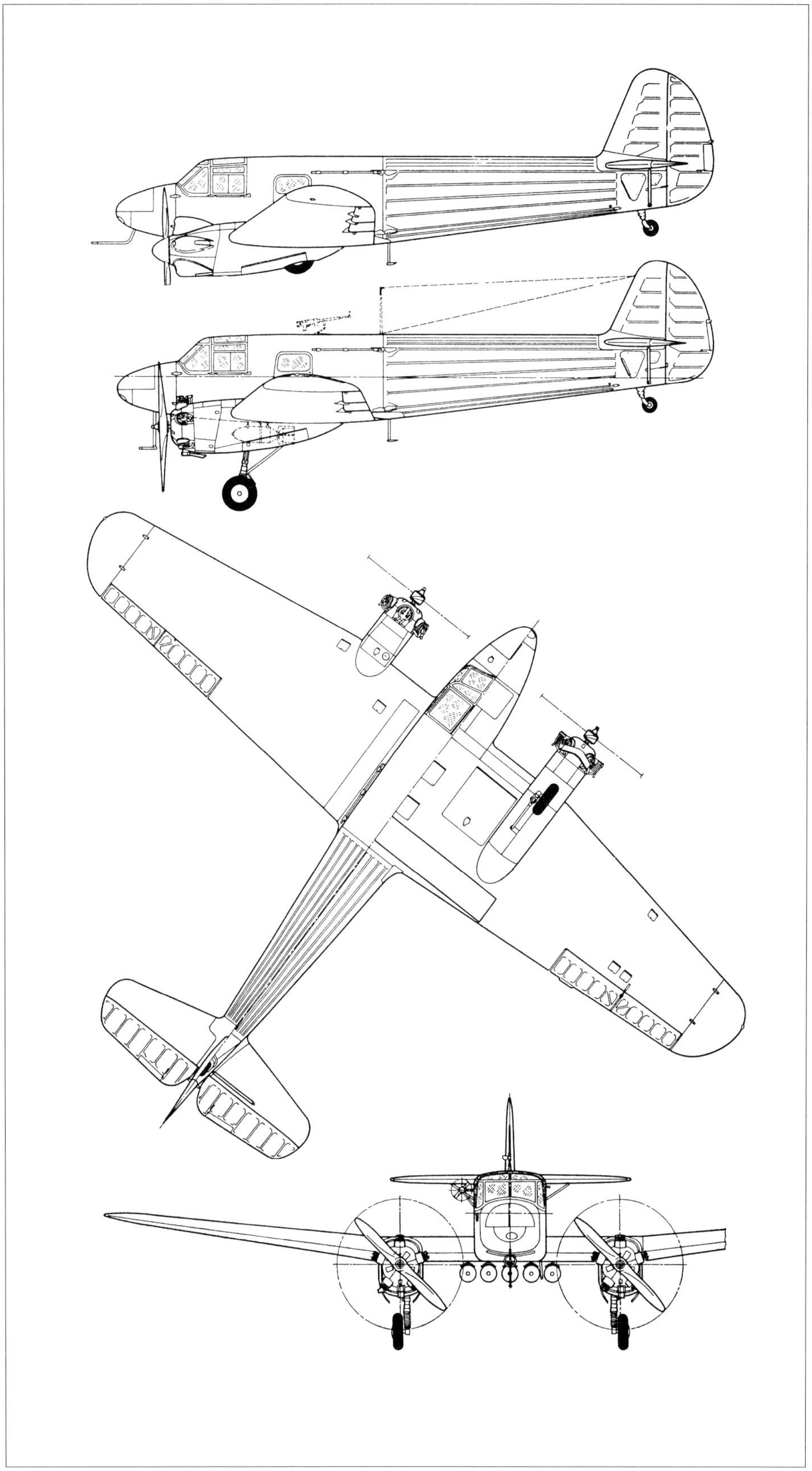

A three-view of the Yak-6 (NBB) light bomber in fixed-gear configuration, with an additional side view of a retractable-gear utility example.

Yakovlev OKB

Yakovlev OKB

Top and above: The first prototype of the Yak-8 transport; its commonality with the retractable-gear Yak-6 is obvious.

Yakovlev OKB

Yakovlev OKB

Two views of the second prototype Yak-8 which had a larger vertical tail. Note the different placement of the national insignia.

or neglected during the tests. It became obvious that the Yak-6 had poor directional stability coupled with excessive lateral stability. Its insufficient longitudinal stability margin harboured the danger of the aircraft entering a spin, which indeed happened in the event of overloading or careless piloting. Among other deficiencies cited was the propensity to propeller overspeeding. Faced with these problems, Yakovlev gave his consent to halting the manufacture of the Yak-6 in Chkalov, where it was superseded on the production line by the Shcherbakov Shche-2 transport. This decision, dated 25th September 1943, prevented the plant from starting the production of a modernised Yak-6, the prototype of which (c/n 014347 – that is, 01st aircraft of Batch 43 built by plant No.47) was to become a standard-setter for batches from 1st October 1943 onwards.

Yak-6 as an attack aircraft

In addition to the Yak-6's night bomber role, the Yakovlev OKB experimented with the idea of turning the light transport into an attack aircraft. In 1945 a Yak-6 was fitted with guide rails for rocket projectiles under the wings (up to ten rockets could be carried). One account states that some Yak-6s modified in this fashion were used operationally during the closing stages of the war.

Yak-6 modified (prototype)

Several examples of the Yak-6 were aerodynamically refined by TsAGI to improve the aircraft's performance and handling. The best results were obtained on the third modified example. It had completely redesigned outer wing panels with a swept leading edge and a straight trailing edge. The wing dihedral was increased to 8°30'. The height of the fin and rudder was increased; the main undercarriage units featured larger wheels which were semi-exposed when retracted. The aircraft was fitted with 145-hp M-11FM engines in helmeted cowlings driving VISh-327 variable-pitch propellers with spinners. The shape of the undernose pitot head was altered.

Yak-8 transport aircraft prototypes

This light passenger aircraft providing accommodation for six passengers and two crew entered manufacturer's tests in early 1944. It was designed by a team led by Oleg K. Antonov who later gained world renown as a designer in his own right. Two prototypes were built, the construction being supervised by Sinitsyn and Adler respectively. The second prototype, refined by Adler, featured a saving of some 250 kg (550 lb) in empty weight compared to the first one.

The Yak-8 resembled a slightly scaled-up Yak-6. It was intended to be powered by

190-hp Kossov M-12 radials, but these engines were not available in time and had to be replaced by M-11FM (M-11M) engines delivering 145 hp – a meagre increase in output over the 140-hp M-11Fs powering the Yak-6. The Yak-8 was of wooden construction with fabric skinning. The centre fuselage housing the cabins was a semi-monocoque structure formed by wooden frames; these were covered both on the outside and the inside with a skin of 2-mm ($0\frac{5}{64}$ in) plywood, resulting in a very rigid box-like structure. The rest of the fuselage was similar to that of the Yak-6. The two-spar wooden wing was made in a single piece. The aircraft had a tailwheel landing gear, the main units retracting aft into the engine nacelles.

The second Yak-8 successfully passed State acceptance trials and was recommended for series production which, however, failed to materialise. This was obviously due to the overall trend towards switching over to metal construction in all aircraft categories.

Specifications of the Yak-8 transport aircraft

Engine type	M-11FM
Engine power, hp	2 x 145
Length	11.35 m (37 ft $2\frac{7}{8}$ in)
Wing span	14.8 m (48 ft $6\frac{5}{8}$ in)
Wing area, m^2 (sq ft)	30.0 (323)
Empty weight, kg (lb)	1,750 (3,860)
All-up weight, kg (lb)	2,700 (5,950)
Maximum speed, km/h (mph)	248 (154)
Landing speed, km/h (mph)	100 (62)
Time to 1,000 m (3,280 ft), minutes	6.4
Service ceiling, m (ft)	3,900 (12,795)
Range, km (miles)	890 (553)
Take-off run, m (ft)	380 (1,250)
Landing run, m (ft)	260 (850)

Yak-16 (Yak-16-I) passenger aircraft prototype

After the end of the Second World War the Civil Air Fleet of the USSR turned its attention to establishing a network of local air services with low-density traffic, where the Li-2P (DC-3) would be too big. A requirement was formulated for a passenger aircraft capable of carrying ten passengers over a distance of 800 km (496 miles) at a cruising speed of 290 km/h (180 mph); it was to be powered by two Shvetsov ASh-21 five-cylinder radials with a nominal rating of 570 hp. A government directive tasking the Yakovlev OKB with the development of this aircraft was issued in February 1946. The first of the two prototypes was to be submitted for State acceptance trials by 1st November 1946. However, the OKB was overburdened with

Yakovlev OKB

Above: This view shows the large passenger/cargo door of the first prototype Yak-8.

Yakovlev OKB

Above: The first prototype Yak-8 in flight with the gear up. Note the strong wing dihedral.

Yakovlev OKB

Yakovlev OKB

Two views of the first prototype Yak-16 (Yak-16-I) serialled '464 Yellow'. A good-looking aeroplane it was.

Yakovlev OKB

Above: The Yak-16-I in flight, showing the retracted position of the main gear units. This aircraft was in airliner configuration with a small entry door to port.

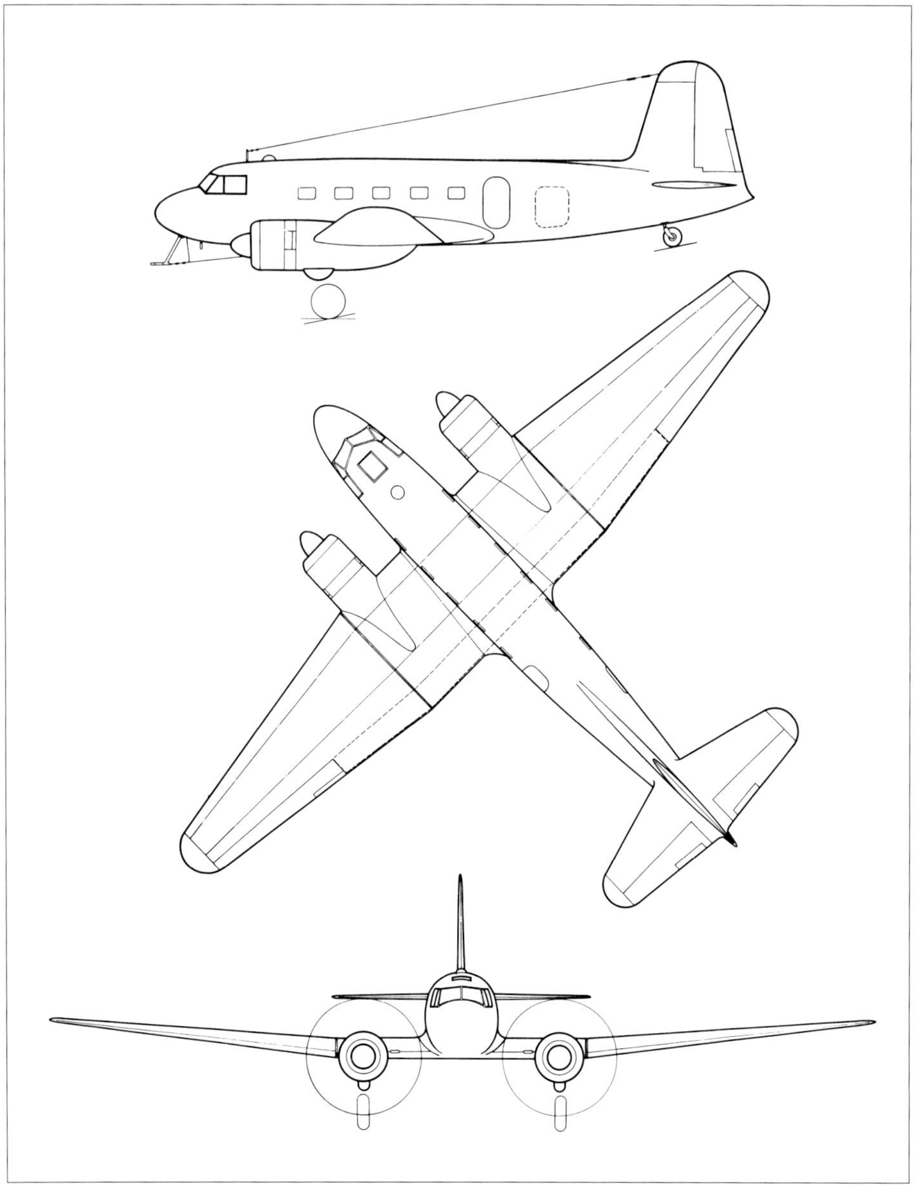

A three-view of the Yak-16-I. The square-shaped baggage door was on the starboard side.

other tasks, and the schedule had to be revised, shifting the State acceptance trials deadline to August 1947. The revised specifications included a maximum speed of no less than 350 km/h (217 mph) at sea level. The military version was to be provided with a dorsal turret armed with a single 20-mm cannon.

Construction of the prototype was undertaken by Plant No.464 which was placed at the disposal of the Yakovlev OKB from July 1946, supplementing the OKB's main plant No.115. The Yak-16 was a low-wing all-metal monoplane with two-spar wings which comprised a horizontal rectangular centre section and outer panels with dihedral and taper. The semi-monocoque circular-section fuselage provided accommodation for a two-crew flightdeck and a 10-passenger cabin supplemented by a toilet and a baggage hold. The conventional tail unit had duralumin frames, the fin and tailplane having a metal skin and the control surfaces being fabric-covered.

The main units of the tailwheel undercarriage retracted forward into the engine nacelles mounted on the wing centre section. When retracted, the main wheels were semi-exposed.

The aircraft had dual controls. The ASh-21 engines enclosed by NACA cowlings drove VISh-111V-20 two-blade variable-pitch propellers provided with large spinners.

Manufacturer's flight tests began in the autumn of 1947. The aircraft performed its maiden flight with test pilot Fyodor L. Abramov at the controls on 24th September. In the course of the testing the aircraft displayed good handling qualities. Test pilots noted its good stability and controllability, moderate stick forces, a good range of speeds, good acceleration at take-off, no tendency to veer off course during take-off and landing run. The Yak-16 displayed excellent single-engine handling, being capable of climb with one engine inoperative, which was an important safety asset. The aircraft could quickly be mastered by average-skilled pilots.

In early 1948 the Yak-16 passed its State acceptance trials at the Civil Air Fleet Research Institute (NII GVF – *Naoochno-issledovatel'skiy institoot Grazhdahnskovo vozdooshnovo flota*). This first prototype was sometimes unofficially referred to as Yak-16-I.

Yak-16-II military transport prototype

In the spring of 1948 the second prototype, also known as the Yak-16-II, joined the test programme. It was configured as a military transport and differed in several respects from the passenger machine. The changes included the installation of V-511 three-bladed feathering propellers with scimitar-shaped blades, a slight increase in the area of the vertical tail and the forward fuselage length, and the strengthening of the cargo cabin floor. The cargo cabin provided accommodation for seven paratroopers or ten fully equipped troops. In an ambulance version the aircraft could carry six stretcher cases and a medical attendant. The loading in this case was facilitated by the provision of a large clamshell side cargo door whose forward half incorporated a smaller inward-opening entry door.

Yakovlev OKB

Yakovlev OKB

Top and above: The Yak-16-II was a military transport featuring a large cargo door and a UTK-1 dorsal turret. Note the new V-511 three-blade propellers with curious scimitar-shaped blades.

Mounted above the front fuselage was a UTK-1 'ball turret' fitted with a 12.7-mm (.50 calibre) Berezin UBT machine-gun. The aircraft was equipped for the carriage of external stores: three TsDDM-120 paradroppable containers could be suspended on racks under the wing centre section. There was also provision for towing an assault glider.

The Yak-16-II passed its manufacturer's tests during April 1948 and was submitted for the State acceptance trials. Albeit the second prototype shared the good handling

Yakovlev OKB

The Yak-16-II ('2 Yellow') in flight. The small entry door incorporated into the cargo door is clearly visible.

Yakovlev OKB

Above: The Yak-16-I in later civilian guise as CCCP-И985 (that is, SSSR-I 985)

Specifications of the Yak-16 passenger and transport aircraft

	Yak-16-I	Yak-16-II
Length	15.6 m (51 ft 2¼ in)	16 m (52 ft 6 in)
Height, tail down	4.66 m (15 ft 3½ in)	n.a.
Wing span	21.5 m (70 ft 6½ in)	21.5 m (70 ft 6½ in)
Wing area, m² (sq ft)	56.25 (605)	56.25 (605)
Empty weight, kg (lb)	4,486 (9,982)	4,560 (10,055)
All-up weight, kg (lb):		
normal	6,021 (13,274)	n.a.
overload	6,230 (13,737)	6,350 (14,002)
Maximum speed, km/h (mph)		
at sea level	350 (217.5)	n.a.
at 2,250 m (7,380 ft)	370 (230)	352 (218.8)
Landing speed, km/h (mph)	95 (59)	105 (65)
Service ceiling, m (ft)	7,750 (25,428)	7,100 (23,295)
Time to 3,000 m (9,840 ft)	6 min	7 min
Range, km (miles)	800 (497)	n.a.
Take-off run, m (ft)	240 (787)	185 (610)
Landing run, m (ft)	250 (820)	240 (790)

qualities of the passenger machine, its military test pilot pointed out that the machine failed to meet some points of the military SOR. To rectify this, several modifications were effected, including an increase in the horizontal tail area and installation of de-icers on the wings and tail surfaces. These and other minor changes led to an increase in the AUW which had to be compensated for by deleting some items of equipment and refining the structure. Thus modified, the Yak-16-II was submitted for a second round of State acceptance tests.

The machine received a positive appraisal, but this proved insufficient to achieve production status. Both the passenger aircraft and the military transport failed to progress beyond prototype stage. This was because a decision was taken in 1948 to order the now famous Antonov An-2 biplane into series production. The decision was motivated by the An-2's greater versatility, albeit the Yak-16 had an edge in passenger comfort. The Yak-16-I (which had carried military markings and the serial '464 Yellow') was registered CCCP-И985 (SSSR-I985) and was used for some time by plant No.464 for transport duties; the 'I' operator designator denoted the People's Commissariat/Ministry of Aircraft Industry. Some efforts were made to promote it for export. In 1948 this machine was shown at an international trade fair in Poznań (Poland) and reportedly also made demonstration visits to Hungary, Finland, Romania and Czechoslovakia. However, foreign air carriers refrained from ordering a machine which was not adopted

Yakovlev OKB

The first prototype of the Yak-14 assault glider, with a very early-production GAZ-51 *sans suffixe* lorry (the 1946 model with a wood/metal cab) parked alongside. Note the lack of a dorsal fin, the car-type door of the glider's cockpit and the landing lights buried in the underside of the nose.

by Aeroflot. Later the Yak-16-I was reregistered CCCP-И1074 and was used by Plant No.115 (the Yakovlev OKB).

Yak-14 assault glider prototype

In 1948 the Yakovlev OKB was tasked with developing a transport and assault glider for the Airborne Troops (VDV – *Vozdooshno-desahntnyye voyska*). The specifications envisaged a machine capable of carrying a lorry or a field gun together with its crew. Initially a payload of 3,000 kg (6,610 lb) was specified. A team headed by project engineers Yevgeniy G. Adler and Leonid L. Selyakov set about the work which seemed to pose some problems due to the lack of previous experience with assault gliders. To overcome this obstacle, the designers carefully studied the structure of a German Gotha Go 242 assault glider captured by Soviet troops and borrowed some structural design features from it (although the two machines were very different externally). The first prototype was built as early as 1948; the testing was conducted by pilots from the VDV.

The Yak-14 featured high-set strut-braced wings with a rectangular centre section and tapering outer panels, and a slab-sided fuselage; the nose ahead of the cockpit and the complete rear fuselage/tail unit could hinge open for loading and unloading. A detachable loading ramp allowed wheeled vehicles to be rolled in. The cockpit seating two pilots side by side was placed above the forward part of the cargo hold and offset to port for better visibility. The cargo cabin could accommodate up to 35 troops (albeit this number was initially limited to 25) or various items of weaponry and vehicles (artillery pieces with their tugs, self-propelled guns, jeeps and the like). The tricycle undercarriage incorporated a 'kneeling' feature, enabling the glider to lie flat on its belly for ease of loading and unloading. This feature also permitted the glider to make belly landings on two wooden skids mounted under the fuselage.

The first prototype could be identified by the lack of a dorsal fin (which appeared on later machines); it had a wing area of 72,98 m^2 (785.6 sq ft, later increased) and an AUW of 6,250 kg (13,780 lb).

The test flights conducted with an Il'yushin IL-12D military transport as a tug showed that the glider carrying a payload of 3,180 kg (7,012 lb) could be towed at a speed of up to 292 km/h (181 mph) at an altitude of 2,000 m (6,560 ft). At the same time the tests revealed a number of deficiencies which had to be rectified before submitting the machine for State acceptance trials. This work resulted in the construction of a second prototype which subsequently formed the basis for the production version.

Sergey and Dmitriy Komissarov archive

Above: The pilots clamber into the cockpit of the Yak-14 prototype as a further crewman enters the cargo cabin. The full-length skids on the bottom of the fuselage are visible.

Sergey and Dmitriy Komissarov archive

Above: This view shows how the landing gear could be semi-retracted, making the Yak-14 rest on the skids to facilitate the loading of vehicles – in this case, a GAZ-67B jeep with a 45-mm anti-tank gun.

Sergey and Dmitriy Komissarov archive

Seen through a whirl of snow from the IL-12D glider tug's propellers, the Yak-14 becomes airborne. The shape of the tail is clearly visible; note the stabiliser bracing wires.

Yakovlev OKB

Above: A production Yak-14 glider painted olive drab overall. The dorsal fin and the upper stabiliser bracing struts are clearly visible. Note that the skids are much shorter and located in line with the main gear units.

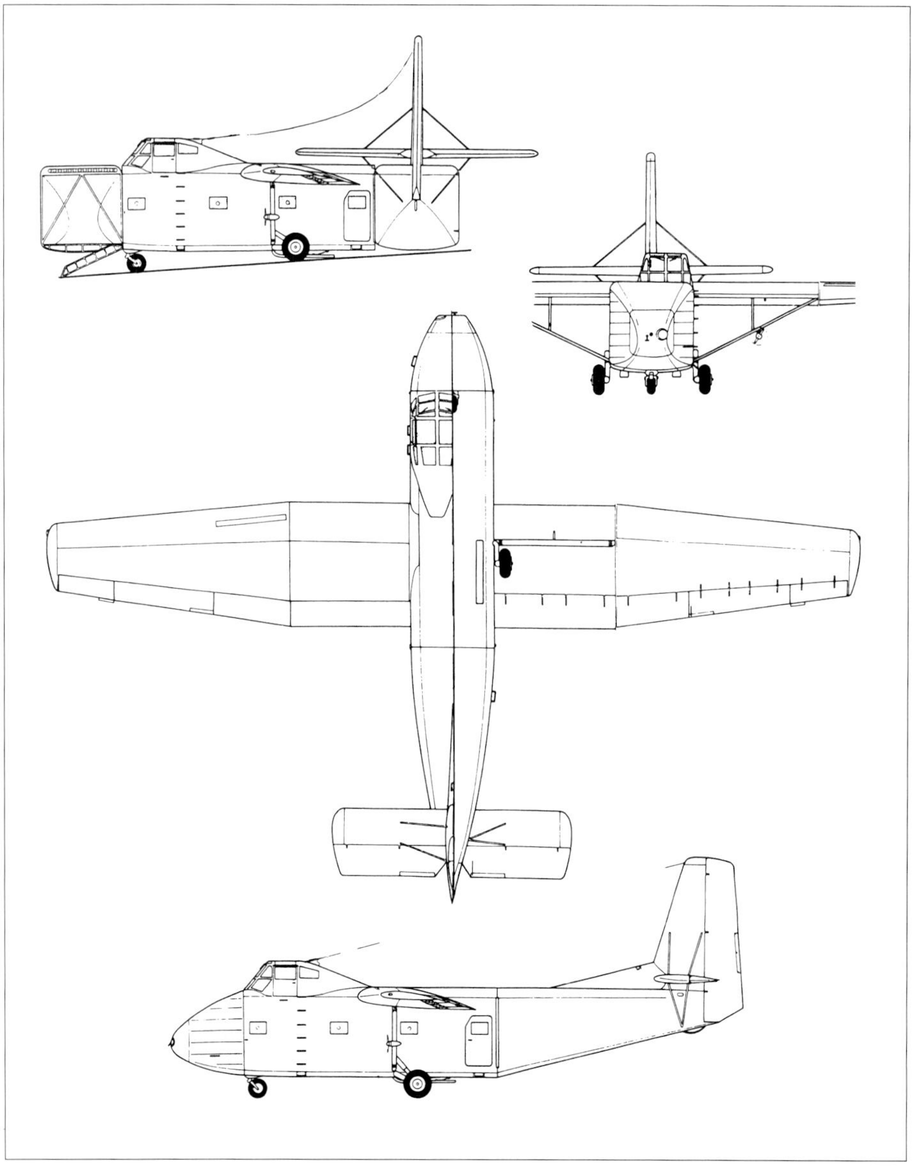

Four views of the Yak-14 glider. Note how the nose and tail sections swing open in opposite directions.

Yak-14 assault glider (production version)

The second prototype incorporated a number of detail improvements; it differed externally from the first prototype in having a prominent dorsal fin which became a standard feature on production machines. During the testing of this second prototype a speed of 315 km/h (196 mph) was attained in free flight, while the speed of flight in tow amounted to 305 km/h (190 mph). In this configuration the glider was placed in series production at Plant No.464.

In August 1949 the OKB-115 presented a modified production example of the Yak-14 (c/n 4640203 – that is, plant No.464, Batch 02, 03rd aircraft in the batch) for State acceptance trials. Its modification was undertaken in response to a directive issued by the Council of Ministers on 13th March 1949. The airframe was strengthened and the payload was increased to 3,500 kg (7,720 lb); the glider was equipped with two skids enabling it to land on unprepared strips and on snow. More efficient spoilers were installed to reduce the landing run. Some improvements were made to the cockpit interior. The glider featured dual controls. By that time the range of combat materiel that could be carried by the glider was increased to include the DP-55 anti-tank gun, the 160-mm mortar with a GAZ-67B towing jeep, pontoon bridges and the 57-mm ASU-57 self-propelled gun (the latter had a weight surpassing by 100 kg (220 lb) the glider's nominal payload). Other items carried by the glider in test flights included a GAZ-51 4 x 2 lorry, a 122-mm howitzer, a 37-mm anti-aircraft gun and other combat materiel. Tests were conducted, as before, with an IL-12D as the towing aircraft; they showed that the glider could be towed at a maximum speed of 277 km/h (172 mph). The State acceptance trials conducted at NII VVS lasted from 2nd August to 17th September 1949. Despite certain shortcomings noted by the test pilots, their general appraisal was positive. In 1949 the Yak-14 entered series production at a further plant – No.168 in Rostov-on-Don.

Specifications of the Yak-14 transport glider

	Prototype, 1948	C/n 4640203, 1949
Length	18.44 m (60 ft 5 63/64 in)	n.a.
Height when parked	7.2 m (23 ft 7 15/32 in)	n.a.
Wing span	26.17 m (85 ft 10 5/16 in)	n.a.
Wing area, m^2 (sq ft)	72.98 (784.73)	82.3 (884.94)
Empty weight, kg (lb)	2,940 (6,480)	3,082 (6,794)
All-up weight, kg (lb)	6,250 (13,780)	6,750 (14,880)
Assault cargo, kg (lb)	3,000 (6,610)	3,500 (7,720)
Number of troops	25	27
Max permissible speed, km/h (mph)	282 (175)	300 (186)
Landing speed, km/h (mph)	83 (51)	88 (54)
Landing run on wheels, m (ft)	255 (840)	270 (885)
Landing run on skids, m (ft)	270 (885)	n.a.

Yak-14M modified production assault glider

In 1950 yet another government directive tasked the Yakovlev OKB with further increasing the load-carrying capacity of the Yak-14. It was stipulated that should be able to deliver 27 troops to a landing site. The new version which was designated Yak-14M (*modifi**tsee**rovannyy*) went into production from March 1950. (According to one account, the Yak-14M was produced from 1951 onwards and featured dual controls).

In all, production plants No.168 and No.464 delivered 413 Yak-14s between 1949 and 1952. During their service with the Soviet Air Force the Yak-14s were deployed over the entire territory of the Soviet Union. At the time when large transport aircraft, such as the An-8 or An-12, were still non-existent, the Yak-14 with its capacious cargo hold was virtually the only means of transporting bulky hardware items by air without disassembling them. This was vividly demonstrated in March 1954 when four Yak-14s and four IL-12Ds performed a long-distance flight to one of the Soviet research stations deployed on the drifting ice of the Arctic Ocean. One of the Yak-14s delivered to the SP-4 (***Sev**ernyy **Pol**yus*, North Pole) station a complete bulldozer, something that no other transport aircraft could do at that time.

NK-14 (Yak-14 in Czechoslovakia)

A certain number of Yak-14 gliders was supplied to the Czechoslovak Air Force where they were designated NK-14 (NK stands for *nákladní klužak*, cargo glider). Some sources assert that Czechoslovakia received several dozen Yak-14s; other sources mention the delivery of only three machines. Yet another source claims that the CzAF operated only two Yak-14s which were subsequently returned to the Soviet Union and went to Seshcha AB near Bryansk, Russia.

Yak-200 (Yak-UTB, Yak-26) bomber trainer prototype

Although the service introduction of the IL-28 jet tactical bomber in 1948 was followed by the appearance of its trainer version (the IL-28U) in 1950, the need was felt for a cheap piston-

Yakovlev OKB

'52 Yellow', the grey-painted prototype of the Yak-200 bomber pilot trainer (alias Yak-UTB or Yak-26), as originally flown. Note the small tail bumper.

Yakovlev OKB

Yakovlev OKB

Top and above: The Yak-200 at a later stage of the trials; note the addition of a dorsal fin and the vertical nose gear strut making an interesting comparison with the Yak-210 on the opposite page.

engined bomber trainer that could be used either as a basic trainer or as an advanced trainer after the Yak-18. To meet this requirement, in February 1951 the Yakovlev OKB started design work on an aircraft which was provisionally designated Yak-UTB (*oo**cheb**no-treni**rov**ochnyy bombardi**rov**shchik*, bomber trainer) and then, for a brief period, was known as the Yak-26; however, this designation was immediately cancelled and was later allocated to a completely different aircraft.

In accordance with the Air Force operational requirement, the Yak-UTB was developed concurrently in two versions: the Yak-200, described here, and the Yak-210 (see below), intended for pilot and navigator training respectively.

The Yak-200 was to be used for the training of pilots at Air Force flying schools. It was to have a crew of two: the trainee and the instructor who sat side by side in the extensively glazed cockpit. The navigator's station in the extreme nose remained empty, being minus seat and equipment.

The following description of the Yak-200 is also basically applicable to the Yak-210.

The Yak-200 was a mid-wing twin-engined monoplane of all-metal construction with a retractable tricycle undercarriage; the nose unit retracted aft and the main units forward into the engine nacelles. The single-spar stressed-skin wings comprised the wing centre section of rectangular planform and detachable outer wing panels of trape-

Yakovlev OKB

This magnificent shot shows how the Yak-200's engine cowlings provided excellent access to the ASh-21 radials for maintenance. Note the cooling shutters.

Yakovlev OKB

Above, below and bottom: The unpainted Yak-210 navigator trainer ('53 Yellow') was almost identical to the Yak-200, differing largely in equipment; it is seen here after the addition of the dorsal fin. Note the shape of the radome and the forward-sloping nose gear unit.

Yakovlev OKB

zoidal planform. The wings featured fabric-covered slot ailerons, metal-skinned flaps on the centre section and fabric-covered flaps on the outer wings. The aircraft was powered by two ASh-21 radials with a nominal rating of 570 hp and a take-off rating of 700 hp, driving VISh-11V-20A two-blade variable-pitch propellers. The engine nacelles were mounted at the outer ends of the wing centre section. The fuselage was a semi-monocoque structure built in three sections: forward, centre and rear. The all-metal tail unit had fabric-covered control surfaces.

The single Yak-200 prototype made its first flight on 10th April 1953 at the hands of test pilot Fyodor L. Abramov. Its performance surpassed the stipulated characteristics. Between 29th July and 10th September

Yakovlev OKB

1953 the Yak-200 underwent State acceptance trials. It was stated that the machine met the Air Force requirements; yet, it was not recommended for service introduction because of difficult handling, insufficient longitudinal stability and considerable trim changes accompanying the changes in engine power. The list of deficiencies also included the lack of anti-icing devices on the windshield and propeller blades. The machine was subjected to various modifications and renewed tests to rectify the faults. External changes included the installation of a dorsal fin, an extension of the fuselage by inserting a 350-mm (1 ft 1²⁵⁄₃₂ in) plug, reduction of the wing dihedral, raising the wings by 10 cm (3¹⁵⁄₁₆ in) and modifying the nose gear unit. The CG was moved forward, the engines were set at a slightly different angle. The aircraft's empty weight rose by 120 kg (265 lb) and the AUW by 195 kg (430 lb), with some increase of the fuel load.

In the course of additional flight testing the OKB succeeded in rectifying virtually all significant drawbacks revealed during the State acceptance trials. The aircraft's handling became simple and pleasant, making it suitable for the training role. After the second round of trials the Yak-200 and the Yak-210 were recommended for production and service, but this never happened. On 26th March 1956 the Council of Ministers issued a directive ordering a halt of all work on the two machines. Possibly, the Air Force came to the conclusion that the IL-28U bomber trainer was sufficient for tackling all the training jobs in hand, after all.

Yak-210 navigator trainer prototype

The Yak-210 was to be used for navigator training at the Air Force's flying schools and proficiency training of pilots and navigators in service units. The crew comprised three persons: the instructor pilot in the upper cockpit, the trainee navigator in the extreme nose and the instructor navigator sitting behind the trainee. In overload configuration the crew could be supplemented by a fourth member – a trainee pilot.

The differences between the Yak-210 and the pilot training machine were confined to features directly connected with navigator training. The forward fuselage was fitted out as a navigator's station equipped with a full set of instruments for this mission, including the PSBN-M 360° search/bomb-aiming radar for bombing under all weather conditions (PSBN = *pri**bor** sle**po**vo bombome**tah**niya i navi**gah**tsii* – blind-bombing and navigational device) – the same model as fitted to the IL-28. The radar antenna enclosed by a dielectric radome under the rear fuselage was the Yak-210's main external recognition feature. Other items of mission equipment included the OPB-6SR optical synchronised bombsight (*op**tich**eskiy pri**tsel** bombardi**rov**ochnyy*), an AP-5 autopilot, and an AFA-BA-40 aerial camera for vertical photography; its axis could be tilted 15° aft to record the bombing results. The weight of equipment on the Yak-210 amounted to 860.5 kg (1,897 lb) versus 390 kg (860 lb) for the Yak-200, which necessitated a reduction of the fuel load by 235 kg (518 lb).

The first flight took place on 1st August 1953, the manufacturer's flight testing was conducted by test pilot F. L. Abramov, LII navigators K. B. Makar'yev and S. D. Dzyuba and project engineer L. A. Mashev. In the course of testing some modifications introduced on the Yak-200 (including the installation of the dorsal fin) were applied to the Yak-210 as well; in addition, the oval-shaped radome was replaced by a teardrop-shaped one. As stated above, the Yak-210 shared the fate of its stablemate when all work on the two aircraft was terminated in March 1956.

Specifications of the Yak-200 and Yak-210 trainers

	Yak-200*	Yak-210
Crew	2	3-4
Engine power (nominal), hp	2 x 570	2 x 570
Length	12.95 m (42 ft 5⅞ in)	12.95 m (42 ft 5⅞ in)
Wing span	17.45 m (57 ft 3¼ in)	17.45 m (57 ft 3¼ in)
Wing area, m² (sq ft)	36.0 (387.5)	36.0 (387.5)
Empty weight, kg (lb)	3,910 (8,620)	4,542 (10,013)
All-up weight, kg (lb)	4,715 (9,206)	5,122 (11,292)
Maximum speed, km/h (mph)	400 (248.5)	400 (248.5)
Landing speed, km/h (mph)	130 (81)	110 (68.3)
Time to 1,000 m (3,280 ft)	2.3 min	n.a.
Service ceiling, m (ft)	7,160 (23,490)	7,350 (24,115)
Range, km (miles)	1,280 (795)	650 (404)
Take-off run, m (ft)	360 (1,180)	300 (950)
Landing run, m (ft)	430 (1,410)	280 (920)

* Presumably prior to modification. The different performance figures cited by some sources may refer to the tests results after modification.

UTB bomber trainer (project)

As early as 1950 the Yakovlev OKB embarked on the design of a turbojet-powered bomber trainer designated simply UTB (not to be confused with the Tupolev UTB, a trainer derivative of the Tu-2 piston-engined bomber). An advanced development project of such an aircraft powered by two RD-500 turbojets delivering a thrust of 1,590 kgp (3,510 lbst) was completed in late 1950. Two versions were envisaged: a two-seat familiarisation (conversion) aircraft and a three-seat trainer.

Basic design parameters included a wing span of 14 m (45 ft 11³⁄₁₆ in), a wing area of 34 m² (366 sq ft), a take-off weight of 7,200 kg (15,880 lb) and a maximum speed of 880 km/h (547 mph). An available source states that the aircraft 'was similar in its geometrical characteristics to the Yak-23 fighter'. Presumably this refers to dimensions, but not to the aerodynamic layout which must have been altogether different. The project did not receive a go-ahead.

Yak-UTB powered by two RD-45 turbojets (project)

Another study of a turbojet-powered bomber trainer made its appearance in 1952. When performing aerodynamic calculations on the Yak-UTB powered by two ASh-21 radials (ultimately known as the Yak-200, see above), the Yakovlev design team also made calculations for a version in which the piston engines were replaced by two RD-45 turbojets. The aircraft was to have a take-off weight of 4,600 kg (10,140 lb) and a maximum speed of 636 km/h (395 mph) at an altitude of 9,000 m (29,530 ft). This project fared no better than the previous one.

Yak-220 transport aircraft (project)

Between October 1953 and March 1954 the Yakovlev OKB was engaged in design work on a small transport aircraft powered by two ASh-21 engines. The aircraft, a derivative of the Yak-200 and Yak-210, was intended to carry eight passengers (military personnel) and two crew. It differed from the mentioned baseline types mainly in having shoulder-mounted wings and a redesigned fuselage housing a passenger cabin. The project did not proceed further than the drawing board.

TRAINERS, UTILITY & SPORTS AIRCRAFT SINCE 1945

7

Yak-11 advanced trainer

Yak-UTI (Yak-3UTI) trainer prototype

In late 1945 the Yakovlev OKB produced a prototype of trainer designated Yak-UTI (called Yak-3UTI in some documents). It was a derivative of the Yak-3U in which the ASh-82FN engine was replaced by the new and less powerful 570-hp ASh-21 seven-cylinder air-cooled radial (essentially half an ASh-82) driving a VISh-11V-20 two-blade constant-speed propeller of 3 m (9 ft 10 7/64 in). The trainer inherited from its progenitor the all-metal wings with a span of 9.4 m (30 ft 10 in) and a mixed-construction fuselage. A second cockpit was added, the canopy being extended aft to feature a separate sliding hood over the rear seat, and the airframe was stressed to an exceptional G-load of 15.4. The armament comprised a synchronised Berezin UBS machine-gun with 100 rounds, plus racks for two FAB-100 bombs or other stores.

Yak-11 production trainer

A production prototype of the Yak-3UTI, tested in 1946, was designated Yak-11. It featured some changes as compared to the original project. The position of the cockpits was slightly altered, the engine was installed on rubber shock-absorbing mounts and enclosed by a long-chord cowling exactly matching with the fuselage downstream, with hinged cooling air exit flaps at the sides. Steerable radial shutters surrounding the propeller hub served for controlling the cooling air flow. The all-metal two-spar wings retained a Clark YH airfoil, with 5° dihedral from the centreline, the duralumin skin having a thickness of 2 mm (0.078 in) back to the front spar and 1 mm (0.039 in) further aft. Outboard of the fuselage the large split flaps were retained, with two settings (0° and 43°).

The fuselage was based on a truss of welded 30KhGSA grade chrome steel tube, which included four longerons and 11 frames, with a light duralumin secondary structure. The skin was mainly stress-bearing duralumin, except for the sides and underside aft of the cockpits, which were fabric-covered over wooden stringers. The fin and tailplane had duralumin skin, and all control surfaces featured a duralumin structure with fabric covering. The ailerons and rudders had ground-adjustable tabs, while the elevators had trim tabs adjustable in flight.

The main undercarriage units were virtually identical to those of the Yak-3, featuring an oleo strut with a half-fork meeting the inner end of the axle carrying a drum-braked wheel with a 600 x 180 mm (23.62 x 7.08 in) tyre. The mainwheels retracted inwards to lie ahead of the front spar, with the leg and part of the wheel faired by two doors. The tailwheel fitted with a 225 x 110 mm (8.85 x 4.33 in) tyre, could be unlocked to castor freely by pushing the control column forward, and retracted aft into a bay with twin doors. The undercarriage, engine cooling flaps, wing flaps and wheel brakes, as well as engine starting, were all actuated by a pneumatic system, with main and emergency air bottles charged by an engine-driven compressor.

Two self-sealing fuel tanks holding 175 litres (38.5 Imp gal) each were accommodated in the inner part of each wing between the spars (173 litres/38 Imp gal on production machines), with a gauge in the wing upper surface. These tanks fed a 13.5-litre (2.97 Imp gal) tank behind the engine, above which was the 38-litre (8.36 Imp gal) oil tank (35 litres/7.7 Imp gal on production machines). The oil cooler was in the port wing root, with an exit chute in the wing

Yakovlev OKB

Above and below: The Yak-11 prototype, designated Yak-3UTI or simply Yak-UTI, on the ground and in flight. Note the oil cooler inlet in the port wing root and the frameless moulded windscreen.

Yakovlev OKB

Yefim Gordon archive

Above: A line-up of production Yak-11s at an unpaved airfield. The aircraft are painted in differing shades of grey, though this may be due to weathering. The star insignia is carried on the fuselage, pre-1955 style. Note the windshield design.

Yefim Gordon archive

Above: A late-production Yak-11 with an S-13 gun camera in a fairing on the cockpit windshield pictured at Chkalovskaya AB during State acceptance trials.

RART

This Yak-11, or possibly C-11 (again equipped with a gun camera), features two additional whip aerials on the fuselage underside.

undersurface. The armament was still a single UBS with 100 rounds in the port side of the nose decking, plus stores up to 100 kg on BD-2-45 racks immediately outboard of the main gear units. Equipment included a PAU-22 gun camera, a landing light in the port wing leading edge and a pitot closer to the port wingtip. Communication with ground control and other aircraft in the air was maintained with the help of an RSIU-3M VHF radio which could be operated from either cockpit. The front (trainee's) cockpit was comprehensively equipped; the rear (instructor's) cockpit was provided with the basic instruments for flight and engine control. The flight, navigation and communication equipment enabled the aircraft to be flown day and night in visual and instrument weather conditions.

Upon completion of manufacturer's tests the prototype was transferred to NII VVS for State acceptance trials, which were satisfactorily completed in October 1946. The aircraft was ordered into production in Saratov (Plant No.292) and later in Leningrad (Plant No.272), with deliveries from mid-1947. Series aircraft were slightly heavier, mainly because of additional equipment.

A non-retractable tailwheel was introduced from the second production batch. Production machines also differed from the prototype in having a cockpit visor with framing instead of the prototype's frameless moulded visor. Late-series Yak-11s had a VISh-111-D-15 propeller with an R-7Ye automatic speed governor; this propeller can be identified by its paddle blades as distinct from the double-taper blades of the earlier VISh-11-V-20 propeller.

Early-series Yak-11s were equipped with a PBP-1A gunsight in the trainee's cockpit and had a PAU-22 gun camera in the port wing panel. Later, these trainers were fitted with the new ASP-1N gunsight and the S-13 gun camera that were used on fighters. A Yak-11 so equipped underwent State acceptance trials in the spring of 1948. The S-13 gun camera was mounted in a characteristic thimble fairing on the cockpit visor. Some machines were equipped with gun cameras of both types. In 1950 a Yak-11 manufactured by the Leningrad plant was fitted with the new ASP-3N gunsight.

In the 1950s an idea cropped up to use the Yak-11 as a fighter intended to destroy enemy helicopters. The idea failed to win support. Yet the Yak-11 was actually used as an *ersatz*-fighter in East Germany against drifting reconnaissance balloons.

The Yak-11 was generally popular with pilots, although one can come across complaints about this aircraft being excessively demanding in piloting. The aircraft's high performance was reflected in several world speed records in its class established on the Yak-11 between 1950 and 1954.

The Soviet plants manufactured a total of 3,859 Yak-11s of all versions between 1946 and 1956. The aircraft served not only with the Soviet AF and DOSAAF, but was also supplied to the Soviet Union's allies in Eastern Europe and some other countries, including Austria. In keeping with the day's practice the Yak-11 received a NATO reporting name in the 'miscellaneous' category, *Moose*.

Let C-11 (Czech-built Yak-11)

Czechoslovakia obtained a licence to manufacture the Yak-11. Production of the type was started at the Let factory in Kunovice, which eventually turned out 707 examples between 1952 and 1956. Initially bearing the factory designation Le-10, they were redesignated C-11 (C = *cvičny* [*letoun*] – trainer) by the Czechoslovak Air Force. The ASh-21 engine was also licence-built as the M-21. The Czechoslovak-built aircraft differed from their Soviet prototype in having additional antennas under the fuselage. They were all fitted with a single ShKAS machine-gun.

The C-11 was produced not only for domestic use, but also for export (in particular, to some countries of the Middle East). Later, after withdrawal from service, many of them received a new lease of life as 'warbirds' amenable to conversions into racing machines (see below).

Yak-11U (C-11U) and Yak-11T trainer prototypes

Taking account of the trend towards the universal use of a tricycle undercarriage in all categories of military and civil aircraft, in 1951 the Yakovlev OKB produced two Yak-11 versions featuring this type of undercarriage. They were designated Yak-11U (*oo**cheb**nyy*, basic trainer) and Yak-11T (*treni**rov**ochnyy*, proficiency trainer) and were externally identical. The Yak-11T differed in its equipment set which came closer to the standard adopted for fighters of that period and included radio navigation aids for landing in adverse weather conditions, a more advanced radio, an IFF transponder and other items. The nose-gear unit retracted aft, the main units inward.

The Yak-11U featured a lengthened forward fuselage increasing the overall length to 8.66 m (28 ft 5 in); the main undercarriage units were redesigned to feature trailing arms and relocated so that the wheels were placed aft of the CG. The empty weight of the nosewheel-gear version was increased by nearly 100 kg (220 lb). To retain the AUW and performance characteristics of the standard Yak-11, the amount of fuel on the U and T versions had to be reduced; the range fell accordingly to 770 km (479 miles).

Testing revealed one more 'weakness' of the machines: with a nosewheel undercarriage they were less suitable for operation from snow-covered airfields or unpaved airfields with a low-strength surface. While the tailwheel-equipped aircraft had no problems with taxying and taking off under these conditions, the nosewheel on the U and T machines tended to sink into the soft ground and could collapse. This was one of the reasons which prevented the nosewheel-equipped Yak-11 versions from being adopted for service.

Production of the Yak-11U was confined to several examples built in Czechoslovakia as the C-11U (some sources claim it was *developed* in Czechoslovakia). The Czech aviation historian Václav Nemeček claims in his book *Sovetska letadla* (Soviet Aircraft) issued in 1969 that in the Soviet Union the C-11U was referred to as the 'Yak-10 with the ASh-21 engine'.

Yak-11 drone director aircraft

A suitably equipped Yak-11 was used for controlling unmanned target drones con-

Yefim Gordon archive

Above: This Yak-11 coded '43' or '45' and sporting a light-coloured spinner and fin cap is unusual in having an extra antenna mast atop the windshield. Note the oil cooler's ventral air outlet.

Yakovlev OKB

Above: The prototype of the tricycle-gear Yak-11U trainer. Note the bomb racks outboard of the main gear units.

Yakovlev OKB

The Yak-11T differed from the Yak-11U only in equipment fit (note the ventral rod aerial). This view accentuates the short landing gear wheelbase and the sharp incline of the nose gear unit.

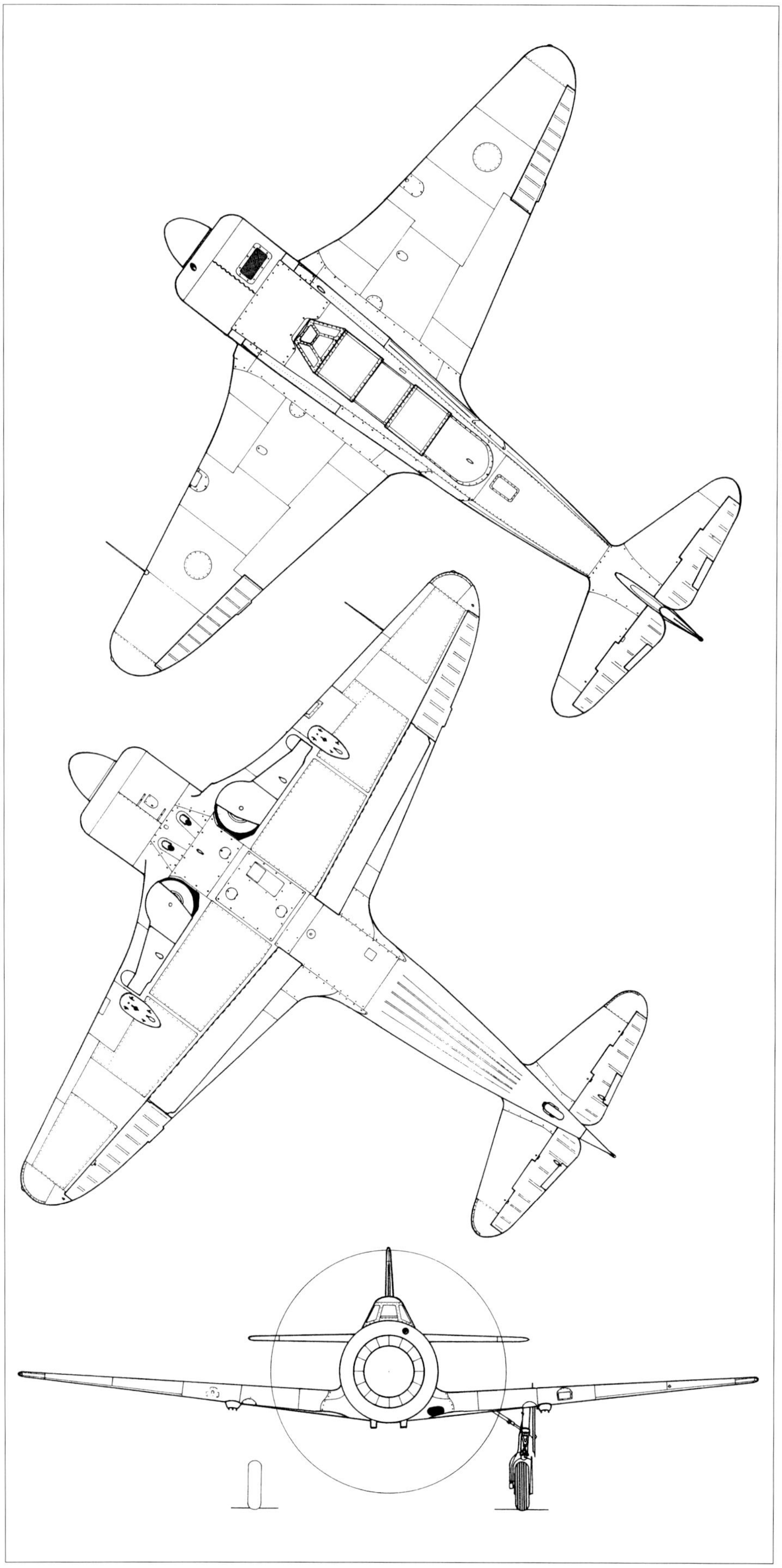

Three views of the production-standard Yak-11 (C-11).

verted from combat aircraft (it can be surmised that the target drone in question was converted from the Tu-16 bomber). Control was effected when the target drone was taking off.

Yak-11 as a target tug

In 1950 a production Yak-11 was tested with a winch for towing a sleeve-type target for air gunnery training.

Yak-11 – Soviet single-seat conversions

Several production Yak-11s were converted into single-seat machines in the Soviet Union by deleting the rear cockpit and its part of the canopy (the rearmost segment of the canopy was moved forward to form a canopy similar to that of the wartime Yak fighters and the rear cockpit opening was faired over). These machines impersonated Yak-3 fighters in the Soviet film *Normandie-Niémen* about the famous French fighter regiment that fought on the Eastern front. The aircraft were flown to provide realistic air combat scenes.

On another occasion the Yak-11s were called upon to represent Polikarpov I-16 fighters in the film *Bal**tiy**skoye **neb**o* (The Baltic Skies). To make them resemble the tubby I-16s, they were subjected to 'surgery' which made them unfit for flying, and they were used only for shooting ground episodes.

Yak-11 single-seat aerobatic aircraft (Romanian conversion)

One of the Yak-11s supplied to Romania was converted into a single-seat aerobatic machine by deleting the front cockpit and retaining only the rear seat, with suitable alterations to the cockpit canopy and the upper decking of the fuselage. This aircraft, sporting the then-current star-type insignia of the Romanian Air Force, was flown by Stefan Calotă and Octavian Bacanu, Romania's leading aerobatic pilots who had been awarded the Paul Tissandier diploma by the FAI.

Yak-11 – European single-seat conversions

Quite a few Yak-11s (to be precise, C-11s) were bought by warbird enthusiasts, museums and aerobatic pilots in Western Europe and the USA from Egypt and other Middle Eastern countries. Some of these aircraft were restored to flying condition and took part in various competitions and air shows where they were appreciated not only for their flying qualities but also for evoking the memory of their progenitor – the Yak-3 fighter. Some of these restored aircraft were repainted in the colour schemes of wartime

Soviet Yak-fighters. To make the resemblance more convincing, some of these machines were converted to a single-seat configuration.

One such conversion was undertaken in France. As distinct from the Soviet single-seat Yak-11 described above, this machine retained the rear cockpit which gave the aircraft greater likeness to the Yak-3; it was painted in the colours of the *Normandie-Niémen* regiment. This machine was restored by Gerry Marchandier together with the French aerobatic pilot Pierre Dague. The original aircraft (the c/n was quoted as '25/III/05') was obtained for conversion from Jean-Baptiste Salis who had recovered 41 Yak airframes from Egypt. The aircraft was completely refurbished, and the conversion job, apart from deleting the front cockpit and suitably modifying the cockpit canopy, included the installation of a Cornelius compressor from a Boeing B-29 in place of the the front cockpit to be used as an emergency air supply, relocating and modifying the cowl flaps, developing and installing a new electric circuit, making the tailwheel retractable, and adding fake gun blisters ahead of the cockpit. Two stainless steel panels were added on the fuselage sides aft of the exhaust pipes. Deviations from the original design included installation of an airspeed indicator and a pitot tube from a Douglas A4D Skyraider, a modern radio and a voltage regulator, and navigation lights from a twin-engined Beechcraft.

The painstaking conversion job took three years to complete, but the end product was a superb-looking machine which was successfully flown. Registered F-AZNN in the French veteran aircraft series (A stands for *ancien*, old), the aircraft wore Soviet Air Force insignia and the code '44 White'.

Yak-11 (C-11) conversions in the USA

In the USA several Czechoslovak-built Yak-11s were specially modified for participation in unlimited air races. Some of these conversions are listed below.

One LET C-11 was modified at Van Nuys for Joseph Kasperoff with a Wright R-3350 Duplex Cyclone two-row radial under a cowling from a DC-7 driving a propeller from a Skyraider. A plug was inserted in the fuselage aft of the wings, presumably to compensate for the CG shift caused by the heavier engine. The tandem cockpits were replaced by a single cockpit which was blended into the contours of the rear fuselage with a raised decking. It was later raced by Darryl Greenamyer as *Mr Awesome* and is now, after further extensive rebuild, with Jack Frost as *Red Heat* with whom it bears the registration N190JF.

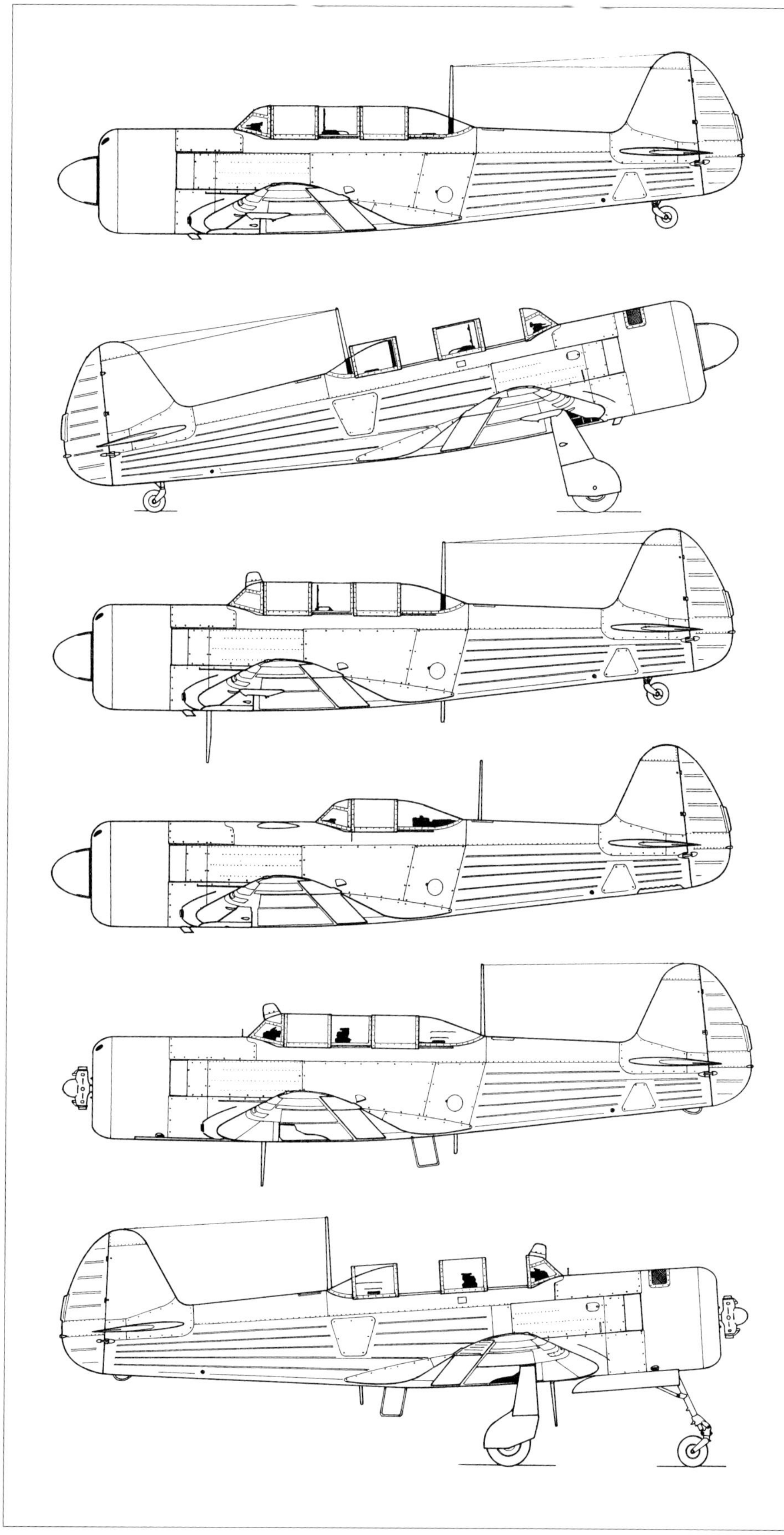

Top to bottom: Port and starboard views of an early-production Yak-11 without a gun camera; a late-production Yak-11 (or C-11) with a gun camera and extra radio equipment; Yak-11 F-AZNN converted by Gerry Marchandier and Pierre Dague; and two views of the experimental Yak-11T. Note the different location of the metal skin panels on the port and starboard sides of the fuselage.

Another Yak-11 (C-11) conversion in the USA, undertaken by Ascher Ward and Ralph Wise, involved installation of a 14-cylinder Pratt & Whitney R-1830 Twin Wasp two-row radial mated to a three-bladed prop sourced from a Douglas DC-3. This powerplant was fitted to a clipped-wing Yak-11 registered N18AW and named *Maniyak*.

An even more ambitious project was undertaken by Bob Yancey (famous for his racing in the modified Vought F4U Corsair *Old Blue*) who decided to transplant the awesome Pratt & Whitney R-2800 Double Wasp 18-cylinder radial engine from the Corsair into a Yak-11 airframe. The engine drove a four-blade propeller with huge paddle-shaped blades. The fuselage was completely stripped down and converted into a single-seat configuration. A turtledeck with a small low-drag racing canopy was fitted and the engine enclosed by a cowling from a Nord 2501 Noratlas transport. The result looked every bit an unlimited racer, and the aircraft made its debut at Reno, Nevada, in 1987 as N5943 (c/n 407). Later Yancey named the beast *Perestroika*, in the 1993 Reno Air Races it achieved an incredible 439.543 mph. It was re-named *Czech Mate* by current owner John Moore, who first raced it at Reno in 1998.

One of the 41 Egyptian Air Force Let C-11s acquired by the Jean Salis Collection in France, Yak-11 N25YK was registered on 5th July 1988 to Robert J. Pond of Palm Springs, California (the c/n was quoted as '25/III/25'). It was re-engined with a Pratt & Whitney R-1830 which drove a propeller sourced from a Douglas SBD Dauntless dive-bomber. Several more rebuilders were considering this route in view of the difficulty in having the Soviet Shvetsov radial overhauled. This aircraft is now an exhibit at the Planes of Fame Air Museum in Eden Prairie, Minnesota.

Another ex-Egyptian C-11 is exhibited at the Planes of Fame Air Museum in Chino, California. Still retaining the R-1830 fitted for racing, it is now painted as the Lavochkin La-7 flown by Soviet ace, Ivan Kozhedoob.

Another example converted by Samuel C. Davis of Corona, California, was entered into the US civil register on 9th October 1998 as N498SD under the spurious designation Yak-3UR-2000. The UR obviously stood for 'unlimited racing' and the numerals referred to the engine, a Pratt & Whitney R-2000 (a bigger-displacement version of the Twin Wasp as used on the Douglas C-54/DC-4 Skymaster).

Other aircraft re-engined with the R-2000 for air racing and exhibition flying in the USA include N11MQ *Czech Ride* and N2124X *Defector*.

Yak-11s converted into Yak-3s or Yak-9s (the USA and Europe)

Being a derivative of the Yak-3, the Yak-11 completed a full circle, returning to the (quasi)original standard at the hands of some enthusiastic rebuilders in the USA and Europe. A technical possibility for such a conversion, involving the replacement of the ASh-21 radial by a liquid-cooled Vee-12 engine, arose thanks to the fact that a certain number of wartime Allison V-1710 engines had been preserved in airworthy condition in the USA. In its dimensions and power rating the V-1710 represented a close approximation of the Klimov VK-105PF engine which, unfortunately, is no longer available (as related in Chapter 2, the Allison engines were also used to produce new-build examples of the Yak-3, Yak-1, Yak-7 and Yak-9 fighters in Russia). Here are some examples of such conversions.

In 1988, a Yak-11 (Let C-11) airframe registered N3YK (the c/n was quoted, obviously in error, as Y337) was being configured as a Yak-3 with an Allison powerplant by John Houston of Harlingen, Texas. This machine was referred to in another source as being converted in Harlingen into a Yak-3 for the Texas Air Museum. Eventually it became one of the museum's exhibits.

Also in 1988, a Yak-11 (C-11) airframe was under rebuild by Jean Garric of Mercedes, Texas. It was to be powered by an Allison engine '*à la* Yak-9', according to the source (albeit the wings of the Yak-9 differed from those of the Yak-3 and, consequently, of the Yak-11 in span and planform).

In the UK, a pair of Yak-3s, obviously rebuilt from Let C-11s, were to be found at the home of the Imperial War Museum in Duxford, Cambridgeshire.

One, a 'Yak-3U' (*sic*) registered as G-BWOE, a rebuild from a former Egyptian Air Force Yak-11 (C-11) c/n 171231, was operated by the Old Flying Machine Company until it went to New Zealand in February 2005. The U suffix in this case has nothing to do with the original Yak-3U (see Chapter 2).

The other, 'Yak-3U' G-BTHD (c/n 170101) was shipped from The Fighter Collection at Duxford to the Planes of Fame Air Museum at Chino in January 2005.

Yak-10 liaison and passenger aircraft (Yak-14 – first use of designation)

This four-seat high-wing strut-braced monoplane was designed in 1944 and entered manufacturer's test in January 1945. The aircraft, similar in concept to Yakovlev's pre-war AIR-5 and AIR-6 machines, was powered by a 145-hp M-11FM engine driving a VISh-327 propeller. It featured a mixed construction with wooden wings, a metal fuselage and metal tail surfaces. The wing had a Clark YH airfoil and a thickness/chord ratio of 11%. Twin bracing struts on each side were made of airfoil-section steel tubes. The fuselage was a welded tubular steel truss; the tail unit featured duralumin frames. Fabric was used for skinning everywhere. The non-retractable undercarriage comprised rubber-sprung pyramid-type main units and a castoring tailwheel.

Originally designated Yak-14, the aircraft was renamed Yak-10 in the course of testing (the Yak-14 designation was re-used for an assault glider). In June 1945 it passed State acceptance trials and was placed into production at plant No.464 in Dolgoproodnyy north of Moscow, where 40 examples were manufactured in 1946. In production the M-11FM gave place to the M-11FR engine rated at 160 hp.

Yak-10V dual-control trainer

The production Yak-10s included a certain number of dual-control machines desig-

Specifications of the Yak-11 and Yak-11U trainers

	Yak-11	Yak-11U
Engine type	ASh-21	ASh-21
Engine power, hp	570	570
Length	8.20 m (26 ft 10$\frac{27}{32}$ in)	8.66 m (28 ft 4$\frac{7}{8}$ in)
Wing span	9.4 m (30 ft 10 in)	9.4 m (30 ft 10 in)
Wing area, m^2 (sq ft)	15.4 (166)	15.4 (166)
Empty weight, kg (lb)	1,900 (4,190)	2,066 (4,555)
All-up weight, kg (lb)	2,440 (5,380)	2,500 (5,510)
Maximum speed, km/h (mph)	465 (289)	460 (286)
Landing speed, km/h (mph)	89 (55)	n.a.
Time to 1,000 m (3,280 ft), min	1.3	n.a.
Service ceiling, m (ft)	7,950 (26,080)	7,200 (23,620)
Range, km (miles)	1,250 (777)	n.a.
Take-off run, m (ft)	395 (1,300)	400 (1,310)
Landing run, m (ft)	500 (1,640)	n.a.

Yakovlev OKB

Above: The Yak-10 prototype, still bearing the original Yak-14 designation. Despite the helmeted engine cowling, the aircraft has an exhaust collector ring and a long exhaust pipe. Note the shape of the main gear struts with sloping forward rods and the rear rods at right angles to the fuselage axis.

nated Yak-10V (*vyvoz**noy***, for familiarisation training).

Yak-10S ambulance aircraft

The Yak-10S (*sani**tar**nyy* – medical or ambulance, used attributively) developed in 1947 featured a long hatch on the port side for loading one or two stretchers and also had a seat for a medical attendant. A small production batch was manufactured (a photo exists of a group flight of Yak-10s carrying the Red Cross emblem on their sides).

Yak-10 equipped with skis

In February 1947 a Yak-10 powered by an M-11FR was tested in the Civil Air Fleet with Canadian-built wooden skis measuring 1,930 x 340 mm (6 ft $3\frac{63}{64}$ in x 1 ft $1\frac{25}{64}$ in) and weighing 20.25 kg (44.7 lb) each, and a tail ski measuring 460 x 120 mm (1 ft $6\frac{7}{64}$ in x $4\frac{47}{64}$ in) and weighing 1.93 kg (4.25 lb). Performance deteriorated noticeably, especially the service ceiling (which was reduced to 2,550 m/8,370 ft), field performance and taxying properties. The ski undercarriage was deemed suitable for use only on rolled-down airfields and was not adopted for the Yak-10s operating on local services in winter.

Yak-10G floatplane

In 1946 a single Yak-10 was fitted with AIR-6 type floats, receiving the designation Yak-10G (***ghid**rosamo**lyot*** – floatplane). It underwent manufacturer's tests but did not go into production. No further particulars are available.

Yak-13 liaison and light passenger aircraft prototype (Yak-12 – first use of designation)

In February 1945 the Yakovlev OKB produced a prototype of a four-seat liaison and light passenger aircraft which eventually received the designation Yak-13 (initially it was known as the Yak-12, or 'passenger low-wing monoplane No.12'; the designation Yak-12 was subsequently re-allocated to a high-wing monoplane described below). The aircraft had been designed since 1944 by a team led by project engineer K. Sinel'shchikov. It was a low-wing monoplane with a cabin accommodating four persons (including the pilot) in two rows of comfortable seats; there was a provision for a baggage compartment. The aircraft had a retractable undercarriage with a castoring tailwheel, the main units retracting inwards. It was equipped with a landing flap on the wing centre section, wheel brakes and a pneumatic engine starter.

Yakovlev OKB

Yefim Gordon archive

Top: The Yak-10 flies over the Moskva River in the suburbs of Moscow. Note the V-shaped windscreen.
Above: The same aircraft at Chkalovskaya AB during State acceptance trials at NII VVS.

Yakovlev OKB

Above: A production-standard Yak-10 (with appropriate titles on the doors) displays different skin tones on the metal-skinned fuselage and the fabric-covered tail. No wheel spats were fitted.

Yakovlev OKB

Above: A fine row of production Yak-10s awaiting delivery at Dolgoproodnyy. The titles on the door appear to be standard. Note the different type of wheels than hitherto.

Yakovlev OKB

Above: A white-painted Yak-10S ambulance aircraft with appropriate markings. The stretcher loading door is barely visible in the area of the red cross.

Yakovlev OKB

The prototype of the Yak-10G floatplane.

The Yak-13 was designed to accept M. Kossov's new M-12 five-cylinder radial rated at 190 hp, but the engine development schedule slipped, and a stock 145-hp M-11FM driving a VISh-327 propeller had to be installed. With this powerplant the Yak-13 passed manufacturer's flight tests, in the course of which it attained a maximum speed of 235 km/h (146 mph) at sea level and 205 km/h (127 mph) at 2,500 m (8,200 ft), the cruising speed being 170 km/h (106 mph). Summing up the results of these tests, the test pilots noted that the Yak-13 was superior to all contemporary Soviet passenger and liaison aircraft powered by the M-11 engine as regards performance, operational qualities and comfort. The aircraft was deemed ready for State acceptance trials. There is, however, no documentary evidence to confirm that it was actually subjected to these trials.

Despite all its virtues, the Yak-13 was not put into series production. The reason was simple: it lost to another contender for the same role of light passenger and liaison aircraft, the high-wing Yak-10, which had certain operational advantages, despite lower performance. Simultaneous production of two closely similar types was considered economically unwise at that time.

Yak-12 multi-purpose aircraft (second use of designation)

The Yak-12 high-wing strut-braced monoplane (second use of designation) appeared in 1947 as a direct development of the Yak-10 described above. Outwardly it differed from its predecessor primarily in having a revised rear fuselage contour with a shallower upper decking. Unlike the Yak-10, this multi-role light utility aircraft saw large-scale production, finding both military and civil applications. A number of versions were developed over the years, incorporating new powerplants and various improvements, as well as changes associated with new roles. The Yak-12 had a successful operational career both at home and abroad, which included licence production in Poland and China. Although the type is long since phased out, some examples of this aircraft are flying to this day.

Described below are the numerous versions of the Yak-12.

Yak-12 initial production version

The prototype and initial production Yak-12 shared the M-11FR-1 engine, mixed construction and the basic wing structure having constant chord with the Yak-10. The wingtips were semi-circular.

The fuselage was a welded truss of KhNZA grade steel tubes with a wooden secondary structure and a mix of ply and fabric covering. The tail surfaces with duralumin frames were also fabric-covered. All controls were cable-operated.

The cabin had two bucket seats for the pilot (on the left) and an observer beside him. Often there was a third seat at the rear.

The M-11FR engine driving a fixed-pitch aluminium propeller had a helmeted cowling and differed from the same installation on the Yak-10 in having short exhaust pipes instead of one long pipe under the fuselage.

The prototype flew in 1947 and production was soon authorised at two plants; most of the initial production machines were used by the Air Force for artillery spotting and liaison. There were also some export deliveries, notably to Poland and China.

An example of the M-11-powered Yak-12 *sans suffixe* registered SP-ASZ is preserved in the Aerospace Museum (*Museum Lotnictwa i Astronautyki*) in Kraków, Poland. The Yak-12 family had the NATO reporting name *Creek* in the 'cargo/commercial' category.

Yak-12UT dual-control trainer

The Yak-12UT (*oo**cheb**no-treni**rov**ochnyy* [*samo**lyot***] – trainer) was a dual-control trainer version of the Yak-12 *sans suffixe* powered by the M-11FR-1 engine. It was intended for practising instrument landing approach and was equipped with appropriate instruments. Electric power for these was provided by a windmill-driven generator under the fuselage.

It is not clear whether the designation Yak-12UT also applied to dual-control examples of the subsequent baseline versions – the Yak-12R, Yak-12M and Yak-12A. See also the familiarisation trainer versions of the Yak-12R and Yak-12M below.

Yak-12G (Yak-12P) floatplane

The Yakovlev OKB produced a floatplane version of the initial production Yak-12 *sans suffixe*. Various sources quote the year of construction as 1947, 1948 or 1949 and identify it as the Yak-12G (***ghid**rosamo**lyot***). Vadim B. Shavrov refers to it as the Yak-12GR but this appears to be a mistake, an OKB source using this designation for a later variant based on the Yak-12R (see below). Shavrov states that the floats were of the same type as those of the AIR-6A floatplane of the 1930s. The designation Yak-12P (*poplav**kov**yy* [*samo**lyot***] – floatplane) is also encountered in documents.

The drag created by the floats caused a serious deterioration of performance. The maximum speed fell to 162 km/h (101 mph), the take-off run rose to 490 m (1,610 ft) and the landing run to 130 m (430 ft). The aircraft was obviously underpowered, which precluded series production.

Yakovlev OKB

Above and below: Two views of the Yak-13 cabin monoplane prototype. Note the strong wing dihedral outboard of the main gear fulcrums.

Yakovlev OKB

Yak-12GM floatplane version

This 'four-seat multipurpose floatplane' produced in 1950 is mentioned by a source without any further details. It remains unclear whether it was based on the Yak-12 *sans suffixe* or on the Yak-12R powered by the AI-14R (described below).

Yak-12S ambulance aircraft

An ambulance version of the M-11FR-1 powered Yak-12 was produced in prototype form in 1948 as the Yak-12S (*sani**tar**nyy*). It accommodated one stretcher case and a medical attendant. The only difference from the basic version was the upward-opening triangular hatch to port for loading the stretcher. This feature became standard on ambulance versions of later Yak-12 variants.

Yak-12SKh agricultural aircraft

The Yak-12 could be used for crop-spraying and crop-dusting, this version being desig-

Yakovlev OKB

'122', one of the Yak-12 *sans suffixe* prototypes, in flight. This version rarely had a propeller spinner. Note the leading-edge slats.

Yakovlev OKB

Above: The Yak-12 prototype – or possibly the same one as on the previous page before the addition of a second '2' and an aerial mast. This view illustrates the constant-chord wings with rounded tips.

Yakovlev OKB

Above: A Yak-12UT trainer equipped for practising instrument approaches; the windmill-driven generator is the giveaway. Note the ski landing gear, the slim rear fuselage and the port side cargo hatch.

Yakovlev OKB

Above: A Yak-12SKh agricultural aircraft (c/n 4640638) configured with spraybars. Note the lack of the rear side windows where the chemical hopper is located and the windmill-driven chemical pump.

Yakovlev OKB

An overall silver doped Yak-12S ambulance aircraft in Soviet Air Force markings. The rear side windows appear to be oversprayed.

nated Yak-12SKh (***sel'**skokho**zyai**stvennyy*, agricultural). It was equipped with a chemical tank aft of the pilot's seat and a dry chemical spreader made of aluminium sheet below the fuselage in almost the same installation as the earlier Polikarpov U-2AP (Po-2A). Spraybars were fitted under the wings for dispensing liquid chemicals. Yet another set of equipment was used for the aerial sowing of plants.

Yak-12 M-11FR – Chinese-built versions

The USSR delivered a certain number of Yak-12s *sans suffixe* in a landplane version to China. Later, a copy of the Yak-12 powered by the M-11FR engine was produced in China, presumably with Soviet assistance. A Chinese publication describes it as the 'Shen Hang-1 multipurpose aircraft developed by the Shenyang Aviation Polytechnic School' and cites 7th October 1958 as the first flight date. Curiously enough, it is asserted that its passenger version could carry 'three passengers and a toilet'. According to some sources, the aircraft received the designation Shenyang Type 5. There is no mention of series manufacture.

An example of the floatplane version of the Yak-12, also powered by the M-11FR, was built by the Shanghai Aircraft Manufacturing Factory under the designation Flying Dragon-1, making its first flight on 15th September 1958. Some sources give it the name Shanghai Hiryu. The aircraft was tested at the mouth of the Yangtze River. Again, there is no evidence of series manufacture.

Interestingly, some indigenous Chinese light aircraft designs of the early 1950s were clearly inspired by the Yak-12 and incorporated some elements of the Soviet aircraft's structure. This was particularly the case with the Heilongjiang-1 multi-purpose aircraft which was developed by the Harbin Aviation Polytechnic School, making its first flight on 16th December 1958. It looked like a Yak-12 *sans suffixe* with the rear fuselage altered to resemble a pod-and-boom configuration and the vertical tail squared off.

Yak-12 M-14 prototype

The output of the M-11FR appeared to be insufficient for the Yak-12, and a more powerful engine was clearly needed. Yet, no production engine in the required power class was available at that time. The designers decided, by way of an experiment, to make use of the new 250-hp M-14 nine-cylinder radial which was then under test and not yet cleared for production. In 1948 this engine was installed in a Yak-12 which, apart from the powerplant and associated cylindrical cowling, apparently retained all the features of the initial production version, including the rounded wingtips and the cabin glazing with two windows on each side.

The new engine produced a marked improvement of field performance and rate of climb. Eventually the M-14 engine developed by Aleksandr G. Ivchenko was renamed AI-14 (the name M-14 came up again at a later stage) and in its AI-14R production version formed the basis for the next baseline version – the Yak-12R described below.

Yak-12R multi-purpose aircraft (liaison and passenger versions)

The Yak-12R powered by the AI-14R engine delivering 260 hp was developed in 1950. The prototype of this version was tested in June 1950; apart from the engine, with its redesigned long-chord cowling and radial shutters round the front, it did not differ from the preceding production version. However, before putting the re-engined version into production, the OKB made a major redesign of the airframe. The new wings retained the Clark YH airfoil but had greater span and area and featured squared-off tips; the plywood leading edge was replaced by metal. The previous plywood skin over the front of the fuselage and rear deck was likewise replaced by D1 duralumin.

With its greater wing area and improved high-lift devices, the Yak-12R became virtually a STOL aircraft. In its baseline two-seat version the Yak-12R had a take-off run of 52 m (170 ft) and a landing run of 81 m (265 ft). To shorten the landing run, the Yak-12R was equipped with a special braking dozer blade (soshnik) which could be extended under the rear fuselage in the event of landing on a grass airstrip.

The cabin of the Yak-12R in its standard production version normally had only one side window (in the door) on each side. Yet, an example supplied to Poland and registered SP-CFO had two windows on each side (as had some Yak-12R prototypes).

Yak-12R glider tug and sports aircraft

The Yak-12 was used by aero clubs as a glider tug and for dropping parachutists. In the former role the Yak-12R (as well as the later Yak-12M) had some advantages over the less powerful Po-2 and the bigger An-2. For this role, in addition to the towing hook, it was equipped with a rear view mirror and with wires running from the lower rear fuselage to the tips of the tailplane to protect the rear fuselage from contact with the towing cable, should the glider take up an overly high position in relation to the aircraft. When used for dropping parachutists, the aircraft was equipped with a large step on the starboard side of the cabin.

Yakovlev OKB

Above: The experimental Yak-12G (alias Yak-12P) floatplane immediately after becoming airborne. As this photo shows, most Yak-12s were manufactured with the port side cargo hatch.

Yakovlev OKB

Above: The experimental Yak-12 M-14 – the first version to have the new M-14 (AI-14) nine-cylinder radial. Note the original tapered cowling, the long exhaust pipes and the braking dozer blade under the tail.

Yakovlev OKB

The prototype of the Yak-12R – the first production version to have the new AI-14R nine-cylinder radial. The redesigned cowling of almost constant diameter is clearly visible.

Yak-12R dual-control familiarisation aircraft

Some Yak-12Rs were provided with dual controls and were used by Aeroflot as familiarisation trainers.

Yak-12R ambulance aircraft version

When provided with a large triangular hatch on port side, like the Yak-12S described above, the Yak-12R could be used in the ambulance role. A Hungarian Yak-12R registered HA-MEB was used as an ambulance aircraft, as evidenced by the Red Cross symbols on the fuselage and under the wings.

Yak-12GR floatplane

The designation Yak-12GR was quoted for a floatplane version of the Yak-12R.

Yak-12M multi-purpose (liaison, sports and passenger) aircraft

In 1954 the Yakovlev OKB produced a new baseline version of the Yak-12 designated Yak-12M (*moderni**zee**rovannyy*). In its pas-

Specifications of the Yak-12 and Yak-12R multi-purpose aircraft

	Yak-12	Yak-12R
Engine type	M-11FR	AI-14R
Engine power, hp	160	260
Length	9.0 m (29 ft 6⅜ in)	9.0 m (29 ft 6⅜ in)
Wing span	12.09 m (39 ft 8 in)	12.60 m (41 ft 4 in)
Wing area, m^2 (sq ft)	21.6 (232.5)	23.8 (265)
Empty weight, kg (lb)	830 (1,830)	912 (2,010)
All-up weight, kg (lb)	1,185 (2,610)	1,172 (2,584)
Maximum speed, km/h (mph)	169 (105)	184 (114)
Cruising speed, km/h (mph)	148 (92)	160 (99.5)
Landing speed, km/h (mph)	64 (40)	60 (37)
Climb time to 1,000 m (3,280 ft)	7.8 min	3.6 min
Service ceiling, m (ft)	3,000 (9,840)	5,800 (19,030)
Range, km (miles)	810 (503)	510 (317)
Take-off run, m (ft)	130 (430)	52 (171)
Landing run, m (ft)	110 (360)	81 (266)

Yakovlev OKB

Above: A further modified prototype of the Yak-12R with the rear windows eliminated and a redesigned cowling. The aircraft entered production in this guise.

senger configuration it had a cabin accommodating a pilot and three passengers, one seated beside the pilot and two behind them on a comfortable bench seat. In the liaison configuration the pilot and one passenger sat on parachutes in bucket seats, the rest of the cabin being used as a cargo hold.

Outwardly the Yak-12M differed from the Yak-12R in having a lengthened rear fuselage and redesigned tail surfaces, the vertical tail featuring a long curved fin fillet. Like its predecessor, it was powered by an AI-14R engine driving a V-530-D-11 two-blade variable-pitch propeller with an R-2 speed governor. The avionics included an R-800 VHF radio and an ARK-5 automatic direction finder.

The Yak-12M was produced in considerable numbers and was operated by the Air Force, the Civil Air Fleet and DOSAAF (a voluntary paramilitary organisation responsible for aero clubs) in a variety of roles; its baseline liaison and passenger/transport version could be adapted to such duties as forestry patrol and even service in support of the Soviet Antarctic expeditions. In winter the aircraft was operated on skis equipped with brakes. Cases are on record when the Yak-12M was used for laying wire communication lines; two reels with cable mounted on the fuselage sides were unwound at a speed of 120 km/h (75 mph) to a distance of up to 8 km (5 miles).

Yak-12M air club aircraft

Yak-12Ms were used in air clubs for glider towing and for the training of parachute jumpers. Their equipment was identical to that of the Yak-12Rs used in a similar role.

Yak-12M dual-control proficiency trainer

The dual-control version of the Yak-12M had the same features as the similar version of the Yak-12R described above and, likewise, was used for the proficiency training of Aeroflot pilots. The second set of controls could be easily removed. Known registrations are CCCP-Ш509 (that is, SSSR-Sh509) and CCCP-Ш511; the Sh operator designator denoted ***shko**ly **Aeroflot**a* – Aeroflot's (flying) schools.

Yak-12M agricultural version

The Yak-12M was successfully adapted to crop-dusting and crop-spraying by installing a hopper in the space normally occupied by the rear passenger seat and fitting spraybars under the wings or a duster under the fuselage. The spraying equipment was actuated by a propeller-like vane under the fuselage. A Yak-12M could be converted to the agricultural version within two hours. Known examples are CCCP-Л4270, CCCP-21093 and CCCP-74077.

Yak-12M ambulance version

The ambulance version of the Yak-12M differed from the baseline Yak-12M only in having a door on port side for loading the stretcher in a fashion similar to the ambulance versions of the Yak-12R and initial production Yak-12. One stretcher case and one sitting patient accompanied by a medical attendant could be carried. Known registrations include CCCP-K175, CCCP-07870 and CCCP-22224.

Yakovlev OKB

A Yak-12M prototype in agricultural configuration with spraybars. The longer rear fuselage and the dorsal fin are clearly visible. The four-blade propeller driving the chemical pump is locked for cruise flight.

Yak-12MM seaplane prototype

A single example of the Yak-12M was converted by the OKB into a floatplane, receiving the designation Yak-12MM. Another source describes the Yak-12MM (without reference to its Yak-12M progenitor) as a four-seat multi-purpose floatplane built in 1957. No further details are available. One might surmise that MM could stand for 'Yak-12M *mor**skoy***' (seaplane).

Yak-12M rebuilds by enthusiasts

Both during the service career of the Yak-12M and long after the type had been phased out, aviation enthusiasts used this aircraft as a basis for various home-built conversions, sometimes altering the baseline machine quite considerably. Some of these conversions are briefly mentioned here.

Danko. Named after an epic character, this was a Yak-12M converted in 1984-85 by a group of enthusiasts from the Donetsk Region in eastern Ukraine. The group was headed by A. A. Balooyev. The most important alteration was the installation of a Czechoslovak Walter M-337 six-cylinder inverted in-line air-cooled engine instead of the AI-14R radial. The aircraft also had a recontoured vertical tail. Internal changes included some revision of the structure involving the use of glassfibre for some of its elements. A dual set of controls was provided. The aircraft was successfully tested and flown very extensively, taking part in various air events. It was progressively modified, including the deletion of leading-edge slats and provision of a dorsal hatch and rails above the cabin for aerial acrobats who performed daring barnstorming flights, perched on top of the machine. In 1996 the Danko reverted to the original AI-14 radial engine with all the associated equipment.

Danko-2. This was a similar conversion effected in 1986-87 in Vinnitsa, western Ukraine, by a group of enthusiasts headed by Aleksandr Kashevskiy. The baseline machine was a Yak-12M (reported in error as a Yak-12A in some sources). As in the previous case, the AI-14 engine was replaced by an M-337. The leading-edge slats were deleted and the wing leading edge was reinforced with a coat of glassfibre with fabric covering. The pneumatic flap actuation system was replaced by a mechanical one. The vertical tail contours were altered. Registered CBC-37001 (that is, SVS-37001; the SVS prefix used in the late 1980s/early 1990s stood for *samo**del'**noye voz**doosh**noye **sood**no* – homebuilt aircraft), in 1991 the Danko-2 took part in a long-range flight of a group of light aircraft spanning over 3,243 km (2,015 miles).

MakrYak. The name of this example of the Yak-12M restored by a group of enthusi-

Yefim Gordon archive

Above: This pristine Soviet Air Force Yak-12M in a civilian-style colour scheme represents the 'four-window' variety. The constant-chord wings and fixed leading-edge slats are clearly visible

Yefim Gordon archive

Above: A line-up of ten Yak-12Ms operated by one of Aeroflot's pilot schools sitting under wraps on a snowy airfield. The nearest aircraft is registered SSSR-Sh511.

Yefim Gordon archive

A ski-equipped production Yak-12M in agricultural configuration with spraybars running along the rear rods of the wing struts is filled up with chemicals; note the lack of the rear windows due to the installation of the chemical hopper. One can only conjecture what agricultural work it was to perform in the winter.

asts is a pun based on a Russian translation of *Scrooge McDuck* (a Walt Disney cartoon character). This machine took part in the SLA-91 ultra-light/homebuilt aircraft show at Chernigov, the Ukraine, in August 1991. Sporting the non-standard three-digit registration CBC-951 (the standard prescribed five digits, the first two of these denoting the region where the aircraft was registered), the aircraft remained almost unchanged externally but featured various internal modifications and improvements.

Yak-12M re-engined with an M-11. On 13th-15th August 2004 a strange hybrid aircraft made its appearance at the Flying Legends air show staged in Monino south of Moscow (it was an international event with the participation of veteran aircraft from several countries). Registered RA-2992K under the current Russian system for aircraft owned by civil air clubs and painted in camouflage colours to represent a wartime aircraft, it was a Yak-12M with the normal AI-14 engine replaced by an M-11 five-cylinder radial, complete with the helmeted cowling characteristic of the Yak-12 *sans suffixe*. In combination with the Yak-12M's longer tail and revised fin shape this cowling gave the aircraft an unfamiliar look. During the show the machine flew with the cowling removed.

Yak-12M – restored examples. Several Yak-12Ms have been restored to flying con-

Yakovlev OKB

Above: A still unregistered Yak-12A in the silver/dark blue colour scheme worn by some Aeroflot examples shows to advantage the tapered outer wing panels with automatic leading-edge slats.

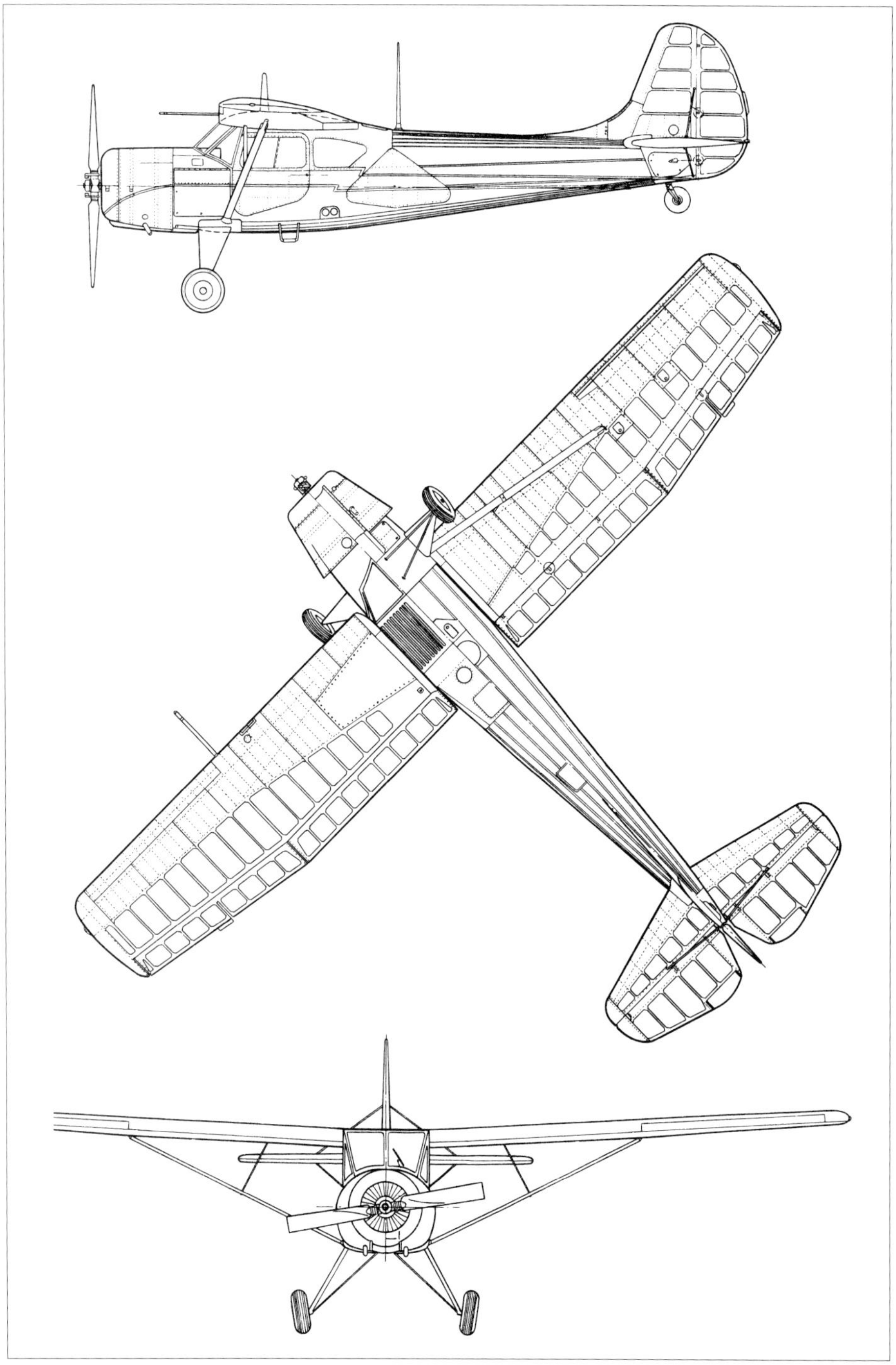

A three-view of the Yak-12A.

dition by members of an aero club in the town of Starodoob, Bryansk Region of Russia. The refurbishment included the installation of new radio equipment, as evidenced by a new array of antennas.

One of these aircraft deserves special mention. Built in 1952 and written off as time-expired in 1973, this machine was restored in 1999 by enthusiasts from Doobna (Moscow region) assisted by the Starodoob air club. Registered ФЛА РФ 01674 under the pre-2003 system (that is, FLA RF 01674; FLA RF = *Fede**rah**tsiya lyu**bit**eley avi**ah**tsiï Ros-**seey**skoy Fede**rah**tsiï* – Aviation Enthusiasts' Federation of Russia) and sporting a blue camouflage scheme, it was publicly demonstrated at the MAKS-2001 airshow in Zhukovskiy with a wheeled undercarriage. Before that, it had flown with twin PA-1700 floats that had been recently developed by specialists in Doobna.

Polish licence-built version of the Yak-12M

In 1955 the Polish authorities took the decision to obtain manufacturing rights for the Yak-12M. The WSK-4 design bureau based at Warsaw-Okęcie airport was tasked with the necessary preparations for production which were conducted in 1956, and licence production began in the same year which saw the roll-out of the first 20 Polish-built Yak-12Ms. Large-scale deliveries started in 1957; the aircraft was built in liaison, passenger, trainer and ambulance versions. It was supplied to the Polish Air Force and Polish air clubs, apart from export deliveries – mainly to the Soviet Union.

Yak-12A multi-purpose aircraft

This was the last baseline version of the Yak-12 which made its appearance in 1957. The design was reworked with a view to enhancing passenger comfort. From the outset the car-like cabin with comfortable seats for a pilot and three passengers featured two windows on each side and improved heating and ventilation. The wings were modified, comprising rectangular inner wing panels and tapered outer panels with forward sweep on the trailing edge. The Vee-shaped bracing struts characteristic of earlier versions gave way to single duralumin struts. The fixed leading-edge slats were replaced with automatic slats neatly fitting flush into the leading edge; this ensured a tangible increase of the flight speeds.

The tailplane halves were tapered, not rectangular, and were braced with only two wires on each side instead of three wires and a strut. Changes were introduced into the elevator and aileron controls, the control stick being replaced by a horned wheel. The fuel load was increased to 166 kg (366 lb), ensuring a more than twofold increase of the range as compared to the Yak-12M. The AI-14R engine drove a V-530 DIS propeller with a speed governor.

Despite the increased AUW, the Yak-12A had considerably improved flight performance as compared to the Yak-12M; it boasted higher maximum and cruising speeds, a greater payload and increased range at the cost of some reduction in service ceiling and rate of climb. Like its predecessors, the Yak-12A served in a variety of roles, as detailed below.

Yak-12A agricultural version

Some Yak-12A aircraft were equipped for crop-dusting and crop-spraying in the same fashion as the Yak-12M.

Yak-12A ambulance version

An ambulance version of the Yak-12A was put into operation; it featured the same kind of loading hatch for a stretcher case as the appropriate versions of the Yak-12R and Yak-12M.

Polish licence-built version of the Yak-12A

In late 1958 the WSK Okęcie factory obtained a licence to manufacture the Yak-12A. The first Polish-built Yak-12A was rolled out in mid-1959. This prototype registered SP-PXA represented a multi-purpose version suitable for use as a passenger, liaison or ambulance aircraft; in fact, the registration clearly indicates it was an ambulance aircraft, as the SP-*X* registration series was reserved for the Polish Flying Ambulance Service (*Lotnicze Pogotowie Ratunkowe*). Series manufacture of the Yak-12A in Poland took place in 1959-60, initially concurrently with the manufacture of the Yak-12M. A total of 1,150 Yak-12Ms and Yak-12As were produced in Poland between 1956 and 1960, mostly for export to the USSR (one source gives the number as 1,516). In Poland this aircraft was especially widely used in the ambulance role.

Yak-12G floatplane version of the Yak-12A (project)

A model of a floatplane version of the Yak-12A, captioned Yak-12G (see this designation above!), was on show in the Aerospace Museum in Moscow. There is no evidence of this version ever existing in hardware form.

Yak-12B prototype

The Yakovlev OKB converted a stock production Yak-12A into an experimental single-bay biplane version designated Yak-12B (*bi**plahn***). The standard wings were augmented by lower wing panels having a shorter span. The diagonal bracing struts already present on the Yak-12A were supplemented by single vertical struts connecting the upper and the lower wings. Both upper and lower wings featured flaps, the upper and lower flap on each side being linked by a strut.

Flight tests showed that the Yak-12B had improved field performance. However, other performance characteristics deteriorated considerably due to the increased weight and drag. Hence the machine remained a one-off.

Yefim Gordon archive

This silver-painted Yak-12A registered in the CCCP-72... block is an ambulance aircraft and carries the registration on the wings only, with a large red cross on the fuselage instead. Note that the Yak-12A's rear side windows are shorter than those of the Yak-12M.

Specifications of the Yak-12M and Yak-12A multi-purpose aircraft

	Yak-12M	Yak-12A
Year of manufacture	1954	1957
Engine type	AI-14R	AI-14R
Engine power, hp	260	260
Length	9.0 m (29 ft 6⅜ in)	9.0 m (29 ft 6⅜ in)
Wing span	12.60 m (41 ft 4 in)	12.60 m (41 ft 4 in)
Wing area, m^2 (sq ft)	23.8 (265)	22.66 (243.9)
Empty weight, kg (lb)	1,026 (2,262)	1,060 (2,340)
Payload, kg (lb)	n.a.	530 (1,169)
All-up weight, kg (lb)	1,450 (3,197)	1,590 (3,506)
Maximum speed, km/h (mph)	182 (113)	220 (137)
Cruising speed, km/h (mph)	n.a.	150-190 (93-118)
Landing speed, km/h (mph)	82 (51)	90 (56)
Climb time to 1,000 m (3,280 ft)	4 min	n.a.
Service ceiling, m (ft)	4,160 (13,648)	4,000 (13,120)
Range, km (miles)	765 (475)	1,070 (665) technical 600-800 (373-497) practical
Take-off run, m (ft)	126 (413)	130-150 (430-490)
Landing run, m (ft)	90 (295)	130-150 (430-490)

Yakovlev OKB

The experimental Yak-12B biplane version in flight. The single upper wing bracing struts indicate clearly this is a derivative of the Yak-12A.

Waclaw Holys

Above: An early-production Bulgarian-registered PZL-101 Gawron in agricultural configuration.

Yak-12B Bulgarian licence-built version (project)

In the early 1950s there were plans for putting the Yak-12 (presumably the Yak-12M) into production in Bulgaria as the Yak-12B (Bulgarian, not to be confused with the Yak-12B biplane). Everything was ready, including manufacturing drawings, jigs and tooling and the necessary materials, when in 1954 the Council for Mutual Economic Assistance (COMECON, the Eastern Bloc's counterpart of the EEC) decided to close down aircraft manufacture in Bulgaria.

Polish derivatives of the Yak-12 (PZL-101 Gawron)

The Yak-12M produced under licence in Poland formed the basis for an indigenous version designated PZL-101 Gawron (Raven) which was produced mainly as a dedicated agricultural aircraft. The design was thoroughly reworked, resulting in a basically new aircraft which initially differed externally in having a bulged top of the fuselage behind the wings to house a larger-capacity chemical hopper. Later versions of the Gawron featured wing panels that were slightly swept back for CG reasons and provided with characteristic oval endplates.

The PZL-101 was built mainly in the agricultural version and, to a lesser extent, for use as ambulance, sports and utility aircraft. The number of the PZL-101s of all versions built before 1972 totalled 325, of which more than 150 were exported to Hungary, Austria, Finland, Turkey, Spain, India, Bulgaria and even the Soviet Union (two examples).

Yakovlev OKB

Above: The first prototype of the original Yak-18 seen during manufacturer's tests. The aircraft is in natural metal/silver dope finish with red/black trim and a red propeller spinner.

Yakovlev OKB

A famous picture of a famous aircraft. This air-to-air of the Yak-18 prototype illustrates the trainer's classic lines and the red/black 'invasion stripes' on the wings and fuselage.

Yak-18 primary trainer

The Yak-18, which became the progenitor of a celebrated family of trainers and competition aerobatic aircraft, was a low-wing two-seat primary trainer developed by the Yakovlev OKB after the war as a direct development of the UT-2 trainer which it was intended to supersede. It owed much to the UT-2's final versions, notably the UT-2L from which the Yak-18 inherited such features as the enclosed cockpits and the helmeted engine cowling. The truss-type fuselage and the two-spar wing with internal wire bracing were also points of commonality with the previous design philosophy. At the same time the Yak-18's design reflected the changes in pilot training requirements associated with the new stage in the development of civil and combat aircraft. Notably, the Yak-18 concept envisaged the possibility of flights at night and in adverse weather.

K. V. Sinel'shchikov, who was appointed project manager for the Yak-18, received instructions from Chief Designer Aleksandr S. Yakovlev on designing this aircraft on 10th December 1945. There were two prototypes differing in powerplant, as described below. Their otherwise common design features included the two-spar wings with a rectangular horizontal centre section and outer panels having equal taper, rounded tips and 6°30' dihedral. The duralumin wing structure had a duralumin skin ahead of the front spar and fabric covering aft of it. The wings were fitted with slotted ailerons and a split flap under the centre section that could be lowered 55° for landing. The fuselage was based on a welded truss of mild steel tube; its covering comprised detachable duralumin panels as far back as the wing trailing edge and fabric further aft. The wire-braced tail had a duralumin structure with fabric covering. The main landing gear struts with braked wheels inboard of the oleos retracted aft, the legs remaining exposed and the wheels projecting to reduce damage in a wheels-up landing; they were supplemented by a fixed castoring tailwheel.

The fuel and oil system was designed to permit prolonged flight under negative G forces and comprised two 65-litre (14.3 Imp gal) fuel tanks accommodated in the wings between the fuselage and the undercarriage. The tandem cockpits were enclosed by a common canopy with aft-sliding sections and fully duplicated instruments and controls. The aircraft had a pneumatic system which served the engine starter, wheel brakes, undercarriage retraction and flap operation. Equipment included the RSI-3M/RSI-4M radio with a twin-wire aerial running from the fin to a mast on the centreline at the aft end of the canopy.

Yefim Gordon archive

Above: The Yak-18 final assembly line at one of the three factories which built the type. The aircraft are early-production examples, as evidenced by the propeller spinners.

Yefim Gordon archive

Row upon row of Yak-18s *sans suffixe* share an airfield with Po-2s, Kiev-built An-2 biplanes, Li-2 transports and a Douglas C-47 in the late 1940s. Note the lack of propeller spinners.

Yak-18 trainer (first prototype)

This prototype was powered by an M-11FM engine rated at 140 hp. It was completed on 19th April 1946 and first flew on 6th May, the manufacturer's tests continuing until 29th May. This configuration was not proceeded with.

Yak-18-2 trainer (second prototype)

The second prototype (known as the Yak-18-2) had an M-11FR-1 engine rated at 160 hp. It was completed on 30th April 1946, passing its manufacturer's tests between 15th May and 6th June 1946, with the first flight on 17th May. It became the main prototype. The test pilots were very impressed; still, along with positive comments, they noted the need for further adjustment work on the V112A/12 propeller.

Between 7th June and 12th September 1946 the Yak-18-2 underwent State acceptance trials at GK NII VVS. Progressively increasing engine vibrations necessitated an engine change; the propeller proved faulty and was replaced by a prototype propeller of the VISh-327YeV-149 model. Balancing the propeller helped eliminate the vibrations. The VISh-327YeV-149 featured automatic pitch control, but had to be used as a two-pitch propeller (fine pitch for take-off and climb and coarse pitch in level flight and during aerobatics) because there was no drive from the engine to the propeller speed governor.

During the State acceptance trials test pilots noted the Yak-18's ease of handling which made it suitable for pilots of less-than-average skill. It displayed indisputable advantages over the Po-2 and UT-2 trainers as regards stability and controllability in all flight modes and general performance (notably, higher speed). Such features as the radio, intercom, wheel brakes, retractable undercarriage, landing flap and pneumatic engine starter allowed the trainees to acquire some experience in handling the equipment of combat aircraft.

The OKB offered the Yak-18 in two versions, both of them two-seaters; each version could be converted into the other under field conditions. One was an aerobatic aircraft with an AUW of 970 kg (2,140 lb); the other was a primary trainer with an AUW increased to 1,060 kg (2,340 lb) on account of the radio, intercom, generator and landing light. Most of the testing was conducted in the primary trainer configuration which was finally endorsed; the aerobatic configuration did not win the approval of the Air Force because of insufficient fuel amount (35 kg/77 lb) and the lack of a radio.

The Yak-18 family received the NATO reporting name *Max*.

Yak-18 initial production trainer

Upon the completion of the State acceptance trials the OKB spent some time rectifying the faults revealed. After check-up tests held between 16th and 27th November 1946 the Yak-18 was formally adopted for service with the Air Force as a primary trainer. In March 1947 the Government ordered this aircraft into production at three plants: No.135 in Khar'kov, No.272 in Leningrad and No.116 in Semyonovsk (now Arsen'yev).

Initial production Yak-18s were basically identical to the second prototype and were fitted with the M-11FR-1 engine driving a VISh-327YeV-149 propeller. Machines from early batches can be identified by their propeller spinners which were deleted later on. In the course of production the aircraft was subjected to numerous detail improvements and minor modifications.

Yak-18 late-production (upgraded) trainer

One of the improvements effected in the course of production was the introduction of the new VISh-501D-81 propeller which had been tested in prototype form in 1948. This propeller set the required pitch automatically, depending on the flight speed. Thanks to this the rate of climb at sea level rose by 0.5 m/sec (98 ft/min), the take-off run was reduced to 80 m (260 ft), the take-off distance was reduced by 180 m (590 ft) and the maximum range was increased by 75 km (47 miles). At the same time (in late 1948) the M-11FR-1 engine was replaced by the 160-hp M-11FR. Among other things, the new engine version featured improved cylinder

Above: The Yak-18 *sans suffixe* could be fitted with skis for winter operations, as illustrated by this example coded '06 White' and sporting the titles of the DOSAAF paramilitary sports society.

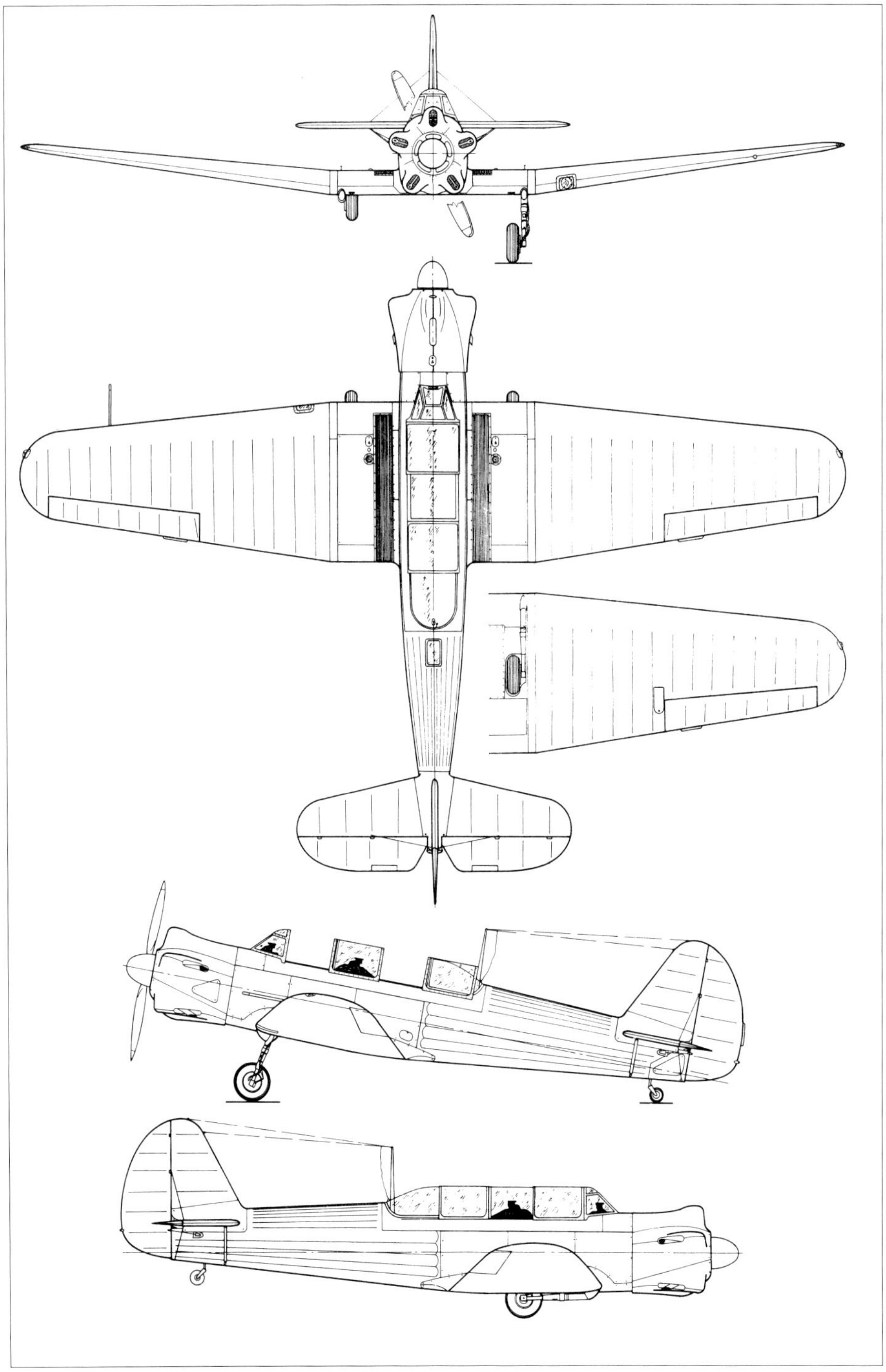

Four views of the Yak-18 *sans suffixe*, with a scrap view of the port wing underside.

cooling fins, an altered cylinder head shape and changes in the oil system. The engine mount received improved shock absorbers. Somewhat later the aircraft received new items of equipment.

M-12-powered Yak-18 trainer prototype

Pursuant to an order issued by MAP on 18th May 1948 the Khar'kov aircraft factory fitted one Yak-18 with a Kossov M-12 radial rated at 190 hp. This was the first aircraft of the Yak-18 family to feature a circular-section engine cowling instead of the helmeted cowling. Development work on this aircraft proceeded for some time and was terminated in early 1951. This was due to the advent of the more powerful Ivchenko AI-14R engine rated at 260 hp, which was used first on the Yak-12R and then on the Yak-12M and the Yak-18A (see below).

Yak-18 with experimental tricycle undercarriage

In 1950 the M-12-powered Yak-18 prototype was converted into a testbed for the experimental installation of a fixed nosewheel undercarriage. The main units were moved aft and fitted with forward bracing struts; the forward-sloping nose unit had a rear bracing strut. The 14 test flights conducted in November 1950 showed that the tricycle undercarriage afforded an improved view during take-off and landing and eliminated the risk of a nose-over when the wheel brakes were applied abruptly. In addition, the take-off and landing techniques became similar to those of jet aircraft which already had a tricycle undercarriage as standard.

Yak-18 on skis

In response to a requirement from the Air Force the OKB developed a fixed ski undercarriage for the Yak-18. The first Khar'kov-built Yak-18 (c/n 01135001 – that is, Batch 01, plant No.135, 001st aircraft in the batch) successfully passed ground and flight tests with skis in December 1947. The skis weighing 19.75 kg (43.5 lb) apiece were made of wood with metal edging. Substitution of the wheels for skis increased the aircraft's AUW by 13.7 kg (30.2 lb). The tailskid was not steerable. Check-up tests conducted in March 1949 in NII VVS confirmed satisfactory characteristics of a production Yak-18 equipped with skis, albeit their use led to a reduction of the maximum speed by 31 km/h (19 mph) and the maximum range by 220 km (137 miles).

Yak-18 as a night bomber

According to some accounts, during the Korean War of 1950-53 the North Korean Air Force used Yak-18 trainers, along with Po-2 biplanes, in the light night bomber role. By the summer of 1951 an Independent Night Bomber Regiment was formed comprising machines of these two types. Reports released by the North Korean command cited successful night air raids against airfields, military depots and other objectives in Seoul, Inchon and Kimpo in 1952-53. A few Yak-18s were reportedly shot down by US Navy F4U Corsairs.

Nanchang CJ-5 (Chinese-built Yak-18 M-11FR)

The initial Yak-18 *sans suffixe* ('taildragger') version powered by the M-11FR engine was built under licence in China. Initially China received Yak-18s from the USSR; in China they were known as Hongzhuan 501. These machines underwent repairs at plant No.302 in Nanchang which started producing some units from 1952 and soon switched to a complete production cycle. One account states that 'the first Yak-18 assembled in China made its first flight' in 1953 (presumably it was a machine assembled largely from Soviet-manufactured parts). In April 1954 the plant received an order for a batch of these aircraft which were allocated the designation CJ-5 (Chuji Jiaolianji-5 or Chujiao-5 – basic trainer, Type 5).

The first CJ-5 performed its maiden flight on 3rd July 1954, piloted by Duan Hiang-lu. This date came to mark the birth of aircraft production in China (as distinct from the mere assembly of foreign aircraft which had taken place earlier). Full-scale production started in late August. Between 1954 and 1958 the plant at Nanchang built a total of 379 CJ-5s.

Hungarian-built Yak-18 M-11FR

A small number of Yak-18s *sans suffixe* was also manufactured under licence in Hungary; production was apparently terminated due to the aforementioned COMECON decision of 1954. Hungarian-built Yak-18s had sequential construction numbers prefixed EM; for instance, a Polish example registered SP-AOP and belonging to the Polish National Air Club (APRL) was c/n EM 005.

Yak-18U production trainer

The Yak-18U was the first Yak-18 production version fitted with a retractable tricycle undercarriage. The nose gear unit retracted aft to lie in the lengthened forward fuselage with the castoring wheel partly exposed. The main units retracted forward; the upper parts of the tyres were recessed in the wings ahead of the front spar, most of the wheel remaining in the slipstream. The axles of the main legs were now mounted on the outer side of the leg which was hinged to a trunnion on the rear spar. Appropriate changes were introduced into the wing centre section, forward fuselage and engine mount.

The cowling of the M-11FR engine was redesigned; it retained individual helmet cowls for the cylinders, but instead of the individual cooling air inlets the cowling was provided with a circular central inlet fitted with controllable radial shutters. Changes were introduced into the oil system, pneumatic system and engine control system. The capacity of each of the two fuel tanks was reduced from 75 to 63 litres (from 16.5 to 13.86 Imp gal). Both cockpits were made 60 mm (2²³⁄₆₄ in) wider by reducing the width of the consoles and moving them closer to the fuselage side skinning. A transparent partition incorporating a small window was installed between the cockpits.

The Yak-18U prototype was converted from a production Yak-18 (c/n 1162509 – that is, plant No.116, Batch 25, 09th aircraft in the batch), passing manufacturer's flight tests between 1st September and 3rd December 1951. Check-up tests at GK NII VVS followed between 24th January and 29th February 1952. Once the defects revealed during the testing had been rectified, the Yak-18U entered production at plants No.116 and No.135.

The Yak-18U, together with the original Yak-18 *sans suffixe*, represented the first basic stage in the development of the Yak-18 family. Overall production of the two types in between 1946 and 1957 amounted to 5,680 machines. In addition to the USSR, the Yak-18 *sans suffixe* saw service in Albania, Bulgaria, East Germany, Poland, Romania, Hungary, China, North Korea, Austria and Afghanistan (of these countries only East Germany also operated the Yak-18U).

Yak-18T advanced trainer prototype (first use of designation)

This aircraft, the first Yak-18 to have the T suffix (T stands for *treni**rovo**chnyy* [*samo**lyot***] – trainer) was a modification of the Yak-18U

Specifications of the Yak-18 trainer, initial versions

	Yak-18 prototype	Yak-18 production	Yak-18 on skis
Year of manufacture	1946	1949	1949
Engine type	M-11FR-1	M-11FR	M-11FR-1
Engine power, hp	160	160	160
Length	8.03 m (26 ft 4⁹⁄₆₄ in)	8.03 m (26 ft 4⁹⁄₆₄ in)	8.03 m (26 ft 4⁹⁄₆₄ in)
Wing span	10.06 m (33 ft 0¹⁄₁₆ in)	10.06 m (33 ft 0¹⁄₁₆ in)	10.06 m (33 ft 0¹⁄₁₆ in)
Wing area, m^2 (sq ft)	17.0 (182.79)	17.0 (182.79)	17.0 (182.79)
Empty weight, kg (lb)	745 (1,640)	769 (1,695)	780 (1,720)
All-up weight, kg (lb)	1,060 (2,340)	1,085 (2,390)	1,096 (2,416)
Maximum speed, km/h (mph)	257 (159)	250 (155)	219 (136)
Landing speed, km/h (mph)	85 (52.79)	84 (52.17)	85 (52.79)
Time to 1,000 m (3,280 ft)	5.0 min	5.4 min	6.0 min
Service ceiling, m (ft)	6,000 (19,685)	n.a.	n.a.
Range, km (miles)	1,050 (652)	1,080 (670)	860 (534)
Take-off run, m (ft)	290 (950)	315 (1,030)	370 (1,210)
Landing run, m (ft)	250 (820)	185 (605)	270 (885)

Khar'kov State Aircraft Manufacturing Co.

The experimental Yak-18 fitted with a Kossov M-12 radial. The cowling design is similar to that of the Yak-11.

Yakovlev OKB

Above: The prototype of the Yak-18U, the first tricycle-gear version. The altered cowling design is clearly visible; note the radial shutters at the front.

described above and shared with it the tricycle undercarriage, differing only in equipment. As distinct from the Yak-18U which was intended for primary training, the Yak-18T was to fill the advanced trainer role, which would enable trainees to practise route flights and flights in adverse weather. Accordingly, its equipment included the OSP-48 ILS comprising the ARK-5 ADF (instead of the Yak-18U's RPKO-10M DF), the RV-2 low-altitude altimeter with transmitter and receiver dipole aerials under the fuselage, the MRP-48 marker beacon receiver with an internal antenna and the DGMK-3 remote-controlled gyro-flux gate compass. The DC battery was located in the wing centre section, replacing the starboard fuel tank; the fuel capacity was reduced to 63 litres (13.86 Imp gal). In addition to the landing light, a special taxi light was fitted.

The Yak-18T prototype converted from a production Khar'kov-built Yak-18U (c/n 06135084) commenced its manufacturer's tests on 1st September 1951; on 24th September they were followed up by State acceptance trials at GK NII VVS. Test pilots noted that the aircraft's equipment enabled Air Force trainee pilots to practise blind landing approaches making use of the OSP-48 system. However, with the AUW increased to 1,204 kg the aircraft was clearly underpowered; the reduced fuel load further diminished the aircraft's usefulness. The Yak-18T was not adopted for series production, but some of its equipment items were later used on production Yak-18A trainers.

The Yak-18T designation was later re-used for a completely different derivative of the Yak-18, described below.

Yak-18A trainer, first prototype (Yak-20, second use of designation)

The need to replace the Yak-18/Yak-18U's 160-hp M-11FR engine with a more potent powerplant had long been recognised; this intention was implemented in 1956 when the aircraft received the Ivchenko AI-14R engine rated at 260 hp. It was a nine-cylinder air-cooled radial with a reduction gearbox and a supercharger; the engine drove a V-530D-35 two-blade hydraulically controlled variable-pitch propeller.

Specifications of the Yak-18U and Yak-18T (1951) trainers

	Yak-18U prototype	Yak-18T prototype	Yak-18U production
Year of manufacture	1951	1951	1955
Engine type	M-11FR	M-11FR	M-11FR
Engine power, hp	160	160	160
Length	8.12 m (26 ft 7 11/16 in)	8.12 m (26 ft 7 11/16 in)	8.13 m (26 ft 8 5/64 in)
Wing span	10.06 m (33 ft 0 1/16 in)	10.06 m (33 ft 0 1/16 in)	10.06 m (33 ft 0 1/16 in)
Wing area, m² (sq ft)	17.0 (182.79)	17.0 (182.79)	17.0 (182.79)
Empty weight, kg (lb)	856 (1,887)	968 (2,134)	882 (1,944)
All-up weight, kg (lb)	1,140 (2,510)	1,204 (2,654)	1,166 (2,570)
Maximum speed, km/h (mph)	235 (146)	225 (139)	230 (142)
Landing speed, km/h (mph)	103 (64)	105 (65)	96 (59)
Time to 1,000 m (3,280 ft)	6.2 min	7.3 min	7.0 min
Service ceiling, m (ft)	4,000 (13,120)	3,500 (11,480)	3,300 (10,830)
Range, km (miles)	825 (512)	295 (183)	750 (465)
Take-off run, m (ft)	235 (770)	255 (840)	260 (850)
Landing run, m (ft)	285 (935)	290 (950)	295 (970)

The first prototype of the new version was a converted Yak-18U (c/n 1161311) built by Plant No.116 in February 1956. In its new guise the machine was initially known simply as a 'modified Yak-18U'; then the designation Yak-20 (second use of designation; see below) came up. However, by the end of 1957 the aircraft was renamed Yak-18A, and this designation was officially endorsed.

The first prototype passed its manufacturer's tests in October 1956, the State acceptance trials following between 26th February and 22nd March 1957. The aircraft was recommended for series production and service as the standard trainer for the Air Force flying schools.

The Yak-18A differed from the Yak-18U externally in having a circular-section engine cowling instead of the helmeted cowling, a deeper cockpit canopy and a vertical tail with a curved fin fillet. Other changes included longer control sticks, greater aerodynamic balance of the ailerons, provisions for cockpit heating and improved ventilation.

The new engine entailed a redesign of the engine mount and of the oil system. Engine cooling was regulated by radial shutters. As distinct from the previous model, the new propeller's pitch was controlled manually by the pilot from his cockpit through a cable linkage and a hydraulic actuator. The R-2 propeller speed governor ensured constant rpm.

The Yak-18A differed from the preceding model in having a completely updated avionics complement. The RSI-6M radio and the RPKO-10M DF were replaced by the RSIU-3M and the ARK-5 ADF respectively. The AGK-47B gyro horizon was replaced by the AGI-1. The 350-watt GS-10-350M generator gave place to a 1.5-kilowatt GS-1500 generator.

Installation of the new engine and new equipment items led to a slight increase of the all-up weight which, in turn, necessitated restressing of some structural members. However, this weight penalty was fully justified by the improvement in performance thanks to the new powerplant.

Yak-18A trainer, second prototype

The second prototype Yak-18A was again converted from a production Yak-18U (c/n 1161003). Its equipment was supplemented by a GIK-1 gyro-flux gate compass and an MRP-48P marker beacon receiver. This enabled the aircraft to operate in adverse weather conditions and perform instrument landings.

The machine passed its manufacturer's tests between 22nd May and 10th June 1957 and check-up tests at GK NII VVS between 6th July and 2nd August 1957.

Yak-18A production trainer

Deliveries of production Yak-18As to military and civil flying schools began in the second half of 1957. In the course of production various improvements were introduced from batch to batch. The maximum permissible diving speed was increased from 310 to 340 km/h (from 193 to 211 mph); the equipment came to include a second 12-A-10 type DC battery, an R-800 radio and an MGP-56P marker beacon receiver. The Yak-18A initially had a service life of just 400 flight hours or two calendar years. This was gradually increased by 1964 to 3,000 hours or 15 years. Production of the Yak-18A lasted until 1960 and totalled 950 machines.

In 1975 most of the Yak-18As serving with the Air Force ran out of service life and were withdrawn from use. At that time only 11 machines remained operational with the DOSAAF air clubs; their designated service life was extended to 3,500 flight hours and 19 calendar years. Some examples logged as many as 4,000 hours and stayed in service for 20 years.

The Yak-18As served also with the East German Air Force and the Air Force of Egypt; some Egyptian examples were later sold to warbird collectors and found their way to aircraft museums and collections in Russia, Great Britain, France, Germany and the USA.

The Yak-18A served as a basis for the development of the Yak-18P, Yak-18PM and Yak-18PS aerobatic aircraft described below.

Yak-18A production trainer on skis

An example of the Yak-18A with a fixed ski landing gear was tested in March 1958.

Nanchang CJ-6 and CJ-6A (Yak-18A derivatives built in China)

An example of the Yak-18A was supplied to China with a view to setting up licence production of this version as a successor to the original Yak-18 *sans suffixe* (CJ-5). The aircraft was carefully studied by Chinese specialists at the Shenyang aircraft factory. They came to the conclusion that direct copying of the Soviet machine was not advisable for reasons associated with production methods. The Yak-18A airframe was based on a steel tube truss; this did not conform to the local production conditions which were determined by the fact that China had built up a large-scale output of rolled aluminium.

A decision was taken to design a new trainer based on the Yak-18A. Design work on this aircraft originally designated Nanchang Type 61 or Hongzhuan 502, and later renamed CJ-6, proceeded in 1957-58. The CJ-6 performed its maiden flight on 27th August 1958. Judging by the shape of the

Yakovlev OKB

Above and below: The Yak-18A prototype during initial tests, showing the cowling design characteristic of all subsequent versions. Note the exposed position of the semi-retractable landing gear.

Yakovlev OKB

Sergey and Dmitriy Komissarov archive

Above: '75 Yellow', the second prototype Yak-18A. All three landing gear struts were inclined forward when extended.

Sergey and Dmitriy Komissarov archive

A production DOSAAF Yak-18A coded '05 White'. Note the white panel ahead of the fin fillet concealing the DF aerial.

Yefim Gordon archive

Above: The Yak-18As delivered to Aeroflot's flying schools wore a civil colour scheme and civil registrations, such as CCCP-81418 here. The registrations were later re-used for Yak-18T four-seaters.

Yefim Gordon

Many people may be excused for believing this is a Yak-18A. It is not; despite its apparent similarity, the Nanchang CJ-6 is a very different aircraft. This example is seen at White Waltham, England in July 2004.

cowling, the first prototype was fitted with a Czechoslovak flat-six engine (possibly a 235-hp Praga M 208C) and a propeller of Czechoslovak manufacture, but in August 1959 the machine was re-engined with the Soviet-produced AI-14R. The second prototype of the CJ-6 completed in 1961 featured a redesigned engine cowling. The machine went into production in January 1962; from 1963 it was fitted with locally produced AI-14R engines and propellers, the engine receiving the designation HS-6A (Huosai – piston engine, Type 6A).

In its general appearance the CJ-6 closely resembled the Yak-18A; especially the engine cowling looked virtually identical to that of the Soviet progenitor. However, it was in reality a new aircraft. Apart from the not very obvious structural changes, the CJ-6 featured some external features instantly differentiating it from the Yak-18A. It received new tail surfaces whose angular shape was quite unlike the rounded tail of the Soviet machine. The main undercarriage units had levered suspension instead of the simple oleo legs of the Yak-18A and retracted inwards, stowing completely in the wing centre section. Finally, the cockpit canopy was altered, incorporating a frameless curved windshield and a new rear cockpit hood without a separate fixed glazing section aft of it.

In 1965 an upgraded version of the HS-6A engine was produced; the trainers equipped with this engine received the designation CJ-6A. There were also versions intended for other roles. The CJ-6B built in ten examples in 1964-66 was a light counter-insurgency (COIN) attack aircraft armed with machine-guns, bombs and rocket projectiles. In all, up to 1998 production of different versions of the CJ-6 totalled more than 2,100 machines for the People's Liberation Air Force (PLAAF), plus some 200 examples for foreign customers, including the air forces of Albania, Bangladesh, Kampuchea, North Korea, Vietnam and Zambia. In recent years, many CJ-6/CJ-6As have been bought by warbird collectors in Australia, Great Britain and the USA. Though technically not a true warbird, the CJ-6 looks and behaves every bit like one – and gives you 'more bangs for the buck'.

In 1985 the Nanchang Aircraft Manufacturing Company converted one CJ-6 into a multi-purpose aircraft for various civil uses, including patrol and observation duties and even an agricultural role. The aircraft, named Petrel, made its first flight on 17th August 1985.

Specifications of the Yak-18A and the CJ-6A

	Yak-18A	CJ-6A
Year of manufacture	1957	1965
Engine type	AI-14R	AI-14RF
Engine power, hp	260	285
Length	8.18 m (26 ft 10 in)	8.47 m (27 ft 9$\frac{15}{32}$ in)
Wing span	10.06 m (33 ft 0$\frac{1}{16}$ in)	10.02 m (32 ft 11$\frac{31}{64}$ in)
Wing area, m^2 (sq ft)	17.0 (182.79)	17.0 (182.79)
Empty weight, kg (lb)	1,025 (2,260)	1,093 (2,409)
All-up weight, kg (lb)	1,316 (2,900)	1,397 (3,080)
Maximum speed, km/h (mph)	263 (163)	297 (184)
Landing speed, km/h (mph)	105 (65)	115 (71)
Climb time to 1,000 m (3,280 ft)	3.4 min	n.a.
Service ceiling, m (ft)	5,000 (16,400)	5,180 (16,990)
Range, km (miles)	725 (450)	690 (428)
Take-off run, m (ft)	215 (705)	280 (920)
Landing run, m (ft)	250 (820)	350 (1,150)

Yak-18P (Yak-18AP) aerobatic aircraft prototype

The Yak-18A two-seat trainer was used as a basis for the development of an aerobatic aircraft for the DOSAAF air clubs. This machine received the designation Yak-18P (*pilo**tazh**nyy* – aerobatic).

According to most accounts, the first prototype of this machine was a straightforward adaptation of the basic Yak-18A with minimum changes, the most important of which was the deletion of the forward seat. Thus, the pilot sat in an aft position under an abbreviated canopy. The aircraft retained the Yak-18A's semi-retractable undercarriage with the main units retracting forward and the wheels semi-recessed in the wing leading edge. Thus configured, the aircraft was usually referred to as the Yak-18P, but some accounts credit it with the Yak-18AP designation distinguishing it from the later, considerably different prototype and production Yak-18Ps.

Like the baseline Yak-18A, this Yak-18P (Yak-18AP) prototype was powered by the

260-hp AI-14R driving a V530-D35 variable-pitch propeller. To suit the machine's new mission, the fuel and oil feed systems were designed to ensure the possibility of using the whole range of power settings and rpm in inverted flight. With the engine running at nominal power (2,000 rpm) and a full service tank, the aircraft could perform inverted flight in the course of five minutes.

The Yak-18AP differed from the Yak-18A in the stabiliser incidence changed from zero to +1° and the elevator deflection angles of the changed to 25° down/20° up instead of 20° and 25° respectively. The CG position was shifted from 24.5% MAC to 29.7% MAC.

This prototype was built in 1958 and underwent testing in 1959, judging by some published accounts. According to one 'unorthodox' account, the Yak-18AP was actually an improvisation resulting from an emergency. Test pilot Boris A. Orlov recalls in his memoirs that shortly before the first World Aerobatics Championship in Bratislava in 1960 a Yak-18P prototype ('true' prototype, a different machine) was damaged beyond repair in a rough landing (test pilot Sergey N. Anokhin, putting up a 'smart' show, miscalculated and touched down so hard that the landing gear collapsed!). *'As a result of the accident* – wrote Orlov – *our team went to Bratislava with a single example of a hastily built new Yak-18P, not the one that had crashed but a machine that was in effect a single-seat version of the Yak-18A and was far inferior to the "real" Yak-18P'.*

Sergey and Dmitriy Komissarov archive

Above: The original aircraft bearing the Yak-18P designation (alias Yak-18AP) was a straightforward adaptation of the Yak-18A. Here it shares an airfield with several Zlin 226s; note the 'Yak-18P' tail titles.

Yak-18P aerobatic aircraft, true prototype

As stated above, there are reasons to believe that the true prototype of the Yak-18P actually preceded the machine bearing the same designation which was shown in Bratislava in 1960. This Yak-18P prototype differed from both the Yak-18A and the Yak-18AP in two respects. Firstly, it was turned into a single-seater by deleting the *rear* cockpit, the pilot sitting up front (experience showed later that this was not the best option). Secondly, the semi-retractable landing gear of the Yak-18A gave place to a fully retractable gear, the main units retracting inward. This prototype featured a fighter-type bubble canopy with a one-piece sliding section aft of the windshield. On production machines it was replaced by a three-piece canopy with the central sliding hood placed between the windshield and a fixed rear section.

Yak-18P production aerobatic aircraft

Apart from the abovementioned modification to the cockpit, the production Yak-18P was identical to the prototype. It differed from the Yak-18A in having the whole structure restressed to ensure adequate strength for performing high-G aerobatic manoeuvres. The fuel and oil systems, as in the case of the Yak-18AP, were redesigned with a view to ensuring the possibility of performing inverted flight for several minutes. The two main fuel tanks were supplemented by a third tank holding 4.5 litres (0.99 Imp gal) for feeding the engine at zero and negative G. To improve controllability when performing aerobatics, considerable alterations were made to the stabilisers, the fin and, to some extent, the wings; the area of all control surfaces was increased. The aircraft was stripped of all superfluous equipment.

The Yak-18P successfully passed manufacturer's tests and State acceptance trials and was built in series in 1960-61, super-

Yakovlev OKB

Above: The ill-starred prototype of the 'real' Yak-18P, a totally different aircraft with a cockpit located well forward and inward-retracting main gear units. Note the canopy design with a one-piece blown hood.

Boris Vdovenko

A production Yak-18P operated by the Soviet national aerobatic team (to be precise, the Central Air Club named after V. P. Chkalov). Note the different three-piece canopy and the club's emblem on the cowling.

Yefim Gordon archive

Above: Four Yak-18Ps, all wearing different colour schemes, at a DOSAAF airfield. The second aircraft in the row is marked 'CCCP', indicating it is flown by the national team; the others are marked 'ДОСААФ' (DOSAAF). Note the bulged covers enclosing the mainwheels.

Specifications of the Yak-18P

	Yak-18P (Yak-18AP) prototype	Yak-18P production
Year of manufacture	1960	1961
Engine type	AI-14R	AI-14R
Engine power, hp	260	260
Length	8.18 m (26 ft 10 in)	8.235 m (27 ft 0⁷⁄₃₂ in)
Wing span	10.6 m (34 ft 9⅜ in)	10.6 m (34 ft 9⅜ in)
Wing area, m² (sq ft)	17.0 (182.79)	17.0 (182.79)
Empty weight, kg (lb)	918 (2,023)	937 (2,065)
All-up weight, kg (lb)	1,065 (2,350)	1,093 (2,409)
Maximum speed, km/h (mph):		
at sea level	275 (170)	306 (190)
at 3,000 m (9,840 ft)	n.a.	280 (174)
Landing speed, km/h (mph)	95 (59)	100 (62)
Service ceiling, m (ft)	6,500 (21,325)	n.a.
Take-off run, m (ft)	120 (390)	163 (534)
Landing run, m (ft)	200 (660)	170 (560)

seding the Zlin 226 aircraft which had been used by DOSAAF air clubs for aerobatics. The Soviet national team flew Yak-18Ps at the second World Aerobatic Championship in Hungary in 1962 and at the third contest of this kind in Spain in September 1964.

Yak-18P with experimental propeller

In June 1962 one example of the Yak-18P was fitted with an experimental propeller 'made of plastic' (in reality probably fibreglass) instead of the standard wooden propeller. The new propeller's greater thrust afforded a marginal increase in performance. However, it was considered too risky to modify in this way the other Yak-18Ps, which thus retained their standard propellers. This proved to be a prudent decision. Exactly a year later, in June 1963, when this particular example of the Yak-18P was flown at Tushino by Boris Chernov, a champion aerobatic pilot, one of the propeller blades broke off. Seconds later, the violent vibrations created by the propeller imbalance caused the engine to break off, too. The pilot bailed out, but, sadly, the altitude was too low for his parachute to deploy...

Yakovlev OKB

The Yak-18PM prototype combined the Yak-18P's landing gear design with a repositioned cockpit and wings featuring reduced dihedral.

Yak-18PM aerobatic aircraft – initial version

The Yak-18P was a generally successful aerobatic aircraft, but it had its faults. The forward position of the cockpit hampered the pilot in assessing the aircraft's attitude and the pronounced wing dihedral made the machine unstable in inverted flight. These faults were remedied in a modified version of the Yak-18P which was developed by a group of young Yakovlev OKB engineers.

Designated Yak-18PM (*pilo**tazh**nyy, modif**itsee**rovannyy* – aerobatic, modified), it featured a number of changes. A more powerful AI-14RF engine rated at 300 hp (F = *for**see**rovannyy* – uprated) was installed instead of the 260-hp AI-14R; the wing dihedral was reduced; some structural members were re-stressed and the aerodynamic balance area of the control surfaces was changed. The cockpit was moved aft, enabling the pilot to see the wings against the horizon and to see the ground behind the wing trailing edge. This gave the pilot indisputable advantages when performing aerobatics over a contest site of limited dimensions. (At such events, pilots must not go outside certain imaginary lines for public safety reasons; offenders are disqualified)

The aircraft was provided with an airbrake on the rear part of the wing centre section. However, experience showed that the use of the airbrake spoiled the visual impression of the aerobatic figure performed, and pilots were reluctant to use it. Eventually it was deleted on production Yak-18PMs.

The Yak-18PM displayed its qualities to advantage during the 4th World Aerobatics Championship which took place at Moscow-Tushino in 1966. In that contest the titles of absolute world champion among men and women went to Soviet pilots V. Martem'-yanov and Galina Korchuganova respectively. The Yak-18PM was operated by Soviet air clubs during the following ten years.

Yak-18PM – late-production aircraft

Late-production Yak-18PMs (as well as Yak-18PSs – see below) had shorter-span wings with cropped tips which were squared-off, not rounded. This modification was aimed at increasing the efficiency of the ailerons by reducing the damping effect of the wings' tip sections. Photographs show these machines also to be fitted with rudders of increased chord.

Yak-18PM turboprop conversion (prototype)

A production Yak-18PM was experimentally re-engined with a 920-ehp Glushenkov TVD-10 turboprop driving a three-blade propeller, the longer nose changing the aircraft's appearance completely. No further details are available.

Yak-18PS aerobatic aircraft

This version owes its birth to some disappointment with the Yak-18PM due to the less-than-successful performance of the Soviet aerobatic team at competitions held in Magdeburg, East Germany, in 1968. On that occasion it was the East German team that scored the best overall result. The criticism levelled at the aircraft was probably not quite fair; the blame could more rightfully have been placed with faulty training procedures. Nevertheless, it was found advisable to provide the Soviet team with an improved aerobatic aircraft for participation in future championships.

In modifying the machine, the OKB placed its emphasis on reducing the airframe weight and hence the wing loading. The weight reduction was achieved by reverting to a tailwheel undercarriage of the type used on the initial Yak-18 *sans suffixe*. Coupled with a simplified equipment fit, this gave a weight saving of some 100 kg (220 lb). At the same time a horn balance was introduced on the rudder to reduce the pedal forces. The resulting aircraft was dubbed Yak-18PS (*pilo**tazh**nyy spetsi**ahl'**-nyy* – aerobatic special).

The mentioned modifications did, in fact, produce a tangible improvement in the machine's handling as compared to the Yak-18PM. After having flown the Yak-18PS, the leading pilots of the Soviet national aerobatic team expressed their unqualified preference for it. The OKB converted three Yak-18PMs into the PS version in time for the 6th World Aerobatics Championship. However, it was decided to back up the new machines with the well-mastered Yak-18PM which was also sent to the championship held in Great Britain in July 1970 in company with Yak-18PSs sporting the side numbers '05 Yellow', '07' Yellow' and '08 Yellow'.

On that occasion the Soviet national team comprising ten pilots ranked second after the US team which won first place; but the title of the absolute world champion went

Yakovlev OKB

Above: The Yak-18PM prototype on the ground. The red/white colour scheme applied to this aircraft became basically standard for later Yakovlev sports and training aircraft.

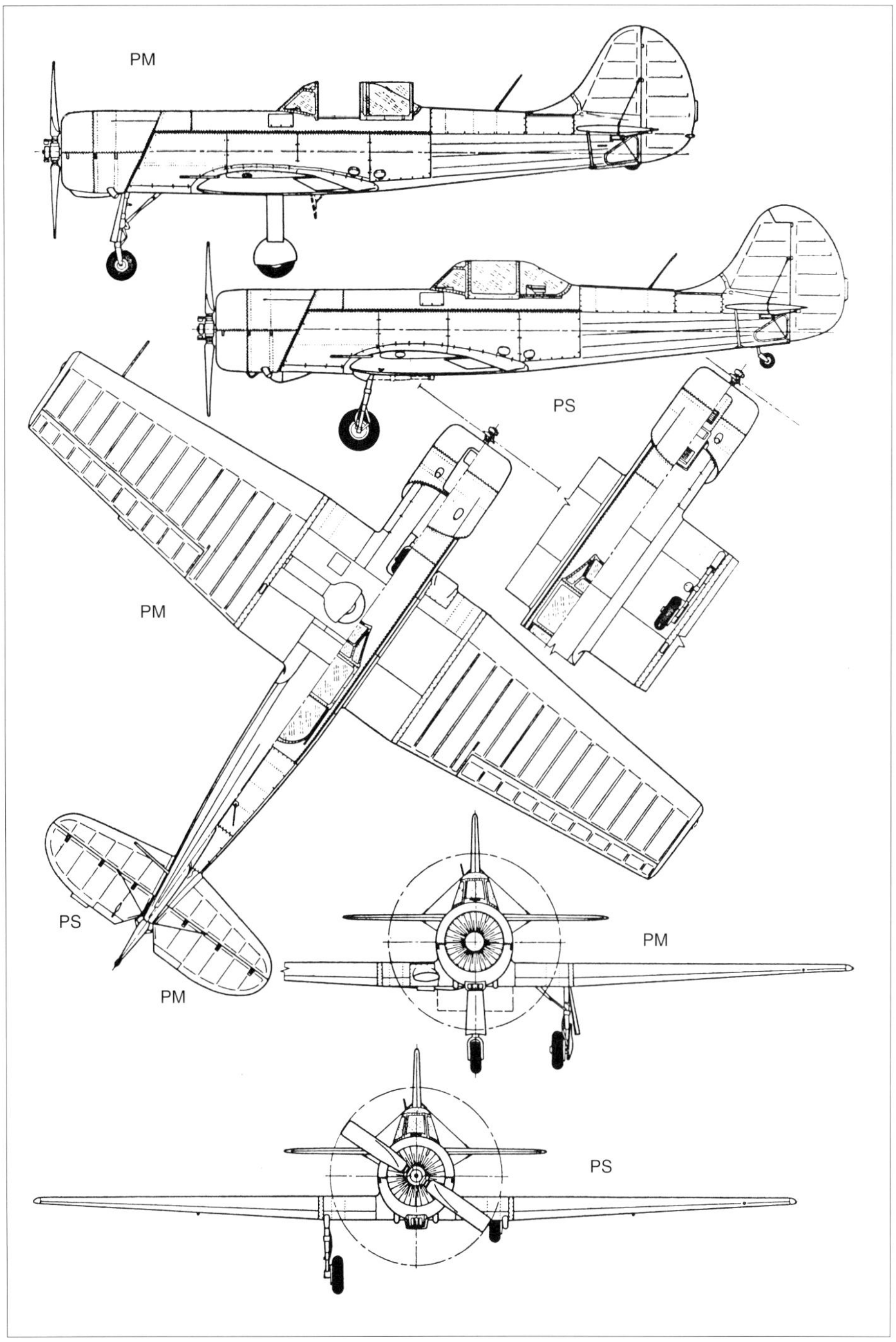

These drawings give a clear comparison of the tricycle-gear Yak-18PM (the late-production cropped-wing version) and the 'taildragger' Yak-18PS.

Sergey and Dmitriy Komissarov archive

'02 Yellow', the prototype of the Yak-18PS. The long forward fuselage, a 'leftover' from the Yak-18PM (where it was a necessity because of the tricycle landing gear), is noteworthy.

Specifications of the Yak-18PM and Yak-18PS

	Yak-18PM	Yak-18PS
Year of manufacture	1965	1970
Engine type	AI-14RF	AI-14RF
Engine power, hp	300	300
Length	8.18 m (26 ft 10 in)	n.a.
Wing span	10.6 m (34 ft 9⅜ in)	10.6 m (34 ft 9⅜ in)
Wing area, m² (sq ft)	17 (182.79); later 16.5 (177.6)	17 (182.79)
Empty weight, kg (lb)	950 (2,094) *	n.a.
All-up weight, kg (lb)	1,100 (2,425)	980 (2,160)
Maximum speed, km/h (mph)	320 (198)	320 (198)
Landing speed, km/h (mph)	115 (71)	n.a.
Initial climb rate, m/s (ft/min)	10 (1,968)	12 (2,361)
Service ceiling, m (ft)	6,700 (21,980)	n.a.
Range, km (miles)	400 (248.5)	n.a.
Take-off run, m (ft)	142 (465)	n.a.
Landing run, m (ft)	133 (436)	n.a.

* Note: Some sources state 825 kg (1,820 lb)

to the Soviet pilot Igor' Yegorov who won it in a neck-and -neck competition with Robert Herendin of the USA.

Yak-18T trainer and multi-purpose aircraft (second use of designation)

This four-seat aircraft intended for a variety of roles, including training was created in 1965-66 as a derivative of the Yak-18PM. However, in most respects it is a completely different aircraft, retaining its Yak-18 designation chiefly as a tribute to tradition. Design work on this machine was conducted by a team of young designers specially formed for the purpose in the Yakovlev OKB. The chief engineers were Sergey A. Yakovlev (A. S. Yakovlev's son) and S. V. Neunylov. The requirement was for a civil aircraft that would serve as a primary and advanced trainer, including blind flying and use of radio, as a transport/aerial taxi for three passengers and baggage or 250 kg (550 lb) of cargo, and as an ambulance with provision for one stretcher case and a medical attendant. The transport role gave rise to the designation Yak-18T, the T in this instance being deciphered as either ***t**rahnsportnyy* (transport, used attributively) or *tak**si*** (taxi).

Initially the project concept envisaged retaining the complete wings, tail unit, undercarriage and other assemblies of the Yak-18PM, the alterations being confined to designing a wider steel tube truss fuselage with a big aft-sliding canopy. In the course of projecting it became clear that more radical changes would be necessary. In the end, a completely new all-metal semi-monocoque fuselage was designed, incorporating a four-seat cabin with car-type doors. Interestingly, the wide fuselage of the Yak-18T weighed a mere 168 kg (370 lb) as compared to 158 kg (348 lb) of the Yak-18A with its considerably smaller dimensions. Provision was made for a baggage hold accessed through a port side door.

The basic wing structure was similar to that of the Yak-18PM. The two-spar outer wing panels were stiffened by bracing wires between the spars and were fabric-covered. The wing centre section was integral with the fuselage and carried a large landing flap. It also provided accommodation for the inward-retracting main undercarriage units. The oleo-pneumatic shock absorber struts of the Yak-18T differed from those of the Yak-18PM in having a two-chamber design ensuring better shock absorption when operating from unprepared airstrips.

The aircraft was provided with dual controls featuring control wheels instead of sticks. New flight and navigation equipment was fitted, comprising an ARK-9 ADF, a compass system, an RV-5 low-altitude radio altimeter, an MRP-65P marker beacon receiver, an Os'-1 (Axis-1) attitude and heading reference system. A full complement of lights was installed for night flying. The communications suite included a Landysh-5 VHF radio and an intercom.

Yak-18T prototype and pre-series aircraft

The first prototype of the Yak-18T was first flown in the spring of 1967 and made its public debut at the 27th Paris Air Show in June of that year. At that time the aircraft was not yet fitted with a full equipment complement.

The first prototype differed outwardly from subsequent examples in a number of respects. For example, it had a panoramic curved windshield; the cabin roof was opaque, except for a narrow 'eyebrow' panel immediately above the windshield. The tip of the fin was not yet crowned with an anti-collision light. This example wore an overall cream colour scheme with red trim and had no titles or registration, except for the 'Yak-18T' titles on the fin. In the course of manufacturer's tests the prototype showed good performance and fine handling qualities, but it took some time before the aircraft reached the production configuration.

Available photos show that the Yak-18T prototype (or prototypes) underwent some detail changes in the course of the flight tests. The changes that can be thus traced included the provision of a new windshield composed of flat panels, extra glazing panels in the cabin roof and a longer carburettor air intake under the nose. The machine so configured retained the same colour scheme and wore no registration. In this configuration the Yak-18T was used to test its spinning characteristics, as witnessed by pictures of the prototype with a spin recovery parachute container attached under the tail. Later this example was turned over to the Soviet Air Force Museum (now Central Russian Air Force Museum) in Monino.

Further modifications resulted in a prototype or pre-series example which was externally identical to the subsequent production machines. It had a windshield consisting of flat panels and a revised framing of

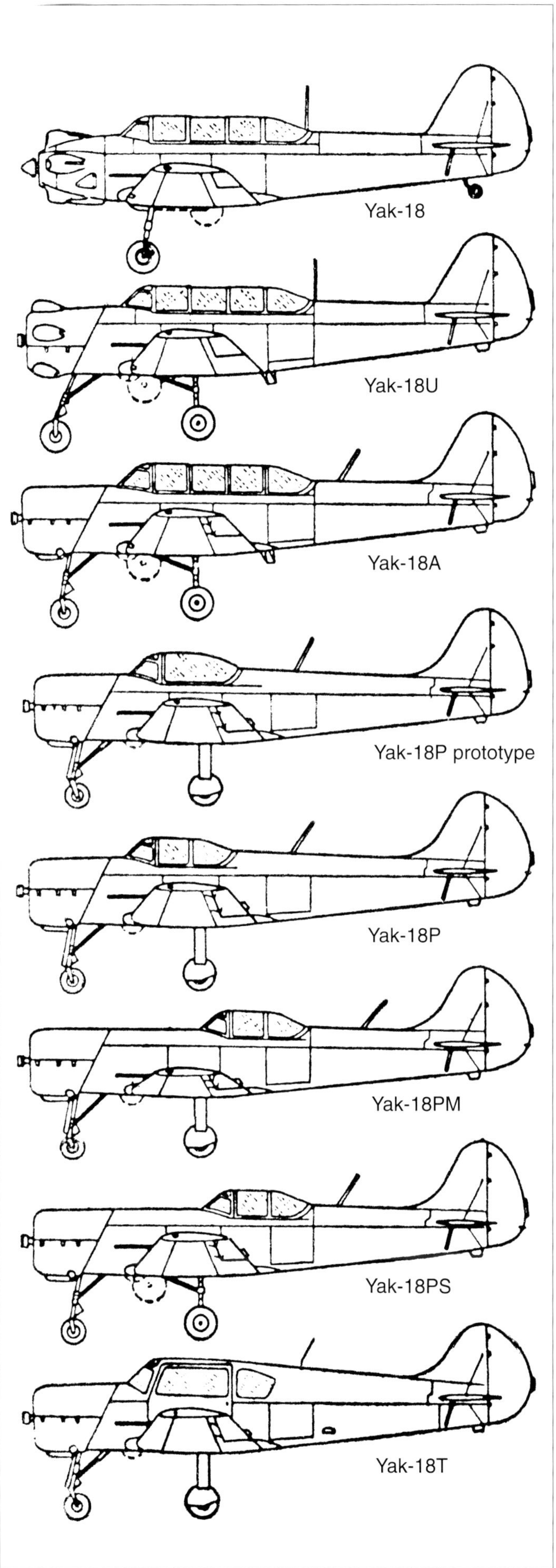

Yakovlev OKB

Yakovlev OKB

Yakovlev OKB

via Yefim Gordon

Top to bottom: The unregistered first prototype of the Yak-18T as originally flown with a curved windshield and a small vertical tail with no root fillet.

The second prototype (or a later example) with the definitive vertical tail and windshield and the original wheel well doors on all three landing gear struts.

CCCP-81400, the one-off Yak-18TS ambulance aircraft prototype. Note how high the Yak-18T sits above the ground.

ФЛАРФ 01265 (that is, FLARF 01265), a privately owned Yak-18T, seen at Moscow-Myachkovo.

Left: These side views show graphically the Yak-18's evolution.

the cabin roof glazing. Directional stability proved unsatisfactory, so the vertical tail was enlarged and provided with a prominent dorsal fin; a revolving red anti-collision light was mounted on the top of the fin. The carburettor air intake under the nose was enlarged. A dipole aerial was mounted on the sides of the fin (production Yak-18Ts lack this aerial). This example, as well as the preceding prototypes, had a wheel well door on the nose gear strut which was later dispensed with on production examples.

Yak-18T production four-seat trainer (*izdeliye* 20)

Production was assigned to the Smolensk aircraft plant No.475, where the aircraft was known as *izdeliye* 20. From 1973 onwards the Yak-18T was manufactured in quantity, initially with the 300-hp Ivchenko AI-14RF engine. On aircraft manufactured from 1975 this engine was superseded by its derivative, the 360-hp Vedeneyev M-14P driving a V-530TA-D35 propeller. Thirty batches (consisting initially of ten but mostly of 20 aircraft each) had been manufactured when production was terminated in 1982. The original production run totalled 537 examples (production was later resumed in post-Soviet Russia – see below).

The machine was produced for Aeroflot's pilot schools in Ul'yanovsk, Sasovo, Buguruslan, Krasnyy Kut, Aktyubinsk and Kirovograd. It was used in two versions (for primary training and for advanced training). In the former case (simple and complex visual flight rules (VFR) training) the aircraft had a crew of two, trainee and instructor (equipped with parachutes). In the second case (circuit flights, route flights and instrument flight rules (IFR) training) the cabin provided accommodation for an instructor and three trainees. In trainer configuration the Yak-18T had pilot seats with dished pans to accommodate the crew's parachutes, while in other cases the cabin was fitted with comfortable car-type seats.

In 1973 four Yak-18Ts underwent operational trials which lasted five months. In 1974 the first 15 Yak-18Ts were delivered to the Sasovo Civil Aviation Flying School; they included CCCP-38335 (c/n 3200103 – that is, year of manufacture 1973, *izdeliye* 20, 01st aircraft in Batch 03) which, despite the c/n commencing with a 3, was manufactured on 14th March 1974. The Yak-18T became a standard trainer in Soviet civil aviation flying schools; it was also exported to Bulgaria and Cuba.

In DOSAAF air clubs, including the Central Air Club named after Valeriy P. Chkalov at Moscow-Tushino, the Yak-18T was used for advanced aerobatics training. Machines pertaining to different batches could have small external differences. For example, the design of the boarding step aft of the port wing changed from an L shape to a U shape, and the lower segments of the main gear doors covering the mainwheels were usually omitted.

On 9th November 1987 the Ministry of Civil Aviation (MGA) issued order No.249 phasing the Yak-18T out of Aeroflot service and requiring the entire fleet of the type to be written off. Nearly half the fleet had been scrapped when, fortunately, someone at the ministry managed to stop this mayhem. Many Yak-18Ts escaped destruction because they were transferred to the DOSAAF. Others were placed in storage and eventually purchased by independent air clubs and private owners in the 1990s; thus, the abovementioned CCCP-38335 became RA-01211.

Yak-18TS ambulance aircraft prototype

An ambulance version of the Yak-18T was developed by the Yakovlev OKB in response to technical requirements issued by the Ministry of Health and the Ministry of Aircraft Industry in 1974. It was intended to transport either one stretcher case and a medical attendant, or three persons of medical personnel for urgent medical help or consultative services. The aircraft was designated Yak-18TS, the S standing for *sani**tar**nyy* (medical).

A prototype converted from a production Yak-18T (CCCP-81400, c/n 4200305) was submitted for evaluation by MGA in 1976. In this version the right-hand set of controls was deleted, the co-pilot's seat and the rear bench seat for passengers were removed to provide space for a stretcher and a seat for medical attendant. The stretcher could be loaded through the usual port side baggage door. Different sets of medical equipment could be carried on a specially installed baggage shelf. The cabin was provided with heating. As was normal for Soviet ambulance aircraft, the Yak-18TS wore an overall white colour scheme with large Red Cross markings.

After successfully passing check-up tests the Yak-18TS prototype underwent operational trials which were conducted in 1977 by ambulance aviation services in the Voronezh and Vladimir Regions. The aircraft proved its worth, flying various medical missions, and was recommended for series production and service. For some reason this recommendation was not implemented, although the need for the ambulance version was estimated at 300-400 machines.

In August 1983 Yak-18TS CCCP-81400 was demonstrated at Moscow-Tushino. The aircraft was last seen at the OKB's flight test facility in Zhukovskiy in September 1993.

Yak-18T modified as a ground engine test rig

A written-off Yak-18T registered CCCP-44310 (c/n 72001313; there was actually one zero too many) was converted into a ground test rig for testing the powerplant of the Molniya-1 six-seat touring aircraft. This machine designed by NPO Molniya ('Lightning' Scientific & Production Association) was fitted with a Vedeneyev M-14PM radial driving a three-blade pusher propeller manufactured by the German company Mühlbauer.

The test rig was intended to simulate the operating conditions of the pusher installation, in particular with regard to the cooling. To this end the aircraft, stripped of its outer wing panels, was fitted with a rather bulky air duct and an electrically-powered impeller; the latter sent a stream of cooling air through the engine in the reverse direction during engine runs on the ground. A test equipment suite, accommodated in a ZiL-131N 6 x 6 lorry with a van body, was hooked up to the aircraft. The installation could be seen at Moscow-Tushino in May 1992.

Yak-18T (new production)

Like the Cessna 172 Skyhawk and Cessna 182 Skylane, the Yak-18T received a new lease of life. In 1993 one of the newly created small design bureaux, the Tekhnoavia Joint-Stock Co., started its activities by reinstating Yak-18T production at the Smolensk aircraft factory.

This step was logical enough. Firstly, a niche existed on the Russian market for small touring and business aircraft, which could not be filled by imported machines because of their high price and the limited purchasing power of the potential customers. The Yak-18T would be considerably cheaper and thus be price-competitive. Secondly, the Tekhnoavia company was headed by Vyacheslav P. Kondrat'yev, who had worked for the Yakovlev OKB and later the Sukhoi OKB; at these bureaux he was responsible for a number of sports aircraft and trainer designs. In consequence, the new firm was in a position to fall back on its leader's previous experience in this field.

Continuing the original Soviet production sequence, new production began with Batch 31. In mid-1978 the Smolensk plant had switched to a new c/n system designed to confuse hypothetical spies; Yak-18T CCCP-44261 was c/n 8201416, while the next aircraft (CCCP-44262) was c/n 22202021517. 222 is a code for plant No.475, 020 means *izdeliye* 20, while the rest has no meaning whatever so as not to reveal the batch number/number of the aircraft in the batch and hence how many had been built; hence there is also a *fuselage number*, 1516 (15th aircraft in Batch 16). The last Soviet-built Yak-18T (CCCP-81595) was c/n 22202055612 and presumably f/n 1330. Now the factory reverted to the rational c/n system; the first 'new' Russian-built Yak-18T, RA-44440, was c/n 01-31.

New-build Yak-18Ts incorporated some minor general upgrading. Since they were intended mostly for business travel, the level of passenger comfort had to be increased. Several new versions of cabin interior were developed, featuring comfortable car-type seats and other amenities. This task was handled not only by the Smolensk plant, but also by some firms specialising in aircraft interior design. Outwardly Russian-built Yak-18Ts could be discerned from the Soviet-built examples by having boarding steps on both sides instead of just one to port; some examples were fitted with large spinners on the propeller hubs, which the original Soviet-built Yak-18Ts never had.

By mid-1999 more than 60 Yak-18Ts had been sold, notably to the Ul'yanovsk Civil Aviation Flying School; more than 10 were delivered to the Russian Federal Border Guard Service. About 30 aircraft went to foreign customers. Recipient countries included the USA, Great Britain, Switzerland, the Philippines, the UAE, Luxembourg and Turkey. The highest known c/n is 05-40 (LY-ATL).

Hopes for a commercial success prompted Tekhnoavia to undertake a further upgrading of the Yak-18T; some of these upgrades are described below.

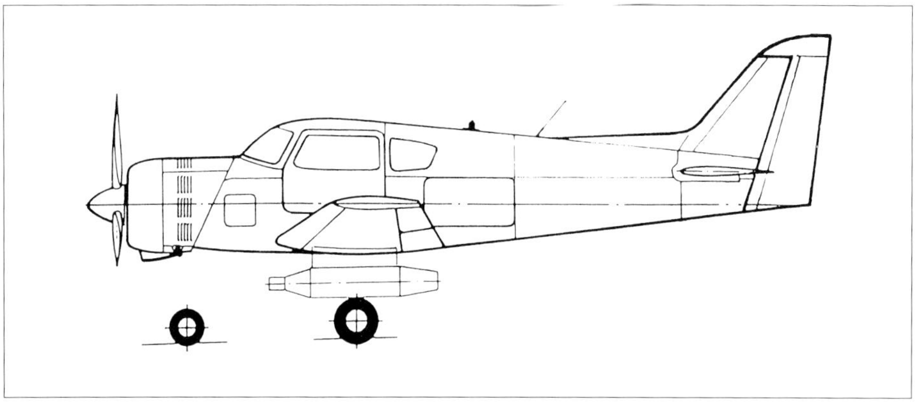

Above: A side view of the proposed Yak-18TM featuring new wings and a new vertical tail. Here it is shown in the armed COIN configuration with rocket pods under the wings.

Yak-18TM trainer and multi-purpose aircraft (project)

The Yak-18TM project was under development by Tekhnoavia in the 1990s. This derivative of the Yak-18T featured all-metal wings and tail surfaces, a more comfortable cabin, an improved powerplant, and updated avionics. Outwardly the Yak-18TM differed from the Yak-18T in having wings and horizontal tail of greater span and area; the planform of wingtips and the tips of the horizontal tail was altered. The vertical tail was swept, as a concession to the modern style (a departure from the traditional Yakovlev vertical tail shape on fighters and light aircraft).

In addition to the usual range of tasks performed by multi-purpose light aircraft, the Yak-18TM was intended to fulfil some military roles as well, including the COIN role. To this end provision was made for wing hardpoints enabling the carriage of armament (bombs and rocket pods). The M-14P engine was to be replaced by the 360-hp M-14N version.

Tekhnoavia SM-94 (Yak-18T upgrade)

In the mid-1990s Tekhnoavia undertook an upgrade of the Yak-18T which, in its new version, received the designation SM-94 (SM stands for *sa**mo**lyot **mno**gotselevoy* – multipurpose aircraft). In its initial version, dubbed SM-94-I, this aircraft was presented for the first time at the MAKS-95 airshow in August 1995. The prototype, RA-44486, differed from the basic Yak-18T in having integral fuel tanks instead of welded tanks and featured aerodynamic refinements, notably a large spinner on the propeller hub and a panoramic frameless windshield. Longer and recontoured rear side cabin windows were provided on some examples. The cabin interior was revised, and the aircraft was fitted with Western avionics. Replacing some bulky equipment items of Russian manufacture by more compact Western counterparts made it possible to provide accommodation for two more passengers. A small number of these aircraft was built (or produced by refurbishing Yak-18Ts built earlier) at the Smolensk aircraft factory and sold to private owners.

A further improvement of the SM-94-I emerged as the SM-94-II with minor differences in equipment. An example registered 02109 ФЛА РФ was converted to this standard in May 1997 at the Smolensk factory from a Yak-18T originally manufactured on 19th July 1975 as CCCP-81431; the modified aircraft was released by the factory on 25th May 1997. In addition to the propeller spinner and more streamlined cabin windshield, the aircraft featured a revised six-seat interior; the avionics suite included a KP-87 ADF, a KY-96A VHF radio, a KA-134 audio panel (all of them Bendix King products), a Garmin GPS-150 satellite navigation set and an AGB-98R artificial horizon with a provision for connecting an autopilot to it. The fuel system was modified, with provisions for installing an extra holding 85-litre (18.7 Imp gal) fuel tank. The aircraft had a cruising speed of 240 km/h (149 mph)

There were reports about Tekhnoavia's plans to produce a yet more radical modification of the SM-94 with a metal skinning on the wings and tail surfaces, a more efficient three-blade propeller and a further upgrade of avionics. An example of the SM-94 with a three-blade propeller (and presumably featuring also the other improvements mentioned above), registered RA-09101, could be seen at Moscow-Tushino in July 2000.

Drawings were published of a prospective armed version of the SM-94 which appears to be identical with the armed version of the Yak-18TM described above.

Sergey Komissarov

RA-09101 seen here at the MAKS-99 airshow illustrates the SM-94-I, a refined derivative of the Yak-18T. Note the characteristic curved windshield and the propeller spinner.

Yakovlev OKB

The first prototype of the Yak-20 basic trainer had an angular vertical tail. Note the ventral Venturi tube and the aerial aft of the wing trailing edge.

Yakovlev OKB

Above: The first prototype in flight. The pilots of the Yak-20 enjoyed an excellent all-round view through the huge blown canopy.

Yakovlev OKB

Yakovlev OKB

Top and above: The second prototype which differed outwardly in having a smaller and rounded vertical tail. Note the cooling air flow adjustment shutters at the front of the engine.

Yak-18T testbeds

The static park of the MAKS-97 airshow (22nd-27th August 1997) included a Yak-18T registered 01001ФЛА РФ (ex-ФЛА СССР-01001) which featured a non-standard propeller. Its blades had unusually shaped tips instead of the normal squared-off tips, resembling an ancient Nepalese sword called *ram dao*. This example (presumably a testbed) had previously flown with a standard propeller. No details are available. The new propeller appears to have remained experimental.

At the MAKS-2003 airshow (19th-24th August 2003) the same aircraft, now with the registration applied as ФЛА-РФ-01001, was displayed statically and in flight as an avionics testbed for testing new navigation system components. Outwardly the aircraft, which was operated by the Moscow-based Fine Instruments Research Institute (NIITP – *Na**ooch**no-is**sled**ovatel'skiy insti**toot** **toch**nykh pri**bor**ov*), differed from the standard configuration only in having a small flat-topped satellite antenna supplanting the anti-collision light on the fin.

Yak-20 trainer and sports aircraft prototype

In 1949 the Yakovlev OKB developed a small two-seater intended for use in air clubs as a pilot trainer and aerobatic sports aircraft. The concept behind it envisaged a cheaper and simpler aircraft than the Yak-18 that had established itself as the basic trainer in military and civil flying schools. Designated Yak-20, the aircraft was a low-wing monoplane with a fixed tailwheel undercarriage; structurally it had much in common with the Yak-18, but featured a simplified design. The fuselage was based on a welded steel tube truss with fabric skinning supported by a light alloy secondary structure. The forward and rear fuselage had a duralumin skin. The single-spar wings comprised two panels attached directly to the fuselage truss. They were covered with duralumin from the leading edge to the spar; with fabric skinning aft of it. The slotted flaps and ailerons were also fabric-covered. The tail surfaces had duralumin frames covered with fabric and were reinforced with bracing wires. The fin and rudder had angular contours. The pyramid-type main undercarriage units were made of steel tubes. Shock absorption was provided by rubber bungees. The steerable tailwheel was controlled by the rudder pedals.

The Yak-20 was powered by the newly developed 80-hp Ivchenko AI-10 five-cylinder radial engine driving a V-515 variable-pitch counterweight-type propeller. Its pitch was controlled automatically, assuming the correct setting under the influence of aerodynamic and centrifugal forces. The 80-hp output was considered sufficient for an aircraft with an AUW of 700 kg (1,540 lb). The long-chord engine cowling was split into upper and lower clamshell halves and had a neat circular cross-section, as distinct from the helmeted cowling on the Yak-18. Two fuel tanks with a total capacity of 70 litres (15,4 Imp gal) were placed in the wing roots ahead of the spar. There was no oil cooler, the oil tank being cooled by the slipstream.

The Yak-20 differed from all other Soviet trainers and sports aircraft of that period in having side-by-side seating for the trainee

Four views of the first prototype Yak-20. The wings' and tail unit's structural members are shown on the plan view for clarity.

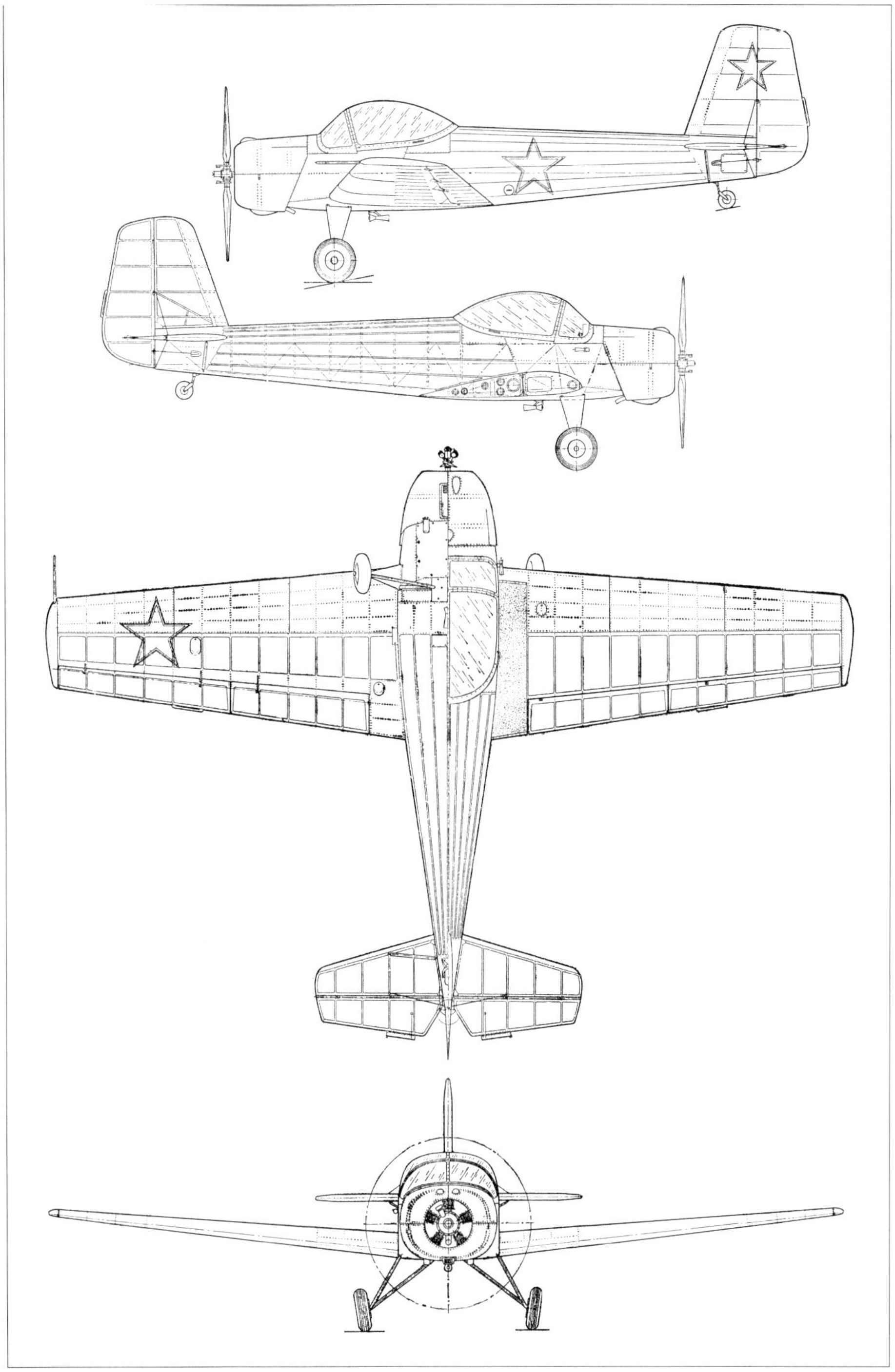

and instructor instead of the usual tandem arrangement. This made the training more effective and helped reduce the airframe weight. A large blown canopy provided an excellent view for the crew. The aircraft had dual controls; all control surfaces were actuated by cable linkages. Both halves of the elevator were provided with trim tabs.

In the course of manufacturer's flight tests conducted by Sergey N. Anokhin and Gheorgiy M. Shiyanov, the Yak-20 displayed exceptionally simple handling and good stability in all flight modes. The aircraft was fully aerobatic and entered a spin only at a very low speed, which obviated the need for slats. Spin recovery posed no problems. Manufacturer's tests were conducted with the participation of several leading aerobatic sports pilots who were unanimous in their praise of the aircraft and urged its adoption as a standard primary trainer for air clubs. Indeed, the Yak-20 appeared to be eminently suitable for this role.

However, the specialists who evaluated the Yak-20 during the State acceptance trials took a different view of the requirements which this aircraft was expected to meet. Their suggestions and critical comments boiled down to raising the performance and capabilities of this trainer to the same level as those of the Yak-18, a more complicated and expensive machine (an approach some specialists later considered to have been erroneous).

The second prototype (sometimes referred to as the Yak-20-2) differed externally in having a smaller vertical tail with more rounded contours. It was built with due regard to the criticism mentioned above; among other things, it was fitted with the necessary equipment for IFR flying. However, the overall result was disappointing. This machine having increased-area wings was considerably heavier than the first prototype; this resulted in a marked deterioration of performance and made the Yak-20 clearly inferior to the Yak-18, which was already in service, so there was no point in putting the new aircraft into production. Thus, the aircraft may be regarded as a victim of ill-advised 'improvement'.

Yak-22 light aircraft (project)

This was a single-seat lightplane resembling the Yak-20 in its general layout. Initial design work on it was conducted in March-April 1950; it envisaged the use of an M-4S engine with a propeller measuring 1.8 m (5 ft 10⅞ in) in diameter. The engine had a take-off rating of 60 hp and a nominal rating of 50 hp.

The aircraft was to have a wing span of 7.4 m (24 ft 3½ in), a length of 5.9 m (19 ft 4 9/32 in), a wing area of 9.0 m^2 (96.88 sq ft) and an all-up weight of 400 kg (882 lb). Its design performance included a maximum speed of 174 km/h (107 mph), a landing speed of 65 km/h (40 mph), a service ceiling of 3,960 m (12,990 ft) and a range of 405 km (252 miles). Climb time to 1,000 m (3,280 ft) was 5.6 minutes; the aircraft would have a take-off run of 115 m (380t) and a landing run of 80 m (260ft).

Yak-30 (Yak-104) advanced trainer

The Yak-30 was the first dedicated jet trainer produced by the Yakovlev OKB, as distinct

Specifications of the first prototype Yak-20

Length	7.06 m (23 ft 2 in)
Wing span	9.56 m (31ft 4⅜ in)
Wing area, m^2 (sq ft)	15.0 (161.5)
Empty weight, kg (lb)	470 (1,040)
All-up weight, kg (lb)	700 (1,540)
Maximum speed, km/h (mph)	160 (99)
Cruising speed, km/h (mph)	142 (88)
Landing speed, km/h (mph)	60 (37)
Service ceiling, m (ft)	3,000 (9,850)
Range, km (miles)	400 (249)
Take-off run, m (ft)	70 (230)
Landing run, m (ft)	80 (260)

Yakovlev OKB

Above: '30 Yellow', the aptly coded first prototype of the Yak-30 advanced trainer, at Zhukovskiy. Note the wire mesh screens closing the engine air intakes buried in the wing roots.

Yakovlev OKB

The second prototype, '50 Yellow', differed in having an aerial to the right of the cockpit windshield. Like the first prototype, it had a huge one-piece transparency in the sliding portion of the canopy.

from trainer versions of fighter aircraft. Yakovlev had long advocated the need for such a machine in the Air Force inventory and made proposals to this effect in 1952 and 1955, which were not accepted but set the ball rolling. In the second part of the 1950s the Soviet Air Force formulated its requirements for a jet trainer stipulating that it should be a totally new machine powered by a specially developed engine. In 1956-57 the Yakovlev OKB set about projecting a jet trainer which received the in-house designation Yak-104. Initial studies of this aircraft (later to be renamed Yak-30) differed considerably from the final configuration.

Yak-104 R5-300 and Yak-104 R5-45 projects

Early design studies of the Yak-104 were based on the use of the R5-300 engine (a derivative of Mikulin's AM-5) as stipulated by government directive dated 28th March 1956; design performance included a maximum speed of 700 km/h (435 mph) at 5,000 m (16,400 ft) at an AUW of 3,000 kg (6,615 lb). On 31st August 1956 this requirement was cancelled in favour of a version of the same aircraft with the R5-45 (presumably another derivative of the AM-5) developed at OKB-45 by Nikolay G. Metskhvarishvili and rated at 1,000 kgp (2,205 lbst). This time the aircraft was to have an AUW of 2,300 kg (5,070 lb), and the design performance envisaged a speed of 600 km/h (373 mph) at 3,000 m (9,840 ft).

The Yak-104, designed around the R5-45, resembled in its general layout the future Czechoslovak Aero L-29 Delfin trainer, sharing with it the powerplant arrangement with the engine nozzle at the aft extremity of the fuselage. It differed from both the L-29 and the future Yak-30 in having lateral air intakes placed above the wing roots rather than in the wing leading edge. A mock-up of this machine was completed in August 1957, but it was not followed by the prototype construction because the development of the R5-45 engine had been discontinued.

In the quest for a new powerplant Aleksandr S. Yakovlev turned to Sergey K. Tumanskiy, the head of the OKB-300 design bureau previously led by Aleksandr A. Mikulin, asking him to develop a compact lightweight engine with a high level of reliability. Tumanskiy grasped the importance of the matter and instructed his staff to start the design work on the engine without waiting for the official sanctioning of this work by a government directive which would follow in due course. The engine, later known as the RU19-300 (*reaktivnyy ooskoritel'* – jet booster), passed manufacturer's tests about two years later, in mid-1959; in February 1961 it successfully passed its State acceptance trials. The new engine had a maximum rating of 900 kgp (1,984 lbst); it weighed a mere 235 kg (518 lb) and measured a modest 1.634 m (5 ft 4⅜ in) in length. The required powerplant was thus available.

Yak-30 (Yak-104) prototypes

In anticipation of this, the Council of Ministers issued a directive on 31st July 1958, tasking the Yakovlev OKB with designing the Yak-104 advanced trainer around the RU19-300 engine. As it turned out, Yakovlev was not alone about this: concurrently the work on jet-powered trainers was initiated in Poland and Czechoslovakia, resulting eventually in the PZL TS-11 Iskra (Spark) and the Aero L-29 Delfin (Dolphin) respectively.

The year of 1960 saw the completion of the first two prototypes of the Yak-104; the first of them, coded '30 Yellow', was rolled out on 15th May and the second ('50 Yellow') followed on 21st July. In the same year the Yak-104 was renamed Yak-30 (third use of the designation, previously applied to a jet fighter and a supersonic reconnaissance aircraft project; see Chapters 3 and 5).

The new machine differed from the original project configuration in having flush wing root air intakes and a jet nozzle placed under the rear fuselage at about two-thirds of the fuselage length, thus resembling the early Soviet fighters with their pod-and-boom configuration. Remarkably, the nozzle was semi-recessed into the rear fuselage underside. The aircraft had low-set straight wings and a cruciform tail unit with the tailplane set at about one-third of the fin's height; the tailplane was unswept, while the vertical tail had fairly strong sweepback.

The near-oval section semi-monocoque fuselage housed a pressurised cockpit fitted with ejection seats. The instructor and trainee were seated in tandem under a common frameless canopy with a fixed windshield (there was also a project version with a canopy without a separate windshield, as on the later General Dynamics F-16B Fighting Falcon). For the first time in the Soviet Union ejection could be performed through the canopy. The tricycle undercarriage featured a forward-retracting nose unit and main units retracting inward into the wing roots; all units had single wheels and levered suspension. Undercarriage operation was by means of a pneumatic system. The aircraft had dual controls, the elevators and ailerons being actuated by push-pull rods, while the rudder was controlled by cable linkages; the elevator trim tabs were remote-controlled electrically.

The RU19-300 turbojet installed amidships was easily accessible and could be removed without difficulty; it was protected against FOD by external wire mesh screens closing the wing root inlets. The fuel system with a total capacity of 870 litres (191.4 Imp gal) comprised a fuselage bag-type tank holding 845 litres (189.5 Imp gal) and a 25-litre (5.5 Imp gal) welded service tank also used for negative-G flight modes. The aircraft was fitted with the most up-to-date flight and navigation equipment of the day, which included a Collins VHF communication radio, an ARK-9 ADF, an MRP-56 marker beacon receiver, a GIK-1 gyro flux-gate compass, a KI-13 magnetic compass, as well as an SPU-7 intercom. The instrument panels featured a design similar to that of fighter aircraft; the Yak-30's avionics suite enabled it to operate day and night in visual and instrument meteorological conditions.

In the course of the manufacturer's tests conducted in the period between 20th May 1960 and March 1961 the two prototypes performed a total of 82 flights logging 43 hours and 36 minutes. One might say, the Yak-30 passed these tests with flying colours: test pilots assessed the aircraft's handling and performance as excellent and recommended it for use as a basic trainer and proficiency trainer. Special note was made of the aircraft's ability to operate from dirt or grass strips.

State acceptance trials started on 30th August 1960, prior to the completion of the manufacturer's tests. A further two machines coded '80 Yellow' and '90 Yellow' joined the trials at this stage. They differed in some respects from the first two prototypes. The wing outer panels were reinforced; the stability margin for G-loads was increased; changes were introduced into the control system, cockpit (the canopy received a frame in the middle) and equipment; vortex generators were installed on the wings to obtain the warning buffet before stall; the radio's whip aerial was transferred from the nose to a place amidships. During the tests the Yak-30 attained a speed of 655 km/h (406 mph) at 5,000 m (16,400 ft). On the machine coded '90 Yellow' tests were made of the ejection seat enabling the pilot to eject through the canopy.

The high performance of the Yak-30 was corroborated by world records established by this aircraft. In September 1961 the second prototype set Class C-1-d world speed and altitude records for light jet aircraft, attaining a speed of 767.308 km/h (476.88 mph) on a 25- to 35-km (15.53 to 18.65-mile) circuit and climbing to 16,128 m (59,916 ft).

The State acceptance trials were completed on 14th August 1961; military test pilots recommended the machine for service introduction. This recommendation was endorsed by the VVS Commander-in-Chief Air Marshal K. A. Vershinin. In February 1962 preparations began for series manufacture of the Yak-30 at plant No.116 in Arsen'yev. However, the machine was not destined to enter production. In November 1960 an idea cropped up of conducting a fly-off between the Soviet Yak-30 and the Czechoslovak L-29. The fly-off took place at Monino near Moscow, starting in August 1961, with the participation of the Yak-30 ('90 Yellow'), the L-29 ('003 Red') and the TS-11 ('03 Red'). Comparative testing revealed the Yak-30's noticeable ascendancy in performance over the other two contenders, which was not surprising, bearing in mind the Yak-30's lower weight and considerably higher thrust/weight ratio.

However, the outcome of this informal contest was determined not so much by technical merits of the contenders as by political considerations. As a result of talks between the leaders of the Soviet Union and Czechoslovakia, a decision was taken to assign to the Czechoslovak aircraft industry the leading role in the development and manufacture of trainers for the Warsaw Pact countries. Accordingly, the L-29 was pronounced the winner of the fly-off and subsequently was produced in large numbers for the air forces of the Soviet Union and its allies. Not surprisingly, the Yakovlev OKB staff's pride was hurt; in their view, this decision failed to do justice to the merits of their machine which deserved a better fate.

Despite this turn of events, development work on the Yak-30 went on for some time, resulting in some modifications which are described below. Again, the Yak-30 received a NATO reporting name in the 'miscellaneous' category, *Magnum*.

Yak-30 '80 Yellow' modified (with rocket and bomb armament)

The third prototype of the Yak-30 ('80 Yellow') underwent some modifications and was tested at LII (Flight Research Institute) and the OKB in November 1961. It was fitted with an ASP-3N computing gunsight, an FKP-2-2 gun camera and two underwing rocket pods (alternatively, two 50- or 100-kg (110- or 220-lb) bombs could be carried). Accordingly, a weapons control button was added to the control stick. Changes to the instrument panel in the rear cockpit improved the forward view for the instructor. Stick forces were reduced by modifying the spring loading device. An airbrake was added under the wing centre section. With these modifications, the Yak-30 was deemed to have achieved ascendancy over its competitors on all counts, as regards performance, serviceability and fuel economy.

Yakovlev OKB

Above: The third prototype Yak-30, '80 Yellow', introduced a transverse member on the sliding portion of the canopy. Note the jet nozzle centrebody showing how the nozzle is semi-recessed into the fuselage.

Yakovlev OKB

'90 Yellow', the final prototype, with the canopy slid all the way back.

Yakovlev OKB

Above: The first prototype Yak-32, again with a very appropriate tactical code, displays the aircraft's clean lines.

Yakovlev OKB

Above: '60 Yellow', the second prototype Yak-32, on the hardstand at Zhukovskiy in its original natural metal finish.

Yefim Gordon archive

Above: The same aircraft in a new colour scheme with orange undersides applied for display purposes.

Yefim Gordon archive

The identically painted third prototype Yak-32, '70 Yellow', in company with the fourth prototype Yak-30 at Moscow-Tushino.

In 1962 an attempt was undertaken to 'save' the Yak-30; it was suggested that the Yak-30 be placed in production for the VVS, while the L-29 could be built for the needs of Czechoslovakia and other countries. Yet, despite the support of such persons as Dmitriy F. Ustinov, Deputy Chairman of the Council of Ministers, and Pyotr V. Dement'yev, head of the State Committee for Aviation Hardware (GKAT – *Gosoo**dar**stvennyy komi**tet** po aviatsi**on**noy **tekh**nike*, as MAP was known in 1957-65), this proposal was rejected.

Yak-30M (Yak-30KP) drone director aircraft

On 24th November 1961 GKAT took a decision to convert two of the Yak-30 prototypes into the Yak-30M (Yak-30KP) drone director aircraft (KP stands for *ko**mand**nyy poonkt* – command post). They were intended for use during the second stage of the Yak-25RV-II target drone's State acceptance trials. One machine underwent conversion in the second quarter of 1962. The front cockpit of the Yak-30M was turned into an operator's workstation. It was fitted with a VHF command radio with a PRD-3S transmitter, served by an aerial installed on the stabiliser leading edge. The Collins radio was replaced by an indigenous RSIU-5, and the associated whip aerial was moved to a place ahead of the cockpit.

The Yak-30M was successfully used both during the State acceptance tests of the Yak-25RV-II and later in routine operations with the target drones of this and other types.

Of the four Yak-30 prototypes, one was lost in an accident, two remain at the OKB and on has become an exhibit of the Central Russian Air Force Museum in Monino.

Yak-30V VTOL aircraft (project)

In late 1960 the OKB prepared an advanced development project of a VTOL aircraft based on the Yak-30. Designated Yak-30V (V presumably stands for *verti**kahl'**nyy vzlyot*, vertical take-off), the aircraft was to be fitted with additional lift engines based on the RU19-300and installed vertically in the fuselage instead of the second cockpit. Some sources refer to this project as a VTOL version of the Yak-104. The work on this machine was terminated when the OKB commenced the work on the Yak-36 VTOL aircraft.

Yak-32 (Yak-104PS) sports aircraft

The Yak-32 (second use of designation) was a single-seat aerobatic derivative of the Yak-30 trainer. Its development was started under the designation Yak-104PS (the suffix presumably denoting *pilo**tazh**nyy samo**lyot*** –

aerobatic aircraft) in response to a Council of Ministers directive issued on 31st July 1958. The first prototype of the Yak-32 (appropriately coded '32 Yellow') was completed on 13th October 1960. The machine differed from the Yak-30 in having no rear cockpit and an accordingly shorter canopy; the service tank's capacity was increased to 45 litres (9.9 Imp gal) and the engine was adapted for inverted flight.

A further two examples coded '60 Yellow' and '70 Yellow' were built in July 1961. The machines differed in some respects. '32 Yellow' was configured for training, while '60 Yellow' was intended as a sporting aerobatic aircraft. The aerobatic configuration featured a 555-kg (1,224-lb) reduction in the AUW achieved by reducing the fuel load and deleting some equipment items. The aerobatic version had greater aileron area and was not fitted with flaps.

Joint manufacturer's and State trials showed the aircraft to be simple and agreeable in handling and capable of performing all normal and inverted aerobatics. The aircraft attained a maximum speed of 663 km/h (412 mph) at 3,000 m (9,840 ft). The Yak-32 was recommended for use by DOSAAF air clubs as a standard sports and aerobatic aircraft. For some reason the single-seater had a separate NATO codename, *Mantis*.

Yak-32Sh attack aircraft (project)

Experience gained in the Vietnam War, where light aircraft were used in the attack role, prompted the Yakovlev OKB to develop a project of a light attack version of the Yak-32. It was designated Yak-32Sh (*shtoormovik* – attack aircraft). The aircraft was intended to carry various external stores under the wings.

In 1961 Yakovlev demonstrated the capabilities of this version to the then Minister of Defence Marshal Rodion Ya. Malinovskiy. The first prototype was fitted with two underwing pylons carrying AAMs and was put on show at Moscow's Central airfield (Khodynka). Placed on the ground near the aircraft was an impressive array of air-to-ground weapons it could carry (guided missiles, unguided rockets, bombs, gun pods and other stores).

Yak-32P sports aircraft (restored Yak-32)

In August 1971 the Ministry of Aircraft Industry issued an order requiring the OKB to restore one of the Yak-32 prototypes in a sports and aerobatic configuration and fit it with the prototype RU19P-300 engine. Rated at the same static thrust of 900 kgp, this engine had a revised oil system permitting a greater duration of inverted flight.

Designated Yak-32P (*pilotazhnyy* – aerobatic), the aircraft was produced by refurbishing one of the Yak-32 prototypes. Changes as compared to the original configuration included the deletion of cockpit pressurisation effected previously by engine bleed air (instead, the cockpit was ventilated by ram air) and installation of new equipment items, such as the Landysh-5 communications radio, the ARK-15 ADF and the GMK-1A gyro flux-gate compass (the latter two items, as well as a marker beacon receiver, were removable and were intended for use only for positioning flights and flights in adverse weather).

In 1972 the Yak-32P passed its manufacturer's check-up tests and put up a display of aerobatics during that year's Aviation Day fest at Moscow-Tushino in August. Again, the aircraft's handling and performance drew praise from pilots.

Yak-52 jet advanced trainer (project, first use of designation)

In 1971 the Yakovlev Design Bureau embarked on a project of a jet-powered trainer which was allocated the designation Yak-52. It was a derivative of the Yak-30 (Yak-104) powered by a 1,500-kgp (3,310-lbst) Ivchenko AI-25 turbofan, which required the air intakes to be greatly enlarged. The

Yakovlev OKB

The first prototype Yak-32 following conversion as the Yak-32Sh attack aircraft with two pylons outboard of the wings. The ordnance array includes (moving from the centre) R-3M AAMs, S-24 unguided rockets, UB-16-57 FFAR pods, FAB-250M-62 low-drag bombs, machine-gun pods, S-21 rockets and napalm tanks.

Yakovlev OKB

Yakovlev OKB

Yakovlev OKB

Upper row, far left: The Yak-32Sh with UB-16-57 FFAR pods. Note the shape of the pylons.

Centre: The Yak-32Sh with S-3K unguided rockets on six-round launchers.

Above: The Yak-32Sh with ZB-360 napalm tanks.

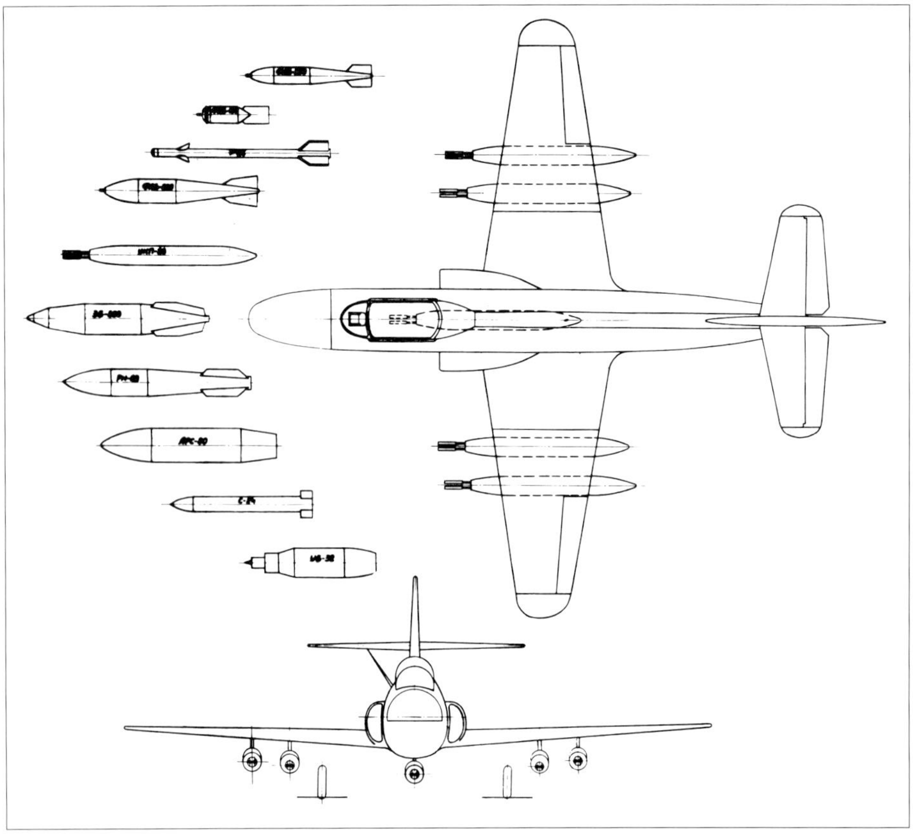

Left: The projected Yak-52Sh attack configuration with cannon pods, high- and low-drag bombs and rockets. Note the enlarged overwing air intakes. The nose gear unit is not depicted.

aircraft was designed in two versions: a two-seat advanced trainer and a single-seat sports (aerobatic) aircraft. It was intended to meet the same specification as the similarly powered Aero L-39 Albatros which was destined to become the successor of the L-29 in the Soviet Air Force. Development of the Yak-52 was discontinued at an early stage, and the designation was re-used for a piston-engined primary trainer.

Yak-52 attack version

A light attack derivative of the single-seat version featuring five external stores hardpoints was also envisaged. The equipment included ASP-PF and PBK-2 sights. Ordnance options included up to four UPK-23-250 cannon pods, four 100- or 250-kg HE bombs, two 500-kg HE bombs or ZB-500 napalm tanks, two S-24 unguided rockets, two UB-32 or B-8M FFAR pods (each with 32-57-mm rockets or six 80-mm rockets respectively), two R-3S or R-13M IR-homing AAMs.

Yak-50 competition aerobatic aircraft prototypes

Despite its apparent similarity to the Yak-18PS, this was effectively a totally new aircraft. The Yak-50 featured an all-metal semi-monocoque fuselage, metal-skinned tail surfaces (except for the rudder and elevators, which were still fabric-covered) and completely new all-metal wings with zero dihedral and squared-off tips. The original project envisaged an angular tail reminiscent of the North American P-51 Mustang, but the actual aircraft retained the rounded tail used hitherto. Finally, the Yak-50 was powered by a 360-hp Vedeneyev M-14P instead of a 300-hp Ivchenko AI-14RF.

Three aircraft coded '01 Yellow', '02 Yellow' and '30 Yellow' can be identified as prototypes of the Yak-50 competition aerobatic aircraft (presumably the first, second and third prototype respectively). All three prototypes had external peculiarities. '01 Yellow' had a retractable undercarriage and a two-piece blown teardrop canopy. '02 Yellow', positively identified as the second prototype, differed from the former aircraft in having a Yak-18PS-style cockpit canopy with fixed forward and aft portions and a sliding hood. It was externally representative of the future production version. Finally, '30 Yellow'

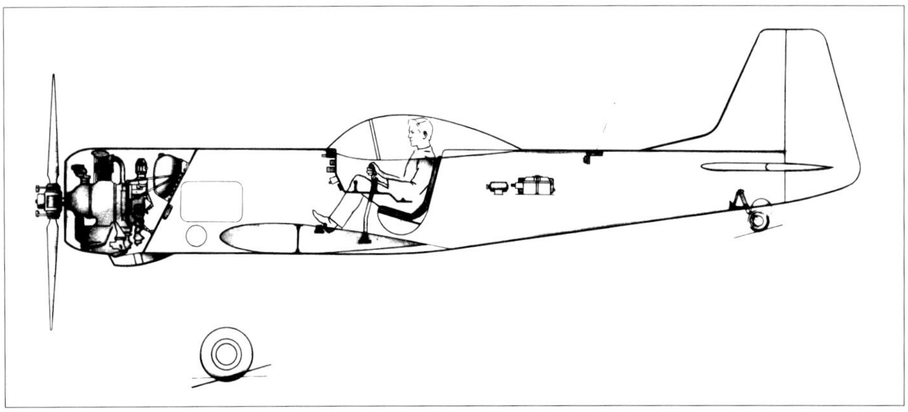

A cutaway drawing of the Yak-50 aerobatic aircraft from the project documents. Note the teardrop canopy, the retractable tailwheel and the angular tail reminiscent of the P-51 Mustang.

Specifications of the Yak-50 aerobatic aircraft

Engine type	M-14P
Engine power, hp	360
Length	7.68 m (25 ft 2¼ in)
Wing span	9.5 m (31 ft 2 in)
Wing area, m^2 (sq ft)	15.0 (161.45)
Empty weight, kg (lb)	780 (1,720)
All-up weight, kg (lb)	915 (2,020)
Maximum speed, km/h (mph)	320 (198)
Initial rate of climb, m/s (ft/min)	16 (3,150)
Landing speed, km/h (mph)	110 (68)
Service ceiling, m (ft)	5,500 (18,045)
Range, km (miles)	390 (242)
Take-off distance, m (ft)	200 (660)
Landing distance, m (ft)	250 (250)

had a fixed spatted undercarriage and a large spinner on the propeller hub; neither of these features was adopted for production. In the course of testing the first prototype had the teardrop canopy replaced by a three-piece canopy of the same type as on the machines coded '02' and '30'.

Yakovlev OKB

Above: '01 Yellow', the first prototype of the Yak-50 aerobatic aircraft, was originally tested with a teardrop canopy featuring a one-piece windshield. The wing and stabiliser planform is clearly visible.

Boris Vdovenko

Above: The same aircraft following installation of a Yak-18PS-style three-piece canopy. Note the wingtip fairings and the Central Air Club badge on the engine cowling.

Yakovlev OKB

'02 Yellow', the second prototype Yak-50, represented the production configuration.

Yak-50 production aircraft

Production Yak-50s were modelled on the pattern set by the second prototype, featuring a retractable undercarriage, a three-piece cockpit canopy and a V-530TA-D35 two-blade wooden propeller with no spinner. In its original production configuration the Yak-50 had small fairings at the squared-off wingtips and inset ailerons right up to these fairings. Late production machines had these fairings removed, and some examples had ailerons fitted with horn balances protruding beyond the tip ribs. At least one DOSAAF Yak-50 had a three-bladed Hoffmann glassfibre propeller with a small spinner (it was seen in this configuration in August 1993).

Between 1973 and 1986 the Arsen'yev plant built 312 Yak-50 aircraft. Their capabilities were displayed to advantage during the aerobatic championships in 1976-1983; on many occasions the Soviet national team flying Yak-50s showed a clear ascendancy over the other competitors.

Due to its limited production run the Yak-50 was used mainly by Soviet sports pilots; export deliveries were made only to Bulgaria and East Germany (seven and ten examples respectively). The East German Yak-50s were inherited by the reunited Germany; one of them (DDR-WQV) became a part of the collection of Flugwerft Schleissheim – Deutsches Museum in Oberschleissheim near Munich. In post-Soviet days many Yak-50s found their way to private owners in other countries – Great Britain, the USA, Sweden, Australia, South Africa and Lithuania (the latter inherited the aircraft from the Soviet Union, of course), where they were either flown or preserved as static exhibits in museums. In Russia a Yak-50 was restored to flying condition by aviation enthusiasts in the town of Starodoob, Bryansk Region, as ФЛА СССР-39008 and was still operational in 1999 with the new registration RA 02120 ФЛА РФ.

In 2000 a written-off Yak-50 airframe was used by Maksim Yegorov, the leader of a Moscow autogyro enthusiast club, for conversion into an autogyro. It was to have a rotor with three blades taken from a Kamov

Yakovlev OKB

Above: The fixed-gear third prototype of the Yak-50 ('30 Yellow') featured wheel spats and a propeller spinner. Note that the test pilot is wearing a 'bone dome' helmet.

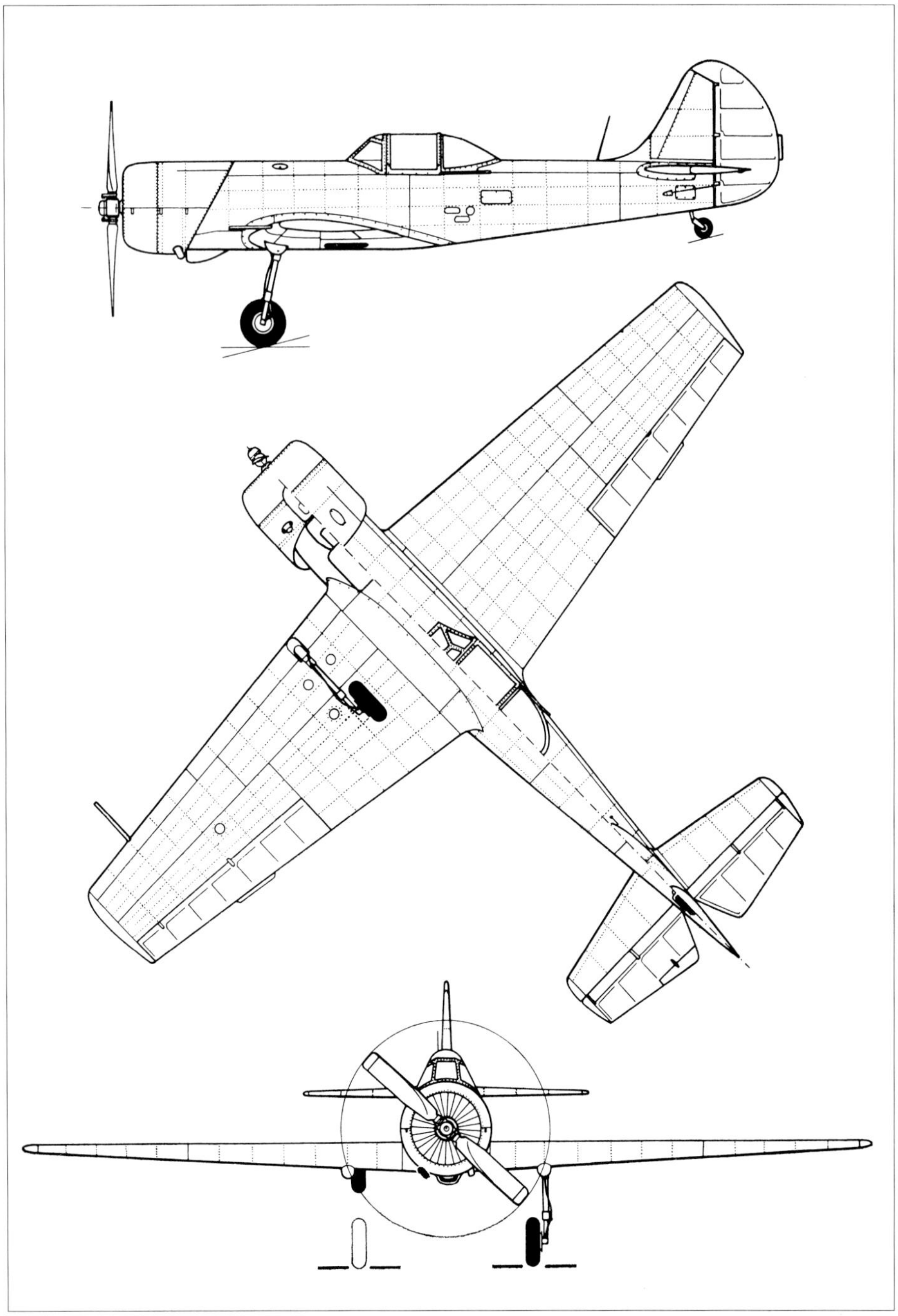

A three-view of the Yak-50.

Ka-26 helicopter, complete with the hub, and installed on a cabane. A long ski under the fuselage would serve as a guard against nose-overs. The aircraft was to be test-flown in 2001, but no confirmation is available.

Yak-52 piston-engined primary trainer (second use of designation, Yak-50U)

Initially designated Yak-50U (*oo**cheb**nyy* – training, used attributively), the Yak-52 basic trainer is a two-seat derivative of the Yak-50. Like its predecessor, it was developed by a team of young Yakovlev OKB designers. It retained the Yak-50's semi-monocoque fuselage in which a second cockpit was added ahead of the existing one, the canopy design resembling that of the Yak-18A.

Unlike the Yak-50, the two-seater has a tricycle undercarriage, also patterned on that of the Yak-18A; the nose unit retracts aft and the main units forward. When retracted, all three wheels lie against the fuselage and the wings, remaining fully exposed. This arrangement has proved its worth as a safety device in the event of a wheels-up landing.

The Yak-52's single-spar wings comprise two detachable panels attached directly to the fuselage, without a centre section. They are fitted with flaps; the root sections house fuel tanks. The use of a nosewheel undercarriage required the oil cooler to be relocated under the starboard wing. Like the Yak-50, the Yak-52 initially had small wingtip fairings which were later deleted, thus slightly reducing the wing span. The tail unit is identical to that of the Yak-50. The airframe is stressed to +7/–5 Gs.

The Yak-52 is powered by a 360-hp M-14P radial; this powerplant gives it a very favourable power/weight ratio. In addition to two 65-litre (14.3 Imp gal) fuel tanks placed in the wing roots, the Yak-52 has an additional service fuel tank in the fuselage, which ensures stable engine operation in inverted flight. The fuel and oil systems enable the engine to run flawlessly under all possible flight modes, including two minutes' flight with negative G-loads.

In addition to a complete standard set of flight and navigation instruments, the Yak-52 is fitted with a GMK-1A gyro compass, an ARK-15M ADF and a Landysh-5 (Lily of the valley) VHF radio. The SPU-9 intercom caters for communication between the instructor and the trainee. An important element of the aircraft's equipment is an audio-visual warning system indicating the onset of critical flight modes, especially when the aircraft approaches its G-load limits.

The Yak-52's flying controls are provided with spring-loading devices as standard. However, these devices were removed on some machines used for aerobatics by skilled sports pilots, making the aircraft more responsive to the controls.

In winter the Yak-52s can be operated on skis. This machine can be used not only for pilot training under all weather conditions day and night, but also for towing gliders. For this purpose the Yak-52 is fitted with a special cable lock.

Series manufacture of the Yak-52 was undertaken in Romania under licence in accordance with a COMECON agreement. The Bacâu aircraft factory (Intreprenderia Avioane Bacâu, renamed Aerostar by 1991) delivered close to 1,700 machines between 1977 and 1991, the Soviet Union being the main customer. The Romanian spelling of the aircraft's name was Iak-52, and Romanian Air Force examples were thus marked on the cowling instead of the usual Russian rendering (Як-52).

Apart from the Soviet Union and Romania, only Hungary is known to have adopted the Yak-52 as a trainer. However, after the break-up of the Soviet Union dozens of Yak-52s from the former DOSAAF air clubs were sold to private owners and museums in Western Europe and the USA. The American and Lithuanian Yak-52 fleets were particularly numerous, although most of the Lithuanian machines appear to have moved to Great Britain now.

Yak-52 late-series production trainer

Late-series Yak-52 manufactured in Romania and supplied to the Soviet Union/Russia featured various minor modifications which, for the most part, were not externally visible. There were, though, at least two new features that could be spotted easily. The first of these was the addition of a spinner to the V-530TA-D35 two-blade propeller. An example featuring a spinner carried the Russian registration RA-44501.

Another feature was the installation of additional elliptical-section fuel tanks at the wingtips. An example fitted with such tanks and registered RF-01055 (the non-standard prefix reveals its ownership by ROSTO – the Russian Defence Sports & Technical Society, the successor of DOSAAF) could be seen at Monino during the Flying Legends air show on 13th-15th August 2004. There were also machines combining the two mentioned features (for instance, RA-44545 No.2; the registration previously appeared on a Yak-18T).

Yak-52W (Iak-52W) trainer

In late 1998 the Aerostar aircraft factory of Romania started development of a new Yak-52 version fitted with Western equipment; accordingly it received the Romanian designation Iak-52W ('Westernised'). The prototype of the new version flew in April

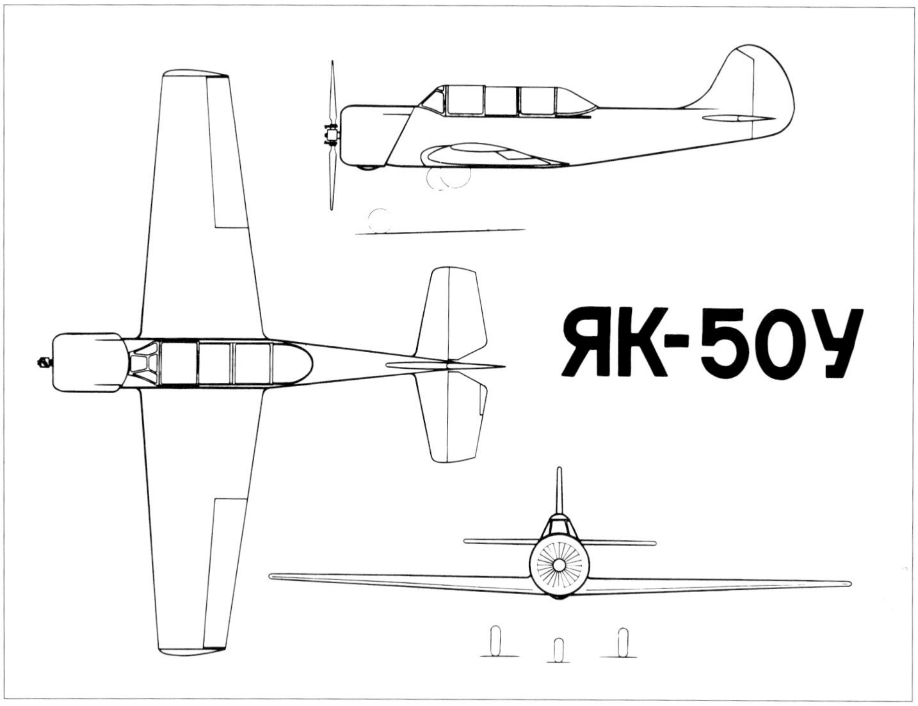

Above: A three-view drawing from the Yak-52's project portfolio bearing the original designation Yak-50U.

Yakovlev OKB

Above: Despite the code '101 Yellow', this is the Yak-52 prototype, as revealed by the tail logo/tail titles. The tactical codes of production aircraft are in fact the sequence number of the Yak-52 built in a given year.

ITAR-TASS

For winter operations the Yak-52 can be fitted with skis, as illustrated by the nearest and the farthest aircraft here. Note the star insignia and the DOSAAF tail titles.

via Yefim Gordon

Above: YR-IKA, a Yak-52 demonstrator displayed by IAv Bacău at various international air shows. The aircraft wears the standard red/white livery but the type is marked on the nose as 'Iak-52' (*their* spelling).

Dmitriy Komissarov

Above: '07 Yellow', the Yak-52Sh COIN aircraft prototype, in the rarely accessible Hangar 8 at Monino. The UB-32A FFAR pods show clearly how small the aircraft is (thank goodness for the tall undercarriage).

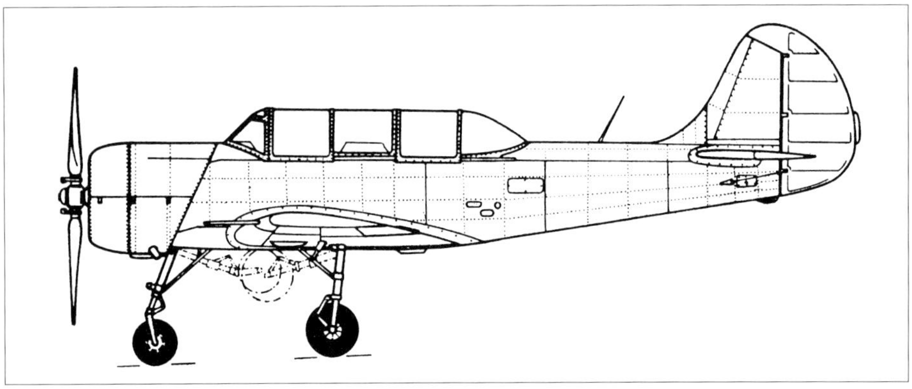

A side view of the Yak-52. Note the position of the wheels when the landing gear is retracted.

Specifications of the Yak-52 trainer

Engine type	M-14P
Engine power, hp	360
Length	7.676 m (25 ft 2 in)
Wing span	9.5 m (31 ft 2 in)
Wing area, m^2 (sq ft)	15.0 (161.5)
Empty weight, kg (lb)	1,000 (2,200)
All-up weight, kg (lb)	1,290 (2,840)
Maximum speed, km/h (mph)	285 (177)
Landing speed, km/h (mph)	110 (68)
Initial rate of climb, m/sec (ft/min)	10 (1,970)
Service ceiling, m (ft)	6,000 (19,685)
Range, km (miles)	550 (342)
Take-off run, m (ft)	170 (560)
Landing run, m (ft)	200 (660)

1999 and was exhibited at the 43rd Paris Air Show in June of the same year.

According to aeronautical press accounts, the Yak-52W offered the same performance as the original version, but gave twice the range due to its two integral fuel tanks in the wings, which doubled the capacity to 280 litres (74 Imp gal). A three-bladed Mühlbauer MTV-9B/C/L250-21 propeller with a spinner was fitted; the wingtips were strengthened for optional tip tanks; the fabric covering on the control surfaces was replaced with aluminium; a retractable cockpit access ladder was added, as were wingtip strobe and navigation lights plus recessed landing/taxi lights. The machine was provided with externally accessible baggage and battery compartments Other new items included hydraulic disc brakes, Cleveland wheels, Western tyres, a Hooker seat harness system and a full set of Western equipment and avionics, including a Garmin IC-A200 radio and GTX transponder with AmeriKing Altitude Encoder and NAT AA80-20 intercom.

In 1999 Aerostar planned to produce 30 Yak-52Ws per year for the next five years and then assess the viability of continuing production. Manufacture of the standard Yak-52 was also expected to continue alongside the new model. Production Yak-52Ws have been sold to the USA and Australia. At least one example (identity unknown) has found its way to a Russian air club based at Moscow-Myachkovo; interestingly, the type is marked on the nose as 'Як-52W'!

Iak-52TW trainer/sports aircraft

In an attempt to add customer appeal by, so to say, 'warbirdising' the Yak-52, in 2000 Aerostar brought out a 'taildragger' version of the Iak-52W designated Iak-52TW (the suffix standing for 'tailwheel undercarriage'). The main gear units are moved forward appreciably, retracting inwards into the wing roots, while the fixed rear strut has a long cantilever spring placing the tailwheel aft of the rudder trailing edge.

The prototype of the new version flew in 2001. At least four have been delivered to private owners in the USA.

Aerostar Condor (Romanian Yak-52 derivative)

The IAv Bacău plant (Aerostar) produced yet another derivative of the Yak-52 dubbed Condor. It was powered by a 300-hp Avco Lycoming AEIO-540-L1B5D flat-six engine driving a three-bladed Hoffmann variable-pitch propeller. The nose contours were suitably restyled. Coupled with the revised vertical tail contours with a squared-off top and the frameless cockpit visor, this gave the new machine a rather different look from the original Yak-52. The machine was clearly intended for Western markets, but presumably remained a prototype.

Yak-52PSh COIN aircraft (Yak-52Sh; Yak-54 – first use of designation)

This light ground attack/COIN version for 'limited conflicts' (Sh stands for *shtoormovik*, attack aircraft) was equipped with two pylons outboard of the main landing gear units for the carriage of various external stores; the instructor's instrument panel was replaced by a weapons selector panel. According to *Jane's All the World's Aircraft 2000-01*, three prototypes of this version were built in Romania where it was designated Yak-54 (this name was latter re-used for a different aircraft). The aircraft was not adopted for series production.

A camouflaged Yak-52PSh prototype coded '07 Yellow' is preserved at the Soviet (now Central Russian) Air Force Museum in Monino, carrying UB-32A unguided rocket pods under the wings. The aircraft bears the c/n 780102 – that is, year of manufacture 1978, Batch 01, 02nd aircraft in the batch; thus it was actually converted from the second production Yak-52.

Curiously, photoproof exists of a Yak-52 registered ФЛА СССР 01087 and coded '14 Red' (ex-DOSAAF '14 Yellow', c/n 822002) with 'Yak-54' titles on the tail! Clearly this is just for show and the aircraft is not representative of the armed COIN version.

Sergey Komissarov

This Ukrainian Yak-52 putting on an aerobatics display at one of the Aviasvit-XXI airshows in Kiev features the wingtip tanks fitted to some late-production examples. The tanks' rear ends are flattened.

Yak-52M upgraded basic trainer

The Yak-52M version (*moderni**zee**rovannyy* – upgraded) was developed in 2002-03 in response to Russian Air Force requirements which stipulated an equipment set enabling night training flights and take-offs and landings in adverse weather conditions. The development of this aircraft was a forced measure. On the one hand, in as-was condition the standard Yak-52 no longer met current requirements concerning operational safety and equipment standards. On the other hand, development of new-generation basic trainers featuring a crew rescue system and state-of-the-art avionics was making slow progress due to funding shortfalls and hence lack of state orders. In these conditions a mid-life update of the existing Yak-52 fleet appeared a more realistic option than waiting for a 'pie-in-the-sky' Yak-152 or Sukhoi Su-49.

The Yak-52M prototype coded '01 White' (c/n 866910) was shown statically at the MAKS-2003 airshow in Zhukovskiy; it had been manufactured as a Yak-52 *sans suffixe* on 14th July 1986 and was probably originally coded '106 Yellow'. This machine differed from the standard Yak-52 in several important respects. Outwardly it was readily identifiable by the new bulged cockpit canopy opening to starboard in the manner of the Yak-54 aerobatic aircraft, which gave the Yak-52M a distinctive humpbacked appearance. The cockpit was equipped with two SKS-94MYa ejection seats developed by NPO *Zvez**da*** ('Star' Science & Production Association). These, together with the new canopy, enabled the pilots to eject through the canopy. The rear seat was placed a little higher than the forward one, improving the instructor's field of view.

Modifications to the wings included the installation of vortex generators in the wing root area; they were designed to ensure that in the event of a stall the airflow separates at the roots first and the aircraft merely drops its nose, without dropping a wing. The M-14P engine was replaced by the M-14X version driving a three-bladed Mühlbauer MTV-9B/C/L250-21 propeller with a spinner. The fuel capacity was increased to 200 litres (44 Imp gal), resulting in a range extended to 900 km (559 miles). Some 40% of the avionics and equipment was replaced; the machine received a 12SAM-18A DC battery and a set of lighting equipment required for night flights. The upgrade was performed by the Russian Air Force's Aircraft Overhaul Plant No.123 in Staraya Roossa.

The changes described above are characteristic for the second stage of the upgrade; the initial stage may be confined to installing a new set of equipment similar to that of the prospective Yak-152 and Su-49 aircraft and endowing the aircraft with night flight capability. In 2002 there were plans to start off with converting some 70 out of the 600-plus Yak-52s currently in service to Stage A standard. In January 2005 Russian Air Force Commander-in-Chief Valeriy Mikhaïlov confirmed the intention to procure the Yak-52M for military flying schools.

Yak-52 with AI-450 turboprop (project)

In 2003 it was revealed that the Yakovlev Design Bureau had signed a contract with the Odessa-based Ukraviaremont aircraft repair facility for upgrading of the Ukrainian Air Force Yak-52s. This work will include fitting a ZKMB Progress AI-450 turboprop and overhauling the airframe. Installation of the AI-450 turboprop will entail lengthening the aircraft's forward fuselage, which will be recontoured and fitted with a redesigned engine cowling. A three- or five-bladed reversible-pitch propeller with automatic feathering is to be used. The nose undercarriage strut will be moved forward and made fully retractable. The existing seats will be replaced with Zvezda SKS-94 ejection seats. Provision will be made for the installation of a retractable periscope in the upper part of the cockpit glazing in order to provide a forward view for the instructor during take-off and landing.

The length and area of the dorsal fin will be increased. A new, taller rudder of increased area will be fitted. It will have an angular top and a two-section trim tab. The ailerons will be modified and fitted with trim tabs. The existing fabric covering of the control surfaces will be replaced by a synthetic material. The avionics suite will be altered to match that of the Aero L-39 advanced trainer. There will also be other equipment changes.

Yak-53 single-seat aerobatic trainer prototype

This was a single-seat aerobatic version of the Yak-52, differing from its progenitor mainly in the deletion of the front cockpit and appropriate changes to the canopy. The Yak-53 dispensed with the spring loading of the controls introduced on the Yak-52, thus reverting to the controls used on the Yak-50. Deletion of some equipment items (including the direction finder and compass system) helped make the aircraft 100 kg (220 lb)

Above: '06 Yellow', the sole prototype of the Yak-53 aerobatic aircraft – basically a tricycle-gear version of the Yak-50.

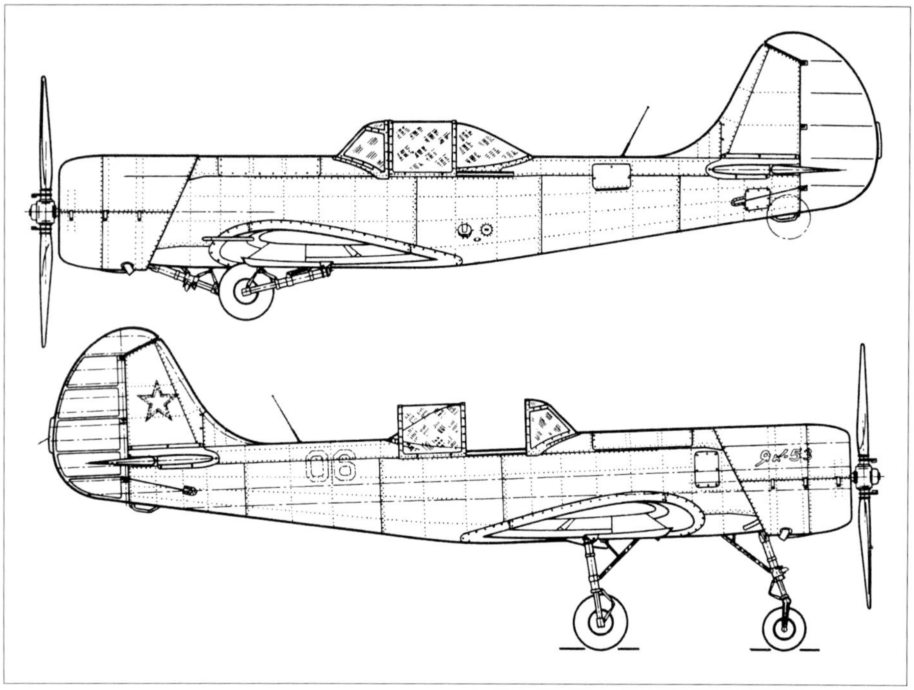

Two views of the Yak-53, showing the altered position of the ,maintenance access panel on the port side.

lighter than the Yak-52, which enhanced its aerobatic performance.

The aircraft was tested in prototype form in 1982 and set a time-to-height record on 15th February 1982, climbing to 3,000 m (9,840 ft) in 5 min 5 sec. On 23rd February 1982 another record achievement was scored when the Yak-53 climbed to an altitude of 6,000 m (19,680 ft) in 13 min 54 sec.

Specifications of the Yak-53 aerobatic aircraft

Engine type	M-14P
Engine power, hp	360
Length	7.68 m (25 ft 2¼ in)
Wing span	9.5 m (31 ft 2 in)
Wing area, m² (sq ft)	15.0 (161.5)
Empty weight, kg (lb)	900 (1,984)
All-up weight, kg (lb)	1,060 (2,340)
Maximum speed, km/h (mph)	300 (186)
Take-off run, m (ft)	150 (490)
Landing run, m (ft)	250 (820)

According to some reports, the record flights were performed with a three-bladed propeller of West German manufacture (presumably a Hoffmann product) instead of the standard two-blade propeller inherited from the Yak-52. Some sources suggest that there were plans for putting the Yak-53 into production at the AAPO Progress plant in Arsen'yev and under licence in Romania; yet the aircraft did not progress beyond the prototype stage.

Yak-55 competition aerobatic aircraft

The story of this aircraft reflects the influence of capriciously fluctuating trends in competition aerobatic flying. The inception of the Yak-55 was due to the experience of the 1976 World Aerobatics Championship in Kiev. Soviet sports pilots emerged as the winners with their Yak-50s, demonstrating a dynamic style of aerobatics with high speeds requiring a sizeable piloting space. However, they were impressed by the style of their foreign colleagues who used slower aircraft and performed aerobatic figures within a more limited space. This style was acknowledged as 'more advanced' at that time, and this prompted Soviet pilots and designers to set about producing a new aerobatic aircraft that would meet the new requirements. It received the designation Yak-55.

Yak-55 prototype and initial production model

The new aircraft was designed around the 360-hp M-14P radial driving a V-530TA-D35 two-blade variable-pitch propeller. It was an all-metal mid-wing monoplane with a fixed tailwheel undercarriage featuring cantilever spring-type main legs with optional spats. The wings were rather thick, having a thickness/chord ratio of 18% as compared to 9-14% on earlier aircraft; they had moderate taper, with negative trailing-edge sweep, and a symmetric airfoil which, coupled with zero dihedral and incidence, facilitated the execution of inverted aerobatic flight. The vertical tail featured a ventral strake intended to improve the aircraft's spin characteristics. The aircraft was fitted with a two-piece blown canopy and had a large spinner on the propeller hub.

Construction of the prototype took quite some time, and the first flight did not take place before May 1981. In August 1982 two prototypes were demonstrated for the first time at Moscow-Tushino, one of them coded '55' and the other sporting the side number '07 Yellow'. By that time the dynamic high-speed piloting style practised by Soviet sports pilots had gained wide acceptance once more, rendering the concept of the Yak-55 outdated at the very start of its career. Its roll rate proved to be rather poor and the aileron stick forces too high.

In its original form the Yak-55 did not find favour with Soviet sports pilots who declined to use the machine during the World Aerobatic Championship of 1982. However, the Yak-55 had its strong points, too. It was eminently suitable for performing inverted aerobatics and displayed excellent spin recovery characteristics. What was needed was a greater speed and higher roll rate. To achieve this, the wing span was reduced, thus decreasing the damping effect of the wingtips. So configured, the Yak-55 attained a maximum speed of 315 km/h (195.8 mph) and a roll rate of 4 radian/sec (229°/sec).

Yak-55 modified (early production version)

The modified Yak-55 featured wings of shorter span (9 m; 29 ft 6 21/64 in), reduced area (14.8 m²/159.3 sq ft), greater taper and a reduced thickness/chord ratio, retaining a symmetrical airfoil. The ailerons had 32% aerodynamic balancing and were provided with aerodynamic balance tabs mounted underneath on twin forward-swept outriggers projecting ahead of the aileron rotation axis. This helped reduce the aileron stick forces. The ailerons had their mass/horn balances neatly faired into the slightly rounded wingtips (the original Yak-55 had horn balances projecting outside the vertically cropped wingtips). An interesting feature of the aircraft was the use of drooping ailerons ('flaperons') interconnected with the elevators, although one of several prototypes under test retained conventional ailerons for comparison purposes.

The aircraft also differed from the initial Yak-55 in having the original two-piece canopy replaced by a three-piece unit, with a fixed aft section behind the aft-sliding centre section. The propeller spinner was usually dispensed with, although some early examples had it. The modified Yak-55 was evaluated with two- and three-blade propellers. To improve the pilot's downward view, the designers made use of a glazed window in a wing root and of a screen in the cockpit, the picture to which was supplied via a fibre-optic cable.

The Yak-55 with the revised wings was publicly shown in 1986 at the VDNKh (National Economy Achievements Exhibition) in Moscow. Prior to that, in 1984, two Yak-55s coded '3 Yellow' (and fitted with a three-blade propeller) and '07 Yellow' took part in the World Aerobatic Championship in Hungary. One of them was flown by woman sports pilot Khalidé Makagonova and brought her the first prize.

In 1985 the Yak-55 entered production in Arsen'yev; eventually some 100 machines were manufactured. In 1986 the Soviet national aerobatic team of woman pilots won the first place at an international contest, Khalidé Makagonova gaining the title of world champion.

Yak-55M aerobatic aircraft

In response to wishes expressed by sports pilots and their DOSAAF leaders, the OKB undertook a further modification of the Yak-55. The new version designated Yak-55M made its first flight in May 1989, the prototype being coded '49 Blue'. The wing span was further reduced to 8.1 m (26 ft $6\frac{29}{32}$ in), the wing area to 12.8 m^2 (137.8 sq ft). The roll rate was increased to 6-7.7 radian/sec (344-441°/sec) at speeds of 300-360 km/h (186.4-223.7 mph).

The new machine won praise from many leading Soviet sports pilots who commended, in particular, the reduced stick forces. The Yak-55M came to be used by the Soviet team in international aerobatic con-

Yakovlev OKB

Above: '07 Yellow', the first prototype of the Yak-55 *sans suffixe*. The thick wing airfoil is readily apparent. Note the canopy design, the aileron horn balances, the wheel spats and the original propeller spinner.

Yakovlev OKB

Above: The same aircraft minus wheel spats and with a shorter propeller spinner. The tailwheel appears to have been replaced with a tailskid.

Yakovlev OKB

Above: The second prototype Yak-55 (aptly coded '55'), again with a two-piece canopy and wheel spats. The oil cooler between the main gear struts appears to be larger.

Yakovlev OKB

'3 Yellow', another Yak-55 *sans suffixe*, was used to test the new, thinner wings with faired aileron horn balances and the new three-piece canopy.

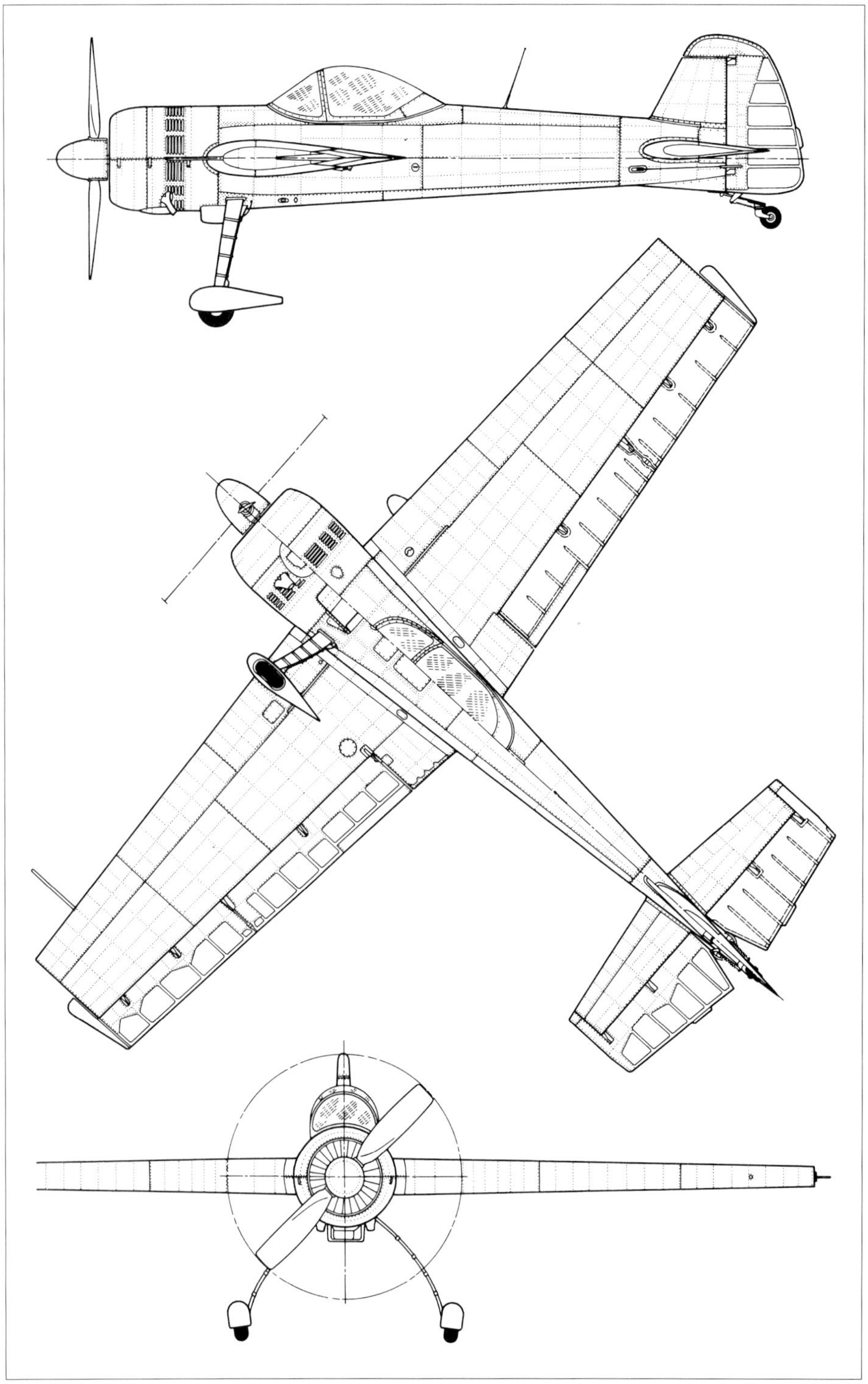

A three-view of the Yak-55 *sans suffixe* as originally flown with large-span wings and protruding elevator horn balances.

tests, starting with a contest in Czechoslovakia in June 1989 where it brought a champion title to the woman sports pilot Svetlana Kabatskaya. In 1990 the Yak-55M superseded the Yak-55 in production at Arsen'yev; more than 100 machines were manufactured for Soviet air clubs.

Tekhnoavia SP-55M Slava aerobatic aircraft

The SP-55 Slava single-seat aerobatic aircraft was developed in the late 1990s by the Tekhnoavia company as a derivative of the Yak-55M (hence the digits in the SP = *samo**lyot** pilo**tazh**nyy*) and built at the Smolensk aircraft plant. As noted in the Yak-18T section, Vyacheslav P. Kondrat'yev, the SP-55M's designer, had previously worked in the Yakovlev OKB and had been the Yak-55's project engineer; thus, in effect, he merely undertook an upgrade of his own progeny. The aircraft's popular name contains a pun; ***sla**va* is Russian for 'glory' or, in some cases, 'reputation' (good or otherwise), but Slava is also the short/familiar form of the name Vyacheslav and a few other Russian men's names (Vladislav, Svyatoslav and so on).

Outwardly the SP-55M differed from the Yak-55M in having a redesigned vertical tail of more angular contours (vaguely reminiscent of a shark's tail), a deeper rear fuselage with a raised upper decking and a suitably redesigned cockpit canopy. The more powerful 400-hp M-14PF engine driving a three-blade MTV-9B/C/L250-21 propeller ensured an improvement in performance compared to the Yak-55M, the maximum speed reaching 450 km/h (279 mph) versus 360 km/h (224 mph).

The SP-55 was shown at the MAKS-2001 airshow (14th-19th August 2001) with no registration on; a different example registered RA44492 (c/n 0104 07 09 00 – that is, Batch 01, 04th aircraft in the batch, presumably manufactured on 7th September 2000) was present at the MAKS-2003. The first production example registered RA44547 (c/n 0101) is based at White Waltham airfield, the seat of the West London Air Club, since 2002 and is presumably sold to a British owner.

Filin primary trainer (project)

In late 1980s a group of young designers of the Yakovlev OKB prepared an advanced development project of a two-seat trainer named ***Fil**in* (Horned Owl) and submitted it to a competition for the best project of a primary trainer arranged by DOSAAF. This was a strut-braced high-wing monoplane with a non-retractable tricycle undercarriage. The instructor and trainee were seated side-by-side in an extensively glazed cabin. The aircraft was to be powered by an unspecified new 150-hp engine with cylinders in X formation. The basic specifications of the Filin were as follows: take-off weight 670 kg (1,480 lb); maximum speed 210 km/h (130 mph); vertical speed at sea level 5.6 m/sec (1,102 ft/min); fuselage width 1.28 m (4 ft 2$\frac{25}{64}$ in); wing area 13.5 m^2 (145.3 sq ft).

This project was short-listed by the organisers of the competition as one of the best projects among those submitted, but it did not receive the go-ahead. In its general contours the project is strikingly similar to the better known Yak-112 general-purpose aircraft (described later in the book), and it stands to reason to surmise that the latter was, in effect, a reworked and scaled-up Filin with four seats instead of two and about twice the weight.

Yak-54 aerobatics trainer/ competition aerobatic aircraft

Faced with the need to keep abreast of the steadily improving Western competitors of the Russian aerobatic aircraft, the Yakovlev

OKB decided to produce a new machine that would be superior to both the Yak-55M and the Su-26M. Aleksandr N. Dondukov, who was the OKB's General Designer at the time, came up with the idea of producing the new machine designated Yak-54 as a derivative of the Yak-55M. The project was handled by Chief designer Dmitriy K. Drach and project engineer Vladimir Popov.

The Yak-55M's design was thoroughly reworked, the end result being essentially a completely new two-seat aircraft with considerably improved performance as compared to its progenitor. The Yak-54 could also score some points on the closely comparable two-seat Su-29 two-seat aerobatic aircraft, possessing better inverted flight performance and stalling characteristics.

The Yak-54 two-seat aerobatic aircraft was intended for training sports pilots, aerobatics training and for participation in various domestic and international competitions. It was a mid-wing monoplane with a fixed tailwheel undercarriage, powered by a 360-hp M-14P engine driving a three-blade high-efficiency propeller. Basic specifications included a take-off weight of 990 kg (2,180 lb), a maximum speed of 460 km/h (286 mph) and a 15-m/sec (2,953 ft/min) rate of climb. The machine's dimensions were: length 6.91 m (22 ft 8 in), wing span 8.16 m (26 ft $9\frac{1}{4}$ in), wing area 12.9 m^2 (138.9 sq ft).

The aircraft boasted a relatively low empty weight; its reduction was achieved thanks to a careful cockpit design, reduction of the equipment weight, the use of parachutes attached to the pilots' backs and the use of hollow main landing gear spring struts. A novel feature of the aircraft was the use of flaperons. In any flight mode the ailerons could be deflected 5° down, and to an angle of either 10° or 15° during take-off and landing. This feature ensured a lower landing speed and precluded loss of altitude during aerobatic manoeuvres. To improve visibility, a glazed area was provided in the lower fuselage skin.

The as-yet unflown Yak-54 prototype, '310 Grey', was shown at the 1993 Paris Air Show; it was flown for the first time on 24th December 1993. Flight testing was conducted by Yakovlev's chief test pilot Andrey Sinitsyn who was enraptured over the aircraft's handling qualities. The Yak-54 was docile and could be easily handled by inexperienced pilots. At the same time it enabled experienced pilots to perform the most intricate sets of aerobatics thanks to its excellent controllability, high structural strength and high power/weight ratio.

Production was launched at the Saratov aircraft factory. Initially the Yak-54 was manufactured exclusively for export to the USA and some other countries. Eventually

Yakovlev OKB

Above: '02 Red', the Yak-55M prototype, with appropriate tail titles.

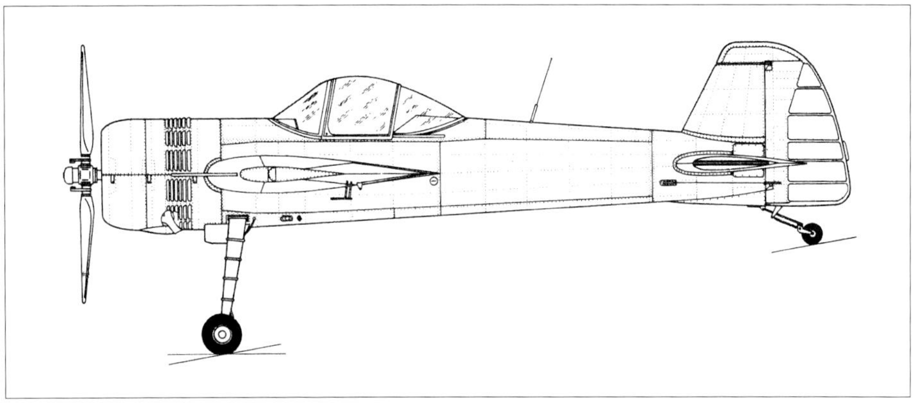

A side view of the Yak-55M.

Specifications of the Yak-55 and Yak-55M

	Yak-55 *sans suffixe*	Yak-55M
Length	7.29 m (23 ft 11 in)	7.29 m (23 ft 11 in)
Height, tail down	2.2 m (7 ft $2\frac{29}{64}$ in)	2.2 m (7 ft $2\frac{29}{64}$ in)
Wing span	9.0 m (29 ft $6\frac{21}{64}$ in)	8.1 m (26 ft $6\frac{3}{4}$ in)
Wing area, m^2 (sq ft)	14.8 (159.3)	12.8 (137.8)
Empty weight, kg (lb)	n.a.	690 (1,520)
All-up weight, kg (lb)	875 (1,930)	855 (1,855)
Never-exceed speed, km/h (mph)	450 (279)	450 (279)
Maximum aerobatics speed, km/h (mph)	360 (224)	360 (224)
Max level airspeed, km/h (mph):		
at take-off power	290 (180)	305 (189.5)
at nominal power	285 (177)	285 (177)
Landing speed, km/h (mph)	130 (81)	130 (81)
Service ceiling, m (ft)		
Ferrying range, km (miles)	595 (370)	595 (370)
Take-off run, m (ft)	150 (490)	150 (490)
Landing run, m (ft)	465 (1,525)	465 (1,525)

via Yefim Gordon

Above: Lithuanian-registered LY-AGL is one of quite a few Yak-55s sold abroad in the 1990s and operated on the European and US airshow circuits.

several examples were supplied to Russian customers as well. By mid-2004 two aircraft were purchased by the Gazprom natural gas concern of Russia. Three machines were acquired by one of ROSTO's air clubs and this organisation placed an option for a further three.

Unfortunately the sports career of the Yak-54 has been marred by at least two crashes. One of these occurred in the USA, the other one took place in Russia on 17th August 1995, when the first production Yak-54 registered 01001 (no prefix, c/n 01 001 – that is, 01st aircraft of Batch 001) crashed near LII's airfield in Zhukovskiy. Nadezhda Sergeyeva, a woman aerobatic pilot who had won many prizes at international aerobatic competitions, lost her life in the crash.

Yak-54M primary trainer (project)

In 1999 the Yakovlev OKB revealed plans for the development of a thoroughly revised and modernised version of the Yak-54 designated Yak-54M. It was to become part of the Yak-UTK comprehensive training system comprising also the Yak-130 trainer and ground simulators. A model of this variant was displayed at the Farnborough Air Show in 2000.

The Yak-54M was, in fact, quite a different aircraft. The radically reworked design incorporated such new features as low-set wings and a tricycle undercarriage with an aft-retracting nose unit and inward-retracting mainwheels. The fuselage was also revised, the cockpit canopy no longer blending into the rear fuselage contours. The new one-piece canopy divided by a frame into front and rear sections (as on the Embraer EMB 312 Tucano) dispensed with the visor, thus providing excellent visibility for the pilot. The aircraft was to be fitted with a 360-hp M-14X engine.

According to an advertising leaflet, the aircraft's dimensions included a length of 7.3 m (24 ft), a wing span of 8.8 m (28 ft 10½ in) and a height of 2.8 m (9 ft 2¼ in). The aircraft was to have a take-off-weight of 1,300 kg (2,870 lb), a maximum speed of 450 km/h (280 mph), a stalling speed of 100 km/h (62 mph), a maximum fuel load of 200 kg (440 lb), a maximum range of 1,000 km (620 miles), and a take-off and landing run of 160 and 340 m (525 and 1,115 ft) respectively.

In 2003 a nearly identical model emerged as the Yak-152, suggesting that the Yak-54M project was developed further under a new designation (see Yak-152 below).

Yak-56 aerobatic trainer (project)

In 1988 Yakovlev announced a new two-seat aerobatic trainer based on the Yak-55M but featuring low-set wings and retractable main undercarriage units. The aircraft was regarded as a prospective replacement for the Yak-52. It was to be powered by a 300-hp Voronezh Engine Design Bureau (VOKBM) M-16 eight-cylinder engine driving an AV-86 three-blade variable-pitch propeller (the engine had an unusual cylinder arrangement comprising two X-shaped rows, one after the other). The wings featured leading-edge root extensions. The cockpit canopy was frameless in the front part, lacking the usual visor, so as to improve the forward view.

Sergey Komissarov

Above: '310 Grey', the Yak-54 prototype, at Le Bourget during the 1993 Paris Air Show. The two-seater looks rather different from the Yak-55 from which it was derived.

Yefim Gordon archive

The wreckage of Yak-54 01001 in the back yard of a country house near Zhukovskiy. Pilot Nadezhda Sergeyeva was killed in this crash on 17th August 1995.

The following basic specifications for the Yak-56, revealed in 1990, included an AUW of 830 kg (1,830 lb), a maximum speed of 460 km/h (286 mph), a stalling speed of 100 km/h (62 mph), a length of 6.32 m (20 ft 8⅘ in), a wing span of 8.3 m (27 ft 2¾ in) and a wing area of 11.53 m² (124.12 sq ft). The aircraft was designed for G-loads of +9/–7. The Yak-56 was to be fitted with full-span flaperons linked to the elevators, a system used by some Western aircraft at that time.

According to some reports in the Western press, the Yakovlev OKB studied the possibility of fitting the Yak-56 with a built-in parachute recovery system designed to bring the whole aircraft safely to the ground in an emergency. Reportedly, the system was to be developed from US-designed recovery devices for light aircraft and remotely piloted vehicles.

The Yak-56 prototype was expected to fly in 1992 or 1993, but delays with the development of the M-16 engine forced the OKB to abandon these plans. Instead, Yakovlev produced the simpler Yak-54 powered by the M-14P engine.

Yak-57 aerobatic aircraft (project)

The Yak-57 was conceived as a single-seat aerobatic version of the Yak-56 and was under development in parallel with the baseline aircraft. This project was mentioned in 1989-91 in Russian magazine articles which referred to the Yakovlev OKB's work on *'the new Yak-56 trainer and the Yak-57 aerobatic aircraft [developed] in response to an order from DOSAAF'*. The single-seat Yak-57, it was stated, would meet all the requirements posed to an aerobatic aircraft.

Some details were revealed in an interview given by Sergey A. Yakovlev, Chief Designer of the Yakovlev OKB, in September 1988 to *Flight International* and later in an article in the Soviet press. He stated that the two aircraft – the two-seat Yak-56 and the single-seat Yak-57 – would share the same airframe. The Yak-57 (which was intended as a replacement for the Yak-55 and, hopefully, might replace the Su-26M in competitions) would be cleared for just +10/–9 Gs (the appropriate figures for the Su-26M being +11/–9 Gs). However, the demise of the Yak-56 project led to the termination of work on the Yak-57 as well.

Yak-130 combat-capable advanced trainer

At the end of the 1980s the Soviet Air Force took the decision to write a new set of general operational requirements (GOR) for a dedicated training aircraft provisionally known simply as the UTS (*oo**cheb**notreni**rov**ochnyy samo**lyot*** – conversion/proficiency trainer). The GOR was issued to a number of design bureaux that were called upon to respond to the requirements.

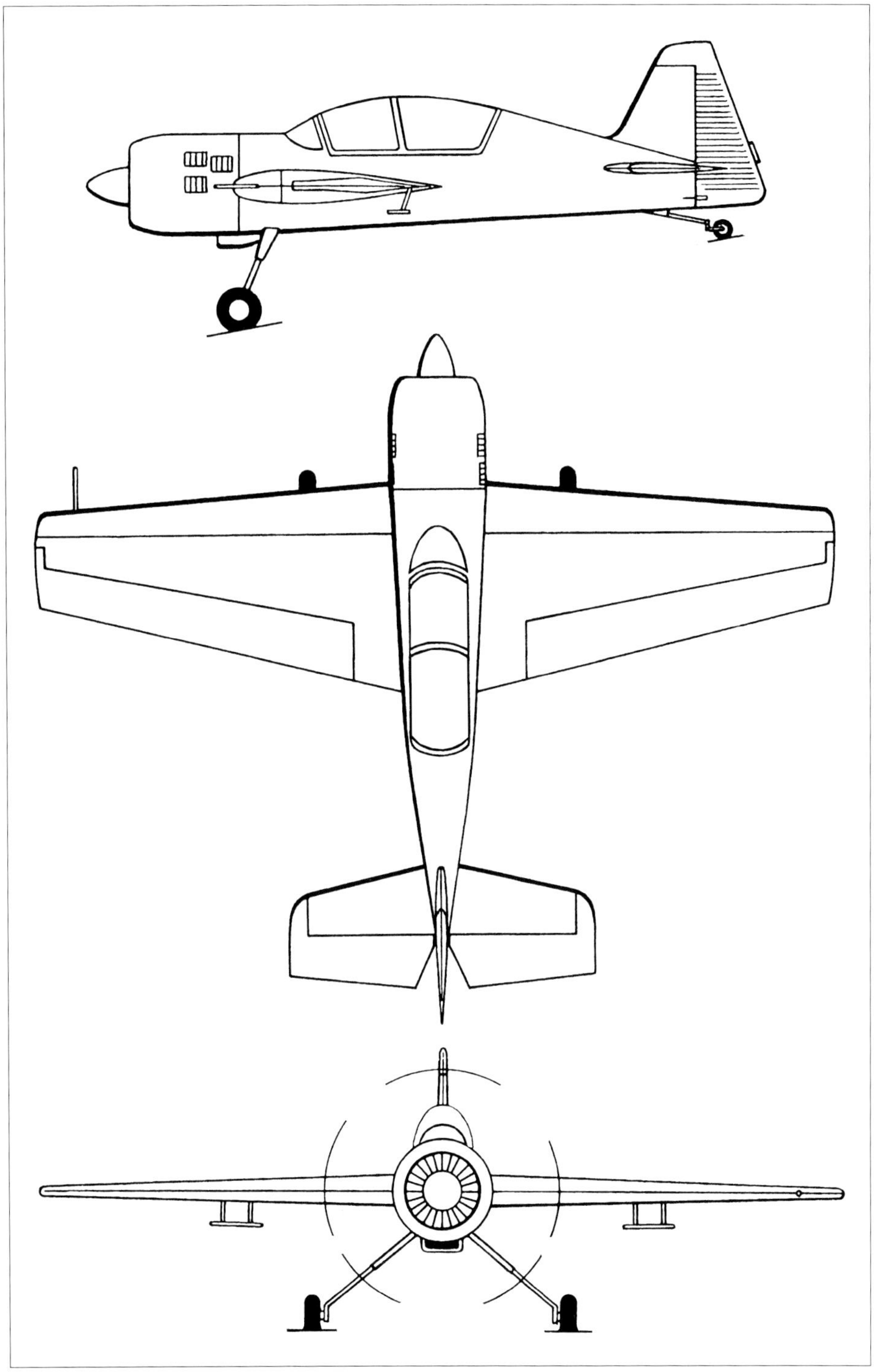

A three-view of the Yak-54 aerobatic trainer.

In the early 1990s, four Russian design bureaux – MiG, Myasishchev, Sukhoi and Yakovlev – took up the challenge and set to work on their individual proposals for a new-generation trainer design. In contrast to the practices of the Soviet period when such undertaking were generously funded by the state, budgetary constraints associated with the *perestroika* compelled the design bureaux to finance the project design work out of their own resources. An important aspect of the competition was its open character, whereby foreign partners were to be invited to participate in a joint project with the winning contender.

The first round of the competition resulted in the Sukhoi and Myasishchev designs being eliminated from the list of contenders, the Yakovlev Yak-130 and Mikoyan MiG-AT projects being chosen for the further work. In both cases, prototypes were built and much effort was spent on winning the support of prospective customers both at home and abroad so as to ensure

Yakovlev OKB

Above: Despite the 'yak-56' nose titles, this model represents the Yak-57, its single-seat derivative. Note the four-bladed propeller and the shape of the cowling concealing an M-16 X-block engine.

Yakovlev OKB

Above: A very provisional display model showing apparently an early project configuration of the Yak-56 aerobatic trainer (note the canopy design and the shape of the tail).

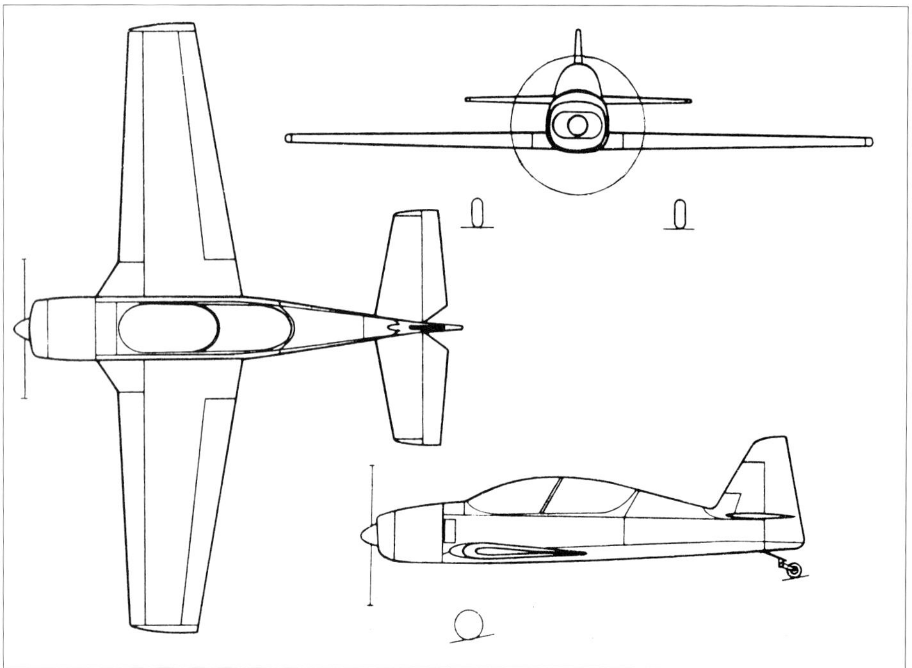

This three-view shows the definitive project configuration of the Yak-56, a much sleeker aircraft.

future sales. Eventually the Russian Air Force made its choice in favour of the Yak-130, but more about it later.

When designing the Yak-130, the project team considered the fact that it would have to train pilots both for existing fighter types and for future fifth-generation fighters with considerably better performance and greater agility than contemporary aircraft. It was decided to utilise an integral (blended wing/body) layout with swept wings and engines offering a comparatively high thrust-to-weight ratio, with the intention to emulate as closely as possible the subsonic manoeuvring characteristics of the newer fighters.

The Yak-130 is a subsonic two-seater with mid-set cropped-delta wings having 31° leading-edge sweep outside of the large curved LERXes which, together with full-span slats, are intended to permit controlled flight at an angle of attack up to 35°. The wings are fitted with tall winglets. The slab stabilisers repeat the wing planform and feature anhedral and a large dogtooth (omitted on the first prototype). The vertical tail is swept. The cockpit is equipped with two Zvezda K-36DM zero-zero ejection seats, which are sharply stepped to ensure good visibility for the instructor. They are enclosed by a large sideways-hinging canopy supplemented with a single-curvature frameless windscreen.

The aircraft was originally intended to be powered by two AI-25 turbofans, but this was soon changed to DV-2 turbofans. The DV-2 was originally developed as a Soviet-Czechoslovak co-operation programme and later as a joint project between ZMKB Progress (the Ukraine) and the Považske Strojarne enterprise (Slovakia); hence the DV initials standing for 'Dnepr-Vltava', the names of the two rivers on which the two enterprises are located. Actually, the model intended for installation in the production Yak-130 was the RD-35M, an NPP Klimov-modified version of the DV-2 rated at 2,200 kgp (4,850 lbst). At a still later stage this engine, in turn, gave place to the ZMKB Progress AI-222-25 turbofans (see below).

A curious feature of the Yak-130, which it shares with the Mikoyan MiG-29 fighter, is the provision of blocker doors which close the air intakes completely when the engines are run on the ground to prevent FOD. When they are closed, the engines draw air through dorsal auxiliary intakes in the LERXes closed by mechanically linked doors; the latter close (and the main blocker doors open) at the moment of rotation on take-off, triggered by weight off the nose gear unit. Unlike the MiG-29, the dorsal intakes are single, not a row of spring-loaded 'shark's gills'.

Originally it was planned to build the Yak-130 as a joint venture with Aermacchi of Italy, with which Yakovlev struck a partnership in 1994. Aermacchi shared the design and production rights and had exclusive marketing rights outside the former Soviet Union. However, as a result of differing timescales and priorities, the Italian company eventually took a decision to develop the concept independently as the Aermacchi M-346, leaving the Russian side to continue development work on the Yak-130 design on its own.

The main Russian participant in the Yak-130 programme, apart from the Yakovlev Corporation itself, is the Sokol (Falcon) aircraft factory in Nizhniy Novgorod; on the engine side, it was the NPO Klimov State Engine Manufacturing Enterprise of St Petersburg, but at present this role has passed to the Moscow-based MMPP Salyut as the prospective supplier of the AI-222-25 engines.

In August 2003, during the MAKS-2003 air show, an agreement was signed on the merger of the Yakovlev Design Bureau and the Irkut Aircraft Corporation. The Yakovlev people hope that this merger will facilitate promotion of the Yak-130 on foreign markets as a useful supplement to its deliveries to the Russian Air Force, which has chosen it in preference to the Mikoyan MiG-AT. The Yak-130 has a chance to win a niche on the Indian market where the Irkut Corporation has a foothold.

The Ukraine has voiced an intention to adopt the Yak-130 as the advanced trainer for its Air Force. On 14th May 2004 Commander-in-Chief of the Ukrainian Air Force Lieutenant-General Yaroslav Skal'ko said in a press interview that there were plans to use the Yak-130 for training Ukrainian military pilots and that assembly of these aircraft would be organised in Odessa. The Ukraine is participating in the Yak-130 programme as a supplier of the engines.

The various versions of the Yak-130, including projects, are detailed below.

Yak-130D (Yak/AEM-130D) prototype

The first prototype of the Yak-130 was designated Yak-130D ('demonstrator') – an appellation which it fully justified by taking part in various air shows in Russia and abroad. The Yak-130D left the assembly shop on 30th November 1994 and made its first flight on 25th April 1996, bearing the number 296 on the nose gear doors. Unlike the prototypes of previous Russian military aircraft, the Yak-130 wore a striking white/bluish grey/red colour scheme in a decidedly civilian style.

In its original configuration the Yak-130D featured large winglets which soon proved to be insufficiently strong and were removed. The extreme nose had a characteristic flattened shape with chines. The slab stabilisers lacked the dogtooth which was envisaged by an early project configuration (and reappeared in the later production version). The machine was demonstrated both with and without weapon pylons (three under each wing).

In December 1998, during routine test flying in Italy from the Aermacchi flight test facility at Venegono, the Yak-130D achieved an angle of attack of 42° in controlled and stabilised flight, thereby demonstrating one of the most important parameters expected of the new trainer; safe handling at critical angles of attack. This was followed

Yakovlev OKB

Above: This desktop model depicts an early project configuration of the Yak-130, then known simply as the Yak-UTS. Note the engine air intakes extending all the way to the nose gear unit.

Yakovlev OKB

Above: A later model depicting the noticeably different design which was 'frozen' before the construction of the prototype. Note the shorter intakes, the new shape of the nose and the altered landing gear.

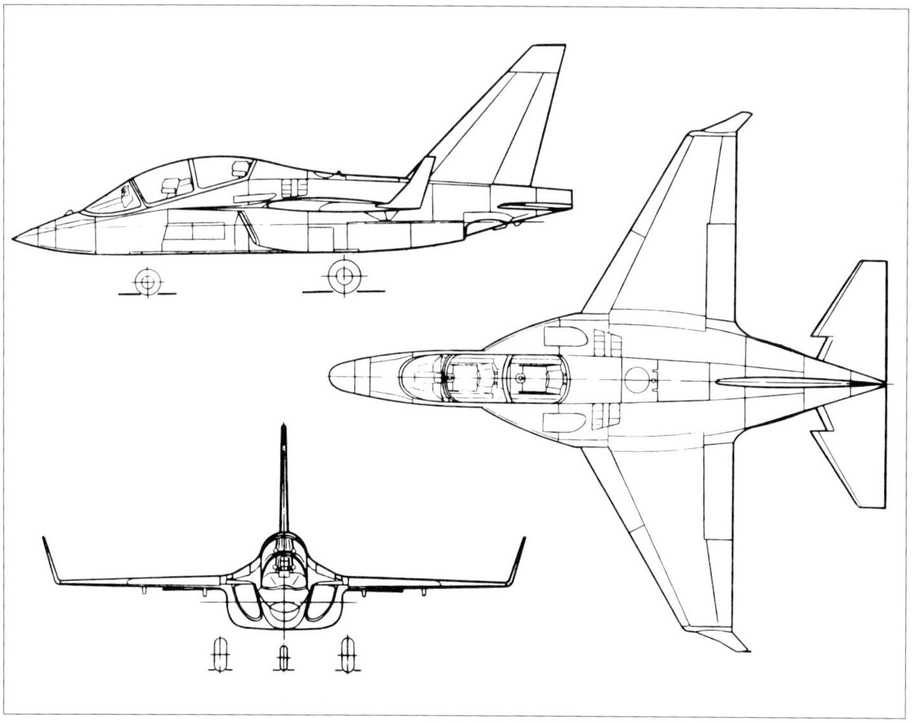

A three-view of the Yak-130D demonstrator following modifications.

Specifications of the Yak-130 combat trainer

	Yak-130D	Production Yak-130
Engine type	DV-2	AI-222-25
Engine thrust at take-off, kg (lb)	2,200 (4,850)	2,500 kg (5,512)
Length	11.245 m (36 ft 10¾ in)	n.a.
Wing span	10.4 m (34 ft 1½ in)	n.a.
Wing area, m^2 (sq ft)	23.5 (253)	n.a.
Empty weight, equipped, kg (lb)	4,410 (9,720)	n.a.
All-up weight, kg (lb)	6,200 (13,670) clean	
	9,000 (19,840) maximum	c. 9,000 (19,850)
Maximum speed, km/h (mph)	1,050 (652)	1,060 (659)
Landing speed, km/h (mph)	175 (109)	n.a.
Service ceiling, m (ft)	12,500 (41,000)	12,000 (39,360)
Range, km (miles)	2,220 (1,379)	2,000 (1,240)
Landing run, m (ft)	485 (1,590)	n.a.

by several more high-alpha flights to verify the flight control system.

In August 1999 the machine was demonstrated with a number of modifications introduced in cooperation with Aermacchi and reflecting the suggestions made by Italian specialists. The winglets were deleted, the wings acquired a dogtooth; small trapezoidal fences were added at the wing/LERX junctions and ahead of the windshield. The demonstrator was now registered RA-43130 and wore a revised colour scheme with Yakovlev OKB and Aermacchi logos. After the dissolution of the partnership it was repainted in camouflage colours and retained only the Yak logo, receiving the tactical code '01 White outline' in addition to the civil registration. Another external change was the addition of a heat shield aft of the APU outlet on the starboard side of the rear fuselage.

Yakovlev OKB

Above: The as-yet unregistered Yak-130 prototype with the number 296 on the nose gear doors and Yak/Aermacchi logos on the tail in the assembly shop of the Yakovlev OKB's experimental plant.

Yakovlev OKB

The Yak-130 makes a low-speed/high-alpha pass, showing the large winglets with which it was initially flown. The winglets soon developed cracks and were removed. Note the APU exhaust.

Yak/AEM-130S production version (joint project)

At a stage when the Yakovlev OKB and the Aermacchi company were still maintaining their partnership, their common plans envisaged starting the manufacture of a production version differing considerably from the Yak-130D (or Yak/AEM-130D) prototype. It would be an aircraft of smaller dimensions, its wing span being reduced from 11.25 m (36 ft 11 in) to 10.4 m (34 ft 1½ in), the fuselage length from 12.4 m (40 ft 8$^9/_{16}$ in) to 11.24 m (36 ft 10½ in) and the wing area reduced to 23.5 m^2 (253 sq ft).

To improve visibility, the nose would be drooped further; the fuselage would be provided with a pointed tailcone and made shallower; the landing gear wheelbase would be reduced. Instead of DV-2 engines, the powerplant would comprise DV-2S turbofans with full-authority digital engine control (FADEC) and operational life improvements developed in conjunction with Klimov and TsIAM in Russia as the RD-35. This version was known as the Yak/AEM-130S (*se**reey**nyy* – production, used attributively).

This project did not reach the hardware stage in its original form, but provided the basis for later production versions developed separately by the Yakovlev OKB and the Aermacchi company (see below).

Yak-130 production combat trainer (Russian version)

The first production version of the Yak-130 embodies a considerable degree of redesign as compared to the Yak-130D prototype. The changes incorporated in this production version reflect the somewhat changed concept of the Yak-130's employment in the Russian Air Force that emerged by the beginning of the 21st century. A new situation in the development of the Russian AF dictated the need to transform the Yak-130 from a pure trainer into a combat trainer capable of being used for practising up to 80% of the entire training programme for service pilots, as well as for live weapons training. Furthermore, the new threats to Russia's security are largely associated with a shift to low-intensity local conflicts in which a modern combat trainer possessing a secondary attack capability is an effective and less costly alternative to using more potent and expensive combat aircraft. The Yak-130 modified to meet the Russian Air Force's additional requirements should cope effectively with both training and combat tasks.

The production version differs externally from the Yak-130D in having a larger-diameter

Yakovlev OKB

Above: The Yak-130 prototype (still with winglets and without 'Yak-130D' titles) deploys its large dorsal airbrake. Note the instrumented probe added to the nose

pointed ogival nose cone housing a radar, reshaped air intakes, a dogtooth stabiliser leading edge and wingtip pylons for short-range air-to-air missiles. A conformal cannon pod is added under the fuselage.

The combat trainer is powered by Ukrainian AI-222-25 turbofans developed by the Ivchenko Progress Engine Design Bureau (ZMKB) and built by Motor-Sich, both of Zaporozhye. The AI-222-25 delivering a 2,500-kg (5,512-lb) thrust will later be built for the Russian Air Force at the MMPP Salyut Moscow Engine Production Association in cooperation with Motor-Sich. This meets the critical requirement posed by the Russian Air Force: its aircraft shall be powered by Russian-made engines (the Yak-130D prototype has Slovak-built DV-2 turbofans).

The production Yak-130's specifications include a take-off weight around nine tons (19,850 lb), a top speed of 1,060 km/h (659 mph), a service ceiling of 12,000 m (39,360 ft) and a range of 2,000 km (1,240 miles). The aircraft can carry up to three tons (6,610 lb) of external stores on its nine hardpoints; these can accommodate essentially the whole range of weapons used by tactical aircraft, including bombs up to 500 kg (1,102 lb), unguided rockets and guided air-to-air and air-to-surface missiles. External stores may include TV- or laser-guided bombs like the KAB-500, Kh-25ML guided air-to-surface missiles, as well as foreign-made missiles like the AGM-65 Maverick or other weapons. Light armour will be provided to protect the cockpit, powerplant and vital equipment items. A mock-up of the production model 's cockpit section exhibited in 2001 featured a retractable in-flight refuelling probe on the starboard side ahead of the windshield.

Yakovlev OKB

Above: This model depicts the proposed light strike/combat trainer version of the Yak-130 but is still based on the initial configuration with winglets and nose chines.

Yefim Gordon

Yak-130D RA-43130 in its current guise – that is, minus winglets and in camouflage colours. It carries UB-32A and B-8M FFAR pods and R-73 air-to-air missiles, and more weapons are arrayed in front of it

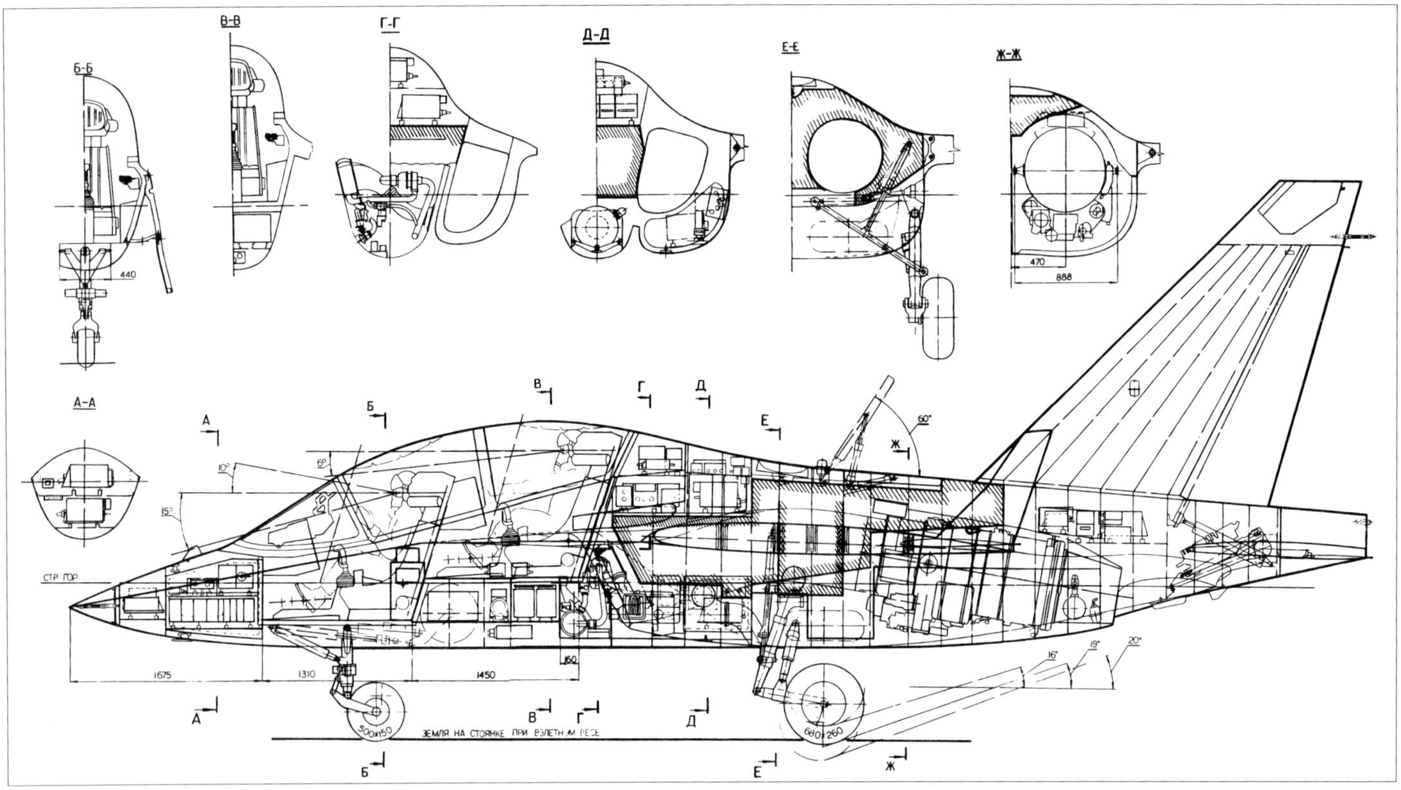

Yakovlev OKB

Above: A cutaway drawing of the Yak-130D demonstrator powered by DV-2S engines.

The first production Yak-130 (confusingly, also coded '01 White outline') made its maiden flight on 30th April 2004 with Yakovlev OKB chief test pilot Roman P. Taskayev at the controls. Thus, this first production model of the Yak-130 powered by two AI-222-25 turbofans started its flight development tests. The integrated test programme will span over two years. According to the present plans, the flight tests will involve four production machines. The official tests of the Yak-130 are to be completed in 2005, the Nizhniy Novgorod-based Sokol plant starting the series production of the aircraft in the same year.

The Russian AF plans to buy about 350 Yak-130s before 2010-2015 to replace the Czech-built L-39 trainers currently forming the backbone of its training fleet. In early 2005 two first production machines were passing their flight test programme. Deliveries to the Russian AF are to begin in 2006.

The prospect of the Yak-130's introduction into the Russian Air Force inventory led the Yakovlev OKB to step up the efforts to promote the aircraft also to the world market. By mid-2004 the Rosoboronexport State Company had held technical demonstrations of the Yak-130 in Algeria, India, Germany, the Ukraine, Chile and Slovakia. With this aircraft, Yakovlev hopes to secure up to 30% of the world's advanced trainer/combat trainer market.

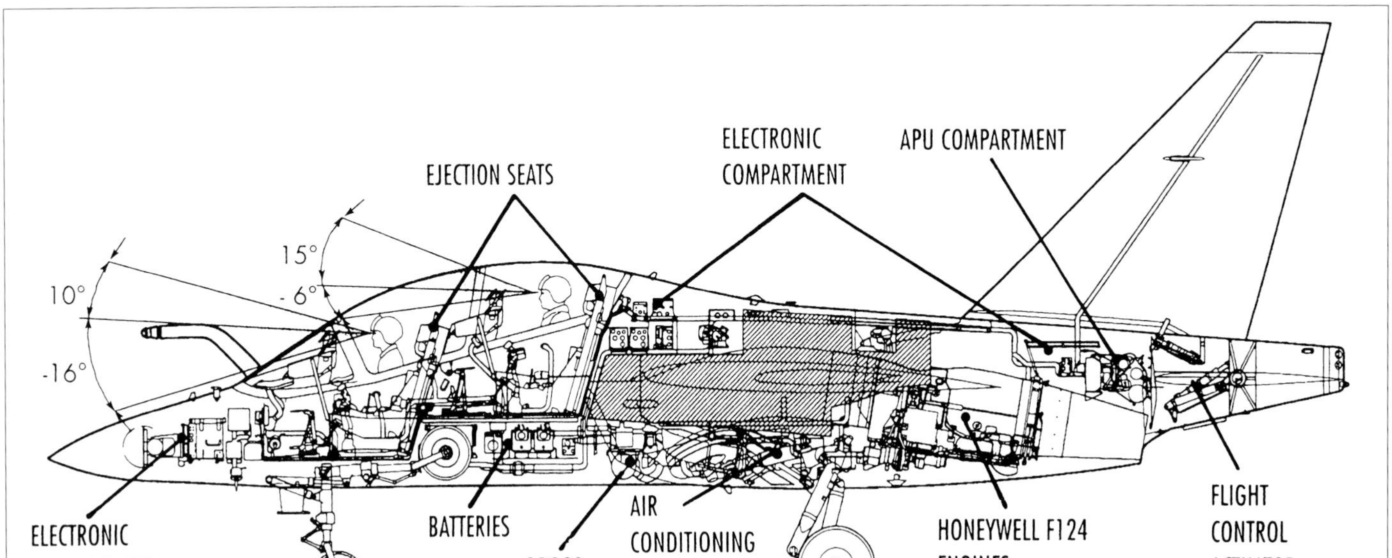

Aermacchi

Here, for the sake of comparison, is a similar cutaway drawing of the Aermacchi M-346.

Yakovlev OKB

Above: '01 White outline', the prototype of the radically redesigned production-standard Yak-130, in a test flight.

The Yak-130 has growth potential that enables the development of its various versions. Plans are in hand to develop a shipboard trainer, a light single-seat fighter, a light attack aircraft, a light reconnaissance plane (operating jointly with the Pchela UAV) and an unmanned attack aircraft.

Yak/AEM-130 with F-124 engines (joint project)

In their effort to boost the export potential of the Yak/AEM-130, the Yakovlev OKB and Aermacchi studied the possibility of re-engining the aircraft with F124 turbofans developed by the International Turbine Engine Co. (ITEC), a subsidiary of the AlliedSignal company. The engines were to be derated to 25 kN if the plan was approved. The dissolution of the partnership between Yakovlev and Aermacchi put an end to these joint studies, but the engines in questions were, in fact, chosen by the Italian company for its own version of the Yak-130 (see below).

Aermacchi M-346 advanced trainer (Italian derivative of the Yak-130)

As mentioned above, in December 1999 the partnership between Yakovlev and Aermacchi was discontinued and the two companies went their separate ways in the further development of the basic design. Aermacchi reportedly received a full set of documentation on the Yak-130 from the Russian side under an agreement between the governments of Russia and Italy; the transfer of documentation was regarded as a part payment of the Russian debt to Italy.

On the basis of this documentation the Italian company reworked the design to suit the requirements of prospective Western customers and produced its own 'Westernised' version of the Yak-130 which received the designation Aermacchi M-346. Externally this lead-in fighter trainer is very similar to the first Yak-130 prototype, but it features a complete structural redesign, a Western set of avionics and Western engines – the aircraft is powered by two ITEC F124-200 turbofans developed jointly by engine manufacturers of the US and Italy. Aermacchi is positioning the aircraft to meet the 12-nation AJEPT (Advanced European Jet Pilot Training) 'Eurotraining' requirement

Coded P.01, the first prototype of the M346 (c/n 6962) was completed on 7th June 2003. It was described as advanced and lead-in fighter trainer. Its first flight took place on 15th July 2004. When presenting it, Aermacchi accentuated such features of the aircraft as vortex aerodynamics, a digital fly-by-wire control system permitting a controlled flight at angles of attack up to 40°, modular avionics, a 'glass' cockpit and so on. The F124-200s engines ensure a thrust/weight ratio close to 1.

Under the terms of the Yakovlev/Aermacchi agreement, Aermacchi has the exclusive right to market its trainer in NATO member countries, while Yakovlev has the exclusive right to the CIS market. In other parts of the world the two companies will engage in free market competition.

Yak-130 in a light fighter version (project)

The use of the Yak-130 in a light fighter role was one of the options considered by the Yakovlev OKB. No details are available.

Yak-130 as a carrier-based trainer (project)

Among the Yak-130 versions studied in the Yakovlev OKB there was a carrier-based trainer intended for replacing the Su-25UTG in the Russian Navy, as well as for export.

Key Publishing

The first prototype Aermacchi M-346 nearing completion at the Venegono plant. The aircraft makes an interesting comparison with the definitive Yak-130 pictured at the top of the page.

Specifications of the Aermacchi M-346 trainer

Engine thrust at take-off, kN (lb)	27.92 (6,277)
Wing span	9.70 m (32 ft 0 in)
Wing area, m^2 (sq ft)	23.52 (253.16)
Maximum external load, kg (lb)	3,000 (6,608)
Maximum speed, km/h (mph)	1,083 (673) Mach 0.86
Initial rate of climb, m/sec (ft/min)	101 (20,000)
Service ceiling, m (ft)	13,635 (45,000)

Yak-130 four-seat trainer derivative (project)

Prospective versions of the Yak-130 include a four-seat version featuring two cabins, each with two seats. It is intended for training pilots and navigators of military transport aircraft and bombers, as well as pilots of civil airliners.

Yak-130 four-seat COD version (project)

A shipboard (carrier on-board delivery) version of the four-seat Yak-130 derivative is intended for quick delivery of people and small cargoes to aircraft carriers. This aircraft is expected to have a maximum cruising speed of 650 km/h (404 mph) and a practical range of 2,000 km (1,243 miles).

Yak-131 single-seat light strike aircraft (project)

A report published in mid-2003 stated that there were plans envisaging the development of a dedicated strike derivative of the Yak-130. This machine designated Yak-131 will be a lightweight single-seat aircraft possessing 90 to 95% commonality with the basic Yak-130. The strike version will have a combat radius extended to 1,000 km (620 miles). It will be provided with a built-in 30-mm cannon. The weapon complement will also include the Vikhr' (Whirlwind) laser-guided anti-tank missiles.

The light strike aircraft version of the Yak-130 will feature armour protection making use of lightweight composite materials; this will afford greater survivability. A new avionics suite will enable the aircraft to tackle its combat tasks day and night under adverse weather conditions. It is to feature the *Moskit-2* (Mosquito) radar (an improved version of the Moskit radar developed by NPO Fazotron). The Yak-131 was envisaged as forming part of a unified reconnaissance and strike system comprising also a reconnaissance/target-designator aircraft, as well as an unmanned reconnaissance drones. The Yak-131 was expected to carry up to 3,000 kg (6,615 lb) of armament externally on nine weapons pylons. External stores options comprised up to eight free-fall bombs weighing up to 250 kg (551 lb) each, up to four guided air bombs weighing up to 500 kg (1,102 lb) each, the Kh-25ML guided missile or (on export versions) the AGM-65 Maverick missile, as well as short- or medium-range air-to-air guided missiles.

L-15 supersonic combat and training aircraft (Chinese project)

As officially acknowledged by the Yakovlev OKB, its members are taking part in the consultative role in design work on the L-15 – a Chinese dual-role (combat/trainer) twin-turbofan aircraft supposed to attain a maximum speed of Mach 1.4. Judging by a published photo of a model, the layout of this aircraft has very much in common with the Yak-130. Commenting on this co-operation, a senior official of the Yakovlev OKB pointed out that the Yak-130 had been expressly designed as a subsonic aircraft and the supersonic capability for the L-15 was due to Chinese requirements, not to Yakovlev's design philosophy.

Yak-152 primary trainer (project)

Drawing upon its long-standing experience in the design of training aircraft, the Yakovlev OKB embarked on design of a new piston-engined two-seat primary trainer intended to supplant the venerable Yak-52. The project was revealed in 2002. Development of the machine proceeded in a direct competition with Sukhoi's Su-49 intended for the same role and possessing a similar configuration and comparable design performance. The OKB's concept envisages using the Yak-152 in conjunction with the Yak-130 jet-powered trainer, the two aircraft forming an integrated training system in the primary and advanced trainer roles respectively. The instrument panel arrangement of the Yak-152 is modelled on that of the Yak-130.

The Yak-152 is being designed on the basis of the Yak-54 sports aircraft and embodies design features incorporated into the Yak-54M project. Indeed, models of the Yak-54M and the Yak-152 displayed at Farnborough International 2000 and MAKS-2003 respectively appear to be nearly identical, featuring only minor differences in the landing gear. As distinct from the basic Yak-54 (but in common with the Yak-52 and Yak-54M), the Yak-152 will have a tricycle undercarriage. A distinctive common feature of the Yak-54M and Yak-152 projects is the blown cockpit canopy covering the two seats. The front part of the cockpit canopy is moulded in one piece, without the usual visor and its framing, and affords excellent visibility. The aircraft is equipped with the SKS-94M ejection seat system.

The aircraft is of all-metal construction and will be powered by the 360-hp M-14X radial engine driving a three-blade propeller. It will have an AUW of 1,320 kg (2,910 lb), an empty weight of 948 kg (2,090 lb), a wing span of 8.82 m (29 ft), a length of 7.72 m (25 ft 4 in), and a wing area of 13 m^2 (140 sq ft). Design performance includes a maximum speed of 500 km/h (311 mph), a flaps-up stalling speed of 110 km/h (68 mph), maximum G-loads of +9/–7, a range of 1,000 km (620 miles) and a take-off/landing run of 175/360 m (575/1,180 ft).

In 2002 the Yak-152 lost to the Su-49 in a tender announced by the Russian Air Force for a primary trainer suited to its needs. However, the future of the Yak-152 seems to have been secured. In May 2004 Commander-in-Chief of the Ukrainian Air Force Lieutenant-General Yaroslav Skal'ko told journalists that there were plans envisaging the use of the Yak-152 for primary training of pilots in the Ukraine. The aircraft is to be manufactured at production facilities of Odessaviaremservice – an aircraft repair enterprise in Odessa belonging to the Ukrainian Ministry of Defence (which, incidentally, will also assemble the Yak-130 combat trainer). The Ukrainian side, according to Skal'ko, has already reached an agreement with the Yakovlev OKB about setting up its branch in Odessa to supervise this production.

Paul Jackson

A model of the projected Yak-152 trainer. A very similar model was displayed earlier as the Yak-54M.

POST-WAR AIRLINERS

8

Yak-40 feederliner

The Yak-40 occupies a special place among the progeny of Aleksandr S. Yakovlev and his design bureau. It was the OKB's first venture into designing relatively large commercial aircraft which became a stepping stone for its subsequent involvement in the creation of big jet airliners. Development of the Yak-40 was initiated by the joint directive passed by the Communist Party Central Committee and the Council of Ministers on 30th April 1965. The design work and prototype construction took just one year, and flight testing of the new machine commenced in October 1966.

The aircraft was intended to supersede the piston-engined Lisunov Li-2, Il'yushin IL-14P/IL-14M and Antonov An-2P airliners on local services, carrying 27-32 passengers over distances up to 1,500 km (930 miles). The designers put an emphasis on the aircraft's reliability, simplicity of construction and operation, safety, and ability to operate into and from unprepared and austerely equipped airfields.

These considerations dictated the choice of layout – a straight-wing monoplane powered by three jet engines. The wings had relatively large area – 70 m^2 (753.5 sq ft), making it possible to do without overly complex high-lift devices. In combination with three specially developed 1,500-kgp (3,310-lbst) Ivchenko AI-25 turbofans providing a high power/weight ratio they ensured that the Yak-40 would have good field performance. The wings had an aspect ratio of 9, a thickness/chord ratio of 15% at the root and 10% at the tip. They were built in port/starboard halves joined at the centreline. The wings featured three-section tracked slotted flaps, with two-section manually actuated ailerons with inset hinges outboard of these.

The powerplant comprised two engines in pylon-mounted nacelles flanking the rear fuselage, with the third engine buried in the fuselage and breathing through a dorsal air intake with an S-duct, the intake trunk blending into the fin. This arrangement made for safety, enabling the aircraft to continue flight confidently, should one of the engines fail. The aft position of the engines, coupled with the unswept wings, resulted in the passenger cabin being placed ahead of the wings. It was accessed via hydraulically powered airstairs in the rear fuselage underside, making the aircraft independent of airport facilities. The same purpose was served by the installation of an Ivchenko AI-9 auxiliary power unit (APU) which supplied compressed air for engine starting and, additionally, could provide air-conditioning for the cabin on the ground. The cabin was pressurised and environmentally controlled by systems using bleed air from the engine compressors. Hot bleed air was also used to de-ice the wing and tail leading edges.

The Yak-40 had a T-tail, with variable-incidence unswept tailplanes placed on top of a swept fin; the sweep was dictated by the need to increase the tailplane arm. The fin/stabiliser junction was enclosed by a bullet fairing projecting beyond the fin leading edge. Control surfaces comprised the manual tabbed rudder and inset-hinge manual elevators. The tricycle undercarriage comprised inward-retracting main units and a forward-retracting nose unit, each unit having a single wheel, levered suspension and an oleo-nitrogen shock strut. The outer faces of the mainwheels remained exposed when retracted, just like on the Boeing 737.

A comprehensive avionics set was provided, enabling night/all-weather operation. It included a *Groza-40* (Thunderstorm-40) weather radar, an MVL-40E autopilot (MVL = ***mes**nyye voz**doosh**nyye **lin**ii* – commuter air routes), an MRP-56P marker receiver, an ARK-9 ADF, GMK-1G gyrocompass, an RV-3M radio altimeter, an SO-70 IFF transponder, R-860-11, ***Lan**dysh-5* (Lily of the valley) and *Bak**lan**-5* (Cormorant) radios and other items.

Low-rate initial production of the Yak-40 was launched at the Saratov aircraft factory No.292 in 1967, when the first orders were placed in anticipation of the certification, which took place in 1968. Revenue services with Aeroflot began on 30th September 1968. Over the years, the Yak-40 firmly established itself on the Soviet Union's local routes, adding a new quality to the short-haul and regional air transport in the country. By 1980 the Aeroflot operated Yak-40 services to 276 domestic destinations.

Production of the Yak-40 was terminated (prematurely, in the opinion of some specialists) in 1981, when a total of more than 1,000 machines had been manufactured. In addition to being widely operated on the Soviet Union's internal routes, the Yak-40 was exported to a number of countries. The recipients included Afghanistan, Angola, Bulgaria, Cuba, Czechoslovakia, Republic of Equatorial Guinee, Ethiopia, Federal Republic of Germany, Hungary, Italy, Cambodia, Laos, Madagascar, Poland, Syria, Vietnam, Yugoslavia, Zambia. France and Bangladesh are also mentioned in some sources. Later, through re-export and leasing agreements, the Yak-40s found their way into the fleets of air carriers of several other countries, Iran being an example (to say nothing of the now independent former Soviet republics). The type soldiers on to this day, albeit in reduced numbers.

Described below are the many versions of the Yak-40, including projects.

Yak-40 prototypes

Bearing the non-standard four-digit 'custom' registration CCCP-1966, the first of a series of prototypes was built at MMZ No.115. It made its first flight on 21st October 1966 at the hands of OKB test pilots A. L. Kolosov and Yu. B. Petrov. It appears that later this machine was reregistered CCCP-19661 – presumably after some modification which included adding a service door on the port side immediately aft of the flightdeck.

The first machine was followed by further prototypes and Saratov-built pre-production examples registered CCCP-1967, -19672, -19673, -19674, -19675, -19676 and -1968. (There is no pictorial evidence for the existence of CCCP-19671 which may have been the former CCCP-1967 renumbered; likewise, CCCP-1968 is listed in some sources as CCCP-19681.) Improvements and alterations were progressively introduced on each of the prototypes, partly reflected in minor external differences. For example, from the third aircraft onwards a longer and more pointed radome was fitted to reduce drag. The abovementioned service door was introduced on CCCP-19661, but was absent on CCCP-1967 and -19672, only to

Yakovlev OKB

Above: Yak-40 CCCP-19661 pictured during tests; it already has the longer radome and port side service door. Note the 'London – Copenhagen – Amsterdam – Paris' legend below the flightdeck windows.

Yefim Gordon archive

Above: Another aspect of CCCP-19661, this time at Moscow-Domodedovo during the 9th July 1967 airshow.

Yefim Gordon archive

Above: CCCP-1967, the second prototype, banks away from the camera ship, showing the exposed mainwheels. Note the original short radome and the hinges of the escape chute door under the nose.

Yefim Gordon archive

Yak-40 CCCP-19672 lacked mainwheel well doors

reappear on subsequent prototypes and production machines. CCCP-1966, -1967 and -19672 had a small hatch closed by a forward-hinged door low on the port side of the nose (immediately aft of the nosewheel well) allowing the crew to bail out in an emergency; this was omitted on subsequent aircraft. CCCP-19672 had no doors on the main landing gear oleos. CCCP-19674 introduced a small fillet between the lower lip of the No.2 engine air intake and the fuselage, while CCCP-19675 featured a raked inlet to the No.2 engine, which was adopted for production machines. Most of the prototypes had eight windows on each side, but CCCP-19675 had nine windows to port, and CCCP-1968 had as many as 11 windows on the starboard side!

Yak-40 production aircraft ('1st series')

Production Yak-40s from different batches differed in the number of windows (eight or nine on each side) and the number and location of emergency exits. They differed also in some other respects, featuring a constant stream of improvements from batch to batch. For the sake of convenience, with regard to these changes, production aircraft are sub-divided into four groups (1st, 2nd, 3rd and 4th series) and accordingly described here, although these are **not** to be regarded as official designations.

The first series comprises batches 00 through 17 of the Yak-40 in its original configuration with a bullet fairing on the fin and no thrust reverser on the centre engine. Initial series aircraft differed from the prototype mainly in having a longer cabin, extended to a length of 7.7 m (23 ft 2$\frac{3}{8}$ in) as compared to 6.7 m (21 ft 11$\frac{25}{32}$ in) on the prototype. Airline-configured Yak-40s of all series could have interiors with various seating arrangements for 24, 27 or 32 passengers three-abreast; the number of seats was reduced in VIP configurations described below.

Initial operation of the Yak-40 was accompanied by an incident in which one of the main undercarriage units broke off as the aircraft lined up for take-off. Investigation showed fatigue cracks in the mainwheel legs of machines that had logged more than 100 landings. To rectify the fault, the leg was redesigned and suitably strengthened.

Yak-40 feederliner ('2nd series')

In 1969 production Yak-40s began to be fitted with a clamshell thrust reverser on the centre engine. This reverser was retrofitted to a Yak-40 bearing the out-of-sequence registration CCCP-87791 (c/n 9920203 – that is, *izdeliye* 9, year of manufacture 1969, 2nd quarter (April-June), 02nd aircraft in Batch 03) which was used by the OKB as a demonstrator and development aircraft. Apparently the thrust reverser was introduced on new-build Yak-40s from Batch 18 onwards. Most initial production machines were retrofitted in a similar way. Initially the reverser was fitted to AI-25 engines rated at 1,500 kgp (3,300 lbst); subsequently reverser-equipped machines were powered by AI-25T engines delivering 1,750 kgp (3,860 lbst). Interestingly, the reverser was designed by the Yakovlev OKB, not by the Ivchenko engine design bureau (OKB-478), because it formed part of the rear fuselage and was not integrated with the engine.

Other modifications included provision of an extra outer wing tank bay increasing fuel capacity to 3,910 litres (860 Imp gal), and strengthened main undercarriage. A ninth row of seats was added to provide a total of 27. CCCP-87791 was displayed at the 28th Paris Air Show in 1971 with a new interior for 40 passengers.

Yakovlev OKB

Above: This air-to-air of a prototype registered CCCP-1968 shows 11 cabin windows to starboard instead of the usual eight or nine and, oddly, no emergency exits.

Yak-40 feederliner ('3rd series')

From 1973 onwards production Yak-40s were cleared to significantly higher weights, and a new standard interior for 32 passengers was introduced. Externally the '3rd series' aircraft could be identified by the deletion of the tailplane bullet fairing and the nib-type nacelle/pylon fairings. (This, however, appears controversial, since the bullet fairing vanished as early as Batch 20, which was mostly manufactured in December 1971, not 1973!)

Yak-40 feederliner ('4th series')

In 1975 the final variety of the Yak-40 *sans suffixe* entered production, with the payload increased to 3,200 kg (7,055 lb) and with interiors arranged for two passenger classes or for rapid conversion from passenger to cargo. Upturned strips were added to the aileron trailing edges and downturned strips to the outboard flaps.

Yak-40E export version

According to the Czech aviation historian Václav Nemeček, the designation Yak-40E (***eks**portnyy*) applied to an export version fitted with Soviet avionics; it is nor clear in what respects it differed from the version intended for Soviet customers, apart from customer-specified interiors. A Western source cites a specific example of the Yak-40E, namely HA-YLR (c/n 9541044) delivered to Hungary and used there as a navaids calibration aircraft. The E suffix was not worn visibly by this machine, nor, for that matter, by any other known example.

Specifications of the Yak-40 ('3rd series')

Crew	2
Length overall	20.36 m (66 ft 9½ in)
Wing span	25.0 m (82 ft 0¼ in)
Wing area, m² (sq ft)	70.0 (753.5)
Powerplant	AI-25
Rating, kgp (lbst)	3 x 1,500 (3 x 3,300)
Empty weight, kg (lb)	8,580 (18,819)
Normal TOW, kg (lb)	16,100 (35,500)
Cruising speed, km/h (mph)	510 (317)
Landing speed, km/h	180 (112)
Service ceiling, m (ft)	8,100 (26,570)
Time to 6,000 m (19,680 ft), minutes	16.1
Range, km (miles)	1,800 (1,118)
Take-off run, m (ft)	715 (2,346)
Landing run, m (ft):	
without thrust reversal	750 (2,460)
with thrust reversal	550 (1,805)

Dmitriy Komissarov

Above: Pictured at Moscow/Vnukovo-1 on 18th February 2002, RA-87625 was one of the last surviving Yak-40s in the version combining the fin top bullet fairing with a thrust reverser on the centre engine.

Yefim Gordon archive

Yak-40 'Salon 1st Class' RF-87659 in 1992, wearing the unofficial RF- registration prefix, 'Yak Aircraft Corporation' titles and the titles of the Russian airline Transaero for which it was then operated.

Dmitriy Komissarov

Above: New-build Yak-40 'Salon 1st Class' aircraft had a window in the service door. RA-87655 of Interprobusiness-M Ltd. (in ex-Vostsibaero colours) is seen at Moscow/Vnukovo-1 on 6th May 2003

Yefim Gordon archive

Above: The other VIP version is called Yak-40 'Salon 2nd Class' and has one window 'plug' on each side. RA-87993 is operated for the Belogor'ye Volleyball Club of Belgorod.

Yefim Gordon archive

The Yak-40K convertible version has a port side cargo door and usually wears appropriate nose titles. Here, CCCP-98110 is seen parked at Zhukovskiy, its home base.

Yak-40P airliner

This unofficial designation was used by the Polish Air Force for those of its Yak-40s which had been delivered in 28-, 30- or 32-seat configuration. The P stood for *pasażerski* (passenger, used attributively) – that is, not executive/VIP.

Yak-40EC export version

In 1970-75 three Yak-40s were delivered to Aertirrena of Italy. This export version, designated Yak-40EC (export, Collins), was fitted with Collins avionics, including the AP-104/FD-108 integrated radio navigation system. The aircraft were registered I-JAKA, I-JAKE and I-JAKI (c/ns 9020409, 9141418 and 9141518). Outwardly the Yak-40EC was identifiable by two 'towel rail' aerials low on the fuselage sides immediately ahead of the wings; these replaced the 'hockey stick' aerials of the domestic version.

Yak-40FG export version

In 1972 the Soviet Union delivered five Yak-40s to General Air of West Germany. In some Western publications they were reported as Yak-40FGs (FG apparently stands for Federal Germany). Their equipment presumably included Western avionics. The aircraft were initially registered D-COBA through D-COBE (later re-registered D-BOBA through D-BOBE; c/ns 9211420, 9211520, 9230122, 9230323 and 9230623). In late 1975 they were returned to the USSR, with the exception of D-BOBD which overshot the runway at Saarbrücken on 9th December 1975 and was damaged beyond repair. The re-export Yak-40FGs were operated by various MAP enterprises.

Yak-40 VIP versions ('Salon 1st Class', 'Salon 2nd Class')

The Yak-40 'Salon 1st Class' and Yak-40 'Salon 2nd Class' are VIP versions identifiable by the white plugs in the foremost cabin windows where coat closets or galley equipment are located. The 'Salon 1st Class' has such plugs (opaque plastic replacing the usual Perspex) in the two foremost windows on each side and usually has an extra emergency exit around the sixth window on the starboard side. Additionally, aircraft built in 'Salon 1st Class' configuration have an extra window in the service door which is normally windowless. CCCP-87792 (c/n 9910303) may be the prototype of this version.

The Yak-40 'Salon 2nd Class' has one plug on each side in the foremost cabin window (some, however, also have plugs in the rearmost window on each side – that is, four plugs). It is usually provided with an extra emergency exit around the fourth window on the starboard side.

These are normally aircraft from production batches without the bullet fairing on top of the fin, but some early production Yak-40s featuring this bullet were also outfitted as VIP aircraft (one such machine with a reverser on the centre engine had a white plug in the rearmost, ninth, window on each side).

Many Yak-40s have been converted from VIP to airline configuration and vice versa, confusing things considerably. For instance, OK-EEC (c/n 9440737) was manufactured as a Yak-40 *sans suffixe* for ČSA Czechoslovak Airlines on 24th October 1974 and refitted as a Yak-40 'Salon 2nd Class' for the Gor'kiy aircraft factory after being re-exported to become CCCP-87211.

Yak-40S

This was the unofficial designation used by the Polish Air Force for the examples delivered in Yak-40 'Salon 2nd Class' configuration; the S stood for 'salon'.

Yak-40K convertible passenger/cargo aircraft

This was a convertible version for cargo and passenger/cargo transportation (K stands for *konverteeruyemyy*, convertible). It was fitted with an upward-opening cargo door on the port side, aft of the service door. The cargo door was 1.6 m (5 ft 3 in) wide and 1.5 m (4 ft 11 in) high.

The Yak-40K prototype (CCCP-87597, c/n 9110117) was converted from a regular passenger example; in 1975 it was reregistered CCCP-87490 for some reason. The aircraft later became an exhibit of the Soviet Air Force Museum (now Central Russian Air Force Museum) in Monino.

Mass production of the Yak-40K began in January 1976, starting with Batch 45; convertible and pure passenger aircraft were interspersed (that is, no batch consisted entirely of Yak-40Ks). In addition, years before new-build Yak-40Ks started rolling off the assembly line, several earlier production machines were converted into this model for use by Aeroflot in accordance with the CofM Presidium's Defence Industry Commission ruling No.18 of 23rd January 1974 and MAP order No.84 of 14th February 1974. Known examples are CCCP-87621 (c/n 9130219), CCCP-87317 (c/n 9331629), CCCP-87321 (c/n 9332029), CCCP-87849 (c/n 9331830), CCCP-87383 (c/n 9311432) and CCCP-87385 (c/n 9411632); the latter was manufactured as a Yak-40 *sans suffixe* on 31st January 1974 and converted on 10th September that year.

In Aeroflot service Yak-40Ks were sometimes operated in passenger configuration; they could also carry 10 to 18 passengers and a cargo of up to 2,150 and 1,150 kg (4,740 and 2,535 lb) respectively. The Yak-40K is sometimes erroneously referred to as the 'Yak-40T'.

Yuriy Kirsanov

Above: Yak-40D RA-88170 (with appropriate nose titles) of AeroTex on finals to Moscow-Sheremet'yevo. This aircraft crashed fatally at Sheremet'yevo on 9th March 2000.

Dmitriy Komissarov

Yak-40KD RA-21506 in the old livery of CNG-Transavia sits at Zhukovskiy on 23rd September 1999.

Yak-40K VIP conversions

In post-Soviet years some Yak-40Ks were refitted to a 'Salon' configuration, making the cargo door non-functional. Known examples include UN 87213, RA-87224, UR-87590 through UR-87592, RA-87906, RA-87912, RA-87952, RA-87953, RA-88287, RA-88298, RA-88306, RA-98109 through RA-98111 and UR-CLA.

Yak-40: recent upgrades

In the course of the recent 10 to 15 years the remaining Yak-40s have been the subject of refurbishment undertaken at the Smolensk aircraft factory with a view to extending their service life. The refurbishment has been accompanied by various modifications dealing primarily with the aircraft's avionics and interior design, often in connection with converting the aircraft into executive/VIP versions. To cite an example, in 1997 Yak-40 RA-88236 (c/n 9640551) operated by the Lipetsk Air Enterprise was leased by the co-located Stinol refrigerator plant and converted into an executive aircraft. This aircraft does not have the window 'plugs' characteristic of the Yak-40 'Salon'.

Avionics upgrades have been effected on some Yak-40s in order to make them suitable for operation on international routes.

Another line in upgrading was associated with the need to extend the Yak-40's range. The result was the emergence of the Yak-40D and Yak-40KD described below.

Yak-40D and Yak-40KD long-range versions

The Smolensk aircraft plant equipped some Yak-40s with two additional integral tanks in the wings increasing the fuel capacity by 1,600 kg (3,530 lb) to 6 tons (13,230 lb). This increases the aircraft's range from 1,200 km (746 miles) to 2,500 km (1,554 miles). After this upgrade the basic (all-passenger) Yak-40 is redesignated Yak-40D, while a similarly modified convertible Yak-40K becomes a Yak-40KD (D stands for ***dahl'**niy*, long-range). In the process of this modification the aircraft is also fitted with a single-point pressure refuelling system.

As a rule, Yak-40Ds are converted from Yak-40 'Salon' VIP jets. Known examples are RA-87807, RA-87311, RA-87334, RA-88170 and RA-87970; the latter is operated by the Rossiya State Transport Co. (the Russian government flight). Only two Yak-40KDs, RA-21506 and EX-87228, are known so far.

Yak-40 navaids calibration aircraft

Several Yak-40s in Russian service and abroad were converted into navaids calibration aircraft. This version sported various additional antennas. An example of this version is RA-87251 (c/n 9310826), which was used by the Siberian Aeronautical Research Institute (SibNIA – *Si**beer**skiy na**ooch**no-is**sled**ovatel'skiy insti**toot** avi**ah**tsii*) in Novosibirsk in the 1980s and 1990s.

One Yak-40 in this version (the above-mentioned Yak-40E HA-YLR) was delivered to Hungary and operated there as a navaids calibration aircraft by Malév Hungarian Airlines; similarly, LZ-DOE (9521441) was operated as a calibrator by Balkan Bulgarian Airlines in a non-standard livery with yellow undersides. The Czechoslovak Airports Authority (ČSSDL – *Československá sprava dopravních letišti*) operated Yak-40 OK-DHA (c/n 9341230) as a calibrator; the aircraft

Yefim Gordon archive

Above: CCCP-87536 was converted into the Yak-40 'Liros' weather research aircraft. It is seen here at Minsk-2 airport, seat of the Aircraft Overhaul Plant No.407, prior to redelivery.

was later reregistered OK-DYA, eventually going to Slovakia as OM-DYA. Yak-40 SP-GEA (c/n 9230224) was used in the same role by the Polish Institute of Aeronautics (*Instytut Lotnictwa*). Two machines served in this role in Yugoslavia. One of them was Yugoslav Air Force '71505', later registered YU-AKV. The other one, serialled 71503 (c/n 9222020), later registered YU-AKT, was used for navaids calibration in 2002 after years of service in the VIP transport role. Both featured a 'towel rail' aerial above the centre engine air intake.

Yak-40 Akva (Aqua) weather research aircraft

This designation was allocated to Yak-40 CCCP-87937 (c/n 9740856) manufactured in standard configuration on 4th January 1978 after it had been converted into a weather research aircraft. The job was performed by Civil Aviation Aircraft Repair Plant No.407 at Minsk-2 airport (ARZ No.407; ARZ = ***a****viare****m****ontnyy* ***z****avo****d***) in 1986. In addition to atmospheric studies, the machine was intended for provoking rainfall by disseminating chemical agents in clouds and for ensuring protection of crops from hail in a similar way, as well as for monitoring air pollution. A special feature of the Yak-40 Akva was the installation of underwing SPVG-40 containers with 26-mm (1-in) PV-26 silver iodide flares. A model of this aircraft shown at an exhibition in Moscow in 1985 featured a nose probe and various sensors, as well as a scabbed-on fuselage dispenser.

The Yak-40 Akva served for several years with the Ukrainian Research Institute under the USSR State Committee on Hydrometeorology. Later it was reconverted to passenger configuration. In the 1990s it was in service with Armenian Airlines as EK-87937 and occasionally leased to various Iranian air carriers. On 17th May 2001 it crashed in Iran as EP-TQP while operated by Queshm Air.

Yak-40 Meteo weather research aircraft

The Yak-40 Meteo (CCCP-87537, c/n 9520242) was a multi-purpose weather research aircraft fitted with various sensors, data recording and processing equipment which enabled the researchers on board to study, among other things, the atmospheric turbulence and the effects of solar radiation on the atmosphere. The equipment also included devices for causing artificial rainfall. Again, it was eventually stripped of mission equipment, serving with Kazakhstan Airlines and subsequently Taraz Wings as UN 87537.

Yak-40 Liros weather research aircraft

One more weather research version, the Yak-40 Liros (CCCP-87536, c/n 9522041; the meaning of the word 'Liros' is unknown), was intended for studying atmospheric conditions and clouds in a broad range of altitudes and for testing the various methods of causing artificial rainfall. The Yak-40 Liros was operated by GosNII GA from Sheremet'yevo-1. This aircraft, too, was reconverted to standard and became EK-87936 with Armenian Airlines.

Yak-40 Shtorm meteorological research aircraft

The Yak-40 Shtorm (Sea Storm) was developed jointly by the High-Mountain Geophysics Institute and the ARZ No.407. The aircraft registered CCCP-87992 (c/n 9541644) was intended for meteorological research and was equipped with various sensors and data recording devices, as well as with devices for actively influencing atmospheric processes. At the beginning of the 1990s the aircraft was reconverted to passenger configuration and transferred to Stavropol' Avia.

Yak-40 Fobos (Phobos) testbed and research aircraft

This research aircraft converted from a production airliner registered CCCP-87304 (c/n 9322028) was developed for the Lavochkin Science and Production Association which, after Lavochkin's demise, had switched to missile and space technology. The Yak-40 Fobos was to be used as a testbed for various equipment items developed by this enterprise for manned and unmanned space vehicles. In addition, it was intended to undertake a broad spectrum of research associated with the study of natural resources, environment, atmospheric phenomena, natural and man-made anomalies, solar radiation, air and water pollution and so on. The results of this research provided the basis for economic analysis of the environment situation with a view to evolving the necessary practical measures. The aircraft filled this role in the late 1980s/early 1990s, whereupon it was reconverted to passenger configuration and was operated since 1993

Yefim Gordon archive

CCCP-88238 in its days as the bizarre Yak-40REO research aircraft, showing the ventral equipment housing, the nose probe and the observation blister.

as RA-87304, first by Aeroflot and then by Belgorod Avia.

Yak-40DTS military transport/ CASEVAC aircraft prototype

A military transport and casualty evacuation (CASEVAC) variant was developed under the designation Yak-40DTS (*de**sahnt**no-**trahns**portnyy i sani**tar**nyy* – assault transport/ambulance, used attributively). According to a Polish source, a single prototype was built in 1971; this may be CCCP-87756 (c/n 9030112) – a picture exists of this aircraft in flight with the airstairs removed, apparently for the purpose of dropping something through the resulting hatch. The suitably modified passenger cabin could accommodate 20 paratroopers.

Yak-40REO avionics testbed

Registered CCCP-88238 (c/n 9640951), the Yak-40REO (*ra**h**dioelek**tron**noye obo**roo**dovaniye* – electronic equipment, or avionics) was one of the numerous avionics testbeds developed by the Moscow-based NPO Vzlyot (Take-off; NPO = *na**ooch**no-proiz**vod**stvennoye obyedi**nen**iye* – Scientific & Production Association). Used for testing such avionics items as inertial navigation systems, altimeters, Doppler systems and so on, the aircraft was readily identifiable by a huge ventral fairing ahead of the wings and a large observation blister with a rear fairing to starboard. By 1992 these excrescences had been removed and CCCP-88238 had been used in a different programme, as evidenced by the flare dispensers attached under the wing roots and to the underside of the airstairs.

Yak-40-25 avionics testbed

A production Yak-40 (identity unknown) was converted into a testbed for verifying the mission avionics of the Mikoyan Ye-155R high-speed reconnaissance aircraft (the future MiG-25R). The nose of the MiG-25R housing an SRS-4A (*izdeliye* 30A) ELINT pack was grafted on the airliner's forward fuselage, replacing the Groza-40 weather radar.

Jak-40LL engine testbed

The first Yak-40 delivered to CSA Czechoslovak Airlines (OK-EEA, c/n 9431436) was transferred to VZLÚ (*Vyzkumní a Zkusební Létecky Útvar* – Flight Test and Development Unit) in Prague, the local equivalent of LII, on 1st February 1982. There the aircraft was converted into a flying testbed for the 1,822-ehp Motorlet M-602 turboprop engine developed by the engine department of the Let (the Czech word for 'flight') enterprise at the end of the 1980s. This engine, designed to drive a V-518 five-bladed propeller, was to power the Let L-610 twin-engined short-haul airliner.

Yakovlev OKB

Above: This desktop model depicts a 'Soviet JetStar' – a proposed four-engined version of the Yak-40. Note the 'pen nib' fairings between the engine nozzles and the shape of the tailcone.

Yakovlev OKB

Another projected spin-off of the Yak-40 was to feature swept wings and stabilisers and be powered by two high-bypass turbofans.

The M-602 was installed in the nose of the Yak-40 instead of the radar – the only possible location (considering its engine testbed role, the aircraft did not really need a weather radar anyway). Initially the aircraft, which was known as the Jak-40LL (Czech spelling; LL = *letajici labor* – 'flying laboratory', that is, testbed), retained its original registration and basic CSA livery but was subsequently reregistered OK-020; the digits after the OK- nationality prefix instead of the usual letters indicated test or development status.

Yak-40 target tug conversion

Perhaps the most unusual role the Yak-40 ever had to fill was target towing for the benefit of anti-aircraft artillery crews. Four of the eighteen Yak-40s delivered to the Polish Air Force's 36th Special Transport Air Regiment (36. *Specjalny Pułk Lotnictwa Transportowego*) were converted to target tugs as a replacement for the obsolete IL-28 bombers used previously. When the IL-28s had been retired, the Polish Air Force was left with no target tugs at all.

The conversion involved removal of the airstairs, installation of a WH-3000-Jak winch (WH = *winda holownicza* – towing

Above: This artist's impression depicts a proposed V/STOL version of the Yak-40 with lift engines in wing pods. The all-too-slim engine nacelles are probably just an artist's error.

winch) with a steel cable 3,000 m (9,840 ft) long in the rear part of the cabin and modification of the rear pressure bulkhead to incorporate an outlet for the winch cable and a viewing window. The winch worked with either a Polish RSS-5C sleeve-type target, which was 5 m (16 ft $4\frac{27}{32}$ in) long and 4 m (13 ft $1\frac{31}{64}$ in) in diameter when fully inflated, or a similar Swedish target. The winch operator's workstation was located in the former baggage compartment at the rear of the cabin. Access to the flightdeck and cabin was now via the port side service door by means of a ladder. The whole system was developed by the Polish Air Force's overhaul plant No.3 (WZL-3, *wojskowe zakłady lotnicze* – Military aircraft factory) in Radom which undertook the conversion in 1978.

The four aircraft thus modified – '031 Red', '032 Red', '034 Red' and '035 Red' (c/ns 9331029 through 9331329) – entered service with the 19th Target-Towing Squadron (19. *Lotnicza Eskadra Holowniczа*) in Słupsk on 12th July 1980, 25th July 1980, 31st December 1979 and 16th April 1980 respectively. Take-off and landing was performed with 35 m (115 ft) of the cable paid out; during the target run the aircraft flew at 200-600 m (660-1,970 ft) and up to 350 km/h (217 mph) with 1,400 m (4,590 ft) of the cable paid out. In addition to target-towing duties, the Yak-40s of the 19. LEH were used for testing new models of parachutes and checking the operation of air defence radars. The unit was disbanded in March 1999, the four aircraft being reconverted to standard and returned to the 36. SPLT.

Yak-40M (project)

Apparently this designation was used for several different projects. According to one source, this was a projected stretched version for forty to forty-four passengers. A model photo showed one of these studies with four engines, slatted wings and twin-wheel main undercarriage units.

A different Yak-40M was mentioned by A. Aksyonov, Deputy Minister of Civil Aviation of the USSR, in his article in the *Grazh**dahn**skaya Avi**ah**tsiya* (Civil Aviation) magazine in 1973, where he claimed that testing had been completed of *'the Yak-40M, a version of the baby jet airliner with a cabin layout for 32-36 passengers'*.

Yak-40 with four engines (project)

Studies were made of a Yak-40 version powered by four jet engines of an unspecified type, as evidenced by a photo of a desktop model from the OKB. The engines were arranged in pairs flanking the rear fuselage in the manner of the Il'yushin IL-62 (or perhaps, considering the size of the aircraft, in the manner of the Lockheed L-1329 JetStar). Due to the deletion of the buried engine the centre air intake and S-duct gave way to a

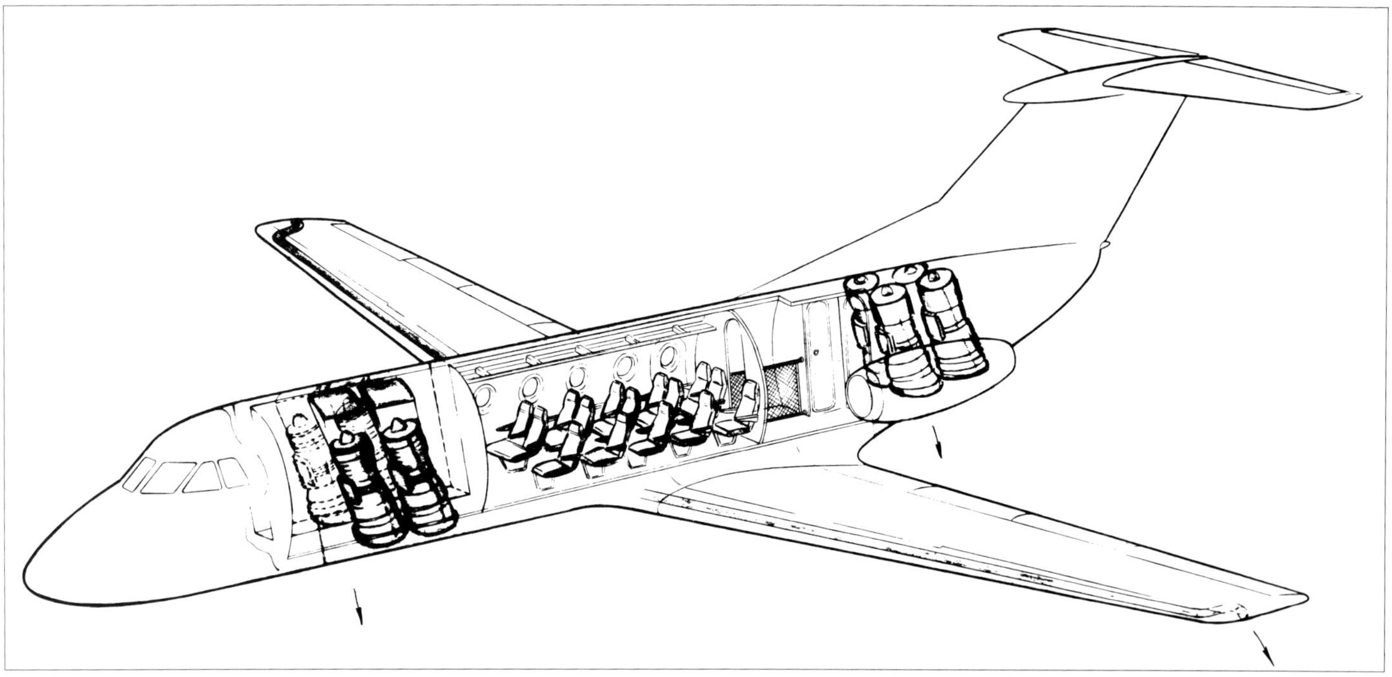

A cutaway drawing depicting an altogether different V/STOL version of the Yak-40, with two cruise engines four pairs of Kolesov RD36-35 lift-jets buried in the fuselage fore and aft of the small cabin. Lateral stability at the hover was ensured by reaction control nozzles under the wingtips. Note the shape of the tail.

dorsal fin, while the nozzle was replaced by a tailcone. The rest of the airframe appears to have remained unchanged. The symbolic registration CCCP-1970 on the model may well hint at the year of the project's inception. It is unclear whether this particular version is connected with the Yak-40M (see above).

Yak-40P long-haul airliner (project)

In 1977 the Yakovlev OKB developed the project of a long-haul version of the Yak-40 airliner dubbed Yak-40P (the meaning of the suffix is unknown). To extend the range, the aircraft was to be fitted with two additional fuel tanks accommodated in fairings ahead of the wing leading edge, each of the tanks holding 1,000 litres (220 Imp gal). No prototype was built.

Yak-40TL (project)

During the Farnborough International '92 airshow the Yakovlev OKB and the US company Textron Lycoming (now Allied Signal) reached an understanding on a joint programme under which the Yak-40 would be fitted with two Lycoming LF507-1N fuel-efficient high-bypass turbofans instead of three thirsty AI-25T engines. The centre engine air intake was replaced by a dorsal fin and the APU was relocated to the tailcone supplanting the thrust reverser.

The new powerplant held a promise of increasing the aircraft's cruising speed by 10% and its range by 25%. A sub-variant with the wing integral tankage of 6 tonnes would have a range of no less than 2,500 km (1,550 miles) with a payload of 1,000 kg (2,205 lb). The re-engined aircraft (provisionally designated Yak-40TL, for Textron Lycoming) would also undergo an avionics upgrade based on the use of state-of-the-art equipment. According to one account, the original project would have had slightly swept wings. The project was not implemented.

Yak-40 with swept wings and two high-bypass engines (project)

This version featured moderate sweep on the wings and horizontal tail (about 20°); the three AI-25s were replaced by two high-bypass turbofans flanking the rear fuselage. No details are known.

Yak-40 VTOL version with buried lift engines (project)

The OKB studied several V/STOL versions of the Yak-40. One of them had two turbofans of an unspecified type for propulsion, with an APU in the tail, and eight Kolesov RD36-35 lift jets in fireproof bays fore and aft of the passenger cabin. The seating capacity was reduced to 15 (five rows of three).

Yakovlev OKB

Above: A model of the first airliner project to be designated Yak-42. A strange beast, to say the least.

Yak-40 VTOL version with lift engines in wing pods (project)

This version appears to have retained the standard three-engine powerplant for forward propulsion. Six RD36-35 lift engines were arranged in two wing-mounted nacelles, the main spar passing between the centre and rear jets on each side. The wing size was optimised for cruise flight.

Yak-42 turboprop airliner (project, first use of designation)

The Yak-42 designation was used by the Yakovlev OKB for the first time in 1967 for a project of a civil transport powered by a single turboprop engine in the nose driving four-bladed contra-rotating propellers. The aircraft featured large-span unswept wings, apparently with anhedral, and a bicycle undercarriage with outrigger struts. Records suggest the engine was a 15,000-ehp Kuznetsov NK-12, but the surviving desktop model, with the pre-1959 style registration CCCP-Л-1970, shows an aircraft much too small for that engine.

Yak-42 medium-haul trijet airliner (second use of designation)

The Yak-42 designation was reallocated in 1972, when the Yakovlev OKB started design work on a short-to-medium-haul jet transport in the 100/120-seat class in response to a specification issued by Aeroflot. The design objectives included simplicity of operation, ability to operate in and out of restricted airfields with poor surfaces and limited facilities, and good economy. This was the largest aircraft designed by the OKB in terms of size, weight and seating capacity. Sergey A. Yakovlev was programme manager at first but was later replaced by S. A. Andreyev.

On the way to the now familiar layout of this airliner the OKB tried and rejected several interim project versions. The first of these, a 114-seat airliner, resembled a scaled-up Yak-40 featuring wings with an unswept trailing edge and very slight leading-edge sweep. The powerplant comprised two Solov'yov D-30K low-bypass turbofans. However, the OKB gave up this version in

Yakovlev OKB

This desktop model bearing the registration CCCP-19671 represents the first step towards the 'real' Yak-42. The aircraft was to have unswept wings and stabilisers and two low-bypass turbofans.

Yakovlev OKB

Above: This model of 1973 (as indicated by the registration) is much closer to the actual aircraft. The shape of the nose and the fin top was to change yet.

Yakovlev OKB

Above: The first prototype Yak-42, CCCP-1974, awaits rollout at MMZ No.115 in Moscow. Note the large window in the starboard-side service door. The window arrangement is door+12+exit+1+exit+2.

favour of a project featuring a three-engine layout and based on the use of high-bypass turbofans (this was dictated by the need to achieve competitiveness with Western airlines powered by fuel-efficient high-bypass engines). The engine in question was the D-36 turbofan developed by V. A. Lotarev and rated at 6,500 kgp (14,330 lbst).

Initial studies of the three-engined version were based on the use of same wings as mentioned above, with slight leading-edge sweep only. The problem of choosing the optimum sweep angle was obviously a tough one, as evidenced by the unusual decision to build the first prototype with wings swept back 11° at quarter-chord and the second with a wing sweep of 25° (this was changed to 23° prior to prototype construction).

Yak-42 first prototype

Bearing the custom registration CCCP-1974, the first prototype (presumably c/n 01001 – Batch 01, 001st aircraft in the batch) made its first flight on 7th March 1975. This aircraft had the 11° wings, a short fuselage with seventeen cabin windows each side, with all four emergency exits above the wings, no leading-edge devices, a bullet fairing at the fin/tailplane junction, twin-wheel main gear units with large wheels (the sidewalls of the outer mainwheels remained exposed when retracted, Yak-40 style) and several systems, such as tail unit de-icing, inoperative. The forward entry door on the port side was positioned rather low, featuring an external hydraulic actuator and hinges to serve as an escape hatch/slipstream deflector during flight tests.

Yakovlev OKB

A beautiful air-to-air of CCCP-1974. Note how low the entry door is located; folding airstairs were to be installed at this location. The first prototype had 11° wing sweep at quarter-chord.

The following description refers both to the prototypes and to production machines.

The Yak-42 is a low-wing monoplane of all-metal construction with moderately swept two-spar wings built in three sections. The two main spars are supplemented by a transverse beam passing unbroken through a cutout in the fuselage. The wing structure incorporates a transverse beam (false spar) carrying the inboard flaps of rectangular shape placed at 90° to the fuselage. Early aircraft had plain flaps preceded by a small fixed slat and with a drooping aft portion. Most production Yak-42s had tracked slotted flaps achieving higher maximum lift. The outer wing panels carry one-piece flaps, with upper-surface spoilers ahead of them, and ailerons which, in most production aircraft, are split into inboard and outboard sections, each with its own actuator. Initial aircraft had no leading-edge devices, but later machines featured full-span leading-edge slats in six sections.

The circular-section semi-monocoque fuselage of 3.8 m (12 ft $5^{39}/_{64}$ in) diameter houses a passenger cabin which extends aft of the wings. The flat rear pressure bulkhead incorporates the main cabin door, leading to the powered airstairs built into the rear fuselage underside.

Two engines in pylon-mounted nacelles flank the rear fuselage; the third engine is buried in the rear fuselage and fed by an S-duct from a dorsal inlet. The engines have separate core and bypass flow nozzles; there are no thrust reversers. The centre air intake trunk blends into the vertical tail with a one-piece rudder. Placed on top of it is the fully powered variable-incidence tailplane

Yakovlev OKB

Above: The second prototype Yak-42, CCCP-1975, introduced the definitive 23° wing sweep. The Yak-42 was the last type to feature the striped rudder characteristic of Yakovlev aircraft since the 1930s.

Yakovlev OKB

Above: The second prototype in flight. The window arrangement is door+13+exit+1+exit+4; there is still a large window in the service door.

Yakovlev OKB

The fourth prototype, CCCP-1977, was used for high-alpha/low-speed trials and was fitted with a spin recovery parachute above the centre engine nozzle. The window arrangement was changed *again* (door+11+exit+1+exit+6)! Note the photo calibration crosses on the fuselage.

Yefim Gordon

Above: Yak-42 CCCP-42302 (possibly ex-CCCP-1977) at the Soviet Air Force Museum in Monino with the paintwork still intact.

carrying one-piece elevators with trim tabs.

The cockpit was designed for two-crew operation down to ICAO Cat II minima initially. Standard equipment included an SAU-42 autopilot, a Groza-42 weather radar, an *Ol'**kha**-1* (Alder) navigation system, a ***Vey**er-M* (Fan) blind navigation aid, a VSP-1-6 speed/altitude autohold module and a Koors-MP-70 (Heading) navigation and blind landing system.

All fuel is housed in integral tanks located in the wing torsion box; their capacity was initially 15,795 litres (3,474.4 Imp gal), increased to 23,175 litres (5,098 Imp gal) from 1981 onwards.

Yak-42 second prototype

The second prototype, again with a non-standard registration (CCCP-1975; c/n 01002?), differed in having a longer cabin with 19 windows on each side. It was fitted with the 23° wings which still had no leading-edge devices, but the flaps were of the area-increasing type, moving on short tracks with external fairings.

Yak-42 third prototype

The third prototype, CCCP-1976 (c/n 01003), had the 23° wings, full de-icing, fairing discs over the outer wheels and other minor improvements. It still retained the bullet fairing and had four overwing emergency exits; however, these were positioned further forward. Some accounts say this machine was used for high-alpha/low-speed tests and was fitted at that stage with a spin recovery parachute in an aft fuselage fairing, but an available picture shows the fourth prototype had this feature (perhaps both aircraft had it?). Eventually this machine was reregistered CCCP-42303 in the series allocated to production Yak-42s. CCCP-42303 was demonstrated at Le Bourget where it sported the same kind of emergency actuator on its front port side door as seen on the first prototype.

Yak-42 fourth prototype

The fourth prototype, CCCP-1977 (the last in the four-digit 'custom' series), was allegedly manufactured by plant No.475 in Smolensk, although this is unconfirmed. It was used for exploring the aircraft's spin characteristics. To this end the aircraft was equipped with an anti-spin parachute housed in a container mounted above the centre engine nozzle. The aircraft bore cruciform photo-theodolite calibration marks on the fuselage.

Yak-42 initial production versions

The Yak-42 was ordered into series production initially at Saratov and Smolensk. The first machines started leaving the assembly lines in Saratov in April 1978; the first production aircraft manufactured on 26th April 1978 (c/n 11820101 – that is, *izdeliye* 11, year of manufacture 1978, 2nd quarter, 01st aircraft in Batch 01) was registered CCCP-42300. At first, the production tempo was very slow, ten machines having flown by mid-1981. The first scheduled passenger service took place on 22nd December 1980 on the Moscow-Krasnodar route.

Production aircraft differed externally from the prototypes and featured minor modifications with each successive series. One of the changes was the introduction of redesigned wingtips having a curved (Küchemann) form, which slightly increased the wing span. Another was the substitution of the twin mainwheels with four-wheel bogies which had wheels of the same diameter as on the nose gear unit and were

Yakovlev OKB

CCCP-42306 was the first Yak-42 to feature four-wheel bogies on the main gear units, a feature that changed the aircraft's looks completely. Note also the modified (penultimate) fin top and the window arrangement (door+6+door+4+exit+1+exit+6); the forward overwing exits appear to be inoperative.

enclosed by large doors when retracted. The latter two changes were first observed on CCCP-42306 (c/n 11840202) manufactured on 3rd December 1978. This aircraft also lacked the fin bullet fairing, resulting in a fin shape akin to late-series Yak-40s. Yet the design proved unsatisfactory, and most later Yak-42s had a kinked fin leading edge with zero sweep at the top to increase the chord ahead of the tailplane.

Saratov-built machines featured two distinctive configurations. The first few aircraft retained the bullet fairing and had two overwing emergency exits and one door on each side; most Yak-42s had no bullet fairing (that is, had the cranked fin leading edge) and one overwing emergency plus two doors on each side of the fuselage. Some early examples, though, were something in between, featuring two doors and two exits on each side; known examples are CCCP-42530 and CCCP-42531 (c/ns 11120204 and 11130304) manufactured on 16th June and 16th September 1981 respectively. Machines built at Smolensk (reportedly series production there was limited to eight examples) matched the first of the three configurations.

Initial Aeroflot aircraft had a 120-passenger seating arrangement with three plus three abreast. It was soon superseded by an arrangement for 104 passengers, comprising a 96-seat tourist-class main cabin and a first-class section for eight.

The Yak-42's service career was nearly wrecked by a fatal accident which took place on 28th June 1982 near Mozyr'. A failure of stabiliser screw jack mechanism caused the horizontal tail of Yak-42 CCCP-42529 (c/n 11040104) to break away and the uncontrollable aircraft plunged into the ground, killing all on board. As a result, the type was grounded. The design fault was identified and remedied, and Aeroflot resumed Yak-42 services in October 1984. However, this accident seriously damaged the Yak-42's reputation and sales prospects, and its production never reached the optimistic target figure of 2,000, being limited to a fraction of that number.

Yak-42ML-1 and Yak-42ML-2 medium-haul airliner

Apart from the purely domestic version, since 1981 the Yak-42 was produced in two modified versions with modified avionics fits enabling operation on international services for use on international services. The two versions, called Yak-42ML-1 and Yak-42ML-2, (the ML suffix stood for *mezhduna**rod**nyye* ***lin**iï* – international services) differ in details of equipment and furnishing, for example, in the languages of notices to passengers and servicing engineers. The Yak-42ML entered service in July 1981 on the Leningrad-Helsinki route.

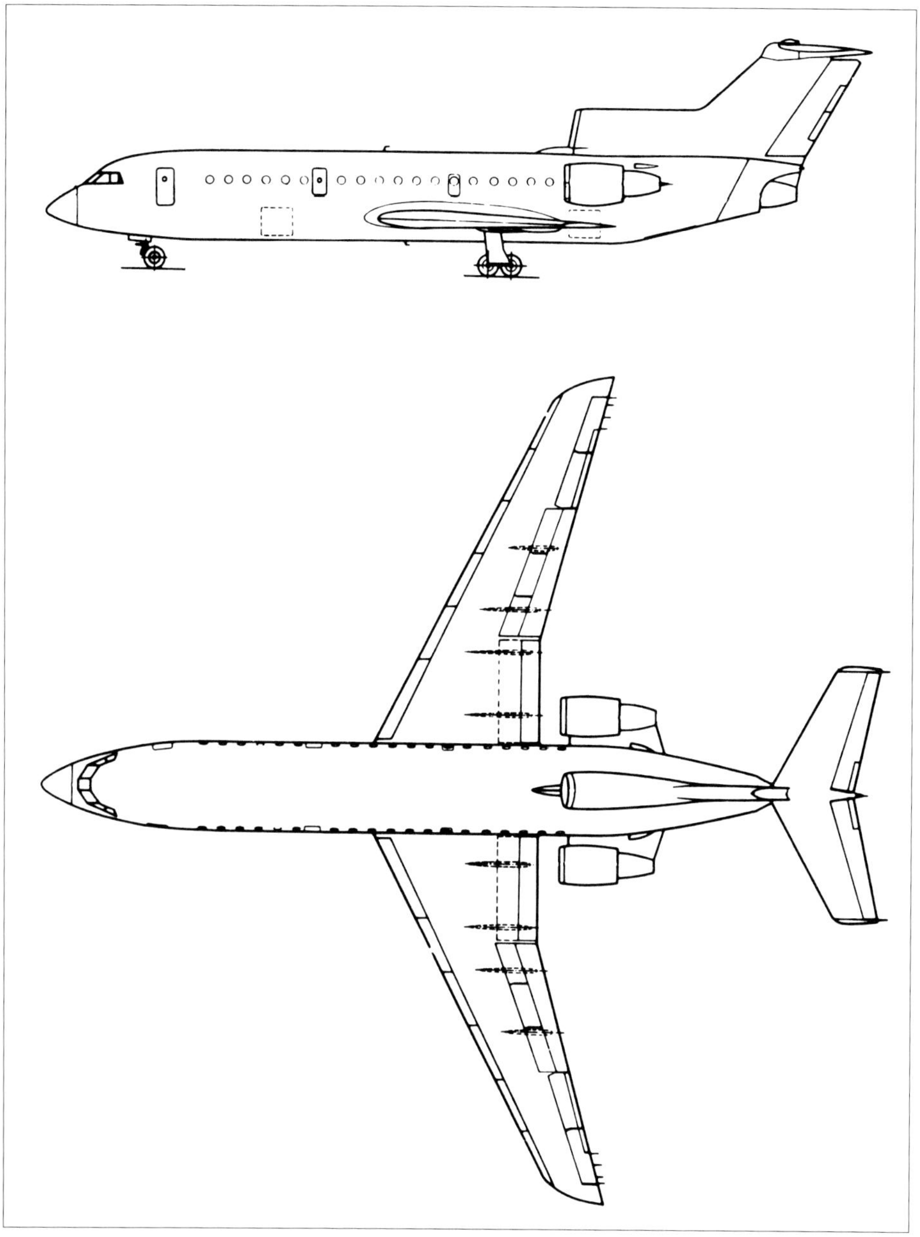

Above: A three-view drawing of a typical production 'four-door' Yak-42.

Yakovlev OKB

Yak-42 CCCP-42313 delivered to the Bykovo United Air Detachment of Aeroflot's Central Regions Civil Aviation Directorate shows the ultimate fin design with the characteristic kink at the top.

Yakovlev OKB

Above: This model depicts the Yak-42A as originally envisaged, with two 'plugs' of unequal size inserted into the fuselage fore and aft of the wings (the window arrangement is door+9+door+6+exit+7).

Yak-42D production airliner

An extended-range version designated Yak-42D (***dahl'****niy*, long-range) was brought out with the fuel tankage increased by 3,100 litres (682 Imp gal). This became a standard version which has been produced at Saratov since 1991, with assistance from the Smolensk plant which supplies wing panels. By 1st January 1995 a total of 185 Yak-42s had been built, including 105 Yak-42Ds. Some Yak-42s *sans suffixe* were upgraded to D standard during overhauls.

The Yak-42/42D achieved a modest degree of success on foreign markets: four machines were sold to Cuba and eight to China. In the 1990s some Yak-42/42Ds came to be operated by Lithuania and other newly independent former Soviet republics; in addition, a few were leased to air carriers in Iran, Pakistan and other countries.

Yak-42A medium-haul airliner

In December 1992 the Yakovlev OKB completed a prototype of this version, converted from an originally standard Yak-42D (RA-42423, c/n 4520424216606, f/n 0913). This version was fitted with Allied Signal avionics. The aircraft was demonstrated at the 40th Paris Air Show in June 1993 and the MAKS-93 airshow in Zhukovskiy on 31st August/5th September 1993. Later that year the aircraft was redesignated Yak-142 (see below). In August 1999 it was shown at MAKS-99 – again as a Yak-42A!

Another Yak-42A, RA-42424 (c/n 4520421502016), was displayed at the MAKS-95 (22nd-27th August 1995); it was fitted with a VIP interior and wore the corporate colours of the LUKoil petroleum company. For a while it was leased by Kazakhstan as UN 42424 for use as the presidential aircraft of Nursultan Nazarbayev.

(Note: In June 1982 the Saratov factory introduced a new c/n system for the Yak-42. In the case of RA-42423, 452 is a code for plant No.292, 042 means *izdeliye* 42 (a zero is added to keep the usual three-digit code format), the seventh and eighth digits (42) mean fourth quarter of 1992, and the remaining five digits are the 'famous last five' meaning nothing at all. Hence there is also a fuselage number; in this case, f/n 0913 means 09th aircraft in Batch 13. From Batch 16 onwards the plant thankfully reverted to the 'rational' approach; thus, the last seven digits of c/n 4520421502016 mean 'first quarter of 1995, 02nd aircraft in Batch 16'.)

Yefim Gordon archive

Yak-42D RA-42435 displays the smart livery of the now-extinct Oryol-Avia. 'Oryol' is Russian for 'eagle', hence the lustily grinning eagle on the tail.

Yak-42B (Yak-42BK) (project)

This was a projected version with Bendix/King avionics. The existence of this version was first revealed at the *Konversiya-93* exhibition in Moscow.

Yak-42K convertible version (project)

The Yakovlev OKB developed a convertible version designated Yak-42K with a large side cargo door. This version was unveiled at the *Konversiya-93* exhibition. The intention was to licence overhaul plants to convert Yak-42s to Yak-42Ks; none have been modified.

Yak-42D VIP/airborne command post version

A Yak-42D registered RA-42441 (c/n 4520421402018) was specially outfitted for the Russian Ministry of Emergency Situations (EMERCOM) as an airborne command post. The aircraft is used for controlling the forces and means engaged in dealing with emergencies, guiding search-and-rescue teams to people in distress and relaying the commands issued by ground guidance posts. In addition, it is used for transporting rescue teams and experts from other ministries and agencies (up to 48 persons) and as a VIP transport. This machine wears full EMERCOM colours and the name *Valeriy Chkalov* (a renowned Soviet test pilot).

Another Yak-42D, RA42446 (*sic* – no hyphen or space after the prefix; c/n 4520423308017) was operated by Vnukovo Airlines for the Federal Agency of Government Communications and Information (FAPSI – *Fede****rahl'****noye* ***a****gh****ent****stvo pra****vi****tel'stvennoy* ***svya****zi i infor****mah****tsii*) in 1999-2000. The aircraft featured special secure communications equipment the presence of which was revealed by a non-standard unswept blade aerial atop the fin. After the demise of Vnukovo Airlines and the dissolution of FAPSI the aircraft joined the EMERCOM fleet in May 2001 and was christened *Vladimir Kokkinaki* after another famous Soviet test pilot, retaining the basically white colours with a blue/red cheatline.

Yak-42D VIP conversions

In recent years several Yak-42Ds have been outfitted with posh VIP interiors, custom-built to suit the requirements of corporate customers. Known examples are RA-42323 (c/n 4520423402116, f/n 0106) operated by Aviamost (the flying division of Most Group that was headed by the notorious tycoon Vladimir Goosinskiy, now a wanted man), RA-42330 (c/n 45204225050122, f/n 0806) belonging to Sirius-Aero, RA-42344 (c/n 4520422708295, f/n 0208) operated by Yak-Service for the IFK-Jet leasing company (the Il'yushin Finance Co.), RA-42368 (c/n 4520422914166, f/n 0610) which was operated by Moscow-based Centre-Avia for the Murmansk Metal Foundry for a while, RA-42442 (c/n 4520422002019) of Gazpromavia and ER-YCA (c/n 4520424306017), the Moldovan presidential jet.

Yak-42D derivative with revised wings (project)

In August 2002 Aleksandr Yermishin, General Director of the Saratov aircraft plant, said in an interview that the plant was ready to start the manufacture of a new airliner based on the Yak-42D. This upgraded version was to receive new 'high-speed' wings having the same area as those of the Tu-334. The advantages of the new wings, he said, had been confirmed by wind tunnel tests at TsAGI. The cruising speed of the rewinged Yak-42D would be raised to 850 km/h (528 mph) and the range with 100 passengers would be 4,120 km (2,560 miles). The new aircraft, similar in performance to the Tu-334, would be offered to customers at a lower price compared to Tupolev's machine. However, there has not been any practical follow-up to this statement.

Yak-42D-90

In August 1999, during the MAKS-99 airshow at Zhukovskiy, the then General Designer of the Yakovlev OKB Aleksandr Dondukov revealed that the OKB was working on a project of converting the Yak-42D and Yak-42A into the Yak-42D-90 model which was intended to carry 90 passengers over 4,000 km (2,486 miles). The new version was to inherit all the improvements introduced on the Yak-42A.

Yak-42-100 and Yak-42D-100

Reports on this version are somewhat confusing. One Russian source describes the Yak-42-100 as a one-off test aircraft with AlliedSignal avionics, in parallel with the similarly equipped Yak-42A and Yak-142. Another Russian source makes a brief mention of 'Yak-42-100 (Yak-142)'.

A Western source cites the Yak-42D-100 as a Yak-42D equipped with a Western avionics suite, 'also dubbed the Yak-142'. This sounds plausible since the features quoted for the Yak-42D-100 (Western navigation, radio, TCAS II and an electronic flight instrumentation system, or EFIS, integrated with the Russian-made autopilot) closely match the description of the Yak-142.

Yak-142 (Yak-42A, Yak-42-100, Yak-42D-100) airliner

As stated above, Yak-42A RA-42423 was redesignated Yak-142 in late 1993. Retaining the AlliedSignal avionics fit introduced on the Yak-42A, the Yak-142 in its final configuration also incorporated a number of modifications to the airframe as compared to the preceding Yak-42D. Airframe changes include flight spoilers/lift dumpers deployed symmetrically for faster letdown, an increased range of flap settings for improved take-off and go-around climb, especially in hot-and-high conditions, and the port side passenger doors enlarged to 1.7 x 0.85 m (5 ft 6$\frac{15}{16}$ in x 2 ft 9$\frac{15}{32}$ in) for compatibility with boarding fingers at major airports.

The two-crew cockpit is totally upgraded with AlliedSignal avionics, including a KNS-660 flight data system, a KAD-480 air data system, an RDR-4A weather radar, an EFIS-10 electronic flight instrumentation system, a KAH-460 attitude/heading reference system, a KHF-950 HF radio, an RTA-44A VHF radio, an RIA-35A ILS, an RVA-36A VOR, DMA-37 DME, a DFA-75A ADF, an ALA-52A radar altimeter, a TRA-67A ATC/SIF

Dmitriy Komissarov

Above: Yak-42D RA-42368 of Tsentr-Avia (c/n 4520422914166, f/n 0610) was briefly operated as an executive jet for the Murmansk Metal Foundry (MMK). Note the satellite communications antenna.

Dmitriy Komissarov

Above: Another VIP-configured Yak-42D, RA-42442 of Gazpromavia (c/n 4520422002019), with a totally different interior. It is seen here at the Civil Aviation-2000 show at Moscow-Domodedovo.

Yakovlev OKB

The cabin of a Yak-142, looking aft.

Yakovlev OKB

Above: A desktop model of the radically redesigned Yak-42T transport aircraft project. Note the main gear units retracting aft into Tupolev-style fairings on the wing trailing edge.

Yakovlev OKB

Above: The same model in modified form as the Yak-44, featuring standard Yak-42 wings and inward-retracting main gear units. Note the forward sweep on the dorsal engine pylon.

transponder and a TPA-81A TCAS-II. The autopilot is the Russian SAU-4201 with an autothrottle, and the dual environmental systems are calibrated in metric and Imperial measures.

Further improvements have been made to furnishings, soundproofing and passenger facilities, and measures have been taken to ensure quicker turn-around and easier maintenance. Standard seating is for 120, or 104 two-class, or 80 three-class. As stated above, the Yak-142 is sometimes referred to as Yak-42-100 or Yak-42D-100.

Yak-42-200 airliner (project)

In this version the passenger capacity of the Yak-42 was to be increased by stretching the fuselage by 6.05 m (19 ft 10$\frac{3}{16}$ in). The project remained on the drawing board.

Yakovlev OKB

An air-to-air of the one-off Yak-42F research aircraft. The machine wore red/white Aeroflot polar colours.

Yak-42R avionics testbed

This aircraft (identity unknown) was a specially converted machine which was used for flight-testing the radar of the Yak-141 V/STOL fighter.

Yak-42T and Yak-44 cargo aircraft projects

This projected cargo version dating back to 1975 was to be a much-modified derivative of the baseline Yak-42. Its distinguishing features included unswept wings, a full-section rear cargo ramp/door, twin fins and rudders at the tips of low-set stabilisers, the centre engine being placed in a nacelle on a short pylon above the rear fuselage. Initial design studies of this version envisaged aft-retracting main gear units styled on those of some Tupolev designs, with four-wheel bogies somersaulting during retraction to lie inverted in fairings extending beyond the wing trailing edge. In its later project configuration, this aircraft had inward-retracting main gear units and the designation was changed to Yak-44 (which was later used for a different aircraft).

Yak-42F research aircraft

A production Yak-42 registered CCCP-42644 (c/n 4520424914090, f/n 0410) was converted into a special research aircraft for earth resource/geophysical survey, as well as for environmental monitoring and atmospheric research. High-definition electro-optical sensors were carried in the passenger cabin and in large cylindrical pods under each wing just outboard of the undercarriage units, the front portions rotating to bring the terrain below into the field of view. This aircraft was shown at the 39th Paris Air Show in June 1991.

Yak-42E (Yak-42LL, Yak-42E-LL) engine testbed

A production Yak-42 (CCCP-42525, c/n 11030703) was converted into a testbed for flight-testing the Lotarev D-236T geared propfan derived from the core of the D-36 turbofan. Developed by the 'Progress' Zaporozh'ye Engine Design Bureau (ZMKB Progress), it was intended both for the Antonov An-70 transport (which has since materialised) and for projected rear-engined propfan airliners (which have not).

The test engine rated at 10,850 shp was mounted in place of the No.3 (right-hand) D-36. The engine drove SV-36 contra-rotating tractor propellers of 4.2 m (13 ft 9$\frac{11}{32}$ in) diameter developed by the Stoopino Machinery Design Bureau. The SV-36 had glassfibre blades with a hollow composite spar and integrated electric de-icing threads. The front and rear rows had eight and six blades respectively, running at 1,100 and 1,000 rpm respectively; the 100-rpm difference was intended to reduce noise and vibration. To ensure the necessary clearance between the propellers and the fuselage, a new pylon was used for mounting the engine.

The aircraft first flew on 15th March 1991; three months later it was exhibited at the 39th Paris Air Show as the Yak-42E-LL. By August 1997 CCCP-42525 had been withdrawn from use at Zhukovskiy.

Rolls-Royce Tay-powered Yak-42 (project)

Studies were made in 1990 for equipping the Yak-42 with Rolls-Royce Tay turbofans featuring thrust reversers. The project was not implemented.

Yakovlev 150-seat airliner (1988 project)

A cryptic reference to this project appeared in *Flight International* on 17th September 1988. The news item ran as follows: *'Tupolev and Yakovlev are to offer rival designs for a new 150-seat airliner destined for Aeroflot service around 1995. The new aircraft would have a range of 1,500-2,000 km, and replace Yak-42s and ageing Tu-134s, says Vladimir Kravchuk, a senior official with the Soviet Ministry of Aviation Industry.'*

Yak-42M airliner (first project)

In 1989 the Ministry of Civil Aviation formulated an SOR for a radical upgrade of the Yak-42D designated Yak-42M (*modern-**izee**rovannyy* – upgraded) – the first project to bear this designation. Retaining the general arrangement of the production version, the Yak-42M featured a 4.62-m (15 ft $1\frac{57}{64}$ in) fuselage stretch; two unequal plugs were inserted fore and aft of the wings, the larger one forward. A desktop model of the Yak-42M showed the following window arrangement on both sides: door+9+large exit+12+two small exits+16; the port side entry door was larger than the service door to starboard, measuring 1.83 x 0.85 m (6 ft 0 in x 2 ft $9\frac{15}{32}$ in) for compatibility with boarding fingers. As on the Boeing 727-200, the rear airstairs doubled as a tail support when extended to prevent the aircraft from falling over on its tail.

The wings swept back 25° at quarter-chord were completely new, featuring winglets, a high aspect ratio (10.5) and a taper of 3.5; they embodied new supercritical airfoils developed jointly by TsAGI and the Yakovlev OKB. The new wings were expected to give the aircraft a lift/drag ratio of up to 18.

The machine was to be powered by three Muravchenko D-436M turbofans – a derivative of the Lotarev D-36 uprated to 7,500 kgp (16,540 lbst) for take-off, with a

Yakovlev OKB

A display model of the Yak-42M as originally envisaged. The stretched fuselage and new high aspect ratio wings with winglets are evident, as are the translating thrust reverser cowls on all three engines. The port side baggage door is obviously painted on in error.

Specifications of the Yak-42 Family

	Yak-42M (SOR data)	Yak-42M (baseline)	Yak-42M (150 seats)	Yak-42D
Powerplant	3 x D-436M	3 x D-436M	3 x D-436M	3 x D-36
Engine thrust, kgp (lbst)	3 x 7,500 (3 x 16,540)	3 x 7,500 (3 x 16,540)	3 x 7,500 (3 x 16,540)	3 x 6,500 (3 x 14,330)
Length overall	–	40.47 m (132 ft $9\frac{5}{16}$ in)	40.47 m (132 ft $9\frac{5}{16}$ in)	36.38 m (119 ft $4\frac{9}{32}$ in)
Wing span	–	35.5 m (116 ft $5\frac{5}{8}$ in)	35.5 m (116 ft 5 $\frac{5}{8}$ in)	34.88 m (114 ft $5\frac{15}{64}$ in)
Wing area, m^2 (sq ft)	–	120 (1,290)	120 (1,290)	150 (1,613)
Seating capacity	156	156	150	120
Take-off weight, kg (lb)	–	63,000 (138,890)	63,500 (139,990)	56,500 (124,560)
Operating empty weight, kg (lb)	–	37,000 (81,570)	38,500 (84,880)	33,700 (74,290)
Payload, kg (lb):				
maximum	16,500 (36,375)	16,500 (36,375)	16,500 (36,375)	13,000 (28,660)
normal	14,820 (32,670)	14,820 (32,670)	14,250 (31,415)	10,500 (23,150)
with maximum fuel	–	9,600 (21,160)	8,600 (18,960)	4,600 (10,140)
Cruise SFC, kg/kgp.hr (lb/lbst.hr)	0.63 *	0.66 *	0.66 *	0.7 †
Fuel burn, g/seat-km (lb/seat-mile)	21 (0.0074)	21 (0.0074)	22 (0.078)	33.4 (0.118)
Thrust/weight ratio	–	0.346	0.346	0.345
Cruise altitude, m (ft)	11,100-11,600 (36,420-38,060)	11,100-11,600 (36,420-38,060)	11,100-11,600 (36,420-38,060)	9,100 (29,860)
Cruising speed, km/h (mph)	800 (496)	800 (496)	800 (496)	750 (465)
Effective range with reserves, km (miles):				
with maximum payload	1,850 (1,149)	1,850 (1,149)	1,500 (931)	1,850 (1,149)
with a normal payload	2,500 (1,552)	2,500 (1,552)	2,300 (1,428)	2,150 (1,335)
with maximum fuel	4,500 (2,795)	4,500 (2,795)	4,320 (2,683)	3,800 (2,360)
Field length, m (ft)	2,200 (7,220)	2,200 (7,220)	2,200 (7,220)	2,200 (7,220)

* At 11,000 m (36,090 ft) and Mach 0.75
† At 8,000 m (26,250 ft) and Mach 0.7

Yakovlev OKB

The Yak-42M soon evolved into a twinjet modelled on the such aircraft as the Tu-204 or the Airbus narrowbody family, receiving the new designation Yak-242.

specific fuel consumption of 0.63 kg/kgp·hr (lb/lbst·hr). All three engines were to be fitted with cascade-type thrust reversers in the bypass flow, the reverser cascades being closed by translating cowls – a necessity, given the higher landing weight.

The Yak-42M was to feature a new TsPNK-42M digital flight avionics/navigation suite (*tsifrovoy pilotazhno-navigatsionnyy kompleks*). The TsPNK-42M made use of state-of-the-art electronic components and was standardised with the flight avionics/ navigation suites of the Tu-204 medium-haul airliner and the IL-114 feederliner. It included a 'glass cockpit' with multi-function displays based on colour cathode-ray tubes and permitted automatic navigation along any of up to 70 pre-programmed routes, as well as ICAO Cat IIIA automatic approach. The OKB took great pains to improve cockpit ergonomics, bringing them on a par with such aircraft as the Boeing 737-300 and the Airbus Industrie A320.

Unlike the Yak-42, which had conventional powered controls, the Yak-42M had a quadruplex fly-by-wire control system. To enhance flight safety the system incorporated an automatic angle-of-attack, bank and G-load limiting feature; an active damping feature was envisaged to give a smoother ride in turbulence. Changes were also made to the hydraulics, with self-contained control surface actuators, which afforded a sizeable weight saving.

The airliner was intended for short-haul routes and was expected to carry 150-168 passengers over 2,500-4,000 km (1,554-2,486 miles) at a speed of 800-830 km/h (496-516 mph). In the baseline tourist-class seating configuration the passenger cabin would accommodate 156 seats (26 rows six-abreast) at 780 mm (30¾ in) pitch. The operational and maintenance procedures system developed for the Yak-42M was to ensure a 60,000-hour service life, with the first major overhaul due after 12,000 to 18,000 flight hours. The aircraft was expected to be 35-40% more fuel-efficient than the production Yak-42 and have 20-25% lower operating costs. Service entry was planned for 1994.

Yak-42M airliner (second project)

By 1992 the Yak-42M project had been completely reworked. In its new guise the airliner featured an altogether different layout from that of the original Yak-42. Emulating the configuration of the McDonnell Douglas DC-10 , it was to have two of the three D-436M engines pylon-mounted under the wings, the third engine being placed on top of the fuselage at the base of the fin so that it passed through the fin torsion box. The seating arrangements and basic design performance remained unchanged.

The Yak-42M did not materialise in this configuration either, but the project was developed further to become the Yak-242 (see below).

Yak-242 airliner (project)

Although derived from the Yak-42M, the Yak-242 actually became a completely new aircraft with even the fuselage diameter altered. This was a short-haul airliner with two turbofan engines in underwing pods (the interim three-engine configuration *à la* DC-10 was abandoned at an early stage).

The advanced development project was completed in the third quarter of 1992, and a full-scale mock-up of the aircraft was ready in 1993. The aircraft's wings featured new supercritical airfoils developed jointly with TsAGI; they had a sweep of 27° at quarter-chord, an area of 120 m^2 (1,290 sq ft) and an aspect ratio of 10.5. The fuselage of 4.0 m (13 ft 1⁴⁷⁄₆₄ in) diameter provided accommodation for 162 passengers in a tourist-class seating arrangement with a 780-mm (30¼ in) seat pitch. In an enhanced comfort configuration the passenger cabin would seat 156 passengers at an 810-mm (31²⁷⁄₃₂ in) seat pitch. There was also a high-density arrangement with a 750-mm (29½ in) pitch. There was no ventral airstair, embarkation and disembarkation taking place through two doors on the port side. The powerplant comprised two Aviadvigatel' (Solov'yov)

Specifications of the Yak-242

Engine type	Solov'yov PS-90A-12
Engine thrust at take-off, kg (lb)	2 x 12,000 (2 x 26,455)
Length	38.0 m (124 ft 8⅛ in)
Wing span	36.25 m (118 ft 11⅛ in)
Wing area, m^2 (sq ft)	120 (1,292)
Empty weight, equipped, kg (lb)	38,400 (84,660)
Payload, max, kg (lb)	18,000 (39,680)
All-up weight, kg (lb)	64,600 (142,420)
Cruising speed, km/h (mph)	800-850 (497-528)
Cruise altitude, m (ft)	11,100-11,600 (36,400-38,000)
Range with CAR reserves, km (miles):	
with maximum payload	1,600 (994)
with normal payload	2,700 (1,678)
with maximum fuel	5,000 (3,107)
Runway length, m (ft)	2,200 (7,220)

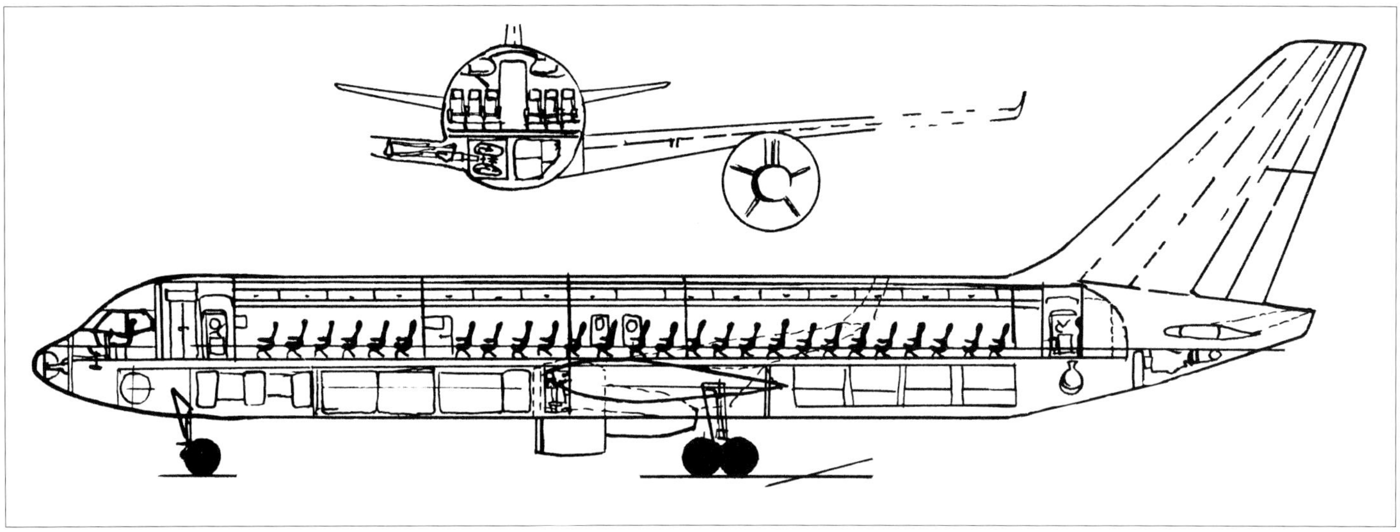

Above: A cutaway drawing of the Yak-46 version powered by ultra-high bypass (UHB) engines.

PS-90A12 turbofans with a take-off rating of 12,000 kgp (26,455 lbst). The engines had cascade-type thrust reversers and the nacelles featured noise-absorbing fan ducts.

Other features included an APU of a new type housed in the tailcone, four-wheel main gear bogies and twin steerable nosewheels, and digital avionics hat were to be certified to ICAO Cat. IIIa. A quadruple-redundant fly-by-wire (FBW) flight control system was to ensure automatic balancing and damping in all three control channels in all flight modes and prevent the aircraft from entering critical flight modes. To enhance passenger comfort and reduce airframe stresses during flight in turbulence, a so-called active damping system was to be used. The avionics set permitted the aircraft to be piloted by a crew of two, albeit there was provision for a third seat in the flightdeck.

The original plans envisaged the construction of prototypes in 1995, with the prospect of completing the flight test phase in 1997 and service introduction in late 1997. These plans turned out to be wishful thinking; financial difficulties prevented the machine from reaching the hardware stage, and the project was temporarily shelved. A few years later, it was resurrected and served as a basis for a project developed by the Yakovlev OKB jointly with the Il'yushin Aviation Complex under the designation MS-21 (see below).

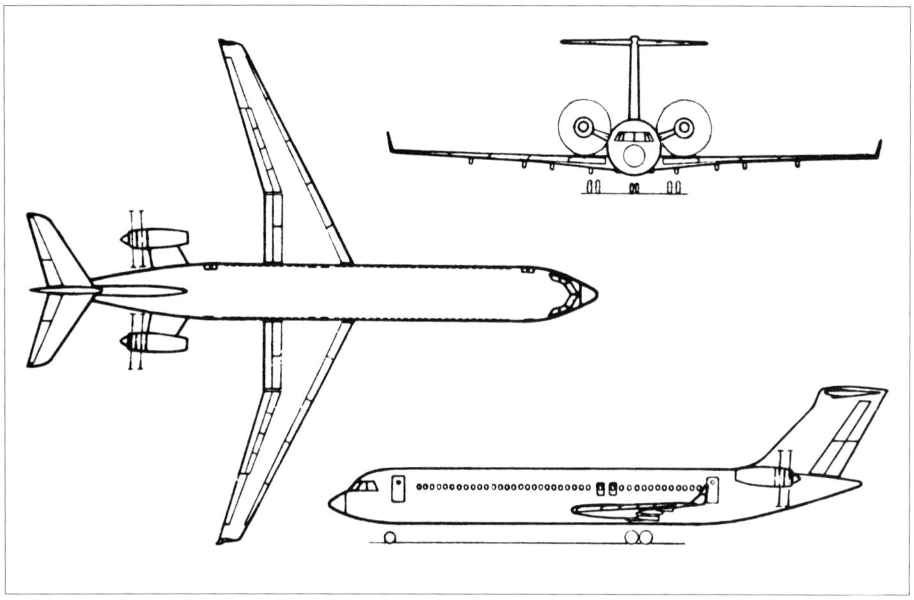

A three-view of the Yak-46 in propfan-powered, T-tail configuration.

Yak-46 turbofan-powered airliner (project)

In early 1990s the Yakovlev OKB embarked on the design of an airliner described as a derivative of the Yak-42M (and Yak-242). Bearing the designation Yak-46, it was projected in two basic versions differing considerably in the powerplant type and aerodynamic layout. One of them (sometimes referred to as the Yak-46-1) was to be powered by two advanced turbofan engines with a thrust rating of 11,000 kgp (24,250 lbst) for take-off. One of the options was the ZMKB Progress D-627 ultra-high bypass (UHB) turbofan then under development. The engines were to be mounted on pylons under the swept wings provided with winglets. The swept tail surfaces had a conventional layout.

The aircraft was intended for carrying 156 passengers at stage lengths up to 4,000 km (2,486 miles) with a maximum speed of 830 to 850 km/h (516 to 528 mph). The seating variants also included a 168-seat all-economy version and a mixed layout with 12 first-class seats and 114 economy-class seats. Provision was made for a convertible version and a transport/CASEVAC version. The aircraft was to be equipped with a fly-by-wire control system. An important aspect of the design was its high fuel efficiency, the fuel burn not exceeding 14-15 g/seat-km (0.049-0.053 lb/seat-mile).

Yak-46 propfan-powered airliner (project)

The other basic version of the Yak-46 (also known as Yak-46-2) was to use two Muravchenko (ZMKB Progress) D-27 unducted propfans with a still higher fuel efficiency, with a fuel burn not exceeding 13-14 g/seat-km (0.046-0.049 lb/seat-mile). The engines rated at about 14,000 ehp, or 11,200 kgp (24,690 lbst) thrust and driving Stoopino SV-27 contra-rotating pusher propellers were to be mounted on the rear fuselage in combination with a T-tail. In terms of seating capacity and range this version was virtually identical to the turbofan-powered variant and differed in having a slightly lower design speed of 800-830 km/h (497-516 mph).

Yakovlev OKB

An artist's impression of the propfan-powered Yak-46 project. So far none of the airliner projects utilising the T-tail/pusher propfan layout have come to fruition.

Specifications of the Yak-46 airliner variants

	Yak-46-I	Yak-46 II
Engine type	Two Kuznetsov turbofans	Two D-27 or D-227 propfans
Engine thrust at take-off, kg (lb)	11,000-kg (24,250-lb) class	–
Engine power, hp	–	c. 14,000 ehp
Length	38.8 m (127 ft 3½ in)	41 m (134 ft 6 in)
Wing span	36.25 m (118 ft 11¼ in)	35.5 m (116 ft 5¾ in)
Empty weight, equipped, kg (lb)	34,840 (76,806)	37,300 (82,230)
Payload, max, kg (lb)	17,500 (38,580 lb)	17,500 (38,580 lb)
All-up weight, kg (lb)	60,200 (132,715)	61,300 (135,140)
Cruising speed, km/h (mph)	830-850 (515-528)	830 (515)
Cruise altitude, m (ft)	11,100 (36,400)	11,100 (36,400)
Range with maximum payload, km (miles)	2,200 (1,367)	1,800 (1,118)
Runway length, m (ft)	2,100 (6,890)	2,100 (6,890)

Yakovlev OKB

A model of an unnamed projected twin-turboprop light aircraft – the Russian answer to the Partenavia P.68RG Victor.

Both variants of the Yak-46 were offered by Yakovlev to Aeroflot in competition with Tupolev designs such as the Tu-204 and Tu-334. This competition, coupled with a sharp decline in state funding of the aircraft industry, prevented the Yakovlev OKB from committing sufficient resources to these projects which appear to have been shelved or, at least, suspended until better days.

Small turboprop transport aircraft (project)

All that has survived from this project dating back to the 1970s is a desktop model; not even the designation is known. The model depicts a small shoulder-wing monoplane strongly reminiscent of the Partenavia P.68RG Victor powered by two turboprop engines (surmised to be in the 500-ehp class) driving three-bladed propellers. It would appear to be in the same category as the An-28 feederliner from which it differs in having a circular-section fuselage, a single-fin tail unit and a retractable undercarriage. The fuselage features extensive glazing and the circular section suggests it was to be pressurised.

Small business jet (project)

During the same period the Yakovlev OKB had developed a project of a business jet looking like a scaled-down Yak-40 with two engines instead of three and a cruciform tail unit instead of a T-tail. The powerplant comprised two turbojets of an unspecified type in slim nacelles. The project was not proceeded with.

Yak-48 business jet (project)

In 1989 the OKB launched development of a long-range business jet intended to carry six to ten passengers over a distance of 4,500 km (2,800 miles) at an AUW of 10,700 kg (23,590 lb). In its overall configuration it was reminiscent of the Dassault Falcon business jets, featuring swept low-set wings with winglets and a swept cruciform tail unit. The powerplant comprised two high-bypass turbofans of Western manufacture in nacelles flanking the rear fuselage. The passenger cabin could be converted into a cargo hold or into an ambulance version capable of accommodating six stretcher cases or three patients and three medical attendants.(see design data in a table below). In the process of design work the project underwent some alterations; in one of the configurations studied it was to weigh 15,175 kg (3,455 lb) and be powered by two rear-mounted Pratt & Whitney Canada PW305 turbofans.

In an effort to secure better prospects for the implementation of the project, Yakovlev established contacts with Israel Aircraft Industries (IAI) which had a similar project, the Galaxy derived from the existing IAI 1125 Westwind Astra SP. In 1993 Yakovlev Aircraft Corporation teamed up with IAI as a risk-sharing partner. Their future common product received the Israeli designation Astra Galaxy, or Astra IV (a model bearing the Yak-48 designation on its side and, additionally, the designation Astra IV on its engines, was exhibited at the MAKS-93 airshow in 1993; it was completely revised as compared to the original project). Other partners included Pratt & Whitney Canada, which was to supply PW306A turbofans rated at 2,580 kg (5,700 lb), and Rockwell Collins, which was responsible for the avionics.

There were plans for series manufacture both in Israel and in Russia at the Orenburg-based Strela enterprise. (Other sources cited plans for starting the manufacture of fuselages and tail units for the Astra Galaxy at Saratov). However, at a certain stage the partnership ran into problems and was eventually dissolved, each of the partners reverting to independent work on their respective projects. For financial reasons the Yak-48 project was shelved until better days; Yakovlev's plans for resurrecting the project remain uncertain.

Yak-77 business jet (project)

In 1992 the Yakovlev OKB commenced work on the project of this large business jet, also intended for use as a passenger aircraft for 'long, thin routes'. Sergey A. Yakovlev was appointed project manager. Two alternative engine types selected for this aircraft included the AE 3010/12, one of the proposed uprated members of the Allison 3000 family of two-shaft turbofans, and an unspecified Russian engine. The chief 'selling point' of the project was the extra-long range (up to 10,000 km/6,200 miles) in combination with a spacious interior for business travel.

Basic features of the aircraft included low-set swept wings with winglets, cruciform swept tail surfaces and engines mounted on the rear fuselage. As a business jet the aircraft would typically seat six to eight persons, and as a commercial transport thirty-two in a four-abreast arrangement with a central aisle (in this case the range would be reduced to 5,000 km/3,100 miles). Stretched versions seating 50 and 70 passengers were studied. There were plans in hand to complete prototype construction, testing and certification not later than 1997, but not even a prototype was flown as a result of financial constraints which afflicted the Russian aircraft industry, as well as the nation's economy as a whole, in the 1990s.

A model of this aircraft was exhibited at the MAKS-93.

Yakovlev OKB

Above: A desktop model of an unnamed business jet project developed by the Yakovlev OKB – probably the company's first.

Yakovlev OKB

Above and below: An artist's impression and a three-view of the Yak-48 business jet as originally conceived. Note the separate core and bypass nozzles of the engines.

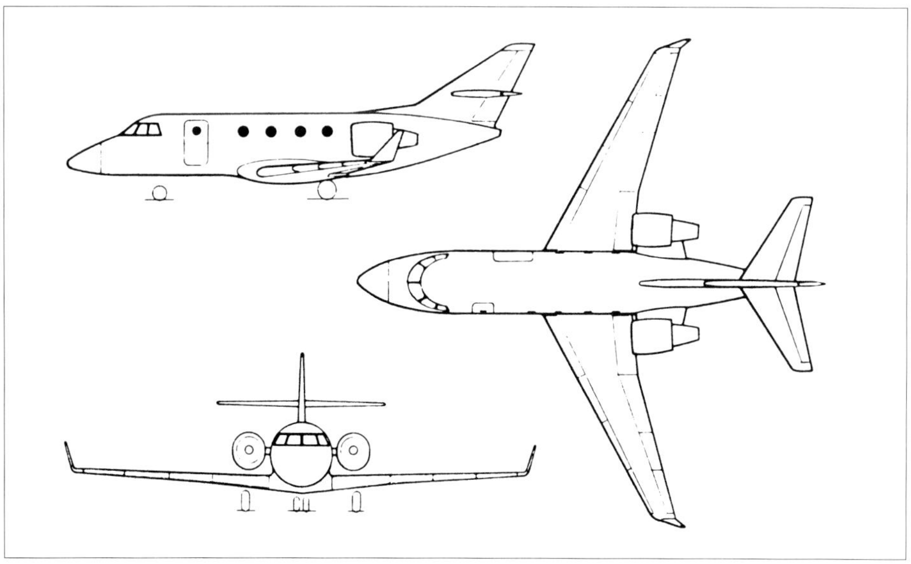

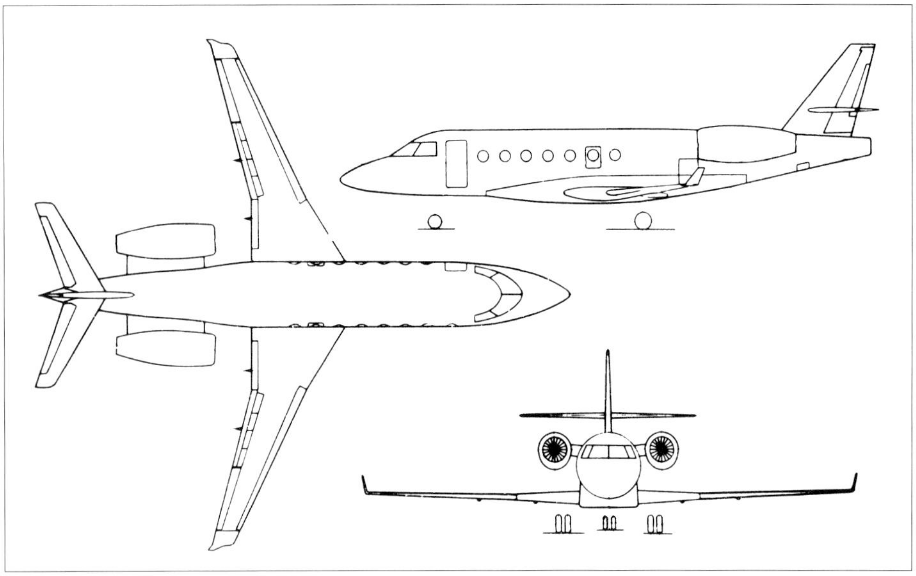

Above: This three-view illustrates the Yak-48 at a later project stage; a totally different and much larger aircraft which is almost identical to the IAI Astra SP as eventually built.

Specifications of the Yak-48 business jet project versions

	Initial Yakovlev project	Later joint project
Engine type	n.a.	PW 306A
Engine thrust at take-off, kg (lb)	2 x 1,850 (2 x 4,080)	2 x 2,586 (2 x 5,700)
Length	15.70 m (51 ft 6 7/64 in)	19.3 m (63 ft 4 in)
Wing span	16.25 m (53 ft 3¾ in)	17.42 m (57 ft 2 in)
Wing area, m^2 (sq ft)	n.a.	30.2 (325)
Empty weight, kg (lb)	n.a.	8,060 (17,769)
Payload, kg (lb)		
(maximum, cargo version)	2,200 (4,850)	1,905 (4,200)
All-up weight, kg (lb)	10,700 (23,590)	15,172 (33,448)
Cruising speed, km/h (mph)	830 (516)	880 (547)
Cruise altitude, m (ft)	11,000 (36,090)	13,715 (45,000)
Range, km (miles)	4,500 (2,800)	6,852 (4,258)
	with 1-hour fuel reserves	four passengers, NBAA reserves
Runway length, m (ft)	1,700 (5,577)	

Yakovlev OKB

A model of the Yak-77 biz-jet, a follow-on to the Yak-48 project

Yak-58 multi-purpose light aircraft

At the beginning of the 1990s the Yakovlev OKB undertook the development of this unorthodox multi-role utility transport and business aircraft which, as the OKB hoped, would be able to win a sizeable share of the light aircraft market and help solve the Design Bureau's financial problems. The aircraft was conceived as a six-seater featuring a twin-boom layout with a pusher powerplant. A full-scale mock-up was exhibited in 1991, followed by the construction of a prototype.

The aircraft had constant-chord wings with downward-curved tips (early studies featured small leading-edge root extensions). The sharply swept fins were mounted directly on the wings (their lower parts forming a sort of fin fillets doubled as tailbooms). The fins were slightly canted inwards and joined at the top by the tailplane. The retractable tricycle undercarriage had single wheels on each of the three struts. The levered-suspension main units retracted inwards in the early project, but were altered to retract outwards on the mock-up and prototypes; the nose unit pivoted forwards. The aircraft was fitted with a 355-hp Vedeneyev M-14PT, a pusher version of the M-14 radial which powers many of Yakovlev's sports and trainer aircraft; it drove a three-blade constant-speed propeller. The engine was mounted at the aft end of the fuselage forming a bullet-like nacelle, the cooling air entering through an annular gap between the fuselage and the cowling; radial shutters were installed at the rear. The cabin provided accommodation for a pilot and five passengers and was accessed through a hinged door on the left and a large sliding door further back on the right. Passenger seats could be removed to provide space for 450 kg (990 lb) of cargo or for mission equipment.

The Yak-58 programme was based at the production aircraft factory in Tbilisi which was assigned the construction of prototypes and, eventually, the series manufacture of the type.

A mock-up of the Yak-58 bearing the registration CCCP-58001 was demonstrated in February 1991; it featured very short tailbooms (fin fillets) which were extended further aft on prototypes, increasing the overall length from 7.8 m (25 ft 7 in) to 8.55 m (28 ft 0 in).

Yak-58 prototypes (with modifications)

A Yak-58 prototype registered RA-58002 (the first prototype?) was shown at the MAKS-93 airshow in Zhukovskiy on 31st August/5th September 1993. In its original configuration it had downward-curved wingtips acting as endplates. The machine was demonstrated before the first prototype's maiden flight, which reportedly took place on 26th December 1993. The first prototype was written off in an accident on 27th May 1994, but the second prototype began flight tests on 10th October 1994. At the beginning of 1996 there were four prototypes on test, plus two static and fatigue test airframes.

In the course of testing the aircraft underwent some modifications. At the MAKS-95 airshow (22nd-27th August 1995) a Yak-58, again registered RA-58002, was presented with small fixed canard foreplanes mounted on the fuselage sides (on the starboard side this foreplane was attached to the door of the passenger cabin). A further modification had been effected by the time this aircraft was demonstrated at the MAKS-97 airshow on 19th-24th August 1997. This time the prototype dispensed with the downward-curved wingtips which were replaced by conventional wingtip fairings. The flightdeck door on the port side was provided with an external cable actuator for opening in an emergency.

The same (or an identical) prototype, repainted and unregistered, was shown Zhukovskiy on 13th-15th August 2004; the only visible difference was the absence of the propeller spinner.

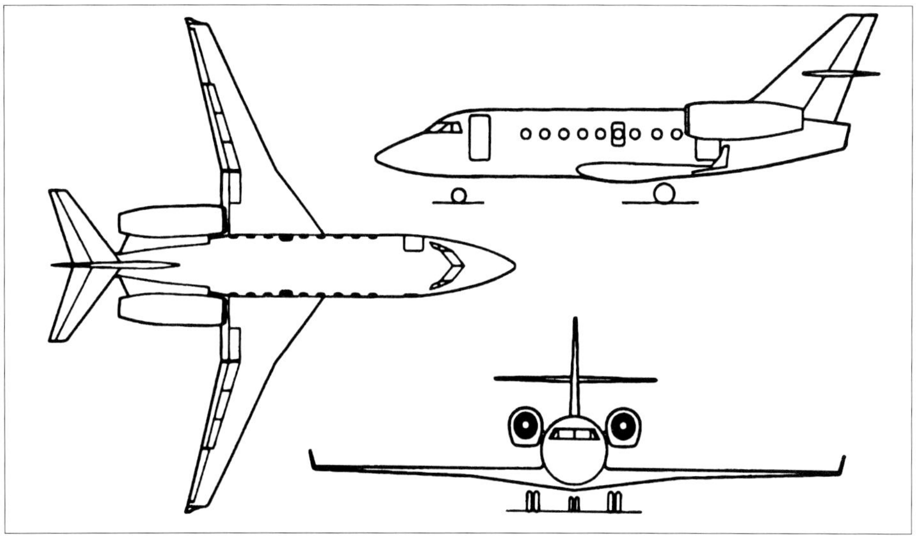

A three-view drawing of the projected Yak-77 business jet. Note the wing planform with the cranked leading edge.

Specifications of the Yak-77 business jet

Configuration	Business jet	Regional airliner
Engine thrust at take-off, kg (lb)	2 x 4,500 (2 x 9.920)	2 x 4,500 (2 x 9,920)
Length	20.45 m (67 ft 1¼ in)	20.45 m (67 ft 1¼ in)
Wing span	21.55 m (70 ft 8½ in)	21.55 m (70 ft 8½ in)
Wing area, m² (sq ft)	c. 55 (592)	c. 55 (592)
Payload	880 (1,940)	3,500 (7,720)
All-up weight, kg (lb)	25,200 (55,555)	25,200 (55,555)
Cruising speed, km/h (mph):		
maximum	850 (528)	850 (528)
economic	800 (496)	800 (496)
Cruise altitude, m (ft)	12,000 (39,370)	12,000 (39,370)
Range, km (miles)		
at maximum speed	9,000 (5,590)	4,500 (2,795)
maximum	10,000 (6,210)	5,000 (3,105)
Balanced runway length, m (ft)	2,200 (7,220)	2,200 (7,220)

Yak-58 ELINT ('radio patrol') version

According to a report in the Russian aeronautical press, the Yakovlev OKB presented a 'radio patrol version' (*sic*) of the Yak-58 at the 42nd Paris Air Show in 1997. The ambiguous term 'radio patrol' may, in actual fact, refer to an ELINT version. Interestingly, another report published in 1998 mentioned an offer by Thomson CSF of France to equip the Yak-58 for ELINT duties (it is not clear whether the offer was accepted and implemented).

Yak-58 ambulance version (project)

An ambulance version of the Yak-58 was under development in 1999.

Yak-58 – prospective versions

The original plans envisaged series manufacture of the Yak-58 at the Tbilisi aircraft plant (Tbilaviamsheni) where all the prototypes were built (this, it was hoped, would facilitate an early start of series production). Optimistically, the OKB predicted the start of deliveries to customers for 1993. However, lack of funding and economic difficulties in Russia, coupled with political unrest within the CIS, prevented these plans from being implemented.

For several years nothing was heard of the Yak-58. Then, in early 2003 steps were taken to resurrect the Yak-58, taking into account the interest for the Yak-58 evinced by potential customers in Kazakhstan and the improving prospects for general aviation in Russia. Participants in this programme were the Yak-Alakon company of Kazakhstan, the Yakovlev OKB and LMP, Inc. of the USA. By August 2003 an investment scheme was evolved envisaging an upgrade of the aircraft and development of a turboprop-powered version.

Advertising materials from Yak-Alakon contained references to a proposed Yak-58 version with a 400-hp Teledyne Continental Motors horizontally-opposed piston engine

Yakovlev OKB

A scale model of the Yak-58. Note the inward-retracting main gear units, a feature that was to change before the project reached the hardware stage.

Yakovlev OKB

Above: A look into the cabin of the full-scale mock-up of the Yak-58 at MMZ Skorost', showing the three rows of seats. Note the sliding passenger door, an uncommon occurrence on such an aircraft.

Yefim Gordon archive

Above: An unregistered Yak-58 prototype with the Tbilaviamsheni badge on the cowling flies over the residential blocks of Tbilisi. Note the small canards on the fuselage sides.

Yefim Gordon

Yak-58 RA-58002 at the Yakovlev flight test facility at Zhukovskiy with the engine uncowled for inspection.

and a version with a 600-hp turboprop from an unspecified manufacturer. The Yakovlev OKB embarked on refurbishing the Yak-58 prototypes in order to restore them to flying condition.

It was surmised that series manufacture of aircraft would be transferred from Tbilisi to the Smolensk aircraft plant due to the deteriorating relations between Russia and Georgia, which was becoming more and more pro-American and NATO-minded. An indirect confirmation of these plans was the participation of a Yak-58 prototype in a small static display arranged at Zhukovskiy in conjunction with an international aerobatics show on 13th-15th August 2004.

Specifications of the Yak-58 multi-purpose aircraft

Length	8.55 m (28 ft 0⅝ in)
Wing span	12.7 m (41 ft 8 in)
Wing area, m² (sq ft)	20 (215.3)
Empty weight, kg (lb)	1,270 (2,800)
Payload, max kg (lb)	450 (990)
All-up weight, max kg (lb)	2,100 (4,630)
Maximum speed, km/h (mph)	300 (186.4)
Landing speed, km/h (mph)	125 (78)
Service ceiling, m (ft)	4,000 (13,123)
Range, km (miles)	
with 45-minute reserves and maximum payload	1,000 (621)
Take-off distance, m (ft)	610 (2,000)
Landing distance, m (ft)	600 (1,968)

Yak-112 multi-purpose light aircraft

The origins of the Yak-112 multi-purpose light aircraft can be traced to a competition held in the Soviet Union in the late 1980s for a two-seat primary trainer for the DOSAAF and aero clubs. The Yakovlev OKB submitted its project of a two-seat strut-braced shoulder-wing monoplane powered by a very fuel-efficient Novikov DN-200 flat-four air-cooled diesel engine then under development at the Rybinsk Engine Design Bureau (RKBM) or a new 150-hp engine with four cylinders in X formation.

The project (designated Filin, Horned Owl, described in Chapter 7) emerged as the winner, but was not implemented in this guise. It was subsequently reworked and developed as a multi-role four-seater. As a consequence of the resulting weight growth, the initial 147-hp rating of the DN-200 engine proved insufficient, and other powerplants had to be sought. In its later version, the DN-200 yielded 200 hp and came into the picture again, but development difficulties prevented the engine from being used. Another powerplant option considered by the Yakovlev OKB was the VOKBM (Voronezh) M-17F with four cylinders in

X-formation, rated (according to different sources) at 175 to 250 hp. Early desktop models of the Yak-112 from the OKB show the aircraft with a rather short and thick nose, a shallow rounded spinner covering the hub of a two-blade propeller, and an elliptical opening for the cooling air in the front area of the engine cowling – presumably reflecting the DN-200 or M-17F-powered versions, or both. The full-size mock-up exhibited at the 39th Paris Air Show in June 1991 was similarly configured. Yet, the M-17F proved to be unavailable, too, and the Yakovlev Design Bureau had to rely on imported engines. These were the Teledyne Continental Motors IO-360-ES flat-four engine rated at 210 hp, driving a two-blade two-pitch propeller, and the Lycoming AIO-360-A1B6 engine, driving a three-blade propeller (in both cases developed by Hartzell).

A Yak-112 was presented in mock-up form in 1991, and a prototype was flown in October 1992 (see details of development versions below). This was a four-seat strut-braced high-wing monoplane of metal construction incorporating extensive use of glassfibre and other composites. Extensively glazed cabin provided excellent visibility. The wing had aerodynamics similar to those of the Yak-58, including the downward-curved wingtips intended to minimise the adverse effect of wingtip vortices. The tail unit comprised rectangular-shaped tailplane and swept vertical tail; the fixed three-unit nosewheel undercarriage had spring-type main units and spats on all the three wheels.

The Yak-112 was intended mainly for the transportation of passengers, small cargoes and mail, as well as for use as a liaison and patrol aircraft; in a dual-control version it could be used for primary training. The cabin could be adapted for ambulance duties.

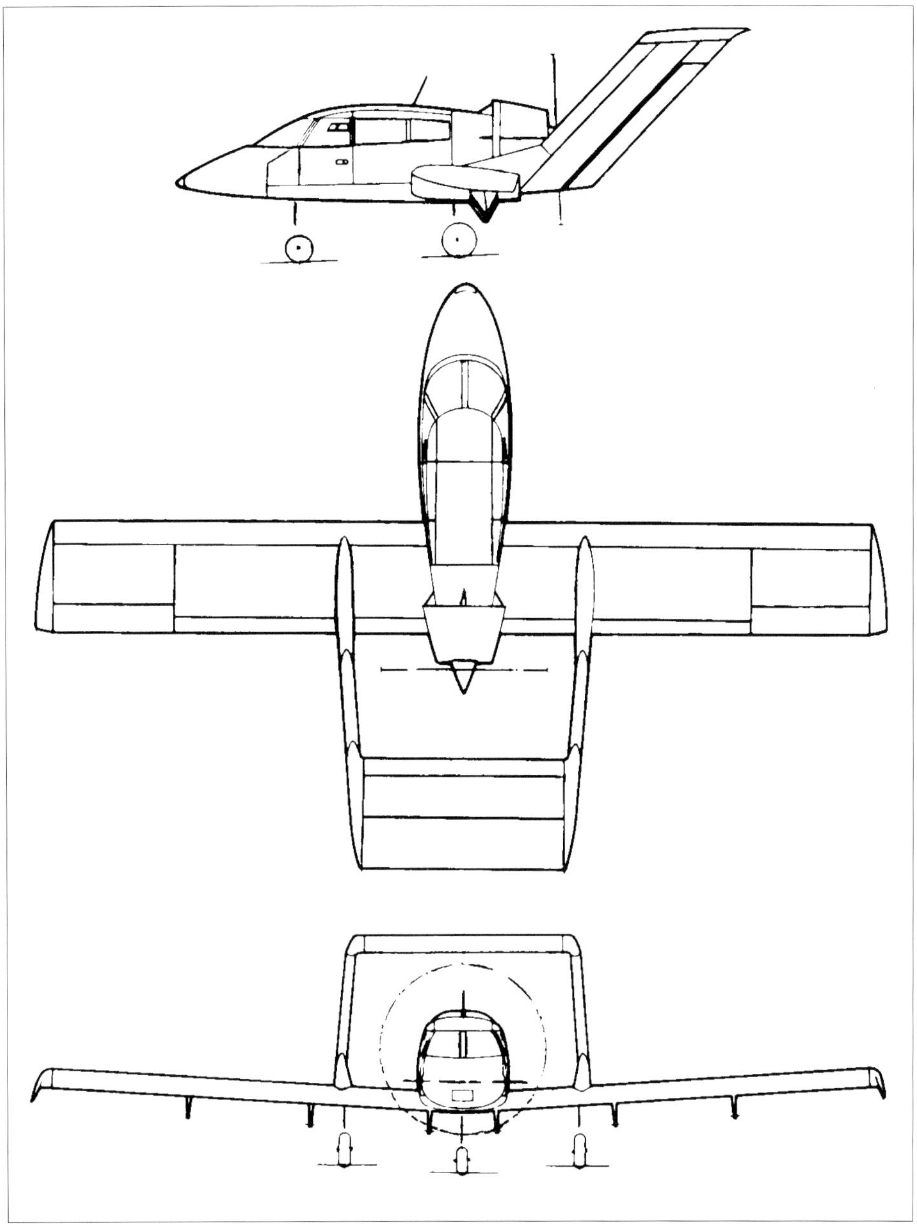

Above: A three-view of the Yak-58 in original form (without the canards on the fuselage sides).

Yakovlev OKB

The first prototype Yak-112, RA-00001 (c/n 00-002), parked in front of the assembly hangar at MMZ Skorost'. Note the spats on the mainwheels and the shape of the wingtips.

Yakovlev OKB

Above: Another view of the first prototype (this time at Zhukovskiy), accentuating the pronounced 'chin' on the engine cowling. Note that the wheel spats have been removed.

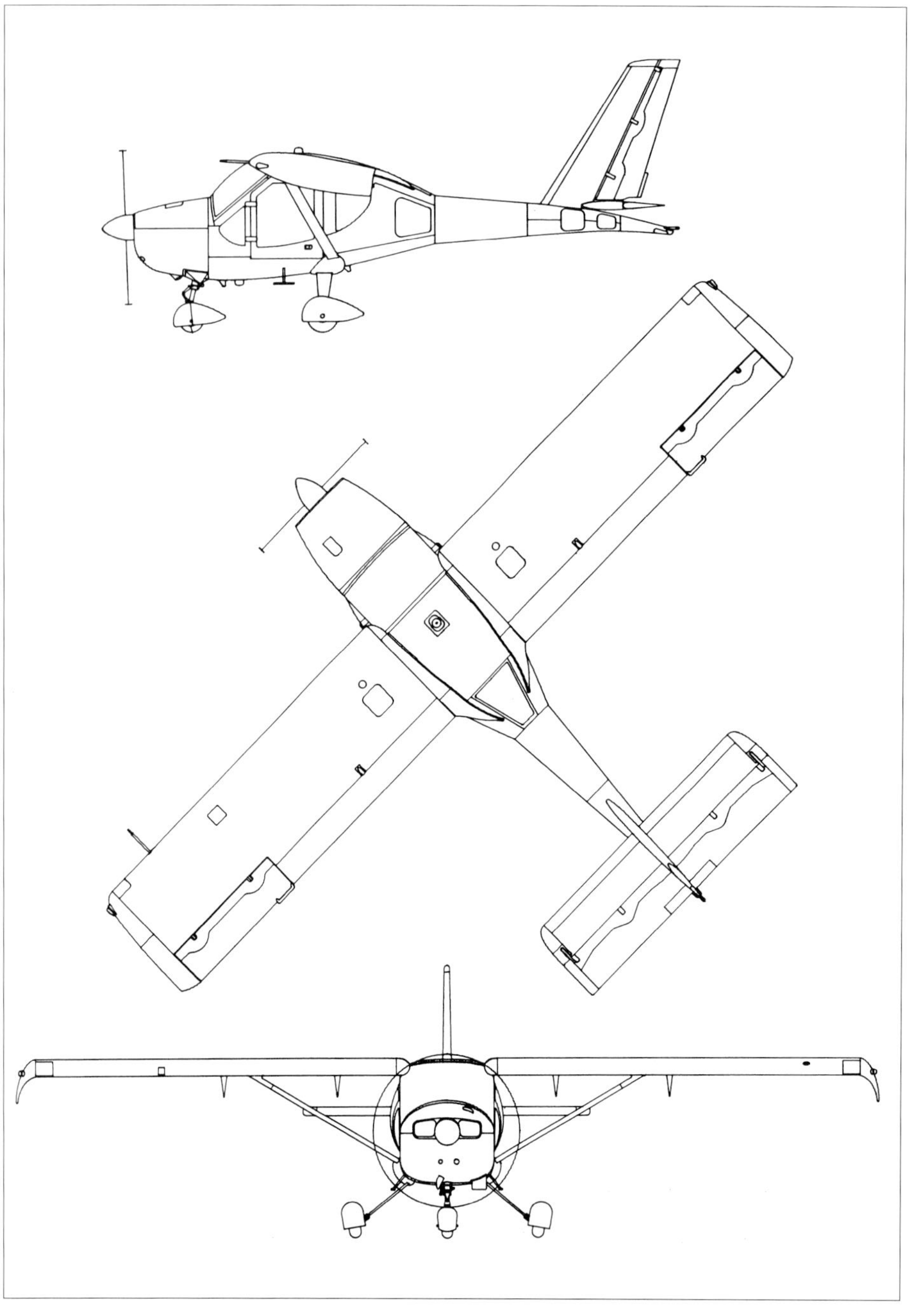

A three-view of the Yak-112 as originally developed to take the M-17 engine (note the cowling shape).

In mid-1994 the Irkutsk Aircraft Production Association (IAPO) had plans for starting series production of the Yak-112 in two versions: a TCM-powered version with Bendix King Avionics and a version with the DN-200 engine and Russian avionics (obviously intended for export and for the home market respectively). Two machines fitted with Bendix/King avionics had been built in 1993, and there were plans for building four machines with Russian avionics in 1994. The IAPO boasted a backlog of orders for the Yak-112 which allegedly stood at 500 aircraft at that time. IAPO's general director Aleksey Fyodorov optimistically spoke at that time about the commencement of series production by the summer of 1995 with a yearly production rate of 200 machines. The Yak-112 was also to be manufactured in series at the Smolensk aircraft plant. However, these optimistic plans proved to be ill-founded and ran into difficulties; the series manufacture never really got started and appears to have been abandoned.

Detailed below are various versions of the Yak-112, including projects.

Yak-112 mock-up configured for the M-17F or DN-200 engines

A full-size mock-up of the Yak-112 bearing the number 0001 on the wings was exhibited at the 39th Paris Air Show in 1991. The shape of its engine cowling was identical to that of the models described above and may be presumed to be adapted to the mentioned engines. In other respects the mock-up was externally representative of the future prototypes, except for the rather large wheel spats on all three undercarriage units and the navigation lights positioned well aft on the downward-curved wingtips.

Yak-112 first prototype with IO-360-ES (or AIO-360-A1B6?) engine

A Yak-112 prototype (presumably the first prototype) was publicly shown for the first time at the MAKS-93 Air Show in Zhukovskiy in August-September 1993. This aircraft, eloquently registered RA-00001, bore the c/n 00-002 (that is, batch 00, 002nd aircraft in the batch; c/n 00-001 presumably was a static test airframe). The RA-00xxx registration batch had been set aside for experimental aircraft.

According to some reports it was powered by a Teledyne Continental Motors IO-360-ES engine (other sources claim it was an Avco Lycoming AIO-360-A1B6 engine) driving a two-bladed Hartzell F8549A-8R propeller. The engine cowling featured a prominent 'chin' and had separate cooling air inlets on both sides of the propeller hub. The navigation lights were moved to the front portions of the wingtip fairings. The boarding steps were attached to the fuselage, not to the main undercarriage legs (as was the case on the models and the full-scale mock-up).

Specifications of the Yak-112 utility aircraft

Engine type	TCM IO-360-ES
Lycoming AIO-360-A1B6	
Engine power, hp	200
Length	6.96 m (22 ft 10 in)
Wing span	10.25 m (33 ft 7½ in)
Wing area, m² (sq ft)	14.0 (150.7)
Empty weight, kg (lb)	775 (1,710)
Payload, max kg (lb)	270 (595)
All-up weight, max kg (lb)	1,260 (2,780)
Cruising speed, km/h (mph):	
normal	190 (118)
maximum	250 (155)
Landing speed, km/h (mph)	125 (78)
Service ceiling, m (ft)	4,000 (13,123)
Range, with 45-min reserve, km (miles):	
with maximum payload	850 (528)
with maximum fuel	2,000 (1,242)
Take-off distance, m (ft)	500 (1,640)
Landing distance, m (ft)	500 (1,640)

Yak-112I production version prototype (with IO-540 engine?)

This version made its appearance in early 1995, when a photo of an unpainted prototype captioned 'Yak-112I' was published in a magazine; the I presumably refers to Irkutsk, the city where the aircraft was to be produced. The engine type is unclear, some reports indicate it as the Textron Lycoming IO-540, but it may well have been the Teledyne Continental IO-360ES.

In its initial configuration, it appeared to be externally identical to the first prototype, featuring the same shape of the engine cowling (with a 'chin'), downward-curved wingtips and cantilever spring main undercarriage legs. In the course of development testing the aircraft underwent a series of modifications. Presumably it was an example of this version that was shown at the MAKS-95 airshow in a green/white colour scheme with black trim, bearing no registration and fitted with a three-blade propeller (the only external difference from the previous configuration).

Further modifications were made to the Yak-112I by the end of that year. According to a press report, a *'thoroughly modernised Yak-112I with the new Lycoming-540 engine and a three-blade Hartzell propeller took to the air on 15th December 1995 with OKB test pilot A. A. Sinitsyn at the controls. [...] In the course of testing the aircraft had been subjected to significant modifications. Among other things, the titanium alloy main undercarriage [spring] struts were replaced by tubular structures.'*

Yak-112 prototypes meeting this description (as regards the modified undercarriage) made their public appearance in 1997. First, an unpainted machine was shown at the LII airfield in Zhukovskiy on 22nd May 1997. In addition to the tubular undercarriage struts, it featured a recontoured (more streamlined) engine cowling without the characteristic 'chin' in combination with a Hartzell F8468A-5R three-bladed propeller. In some reports the absence of the 'chin' is associated with the installation of the Lycoming engine instead of the TCM model. A notable new feature was the redesign of the wingtips which were now ordinary, without the downturned endplates.

This machine appears to have remained unpainted, since it could be seen in its primer colours in the flying display programme of the MAKS-99 air show two years later (or perhaps it was another example). But at the MAKS-97 air show in August 1997 the static exhibits included a Yak-112 registered RA-00012 painted white and blue. It was externally almost identical to the example described above, differing only in having airfoil-shaped fairings on the tubular undercarriage legs and small rudimentary fairings ('demi-spats', so to say) on the mainwheels.

Yak-112P floatplane (project)

Back in 1999 studies were made at the Yakovlev OKB of a floatplane version of the Yak-112 designated Yak-112P (*poplav**kovyy*** – floatplane, used attributively). Still earlier, at the initial stage of projecting, a model was built at the OKB showing a Yak-112 with the initial powerplant (M-17F or DN-200 engine) on twin floats.

Yak-112A six-seat version (project)

In 1999 a brief reference appeared in the Russian aeronautical press to design work being conducted on a more spacious variant of the Yak-112. This version, dubbed Yak-112A, was intended to seat six persons. Nothing was heard of it subsequently.

MS-21 airliner (project)

The designation MS-21 (*maghi**strahl'**nyy samo**lyot dva**dtsat' **per**vovo **vek**a* – 21st Century Mainliner) covers a family of short- and medium-haul airliners which are to be developed jointly by the Il'yushin Aircraft Complex and the Yakovlev OKB. This family of aircraft is based on the Yak-242 project and is intended to replace the Tu-154 medium-haul airliner.

The advanced development project of this airliner was prepared by a special team set up jointly by the Il'yushin and Yakovlev concerns as their submission to a tender for

Yefim Gordon

Yefim Gordon

Two views of a model of the MS21 demonstrated at the 2003 Paris Air Show. The titles on the tail are in Russian.

Dmitry Komissarov

Above and below: This flightdeck mock-up was displayed at the MAKS-2003 . The MS-21 is to have a 'glass cockpit', with only three back-up electromechanical instruments (artificial horizon, ASI and altimeter).

Dmitry Komissarov

Design performance of the MS-21 with Russian-built engines

	MS-21-100	MS-21-200	MS-21-300
Take-off weight, kg (lb)	65,800 (145,060)	71,100 (156,750)	72,000 (158,730)
Engine thrust, kgp (lbst)	2 x 11,800 (2 x 26,010)	2 x 11,800 (2 x 26,010)	2 x 11,800 (2 x 26,010)
Cruise specific fuel consumption, kg/kgp·h (lb/lbst·h)	0.545	0.545	0.545
Thrust/weight ratio	0.36	0.33	0.328
Seating capacity (maximum)	132	156	174
Empty operating weight, kg (lb)	37,000 (81,570)	38,600 (85,100)	39,900 (87,960)
Airframe weight per passenger, kg (lb)	280 (617)	247 (544)	229 (504)
Payload at 100% load factor, kg (lb)	12,540 (27,645)	14,820 (32,670)	16,530 (36,440)
Maximum fuel load, kg (lb)	22,000 (48,500)	22,000 (48,500)	22,000 (48,500)
Cruise altitude, m (ft)	11,600 (38,060)	11,600 (38,060)	11,600 (38,060)
Cruising speed, km/h (mph)	850 (527)	850 (527)	850 (527)
Operational range at 100% load factor, km (miles)	4,700 (2,919)	5,500 (3,416)	4,500 (2,795)
Fuel consumption, g/seat-km (lb/seat-mile)	21.5 (0.076)	17.5 (0.062)	16.6 (0.059)
Runway length, m (ft)	2,100 (6,890)	2,400 (7,870)	2,400 (7,870)

a short/medium-haul airliner announced by Rosaviakosmos (the Russian Aerospace Agency). In mid-2003 this proposal was declared the winner of this tender, and the two concerns took a decision to keep this joint team numbering some 60 persons. The project was advertised at the MAKS-2003 airshow (19th-24th August 2003) where a mock-up of the flightdeck was shown; a complete model of the MS-21 was displayed at Farnborough International 2004 airshow. In its basic layout the aircraft is virtually identical to the Yak-242.

The merger of the Yakovlev OKB with the Irkut Aircraft Corporation (formerly IAPO) introduced new elements into the picture. Now the Irkut Corporation has overall responsibility for the project on behalf of the Yakovlev team. The management of the Irkut Corporation is weighing the financial aspects of this venture; it keeps various options open, including termination of the project in the case of more promising cooperation proposals, such as, for example, an eventual proposal from the Bombardier Company (Canada) for participation in the development of its C-series airliners. Well, politics are always intruding into aircraft design affairs, what more can we say.

In late December 2004 it became known that the leading role in the partnership around the MS-21 was transferred from the Il'yushin Aircraft Complex (which was overburdened with other programmes) to the Yakovlev OKB, Il'yushin becoming the latter's subcontractor. At present an advanced development project is under preparation.

New versions of the MS-21 family have been announced recently. These are the MS-21-100 seating 132 or 116 passengers (in all-economy or two-class configuration respectively), the MS-21-200 configured for 156 or 140, the MS-21-300 seating 178 or 154, the all-cargo MS-21K and the convertible/combi MS-21KP. The engines envisaged for the MS-21 are the Aviadvigatel' PS-90A-12 or PS-12 in the 11,000 to 12,000-kgp (24,250 to 26,455-lbst) thrust class.

RRJ (Russian Regional Jet) airliner (project)

RRJ is a family of regional airliners which is being developed by the Sukhoi Aircraft Holding Company in cooperation with the Il'yushin Aircraft Complex and the Boeing Commercial Airplane Group. These three companies jointly took part in a tender for the development of a regional jet to replace the Tu-134 short/medium-haul airliner which was announced by Rosaviakosmos in 2002. At a certain stage the Yakovlev OKB was also invited to join the project. The present role of the Yakovlev OKB in the programme is unknown.

V/STOL AND SHIPBOARD AIRCRAFT

9

Yak-V VTOL fighter (project)

This project marks the beginning of work undertaken by the Yakovlev OKB on vertical take-off and landing aircraft for naval use which eventually resulted in the emergence of the Yak-38 VTOL light attack aircraft. The first steps in this direction date back to 1960 when Yakovlev, under the impression of successful flights of the Short SC.1 experimental jet-lift aircraft, took a decision to build something similar and formed a special design team to tackle the task. Led by Stanislav G. Mordovin and Leon M. Shekhter, this team undertook studies in the field of jet-engined VTOL aircraft in close co-operation with LII engineers who, under the guidance of Prof V. N. Matveyev and Aram N. Rafaelyants, had earlier produced and tested the *Toorbo**lyot*** ('Turbo-flyer') jet-lift vehicle.

Studying possible layouts, members of the team began with the idea of converting the Yak-30 (Yak-104) jet trainer, then under development, into a VTOL aircraft (Yak-30V) by installing two additional RU19-300 turbojets vertically between the ducts to the propulsion engine. This was abandoned in favour of the project designated Yak-V (*verti**kal'**nyy* [*vzlyot i po**sahd**ka*] – VTOL) which shared the concept of the Hawker P.1127 Kestrel. The aircraft was to be designed around a turbofan in the 10,000-kgp (22,045-lbst) thrust class with four vectoring nozzles. No such engine existed in the Soviet Union; moreover, it proved difficult to find an engine with a suitable by-pass ratio for fitting it with such nozzles. Yakovlev's efforts to obtain a government decision tasking engine design bureaux with creating such an engine were to no avail. A different solution had to be sought.

Yak-36 experimental VTOL aircraft (*izdeliye* V)

An alternative approach consisted in fitting an existing jet engine with a vectoring nozzle – a less demanding job, which the Ministry of Aircraft Industry was prepared to fund. Yakovlev had to put up with this second-best solution. Its practical application was based on adapting the R27-300 engine developed in the OKB-300 engine design bureau led by Sergey K. Tumanskiy. Two such engines rated at 6,350 kgp (14,000 lb) were to be used in the experimental VTOL machine designated Yak-36. Each engine was fitted with a jetpipe and a nozzle of flattened oval section. This was joined to the engine by a 'lobster shell' connection, so that it could be rotated about an inclined axis through an angular range of 95° to deflect the jet downwards for lift, or to the rear for forward flight, or 10° forward of the vertical for in-flight deceleration to the hover. The engines were handed, one with the nozzle on the left side and the other on the right. Of course, both nozzles had to be driven in unison.

The basic configuration of the Yak-36 (initially known in-house as *izdeliye* V, again standing for *verti**kal'**nyy* [*vzlyot i po**sahd**ka*]) was dictated by the need to place the engine nozzles near the aircraft's CG. This meant that the engines had to be accommodated in the nose, with the nozzles under the wings, a layout reminiscent of Yakovlev's early jet fighters. The engines were fed by a plain oval-section nose inlet divided by a vertical splitter.

The aircraft had short mid-set cropped-delta wings of small area optimised for cruise flight, with a leading-edge sweep of 37° and 5° anhedral. Placed above the wing leading edge was the pressurised cockpit; the upper part of the fuselage ahead of it was

Yakovlev OKB

Above: An early display model depicting the Yak-36 VTOL fighter as originally envisaged with a short nosecone.

Yakovlev OKB

This model is much closer to the real thing, featuring a long nose boom carrying a reaction control nozzle. Note the underwing gun pods.

Yakovlev OKB

Above: Aptly coded '36 Yellow', this was the first real, but non-flyable, example of the Yak-36. Note the main gear door design, the lack of recirculation dams and the landing light built into the air intake splitter.

Yakovlev OKB

Above: '37 Yellow', the first flying prototype, is prepared for a mission. Note the recirculation dams; the hinged forward one incorporates the nose gear doors.

Yakovlev OKB

'38 Yellow', the other flying prototype. Note the enlarged air scoop just aft of the tactical code (compare with '36 Yellow').

occupied by an equipment bay. The sharply tapering rear fuselage carried a cruciform swept tail unit. The bicycle undercarriage comprised a castoring single-wheel nose unit retracting forwards into a bay between the inlet ducts, a braked twin-wheel main unit retracting aft into a bay with twin doors, and castoring outriggers retracting forwards into long probe-carrying wingtip fairings.

To ensure controllability during the hover, the machine was fitted with reaction-control jet nozzles (air bleed ejectors, or puffer pipes) placed under each wingtip, in a short fairing on the tailcone and at the end of a large boom projecting ahead of the nose. The nozzles used bleed air from the engines.

Authorisation was given to build four airframes, one of which served as a structural test article. The other three were built as complete aircraft. The first of them, coded '36 Yellow', was rolled out in the autumn of 1962. It was used for static testing and for wind tunnel tests, including tests with the engines running. The others, '37 Yellow' and '38 Yellow', were flight-test articles. Outdoor testing of the tethered aircraft, which started later in 1962 at LII in Zhukovskiy, revealed some problems with re-ingestion of hot gases as a result of re-circulation; this was compounded by the appearance of suck-down forces acting on the airframe. The problem was cured in part by fitting two parallel strakes to the underside of '37 Yellow'.

In January 1963 the Yak-36 was ready for flight testing. The assigned test pilot was Yuriy A. Garnayev from LII who had previous experience with testing the Toorbolyot VTOL machine. To begin with, Garnayev performed taxying runs and small hops in vertical take-off mode. But before the tests could get into full stride, Garnayev was withdrawn from the programme and sent on a mission to France as an experienced helicopter pilot; tragically, he was killed there in the crash of a Mil' Mi-6PZh fire-fighting helicopter. Garnayev was superseded by Yakovlev OKB test pilot Valentin G. Mookhin. It took him some time to learn all the intricacies of this aircraft. Mookhin performed the Yak-36's maiden flight on 27th July 1964.

The test programme between April and August 1965 was devoted to studying and perfecting the behaviour of the aircraft at the hover. During vertical take-offs and landings the machine was controlled both manually and with the help of an automatic control system. It turned out that if the latter failed, the pilot was capable of balancing the machine through manual control.

Full-scale testing of the Yak-36 lasted nine months. During that period the aircraft was repeatedly subjected to modifications. To prevent ingestion of hot gases into the air intake, a large forward-retracting two-segment flap (recirculation dam) incorporating the nosewheel well doors was installed under the lower air intake lip; it could be deflected during take-off and landing. Still, the exhaust ingestion problem remained a source of vexation later, when it came to testing VTOL machines of other types.

The third prototype Yak-36 ('38 Yellow') caused something of a sensation when it was publicly demonstrated at an air display at Moscow-Domodedovo airport on 9th July 1967. Mookhin performed a vertical take-off followed by a brief circular flight in cruise mode and then accomplished a gentle touchdown after a vertical descent, evoking much interest on the part of both the Soviet spectators and the foreign guests. Few of the spectators knew that on the previous day, during a dress rehearsal for the flypast, Mookhin had suffered a mishap when flying the other flight test aircraft, '37 Yellow', and damaged it. Fortunately, '38 Yellow' was at the ready as a back-up, thus saving the day.

During the Domodedovo air display the Yak-36 carried two UB-16-57 pods with unguided rockets on underwing pylons. Hence after the event the aircraft received the NATO reporting name *Freehand* in the misguided belief that it was intended for production and service. The project also contained a provision for installing the GSh-23 twin-barrel cannon. However, all this was no more than token armament. The Yak-36 was a purely experimental machine unsuitable for operational combat service. The aircraft's performance was not particularly impressive, and the available engine thrust did not permit the carriage of a normal weapon load. At an AUW of 11,700 kg (25,790 lb) without external stores, the aircraft attained a maximum speed of 1,009 km/h (627 mph), a service ceiling of 12,000 m (39,370 ft) and a range of 370 km (230 miles).

Flight testing of the Yak-36 demonstrated that the chosen powerplant layout made balancing the aircraft in the VTOL and transition modes too complicated. For this reason further work on the machine was discontinued after the Domodedovo show. The static test aircraft ('36 Yellow') was turned over to the Soviet Air Force museum in Monino. From 1968 the OKB started design work on a VTOL machine featuring a completely different powerplant layout, namely, a combination of lift/cruise and lift engines.

Interestingly, Aleksandr S. Yakovlev was initially very sceptical about this layout and placed more confidence in lift/cruise engines unaided by lift engines. Accordingly, there were some initial project studies featuring this arrangement. A model of one of these projects shows a machine very similar in concept to the experimental Yak-36 but

Boris Vdovenko

Above: Yak-36 '37 Yellow' lifts off in a cloud of dust, showing the downward-rotated vectoring nozzles. Note the small port side cooling air scoop.

ITAR-TASS

Above: The second flying prototype hovers at Moscow-Domodedovo on 9th July 1967, carrying two UB-16-57 rocket pods on underwing pylons. These were just for show purposes.

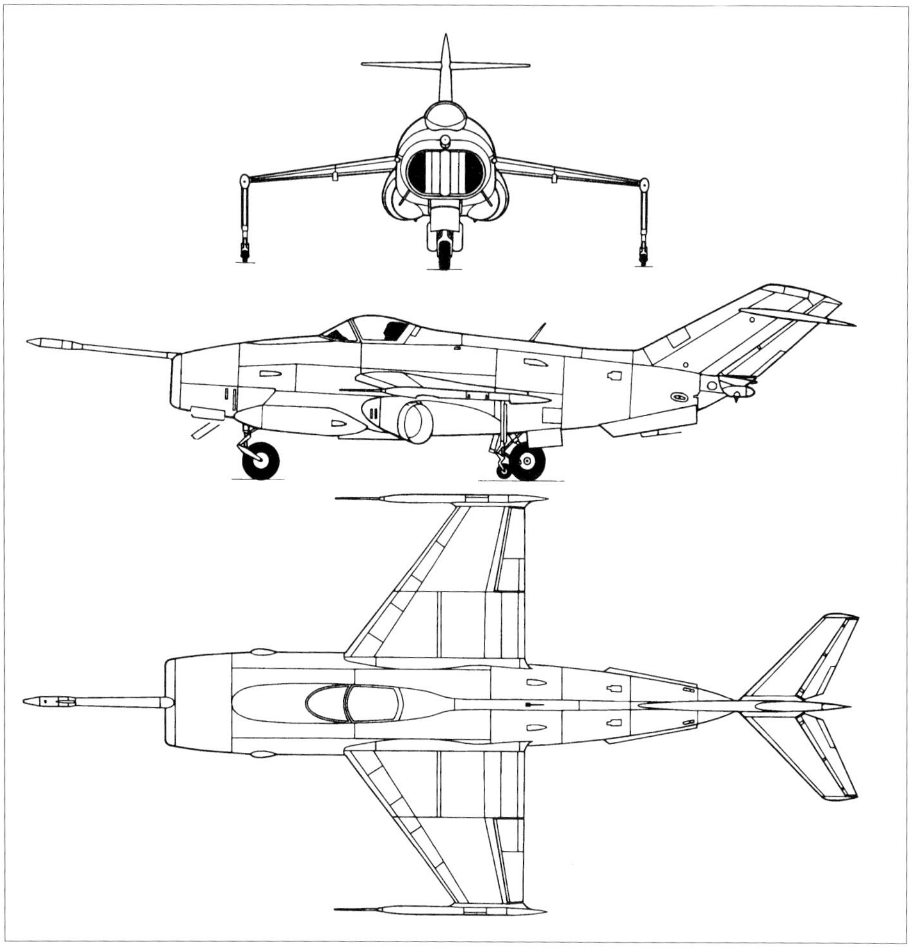

A three-view of the Yak-36.

having a slimmer fuselage, a chin air intake topped by a pointed radar nose cone and a more forward position of the cockpit. The other approach was advocated by S. G. Mordovin who was responsible for the VTOL projects in the OKB. A dispute between the partisans of the two alternative powerplant layouts was finally settled by the Ministry of Aircraft Industry, which supported the composite engine layout.

Specifications of the Yak-36 experimental VTOL aircraft

Length with nose probe	17.00 m (55 ft 9⅜ in)
Height	4.50 m (14 ft 9⅛ in)
Wing span	10.00 m (38 ft 9¾ in)
Wing area, m^2 (sq ft)	17.00 (183.0)
Engine type	R27-300
Engine thrust, kgp (lbst)	2 x 5,300 (2 x 11,690)
Empty weight, equipped, kg (lb)	5,300 (11,690)
All-up weight, kg (lb)	8,900 (19,625)
Maximum speed, km/h (mph)	900 (560)
Service ceiling, m (ft)	12,000 (39,370)
Range, km (miles)	370 (230)

Yak-36M VTOL shipboard attack aircraft prototypes (*izdeliye* VM)

Development of this machine was undertaken with a view to meeting a very specific requirement. The Soviet leaders came to the conclusion that the country's armed forces were in need of aircraft carriers (which the Soviet Union lacked at that time), and several ships of that class were laid down at shipyards. By the time the first of these ships, which were termed 'heavy aircraft-carrying cruisers' for political reasons (to circumvent an international treaty prohibiting the passage of aircraft carriers through the Bosporus and Dardanelles), was launched, a test batch of shipboard attack aircraft was to be built. The Yakovlev OKB was entrusted with this task; the initial requirement issued in 1967 called for the development of a supersonic V/STOL aircraft capable of both carrier-borne and land-based operations, with a maximum speed of 1,400 km/h (870 mph) at high altitude.

A project intended to meet this requirement was prepared in 1968. The aircraft was to be powered by two R27VM-300 lift/cruise engines with thrust-vectoring nozzles placed under the wing roots, which were similar to those of the Yak-36; it had an elongated fuselage with a pointed nose and lateral intakes. This project was abandoned in favour of a machine with a composite powerplant and subsonic performance. The new aircraft was allocated the in-house designation *izdeliye* VM. Its development was conducted under the direction of S. Mordovin as project manager.

Officially the new aircraft was designated Yak-36M (despite the fact that structurally it had nothing in common with the Yak-36 described above). Possibly this was a ploy aimed at securing state funding for the new aircraft. It would have been hard to persuade MAP to finance development of a 'clean sheet of paper' VTOL fighter if the ministry got the idea that the Yak-36 had been a failure and a waste of money. Therefore, quite possibly the OKB wanted to pass the new machine off as a 'modification' of the existing design. The meaning of the M suffix in this case is ambiguous. According to the prevailing opinion among the OKB's staff, this letter stands for *mor**skoy*** (naval), but it can also be interpreted more traditionally as *moderni**zee**rovannyy* (upgraded).

The powerplant of the *izdeliye* VM was radically different from that of the Yak-36. It comprised two types of engines differing in the direction of their thrust. The main engine was a lift/cruise engine; it contributed to providing lift in the take-off and landing mode by means of vectoring nozzles. In these flight modes additional lift was provided by two special lift engines placed almost vertically in a tandem arrangement behind the cockpit; they were slightly inclined forward. After lifting off vertically, the aircraft would perform a transition to horizontal flight, gradually reducing the thrust of the lift engines to complete shutdown in the course of this transition. At the same time the vectoring nozzles of the main engine would be gradually rotated into horizontal position. It was clear that precise balancing of the lift engines' thrust during take-off and landing would be hard to achieve manually; therefore, it was considered necessary to automate fully this process with the help of a specially designed SAU-36 automatic control system (*sis**tema** avtoma**tich**eskovo oopravl**en**iya*).

The main lift/cruise engine was again obtained by adapting the R27-300 turbojet which in its modified version received the designation R27V-300 (*izdeliye* F). It was a two-spool engine with an axial compressor comprising five LP stages and six HP stages, an annular combustion chamber, a two-stage turbine with cooled guide vanes and blades of the HP stage, and a curved bifurcated jetpipe with two tapering rotating nozzles actuated by hydraulic motors. During the bench tests the thrust was initially slightly in excess of 6,000 kgp (13,230 lbst); later, in the course of series production of the Yak-38, the thrust rating was raised (in a modified version) to 6,800 kgp (14,990 lbst).

The lift engines selected for the Yak-36M were RD36-35 turbojets developed at the Rybinsk Engine Design Bureau (RKBM) under the direction of Pyotr A. Kolesov. They had already been tested in Sukhoi's and Mikoyan's experimental aircraft with mixed powerplants. The RD36-35 had a six-stage compressor and a single-stage turbine; during bench tests the engine weighing 176 kg (388 lb) delivered a thrust of 2,350 kg (5,182 lb). The uprated version used on the Yak-36M was designated RD36-35FV (the suffix is sometimes given as VF); it had a thrust rating of 2,900 kgp (6,395 lbst).

The chosen layout with separate lift engines had one inherent shortcoming: the nozzles of the lift engines and the main engine were placed rather far apart, so that the failure of any one engine at the hover would cause a serious problem with balancing the aircraft. Incidentally, one of Yakovlev's associates, Yevgeniy G. Adler, developed his own basic layout of this aircraft which had the merit of placing the vertical jet efflux of all three engines very close to the CG; this eliminated the possibility of a dangerous imbalance in the event of one of the engines becoming inoperative during take-off or landing. However, the proposal came too late to be incorporated into the prototype already under construction.

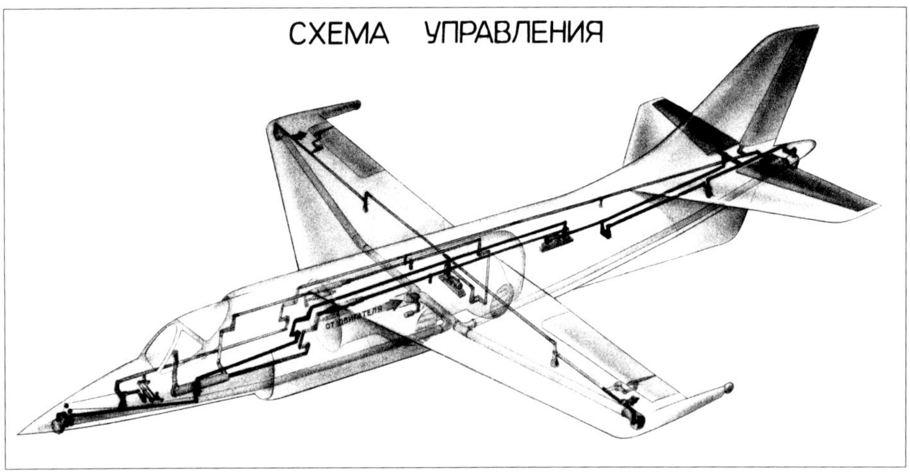

This Yakovlev OKB drawing depicting an early project version of the Yak-36M and marked 'Control system layout' shows an aircraft remarkably similar to the Hawker Siddeley P.1127 Kestrel.

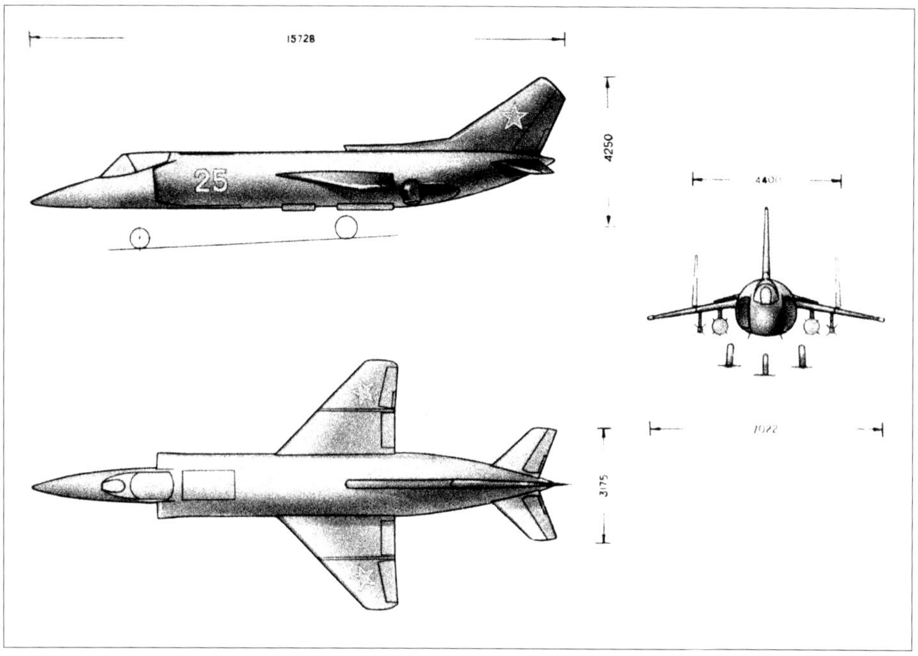

Above: An early general arrangement drawing of the Yak-36M/*izdeliye* VM (the future Yak-38). The all-important dimensions with the wings folded for below-deck stowage are shown.

The machine that thus emerged again featured small mid-set cropped-delta wings with a low aspect ratio; the outer sections folded vertically for carrier stowage. The swept tail surfaces were conventional; both the wings and the tailplane featured strong anhedral.

The fuselage was totally different from the Yak-36's, with a much higher fineness ratio and a pointed nose. The main engine drew air through semi-oval cross-section lateral air intakes just aft of the cockpit (provided at a later stage with a row of auxiliary blow-in doors), exhausting through a bifurcated jetpipe with two small vectoring nozzles. The latter were joined by a transverse synchronising shaft. The two lift engines were enclosed in cruise flight by an aft-hinged cover with spring-loaded blow-in doors; it opened during take-off and landing to act as an air scoop. The jet orifices were closed in cruise flight by twin doors. The fuel was housed in two integral fuselage tanks; one of these was placed between the lift engines and the main engine; the other was accommodated in the rear fuselage.

The tricycle undercarriage comprised single-wheel main units retracting forward into the fuselage, the wheels rotating around the oleos in so doing, and an aft-retracting castoring single-wheel nose unit. The wheel track on the prototypes and the first ten production machines was 2.2 m (7 ft 2⅝ in); on subsequent aircraft it was increased to 2.75 m (9 ft 0¼ in), the main gear units being redesigned to feature a fairly complex retraction sequence.

The aircraft's control system, in addition to the usual control surfaces, included air bleed ejectors placed under the nose and the tail cone and under the wingtips. The Yak-36M featured a unique crew rescue system. It was intended to tackle the possibility of an acute emergency arising instantaneously. To guarantee the pilot's safety in such cases, the aircraft was fitted with the KYa-1 ejection seat and the SK-EM system ejecting the pilot automatically, regardless of his will, the action being triggered by the attainment of definite bank or pitch angles, as well as by certain combinations of angle speeds (roll and pitch rates).

It took about a year to prepare the project and to issue the first detail drawings. On 10th January 1969 the OKB's prototype production facility (MMZ No.115) began construction of a flying testbed dubbed DLL which was intended for flight testing the powerplant. A complete *izdeliye* VM fuselage with the engines installed was suspended under the fuselage of the specially equipped Tu-16LL engine testbed. A little later, on 23rd January, the fuselage of the first VM prototype began to take shape in the assembly jig. In the OKB the first prototype of the VM was known as EVM or VM-01.

The DLL flying testbed was completed in late May 1969 and then transferred to TsIAM (Central Aero Engine Institute) where it was subjected to ground tests in the period between the end of 1969 and June 1970. In July 1970 the DLL was turned over to LII for flight testing. Two of LII's nine Tu-16LLs – '02 Blue' (c/n 4201002) and '10 Red' (c/n 1881110) – were fitted with this installation at one stage of their test career.

Coded '05 Yellow', the first prototype Yak-36M was completed on 14th April 1971. It was moved to the OKB's flight test facility in Zhukovskiy and underwent lengthy ground development and adjustment which began in mid-1970 and lasted for almost a year. In September 1970 the VM-01 piloted by Valentin G. Mookhin twice performed short vertical lift-offs under its own power.

The second prototype, designated VM-02 and coded '25 Yellow', was completed on 5th October 1970. In late Novem-

Yakovlev OKB

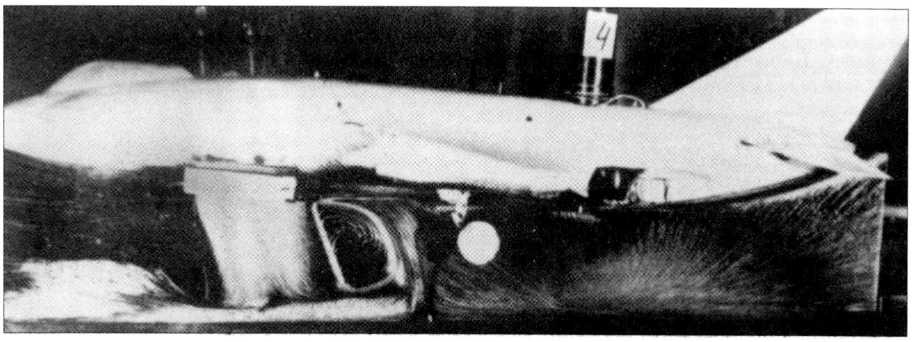

Yakovlev OKB

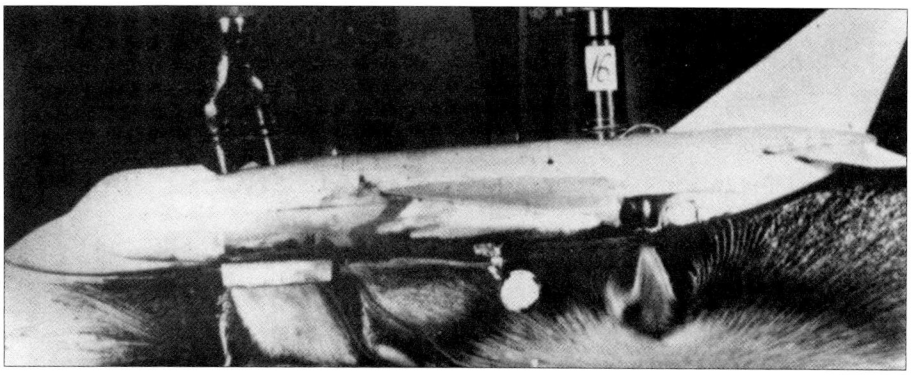

Two views of a model of the Yak-36M/*izdeliye* VM built to visualise the exhaust flow pattern of the main and lift engines by means of water jets.

Yakovlev OKB

Above: Tu-16LL '02 Blue' flies with the DLL full-scale test article lowered clear of the bomb bay for engine running

Yefim Gordon archive

Above: The first prototype Yak-36M (VM-01) with dummy Kh-23 air-to-surface missiles under the wings; it appears to be coded '01 Yellow'. Note the very short ogival nose of identical shape to the DLL test rig.

ber Mookhin put it through the first high-speed taxying runs at the LII airfield, and 20 days later the machine made its first lift-off (a hop). Construction of the third Yak-36M prototype ('55 Yellow') began in the same year; the VM-03 was completed on 29th March 1971. In the meantime the first two machines underwent further development. The VM-01 performed its first horizontal flight on 25th May 1971. On 16th June the VM-03, piloted by Shevyakov, also performed a horizontal flight, but rolled over on its back during a landing and had to undergo repairs which lasted until June 1972.

Yakovlev OKB

The first prototype (now with its proper code '05 Yellow') suspended from a dynamometric gantry during tethered tests to measure the engine thrust. Note the lift engines' open air intake and exhaust doors.

The first half of 1972 saw intensive manufacturer's flight testing of the Yak-36M. On 25th February the second prototype performed the first flight featuring the full flight envelope – that is, vertical take-off, horizontal flight and vertical landing. On 20th March the VM-01 performed a similar flight. From the late spring of 1972 the VM-01 was subjected to air intake modifications which necessitated a new round of testing the aircraft's control system.

By the summer of 1972 the VM-03 was fully repaired; on 19th June it made its first vertical take-off, and on 1st August, a flight over the full envelope. In late February 1972 construction began of the fourth prototype (VM-04, '45 Yellow').

In the summer of 1972 two prototypes, the VM-02 and VM-03, were submitted for joint State acceptance trials which were conducted by the Naval Air Arm' (AVMF – *Avi-**ah**tsiya Vo**yen**no-mor**sko**vo **flot**a*) together with MAP and the Yakovlev OKB. The tests were divided into stages A and B, the former being conducted with a simplified set of equipment. Both machines were to pass both stages of the tests. Later the VM-01 and VM-04 also joined the test programme.

The most important event of the stage A of the tests was the first landing of the Yak-36M on the deck of the helicopter carrier SNS (Soviet Navy ship) *Moskva* in the open sea. The landing was performed by test pilot Mikhail Deksbakh flying the VM-02 on 18th November 1972. On 22nd November he performed a full envelope flight comprising a vertical take-off from the deck and a vertical landing on the deck.

On 30th September 1974 all four prototypes completed Stage B of the State acceptance trials. A preliminary conclusion recommending the type to be put into series production had been signed in 1973, and tooling up for production, in anticipation of this decision, had been started at the Saratov aircraft factory No.292 as far back as 1970-71. Three production Yak-36Ms from the first batch (fuselage numbers 0101, 0201 and 0301) were built by the end of 1974. The second batch comprised five machines (f/ns 0102 through 0502) and all subsequent batches consisted of ten aircraft. They were equipped with RD36-35VF type (*izdeliye* 24) lift engines. In some accounts the suffix of this designation is spelled FV; accordingly, the later VFR version with a higher rotor speed is sometimes spelled FVR (R = *ras**kroo**chennyy* – lit. spun-up).

In 1975-76 the early-production Yak-36Ms were used mainly for ground tests in the course of which development work was performed with instruments, gunsights and various items of onboard equipment, as well as with the aircraft's armament.

Yakovlev OKB

Above: The second prototype Yak-36M (VM-02), '25 Yellow', had a longer and more pointed nose. Here it is again seen with Kh-23 ASMs. Note the recirculation dams under the fuselage.

Yakovlev OKB

Above: The VM-02 in cruise flight. Like the first prototype, it had a horizontally cropped fin.

Production Yak-36Ms manufactured by the Saratov plant were used either for various kinds of special tests or for the training of AVMF pilots. A shipboard aviation regiment was formed with Feoktist Matkov-skiy as its first commander. By the spring of 1975 the first Soviet aircraft-carrying cruiser SNS *Kiev* was ready for shipboard testing of the Yak-36M attack aircraft.

The test programme was started with the VM-02 in which factory test pilots practised landings and take-offs on the cruiser's deck. This training lasted from March until October; then came the turn of service pilots. On 15th December 1975 the regiment's commander Matkovskiy made the first landing on the deck of the *Kiev*. This marked the beginning of endowing the carrier with operational capability.

Yak-38 production shipboard attack aircraft

In the summer of 1976 the first squadron of Yak-36M shipboard attack aircraft was stationed aboard the *Kiev*. That same year the aircraft was introduced into squadron service under the designation Yak-38, thus acknowledging the obvious fact that it was a different design; its dual-control training version (described below) was designated Yak-38U. One account states, though, that the formal decision on service introduction and allocation of the Yak-38 designation was adopted by high authorities on 11th August 1977 pursuant to Communist Party Central Committee and Council of Ministers directive No.644-210, and the designation Yak-38U was formalised even later in accordance with MoD order No.196 on 15th October 1978. The aircraft received the NATO codename *Forger*.

The carrier could house 20 Yak-38s in the hangar below her deck. Production machines were sent to two airbases: Novofyodorovka AB in Saki on the Crimea Peninsula and Severomorsk AB near Murmansk on the Kola Peninsula. The choice of the latter location was due to the fact that an aircraft-carrying cruiser was to be included into the complement of the Northern Fleet. Furthermore, the Yak-38's VTOL capability made it amenable for operations from small

Yakovlev OKB

The VM-02 with four pylons and an impressive array of weapons in front of it, comprising S-24 unguided rockets, S-5 FFARs (and UB-16-57 and UB-32-57 pods for the latter), UPK-23-250 gun pods, Kh-23 air-to-surface missiles, R-3S air-to-air missiles and 250-kg (551-lb) 'iron' bombs.

Yakovlev OKB

Above: The third prototype, VM-03, had an out-of-sequence code ('55 Yellow'). Outwardly it was identical to the second prototype, retaining the old nose and fin shape and the old main gear design.

Yakovlev OKB

Above: The fourth prototype, VM-04, introduced a reshaped nose, a larger fin with a raked tip, modified cruise engine air intakes with splitter plates and extra blow-in doors, and a new wider-track landing gear.

Yakovlev OKB

The VM-4 performs a vertical take-off from the deck of the SNS *Kiev*, carrying napalm tanks on the outboard pylons. The aircraft has been updated to feature a nose pitot; note the blow-in doors.

land airstrips for the purpose of coastal defence.

The service introduction of the Yak-38 was no simple matter; inevitably, it was accompanied by incidents and crashes. Many such incidents happened from October 1978 on the aircraft-carrying cruiser SNS *Minsk*, a sister ship of the *Kiev*. Seven machines were lost between January 1979 and September 1980. In some cases the cause of the crashes was traced to the failure of the thrust-vectoring nozzles of the lift/cruise engine. However, the complex and untried VTOL technology made such losses inevitable. Anyway, available statistics show that the Yak-38's attrition rate was not higher than that of its British counterpart, the British Aerospace Harrier.

Interestingly, the Yak-38 was tested operationally not only in the northern climate, but in a hot-and-high environment as well. In April 1980 four such machines were sent to Afghanistan and remained there until mid-summer, receiving desert camouflage for the occasion instead of the normal deep blue colour scheme with green undersides. The tests resulted in the conclusion that with the powerplant available the VTOL attack aircraft was unsuitable for use in hot-and high conditions.

The Yak-38 underwent numerous modifications in the course of series production. For example, the original RD36-35VF lift engines were superseded by the more powerful RD36-35VFR (*izdeliye* 28), and the R27V-300 main engine was uprated (with no change to the designation).

The problem of preventing exhaust gas ingestion during take-off proved a major challenge. Several production Yak-38s were fitted experimentally with special strakes atop the fuselage on the sides of the lift engines' air intake and with strakes placed below the fuselage, running to the middle of its length, to act as recirculation dams. After testing this modification was introduced into production, and earlier series machines were retrofitted with these strakes.

The safety devices were also modernised in the course of production. The KYa-1 ejection seat designed in-house and the SK-EM automatic ejection system gave place to the Zvezda K-36VM seat (a version of the standardised K-36 model adapted for the Yak-38) and the SK-EMP system which could be used within a greater range of speeds and altitudes.

The armament carried by the Yak-38 was also subjected to updates and improvement. The weapons load was carried on four BD3-60-23F1 racks mounted on pylons under the inner wings, two on each side. The complement of external stores was intended for use against land and sea surface targets in the daytime and at night and, in case of need, against aerial targets in the daytime. Surface targets could be dealt with by using Kh-23 guided missiles in conjunction with the Del'ta-NG radio guidance equipment; to supplement this, there were unguided missiles, free-fall bombs weighing up to 500 kg (1,102 lb) apiece, ZB-500 napalm tanks and 'special stores' (almost certainly small nuclear bombs). Aerial targets could be hit by R-60 or R-60M homing missiles. The ordnance load could reach 1,000 kg (2,205 lb) in VTOL mode and 1,500 kg (3,315 lb) in STOL mode.

Attempts were made to provide the Yak-38 with a built-in GSh-23 twin-barrel cannon. However, as revealed during tests, firing this cannon often caused the engine to surge, and the designers had to give up the idea of fitting the cannon to the fuselage. The only viable option was the use of UPK-23-250 cannon pods suspended under the wings of the Yak-38.

A serious problem was the need to limit the weapons load in the event of operations in high ambient temperatures; retaining a normal ordnance load was possible only at the expense of reducing the fuel load and range. To tackle this problem, trials were conducted to determine the possibility of operating the Yak-38 in STOL mode which promised an increase in weapon load and range. The tests were conducted initially on dry land and then, in 1979, aboard the carrier SNS *Minsk*. This operating mode was introduced into service use.

Albeit the Yak-38 was primarily a shipboard aircraft, test were made to ascertain its suitability for land-based operations, in particular, from special mobile take-off pads. One more mode of employment was the operation of the Yak-38 from civil cargo vessels (container ships). A container ship was fitted with a take-off and landing pad measuring 18 x 23 m (59 x 75 ft) and covered with metal plating. The tests were successful, confirming the possibility of using container ships for delivering the Yak-38s to the aircraft-carrying cruisers stationed in far-off ocean regions.

Yak-38U (*izdeliye* VMU) trainer

This dual-control version of the Yak-38 (inhouse designation *izdeliye* VMU) was developed pursuant to a government directive issued on 28th December 1967. The second cockpit for the trainee was added ahead of the existing one which was occupied by the instructor. To provide sufficient forward view for the latter, the nose had to be drooped, producing an eerie 'broken back' impression. To preserve the CG position, a 'plug' was inserted aft of the wings. The two cockpits were unpressurised and were fitted with

ITAR-TASS

Above: Soviet Navy pilots sprint to their Yak-38s during a practice alert. Note the Soviet navy flag on the air intakes of these operational aircraft and the circles on the deck marking the jets' parking spaces.

Yakovlev OKB

Above: '05 Yellow', the prototype of the Yak-38U (*izdeliye* VMU) conversion trainer, displays its curious banana-like fuselage shape meant to improve the view forward for the instructor.

via Yefim Gordon

This weathered Yak-38U coded '24 Yellow' is seen languishing at Zhukovskiy in the summer of 2003.

Yakovlev OKB

Above: '82 Yellow', the prototype of the re-engined Yak-38M. The tactical code matches the product code, *izdeliye* 82.

Yakovlev OKB

A production Yak-38M coded '48 Yellow'. Outwardly the Yak-38M can be discerned from the original Yak-38 *sans suffixe* by the dorsal strakes flanking the lift engines' air intake.

Specifications of the Yak-38 and Yak-38M shipboard attack aircraft

	Yak-38	Yak-38M	Yak-38U
Lift/cruise engine, type	R27V-300	R28-300	R27V-300
thrust, kgp (lbst)	1 x 6,100 (1 x 13,450)	1 x 6,700 (1 x 14,770)	1 x 6,100 (1 x 13,450)
Lift engines, type	RD-36-35FVR	RD-38	RD-36-35FVR
thrust, kgp (lbst)	2 x 3,050 (2 x 6,725)	2 x 3,250 (2 x 7,165)	2 x 3,050 (2 x 6,725)
Length with nose probe	16.37 m (53 ft 8 47/64 in)	16.37 m (53 ft 8 47/64 in)	17.76 m (58 ft 3 7/32 in)
Wing span:			
full	7.022 m (23 ft 0 29/64 in)	7.022 m (23 ft 0 29/64 in)	7.022 m (23 ft 0 29/64 in)
with folded outer panels	4.45 m (14 ft 7 13/64 in)	4.45 m (14 ft 7 13/64 in)	4.45 m (14 ft 7 13/64 in)
Wing area, m^2 (sq ft)	18.41 (197.95)	18.41 (197.95)	18.41 (197.95)
Empty weight, equipped, kg (lb)	7,020 (15,480)	7,500 (16,530)	n.a.
All-up weight, kg (lb):			
in short rolling take-off mode	n.a.	11,800 (26,010)	n.a.
in vertical take-off mode	10,300 (22,710)	10,800 (23,810)	10,000 (22,045)
Fuel load, kg (lb):			
internal	2,750 (6,060)	2,750 (6,060)	2,750 (6,060)
external	none	800 (1,760)	none
Ordnance load, kg (lb):			
in short rolling take-off mode	1,500 (3,310)*	2,000 (4,410)	–
in vertical take-off mode	1,000 (2,205)*	1,000 (2,205)	–
Speed at sea level, km/h (mph)	1,210 (751)	1,210 (751)	850 (528)
Speed at altitude, km/h (mph)	1,100 (683)	n.a.	n.a.
Service ceiling, m (ft)	11,000 (36,090)	11,000 (36,090)	5,000 (16,400)
Range in short rolling take-off mode with 1,000 kg of weapons, km (miles)	600 (373)	600 (373)	–

*early production version (Yak-36M)

K-36VMU seats linked to the two-pilot SK-EM automatic system which ensured automatic simultaneous ejection on diverging trajectories, precluding a mid-air collision.

The first prototype (VMU-01) was built in the spring of 1972 and, after a protracted ground test cycle, made its first flight in late March 1973. It took another year for the machine to fly its first complete flight envelope. Series aircraft were ordered as the Yak-36U (some sources say Yak-36MU), but this designation was changed to Yak-38U in AVMF service. The type entered limited production, 38 machines being manufactured between 1975 and 1981. Machines from early batches had no dorsal strakes but later-production examples sported them, just like the late-production single-seaters.

Yak-38M (*izdeliye* VMM, *izdeliye* 82) upgraded light attack aircraft

The designers of the Yak-38 were well aware of the fact that this shipboard attack aircraft had a rather limited combat potential because of its short combat radius, insufficient weapons complement and inadequate equipment. All this prompted them to seek ways of radically upgrading the aircraft. Work on several projects commenced in the late 1970s. One of these was known in-house as the *izdeliye* VMM (VM *modern-**izee**rovannyy*, modernised). It envisaged equipping the aircraft with more powerful engines, revising the design of the air intakes, the wings and the tailplane, making the nose gear unit steerable and, importantly, adding a provision for carrying external fuel tanks.

The possibility of putting these plans into effect was associated with the new lift/cruise engine developed by NPO *So**yooz*** ('Union' Research & Production Association) under the direction of O. Favorskiy. This was the R28-300 (*izdeliye* 59) engine with a thrust rating of 6,700 kgp (14,770 lbst) in the vertical thrust mode which was, in effect, a derivative of the R27V-300 with a new LP rotor and a new nozzle. The HP rotor, combustion chamber and turbine were taken from the preceding model. The specialists of the Rybinsk engine design bureau also succeeded in improving the parameters of the lift engines. The new RD-38 lift engine had a thrust rating of 3,285 kg (7,240 lb). These engines were to be used in the new Yak-38 version under development.

In the course of projecting, the upgraded version was allocated a new in-house designation – *izdeliye* 82. Officially the new machine was designated Yak-38M ('modernised'). There were two flying prototypes ('82-1' and '82-2') which were completed in 1982. In addition, one example was built for static tests and yet another, dubbed LL-82 (possibly '82 Yellow', c/n unknown, f/n 0413), was to be used as a flying testbed for the new powerplant.

Outwardly the Yak-38M did not differ in any significant way from its predecessor. Apart from the engines, the changes included revised air intakes, some alterations in the fuselage and wing structure, a steerable nose gear unit instead of a castoring one and 'wet' pylons for the carriage of drop tanks. There were also changes in the armament and equipment complement. At the end of 1982, prior to the commencement of testing, a decision was taken to launch production of *izdeliye* 82.

The tests began in 1983 and lasted several years. The Yak-38M displayed an improvement in performance as compared to the Yak-38 *sans suffixe*. The AUW in the STOL mode rose to 11,800 kg (26,020 lb) and the maximum weight of external stores to 2,000 kg (4,410 lb). The range increased to 410 km (255 miles) in VTOL mode with an ordnance load of 750 kg (1,654 lb), or 600 km (373 miles) in STOL mode with an ordnance load of 1,000 kg (2,205 lb).

The new model superseded the Yak-38 at the production line in Saratov. From the mid-1980s onwards the Yak-38M was delivered to ship-based squadrons. Yet, the original plans for raising the efficiency of the shipboard attack aircraft were not fully implemented. The majority of Yak-38Ms in squadron service failed to receive the drop tanks, while the new powerplant consumed more fuel. This entailed a further reduction in the machine's combat radius. The number of Yak-38s and Yak-38Ms manufactured for the Soviet Navy totalled 231.

The service career of the Yak-38 tapered off in the early 1990s, not least because of the problems associated with the demise of the Soviet Union which had their repercussions for the state of affairs in the Russian Navy. By that time many Yak-38s were approaching the end of their service life. Most of the machines were sent to coastal bases. The Saratov aircraft plant proved unable to launch series manufacture of external fuel tanks for the Yak-38Ms, without which the aircraft's operational capabilities were severely restricted. The government of Russia failed to provide the necessary funding for extending the service life of these aircraft. At present all of them are placed in storage and their future fate is uncertain.

Examples of the Yak-38 have been preserved in the Yakovlev OKB museum (the VM-2 prototype), in the Central Russian Air Force Museum in Monino ('14 Yellow', c/n 7977864401137; 797 is a code for plant No.292, 786 is a product code for the Yak-38 *sans suffixe*, 44 means fourth quarter of 1974, while the rest does not mean anything

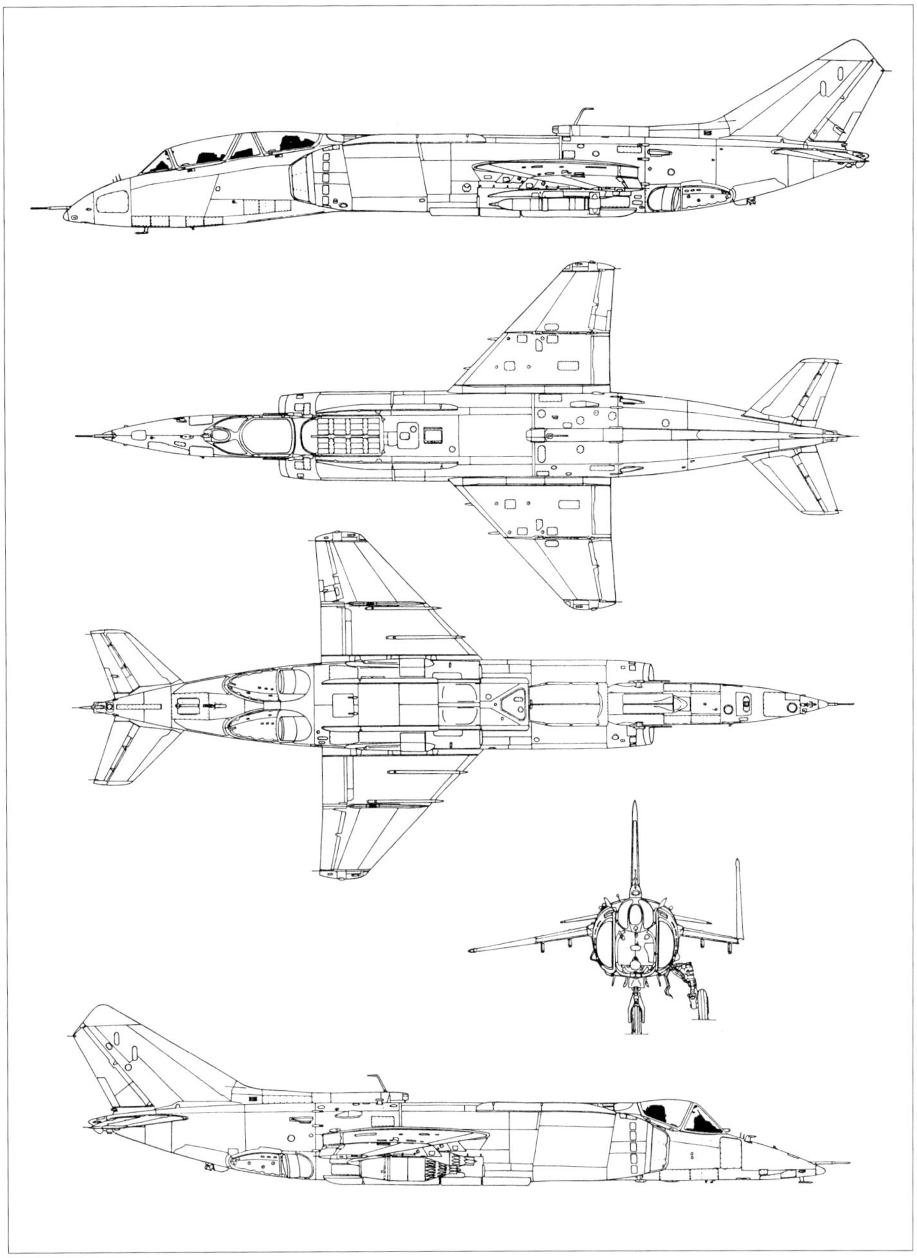

Above: Four views of the Yak-38M, with an additional side view of the Yak-38U trainer.

Victor Drushlyakov

This model depicts a proposed supersonic derivative of the Yak-38, an early step towards the future Yak-41. Note the single afterburner nozzle, the lift nozzles aft of the fuselage break point and the boxy air intakes.

at all so as not to reveal the batch number/number of the aircraft in the batch and hence how many have been built) and in the Ukrainian National Aviation Museum at Kiev-Zhulyany. Yak-38 *sans suffixe* '60 Yellow' (c/n 7977864060699) was preserved in the open-air museum at Moscow-Khodynka, but its fate after the closure of the museum is uncertain. The fourth prototype (VM-4) is now a cutaway instructional airframe at the Moscow Aviation Institute.

Yak-36A attack aircraft (project)

This project may be regarded as a development of the Yak-36M (Yak-38) with an R-49V lift/cruise engine and two lift jets. The work on the project went as far as the construction of the fuselage which was configured as a test vehicle to be suspended under a Tu-16LL engine testbed. However, the project was not proceeded with.

Yak-36O shipboard attack aircraft (project)

One account states that 'this development of the Yak-36M was to be an improved ship-based operational aircraft. It was to have had a redesigned fuselage housing a Type 55 engine in the 15,000-kg (33,070-lb) thrust class. Later this engine was to be replaced by the R-61V. These engines have not yet been identified.' The meaning of the O suffix is likewise unknown.

Yak-38L attack aircraft (project)

The designation Yak-38L applied to a projected variant of the Yak-38 with a Lyul'ka AL-21F turbojet appropriately modified as a lift/cruise engine, hence the L.

Yak-38MP fighter/attack aircraft (project)

This projected upgrade of the Yak-38 was to feature a Fazotron N019 fire control radar (the same model as fitted to the MiG-29 fighter) in a redesigned and fatter nose featuring an ogival radome, plus advanced navigation/attack systems.

Yak-39 ('39') fighter/attack aircraft (project)

One of the projected upgraded derivatives of the Yak-38 was designated '39' and was sometimes referred to as Yak-39. It was to be a multi-role aircraft that would be built in several versions, including a fighter and an attack aircraft. It would differ from the Yak-38 in having more powerful engines, greater wing area and a multi-mode radar forming part of the PrNK-39 targeting and navigation suite (*pri**tsel'**no-navigatsi**on**nyy **kom**pleks*). The aircraft was to have greater fuel capacity and increased spectrum of weapons. The project was not given the go-ahead and remained on the drawing board due to 'lack of interest on the part of MAP and the AVMF'.

Yak-36P interceptor (project)

This was to be a pure air defence version of the Yak-36M (hence P for *pere**khvaht**-chik* – interceptor) equipped with a radar and armed with medium-range AAMs to protect carrier groups against attack by enemy aircraft. The project was under consideration at an early stage of the work on the Yak-36M. Design features included the installation of three RD36-35 lift engines (versus two on the Yak-36M), apparently in connection with a higher AUW.

Yak-36-70F VTOL supersonic fighter (project)

This supersonic light fighter was studied by the OKB in 1970. It was to be powered by two afterburning lift/cruise engines with variable inlets; there were no lift engines. The aircraft featured a bicycle undercarriage.

Yakovlev OKB

Yakovlev OKB

Above: These desktop models represent some of the earliest project studies that led to the Yak-41 and have a downright cartoon-like appearance. Note that the one on the right has a supersonic air intake with a movable half-cone and main gear units retracting aft into fairings projecting beyond the wing trailing edge.

Yakovlev OKB

Yakovlev OKB

Two more projects leading to the Yak-41. The one on the left has lift engines in the nose ahead of the cockpit, an idea soon abandoned, and a two-dimensional main engine nozzle. The model on the right is virtually the final project configuration.

Despite the '36' designation, it is presumably structurally unrelated either to the original Yak-36, or to the Yak-36M (Yak-38).

Yak-41 (*izdeliye* 48) supersonic VTOL fighter project

The work on the project of a supersonic VTOL aircraft designated Yak-41 (*izdeliye* 48) was initiated by the Yakovlev OKB in 1975. It was formally endorsed by a government directive issued in November 1977 which tasked Yakovlev with creating a supersonic VTOL fighter; the latter was to be submitted for State acceptance trials in 1982. The work was conducted initially under the guidance of Sergey A. Yakovlev (the General Designer's elder son) as project manager.

As is usual at an early project stage, several widely differing aerodynamic configurations and powerplant layouts were considered. Initially preference was given to a layout based on the use of a single lift/cruise engine. Gradually the emphasis shifted to studying alternative layouts based on the use of a powerplant comprising two types of engines *à la* Yak-38, albeit the work on the single-engine layout continued for a while.

The earliest project version of the Yak-41 was an aircraft featuring a single lift/cruise engine with a single large vectoring nozzle. A model shows it with a fuselage and tail unit similar to those of the Yak-38, but featuring a pointed nose cone (possibly intended to house a radar) and a chin air intake with a movable half-cone centrebody beneath this cone; the moderately swept wings had small leading edge root extensions (LERXes) and a single vectoring engine nozzle was placed under the fuselage right aft of the wings.

Another proposed version made use of an aft-mounted lift/cruise turbofan with a two-dimensional thrust-vectoring nozzle. It used separate outlets for the core and bypass flows, the bypass flow being ducted to a small afterburner chamber in the forward fuselage; it was intended to balance the thrust created by the flat vectoring nozzle. The layout was, however, considered too complicated to be viable.

Yet another, completely different Yak-41 configuration studied in 1979 was based on the Yak-45 twinjet fighter project that earlier had lost out to the Su-27 and MiG-29. In its original form, the Yak-45 had a normal tail unit in combination with composite-sweep (double-delta) wings carrying two engines underneath. The project of the VTOL version inherited the cranked wings of the Yak-45, replacing the normal tail unit with canard foreplanes and adding thrust-vectoring nozzles to the underwing-mounted main engines, which were moved forward to place the nozzles under the wings. The machine would also be fitted with two widely spaced RD-38 lift engines in the centre fuselage. The rather exotic layout was not proceeded with. This list of project variants is by no means exhaustive.

Victor Drushlyakov

A model of the proposed production version of the Yak-41M with the wings folded. Note the fuselage break point and the ECM pods at the wingtips.

Another series of studies, eventually leading to the final prototype configuration, was based on the use of a composite powerplant, wings with prominent LERXes and a twin-boom tail unit with twin fins and rudders. The latter feature was due to the need to place the vectoring nozzle of the main (lift/cruise) engine closer to the CG so as to shorten its arm and ensure balance with the lesser thrust of the two lift engines placed in the forward fuselage. In one of the studies, the lift engines were placed in the extreme nose, between the cockpit and the radar, in order to widen the distance between these engines and the vectoring nozzle of the main engine. This layout proved unacceptable for a number of reasons. The lift engines had to be placed aft of the cockpit. An interim design study coming close to the final shape differed from it only in having Dassault Mirage-style semi-circular air intakes of the lift/cruise engine; they gave way to the MiG-31-style raked rectangular intakes on another study coming still closer to the future prototype.

In March 1979 the OKB completed the advanced development project and a full-scale mock-up of an aircraft powered by a single lift/cruise engine – a Khachatoorov R79V-300 afterburning turbofan. At the same time the OKB submitted to the Ministry of Defence a proposal for a multi-purpose fighter capable of carrying a wider weapons complement and equipped with a composite powerplant. The latter was to comprise two RD-41 lift engines, each delivering a thrust of 4,100 kg (9,040 lb), and a single R79V-300 lift/cruise engine rated at 15,500 kg (34,180 lb). That was deemed sufficient to enable the aircraft to lift off vertically or with a short take-off run from a carrier deck at an AUW of 19,500 kg (43,000 lb).

Yet, the design could not be 'frozen' because of a series of changes successively introduced into the SOR which initially envisaged a fighter aircraft, then, additionally, an attack aircraft based on the latter, and finally, a multi-purpose aircraft which was intended to perform missions ranging from beyond visual range (BVR) intercept and dogfighting to attacks against surface ships and land targets. Accordingly, the construction of prototypes was repeatedly postponed, not least because of delays in the development of the new engines. It was not before the end of 1984 that the R79V-300 lift/cruise engine was ready for bench testing. The Yak-41's submission for tests was postponed to 1985, but that proved to be unrealistic, too. Yakovlev's retirement in 1984 was one of the circumstances impeding the work on the VTOL machine; as a result, the directive of 1977 was to remain unfulfilled.

Yak-41M (Yak-141, *izdeliye* 48M) supersonic V/STOL fighter

In 1986 one more in the series of government directives was issued, tasking the Yakovlev OKB with creating the Yak-41M shipboard multipurpose fighter based on the Yak-41 design studies. Submission of the Yak-41M for State acceptance trials was slated for 1988, to be followed by the Yak-41UT trainer (*oo**chebno**-treni**rov**ochnyy*) in 1989. At the same time the task for the development of an attack aircraft based on the Yak-41 was cancelled (the multi-purpose role of the Yak-41M, also known as *izdeliye* 48M, included a ground-attack capability).

The OKB built a small batch of Yak-41Ms comprising four machines. Of these, one was intended for static tests, another one (coded '48 Yellow') was to be used for evaluating the forces and loads acting on the aircraft in different flight modes, and checking the functioning of the powerplant. The other two examples coded '75 White' and '77 White' (known at the OKB as the '48-2' and

Yakovlev OKB

Above: The non-flyable first Yak-41M (Yak-141), '48 Yellow', at the Yakovlev OKB's flight test facility in Zhukovskiy. Note the light grey colour scheme, the LERXes and the large nosewheel.

'48-3' respectively) were subjected to flight testing both at shore bases and in shipboard operations from the deck of the aircraft carrier ('heavy aircraft-carrying cruiser') SNS *Fleet Admiral Kuznetsov*.

The Yak-41M was a single-seat aircraft with shoulder-mounted moderately swept wings featuring full-span leading edge slats outboard of large curved leading-edge root extensions (LERXes). The trailing edge was kinked at the wing folding joints, the inboard parts running at right angles to the fuselage; these parts carried the flaps, while the outboard parts were occupied by ailerons. The slab-sided fuselage was area-ruled. The nose incorporated a fire control radar in an ogival radome; placed behind the cockpit in tandem were two RD-41 lift engines, their common air intake and jet nozzles protected by doors which opened for take-off and landing in the same fashion as on the Yak-38. The lift engines were flanked by the ducts of the R79V-300 lift/cruise engine's two-dimensional air intakes featuring horizontal flow control ramps. The engine itself had a unique thrust-vectoring nozzle comprising three wedge-shaped segments rotating in opposite directions. Their rotation in skewed planes caused the jet pipe to bend within an angular range of 95°, directing the jet efflux downwards; the rearmost segment featured controllable nozzle petals.

The jet nozzle was flanked by two tail booms carrying the twin fins and slab stabilisers mounted on the outboard sides of the booms. The use of this arrangement aligned the Yak-41M's vectored thrust as close as possible to the CG, thus minimising the power balancing requirements from the lift engines. A large two-segment snap-action door in the rear fuselage underside enclosed the nozzle of the R79, hinging downwards like an air brake in the vertical thrust mode to act as a recirculation dam.

A special automated system catered for the control of the three engines' nozzle vectoring in coordination with the deflection of the tailplane. The aerodynamic control surfaces were supplemented by reaction control nozzles (air bleed ejectors) which were arranged differently on the two flyable examples: '75 White' had air bleed ejectors for directional control in the tail booms, while '77 White' had jet control nozzles in the nose.

The aircraft were fitted with a comprehensive avionics suite for navigation and flight control, as well as for target seeking and weapon aiming. The armament included a 30-mm (1.18 calibre) GSh-301 cannon with 120 rounds in a semi-buried installation under the fuselage, offset to port. External stores with a total weight of 2,600 kg (5,730 lb) could be carried on four underwing pylons and comprised various types of missiles of three main categories: AAMs, air-to-surface missiles for use against land targets and anti-shipping missiles.

The first test flight involving a conventional take-off and landing took place on 9th March 1987. The manufacturer's testing took more time than anticipated, and the State acceptance trials (originally slated for 1988) had to be postponed. This decision was accompanied by a change in the designation of the aircraft which forthwith bore the name Yak-141. (Some reports, however, associate the change of designation with the events of 1991, when the Yak-41M piloted by A. Sinitsyn established a series of world records. The Yak-141 designation was used in the application to FAI for ratification of these records, allegedly because the Yak-41M designation was still classified at the time).

It was not before late 1989 that tests of this aircraft in VTOL mode were started. The first flight with a vertical take-off and landing was performed by test pilot A. Sinitsyn on 13th June 1990. In September 1991 the aircraft entered the phase of active testing in shipboard conditions. Initially everything

Yakovlev OKB

Above: '75 White', the first flying prototype of the Yak-41M (Yak-141), in low-speed forward flight with the flaps fully deployed. Note the dark grey finish and the smaller nosewheel.

Yakovlev OKB

The same aircraft hovers above the aircraft carrier SNS *Admiral Gorshkov*, showing the open lift engine intake and exhaust doors, the auxiliary blow-in doors and the door ahead of the main engine nozzle.

proceeded normally, both aircraft having performed successful vertical landings on the deck of the heavy aircraft-carrying cruiser SNS *Admiral Gorshkov* (formerly *Baku*). However, on 3rd October 1989 test pilot Yakimov flying Yak-141 '77 White' exceeded the admissible descent speed during a vertical landing on the *Admiral Gorshkov*, which resulted in a crash. On impact the mainwheel struts punched through the fuel tanks and a massive fire ensued; the pilot ejected to safety but the aircraft was seriously damaged, never to fly again. After the incident, flight tests of the remaining airworthy Yak-141 were suspended.

In September 1992 Yak-141 '75 White' was publicly demonstrated for the first time at the Farnborough International '92 airshow and later at various other shows. For the FI'92 appearance the aircraft, previously painted flat grey overall, received a two-tone grey camouflage scheme and the symbolic new code '141 White'. The written-off Yak-141 '77 White' was rebuilt as a static exhibit, confusingly receiving the same colour scheme and the same code '141 White'. It is now a museum piece at the Yakovlev OKB.

The accident described above served as a formal pretext for closing down the supersonic VTOL fighter programme. The real reason was the political and economic situation in the country which was lapsing into a state of disarray. The lack of funding caused termination of the tooling up for series production of the Yak-141, which had started at the Saratov aircraft factory. For some years thereafter the Yakovlev OKB continued on its own the work on a further development of the VTOL supersonic aircraft in the hope of obtaining orders at home and abroad.

Yak-141M VTOL supersonic fighter (project)

One of the projects representing the result of these efforts was the Yak-141M. As follows from the published drawing of one of the studies of this project, it retained the fuselage and tail unit of the baseline version, combining them with new wings and altered air intakes of the lift/cruise engine. The new wings featured a completely different planform; basically they were delta wings with increased sweep on the leading edge and without LERXes. Instead of the latter, the sides of the box-like air intakes were canted outwards.

In comparison with the baseline Yak-141, the modified version retained the same dimensions, including wing area. Its design performance figures were superior to those of its predecessor an almost all counts (range, service ceiling, ordnance load), with the exception of the maximum speed which

Yakovlev OKB

Above: The ill-fated second prototype of the Yak-141, '77 White', also had a small nosewheel. No pylons are fitted, although the attachment points are clearly visible under the port wing.

Yakovlev OKB

A fine air-to-air of '77 White' climbing above thick overcast.

Design specifications of the Yak-141 (Yak-41M) and Yak-141M

	Yak-141 (Yak-41M)	Yak-141M
Length	18.3 m (60 ft 0 in)	18.3 m (60 ft 0 in)
Height	5.0 m (16 ft 4⅞ in)	5.0 m (16 ft 4⅞ in)
Wing span, full	10.1 m (33 ft 1⅝ in)	10.1 m (33 ft 1⅝ in)
with outer panels folded	5.9 m (19 ft 4⅜ in)	5.9 m (19 ft 4⅜ in)
Wing area, m² (sq ft)	31.7 (341.24)	31.7 (341.24)
All-up weight, kg (lb):		
in vertical take-off mode	15,800 (34,840)	15,800 (34,840)
in short rolling take-off mode	19,500 (43,000)	21,500 (47,400)
Maximum speed, km/h (mph)		
at sea level	1,250 (777)	1,250 (777)
at 11,000 m (36,090ft)	1,800 (1,119)	1,600 (995)
Service ceiling, m (ft)	15,000 (49,215)	15,500 (50,855)
Combat action radius with a 2,000-kg (4,410-lb) warload and a 120-m (390-ft) take-off run, km (miles)	690 (429)	900 (559)
Service range, km (miles):		
vertical take-off, no payload:		
at sea level	650 (403)	1,100 (684)
at 10,000 m (32,800 ft)	1,400 (870)	2,400 (1,490)
rolling take-off, 1-tonne payload:		
at sea level	1,010 (628)	1,100 (684)
at 10,000 m	1,400 (870)	2,400 (1,490)
at 10,000 m with external tanks	2,100 (1,305)	n.a.

Yakovlev OKB

Above: Yak-141 '141 White' hovers during a demonstration flight at the Farnborough International '92 airshow. Note the recirculation dams deployed under the air intakes.

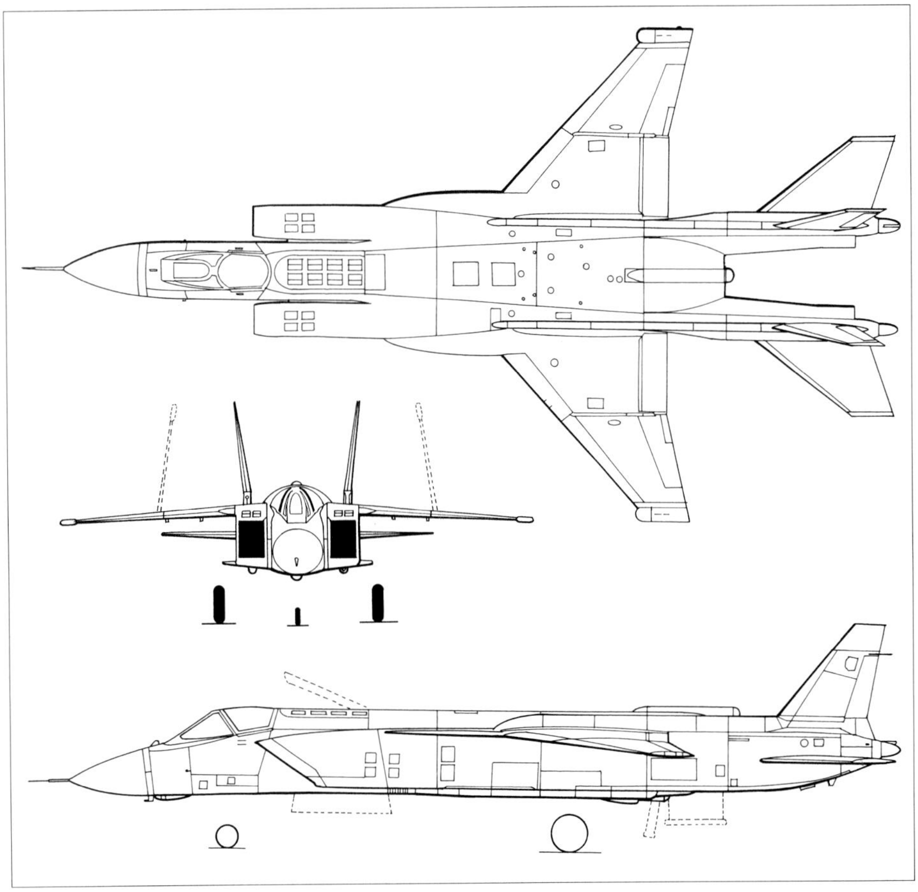

remained the same at sea level and was slightly lower at altitude.

Yak-43 multi-role V/STOL fighter (project)

In 1983-84 the Yakovlev OKB embarked on the design of a next-generation multi-role V/STOL fighter intended to follow the Yak-41M. In its general layout it had a marked resemblance to the Lockheed Martin F-22 Raptor (albeit the latter was designed to a somewhat later time-scale and had no VTOL capability). Like its predecessor, the Yak-43 was to have a cruise engine and separate lift jets; these were to be supplemented by an auxiliary combustion chamber in the nose provided with air bleed piped from the main engine. The latter was to be a derivative of the Kuznetsov NK-321 three-shaft afterburning turbofan with a take-off rating of 24,980 kgp (55,070 lbst; four such engines power the Tupolev Tu-160 strategic bomber).

The Yak-43 promised a considerable improvement on the Yak-41M in performance and combat capabilities. The wings of bigger area ensured greater agility; the increased fuel tankage afforded a longer combat mission radius, the aircraft would be capable of carrying a bigger weapons load and would have a reduced radar signature. However, with the Yak-41M (Yak-141) shelved, the new project had no chances of reaching the hardware stage.

Joint work with Lockheed Martin on a VTOL aircraft project

In 1995 the Yakovlev OKB received permission from the Russian Ministry of Defence to conduct joint work with the Lockheed Martin Company (USA) on the development of a new-generation VTOL combat aircraft. The US partners estimated the cost of research and development work under this programme intended to last until 1998 at some US$ 4 billion. The Yakovlev OKB promptly rendered the necessary assistance to the Lockheed Martin Company in solving many

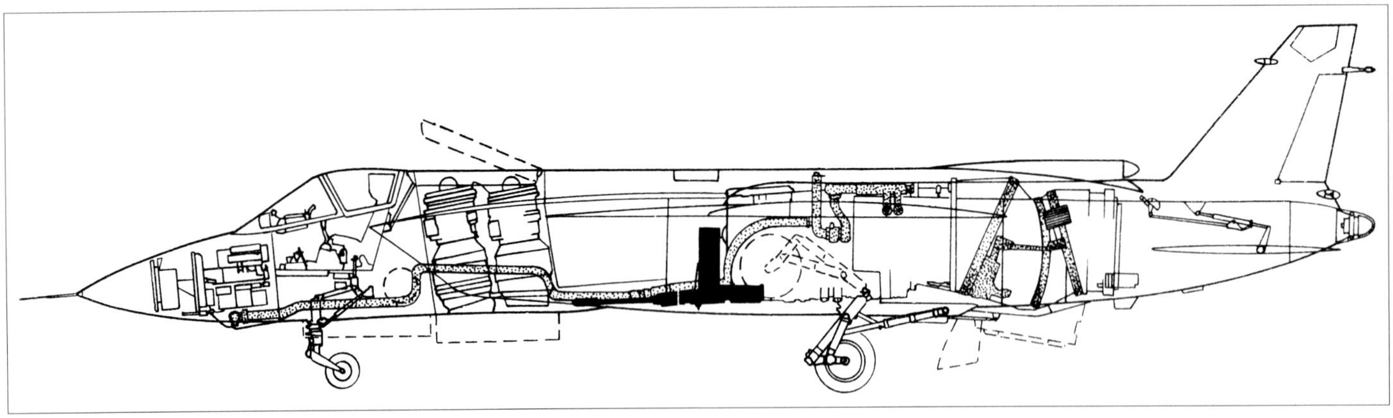

A three-view drawing and a cutaway drawing of the Yak-141.

basic design problems on a concept level, associated with the choice of the main parameters of a VTOL fighter as a whole and of its powerplant, as well as operational problems concerned with exhaust gas recirculation and flight control during take-off and landing.

Upon completion of this stage the US partner took a decision to conduct further work on the project entirely on its own, albeit the potential of co-operation had by no means been exhausted. In late 2001 Lockheed Martin rolled out its X-35B technology demonstrator which, in the opinion of some specialists, had much in common with the Yak-141. As is well known, the Lockheed Martin project became the winner of the Joint Strike Fighter (JSF) contest and the company received an order for the development and production of the F-35 aircraft under this programme.

Yefim Gordon archive

A model of the projected aircraft carrier SNS *Ul'yanovsk* showing a Yak-44 AEW aircraft sharing the flightdeck with Sukhoi Su-33 (Su-27K) fighters.

Yak-44E AEW aircraft (project)

The Yak-44E project owes its inception to an ambitious programme aimed at building a fleet of conventional take-off and landing (CTOL) aircraft carriers which had been pursued by the USSR since the mid-1960s. It was fairly obvious that operations of carrier-borne combat jets would not be effective unless they were backed up by adequate and timely information about the adversary forces in operational areas. This information could be provided only by a shipboard airborne early warning (AEW) aircraft.

The Yakovlev OKB was tasked with the development of such a machine as early as 1979. In addition to the baseline AEW version, Yak-44 variants for other roles, including anti-submarine warfare (ASW), were to be developed in due course of time. By November 1979 the OKB submitted a technical proposal envisaging two alternative types of mission equipment for the Yak-44E. The first of them, dubbed ***Fakel*** (Torch), comprised two radar antennas housed inside the fuselage and accommodated respectively in the fuselage nose and tail cone. The other option, called E-700, featured a radar antenna housed in a rotodome mounted on a pylon above the fuselage. Initially the go-ahead for further development was given to the Fakel-equipped version of the Yak-44E; its powerplant was to comprise two turboprops in underwing nacelles for cruise and four lift engines in the fuselage, to be used for STOL carrier-deck operations. However, placing the lift engines in the fuselage made it very difficult to provide enough space for the mission avionics. Difficulties encountered in the projecting of the Fakel radar system itself brought this work to a standstill, putting the Yak-44E project as a whole in jeopardy.

Concurrently, the Antonov OKB was working on an alternative AEW aircraft project known as the An-71 and intended for the Air Force. Attempts were made to adapt this project for naval use, but eventually it was recognised that the size and AUW of the An-71 made it unsuitable for carrier-borne operations. As a result, the Yakovlev OKB was called upon to resume the work on the Yak-44E, this time on the basis of a new concept which dispensed with the lift engines. The new version of the Yak-44E was to be powered by two D-27 propfan engines driving Aerosila SV-27 contraprops and be equipped with the E-700 AEW mission suite, featuring a revolving disc-shaped radome (rotodome) above the fuselage.

The Yak-44E was an AEW and airborne command post aircraft; its mission included monitoring the airspace and surface situation in designated areas, supplying information about the adversary's actions and target designation for combat aircraft. In its general configuration this shoulder-wing aircraft closely followed the Grumman E-2C Hawkeye; however, it was by no means a direct 'copy', incorporating some different design features.

Unlike the E-2, whose rotodome was supported by a rigid tripod-like structure, the Yak-44's rotodome was mounted on a telescopic pylon and could be lowered by more that 1.8 m (5 ft 10⅞ in). This, coupled with the twin-fin tail unit, was intended to reduce the aircraft's height for below-deck stowage. There were only two vertical tails (the Hawkeye had additional dorsal 'half-fins' at about half-span of each stabiliser); on the other hand, the rudders were double-hinged, whereas the E-2 had simple rudders. The Yak-44 had unswept high aspect ratio wings equipped with winglets (a feature absent on the E-2), powerful trailing-edge flaps and slats. The wing panels outboard of the engines folded upwards about skewed hinges, intersecting when folded; in contrast, the E-2's wings folded aft, rotating nose-down in the process.

The powerplant comprised two Muravchenko D-27 propfans designed by ZMKB 'Progress' (*Zaporozhskoye motorno-konstrooktorskoye byuro* – 'Progress' Zaporozh'ye Engine Design Bureau); D-227s are also mentioned in some sources. The engines were rated at 14,000 eshp (10,290 kW), driving Aerosila SV-27 contraprops of 4.49 m (14 ft 8¾ in) diameter. The front and rear rows had eight and six scimitar-shaped blades respectively, running at different speeds to reduce noise and vibration. Thus, being only marginally bigger than the Hawkeye, the Yak-44 had 2.5 times the engine power. Accordingly, it had almost twice the weight and a far greater fuel capacity.

The mission equipment included a powerful pulse-Doppler search radar developed by NPO Vega-M. In its characteristics it was expected to be considerably superior to the Westinghouse AN/APS-138 radar installed on the E-2C Hawkeye. The heat exchanger for the radar set was served by a ram air inlet ahead of the wings, just like on the Hawkeye. The aircraft featured a fly-by-wire flight control system. The pressurised fuselage provided accommodation for a crew of five with a high degree of comfort for long-duration flights (a toilet, a galley and a rest area with bunks were provided). The tricycle undercarriage had a track of 7.929 m (26 ft) and a wheelbase of 9.373 m (30 ft 9 in); unlike the E-2, the main units had twin wheels. The airframe would be fully ice-protected, with pneumatic de-icer boots on the tail surfaces and hot-air de-icing on the wings, and stressed for catapult launch and arrested landings. At the same time the aircraft was suitable also for take-offs from the ski-jump deck incorporated on the Russian Navy's first (and so far only) CTOL carrier.

The Yak-44E was intended to enter Soviet Navy service for operations from *Kuznetsov* class aircraft carriers. More exactly, it was to be stationed on board the carriers *Fleet Admiral Kuznetsov* and *Ul'yanovsk*, both officially termed heavy aircraft-carrying cruisers.

The former ship (Project 1143.5) was a carrier intended originally for catapult launching of aircraft but completed without deck catapults and with a ski-jump nose deck. The ship was laid down in 1982 as the SNS *Leonid I. Brezhnev*; she was launched as the SNS Tbilisi and bore this name during the tests before finally entering service with the Navy of a different state – the Russian Federation – as the RNS *Fleet Admiral Kuznetsov*.

The SNS *Ul'yanovsk* (project 1143.7) was to be the first Soviet fully-fledged CTOL aircraft carrier with a nuclear powerplant; for the first time she was going to be fitted with deck catapults for launching aircraft. On 4th October 1988 the as-yet non-existent ship was officially included into the inventory of the Soviet Navy. Construction started on 25th November 1988 at the Black Sea Shipyard in Nikolayev, the Ukraine, where all other Soviet aircraft carriers had been built. However, on 1st November 1991, when the ship was 20% complete, construction was discontinued and the ship was excluded from the inventory; on 4th February 1992 she was scrapped.

An advanced development project of the new version of the Yak-44E was completed by September 1988. It was officially endorsed by an appropriate directive of the Communist party Central Committee and the USSR Council of Ministers in January 1989; the document also called for the development of other carrier-based derivatives of the Yak-44E. The aircraft was also intended for introduction into Air Force service.

From June 1989 onwards the Yakovlev OKB commenced detail design of the Yak-44E; a full-scale mock-up was built and submitted to the mock-up review commission in January 1990. However, further work on the project was seriously hampered by turmoil following the collapse of the Soviet Union and the ensuing drastically reduced funding. In 1992, following a decision to cancel the construction of the aircraft carrier *Ul'yanovsk*, the work on the project was terminated at a stage when preparations were in hand for the construction of a prototype. For some time the Yakovlev OKB kept the project afloat and continued at least until 1995 to promote the aircraft in the export market. These efforts failed to bring tangible results, nor did expectations of a renewed interest from the Russian Navy materialise. In the opinion of Russian specialists, the Yak-44E project, despite a considerable lapse of time since then, retains its value and competitiveness; if need be, it can be resurrected to meet present-day requirements.

Design specifications for the Yak-44E AEW aircraft

Length	20.39 m (56 ft 10¾ in)
Wing span	25.7 m (84 ft 3¾ in)
All-up weight, kg (lb)	40,000 kg (88,180 lb)
Maximum speed, km/h (mph)	740 (460)
Service ceiling, m (ft)	13,000 (42,650)
Ferry range, km (miles)	4,000 (2,486)
Runway length for free take-off, m (ft)	1,300 (4,265)

Yakovlev OKB

The full-scale mock-up of the Yak-44E with wings folded and rotodome lowered on the deck of one of the Soviet aircraft carriers; the rotodome is covered by camouflage netting. The similarity to the Grumman E-2 Hawkeye and the difference from it are equally obvious.

10

HELICOPTERS OF THE YAKOVLEV OKB

EG (Sh, Yak-M-11FR-1) experimental helicopter

In 1946 the Yakovlev Design Bureau embarked on the development of an experimental helicopter which was designated EG (*Eksperimen**tahl'**nyy gheli**kop**ter* – experimental helicopter); the designation is sometimes altered to Yak-EG in recent accounts. It was also referred to as Sh (from ***shoot**ka* – joke, a name coined from a surprised exclamation of a design staff member whose reaction to the announcement of the new task was 'are you kidding?'). Yet another name for it that gained some currency was Yak-M-11FR-1; this designation included a reference to the engine around which the machine was designed. It was a 140-hp five-cylinder radial used mainly on light utility, sports and training aircraft (including many of Yakovlev's earlier designs, as described in the preceding chapters). The engine was mounted in the normal attitude (with its shaft horizontal) and drove co-axial two-blade rotors via a transmission comprising a cooling fan, a centrifugal clutch and a 90° bevel gearbox. The rotors turned at 233 rpm in opposite directions and featured blades of laminated pine and hardwood, covered with glued fabric. The rotors were mounted on a fully articulated hub with swashplates giving collective and cyclic pitch control.

The fuselage was a simple truss of welded steel tube with duralumin skinning as far back as the rear of the engine compartment. The rearmost part of the fuselage tapered off to form a fin supplemented by two endplates mounted at the tips of a horizontal tail. This arrangement was expected to provide better stability in cruising flight. The aft fuselage was fabric-covered. The tricycle undercarriage comprised main units with vertical shock struts mounted on trusses attached to the sides of the engine compartment, and a nose unit with a castoring wheel mounted on a vertical shock strut. The nosewheel featured a levered suspension. The extensively glazed cockpit accommodated a crew of two seated side by side and was accessed through a door on each side. Prior to building the prototype, the design team headed by S. A. Bemov as project engineer built a subscale flying model with coaxial rotors which was used for verifying the layout of the future 'real thing'.

The EG (Sh) was completed in the summer of 1947, undergoing flight tests at the hands of the OKB's chief test pilot V. V. Tezavrovskiy who, after a number of tethered tests, performed the first free flight on 20th December 1947. In the course of testing the horizontal tail together with the endplates was removed. In early 1948 the M-11FR-1 engine was replaced by an M-12 five-cylinder radial, with which the machine made the first flight on 9th April. However, the engine

Specifications of the EG helicopter (based on test results)

Empty weight, kg (lb)	878 (1,935)
All-up weight, kg (lb)	1,020 (2,248)
Maximum speed, km/h (mph)	150 (93)
Rate of climb, m/sec (ft/min)	3.1 (610)
Dynamic ceiling, m (ft)	2,700 (8,860)
Range, km (miles)	235 (146)

Yakovlev OKB

Above: A flying scale model of the Yakovlev EG helicopter marked ED 115; the digits refer to OKB-115. The model is powered by a piston engine.

Yakovlev OKB

A desktop model of the EG. The shape of the nose is much closer to the real thing

Yakovlev OKB

Yakovlev OKB

Yakovlev OKB

Yakovlev OKB

Top left: The Yak EG sits on the premises of the Yakovlev OKB in Moscow with the rotor blades folded. Note the cooling gills and the landing gear design.
Above left: The same machine at the same location with the rotors deployed.
Top right and above right: The EG as originally flown with a horizontal tail, endplate fins and tail bumper. The extensive cockpit glazing is evident.

proved troublesome and the designers had to revert to the M-11FR-1 with which the testing continued until its completion on 8th July 1948. The test results were generally positive although some vibration was noted at forward speeds in excess of 20-30 km/h (12.4-18.6 mph).

Eventually the testing was discontinued and the co-axial configuration was abandoned in favour of other layouts. A second prototype was left uncompleted. The first prototype was preserved at the Moscow Aviation Institute; in 1954-55 a team of students converted it into an aero-sleigh which was donated to a state farm in Kazakhstan.

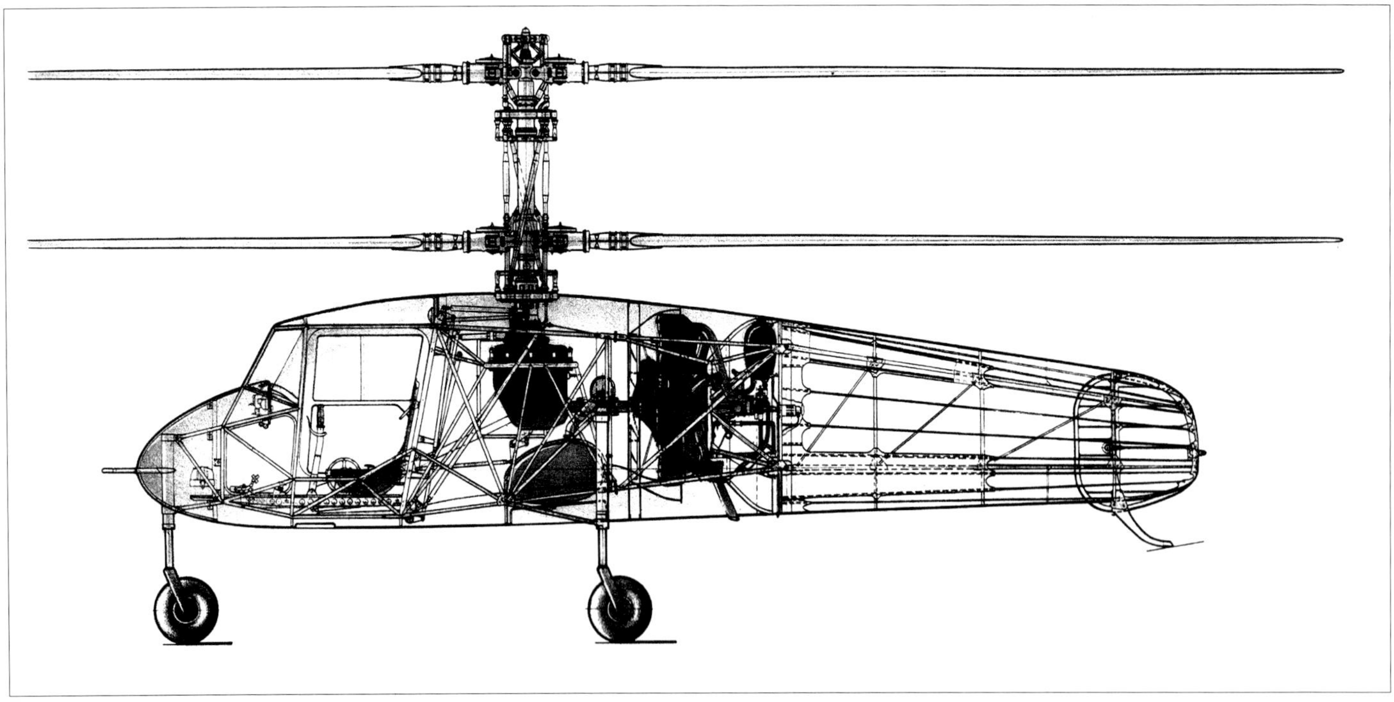

A cutaway drawing of the Yakovlev EG helicopter, showing the aft location of the engine.

Yak-100 (Yak-22) light helicopter

In 1948 the Yakovlev OKB built two prototypes of its second helicopter which was initially designated Yak-22 and then renamed Yak-100 (an out-of-sequence number). As distinct from the co-axial EG helicopter, the new machine featured the classic single-rotor layout. In its overall appearance the new helicopter bore a striking resemblance to the Sikorsky S-51, the main external difference being the Soviet machine's cranked tail boom. However, Yakovlev insisted the resemblance was purely superficial.

The Yak-100 was being developed in direct competition with a similar helicopter from OKB-329 headed by Mikhail L. Mil', which came to be known as the Mi-1. Both machines featured the same basic layout, had roughly the same dimensions and were designed around the same Ivchenko AI-26 seven-cylinder radial (in the case of the Yak-100, an AI-26GRFL with a nominal rating of 420 hp). The flight testing of both machines began almost simultaneously in the autumn of 1948, the Yak-100 making its first flight in November.

The Yak-100 had a three-blade main rotor of 14.5 m (47 ft 7 in) diameter turning anti-clockwise when seen from above (as on most Western helicopters but not on Soviet ones) and a three-blade tail rotor with a diameter of 2.6 m (8 ft 6$^{23}/_{64}$ in) located on the port side in pusher configuration, the forward blade moving against the main rotor downwash. The rotor blades were of wooden construction in both cases. The all-metal fuselage comprised three major subassemblies. The centre fuselage housing the engine and transmission was a steel-tube truss covered with detachable skin panels; the extensively glazed forward fuselage providing accommodation for the crew was formed by duralumin frames and stringers, while the rear fuselage was a semi-monocoque tailboom. The cockpit was fitted with two seats in tandem for a dual-control trainer version and with three seats (a pilot's seat and a bench for two behind it) in a liaison version. The helicopter had a fixed tricycle undercarriage with a castoring levered-suspension nosewheel and triangular-braced main units. The AI-26GRFL engine was mounted with its shaft horizontal and transmitted its power to the rotor shaft through a bevel gearbox.

The first prototype (c/n 01), flown by test pilots Mark L. Gallai and G. I. Komarov, was configured as a dual-control trainer. Initially it suffered from severe vibration and apparent blade flutter. To cure this problem, the rotor blades were modified with the CG moved further aft, behind the flexural axis. These blades were fitted to the second prototype (c/n 02) which joined the test programme in July 1949. This machine, flown by M. D. Goorov and G. I. Komarov, had a three-seat cabin and differed from the first prototype in some other minor details.

Manufacturer's tests were completed in June 1950 and the machine successfully passed its State acceptance tests later that year. Test pilots gave the Yak-100 a very favourable appraisal, noting its good stability and controllability and pleasant handling; the helicopter could be flown without problems by average-skilled pilots. The excellent view from the cockpit and well-arranged instruments and control levers and switches also earned praise. The Yak-100 fully met the requirements but was not put into production, having lost out to the Mi-1; the Mil' machine was already adopted for series production by the time the Yak-100 completed its State acceptance tests.

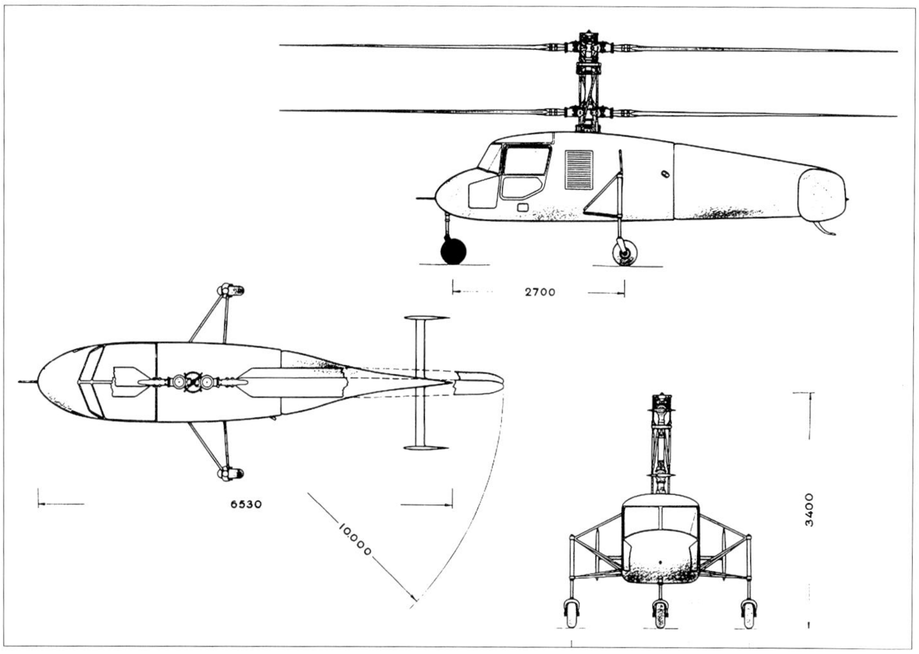

Yakovlev OKB

Above: A three-view of the Yak EG
Below: The EG in flight with the horizontal tail removed. Note the joint of the 'fish tail'.

Specifications of the Yak-100 helicopter

Main rotor diameter	14.5 m (47 ft 6⅞ in)
Length of fuselage (incl. tail rotor)	13.91 m (45 ft 7⅝ in)
Empty weight, kg (lb)	first prototype, 1,690 (3,726); second prototype, 1,805 (3,979)
Maximum all-up weight, kg (lb)	first prototype, 2,090 (4,607); second prototype, 2,180 (4,806)
Maximum speed, km/h (mph)	170 (105.6)
Hovering ceiling, m (ft)	2,720 (8,924)
Ceiling (forward flight), m (ft)	5,250 (17,224)
Range, km (miles)	325 (202)

Yakovlev OKB

Yakovlev OKB

Yakovlev OKB

Yakovlev OKB

Top and centre left: The unmarked first prototype of the Yak-100 during manufacturer's tests. While the helicopter bore a striking resemblance to the Sikorsky S-51, it was by no means a copy.

Centre right: The first prototype takes off, with the engine cowling panels on the port side removed to improve cooling. Note that the tail rotor drive fairing is also removed.

Above: The second prototype Yak-100 wore Soviet Air Force insignia.

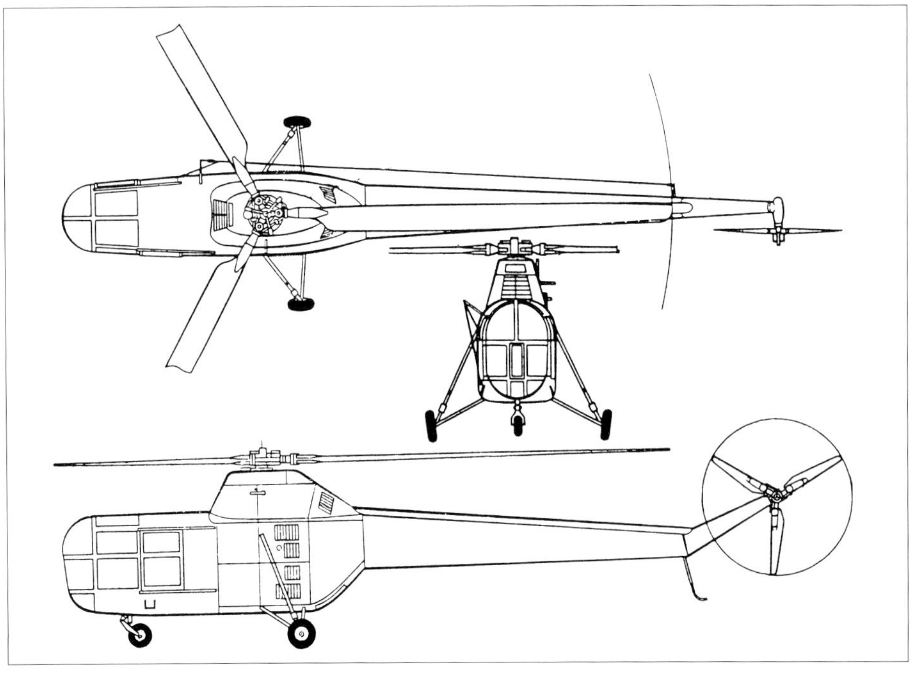

Left: A three-view of the Yak-100.

Above: The first prototype Yak-24 transport helicopter, showing the tail unit design and the large trim tab on the fin.

Yakovlev OKB

Yak-24 twin-rotor transport helicopter

The origins of this helicopter date back to September 1951 when the Yakovlev OKB became involved in a 'crash programme' aiming at the creation of two helicopters: a machine with a payload of 1,200 kg (2,650 lb) and a helicopter capable of lifting twice as much. The question was discussed at a special meeting between Iosif V. Stalin and leading Soviet aircraft designers. Development of the first of these machines was assigned to OKB-329 headed by Mikhail L. Mil' and eventually led to the emergence of the Mi-4 single-rotor helicopter; Yakovlev ventured to embark on the development of the larger machine featuring a tandem twin-rotor layout which became the Yak-24.

To speed up the work, the helicopter section of the Yakovlev OKB headed by Igor' A. Erlikh decided to make use of the main rotor/powerplant installation developed by the Mil' Design Bureau which comprised the Shvetsov ASh-82V 14-cylinder two-row radial engine specially adapted for helicopters, a four-blade rotor with a diameter of 21 m (68 ft 10⁴⁹⁄₆₄ in) and an appropriate transmission and rotor control system (swashplate). Thus, the Yak-24 was regarded as a kind of 'coupled Mi-4'.

The two rotors with their powerplants were installed at the front and rear ends of a large boxcar-like fuselage having a nearly square cross-section. To reduce the harmful influence of the front rotor on the rear rotor, the latter was placed higher up in a staggered arrangement, for which purpose the rear fuselage was provided with a fin-shaped pylon of considerable size. The fuselage was based on a truss of welded steel tubes and featured mixed-type skinning. The forward fuselage, including the engine bay, was skinned with duralumin; so was the rear engine bay and the rotor pylon. The centre fuselage between the engine bays was covered with doped fabric. It rested upon four undercarriage units formed by triangulated trusses and incorporating vertical shock struts with levered-suspension wheels (the forward ones were castoring). The cargo hold measured 9.37 x 1.95 x 1.92 m (30 ft 9 in x 6 ft 4¾ in x 6 ft 3¹⁹⁄₃₂ in) and initially provided accommodation for 19 troops on folding seats along the sides or space for combat vehicles or cargoes weighing up to 2 tonnes (4,410 lb), increased to 3 tonnes (6,615 lb) in overload configuration. In ambulance configuration it accommodated 12 stretchers arranged in two tiers along the cabin walls; later a three-tier arrangement was adopted, increasing the number of stretchers to 18. There were six windows on each side. The cabin was accessed through a side door. In addition, the loading and unloading of cargoes could be effected through a rear door with a full-width ramp enabling wheeled vehicles to be driven in.

The well-glazed flightdeck at the forward extremity of the fuselage provided side-by-side seating for two pilots; between them was a work station for a flight engineer/gunner who operated a 12.7-mm (.50 calibre) TKB-481 (Afanas'yev A-12.7) machine-gun on an NUV-3 flexible mount. The machine-gun could be used against ground targets in the front hemisphere.

The ASh-82 engines rated at 1,430 hp and delivering 1,700 hp for take-off were cleared to operate in any attitude. The front engine was installed between the flightdeck and the cargo hold at an angle of 60° to drive the gearbox under the front rotor. The rear engine was installed in the normal attitude in the base of the rear pylon and drove the rear rotor through a 90° bevel gearbox. The two rotors driven by these engines rotated in opposite directions (front, anti-clockwise; rear, clockwise), being a mirror reflection of one another, and the rear rotor oversailed the front one. The gearboxes were linked by a torque shaft ensuring synchronisation of their rotation and the possibility of single-engine operation in an emergency.

Yefim Gordon archive

A Yakovlev OKB test crew in front of a Yak-24, with a sister ship parked behind it. Note the design of the vertically paired pitot heads.

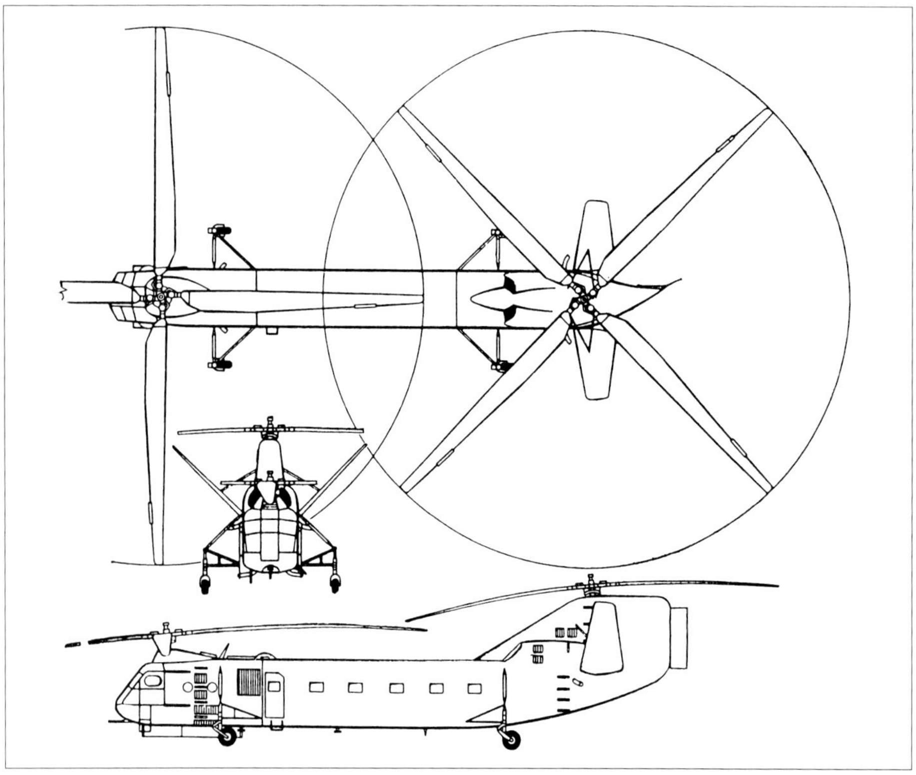

Above: A three-view of the initial production version of the Yak-24.

Yefim Gordon archive

Above: A typical early-production Yak-24 coded '81 Yellow' in flight.

The fin performing the role of a rotor pylon was asymmetrically shaped to provide side thrust that was necessary to compensate the torque produced by the rotor system. In the course of testing the asymmetrical fin was supplemented by a vertical rudder, or 'trim tab' as it was termed in Russian technical descriptions. Flanking the pylon were two variable-incidence stabilisers in a V arrangement, canted 45º and braced by V-struts.

The helicopter's control system incorporated hydraulic actuators (main and backup), inertial dampers and spring tabs.

Yak-24 prototypes

Four prototypes of the Yak-24 were built simultaneously at aircraft plant No.272 in Leningrad, two of them – c/ns 01272301 (that is, Batch 01, plant No.272, third type produced, 01st aircraft in the batch) and 01272302 – being intended for flight testing and the other two serving as static and fatigue test airframes. The first lift-off on Yak-24 c/n 01272301 was performed by test pilot Sergey G. Brovtsev on 1st July 1952, followed by the first free flight two days later. At that time the Yak-24 was the biggest helicopter in the world in terms of dimensions and weight; this was borne out by the world records set on this helicopter by Ye. F. Milyutichev and G. A. Tinyakov on 17th December 1955, when this machine lifted 4,000 kg (8,820 lb) to an altitude of 2,000 m (6,560 ft) and reached an altitude of 5,082 m (16,674 ft) with a load of 2,000 kg (4,410 lb).

However, the flight development of the helicopter was beset by numerous problems, the most serious of them being severe vibration in flight. In some flight modes this vibration posed a real safety threat. According to Aleksandr S. Yakovlev's reminiscences, this vibration led to the destruction of the fatigue test airframe which was equipped with a complete powerplant and transmission but had special cropped rotors with weights. As the engines were ground-run, the rear engine mount suffered a fatigue failure die to vibration; the rear engine and

Yefim Gordon archive

Three early-production Yak-24s coded '22 Yellow', '40 Yellow' and '25 Yellow' at an airfield. Note how the tactical code is repeated on the fin

Yefim Gordon archive

Above: Seen over the nose of a parked sister ship, two Yak-24s *sans suffixe* representing the upgraded version with 24° stabiliser dihedral and vertical endplate fins are just about to take off. Note the bulged blister windows of the flightdeck doors.

rotor tipped over forward, snapping the synchronisation shaft and chopping through fuel lines, and the ensuing massive fire destroyed the aircraft. The machine had by then logged 178 hours' total time – little more than half the target figure of 300 hours.

To tackle this problem, the designers stiffened the fuselage truss, reduced the rotor diameter by 1 m (3 ft) and introduced changes into the control system linkages. Manufacturer's testing lasted until 15th November 1952, the two flying prototypes logging 36 hours 06 minutes in 141 flights; nearly half of these (63 flights totalling 14 hours 48 minutes) were devoted to investigating the vibration problem.

The machine was then submitted for State acceptance trials on 31st December 1952, but the trials had to be suspended on 13th January 1953 due to an in-flight fire caused by a transmission bearing failure on the second prototype. The first prototype crashed on 19th February 1953 during tethered tests; as the engines were run, the tethers snapped and the helicopter took off. Only a mechanic was on the flightdeck at that moment; not knowing how to fly the machine, he throttled back the engines and the Yak-24 came down from an altitude of 3-6 m (10-20 ft), rolling over. The trials were finally completed in April 1955.

Yak-24 initial production version

Tooling up for series manufacture of the Yak-24 started at plant No.272 in early 1953, two years before the completion of the State acceptance trials; the first production example has been quoted as c/n 2720201. At present this plant bears the name 'Northern Plant', St. Petersburg. To supervise the production and solve the inevitable problems, a branch office of the Yakovlev OKB headed by Igor' A. Erlikh was set up at plant No.272. The first production Yak-24s started reaching the service units in 1956.

Yefim Gordon archive

Smartly painted orange overall with a black cheatline, '35 Yellow' was the prototype of the Yak-24U with zero-dihedral stabilisers and endplate fins. It is seen here carrying a 1949 model GAZ-51 lorry (number plate PD 13-10) as a slung load. The flightdeck glazing area appears to be reduced.

In its early production version the Yak-24 was outwardly identical to the prototypes and the four pre-production examples that flew at the Tushino Aviation Day in August 1955, their common feature being the V-shaped stabilisers. Only a small number were produced in this configuration.

Yak-24 production helicopter with endplate fins

From the outset the OKB branch at plant No.272 began introducing various improvements and modifications into production machines, in parallel with the development of specialised versions. In 1955 certain structural elements of the airframe were strengthened and the empennage was redesigned: the stabiliser dihedral was reduced to 24° and square-shaped endplate fins were mounted at the stabiliser tips. In 1956 the reversible hydraulic actuators incorporated into the control systems gave place to irreversible actuators; the cargo cabin was equipped with additional attachment points enabling the helicopter to transport 30 troopers or 18 stretcher cases.

Yak-24U upgraded production helicopter

In late 1957 the Leningrad branch of the Yakovlev OKB developed an improved version of the Yak-24 possessing greater load-carrying capacity. The helicopter was designated Yak-24U; the suffix is sometimes presumed to mean ***ulooch****shennyy* (improved), but other sources interpret it as *u****shee****rennyy* (widened) – a reference to the increased width of its fuselage. Indeed, the width of the fuselage was increased by 0.4 m (1 ft 3¼ in), which imparted it both greater volume and greater strength, enabling the helicopter to carry up to 37 troopers and some new kinds of combat vehicles, including the 57-mm (2.24-in) ASU-57 self-propelled gun used by the Airborne Troops. The maximum load carried internally was

Yakovlev OKB

Above: The first prototype of the Yak-24K. The VIP version was immediately recognisable by the large square cabin windows, four on each side. Note the 24° stabiliser dihedral and square-shaped endplate fins.

Yakovlev OKB

Above: The smart-looking prototype of the Yak-24A 'helibus' had zero-dihedral tailplanes, being based on the Yak-24U, and enlarged (differently) cabin windows. Alas, it remained a prototype.

Yefim Gordon archive

Yak-24 '62 Yellow' was equipped with a special panoramic cine camera installation for 360° filming.

Yefim Gordon archive

A close-up of the installation.

Yefim Gordon archive

Yak-24 '79 Yellow' was converted into the one-off Yak-24T (Yak-24 'Nerpa') pipelayer for constructing POL lines in the field. The lateral cassettes holding forty-four 40-mm pipes each are clearly visible.

increased to 4 tonnes (8,820 lb), and the external slung load rating was increased to 3,500 kg (7,716 lb).

Other significant changes included restoring the rotor diameter to the original 21 m (68 ft 10 49/64 in) value and increasing the tilt angle of the canted rotor axes to 2°30' (the front hub to starboard, the rear to port). The latter feature made it possible to dispense with the trim tab on the fin and redesign the rear fuselage for minimum drag because it was no longer expected to create a side force. The horizontal tail was redesigned to feature reduced area and zero dihedral. Various structural elements were reinforced, and the rear landing gear oleos were modified to eliminate any tendency to ground resonance.

The prototype was converted from a production Yak-24 *sans suffixe* (c/n 27201304 – that is, plant No.272, batch 01, Type 3, 04th helicopter in the batch). After conversion the machine received a new c/n, 0104.

Much effort was put into improving the helicopter's autostabilisation system. From early 1958 onwards its control system included AP-102M automatic damping devices in the lateral, longitudinal and directional stability channels. Subsequently they were introduced on all production Yak-24s. In 1959 the AP-31 autopilot was successfully tested on the Yak-24U prototype. That same year this machine, wearing a non-standard colour scheme with a red cheatline and the tactical code '35 Yellow', was used as a flying crane during the restoration of the Yekaterininskiy Palace in Pushkin, a suburb of Leningrad. Flown by test pilot I. S. Grigor'yev, the helicopter installed 30 wooden roof trusses weighing 3,200 kg (7,050 lb) each and 18 metal trusses weighing 2,000 kg (4,4510 lb) each.

Specifications of Yak-24 helicopter versions

	Yak-24	Yak-24U
Rotor diameter (both)	20 m (65 ft 7⅜ in)	21.0 m (68 ft 10¾ in)
Length overall	33.03 m (108 ft 4⅜ in)	33.03 m (108 ft 4⅜ in)
Fuselage length	21.34 m (70 ft 0⅛ in)	21.34 m (70 ft 0⅛ in)
Height on ground	6.5 m (21 ft 3⅞ in)	6.5 m (21 ft 3⅞ in)
Empty weight, kg (lb)	10,607 (23,384)	11,000 (24,250)
Maximum all-up weight, kg (lb)	14,270 (31,460)	15,830 (34,900)
Same, with additional tankage	–	16,800 (37,037)
Maximum speed, km/h (mph)	175 (109)	173 (107.5)
Cruising speed, km/h (mph)	156 (97)	n.a.
Hovering ceiling, m (ft)	2,000 (6,562)	1,500 (4,920)
Normal dynamic ceiling, m (ft)	4,200 (13,780)	2,700 (8,858)
Range, km (miles)	265 (165)	255 (158)
Range with increased tankage, km (miles)	–	300 (186)*

* with a 4,000-kg (8,820-lb) payload

Yak-24K VIP helicopter prototypes

In late 1959 the Leningrad branch of the Yakovlev OKB at Plant No.272 developed and built a VIP version designated Yak-24K, the K standing for *kom**fort*** – (enhanced) comfort. Two examples (c/ns 27208303 and 27208304) were converted to this configuration.

Outwardly the Yak-24K was identifiable by the extra large square windows (four on each side of the fuselage); the second prototype even carried appropriate 'Yak-24K' nose titles. The luxuriously furnished cabin provided accommodation for eight passengers and featured eight swivelling armchairs, soundproofing, heating, ventilation, telephone communications equip- ment and other 'adjuncts of civilisation'. The two machines were fitted with an electrically operated airstair door.

Yak-24A passenger helicopter

In 1960 the Moscow-based head office of the Yakovlev OKB developed and built a single prototype of the Yak-24A passenger version seating 30 passengers in ten rows of three abreast, with an aisle offset to starboard. The Yak-24A 'helibus' was a derivative of the Yak-24K VIP helicopter. There were six cabin windows on each side (smaller than those of the Yak-24K but larger than on the standard version). The rear cargo ramp/door was eliminated, the passenger door on the port side was fitted with fold-down steps, and the cabin was fitted with larger windows. This version was based on the Yak-24U, as evidenced by the zero-dihedral horizontal tail.

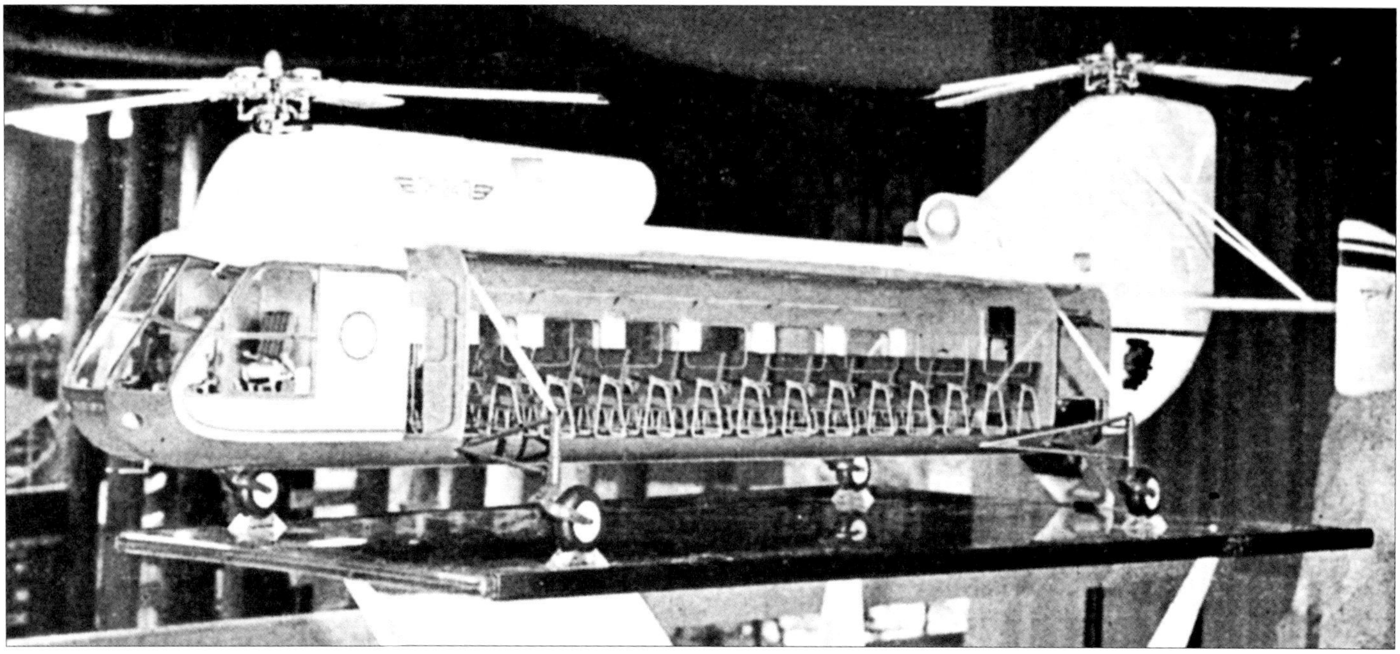

Sergey and Dmitriy Komissarov archive

Above: What might have been the Soviet counterpart of the Boeing Vertol 234C Commercial Chinook. This model of the projected Yak-24P passenger helicopter was displayed at an exhibition in London. The turboshaft engines and rear airstairs are very apparent.

Yakovlev OKB

Above: The projected Yak-32 four-turboshaft heavy-lift helicopter looked like a Yak-24 on anabolic steroids (the models are to the same scale). The flightdeck section appears to have been borrowed wholesale.

Yakovlev OKB

An artist's impression of the Yak-32 carrying a T-34 tank as a slung load.

Yak-24P passenger helicopter (project)

In 1961 a model of a projected version of the Yak-24 was demonstrated at the Soviet Trade Exhibition in Earl's Court in London. Designated Yak-24P (*passa**zheer**skiy* – passenger, used attributively), it sported the installation of two 2,700-eshp AI-24V turboshaft engines in place of the earlier ASh-82V piston engines. The new powerplants were mounted above the fuselage to maximise the useful volume, with the air intake of the forward engine facing aft. The new engine arrangement made more space available for the passenger cabin which provided comfortable accommodation for 39 passengers, with baggage space and a toilet. The machine was expected to have a cruising speed of 210 km/h (130.5 mph). The project failed to reach the hardware stage but, as a point of interest, it was immortalised in a 1/50th scale model released by the East German model kit manufacturer Plastikart.

Yak-24 refuelling tanker

In 1957 a single production Yak-24 was converted by Plant No.272 into a refuelling tanker. It was intended for refuelling, with petrol or kerosene, the different kinds of combat materiel (tanks, trucks, aircraft) deployed near the frontline. Three large fuel tanks holding 3,156 litres (694.3 Imp gal) of fuel in all were installed in the helicopter's cargo hold; they were supplemented by a panel for dispensing the fuel. A team of specialists engaged in testing the system developed devices enabling the helicopter to refuel various combat vehicles while hovering close to the ground. The refuelling system bore the name 'Looch' (Ray, or Beam).

The prototype was converted from Yak-24 c/n 27203301. The helicopter's crew, captained by S. G. Brovtsev, developed and tested procedures for refuelling tanks sheltered in a wood or amphibious tanks afloat.

Yak-24T (Yak-24 'Nerpa') pipeline-laying helicopter

In 1957 one production Yak-24 was converted by Plant No.272 into a version intended to be used in the construction of PMT-100 field petrol, oil and lubricant (POL) pipelines (*pole**voy** maghi**strahl'**nyy **troo**bo-pro**vod***) for supplying tank bases in the field with fuel (!). The system bore the name Nerpa (a kind of fur seal inhabiting Lake Baikal in Russia); hence the helicopter is sometimes referred to as the Yak-24 'Nerpa'. An alternative designation was Yak-24T (***troo**bo'u**klahd**chik* – pipelayer).

Coded '79 Yellow' (c/n 27203301), the helicopter was equipped with cassettes strapped on to the centre fuselage sides, holding 44 pipes each of 40 mm (1.57 in) diameter. The pipes were dropped one by one from low altitude, whereupon the ground construction teams could proceed with their job. The system was tested but not accepted because someone apparently realised such pipelines would be too vulnerable and impractical.

Yak-24 with panoramic cine camera installation

In 1959 the Yak-24 took part in the creation of a unique panoramic documentary film providing full 360° coverage (for which

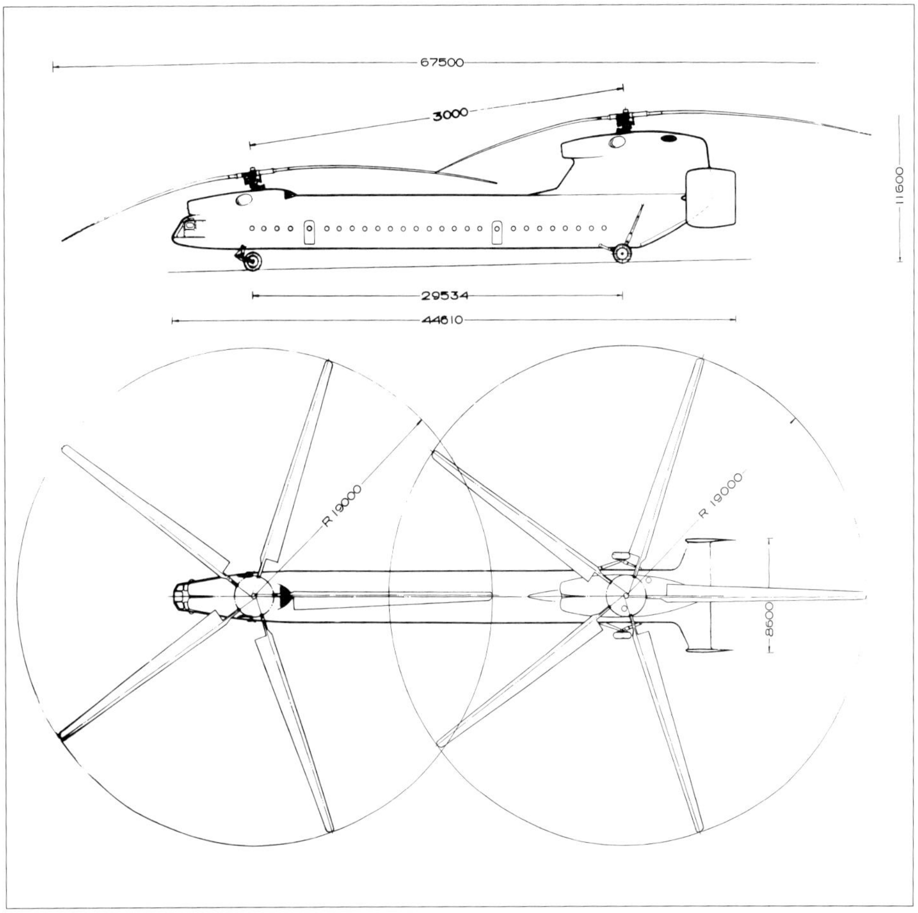

Above: Two views of the Yak-32 helicopter. The dimensions were to be as follows: Length overall, rotors turning, 67.5 m (221 ft 5 47/64 in), height 11.6 m (38 ft 0 43/64 in), rotor diameter 38 m (124 ft 8 in), fuselage length 44.51 m (146 ft 0 5/8 in), landing gear wheelbase 29.534 m (96 ft 10 3/4 in), distance between rotor axes 30 m (98 ft 5 7/64 in), stabiliser span 8.5 m (27 ft 10 21/32 in).

Below: The freight hold of the Yak-32 was to house three infantry fighting vehicles and their complement of troops.

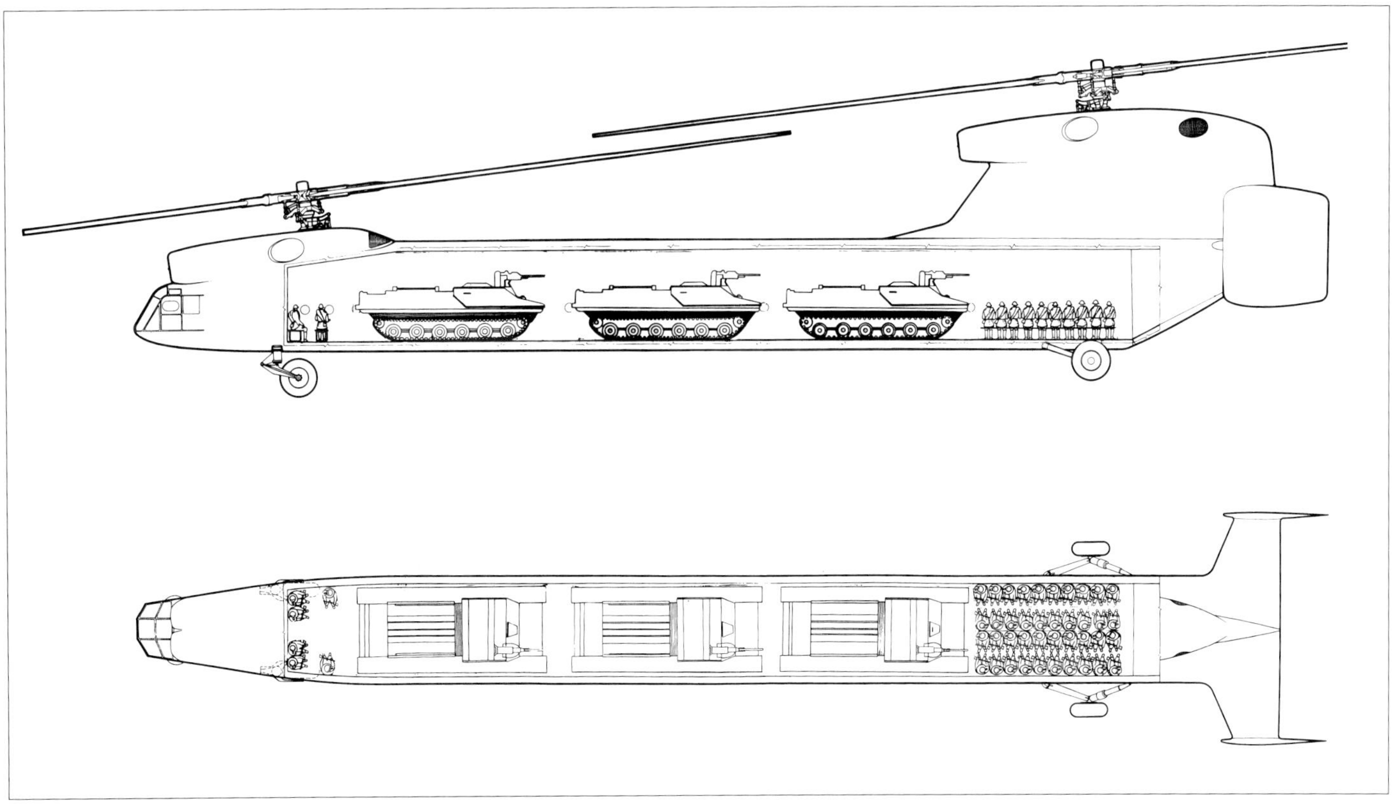

reason it was dubbed 'circorama'). To provide aerial, one Yak-24 ('62 Yellow', c/n 27206304) was equipped with a special cine camera installation in which a number of cameras were mounted around a circle so as to obtain, collectively, the 360° field of view. This installation was suspended under the helicopter on a truss hinged at the front that was stowed against the belly when on the ground, the bulky camera pod fitting into a recess in the belly. It could be lowered by a GL-200 winch by means of cables and pulleys during filming and retracted for take-off and landing.

Yak-32 heavy transport helicopter (project)

The Yakovlev OKB pursued for some time its helicopter design activities along the lines marked by the Yak-24 with its tandem twin-rotor layout. Design studies were conducted on a giant four-engined turboshaft-powered helicopter featuring the same basic layout as the Yak-24. It was to have an impressive load-carrying capacity of 40 tonnes (88,200 lb), being in the same class as the Mil' V-12 which featured a side-by-side twin rotor layout. This proposal from the Yakovlev OKB failed to gain acceptance, and no detail design was undertaken.

VVP-6 heavy multi-rotor helicopter

The development of V/STOL aircraft made it desirable also to develop supporting vehicles able to supply their dispersed and frequently relocated operating bases with food, fuel and munitions, and also to provide defence. Hence the Yakovlev OKB proposed a truly unusual project of a complete flying surface-to-air missile (SAM) system. The vehicle bore the designation VVP-6 (*ver-ti**kal'**no vzle**ta**yushchaya plat**for**ma* – vertical take-off platform), the 6 referring to the number of rotors.

A desktop model of the VVP-6 showed a huge boxy fuselage (about 49 m/160 ft long!), with three sets of unswept high-set wings in tandem on each side. Each wing carried a six-blade rotor of comparatively small diameter close to the tip; each rotor was driven by four turboshaft engines mounted in twin pylon-mounted nacelles under the wing on either side of the rotor shaft. The engines exhausted through downward-angled lateral jetpipes, the free turbine shafts of each pair were connected to an intermediate (coupling) gearbox from which a transmission shaft ran up through the pylon into the wing and then, via bevel gears, to the final drive gearbox. The cockpit was located at the forward extremity of the fuselage which was cut away from below, as was the rear extremity; apparently one of them (or both) incorporated loading ramps.

Six V-750 (11D) or V-775 (20D) SAMs (NATO reporting name SA-1 *Guideline*) were positioned in pairs on top of the fuselage; apparently they would have to be enclosed by some form of covers or fairing to prevent damage in flight. One source describes them as 'ready to fire on their launchers', but this is utterly impossible for the simple reason that the launchers of the V-750 and V-775 SAMs rotated through 360° in order to track the target before launch, and the huge flames belched by the missiles at the moment of launch would certainly damage the vehicle's rotors! Reloads, as well as the search/target illumination radars and auxiliary power supplies, were stowed at lower deck level.

The VVP-6 never got off the drawing board, as the requirement for this monster evaporated with the failure to put land-based V/STOL jets into service.

Yakovlev OKB

A display model of the bizarre VVP-6 SAM system transporter, showing the six pairs of turboshaft engines under the wings and the paired V-75 missiles on the 'upper deck'.

THE PILOTLESS AIRCRAFT

11

DPLA-60 Pchela-1 (*izdeliye* 60) tactical reconnaissance RPV

In 1982 the Soviet Ministry of Defence commissioned the development of a tactical unmanned aerial reconnaissance system to be used at the regimental level. On 12th July that year the Council of Ministers issued a directive to this effect, followed a few days later by the appropriate Ministry of Aircraft Industry, Ministry of Electronics Industry (MRP – *Ministerstvo rahdioelektronnoy promyshlennosti*) and MoD orders. The MRP's *Koolon* (Coulomb) Research Institute had overall responsibility for the programme, while OKB-115 headed by Aleksandr S. Yakovlev was tasked with designing the UAV as such. Sergey A. Yakovlev, the General Designer's elder son, and Yuriy I. Yankevich headed the development effort.

The system's prospective operator was the Soviet Airborne Troops (VDV – *Vozdooshno-desahntnyye voyskah*) which needed an autonomous reconnaissance asset to support the operations of a commando or reconnaissance group inserted behind enemy lines and having virtually no outside information channels to rely on. This operational aspect accounted for some of the system's design features. Specifically, the UAV had to be as simple and cheap as possible; the system had to be air-droppable and capable of fully autonomous operation. Finally, the ground part of the system was to be based on the standard wheeled and tracked vehicles used by the VDV.

In accordance with the specifications drawn up by the three ministries the tactical unmanned aerial reconnaissance system was to feature a re-usable compact remotely piloted vehicle. The deadline for the beginning of the tests was set at February 1983.

The Yakovlev OKB contemplated several alternative general arrangements, including a biplane layout. The version selected eventually featured shoulder-mounted unswept wings folding along the fuselage for transportation and storage.

Besides the aircraft itself, the system – the first of its kind in the Soviet Union – included an integrated launch/control vehicle (LCV) based on the BTR-D paradroppable armoured personnel carrier, a set of test equipment and an evacuation/support vehicle based on the GAZ-66 4 x 4 army lorry (the Soviet counterpart of the Bedford RL and Bedford MJ). The system enabled pre-launch servicing of the RPV, launch and radio-controlled flight, reception of the TV image generated by the machine and its safe recovery for further use. The RPV, which received the manufacturer's designation *izdeliye* 60, was primarily a daytime TV reconnaissance platform but could also be fitted out as an electronic countermeasures (ECM) aircraft for jamming enemy radio communications. Other designations used are DPLA-60 (*distantsionno-piloteeruyemyy letahtel'nyy apparaht* – RPV) and *Pchela-1* (Honeybee-1).

The *izdeliye* 60 was a high-wing monoplane with an annular empennage; the wings of rectangular planform folded forward along the fuselage to save space during storage. The fuselage had a modular design, being divided into a forward section housing the mission equipment, a centre section accommodating the fuel tank, the wing folding/locking mechanism and the basic aircraft systems and equipment, and a rear section housing the engine with its extension shaft. The rear fuselage section carried the tail unit – a wide-chord ring-shaped stabiliser attached by four struts, with one-piece control surfaces (rudder and elevator) at 90° to each other inside the ring. Additionally, cruciform fins of trapezoidal

Yakovlev OKB

A mock-up of the Pchela-1 (*izdeliye* 60) RPV on a very provisional launcher at the Yakovlev OKB's test facility in Zhukovskiy in 1982, showing the cruciform fins and the lack of a parachute recovery system. Note the Tupolev Tu-104 radar testbed in the background on the left.

Yakovlev OKB

Above: The first *izdeliye* 60 (Pchela-1) prototype on the same launcher; the machine is painted silver overall and again has no parachute recovery system. This shot illustrates the tail unit design with cruciform control surfaces inside the annular stabiliser/propeller duct and the launch sled with the two rocket boosters.

Yakovlev OKB

A night-time test launch of the same prototype. A huge net was used for recovery.

planform set at 45° to the horizontal plane were mounted immediately aft of the wings.

The RPV was powered by a P-020 air-cooled two-cylinder horizontally opposed engine delivering 20 hp at 7,300 rpm; this two-stroke engine with a displacement of 274 cc was developed by the design bureau of the Kuibyshev engine plant named after Mikhail V. Frunze. The plant is now known as the *Mo**tor**ostroi**tel**'* (Engine Manufacturer) Joint-Stock Co., while the design office has become a separate enterprise called SKBM, *Sa**mar**skoye kon**strook**torskoye byu**ro** mo**tor**ostroyeniya* – Samara Engine Design Bureau. The engine drove an AV-23 two-bladed fixed-pitch ducted pusher propeller of 600 mm (1 ft 11⅝ in) diameter via an extension shaft; the annular stabiliser doubled as the propeller duct, increasing the propeller thrust and control surface efficiency.

The *izdeliye* 60 was launched from a vehicle-mounted sloping ramp, using two solid-fuel rocket boosters for take-off. A parachute recovery system installed in a fairing atop the centre fuselage made for a safe landing; a smooth touchdown was ensured by an inflatable bumper made of rubberised nylon fabric under the centre fuselage.

As already mentioned, overall responsibility for the programme was assigned to the Koolon research institute headed by Anatoliy S. Novosyolov. The Moscow-based Gorizont (Horizon) OKB, then headed by V. V. Il'yichov, was tasked with designing the launch system. OKB-115 developed the UAV and performed integration of all systems.

The work proceeded fast. The first test 'launch' of a dummy equal in weight to the *izdeliye* 60 from a test rig took place at LII's test ground as early as 21st October 1982; eleven days later, on 3rd November, the dummy was 'fired' for the first time from the actual LCV. On 24th January 1983 the Yakovlev OKB took delivery of the first P-020 engine from the Kuibyshev engine design bureau; the same afternoon it was fitted to the first flying prototype of *izdeliye* 60 in the presence of Minister of Aircraft Industry Ivan S. Silayev who was visiting the OKB to check up on progress. Meanwhile, the Koolon research institute was going full steam ahead with assembly and debugging of the radio control system and other avionics.

By then the Yakovlev OKB had built up a close-knit team of specialists which took all the problems associated with the RPV's development and operation in stride. The design and production technology team included first and foremost V. I. Sergin, Ye. F. Shatalina, S. S. Sharin, Yu. N. Ivanov, Yu. N. Viryachev, B. N. Dedkov, N. V. Mel'nikova, A. L. Toobin, Yu. P. Salykov, V. N. Ganyushkin, A. A. Yakoobov *et al*. Test engineers Ye. P. Golubkov, A. I. Goortovoy, V. I. Palyonov, V. V. Tsaryov, V. I. Zasypkin, M. N. Veselov, V. N. Myasnikov, V. N. Pirogov, A. T. Volkov, S. A. Moorashov, B. B. Korochkin, S. D. Terent'yev and Ya. M. Galinskiy were also involved. A special UAV production department was set up at the OKB's experimental plant, MMZ No.115 'Strela'; it included a highly skilled manufacturing workforce headed by A. F. Vetrov, V. N. Boormistrov, I. S. Ovsyankin and A. G. Korolyov. At various times the group of UAV project engineers included V. F. Kulichenko, N. N. Dolzhenkov, V. A. Mit'kin, V. I. Baranov, A. I. Ovchinnikov, S. K. Koozin, Yu. V. Verkin, P. A. Rips, S. V. Dolinskiy, V. O. Zavarykin and V. K. Sharapov.

A special team of high-class specialists was formed at MMZ No.115 for promptly resolving the issues arising during construction of the *izdeliye* 60 prototypes; it was headed by the plant's Director, S. D. Savin. For safety (and security) reasons the testing of the system as a whole was to take place at a specially equipped and instrumented test range well away from residential areas and 'nosy neighbours'.

On 22nd February 1983 representatives of the principal OKBs and research institutions participating in the programme arrived in force at one of the test ranges operated by GNIKI VVS (the new name of GK NII VVS) together with the hardware to be tested. The commanders of the test ranges had long since allocated a stretch of wasteland with several auxiliary buildings in the middle of nowhere, about 40 km (25 miles) from the range's main facility, for the UAV test programme.

At the crack of dawn on 5th March 1983 the whole team was in position at the range. It was a bright and sunny morning, and the temperature was a crisp –15°C (+5°F). The men put on whatever clothes they had brought with them to keep warm, but the

Yakovlev OKB

Above: Aptly coded '601 Black' (that is, *izdeliye* 60 No.1), the first prototype is shown here on its launch and control vehicle based on the BTR-D paradroppable APC at the Yakovlev OKB's flight test facility (note the Yak-42 in the background). The RPV appears to have an optically flat forward-looking camera port.

adjustment operations on the aircraft and inside the ground control station (GCS) could only be performed with bare hands, and eventually nature won. Still, the men held on and, at 13.07 Moscow time, the RPV was finally installed on the launch rail. Imagine the consternation of the entire team when it turned out that the rocket boosters had been taken away to the main facility early in the morning for warming up to make sure they would fire properly – and were still there, 40 km away. An hour later, at 14.05, the boosters had been fitted and everything was set for launch – and then the autopilot's pitch control channel failed.

Thus, instead of a first flight the tests began with an incident investigation – not the most encouraging start. At an MAP conference dealing with the matter it was decided to change the launch technique, dispensing with the troublesome rocket boosters; the RPV would be catapult-launched instead. Eventually, however, the boosters were retained, albeit they were much improved. The effort to redesign the launcher, modify the aircraft accordingly and perform bench tests of the new system lasted nearly three months.

On 17th June 1983 the efforts finally paid off: at 16.00 Moscow time the *izdeliye* 60 prototype carrying the serial '601 Black' made its successful catapult-launched maiden flight. After a five-minute flight the RPV was guided to a safe landing which involved the use of a recovery net similar to the crash barriers used on military airfields.

The second launch that followed on 28th June was also successful, albeit accompanied by a hair-raising experience. After clearing the catapult rail the RPV did not proceed on its assigned course immediately; instead, it made a round of the launch pad, descending gradually to 100 m (330 ft). Everyone standing in the open ran for cover, trying to find a ditch or a hole in the ground in case the runaway UAV should come their way. Imagine how the crew of a Mi-8 helicopter standing 'with steam up' right beside the launch pad felt. The chopper was to fly chase and pinpoint the location of the RPV's emergency landing, should anything go wrong; now it found itself in the rather unlikely position of being 'chased', and possibly even 'attacked', by the test article! Luckily, 45 seconds later the RPV recovered and the rest of the mission went as planned, the prototype landing in the designated area 15 minutes after launch.

Gradually, as the creators of the system built up experience, they became more and more confident that the system had a future – even though the next test mission on 26th July again proved abortive. By the end of October 1983 the 'score' (the success/failure ratio) after 15 test launches was 10:5 in favour of the designers and the system's basic performance was more or less clear, making it possible to draw up a so-called preliminary conclusion recommending the RPV and the system as a whole for production. At this point, however, the *izdeliye* 60 was retrofitted with a more sophisticated TV camera at the customer's insistence, and a while later an active jammer became available for installation in the RPV. The resulting

Yakovlev OKB

The second prototype *izdeliye* 60 ('02 Red'), shown here inside a hangar at the OKB's flight test facility, was the first to feature the intended parachute recovery system (its bay is open) and inflatable bumper. Judging by the hemispherical dielectric radome, this example was completed in ECM configuration.

additional series of tests dragged on for a long time because the winter/spring season, which had set in by then, was a rotten time for testing RPVs in the field.

The final launch of an *izdeliye* 60 under the special trials programme held jointly by the designers and the customer took place on 28th May 1984. A month later the State commission endorsed the trials report. A total of 23 flying prototypes of the *izdeliye* 60 with different equipment fits and a further nine examples for ground tests were manufactured by MMZ No.115 between June 1982 and July 1983. The trials programme included a total of 25 test flights, 20 of these being accepted 'for the record'. The concluding part of the trials report went:

'1. The system basically conforms to the technical outlook (that is, operational requirement – *Auth.*). *The feasibility of such systems has been confirmed.*

2. It is the recommendation of the undersigned that an initial production batch be manufactured for service trials.'

Basic specifications of the tactical unmanned aerial reconnaissance system based on the Pchela-1 UAV (*izdeliye* 60)

Length overall	2.7 m (8 ft 10 19/64 in)
Wing span	2.4 m (7 ft 10 31/64 in)
Wing area, m^2 (sq ft)	1.4 (15.0)
Launch weight, kg (lb)	98 (216)
Speed, km/h (mph)	120-180 (74.5-112)
Flight altitude, m (ft)	100-1,000 (330-3,280)
Rate of climb, m/sec (ft/min)	3.5 (689)
Endurance	2 hrs
Combat radius, km (miles)	30 (18.6)
Maximum number of RPVs airborne at any one time	1

Pchela-1M (Pchela-1TM, *izdeliye* 60MS) tactical reconnaissance RPV

For another year or so virtually all of the Soviet Armed Forces' arms and services scrutinised Yakovlev's tactical unmanned aerial reconnaissance system, assessing its capabilities, before eventually giving it a thumbs-up. On 25th September 1985 the Council of Ministers issued a directive ordering an initial batch of 50 RPVs to be manufactured by aircraft factory No.475 in Smolensk. The production version was designated *izdeliye* 60MS (*moderni**zee**rovannoye, se**reey**noye* – upgraded, production) or Pchela-1M; the designation Pchela-1TM is also used, the T referring to the TV reconnaissance system in order to distinguish it from the ECM version described below.

The *izdeliye* 60MS differed from the prototypes in having a parachute recovery system which included an inflatable bumper made of rubberised nylon fabric to ensure a smooth touchdown; the parachute and the bumper were housed in prominent angular fairings above and below the centre fuselage. Another obvious recognition feature was that the upper pair of cruciform fins between the wings and the propeller duct was deleted.

Dozens of enterprises within the frameworks of several ministries were involved in the production of the RPV. As it did with all of its aircraft achieving series production, OKB-115 assigned a team of specialists (headed in this instance by project engineer S. K. Koozin) to monitor and support the production process, extending aid to the factory if necessary. Much attention was paid to the manufacturing standards of production RPVs and, even more importantly, to observing the delivery schedule. The OKB's new General Designer A. A. Levinskikh, who had succeeded Aleksandr S. Yakovlev after his retirement in 1984, exercised overall supervision of the programme.

A series of checkout and refinement tests involving six launches was held between May 1987 and 28th December 1987 to see if the production version of the system met the specs. In April 1988 the new tactical unmanned aerial reconnaissance system based on the Pchela-1M (*izdeliye* 60MS) RPV was delivered to first-line units of various arms and services for evaluation and operational tactics development.

Yakovlev OKB

No, this is not a skid landing gear, just a display stand. The production *izdeliye* 60MS (Pchela-1M) lacks the upper fins, the lower ones being retained for attachment to the launch sled. Note the 'fishbowl' nose housing a TV camera and the recovery system housings; the Pchela-1M would win no prizes at a beauty contest.

Pchela-1PM unmanned ECM aircraft

A version of the Pchela-1M equipped with an active jammer was designated Pchela-1PM, the P denoting *posta**nov**shchik po**mekh*** – ECM aircraft. Outwardly it differed from the basic reconnaissance version in having the transparent 'fishbowl' nose housing the movable TV camera replaced with a hemispherical dielectric radome enclosing emitter antennas.

Sterkh tactical unmanned aerial reconnaissance system

DPLA-61 Shmel'-1 (*izdeliye* 61) tactical reconnaissance RPV

From 9th June 1984 onwards the tactical unmanned aerial reconnaissance system based on the Pchela-1M (*izdeliye* 60MS) was officially assigned experimental status. The reason was that the ADP of a much-improved version of the system (regarded as the definitive version which was to enter full-scale production and service) passed its in-house project review at the Koolon research institute that very day. Designed to meet an SOR issued by the Ministry of Defence, the new system, which was subsequently named Sterkh (Japanese Crane), included the new Shmel'-1 (Bumblebee-1) RPV, alias *izdeliye* 61; some sources refer to it as the DPLA-61, Yak-061 or Yak-61. While the basic architecture of the system and the RPV's general arrangement were retained, the aircraft itself was a new design. In particular, the wings had a higher aspect ratio and specially profiled downward-canted wingtips instead of simple squared-off tips. The remaining ventral fins were deleted, as was the inflatable bumper; the *izdeliye* 61 had a landing gear comprising two pairs of cantilever spring struts terminating in dished pads. The recovery parachute housing ahead of the wings was given an airfoil shape. The powerplant was also new – a 440-cc P-032 air-cooled flat-twin two-stroke engine developed by the same Kuibyshev Engine Design Bureau and delivering 35.5 hp at 6,600 rpm.

Ground testing of the system's components began on 1st December 1984. Nearly seventeen months later, on 26th April 1986, the first prototype *izdeliye* 61 bearing the serial '001 Black' performed its maiden flight at the same GNIKI VVS test range near Akhtoobinsk where the *izdeliye* 60 had been first launched in 1983. Continuing the 'tradition' established with the predecessor, the RPV again went haywire during the first launch – this time due to a software glitch in the control system computer.

After a debugging effort and a series of successful launches it was decided to hold full-scale manufacturer's tests and State acceptance trials of the Sterkh tactical unmanned aerial reconnaissance system based on the *izdeliye* 61 RPV fitted with TV reconnaissance equipment.

Basic specifications of the Sterkh system based on the Shmel'-1 (*izdeliye* 61)

Wing span	3.2 m (10 ft 5 63/64 in)
Wing area, m^2 (sq ft)	1.8 (19.35)
Launch weight, kg (lb)	130 (286)
Speed, km/h (mph)	110-180 (68-112)
Flight altitude, m (ft)	100-2,500 (330-8,200)
Rate of climb, m/sec (ft/min)	4.0 (787)
Endurance	2 hrs
Combat radius, km (miles)	60 (37.2)
Maximum number of RPVs airborne at any one time	2

Shmel' RPV as part of the Malakhit system

In 1997 the Russian press carried a report stating that a group of designers had been awarded a Government prize for that year for the development of a new RPV reconnaissance system. Dubbed *Malakhit* (Malachite), this system is equipped with Shmel' RPVs (the version is not identified) and is intended for detecting single or group objects on the move in day-time and at night and pinpointing their location, for radiation and chemical reconnaissance, as well as for various surveillance and monitoring missions.

The whole system, including the Shmel' vehicles, is air-droppable and can be delivered to any site by air. No mention was made of the system's special features as compared to the Sterkh system.

Shmel'-2 tactical reconnaissance RPV (project)

A retractable-gear version of the Shmel'-1 RPV was also under development. The landing gear comprised aft-folding tubular skids.

Shmel II tactical reconnaissance RPV (export version, project)

Despite its name, this RPV looks totally different from the original Shmel-1, although it makes use of a fuselage and outer wings

Yakovlev OKB

An *izdeliye* 61 UAV on a *very* provisional launcher. Interestingly, this example has neither a landing gear nor a parachute recovery system, presumably being a full-scale mock-up. Note that the variable-sweep wings are set to forward sweep in this instance.

derived from its predecessor. This project developed by the Yakovlev OKB was initially known as 'DPLA 250 kg' (referring to the all-up weight of 250 kg/551 lb), or as 'DPLA civil/export'.

The vehicle has been configured as a mini-aeroplane featuring a low-set wing and an inverted-V tail carried by two rear booms. Propulsion is provided by the same P-032 engine with an unshrouded pusher propeller. A tricycle wheeled undercarriage with cantilever-spring struts is fitted, enabling the vehicle to take-off and land in aeroplane fashion, but it can also be launched by a catapult and recovered by parachute.

The RPV was intended for various civil reconnaissance or surveillance missions, providing a platform for different mission payloads, such as TV or IR equipment and the like.

Stroy-P tactical unmanned aerial reconnaissance system

Pchela-1T (*izdeliye* 61T) tactical reconnaissance RPV

The moment of truth during the launch of the RPV came 15 seconds before the actual launch, when ground power was disconnected and the machine switched to autonomous mode. During this final countdown it was beyond human power to do anything – it was all up to the RPV's launch computer to 'take the right decision'. At the closing stage of the Sterkh system's flight tests/State acceptance trials (which proceeded in parallel) the 13 RPVs involved were 'given a chance to think' 68 times. On 52 occasions the flights were successful, being accepted 'for the record'.

The final stage of the State acceptance trials serving to verify the system's perform-

Sergey Komissarov

This scale model of the Shmel'-1 displayed at one of the Moscow airshows features a large ventral fairing which could be an inflatable bumper or an extra fuel tank.

Yakovlev OKB

Above and below: Two views of '023 Black', a production-standard Shmel' UAV in red/blue/white demonstrator colours. Note the arms in between the landing gear struts; they swing down to engage fittings on the launcher's sled.

Yakovlev OKB

ance in mountainous areas was concluded on 28th September 1989. In the course of the trials the 13 *izdeliye* 61T RPVs, as the version of the Shmel'-1 equipped with a TV reconnaissance system was called (T = *televizi**onn**oye obo**roo**dovaniye* – TV equipment), logged a total of 50 flight hours between them. The early prototypes, such as '005 Black', had a 'goldfish bowl' nose patterned on that of the *izdeliye* 60 and simple (that is, not downward-canted) wingtip fairings. In due course the forward fuselage was redesigned; the new nose had a parabolic shape cut away (that is, flattened) from below to provide a mounting platform for a revolving hemispherical camera turret.

The test flights did not always go smoothly; six of the prototypes were lost in crashes, which equals an attrition rate of 46%. Despite this, the State commission's report said: *'...the system's performance renders it suitable for series production and service'*.

Some components of the new tactical unmanned aerial reconnaissance system were first shown publicly at the Moscow Aerospace '90 trade fair held at the VDNKh fairground (***Vys**tavka dosti**zhen**iy na**rod**novo kho**ziay**stva* – National Economy Achievements Exhibition) in December 1990. The system's international debut followed in July 1991 at the 39th Paris Air Show. The complete system was demonstrated for the first time at the MosAeroShow '92 at Zhukovskiy on 11th-16th August 1992 under the new name Stroy-P (Formation-P); the export version was marketed as the Sterkh.

Five production Stroy-P systems, each with a complement of ten Pchela-1T RPVs, had been completed and delivered to various arms and services by 1991; the first of these entered service in the Leningrad Defence District. On 16th June 1997 the Stroy-P tactical unmanned aerial reconnaissance system was finally formally included into the Russian Army inventory pursuant to Russian Government directive No.730.

The Pchela-1T (*izdeliye* 61T) RPV has a designated service life of ten cycles. Its modular design makes sure that the principal airframe subassemblies (the fuselage sections, wings, empennage and landing gear) are interchangeable with the other versions of *izdeliye* 61, allowing the RPV to be easily assembled from standard components in the field.

The mission equipment of the Pchela-1T is a TV camera with a zoom lens mounted on a gyrostabilised platform under the nose;

Yakovlev OKB

Another view of the prototype LCV with a mock-up simulating one of the Yakovlev UAVs as regards weight, but not appearance.

however, the modular airframe design makes it possible to reconfigure the RPV quickly for missions other than TV reconnaissance. For instance, fitting an active jammer turns it into an ECM aircraft capable of suppressing enemy UHF radio communications within a radius of 10-20 km (6.2-12.4 miles). A time-expired RPV can be converted into a target drone by installing equipment increasing its radar and heat signature (including transponders and tracers); this allows the Pchela-1 to emulate light aircraft flying at 120-180 km/h (75-112 mph) and 100-2,500 m (330-8,200 ft), such as might be used to insert or extract spies or saboteurs. The Russian military sure learned their lesson after Mathias Rust's famous landing in Moscow's Red Square in a Cessna 150! For night operations the TV camera can be replaced with infra-red linescan or three-band scanning equipment.

The Stroy-P system comprises ten Pchela-1T RPVs, an integrated LCV based on the BTR-D paradroppable amphibious tracked armoured personnel carrier (a design related to the BMD-2 paradroppable infantry fighting vehicle) and an evacuation/maintenance vehicle on a GAZ-66 chassis. All components of the system can be airlifted by Antonov An-12 and Il'yushin IL-76 transport aircraft or Mil' Mi-26 transport helicopters and paradropped on P-7 pallets with a multi-canopy parachute/retro-rocket system. After launch the RPV normally follows a pre-programmed route with up to 24 waypoints; if necessary the operator sitting in the LCV can take over and 'fly' the machine via radio control. The LCV's mission equipment makes it possible to control up to two RPVs at a time.

Starting in the late 1980s, the Stroy-P tactical unmanned aerial reconnaissance system has been used successfully in nine major military exercises. For instance, a special tactical exercise involving reconnaissance RPV launches was held in the North Caucasus Defence District in 1999; the flights of the RPVs were monitored by two Mikoyan MiG-29 fighters which tracked them with their radars. This was the first exercise of its kind in the history of the North Caucasus DD. The squadron of Stroy-P reconnaissance systems was commanded by Lt. Col. A. Koshootin.

In the course of the exercise two RPVs penetrated deep into Blue Force territory. (In Western exercises, Blue Force is the 'good guys'; in Russian ones, it's the 'bad guys'.) The first one turned back after travelling about 70 km (43.5 miles) from the place where it was launched. The mission of the second machine was more eventful; after covering more than 60 km (37.3 miles) in a straight line it made a 90° turn and flew another 40 km (25 miles) before making for home. Each aircraft had by then eight missions on its tail but had been performing flawlessly so far.

The first sortie that day ended in an 'oops': the recovery parachute broke away after deploying and the RPV stalled and spun in, slamming into the ground and keeping the fire and rescue crews busy for a while. Yet, incredibly, the mission equipment (including the all-important intelligence data recorder) was retrieved intact from the wreckage, so the mission was a success after all. The second RPV landed normally. North Caucasus DD Aviation Commander Maj. Gen. Aleksey Bazarov, who was among the 'brass hats' attending the exercise, congratulated the squadron's personnel on the good results and on the fact that the unit had become fully operational. The most recent exercise in which the Stroy-P system was involved took place near the Caspian Sea in the summer of 2002.

The system also saw actual combat – sadly enough, on its home ground, being used by the VDV and the Army alike during

Yakovlev OKB

Above: Pchela-1T '209 Red' on the LCV a few minutes before launch at a soggy practice range. Note the ground power supply arm attached to the UAV (it is disengaged 15 seconds before launch) and the lowered bars linking the centre fuselage with the launcher carriage.

Yakovlev OKB

The same UAV after landing, with the recovery parachute lying on the ground beside it. Note the open cover of the parachute housing and the position of the landing gear struts after landing.

Dmitriy Komissarov

The modular design of the Pchela-1T sometimes leads to peculiar results. This example ('777 Red') displayed at the MAKS-2003 airshow carried the c/n 5950611207408 on the wings and the c/n 5950611207340 on the stabiliser/propeller duct. Curiously, both ailerons are drooped.

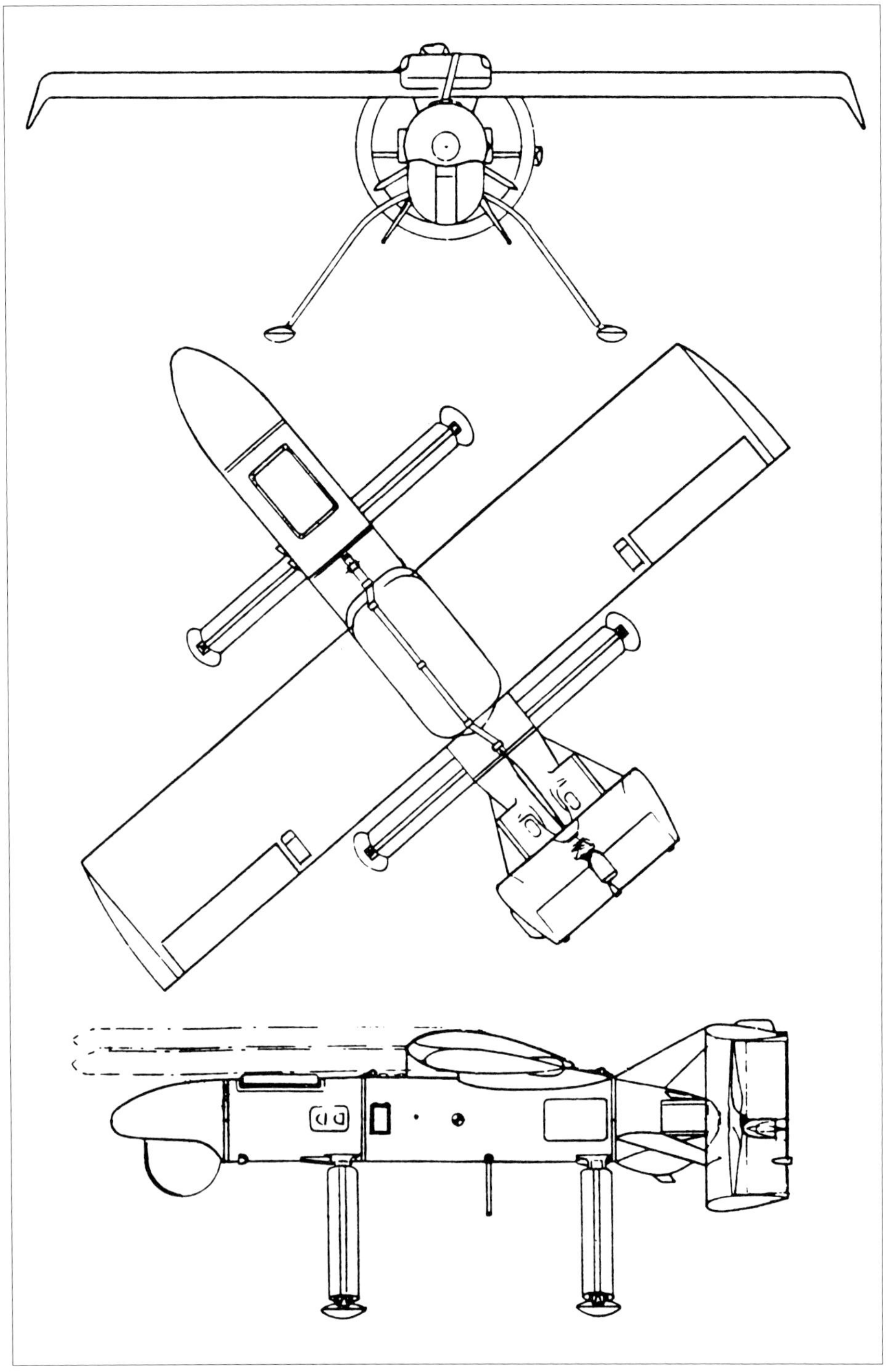

A three-view of the Shmel'-1 (alias Pchela-1T) UAV; the folded position of the wings is shown by hatched lines. The stabiliser is shown in cutaway form to show the propeller and control surfaces.

both Chechen Wars. Actually this was more like combat evaluation. The first combat sorties date back to May 1995, six months after the outbreak of the First Chechen War, when the Stroy-P aerial reconnaissance system was used by North Caucasian DD forces in support of Airborne Troops operations. Five Pchela-1T RPVs were in action on the Caucasian theatre of operations, flying ten sorties, including eight combat ones; total time was 7 hours 25 minutes.

The machines flew their missions at 600-2,200 m (1,970-7,220 ft), operating up to 55 km (34 miles) from the LCV. In so doing, two of the 'bees' were shot down by the Chechen guerrillas who put up field AA guns and concentrated small arms fire along the RPVs' route. On the other hand, there were no casualties among the personnel operating the system.

The 'bees' furnished unique information on enemy movements that would have been unobtainable by other means – at least not without high losses; this information was vital for Russian artillery units pounding guerrilla strongholds with rocket projectiles and for VDV commando groups on seek-and-destroy missions.

After a long pause caused by funding shortfalls, when the Russian Armed Forces were unable to re-equip for lack of money, production of the Stroy-P system resumed in the late 1990s. The ongoing hostilities in the North Caucasus were the primary reason for this; with the enemy cunningly hiding in residential areas and woodland, the Joint Task Force in the North Caucasus, or OGV (*Obyedi***nyon***naya grooppi***rov***ka voysk na* ***Sev****ernom Kav***kah***ze*), needed highly effective pilotless reconnaissance assets.

Trying to make the system more cost-effective, the Yakovlev OKB worked on increasing its service life; among other things, this was to be achieved by resorting to conventional landing, which imposed less stress and strain on the airframe than the parachute recovery technique. Back in the 1980s, when the Stroy-P system was created, no reliable means existed of giving a UAV automatic landing capability. In the meantime, however, aviation technology has advanced, and now such means can be developed in Russia fairly quickly and at an affordable price. Two versions of the upgraded system were envisaged; one would be mobile, like the existing model, while the other version would be permanently based at airfields, the RPVs using runways for landing. The latter version offered the advantage of lower operating costs; also, it would allow 'real-life' crew training to take place much more often than once a month, the way it is now.

Another thrust of the improvement effort was directed at giving the RPV round-the-clock capability. In particular, the upgraded Pchela was to feature a thermal imager linked to a laser rangefinder/marked target seeker; such an aircraft would be capable of not only reconnaissance but would act as a target designator for ground forces or manned strike aircraft (including those carrying laser-guided munitions). Also, a satellite navigation system receiver was to be fitted, allowing the target co-ordinates to be determined much more accurately.

Another possibility now under consideration is that of turning the Stroy-P system into a real-time artillery spotter. The Koolon research institute has developed an algorithm enabling the RPV to work in concert with the Smerch (Tornado) and Grad (Hail) multiple launcher rocket systems and the 152-mm 2S19M1 self-propelled howitzer. Tests of a reconnaissance/strike weapons system combining the Stroy-P system and a modified Smerch MLRS gave impressive results; the MLRS's deployment/target engagement time was reduced by a factor of five to six, the expenditure of rockets was cut by 33%, and the total time from detection to destruction was no more than two or three minutes. The possibility of target designating for attack helicopters was also contemplated.

In 1997 a large group of Russian defence industry and MoD specialists was nominated for the Russian Government Prize for Science and Technology Development in recognition of their part in developing and putting into production the Stroy-P tactical unmanned aerial reconnaissance system based on the Pchela-1 RPV. Anatoliy S. Novosyolov, General Designer at the Koolon research institute, was the first in the list of nominees.

The operational experience with the Stroy-P gained at the Chechen TO is being used to further refine the system. The latest version utilises only cross-country wheeled vehicles – specifically, the GAZ-66 and the Ural-4320 6 x 6 army lorry; the elimination of tracked vehicles has cut operating costs and made the system more mobile.

Pchela-1 IK RPV

A new version of the Pchela-1 RPV, capable of flying night surveillance missions, successfully passed its State acceptance trials in March 2001. The vehicle was designated Pchela-1IK, the IK suffix obviously denoting *infra**kras**nyy*, infra-red. Since infra-red sensors also made part of the equipment of the previous Pchela-1 versions, it may be assumed that this one is a dedicated night-time version.

Colibri unmanned aerial reconnaissance system project

In the early 1990s the Yakovlev OKB teamed up with one of the research institutes in Zelenograd (a town north of Moscow which is the Russian answer to Silicone Valley) to form the NPO AVICS research and production association. This consortium is responsible for a joint project of a successor to the Stroy-P tactical unmanned aerial reconnaissance system. Designated Colibri, it is based on an advanced UAV embodying stealth technology; the aircraft is slightly larger than the Pchela-1T and is to have an endurance of about 12 hours.

NPO AVICS paid much attention to improving the system's operating economics, including intelligence gathering costs. As distinct from the Pchela-1T, which was designed for use at the regimental level, the Colibri was intended for larger Army units (divisions or armies). The envisaged long range was to allow the UAV to be based up to 130 km (80 miles) from the FLOT (Forward Line of Own Troops), reducing its vulnerability to enemy action.

Intelligence gathered by the UAV (which was to have day/night, all-weather capabil-

Basic specifications of the Stroy-P tactical unmanned aerial reconnaissance system and the Pchela-1T (*izdeliye* 61)

Length overall	2.78m (8 ft 10 19/64 in)
Wing span	3.25 m (10 ft 7 31/32 in)
Height on ground	1.1 m (3 ft 7 19/64 in)
Combat radius (TV signal/control command reception range), km (miles)	60 (37.2)
Preparation time for a repeat sortie, hours	20
Flight altitude above sea level, m (ft)	100-2,500-3,000 (330-8,200-9,840)
Speed, km/h (mph)	120-180 (74.5-112)
Launch weight, kg (lb)	Up to 138 (304)
Rate of climb at sea level, m/sec (ft/min)	At least 5.0 (984)
Maximum launch altitude above sea level, m (ft):	2,500 (8,200)
Flight altitude above ground level, m (ft)	100-1,000 (330-3,280)
Turn rate, deg/sec (ft/min)	At least 3
TV camera field of view	3-30°
TV camera movement limits:	
in elevation	+5°/–65°
in azimuth	±170°
Target coordinate measurement error margin:	
in range, m (ft)	150 (490)
in azimuth	1°
Endurance	2 hrs
Service life, cycles (each RPV)	At least 5
Operating temperatures	–30/+50°C (–22/+122°F)
Air humidity limit at +30°C (+86°F)	95%
Maximum wind speed, m/sec (kts):	
launch	10 (20)
landing	8 (16)
Minimum parachute deployment altitude, m (ft)	150 (490)
Maximum number of RPVs airborne at any one time	2
Control crew	2

Sergey Komissarov

This overall black model of the Colibri was displayed at the MAKS-93 airshow; curiously, it does not feature a reconnaissance system turret.

ity) was to be downloaded via data link to mobile user terminals situated at the forward line of the troops the UAV was flying its mission for. Alternatively, it could be relayed to combat aircraft or helicopters to give their crews an indication of the situation near the target in advance, or to combat vehicles on the ground.

The use of standard laser designators made it possible to use a single UAV type for guiding laser-guided munitions of all types (cannon shells, missiles and 'smart' bombs). The result was expected to change the philosophy of ground combat completely; friendly forces would have constant information on the enemy's position, which would enable 'stand-off' engagement by means of precision-guided munitions. For instance, the Msta SP howitzer now in Russian Army service can fire guided shells at targets up to 20 km (12.4 miles) away; advanced 152-mm artillery systems should have twice the 'kill' range. Thus, the use of UAVs (among other things, in anti-terrorist operations) will allow Russian Army units to locate and destroy targets effectively over huge areas without changing their own position.

The Colibri is a low-wing aircraft with gently swept wings and a T-tail powered by a 75-hp engine driving a pusher propeller; it is to carry a mission load of 70 kg (154 lb). The maximum operating radius is 180 km (112 miles), increasing to 700 km (434 miles) if a second UAV acting as a comms relay platform is used.

A model of the Colibri was displayed at the MAKS-93 airshow on 31st August/5th September 1993.

Yak-133BR unmanned combat/aerial reconnaissance system project

Another area of UAV design which is receiving much attention lately is UCAVs, and the Yakovlev OKB has pursued this subject at its own initiative. There have been press reports of Yakovlev UCAVs, including the Yak-133BR derived from the Yak-130 twin-turbofan advanced trainer. Actually the two have little in common, as the Yak-133 is a tailless delta. At an all-up weight of some 6,000 kg (13,230 lb), it compares closely to the US X-45 and X-47 UCAVs.

Preliminary development work performed by the Yakovlev OKB shows that Russian unmanned reconnaissance and strike systems could be developed in time to keep up with the Joneses (that is, the NATO) and at an acceptable price. Yet, so far the Russian MoD has not clear-cut operational requirements for such vehicles, which hampers large-scale research in this field.

Now, however, the situation is slowly changing as the decision makers realise that the future of military aviation lies with unmanned aerial systems of various classes, including special mission UAVs. Russia's late start in the race could jeopardise both its defensive capability and the export potential of Russian aircraft.

Basic specifications of the Colibri unmanned aerial reconnaissance system

Length overall	4.25 m (13 ft 11 21/64 in)
Wing span	5.9 m (19 ft 4 9/32 in)
Stabiliser span	1.7 m (5 ft 6 59/64 in)
Launch weight, kg (lb)	280 (620)
Fuel load, kg (lb)	72 (158)
Payload, kg (lb)	70 (154)
Flight altitude above S/L, m (ft)	50-3,500 (164-9,840)
Speed, km/h (mph)	250 (74.5-112)

Albatross VTOL UAV project

Starting with the MAKS-95, the Yakovlev OKB has been displaying models of its advanced compact UAV projects. These include a very intriguing project of a vertical take-off and landing (VTOL) UAV designated Albatross. The 450-kg (990-lb) vehicle is a convertiplane featuring tilt-rotors at the tips of its unswept low aspect ratio wings. The machine has a blended wing/body design with an oval-section fuselage, an inverted-V tail unit and a monowheel undercarriage (the vertical tails double as supports on the ground). A single turboshaft engine is buried in the fuselage, breathing through two air intakes in the wing roots; the jetpipe is located ventrally and is screened by the vertical tails in side view to reduce the IR signature.

A hemispherical gyrostabilised turret with reconnaissance system sensors is located ahead of the monowheel. The mission equipment is to enable round-the-clock, all-weather operations.

This vehicle has been advertised as being intended for TV or IR air reconnaissance and for environment monitoring. It can be used for such missions as reconnaissance, forest fire spotting and mapping, search and rescue operations, fishery reconnaissance, ice reconnaissance, finding narcotic crops, patrolling gas and oil pipelines, coastal patrol and meteorological observation. The VTOL capability enables the Albatross to be operated from small-tonnage sea-going ships and river boats and from small sites in hard-to-access areas (a landing site of 7 x 7 m/23 x 23 ft is required). The system, comprising the RPV proper, ground control station and servicing equipment, may be based on a ship or two KamAZ-4310 6 x 6 trucks.

Basic specifications of the Albatross unmanned aerial reconnaissance system

Length overall	4.80 m (15 ft 9 in)
Wing span	8.0 m (26 ft 3 in)
Engine power, eshp	160
Take-off weight, kg (lb)	450 (992)
Speed range, km/h (mph)	0 to 300 (0 to 186)
Flight altitude range, m (ft)	0 to 350 (0 to 1,150)
Information receive/transmit range, km (miles)	50/100 (31/62)
Rate of climb, m/sec (ft/min)	9 (1,772)
Endurance	7 hrs

Zhavoronok RPV project

Another product of the NPO AVICS consortium is an interesting UAV project called ***Zha**voronok* (Skylark). This is nothing other than an autogyro based on the airframe and powerplant of the Shmel'-1 UAV. The wings and parachute recovery system have given way to a two-blade rotor driven by the slipstream; a skid landing gear replaces the four cantilever struts. The Zhavoronok is to be powered by either the same P-032 engine or a Voronezh Engine Design Bureau (VOKBM – *Voro**n**ezhskoye **o**pytno-kon**s**t**r**oo**k**tor-skoye **b**yuro **m**otorostro**y**eniya*) M18-02 air-cooled flat-twin two-stroke engine delivering 55 hp at 2,200 rpm. The mission equipment placement is unchanged.

Two project versions differing in engine type, rotor diameter, landing gear design and payload are known. A model of one version was displayed at the MAKS-93 airshow on 31st August/5th September 1993.

Aist captive RPV project

The Aist (Stork) RPV developed, again, by the NPO AVICS consortium, was intended to become the third component of a UAV-based environmental monitoring system, the other two elements being provided by the Colibri and Zhavoronok described above (in this case converted for civil uses). These three vehicles were assigned different roles in the project. The Colibris were expected to ensure surveillance of large areas at relatively high speeds. The Zhavoronok RPVs were intended to effect a more detailed inspection of certain areas at closer range and slower speeds. Finally, the Aist was a device intended for use in emergency situations at industrial sites, such as nuclear power stations and chemical plants, involving danger for human health. The Aist comprised a captive VTOL platform, a ground station for remote control of the vehicle and information acquisition/processing, and a transportation container doubling as a launch tube, installed on an armoured personnel carrier. As distinct from the other two vehicles, remote control was effected in this

Sergey Komissarov

Yefim Gordon archive

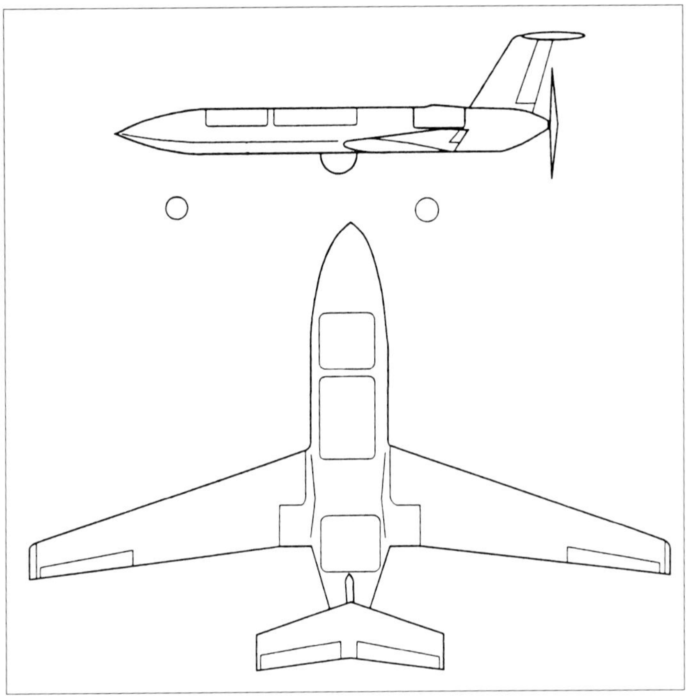

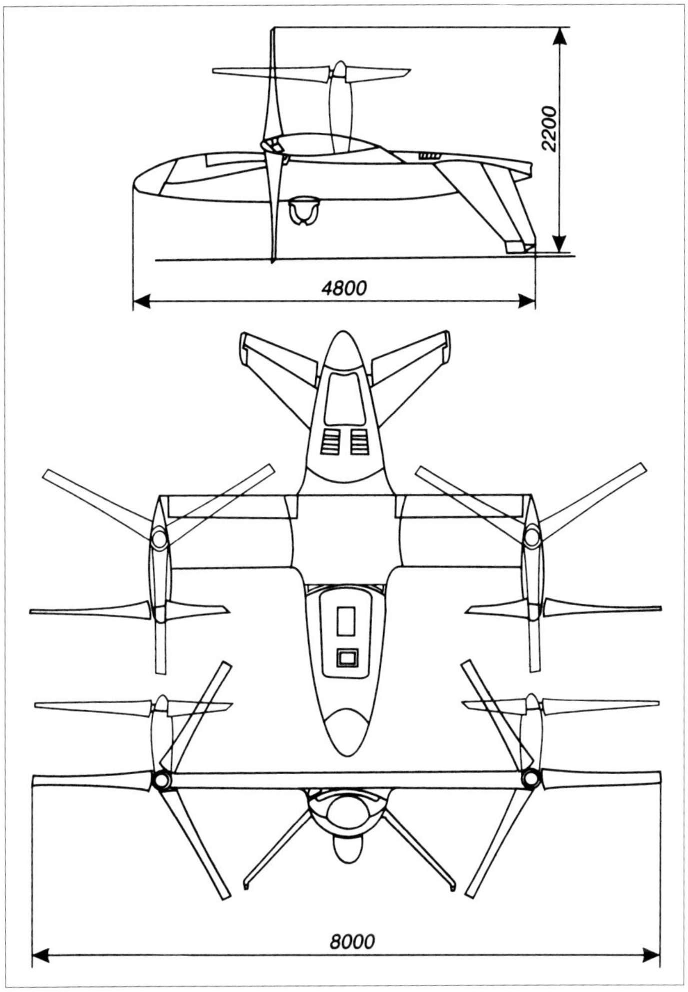

Top and above: Another view of the model and two drawings of the projected Colibri reconnaissance UAV.

Top right: A model of the Albatross VTUAV with the final drive fairings/rotors tilted up. Note how the machine sits on its monowheel and inverted-Vee tail tips.

Right: A three-view of the Albatross vertical take-off and landing UAV. The aircraft is 4.8 m (15 ft 8$\frac{63}{64}$ in) long and 2.2 m (7 ft 2$\frac{39}{64}$ in) high in cruise configuration; the 'wing span' with rotors turning is 8.0 m (26 ft 2$\frac{31}{32}$ in).

case not by radio, but through a fibre-optic cable. In 1993 the NPO AVICS advertised this programme in the hope of attracting investors that would provide the necessary funding. Presumably financial problems remained unresolved and the project was not proceeded with.

Malinovka mini-RPV project

Another Yakovlev mini-RPV optimised for lengthy reconnaissance or surveillance missions is called Malinovka (Robin, the bird). The aircraft has highly unusual looks, featuring an integral (blended wing/body) layout with a flattened fuselage and forward-swept wings located well aft; the wings have downward-angled tips and fold forward for storage. Two tailbooms attached to the wing roots carry an inverted-Vee tail unit. A piston engine located atop the rear fuselage in a bulged cowling drives a two-blade pusher propeller. A hemispherical gyrostabilised turret with reconnaissance system sensors is located ventrally amidships. Like the Albatross, the Malinovka clearly embodies stealth technology.

The operational concept is rather unusual as well; the aircraft comes in a transport container which doubles as a launch tube! The Malinovka is extracted from the tube by a squib attached to the aircraft by a long line, the wings deploying as the aircraft clears the opening; the squib then falls away as the aircraft climbs under its own power. A parachute recovery system is used for landing.

A model of the Malinovka was first displayed publicly at the MAKS-95 airshow.

Expert mini-RPV project

Yet another unmanned aerial system the Yakovlev OKB has in the making – the Bars (Snow Leopard) – is to feature an RPV with an all-up weight up to 60 kg (132 lb). The system makes use of the latest technologies, including SATNAV systems, electronic maps and electronic components borrowed from spacecraft production. The operator's workstation is based on a laptop computer.

One of the UAVs that could be part of the Bars system is the Expert mini-RPV with an

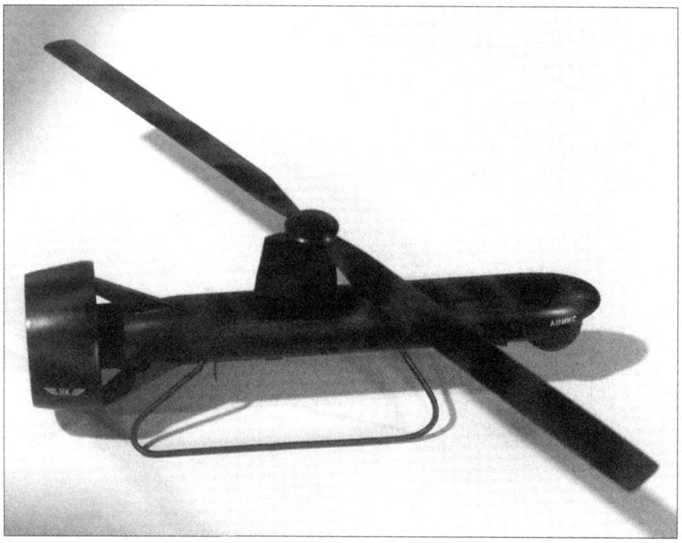

Sergey Komissarov

A model of the Zhavoronok-1 autogyro UAV; its Shmel'-1 lineage is obvious.

Sergey Komissarov

Above: A model of the stealthy, tube-launched, forward-swept-wing Malinovka UAV. The reconnaissance system turret is just visible under the fuselage.

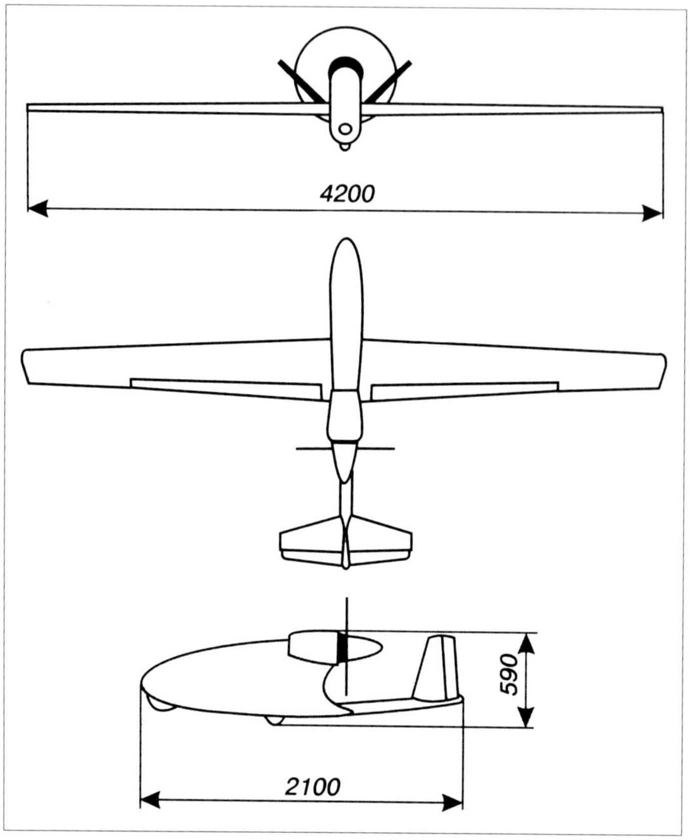

Above: A three-view drawing of the Expert compact UAV. It is 2.1 m (6 ft $10\frac{11}{16}$ in) long and 0.59 m (1 ft $11\frac{15}{64}$) in high, with a wing span of 4.2 m (13 ft $9\frac{23}{64}$ in).

Below left: A model of the Dyatel UAV. The machine has an inverted-V tail carried on twin booms and a pusher propeller; the sharp-chined fuselage blended with the wings appears to have an internal payload bay.

Below right: The broadly similar but smaller Klyost UAV, showing the ventral sensor turret.

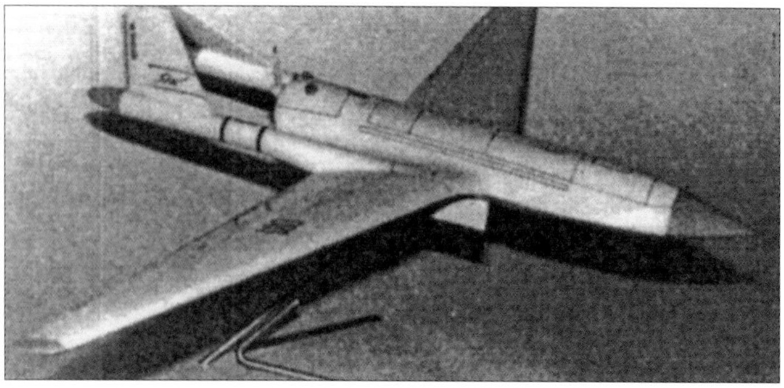

Yefim Gordon archive

Yefim Gordon archive

Basic specifications of the Expert unmanned aerial reconnaissance system

Length overall	2.10 m (6 ft $10\frac{43}{64}$ in)
Wing span	4.20 m (13 ft $9\frac{23}{64}$ in)
Speed range, km/h (mph)	65-110 (40-68)
Flight altitude range, m (ft)	50-450 (164-1,476)
Information receive/transmit	
Range, km (miles)	50/100 (31/62)
Rate of climb, m/sec (ft/min)	5 (984)
Endurance	6 hrs

AUW of 40 kg (88 lb) intended for civilian applications. This diminutive vehicle is driven by a 10-hp engine with a two-blade pusher propeller behind a straight mid-set wing and features a V-type empennage on a tail boom. It is advertised as being capable of the same type of missions as those listed for the Albatros UAV (see above).

The Expert is used in an integrated system comprising three RPVs, ground control station, launcher and servicing equipment. The system is accommodated in a minivan-type car.

DPLA-110 remotely piloted glider prototypes

This remotely piloted glider is equipped with day-time TV cameras and is intended to send back video signals after being launched by a special rocket vehicle and released over a target. Two prototypes have been tested.

Electrically-powered mini-RPV (project)

Plans were in hand at Yakovlev OKB to produce a small vehicle powered by a quiet electric motor, with a night TV camera.

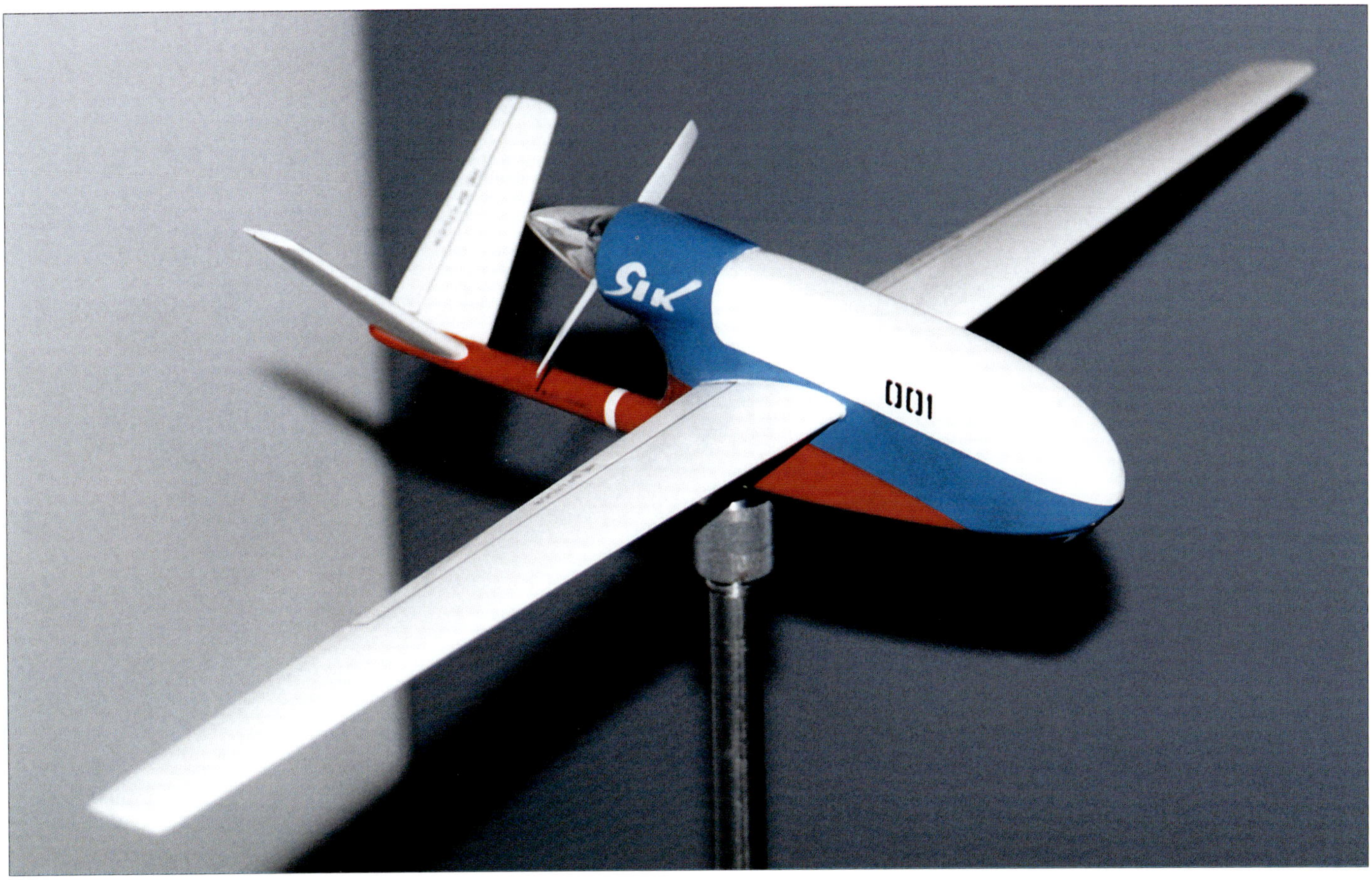

Above: A model of the Yakovlev OKB's Expert UAV designed for commercial applications. The aircraft is much smaller than one might think. The white band on the tailboom indicates the propeller rotation plane.

This view shows a Pchela-1T (*izdeliye* 61T) blasting off.

Sergey Komissarov

Above: This pristine UT-2 *sans suffixe* is preserved at the Yakovlev OKB Museum. It is shown here at Moscow-Tushino during one of the annual air displays.

Yefim Gordon

This Yak-9P is preserved at the Muzeum Wojska Polskiego (Polish Army Museum) in Warsaw.

Yefim Gordon

Above: This Yak-3 bearing the legend 'From Ferapont Petrovich Golovatyy – the second aircraft for the terminal rout of the enemy!' and 14 'kill' stars was also preserved at the OKB Museum and is also shown here at Moscow-Tushino.

Yefim Gordon

Yak-15 '37 Yellow' is probably the sole survivng example and is seen here at Moscow-Khodynka in August 1989 during a brief outing from the OKB Museum.

Marcus Fülber

Above: Yak-23 '15 Red' sits amid deep snow in the open-air display at the Central Russian Air Force Museum in Monino. A Yak-25M is visible on the right.

Yefim Gordon

The Yak-23UTI in its current guise, another Yakovlev OKB Museum exhibit, is seen at Moscow-Khodynka during one of the Aviation Day displays.

Yefim Gordon archive

Above: '22 Yellow', an early-model Yak-24 *sans suffixe* with dihedral tailplanes, in flight. The red fin trim tab is noteworthy.

Yakovlev OKB

The sleek Yak-140 prototype at the OKB's flight test facility in Zhukovskiy.

Yakovlev OKB

Above: A typical production Yak-25M (c/n 1021) awaiting delivery at Saratov-Yoozhnyy, the factory airfield of plant No.292. The light grey nose was standard.

Yakovlev OKB

'54 Yellow', the second prototype of the Yak-26 bomber, as originally flown (with twin boundary layer fences and no wing dogtooth).

Yakovlev OKB

Above: A three-quarters rear view of the prototype of the Yak-25RV-I target drone ('76 Yellow'). Note the flares on the tailcone assisting visual tracking of the aircraft at night.

Yakovlev OKB

'58 Yellow', the prototype of the Yak-27K-8 missile-armed interceptor. Note the shape of the pylons.

Yakovlev OKB

Above: The prototype of the Yak-28P interceptor in an intermediate configuration as regards the powerplant, fitted with dummy K-8 missiles and sporting four 'kill' stars to mark test launches.

Yakovlev OKB

The first production Yak-28U trainer cruises under dark blue stratospheric skies.

Yakovlev OKB

Above: '30 Yellow', the first prototype of the Yak-30 trainer.

Yefim Gordon archive

'38 Yellow', the second flying prototype of the Yak-36 technology demonstrator, is about to take off. The orange-painted UB-16-57 FFAR pods are noteworthy.

Yefim Gordon archive

Above: Late-production Yak-38 attack aircraft are lined up on the angled flightdeck of a *Kiev* class aircraft carrier. Note the strakes flanking the lift engines' dorsal air intake (they are taller on the Yak-38M) and the surreal green undersurfaces characteristic of this type.

Yakovlev OKB

'77 White', the ill-fated second prototype of the Yak-141, in cruise configuration. The four pylons are in place but no missile launch rails are attached.

Yuriy Kirsanov

Above: Yak-40 'Salon 2nd Class' RA-87970 in the current livery of the Rossiya State Transport Co. (the Russian government flight) approaches runway 24 at Moscow-Vnukovo on 17th March 2002.

Dmitriy Komissarov

Yak-42 RA-42326 in the full livery of Saravia (Saratov Airlines) is caught by the camera a few seconds before touching down on runway 32R at Moscow-Domodedovo.

Yakovlev OKB

Above: The Yak-55M prototype, '02 Red', in a decidedly non-standard colour scheme makes a knife-edge pass (note the position of the rudder and the attitude of the pilot).

Yakovlev OKB

'01 White', the prototype of the production-standard Yak-130 combat trainer, displays its distinctive planform in a banked turn.